D0142207

A First Course in Digital Communications

Communication technology has become pervasive in the modern world, and ever more complex. This introductory textbook allows students to understand the principles underlying the reliable transmission of information by focusing on the most basic ideas of digital communications. Carefully paced, logically structured, and packed with insights and illustrative examples, this is an ideal textbook for anyone who wants to learn about modern digital communications.

- The simple, logical presentation covers the fundamentals in sufficient detail to allow students to master the basic concepts in digital communications without being overwhelmed.
- Step-by-step examples and extensive problem sets, including MATLAB® exercises, allow hands on use of the key concepts, to reinforce learning.
- Mathematical derivations are complete, but kept clear and simple to ensure a thorough understanding.
- Initial chapters provide background material to remind students of the basics.
- Advanced topics and applications are also covered to stimulate the reader's interest and appreciation.
- Extensive illustrations provide an intuitive understanding of the derivations and concepts.

Ha H. Nguyen is a Professor at the Department of Electrical and Computer Engineering, University of Saskatchewan, Canada, where he has developed and taught undergraduate and graduate courses in signals and systems, digital communications, error control coding, and wireless communications. He currently serves as an Associate Editor for the *IEEE Transactions on Wireless Communications* and the *IEEE Transactions on Vehicular Technology.*

Ed Shwedyk is a Senior Scholar at the Department of Electrical and Computer Engineering, University of Manitoba, Canada. Since joining the Department in 1974, Dr. Shwedyk has developed and taught many courses, including signals and systems, digital communications, information theory and coding, at both undergraduate and graduate levels. In 1979 he received the Outstanding IEEE Student Counsellor Award for his role in setting up the first IEEE McNaughton Centre. In 2000 he won the University of Manitoba's Dr. and Mrs. H. H. Saunderson Award for Teaching Excellence.

“The use of a step-by-step approach to the design of signal transmission and reception techniques with the aid of examples, illustrations, and problem solving linked to practical systems is very useful.”

Falah Ali, Senior Lecturer in Electrical Engineering, University of Sussex

“What makes this book different from the existing books is an elaboration of simple concepts and more examples.”

Hsuan-Jung Su, Associate Professor of Electrical Engineering, National Taiwan University

A First Course in Digital Communications

HA H. NGUYEN

Department of Electrical & Computer Engineering

University of Saskatchewan

ED SHWEDYK

Department of Electrical & Computer Engineering

University of Manitoba

CAMBRIDGE UNIVERSITY PRESS
Cambridge, New York, Melbourne, Madrid, Cape Town, Singapore, São Paulo, Delhi

Cambridge University Press
The Edinburgh Building, Cambridge CB2 8RU, UK

Published in the United States of America by Cambridge University Press, New York

www.cambridge.org
Information on this title: www.cambridge.org/9780521876131

First published 2009

Printed in the United Kingdom at the University Press, Cambridge

A catalog record for this publication is available from the British Library

ISBN 978-0-521-87613-1 hardback

Additional resources for this publication at www.cambridge.org/9780521876131

Contents

Preface

As the title indicates, the text is intended for persons who are undertaking a study of digital communications for the first time. Though it can be used for self-study the orientation is towards the classroom for students at the fourth-year (senior) level. The text can also serve readily for a beginning-level graduate course. The basic background assumed of the reader is: (i) introductory linear circuit and systems concepts, (ii) basic signal theory and analysis, and (iii) elementary probability concepts. Though most undergraduate electrical and computer engineering students have this background by their final year, the text does include two review chapters which the reader is strongly encouraged to read.

By reading these chapters she/he will obtain a sense of the authors' pedagogical style and the notation used. The notation used is quite standard except (perhaps) in the case of *random* variables or events. They are denoted (faithfully and slavishly) by boldface. As importantly, because of their importance in digital communications, several topics that may or may not be covered in typical introductory courses, are explained in detail in these chapters. The primary topic is random signals which, after a treatment of random variables and probability concepts, are explained in the necessary depth in Chapter 3. Another topic of importance that typically is not touched on or is treated in only a cursory fashion in an introductory signal course is auto- and crosscorrelation and the corresponding energy and power spectral densities. These are explained in Chapter 2 for deterministic signals and Chapter 3 for random processes.

The text material is a reflection of many years (over two decades for the second author) of teaching digital (and analog) communications at the undergraduate level at three universities. At all three universities the course was one term or semester in duration with approximately 36 hours of lectures. The students had diverse backgrounds: at one university they had a prerequisite course where analog communications and random processes were covered, at the second university the students had no background in random concepts even at the elementary level. In both universities the students were in a quite general electrical and engineering program. The third university was one where the students specialized in telecommunications. Except perhaps for the material in the advanced modulation chapter (Chapter 11), the text material can be quite comfortably covered in the allotted time.

The presentation of the material has been strongly influenced by two classic texts: *Principles of Communication Engineering* (John Wiley & Sons, 1965) by J. M. Wozencraft and I. M. Jacobs, and *Detection, Estimation, and Modulation Theory* – Part I (Wiley & Sons, 1968) by H. L. van Trees. Both texts were written at a graduate level but the authors' experience is that their approach to the fundamental concepts of digital communications can be, and in the authors' opinion should be, readily mastered by the target undergraduate audience. The approach taken in the text is to introduce the signal space approach and the concept of sufficient statistics and the likelihood ratio test at the outset. Most introductory texts present these much later. In our experience, the geometrical interpretation offered by the signal space approach offsets any perceived abstraction. Certainly the students responded very favorably to it. Another major and unique difference from standard texts at this level is the introduction of the dynamic search algorithm known as Viterbi's algorithm. This occurs early in the text (in Chapter 6 on baseband modulation) and the algorithm is then used in later chapters. The search algorithm is paramount for modulation/coding techniques that exploit "memory." Almost invariably, modern digital communication systems use this type of modulation/coding. The concept is too important to be left for advanced studies.

Though it would be presumptuous of the authors to tell an instructor how to teach the material in the text, a few guidelines are appropriate. The core material is Chapters 4–8. The presentation is such that these chapters should be taught in the order in which they appear and that the material covered should be treated in some detail since it is very basic to digital communications. After these chapters the authors typically cover Chapter 12, synchronization. This at least introduces the student to this important aspect of communications engineering. Chapters 9 and 10 can be taught independently, the first one deals with bandlimited channels and additive white Gaussian noise, while Chapter 10 considers the fading channel. Chapter 11, which examines advanced modulation techniques has three major sections. The first, on trellis-coded modulation, can be taught after Chapter 8 while the other two, on code-division multiple access and space-time transmission, require the knowledge of the Rayleigh fading channel model developed in Section 10.4.1. Within any of these three chapters, the instructor is free to highlight different topics and assign the rest as reading material. Pedagogically, they build on the concepts of the core chapters so that a reader should be able to read them without much difficulty.

Acknowledgements

The first author would like to thank the undergraduate and graduate students at the University of Saskatchewan who took his courses. Their questions, comments, and mistakes helped to improve the author's teaching and writing of this book. The professional assistance and encouragement that the author received from his colleagues in the Communications Systems Research Group at the Department of Electrical and Computer Engineering, University of Saskatchewan are very much appreciated. The author is pleased to acknowledge the financial support from the University of Saskatchewan in the form of a Publication Fund toward the completion of this book. He is extremely indebted to his wife, Hạnh, for her hard work and devotion in raising our three lovely children, Namca, Nammi, and Namvinh. Without the love, support, and happiness they gave, this book would not have been completed.

The second author would like to acknowledge all the graduate students he has had the good fortune to work with over the years. From them much was learned. Institutionally, the Department of Electrical and Computer Engineering, University of Manitoba, provided an excellent environment for professional growth. The support of Canada's Natural Sciences and Engineering Research Council (NSERC) for the author's research was also crucial to this process. To his wife, Gail, and children, Edwina and Dean, a heartfelt thank you for the support and love for what is approaching half a century. Finally the text is dedicated to both the past and the future. The past is represented by the author's father, George, who instilled in him a fierce love of learning. The future is two lovely grandchildren, Miriam and Silas.

Abbreviations

A/D	Analog to Digital
AC	Alternating Current
AM-SC	Amplitude Modulation – Suppressed Carrier
AM	Amplitude Modulation
AMI-NRZ	Alternate Mark Invert – Nonreturn to Zero
ASCII	American Standard Code for Information Interchange
ASK	Amplitude Shift Keying
AT&T	American Telephone & Telegraph Company
AWGN	Additive White Gaussian Noise
BASK	Binary Amplitude Shift Keying
BER	Bit Error Rate
BFSK	Binary Frequency Shift Keying
Biϕ-L	Biphase-Level
Biϕ:	Biphase
BPF	Bandpass Filter
BPSK	Binary Phase Shift Keying
BSC	Binary Symmetric Channel
cdf	cumulative distribution function
CDMA	Code Division Multiple Access
CMI	Coded Mode Inversion
CPFSK	Continuous Phase Frequency Shift Keying
CPM	Continuous-Phase Modulation
DBPSK	Differential Binary Phase Shift Keying
DA	Data-Aided
DC	Direct Current
DMI	Differential Mode Inversion
DPCM	Differential Pulse Code Modulation
DSB-SC	Double Sideband – Suppressed Carrier
DSSS	Direct Sequence Spread Spectrum
FDM	Frequency Divison Multiplexing
FM	Frequency Modulation
FSK	Frequency Shift Keying
i.i.d.	independent and identically distributed
ISI	InterSymbol Interference

LAN	Local Area Network
LHS	Left-Hand Side
LOS	Line Of Sight
LDPC	Low-Density Parity Check
LPF	LowPass Filter
LSSB	Lower Single Sideband
LTI	Linear Time Invariant
MAC	Medium Access Control
MAI	Multiple Access Interference
MLSD	Maximum Likehood Sequence Demodulation
MLSE	Maximum Likehood Sequence Estimator (or Estimate)
MMSE	Minimum Mean-Square Error
MSK	Minimum Shift Keying
NDA	NonData-Aided
NRZ-L	Nonreturn to Zero – Level
NRZ	Nonreturn to Zero
NRZI	Nonreturn to Zero Inverse
OFDM	Orthogonal Frequency Division Multiplexing
OOK	On Off Keying
OQPSK	Offset Quadrature Phase-Shift Keying
OSI	Open System Interconnection
PAM	Pulse Amplitude Modulation
PCM	Pulse Code Modulation
pdf	probability density function
PLL	Phase Lock Loop
pmf	probability mass function
PPM	Pulse Position Modulation
PRS	Partial Response Signalling
PSD	Power Spectral Density
PSK	Phase Shift Keying
PSTN	Public Switched Telephone Network
PWM	Pulse Width Modulation
QAM	Quadrature Amplitude Modulation
QoS	Quality of Service
QPSK	Quadrature Phase-Shift Keying
RHS	Right-Hand Side
RMS	Root Mean Squared
RZ-L	Return to Zero – Level
RZ	Return to Zero
SIR	Signal to Interference Ratio
SNR	Signal to Noise Ratio
SNR_q	Signal to Noise Ratio quantization
SRRC	Square Root Raised Cosine
SSB-ASK	Single Sideband – Amplitude Shift Keying

SSB	Single Sideband
STBC	Space Time Block Code
TCM	Trellis Coded Modulation
USB	Universal Serial Bus
USSB	Upper Single Sideband
VCO	Voltage Controlled Oscillator
WSS	Wide-Sense Stationary
XOR	Exclusive Or

1 Introduction

Anytime, *anywhere*, *anything* can be taken as the motto and objective of digital communications. *Anytime* means that one can communicate on a 24/7 basis; *anywhere* states this communication can take place in any geographical location, at minimum one no longer is tied to being close to one's land line; *anything* implies that not only traditional voice and video but also other messages can be transmitted over the same channel, principally text, and not only individually but in combination. In large part digital communication systems over the past three decades have achieved the three objectives. Text messaging in all its forms, such as email, internet access, etc., is a reality. Webcasting and podcasting are becoming common. All parts of the globe are connected to the world wide communication system provided by the Internet.

Perhaps, and arguably just as important, a fourth "any" can be added, *anybody*. Though perhaps not as well developed as the first three, digital communication has the potential to make communication affordable to everyone. One feature of digital circuitry is that its cost, relative to its capability, keeps dropping dramatically. Thus though analog communication could achieve the above objectives, *digital communications*, due to this increasingly low cost, flexibility, robustness, and ease of implementation, has become the preferred technology.

This text is an introduction to the basics of digital communication and is meant for those who wish a fundamental understanding of important aspects of digital communication design. Before detailing the text's content and organization, a brief history of digital communication and general comments about it are in order.

Though it does not appear on Maslow's hierarchy [1] of human needs, communication is crucial for any living organism, human or otherwise. To aid the process of communication, humans and human society have developed various techniques and technologies. The impetuses for the developments have been the need to increase the distances over which communication takes place, to increase the rate of communication, and to maintain the reliability of communication. It was only in the last half of the twentieth century that communication systems started to achieve reliable, universal, global communication. Many factors have contributed to this but at the core is the rapid adoption of digital communications.

So what is digital communication? Definitions vary, but the simplest one is that it is the communication (or transmission) of a message using a *finite alphabet* (symbol set) during a *finite time interval* (symbol interval). As such the raising of an eyebrow, a wink, the nod of one's head or a shoulder shrug may be considered to be digital communications.[1] However, both the message set and the distances are limited in these examples. More meaningful digital communication systems were developed early: the Roman army used shields and the sun to flash signals over line-of-sight distances; similarly, North American natives used smoke signals, to list just two preindustrial digital communication systems. But communications as we know today and experience daily was ushered in during the nineteenth century. It started with the "harnessing" of electricity. This harnessing, which began in the mid-eighteenth century, meant that communication at distances further than one could see or hear became feasible.

In 1729, the English scientist Stephan Gray demonstrated that static electricity could be conducted by some materials, e.g., wet twine. The first idea for an electric telegraph was outlined in 1753 in a letter to the *Scots Magazine*. The letter was signed "C.M." and the anonymous author is believed to be Charles Morrison, a surgeon. He proposed the use of as many wires as there are letters in the alphabet. A message was sent sequentially in time with the specific letter indicated by sending an electrostatic charge over the appropriate wire. Dispatches could be sent at distances of 1 or 2 miles with considerable speed. Though there followed various variants and improvements on this idea, it was not until 80 years later that a practical digital communication system, the electric telegraph, was developed and patented. Indeed two patents were granted. One to Charles Wheatstone (of Wheatstone bridge fame) in 1837 in London; the other to Samuel Morse,[2] applied for in 1840, based on his 1837 caveat, with a US patent granted in 1849.

Until 1877, all rapid long distance communication depended upon the telegraph, in essence digital communication with only text messages being transmitted. But in 1877 the telephone was invented by Alexander Graham Bell and this heralded the arrival of long distance analog communications. Coupled with Hertz's discovery of the propagation of electromagnetic waves and Marconi's subsequent exploitation of this phenomenon to greatly increase communication distances, analog communication was ascendent for most of the twentieth century.

However, the second half of the twentieth century, particularly the last two decades, saw a resurgence in digital communications. In 1948 Claude Shannon published a landmark paper [2] in the annals of science in which he showed that by using digital communications it was possible even in the presence of noise to achieve a vanishingly small error probability at a finite communication rate (or finite bandwidth) and with finite signal power. His was a theoretical result and promised the Holy Grail that communication engineers have pursued since. At approximately the same time, R. W. Hamming proposed the Hamming codes [3] for error detection and correction of digital data. The invention of the transistor, also in 1948, and subsequent development of integrated circuitry provided the last component for a digital communications resurrection.

Initially digital communication systems were developed for deep space communications where data reliability was paramount and cost of lesser consideration. The first commercial

[1] In the context of a typical digital communication system block diagram presented later, it is left to the reader to identify the source alphabet, transmitter, receiver, channel, possible impairments, etc., in the described scenarios.

[2] Samuel Morse also devised the dot-dash system known as Morse code.

application started in 1962 when Bell Systems introduced the T1 transmission system for telephone networks.

However, analog communication was still dominant and the first mobile telephone system introduced in North America in the 1980s was analog based. But the ever increasing integrated chip densities and the concomitant decrease in cost meant that the intensive signal processing required by digital communications became feasible. And indeed the late 1980s and the last decade of the twentieth century saw several digital communication systems developed. The GSM mobile phone system[3] was introduced in Europe and DARPA's (Defense Advanced Research Projects Agency) sponsor of a computer communication network, led to the establishment of the Internet. At the end of the millennium one could reasonably state that digital communication systems were dominant again.

So what is a digital communication system and what does it involve? Let us start to answer these questions by considering two general paradigms.

1.1 Open system interconnection (OSI) model

The OSI model was developed explicitly for computer communications over public communication networks. However, it may also serve as a model for other communications. Communication networks such as the public switched telephone network (PSTN) and the Internet are very complex systems which provide transparent and seamless communication between cities, different countries, different languages, and cultures. The OSI model serves to abstract the fundamentals of such a system. It is a seven-layer model for the functions that occur in a communication process.

Figure 1.1 illustrates the different layers. Each layer performs one or a number of related functions in the communication process. Using terminology that arose from computer communications, the seven layers, or any subset of them, are often called a protocol stack. Layers at the same level are known as peer processes. The following description of the functionality of each layer is taken from [4].

(1) *Physical layer* This first layer provides a physical mechanism for transmitting bits between any pair of nodes. The module for performing this function is often called a *modem* (*mo*dulator and *dem*odulator).
(2) *Data link layer* This second layer performs error correction or detection in order to provide a reliable error-free link to the higher layers. Often, the data link layer will retransmit packets that are received in error, but in some implementations it discards them and relies on higher layers to do the retransmission. The data link layer is also responsible for the ordering of packets, to make sure that all packets are presented to the higher layers in the proper order. The role of the data link layer is more complicated when multiple nodes share the same media, as usually occurs in wireless systems. The

[3] The abbreviation "GSM" originally came from the French phrase *Groupe Spécial Mobile*. It is also interpreted as the abbreviation for "Global System for Mobile communications."

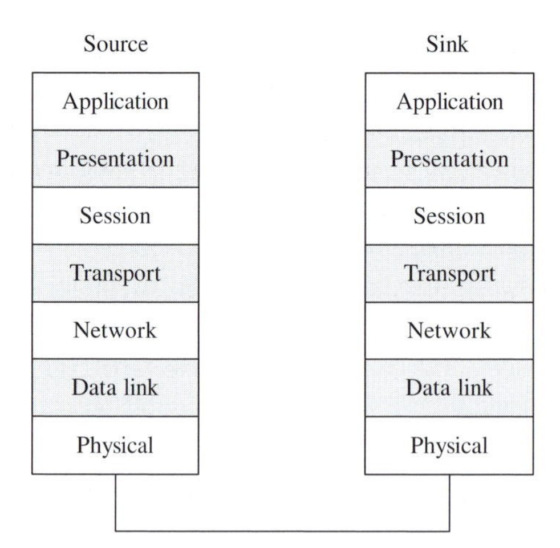

Fig. 1.1 Illustration of different layers in the OSI model.

component of the data link layer that controls *multiple-access communications* is the *medium access control* (MAC) *sublayer*, the purpose of which is to allow frames to be sent over the shared media without undue interference from other nodes.

(3) *Network layer* This third layer has many functions. One function is to determine the *routing* of the packet. A second is to determine the *quality of service* (QoS), which is often controlled by the choice of a connectionless or connection-oriented service. A third function is *flow control*, which ensures that the network does not become congested. Note that the network layer can generate its own packets for control purposes. Often, there is a need to connect different subnetworks together. The connectivity is accomplished by adding an *internet sublayer* to the network layer – a sublayer that provides the necessary translation facilities between the two networks.

(4) *Transport layer* The fourth layer separates messages into packets for transmission and reassembles the packets at the other end. If the network layer is unreliable, the transport layer provides reliable end-to-end communications by retransmitting incomplete or erroneous messages; it also restarts transmissions after a connection failure. In addition, the transport layer may provide a multiplexing function by combining sessions with the same source and destination; an example is parallel sessions consisting of a gaming application and a messaging application. The other peer transport layer would separate the two sessions and deliver them to the respective peer applications.

(5) *Session layer* This fifth layer finds the right delivery service and determines access rights.

(6) *Presentation layer* The major functions of the presentation layer are data encryption, data compression, and code conversion.

(7) *Application layer* This final layer provides the interface to the user.

Though the OSI model has conceptual utility, one should realize that in practice, for many (if not most) digital communication systems, use of the OSI model becomes quite ambiguous if not downright misleading. To give a simple but important example consider error correction or detection. Though assigned to the data link layer, it is just as readily found at the physical layer when hardware error correction/detection integrated circuits are designed. Even more dramatically, data encryption/decryption, data compression, and code conversion can be performed at the physical layer level. Therefore, though the OSI model is of use to elucidate the different functions, in practical implementations of a digital communication system the levels are blurred considerably. For engineering purposes a more pertinent block diagram of a digital communication system is discussed in the next section. In essence, such a block diagram is mainly concerned with the functionality and issues in the *physical layer* of the OSI seven-layer model.

1.2 Block diagram of a typical digital communication system

Figure 1.2(a) shows a block diagram applicable to either an analog or a digital communication system (to be precise, the "synchronization" block is typically only needed in a digital system). It is rare for the system designer to have control over the channel and even rarer, if ever, for the designer to have control over the source. Therefore design and analysis are focused on the transmitter and receiver blocks. The design, of course, must take the source and channel characteristics into consideration. With regard to the transmitter and receiver blocks, for a digital communication system they can be further subdivided for our purposes as shown in Figure 1.2(b).

Consider the transmitter block. The source is typically first passed through a source encoder which prepares the source messages. The encoder details depend on the source and may be further subdivided. For a voice or video source it would consist of, at minimum, an analog-to-digital converter, for a keyboard it could be an ASCII code mapper, etc. Data encryption may be considered to be part of it. Ideally what the source encoder should do is remove all redundancy from the source message, represent it by a symbol drawn from a *finite alphabet*, and transmit this symbol every T_s seconds. Typically the alphabet is binary and the source encoder output is a bit stream or sequence. Removal of the redundancy implies that the source rate (typically measured in bits/second) is reduced, which in turn, as shall be seen, means that the required frequency spectrum bandwidth is reduced.

Having removed at least some of the redundancy by means of the source encoder, it may appear to be counterintuitive to have redundancy added back in by the channel encoder. This redundancy, however, is added back in a *controlled* fashion for error detection/correction purposes.[4] The channel encoder therefore maps the input symbol sequence into an output symbol sequence. To transmit this symbol sequence across a physical channel requires energy and this is the function of the modulator block. It takes the symbol occurring in each T_s (symbol interval) and maps it onto a *continuous-time* waveform which is then sent across the channel. It is important to emphasize that, over any finite

[4] The reader may be familiar with odd/even parity check bits which add an extra binary digit for error detection purposes at the receiver.

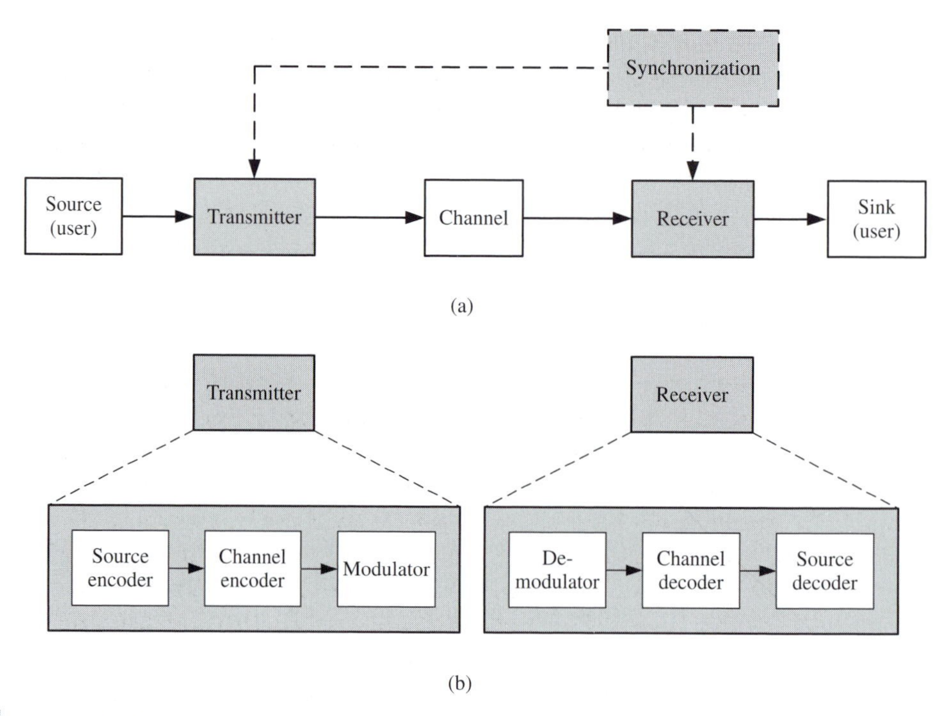

Fig. 1.2 (a) General block diagram of a communication system. The "synchronization" block is only present in a digital system. (b) More detailed block diagram of the transmitter and receiver blocks of a digital communication system. Note that the outputs of the source and channel encoders are *digital* data streams, while that of the modulator block is a *continuous-time* signal.

time interval, the continuous-time waveform at the output of a digital communication system belongs to a *finite set* of possible waveforms. This is in sharp contrast to the analog communication system, in which the modulator directly maps the analog message to a continuous-time signal for transmission. Since analog messages are characterized by data whose values vary over a continuous range, the output of an analog communication system can assume an *infinite* number of possible waveforms. As shall be seen throughout the text, it is the finite property of the sets of the digital messages and modulated waveforms that makes the digital communication system more reliable than its counterpart by applying advanced signal processing algorithms at both the transmitter and the receiver.

The subdivision of the transmitter block, though still relevant, is somewhat classical. The channel encoder/modulator blocks can be, and are, combined to produce a coded modulation paradigm. Indeed all three blocks, source encoder, channel encoder, and modulator, can be combined though this approach is still in the research stage.

At the receiver, one simply passes the received signal through the inverse of the operations at the transmitter; simply, except it is not that simple due to the influence of the channel. If the channel did not filter or distort the signal, did not add noise to it, and if there was no interference from other users, then it would be simple. However, some or all of these degradations are present in physical channels. Therefore one must attempt to overcome the degradations with one's design. For the design or analysis one needs reasonable engineering models of the channel. The channel model depends on the media over which the transmission of modulated signals takes place. Consider guided media such as twisted-pair wire, coaxial cable, and optical fiber. In these the background noise is mainly Gaussian. However, they may exhibit a signal distortion where the transmitted signal is "smeared" out and causes intersymbol interference, i.e., signals in adjacent or nearby time slots interfere with each other. In twisted-pair wire and coaxial cable the smearing occurs due to the finite bandwidth property, while in optical fiber it is due to the dispersion of the light as it travels down the fiber.[5]

A space channel, i.e., satellite communications, typically only adds Gaussian noise to the received signal. The terrestrial microwave channel is similar but the transmitted signal may also be subjected to reflection, diffraction, and refraction, which leads to fading. Fading is a predominant degradation in mobile communications where the signal path from the transmitter to the receiver changes rapidly. In mobile communications, because a great many users must be accommodated in a given geographic area, interference from other users becomes a factor and the chosen modulation/demodulation must reflect this.

As mentioned above, the designer of a communication system typically has little or no control over the source or channel. One is primarily concerned with the transmitter/receiver design. In this introductory text, the major concern is with the *mo*dulator/*dem*odulator blocks, commonly called the *modem*, which are part of every digital communication system. The introduction is concluded by an outline of the text contents.

1.3 Text outline

Since digital communications, particularly at the physical layer and particularly the modem block, involves the transmission of continuous-time waveforms with their reception corrupted by continuous-time noise signals, the next two chapters are concerned with continuous-time signals and systems. Chapter 2 deals with deterministic signals and different methods of characterizing and analyzing them. The material covered is that found in a typical undergraduate signals/systems course. As such it is meant primarily for review purposes but since it is quite self-contained it may be used by a reader with little or no background in the subject. The reader is encouraged to read it and also to attempt the problems at the chapter's end, or at least to read them.

Similar comments apply to Chapter 3 which deals with random signals, or random processes as common terminology terms them (stochastic processes is another name). First, a review of probability theory and random variables is provided. A random process is in essence nothing more than a random variable that evolves in time. Though one initially encounters a random process as additive noise in the channel model it should be pointed

[5] The bandwidth of twisted-pair wire is on the order of kilohertz, that of coaxial cable on the order of megahertz, and for fiber optic it is gigahertz.

out that basically any message source is best modeled as a random signal. Furthermore, the time-varying characteristic of a wireless channel can also be modeled by a random process.

Chapter 4 starts the reader on the road of digital communication. It deals with what, arguably, are the two most important sources, audio and video. These are sources whose outputs are inherently analog in nature. The chapter investigates fundamental principles of converting the analog signal into a digital format. The techniques are applicable to any analog source, e.g., a temperature measurement system. Converting the analog signal to a digital format can be considered to be an elementary form of source encoding.

Chapter 5 is where the study of digital communication systems starts in earnest. Since digital communication systems are found in diverse configurations and applications, the chapter develops a general approach to analyze and eventually design modems. The underlying philosophy is based on the phrase "*short-term pain for long-term gain*".[6] First the modulator output and channel noise are characterized. Then the demodulator is designed to satisfy a chosen performance criterion, namely to minimize the probability of (symbol) error. This is in contrast to analog communications where the criterion is typically the power signal-to-noise ratio (SNR) at the receiver output. The two criteria are related but are not synonymous. Though the symbols are assumed to be *bi*nary digi*ts* (bits) the concepts are readily extended to nonbinary symbols, and they are in later chapters. The chapter concludes with the derivation of the power spectral density (watts/hertz) for general memoryless binary modulation.

Chapters 6 and 7 deal with modem design. Chapter 6 considers baseband signaling. Loosely speaking, in baseband modulation the transmitted signal energy is concentrated around 0 hertz in the frequency band. However, the occupied band can be quite large, ranging from kilohertz (telephone line) to megahertz (cable) and gigahertz (fiber). Since using electromagnetic radiation is not feasible, baseband communication is over guided media. Chapter 7 considers modulation where the modulated signal is shifted to an appropriate frequency band (passband modulation) and then typically transmitted via antennas. Typically the channels are those found in space communications, terrestrial microwave communications, and mobile communications. However, passband modulation may also be found in cable channels for instance.

The analysis/design is concerned with the bandwidth and the average power (or equivalently energy) required by a given or proposed modulation format to achieve a certain performance level, as measured by the bit error probability. Bandwidth and power are viewed as the two natural resources that one has to utilize in the design. To reduce the bandwidth requirement at a given average power-per-symbol level (and hence error performance) one can resort to a nonbinary or M-ary modulation format. This is the topic of Chapter 8.

[6] Though the phrase sounds good, caution is advised. Excluding its application to dental visits, the second author is reminded of a political scenario where a federal minister proposed a budget that included a 25 cents tax increase on a liter of petrol and invoked the phrase. Needless to say the budget and government were defeated on a motion of confidence. In these days of global warming perhaps the minister was correct, though that was hardly his motivation. Furthermore, as subsequent events unfolded what resulted was the nation went through long-term pain for no gain (but this is only a personal opinion).

Up to Chapter 8 the channel model is that of additive white Gaussian noise (thermal noise) with a bandwidth that is infinite or at least sufficiently large so that any effect it has on the transmitted signal can be ignored. Chapter 9 studies the effect of bandlimitation. Bandlimitation results in interference where the transmitted signal in a given signaling interval interferes with the signals in adjacent intervals. Therefore, besides random thermal noise, one has to contend with intersymbol interference. Methods to eliminate or mitigate this interference are investigated in this chapter.

Wireless communications involves a transmitted signal being received over many paths. This results in a phenomenon called fading where the received signal strength varies with time. Chapter 10 develops this important channel model and the various challenges and solutions associated with it.

The modulation/demodulation aspect concludes with Chapter 11 where three modern modulation approaches are described and analyzed. The first modulation, known as trellis-coded modulation (TCM) was discovered in the late 1970s and developed during the 1980s. Its main feature is that a significant bandwidth reduction is achieved with no power or error penalty. The second modulation paradigm presented is code-division multiple access (CDMA) used in the so-called third (and future) generations of wireless communication networks. Finally space-time coding, introduced in the late 1990s, concludes the chapter. Space-time coding provides a significant improvement for wireless communication where fading is present. Though advanced, the modulations are quite readily understood in terms of the fundamental background material presented in the text up to this chapter. In fact all three modulations can be classed under the rubric of coded modulation where coding/modulation is viewed as a single entity. The coding (i.e., adding redundancy), however, is performed differently for the three modulations: in TCM it is done in the time domain, in CDMA in the frequency domain, and in space-time modulation the redundancy is accomplished by having a symbol and its variants transmitted over multiple antennas.

The book concludes with a chapter on synchronization, which appears last not because it is any less important than the other topics covered but simply because one has to start and end somewhere. Synchronization provides all the timing necessary in a digital communication system and therefore is of equal importance to modem design. Chapter 12 presents two very common circuits used in digital communications: the phase-locked loop and the early–late gate.

References

[1] A. H. Maslow, "A theory of human motivation," *Psychological Review*, vol. 50, pp. 370–396, 1943.

[2] C. E. Shannon, "A mathematical theory of communications," *Bell System Technical Journal*, vol. 27, pp. 379–423, 623–657, 1948.

[3] R. W. Hamming, "Error detecting and error correcting codes," *Bell System Technical Journal*, vol. 29, pp. 147–160, Apr. 1950.

[4] S. Haykin and M. Moher, *Modern Wireless Communications*. Prentice Hall, 2005.

2 Deterministic signal characterization and analysis

2.1 Introduction

The main objective of a communication system is the reliable transfer of information over a channel. Typically the information is represented by audio or video signals, though one may easily postulate other signals, e.g., chemical, temperature, and of course text, i.e., the written word which you are now reading. Regardless of how the message signals originate they are, by their nature, best modeled as a random signal (or process). This is due to the fact that any signal that conveys information must have some uncertainty in it. Otherwise its transmission would be of no interest to the receiver, indeed the message would be quite boring (known knowns so to speak[1]). Further, when a message signal is transmitted through a channel it is inevitably distorted or corrupted due to channel imperfections. Again the corrupting influences such as the addition of the ever present thermal noise in electronic components, the multipath fading experienced in wireless communications, are unpredictable in nature and again best modeled as nondeterministic signals or random processes.

However, in communication systems one also utilizes signals that are deterministic, i.e., completely determined and therefore predictable or nonrandom. The simplest example is perhaps the carrier used by AM or FM analog modulation. Another common example is the use of test signals to probe a channel's characteristics. Channel imperfections can also be modeled as deterministic phenomena: these include linear and nonlinear distortion, intersymbol interference in bandlimited channels, etc.

This chapter looks at the characterization and analysis of deterministic signals. Random signals are treated in the next chapter. It should be mentioned at the outset that in certain situations whether a signal is considered deterministic or random is in the eyes of the viewer. Only experience and application can answer this question. The chapter reviews deterministic signal concepts such as Fourier series, Fourier transform, energy/power spectra, auto and crosscorrelation operations. Most of the concepts discussed carry over to random signals. Though intended to be self-contained it is expected that the reader has been exposed to these concepts in other undergraduate courses, hence the treatment is presented succinctly.

[1] For a discussion of knowns and unknowns see "The Secret Poetry of Donald Rumsfeld" website http://maisonbisson.com/blog/post/10086/.

2.2 Deterministic signals

A deterministic signals is a defined, or completely specified, function of an independent variable. The independent variable is invariably taken to be time (seconds) but it may be some other independent variable, e.g., position, heat input (joules), etc. Typically the signal is represented by a graph or analytically by an equation; but other possible representations are a table of values, a statement in English, an algebraic or differential equation, or some combination of the above. The most common representation is that of an equation, several examples of which are given below along with their graphical representations.

(a) $s(t) = \begin{cases} 1, & t \geq 0 \\ 0, & t < 1 \end{cases}$. Known as the unit step function and often denoted by $u(t)$.
(b) $s(t) = \mathrm{e}^{-at}u(t)$, $a > 0$. A *decaying* exponential function.
(c) $s(t) = \mathrm{e}^{-a|t|}$, $a > 0$. Also a decaying exponential. It is an even function (also known as a tent function).
(d) $s(t) = V\cos(2\pi f_c t)$. A periodic sinusoid, whose fundamental period is $1/f_c$ (seconds).
(e) $s(t) = V\cos(2\pi f_c t)\left[u\left(t + T/2\right) - u\left(t - T/2\right)\right]$. A tone burst of duration T (seconds).
(f) $s(t) = \sum_{k=-\infty}^{\infty} p(t - kT)$, where $p(t)$ is an arbitrary signal (pulse) for $0 < t \leq T$ and equal to zero elsewhere. This is a train of delayed pulses.
(g) $s(t) = m(t)\cos\left(2\pi f_c t + \mathrm{d}m(t)/\mathrm{d}t\right)$, where $m(t) = A\sin(2\pi f_m t)$, $f_m \ll f_c$. An amplitude/phase modulated signal.
(h) $s(t) = g(t - t_0)$, where $g(t)$ is any defined signal, perhaps one of the above. This is simply a time shift of $g(t)$ by t_0 seconds – to the right if t_0 is positive and to the left if negative.

Signals (a)–(f) are represented graphically in Figure 2.1.

Though signals can be represented graphically or analytically in the time domain, it is common to seek other representations which allow one to gain insight into the signal characteristics and/or to deal with them more successfully. The most ubiquitous representation is a frequency domain one via either the Fourier series (for periodic signals) or the Fourier transform (for aperiodic signals). Periodic signals have more structure than aperiodic signals and since their representation provides the stepping stone to aperiodic signal representation, they are discussed next.

2.3 Periodic signals

2.3.1 Representation in a (Fourier) series expansion

A periodic signal is simply one where a "pulse," $p(t)$, is defined over a finite time interval of $[0, T_1]$ seconds and repeated every T seconds, where $T \geq T_1$. Implicitly it is assumed $p(t)$ is zero outside the interval $[0, T_1]$. Figure 2.2 illustrates various examples of periodic signals. Analytically a periodic signal can be expressed as

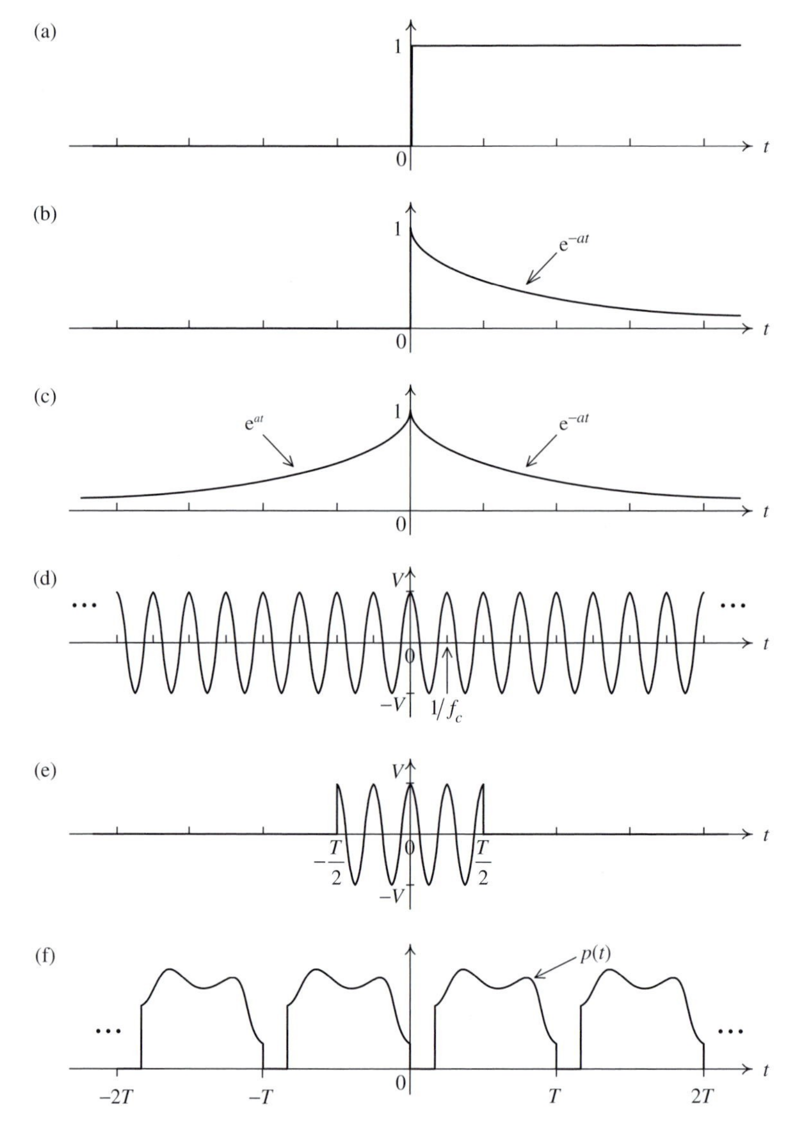

Fig. 2.1 Graphs of example signals: (a) $s(t) = u(t)$; (b) $s(t) = e^{-at}u(t)$, $a > 0$; (c) $s(t) = e^{-a|t|}u(t), a > 0$; (d) $s(t) = V\cos(2\pi f_c t)$; (e) $s(t) = V\cos(2\pi f_c t)[u(t + T/2) - u(t - T/2)]$; (f) $s(t) = \sum_{k=-\infty}^{\infty} p(t - kT)$.

$$s(t) = \sum_{k=-\infty}^{\infty} p(t - kT), \text{ where } p(t) = \begin{cases} \text{arbitrary}, & t \in [0, T_1],\ T_1 \leq T, \\ 0, & \text{elsewhere} \end{cases} \tag{2.1}$$

which is simply example (f) of the previous section, written slightly differently. It is worthwhile to make two observations: (i) if a periodic signal is shifted to the left or right by T or any integer multiple of T seconds, one sees graphically exactly the same signal, (ii) a periodic signal has to last from minus infinity to plus infinity in time.

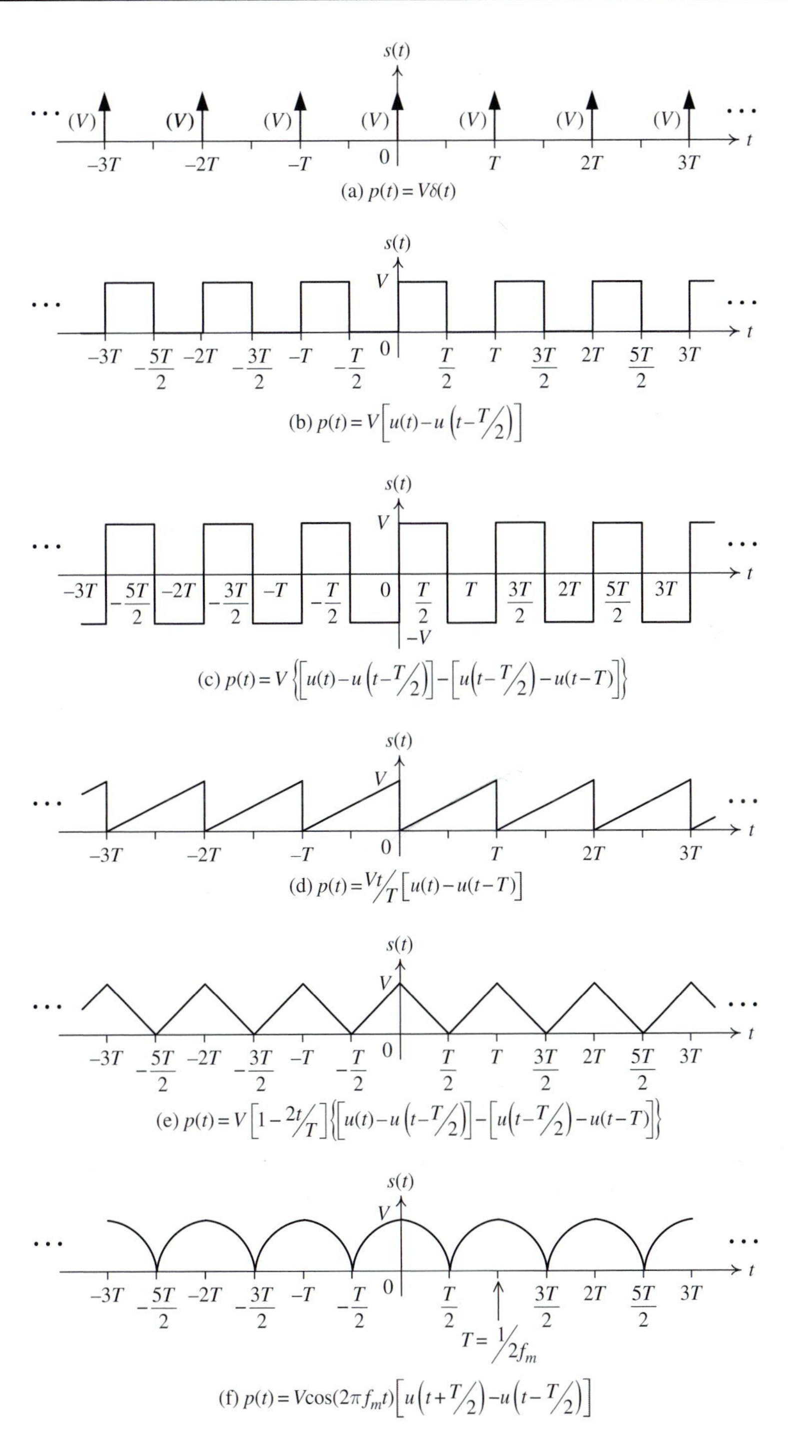

Fig. 2.2 Examples of commonly encountered periodic signals $s(t) = \sum_{k=-\infty}^{\infty} p(t - kT)$: (a) impulse train; (b) square pulse train; (c) square wave; (d) sawtooth; (e) triangular wave; (f) full-wave rectified sinusoid.

Though (2.1) describes the signal $s(t)$, indeed one could even say that it does so concisely, it turns out that when one considers the processing/filtering/generation of the signal it is more convenient to expand $s(t)$ in a series of sinusoidal functions as follows:[2]

$$s(t) = \sum_{k=-\infty}^{\infty} \left[A_k \cos(2\pi k f_r t) + B_k \sin(2\pi k f_r t)\right], \tag{2.2}$$

with $f_r = 1/T$ cycles/second or hertz.

The quantity f_r is the repetition frequency,[3] usually called the fundamental frequency, and kf_r, k a positive integer, is the kth harmonic frequency. The special case of $k = 0$ corresponds to the average or DC value of the signal.

Now elementary trigonometry establishes that

$$a\cos(x) + b\sin(x) = \sqrt{a^2+b^2}\cos\left(x - \tan^{-1}\left(\frac{b}{a}\right)\right). \tag{2.3}$$

Applying this to (2.2) yields an equivalent representation for $s(t)$ as follows:

$$\begin{aligned} s(t) &= \sum_{k=-\infty}^{\infty} \sqrt{A_k^2+B_k^2}\cos\left(2\pi k f_r t - \tan^{-1}\left(\frac{B_k}{A_k}\right)\right) \\ &= \sum_{k=-\infty}^{\infty} C_k \cos\left(2\pi k f_r t - \theta_k\right), \end{aligned} \tag{2.4}$$

where $C_k = \sqrt{A_k^2+B_k^2}$ is the amplitude of the kth harmonic with units of $s(t)$ and $\theta_k = \tan^{-1}(B_k/A_k)$ (radians) is the phase.

Yet another equivalent representation is obtained by using the complex exponential forms to express the cosine and sine functions, namely $\cos(x) = \frac{1}{2}\left(e^{jx} + e^{-jx}\right)$ and $\sin(x) = (1/2j)\left(e^{jx} - e^{-jx}\right)$. Using these identities in (2.2) and performing straightforward algebra gives:

$$s(t) = \sum_{k=0}^{\infty}\left[\left(\frac{A_k - jB_k}{2}\right)e^{j2\pi k f_r t} + \left(\frac{A_k + jB_k}{2}\right)e^{-j2\pi k f_r t}\right]. \tag{2.5}$$

Denote $(A_k - jB_k)/2$ as D_k and observe that $D_k^* = (A_k + jB_k)/2$. With this notation, (2.5) can be written as

$$s(t) = \sum_{k=0}^{\infty}\left[D_k e^{j2\pi k f_r t} + D_k^* e^{-j2\pi k f_r t}\right]. \tag{2.6}$$

Moreover, it can be shown (left as an exercise) that $D_k^* = D_{-k}$ and the final representation for $s(t)$ in terms of complex exponentials is then

$$s(t) = \sum_{k=-\infty}^{\infty} D_k e^{j2\pi k f_r t}. \tag{2.7}$$

[2] The series expansion is analogous to the Taylor/MacLaurin series expansion of a function.

[3] In signal theory it is customary to use radian frequency $\omega = 2\pi f$, with a unit of 2π(radian/cycle) (cycle/second) = radian/second. We prefer the use of f and it will be used almost exclusively in this book.

To proceed it seems advisable to establish a procedure to determine the coefficients of the series expansion. If the coefficients A_k, B_k can be found then C_k, θ_k, and D_k are known. Similarly, knowledge of D_k allows one to determine A_k, B_k or C_k, θ_k. The coefficients A_k, B_k, D_k are readily determined by using the *orthogonality* property of harmonic sinusoids; indeed that is why the frequencies of the sinusoids were chosen as they were. Orthogonality is an important enough property to warrant a more general definition and discussion.

Orthogonality Two signals $s_1(t)$ and $s_2(t)$ are said to be orthogonal over a time interval, $t \in (t_0, t_0 + T)$, of duration T if and only if

$$\int_{t_0}^{t_0+T} s_1(t)s_2(t)\mathrm{d}t = 0. \tag{2.8}$$

Note that two aspects determine whether two signals are orthogonal: (i) the shapes of the two signals and (ii) the time interval $(t_0, t_0 + T)$. Orthogonality of two signals has the connotation that the two signals are at "right angles" and indeed the mathematical operation expressed by (2.8) is directly analogous to the dot product of vectors encountered in mechanics. This geometrical implication is exploited later to give a visual interpretation of many concepts.

Going back to the problem at hand, that of determining the coefficients, straightforward integration establishes the following:

$$\text{(a)} \quad \int_{t_0}^{t_0+T} \cos(2\pi k f_r t)\cos(2\pi l f_r t)\mathrm{d}t = \begin{cases} 0, & k \neq l \\ T/2, & k = l \neq 0 \\ T, & k = l = 0 \end{cases}, \tag{2.9a}$$

$$\text{(b)} \quad \int_{t_0}^{t_0+T} \sin(2\pi k f_r t)\sin(2\pi l f_r t)\mathrm{d}t = \begin{cases} 0, & k \neq l \\ T/2, & k = l \neq 0 \\ 0, & k = l = 0 \end{cases}, \tag{2.9b}$$

$$\text{(c)} \quad \int_{t_0}^{t_0+T} \cos(2\pi k f_r t)\sin(2\pi l f_r t)\mathrm{d}t = 0, \tag{2.9c}$$

$$\text{(d)} \quad \int_{t_0}^{t_0+T} \mathrm{e}^{\mathrm{j}2\pi k f_r t}\mathrm{e}^{-\mathrm{j}2\pi l f_r t}\mathrm{d}t = \begin{cases} 0, & k \neq l \\ T, & k = l \end{cases}, \tag{2.9d}$$

where t_0 is arbitrary and $f_r = 1/T$. Therefore the sinusoids (or complex exponentials) of harmonic frequency, kf_r, are orthogonal over a time interval of T seconds.

The fact that two sinusoids of different harmonic frequencies are orthogonal over the interval $T = 1/f_r$ means that the coefficients A_k, B_k, D_k in the Fourier series expansion can be determined "individually." Simply multiply both sides of (2.2) by $(2/T)\cos(2\pi m f_r t)$ or $(2/T)\sin(2\pi m f_r t)$ and integrate over a time interval of T seconds. Since all terms on the right-hand side (RHS) of (2.2), except for the $k = m$ term, integrate to zero one has:

$$A_m = \begin{cases} \dfrac{2}{T}\displaystyle\int_{t\in T} s(t)\cos(2\pi m f_r t)\mathrm{d}t, & m = 1, 2, \cdots \\ \dfrac{1}{T}\displaystyle\int_{t\in T} s(t)\mathrm{d}t, & m = 0 \quad \text{(average or DC of } s(t)) \end{cases} \tag{2.10a}$$

$$B_m = \frac{2}{T}\int_{t\in T} s(t)\sin(2\pi m f_r t)\mathrm{d}t, \quad m = 1, 2, \ldots \tag{2.10b}$$

The factor $2/T$ (or $1/T$ for $m = 0$) is simply a normalizing constant used for convenience. Similarly, the complex coefficients D_k are determined by

$$D_m = \frac{1}{T}\int_{t\in T} s(t)\mathrm{e}^{-\mathrm{j}2\pi m f_r t}\mathrm{d}t, \quad m = 0, \pm 1, \pm 2, \ldots \tag{2.11}$$

Note that D_m needs to be calculated only for $m = 0, 1, 2, \ldots$. Then the fact that $D_{-m} = D_m^*$ is used for negative m.

Though simple enough in principle the integrations of (2.10), (2.11) and the resultant algebra can be quite tedious in practice. Some of this tedium can be alleviated, but alas never completely eliminated, by using the six properties that are presented and discussed next. Judicious application of these properties along with exploitation of signal symmetries can in some instances significantly reduce the effort. Even if the straightforward approach is taken to determine the coefficients, the properties may be used to provide a check on one's work.

2.3.2 Properties of the Fourier series

1. **Superposition** Suppose that the function, $s(t)$, is expressed as, or can be expressed as, the weighted sum of two functions, i.e.,

$$s(t) = \alpha s_1(t) + \beta s_2(t), \tag{2.12}$$

where α, β are arbitrary constants and $s_1(t)$, $s_2(t)$ are periodic functions. The following three cases are possible.

(i) The fundamental periods of $s_1(t)$ and $s_2(t)$ are the same and equal T seconds. Then $s(t)$ is periodic with fundamental period an integer multiple of T, typically T, and

$$s(t) = \sum_{k=-\infty}^{\infty}\left[\alpha D_k^{[s_1(t)]} + \beta D_k^{[s_2(t)]}\right]\mathrm{e}^{\mathrm{j}2\pi k f_r t}$$

$$\Rightarrow D_k^{[s(t)]} = \alpha D_k^{[s_1(t)]} + \beta D_k^{[s_2(t)]}, \tag{2.13}$$

where hopefully the notation $D_k^{[\cdot]}$ is self-evident.

(ii) The functions $s_1(t)$ and $s_2(t)$ are periodic with fundamental periods T_1 and T_2, respectively. Moreover, the ratio T_1/T_2 is a rational number. Then $s(t)$ is periodic with fundamental period[4] $T = \text{LCM}\{T_1, T_2\}$. The Fourier series coefficients are related again to those of $s_1(t)$, $s_2(t)$, but not as simply as in (i).

(iii) The functions $s_1(t)$ and $s_2(t)$ are periodic with fundamental periods T_1 and T_2, respectively. However, the ratio T_1/T_2 is an irrational number. Then there is no finite interval T that contains simultaneously an integer number of periods of both signals $s_1(t)$ and $s_2(t)$. The signal $s(t)$ is therefore not periodic (i.e., it is aperiodic).

2. **Change of interval** The integration to determine A_k, B_k, D_k can be performed over any time interval $(t_0, t_0 + T)$, where t_0 is arbitrary. Typically t_0 is chosen to be $-T/2$ but other choices may simplify the integration.

3. **Time scaling** Let $s(t)$ be a periodic function with fundamental period T. Define $s_1(t) = s(\gamma t)$, where $\gamma > 0$ is the time scaling factor.

At $t = 0$, $s_1(0) = s(0)$ and at $t = T/\gamma$, $s_1(T/\gamma) = s(T)$. This implies that $s_1(t)$ is periodic with fundamental period $T_1 = T/\gamma$, which means that $f_r^{[s_1(t)]} = 1/T_1 = \gamma/T = \gamma f_r$ (Hz). Note that $s_1(t/\gamma) = s(t)$. Now the coefficient $D_k^{[s_1(t)]}$ in the Fourier series expansion of $s_1(t)$ is given by

$$D_k^{[s_1(t)]} = \frac{1}{T_1}\int_{-T_1/2}^{T_1/2} s_1(t)\mathrm{e}^{-\mathrm{j}2\pi k f_r^{[s_1(t)]} t}\mathrm{d}t = \frac{\gamma}{T}\int_{-T/2\gamma}^{T/2\gamma} s_1(t)\mathrm{e}^{-\mathrm{j}2\pi k\gamma f_r t}\mathrm{d}t. \tag{2.14}$$

Changing the integration variable to $\lambda = \gamma t$ gives

$$\begin{aligned} D_k^{[s_1(t)]} &= \frac{\gamma}{T}\int_{\lambda=-T/2}^{T/2} s_1\left(\frac{\lambda}{\gamma}\right)\mathrm{e}^{-\mathrm{j}2\pi k\gamma f_r(\lambda/\gamma)}\frac{\mathrm{d}\lambda}{\gamma} \\ &= \frac{1}{T}\int_{\lambda=-T/2}^{T/2} s(\lambda)\mathrm{e}^{-\mathrm{j}2\pi k f_r \lambda}\mathrm{d}\lambda = D_k^{[s(t)]}, \end{aligned} \tag{2.15}$$

which shows that the Fourier series coefficients are unchanged in magnitude or phase. But the frequencies they are now associated with are changed from kf_r to $k\gamma f_r$.

4. **Time displacement** Consider a signal that is a time shifted version of a periodic signal, $s(t)$, i.e., $s_1(t) = s(t - \tau)$, where τ is the time shift. It is easy to see graphically that $s_1(t)$ is also periodic with the same fundamental period T. To see the relationship between the Fourier series coefficients of $s(t)$ and those of $s_1(t)$, start with $s(t) = \sum_{k=-\infty}^{\infty} D_k \mathrm{e}^{\mathrm{j}2\pi k f_r t}$. Then $s_1(t) = s(t - \tau) = \sum_{k=-\infty}^{\infty} D_k \mathrm{e}^{\mathrm{j}2\pi k f_r (t-\tau)} = \sum_{k=-\infty}^{\infty} \left[D_k \mathrm{e}^{-\mathrm{j}2\pi k f_r \tau}\right]\mathrm{e}^{\mathrm{j}2\pi k f_r t}$. But the RHS is simply the Fourier series of $s_1(t)$ and therefore the Fourier series coefficients of $s_1(t)$ are

$$D_k^{[s_1(t)]} = D_k^{[s(t)]}\mathrm{e}^{-\mathrm{j}2\pi k f_r \tau}. \tag{2.16}$$

5. **Integration** Start with $s(t)$, a periodic function with fundamental period T and define $s_1(t) = \int^t s(\lambda)\mathrm{d}\lambda$. The first observation is that if the DC value of $s(t)$ is nonzero,

[4] LCM means least common multiple. For example, $\text{LCM}\{3, 4\} = 12$.

i.e., $A_0 = C_0 = D_0 \neq 0$, then $s_1(t)$ is not periodic because it will have a linear component $D_0 t$ which is not periodic. However, if the DC value is zero then

$$\begin{aligned} s_1(t) &= \int^t s(\lambda)\mathrm{d}\lambda = \int^t \sum_{k=-\infty}^{\infty} D_k \mathrm{e}^{\mathrm{j}2\pi k f_r \lambda}\mathrm{d}\lambda \\ &= \sum_{k=-\infty}^{\infty} D_k \int^t \mathrm{e}^{\mathrm{j}2\pi k f_r \lambda}\mathrm{d}\lambda \\ &= K + \sum_{k=-\infty}^{\infty} \frac{D_k}{\mathrm{j}2\pi k f_r}\mathrm{e}^{\mathrm{j}2\pi k f_r t}, \end{aligned} \tag{2.17}$$

where K is a constant of integration. This shows that

$$D_k^{[s_1(t)]} = \begin{cases} \frac{D_k^{[s(t)]}}{\mathrm{j}2\pi k f_r}, & k = \pm 1, \pm 2, \ldots \\ K, & k = 0. \end{cases} \tag{2.18}$$

The constant of integration represents the DC value of $s_1(t)$ and depends on how the integration is performed. This will be illustrated in an example later.

6. **Differentiation** Let $s_1(t) = \mathrm{d}s(t)/\mathrm{d}t$ where, as usual, $s(t) = \sum_{k=-\infty}^{\infty} D_k \mathrm{e}^{\mathrm{j}2\pi k f_r t}$ and $f_r = 1/T$. Direct differentiation of the RHS gives $s_1(t) = \sum_{k=-\infty}^{\infty} \left[D_k(\mathrm{j}2\pi k f_r)\right] \mathrm{e}^{\mathrm{j}2\pi k f_r t}$, which simply implies that

$$D_k^{[s_1(t)]} = \mathrm{j}2\pi k f_r D_k^{[s(t)]}. \tag{2.19}$$

2.3.3 Examples of Fourier series

When one models a physical signal, one looks for a model that both captures the essence of the signal and also simplifies subsequent analysis. One consequence is that discontinuities are introduced. Perhaps the square pulse is the simplest example of this. Differentiation of discontinuities leads to the concept of the impulse or delta function. The impulse function, though strictly speaking a mathematical fiction, is of considerable importance in signal and system analysis. Before presenting the examples, the impulse function is discussed.

Impulse function The impulse function, also known as the delta function, is typically denoted by $\delta(t)$ and is a function that is zero for $t \neq 0$, infinite (i.e., mathematically undefined) at $t = 0$, and has an area under it that is unity, i.e., $\int_{-\infty}^{\infty} \delta(t)\mathrm{d}t = 1$. It is commonly defined as the limit of the square pulse shown in Figure 2.3(a). As Δ (hence one of the names) tends to zero, it is easy to see that the area of the pulse remains 1, that $s_\Delta(t)$ becomes zero for nonzero t, and that at $t = 0$ the amplitude goes to infinity which corresponds to all the characteristics ascribed to an impulse function. Therefore $\delta(t) = \lim_{\Delta \to 0} s_\Delta(t)$.

One can generalize the impulse by considering the square pulse of Figure 2.3(b). Now in the limit as $\Delta \to 0$, the following happens. The function is zero for $t \neq t_0$, its amplitude approaches infinity at $t = t_0$, and the area of the pulse is always V. The quantity V is called the strength of the impulse. Notationally, in the limit the square pulse in Figure 2.3(b) results in $V\delta(t - t_0)$.

Two parameters define an impulse function: its strength V and its location t_0. A very important property of the impulse function, particularly for our purposes, is the sifting or sampling property which states that whenever a unit ($V = 1$) impulse is "embedded" within an integrand the value of the integral is simply the value of whatever else is keeping it company in the integrand evaluated at the time the impulse occurs, i.e.,

$$\int_{-\infty}^{\infty} \delta(t - t_0)s(t)dt = s(t)\Big|_{t=t_0} = s(t_0), \tag{2.20}$$

where it is assumed $s(t)$ is continuous at $t = t_0$.

The proof relies on the characteristics of the impulse function, is quite straightforward, and is left to the reader. More specifically, it relies on the fact that the strength of the impulse when multiplied by a function is the value of the function evaluated at the location of the impulse,[5] i.e.,

$$s(t)\delta(t - t_0) = s(t_0)\delta(t - t_0). \tag{2.21}$$

Other useful relationships regarding the impulse function are:

$$\delta(t) = \delta(-t), \tag{2.22}$$

and

$$\delta(\gamma t) = \frac{1}{\gamma}\delta(t), \; \gamma > 0. \tag{2.23}$$

Having presented a number of properties and introduced the impulse function it is time to consider examples. Because of properties 2 (change of interval) and 4 (time shift), in the examples the integration period is taken to be from $-T/2$ to $T/2$ with the signal "centered" at $t = 0$ to exploit any symmetry. Only the coefficients, D_k, are determined. As mentioned, it is straightforward to obtain A_k, B_k or C_k, θ_k coefficients from D_k.

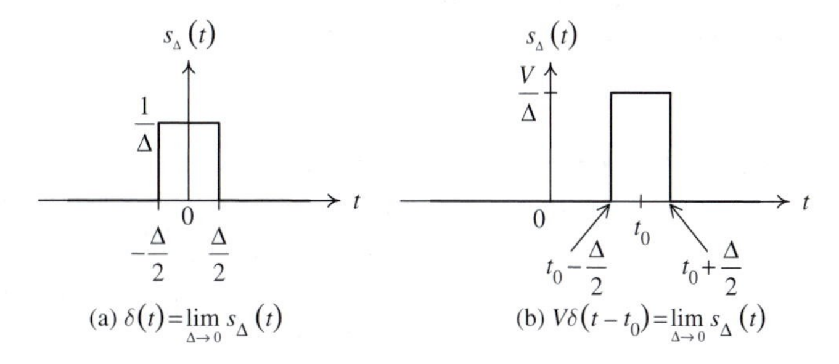

Fig. 2.3 Definition of the impulse function as the limit of a square pulse: (a) unit strength impulse at $t = 0$; (b) impulse of strength V at $t = t_0$.

[5] Therefore in terms of terminology one can say that in (2.21) the impulse samples the value of $s(t)$ whereas in (2.20) it sifts out the value of $s(t)$.

Example 2.1 impulse train Because a periodic signal with discontinuities when differentiated results in a periodic train of impulses, the first example considers the periodic train of impulses shown in Figure 2.2(a). The Fourier coefficients are

$$D_k = \frac{1}{T}\int_{-T/2}^{T/2} s(t)\mathrm{e}^{-\mathrm{j}2\pi k f_r t}\mathrm{d}t = \frac{1}{T}\int_{-T/2}^{T/2} V\delta(t)\mathrm{e}^{-\mathrm{j}2\pi k f_r t}\mathrm{d}t = \frac{V}{T}. \tag{2.24}$$

Observe that a periodic train of impulses results in a Fourier series where the coefficient of each frequency is the same. Since the average power of a sinusoid, $V\cos(2\pi f t)$, is $V^2/2$ watts, one concludes that a periodic sequence of impulses has equal power at each harmonic frequency. ■

Example 2.2 square pulse train For the square pulse train shown in Figure 2.2(b), it is straightforward to determine D_k directly as follows:

$$D_k = \frac{1}{T}\int_{-T/2}^{T/2} s(t)\mathrm{e}^{-\mathrm{j}2\pi k f_r t}\mathrm{d}t = \frac{V}{T}\int_{0}^{T/2} \mathrm{e}^{-\mathrm{j}2\pi \frac{k}{T} t}\mathrm{d}t = \frac{V}{\mathrm{j}2\pi k}\left(1 - \mathrm{e}^{-\mathrm{j}\pi k}\right). \tag{2.25}$$

Since $\mathrm{e}^{-\mathrm{j}\pi k} = 1$ when k is even and $\mathrm{e}^{-\mathrm{j}\pi k} = -1$ when k is odd, one has

$$D_k = \begin{cases} V/\mathrm{j}\pi k, & k \text{ odd} \\ 0, & k \text{ even} \\ V/2, & k = 0 \text{ (i.e., } D_0) \end{cases}. \tag{2.26}$$

Note that to determine D_0 by setting $k = 0$ in the general expression for D_k results in $0/0$ and one must resort to l'Hospitale's rule to determine the value of this indeterminate expression. However, recall that D_0 is the average or DC value of the signal over T seconds. Therefore D_0 = area under the pulse/period $= V(T/2)/T = V/2$. ■

Example 2.3 square wave For the square wave in Figure 2.2(c), again though it is straightforward enough to determine $D_k = D_k^{[s(t)]}$ directly, to illustrate the use of the properties "decompose" $s(t)$ as shown in Figure 2.4 into $s_1(t)$ and $s_2(t)$. Since $s(t) = s_1(t) + s_2(t)$, it follows from the superposition property that $D_k^{[s(t)]} = D_k^{[s_1(t)]} + D_k^{[s_2(t)]}$. Note that $D_k^{[s_1(t)]}$ has been already determined in the last example. Furthermore, $s_2(t) = -s_1(t - T/2)$. Applying the amplitude scaling and time shifting ($\tau = T/2$) properties yields $D_k^{[s_2(t)]} = -D_k^{[s_1(t)]}\mathrm{e}^{-\mathrm{j}\pi k}$. Finally,

$$\begin{aligned} D_k^{[s(t)]} &= D_k^{[s_1(t)]} + D_k^{[s_2(t)]} = D_k^{[s_1(t)]}\left(1 - \mathrm{e}^{-\mathrm{j}k\pi}\right) \\ &= \begin{cases} 2V/\mathrm{j}\pi k, & k \text{ odd} \\ 0, & k \text{ even} \end{cases}. \end{aligned} \tag{2.27}$$

Note that $D_0^{[s(t)]} = 0$ as is easily seen from the graph of $s(t)$. ■

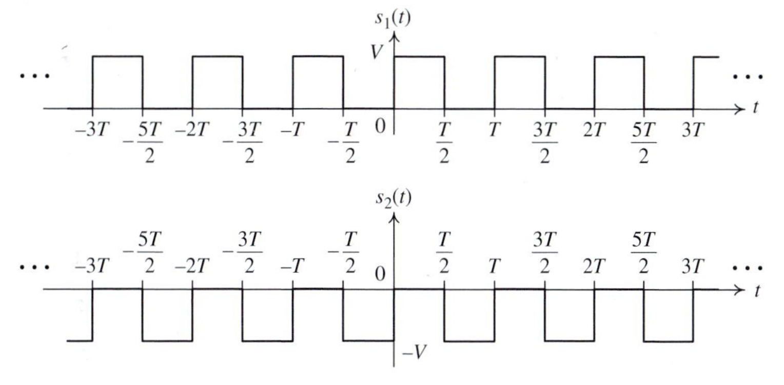

Fig. 2.4 Representing the square wave of Figure 2.2(c) as the sum of $s_1(t)$ and $s_2(t)$.

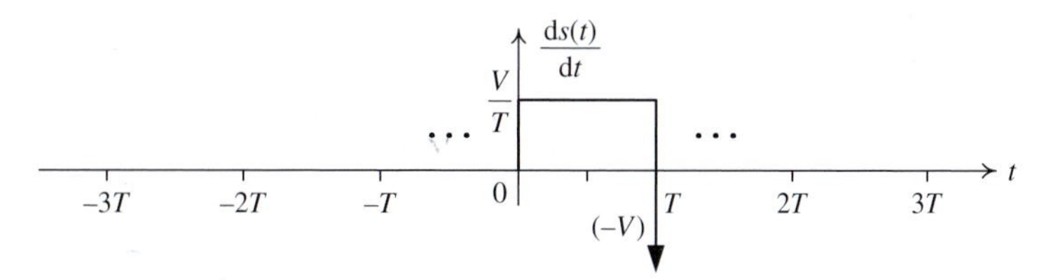

Fig. 2.5 Derivative of the sawtooth wave of Figure 2.2(d) over interval 0^+ to T^+.

Example 2.4 sawtooth wave Consider the sawtooth wave in Figure 2.2(d). The DC value is easily determined to be $D_0^{[s(t)]} = V/2$. To determine $D_k^{[s(t)]}$ for $k \neq 0$, differentiate $s(t)$ to obtain the waveform $s_1(t)$ shown in Figure 2.5.

From the differentiation property one has $D_k^{[s_1(t)]} = (\mathrm{j}2\pi k/T)D_k^{[s(t)]}$ or $D_k^{[s(t)]} = D_k^{[s_1(t)]}/(\mathrm{j}2\pi k/T)$. Next, determine $D_k^{[s_1(t)]}$ by direct integration over the interval 0^+ to T^+ as follows:

$$\begin{aligned} D_k^{[s_1(t)]} &= \frac{1}{T}\int_{0^+}^{T^+} \frac{V}{T}\mathrm{e}^{-\mathrm{j}2\pi\frac{k}{T}t}\mathrm{d}t + \frac{1}{T}\int_{0^+}^{T^+} -V\delta(t-T)\mathrm{e}^{-\mathrm{j}2\pi\frac{k}{T}t}\mathrm{d}t \\ &= \frac{V}{T}\frac{1-\mathrm{e}^{-\mathrm{j}2\pi k}}{\mathrm{j}2\pi k} - \frac{V}{T}\mathrm{e}^{-\mathrm{j}2\pi k} = -\frac{V}{T}. \end{aligned} \tag{2.28}$$

Therefore

$$D_k^{[s(t)]} = \frac{D_k^{[s_1(t)]}}{\mathrm{j}2\pi k/T} = \begin{cases} -V/\mathrm{j}2\pi k, & k \neq 0 \\ V/2, & k = 0 \end{cases}. \tag{2.29}$$

■

Example 2.5 triangular wave For the triangular wave in Figure 2.2(e), differentiation of $s(t)$ results in the square wave of Figure 2.6, which is the square wave considered in Example 2.3 scaled by $-2/T$. As seen in the previous example the DC term vanishes due to the differentiation. But $D_0^{[s(t)]}$ is easily seen to be $(VT/2)/T = V/2$. Next, to apply the

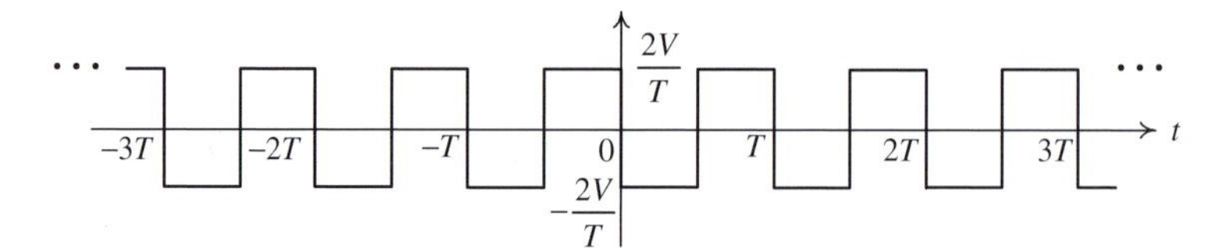

Fig. 2.6 Derivative of the triangular wave of Figure 2.2(e).

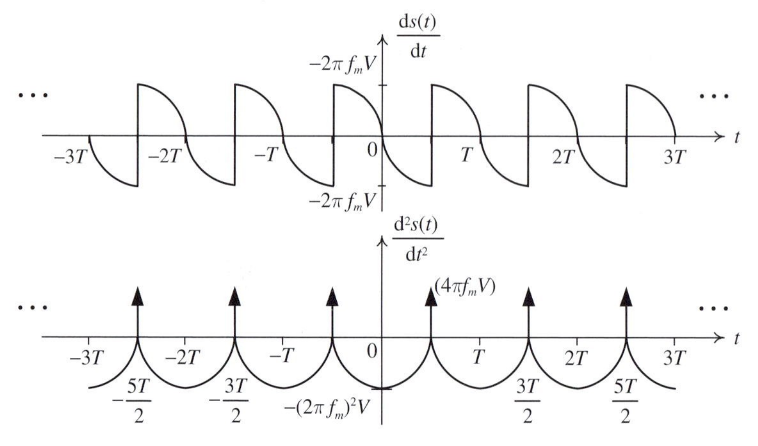

Fig. 2.7 First and second derivatives of the full-wave rectified sinusoid of Figure 2.2(f).

differentiation property $D_k^{[s(t)]} = D_k^{[s_1(t)]}/(\mathrm{j}2\pi k/T)$, determine $D_k^{[s_1(t)]}$ by applying amplitude scaling to the results of Example 2.3 to obtain $D_k^{[s_1(t)]} = (-2/T)\,(2V/\mathrm{j}\pi k)$ for k odd, and $D_k^{[s_1(t)]} = 0$ for k even. Therefore

$$D_k^{[s(t)]} = \begin{cases} 2V/\pi^2 k^2, & k \text{ odd} \\ 0, & k \text{ even}, k \neq 0 \, . \\ V/2, & k = 0 \end{cases} \tag{2.30}$$

■

Example 2.6 rectified sinusoid The rectified sinusoid is shown in Figure 2.2(f). After two differentiations (see Figure 2.7) one obtains the following signal in the interval $-T^+/2$ to $T^+/2$:

$$\begin{aligned} \frac{\mathrm{d}^2 s(t)}{\mathrm{d}t^2} &= -(2\pi f_m)^2 V \cos(2\pi f_m t) + 4\pi f_m V \delta\left(t - \frac{T}{2}\right) \\ &= -(2\pi f_m)^2 s(t) + 4\pi f_m V \delta\left(t - \frac{T}{2}\right). \end{aligned} \tag{2.31}$$

Equivalently,

$$\frac{\mathrm{d}^2 s(t)}{\mathrm{d}t^2} + (2\pi f_m)^2 s(t) = 4\pi f_m V\delta\left(t - \frac{T}{2}\right), \tag{2.32}$$

which implies that

$$(\mathrm{j}2\pi k f_r)^2 D_k^{[s(t)]} + (2\pi f_m)^2 D_k^{[s(t)]} = 4\pi f_m V D_k^{[\delta(t-T/2)]} \tag{2.33}$$

where $f_r = 2f_m = 1/T$. Now

$$D_k^{[\delta(t-T/2)]} = \frac{1}{T}\int_{-T^+/2}^{T^+/2} \delta\left(t - \frac{T}{2}\right) \mathrm{e}^{-\mathrm{j}2\pi \frac{k}{T} t} \mathrm{d}t = \frac{\mathrm{e}^{-\mathrm{j}k\pi}}{T} = \frac{(-1)^k}{T}. \tag{2.34}$$

Therefore,

$$\begin{aligned} \left[-4\pi^2 k^2 f_r^2 + \pi^2 f_r^2\right] D_k^{[s(t)]} &= 2\pi f_r^2 V(-1)^k \\ \Rightarrow D_k^{[s(t)]} &= \frac{2V(-1)^k}{\pi(1-4k^2)}, \ \forall k. \end{aligned} \tag{2.35}$$

■

Example 2.7 integration The last example to be considered illustrates the integration property. Let the signal of Figure 2.2(c) be passed through an integrator as shown in Figure 2.8, where $1/T$ is a scaling factor which makes the output amplitude independent of T.

Using the results of Example 2.3 and the integration property, one has

$$D_k^{[s_0(t)]} = \frac{1}{T}\frac{D_k^{[s(t)]}}{\mathrm{j}2\pi k/T} = \frac{2V}{(\mathrm{j}2\pi k)(\mathrm{j}\pi k)} = \begin{cases} -V/\pi^2 k^2, & k \text{ odd} \\ 0, & k \text{ even}, k \neq 0 \end{cases}. \tag{2.36}$$

The DC value, $D_0^{[s_0(t)]}$, depends on when we visualize the integrator as having "started." This is shown in Figure 2.9, where the input signal is shown along with the output signal for three different starting times. Note that one should visualize the integrator as acting both before and after the starting time. Clearly, $D_0^{[s_0(t)]} = -V/4$ for t_1, $D_0^{[s_0(t)]} = 0$ for t_2, and $D_0^{[s_0(t)]} = V/4$ for t_3. ■

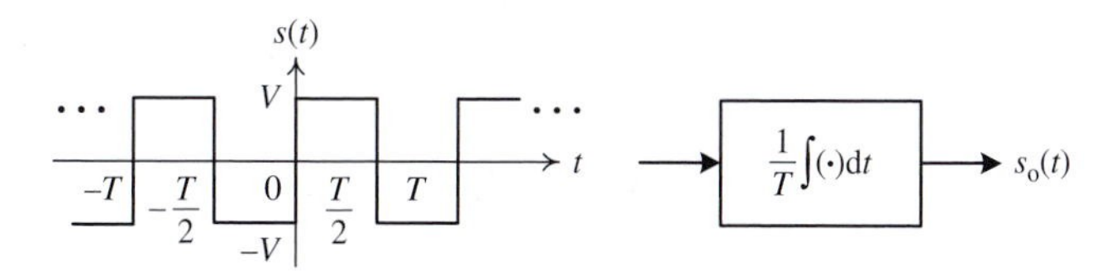

Fig. 2.8 Integration of a periodic signal.

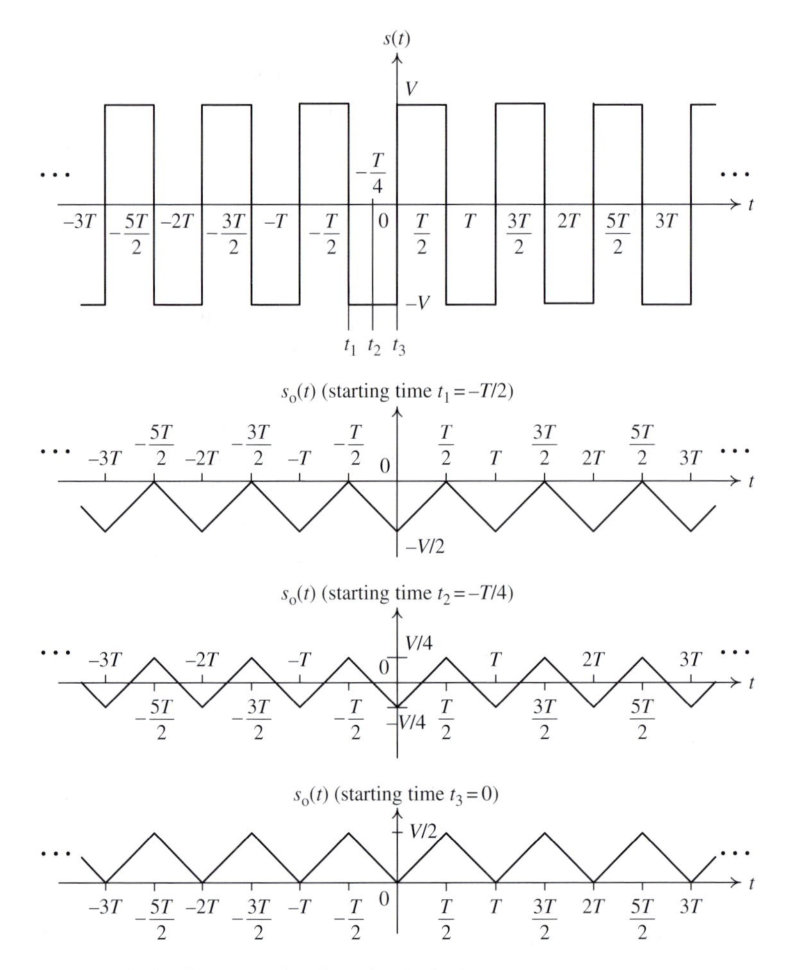

Fig. 2.9 Effect on output signal of different starting times for the integrator.

2.3.4 Discussion of examples

Having determined the Fourier series for a number of periodic signals several statements can be made from the obtained coefficients. Coefficients D_k for the square wave of Figure 2.2(c) have only an imaginary component, which means that $A_k = 0$, or that only sine terms are present in the signal. Note that $s(t)$ is an odd function about $t = 0$, i.e., $s(t) = -s(-t)$ which implies that only the harmonics $\sin(2\pi k f_r t)$ (which are odd time functions) are needed to represent it. This is true in general: if (and only if) $s(t)$ is an odd periodic function, then the real component of D_k is zero, i.e., $A_k = 0$ or $\theta_k = \pm\pi/2$. Though the signal of Figure 2.2(b) is not odd, the coefficients have $A_k = 0$ for all nonzero k, which appears to contradict what has just been said. However, once the DC value of $V/2$ is subtracted, one obtains an odd function. The same remark applies to the signal of Figure 2.2(d).

The signals of Figures 2.2(e), 2.2(f) are even functions about $t = 0$, i.e., $s(t) = s(-t)$. The Fourier coefficients, D_k, have only real components since $B_k = 0$, implying that only the harmonics $\cos(2\pi k f_r t)$ (even time functions) are needed for the representation. Again this is true in general: if (and only if) $s(t)$ is an even periodic function, then the imaginary component of D_k is zero, i.e., $B_k = 0$ for all k or $\theta_k = 0$ (or π if A_k is negative).

The signals of Examples 2.2, 2.3, 2.5 have nonzero harmonics only for odd k. This is not happenstance. It is because these signals possess what is called half-wave symmetry. If one shifts a signal by $\pm T/2$ seconds and upon flipping it over obtains the same signal, then the signal has half-wave symmetry, i.e., $s(t) = -s(t \pm T/2)$. In general, if a periodic signal has half-wave symmetry (after the DC value is subtracted out) it will have only nonzero odd harmonics., i.e., excluding D_0, $D_k = 0$ for all even k. It should be pointed out that $s(t)$ can exhibit half-wave symmetry and be neither an odd nor even function. If in addition to having half-wave symmetry, $s(t)$ is also even or odd, then it is said to exhibit quarter-wave symmetry, even or odd quarter-wave symmetry, respectively.

Finally consider the behavior of the coefficients with k keeping in mind that k indexes the frequency of the harmonics. In particular, consider the magnitude, $|D_k|$. The harmonics of the first example have the same magnitude at all k. The second, third, and fourth examples have harmonics whose magnitudes decay as $1/k$, while those of the fifth and sixth examples decay as $1/k^2$, considerably faster. The behavior of the harmonic magnitudes is related directly to how many times $s(t)$ can be differentiated before an impulse (or equivalently a discontinuity) first appears. The rule is that if an impulse first appears in $\mathrm{d}^n s(t)/\mathrm{d}t^n$ then $|D_k| \propto 1/k^n$.

2.3.5 Frequency spectrum of periodic signals

The expansion of a periodic signal into an infinite series of orthonormal sinusoidal functions is done to obtain a representation in the frequency domain. At each harmonic frequency the signal has a magnitude and phase which can be obtained from one of the sets $\{A_k, B_k\}$, $\{C_k, \theta_k\}$, or $\{D_k\}$. It is customary to plot $|D_k|$, $\angle D_k$, $k = 0, \pm 1, \pm 2, \ldots$ to obtain a two-sided spectrum where negative ks reflect the $\mathrm{e}^{-\mathrm{j}2\pi k f_r t}$ terms and are called negative frequencies. The same information is presented by $\{C_k, \theta_k\}$ except that since $k = 0, 1, 2, \ldots$, only the positive frequency axis is needed. Regarding C_k, a distinction should be made between C_k, called the *amplitude*, and the *magnitude* of C_k, namely $|C_k|$. The coefficient C_k is always a real number (for a real signal, $s(t)$), but it could be positive or negative. If negative, then in addition to the phase θ_k the harmonic has a further phase of π radians which should be taken into account if plots of $\{|C_k|, \theta_k\}$ versus kf_r are presented. Finally note that $|C_k| = 2|D_k|$ and $\theta_k = \angle D_k$ so that only plots of $\{|D_k|, \angle D_k\}$ need to be considered and that is what is done here.

Figure 2.10 shows the magnitude and phase plots for Examples 2.3, 2.4, 2.5. Because a periodic signal contains only discrete frequency components it is logical to call this a discrete frequency spectrum. Though the discrete frequency spectrum gives an alternative representation of a periodic signal, the motivation for using this representation still needs to be addressed. There are at least three primary motivations: (i) to synthesize the signal, (ii) to filter the signals, which typically involves the filtering of undesired signal components (usually called noise) while leaving the desired signal as undistorted as possible,

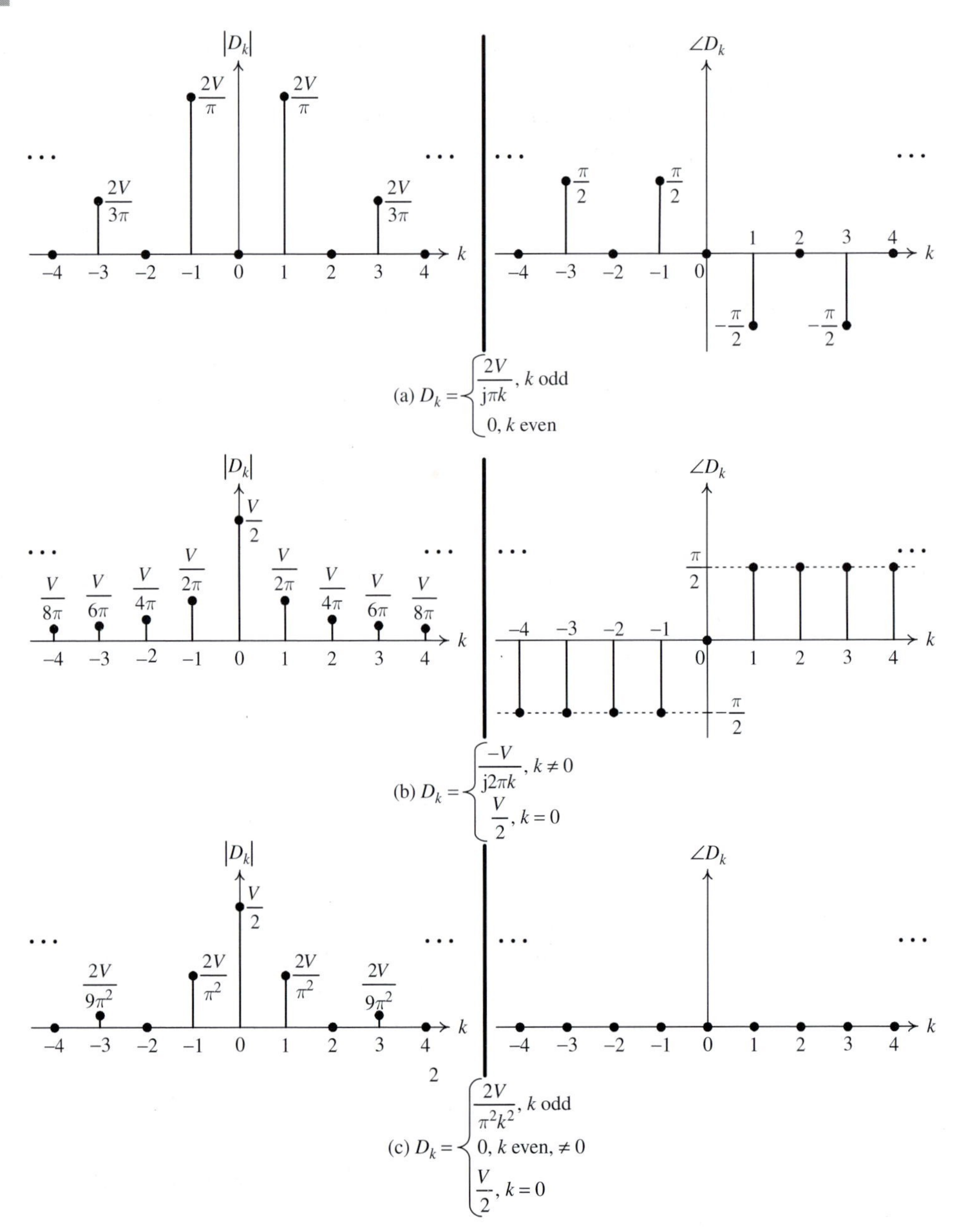

Fig. 2.10 Magnitude and phase plots for: (a) Example 2.3, (b) Example 2.4, and (c) Example 2.5. Note that the axes' units are: (i) for the vertical axis $|D_k|$ has the unit of the signal, typically volts, $\angle D_k$ has the unit of radians, (ii) the horizontal axis is a normalized frequency, $k = f/f_r$, where f_r is the fundamental frequency.

(iii) to determine the bandwidth required to pass the signal through a channel relatively undistorted.

All of the above require a measure of performance and the typical measure chosen is average power. The average power of a periodic signal is[6]

$$P_{\text{av}} = \frac{1}{T}\int_{-T/2}^{T/2} |s(t)|^2 \text{d}t = \frac{1}{T}\int_{-T/2}^{T/2} s(t)s^*(t)\text{d}t \quad \text{(watts)}. \tag{2.37}$$

In terms of the Fourier series representation, it can be calculated as

$$\begin{aligned} P_{\text{av}} &= \frac{1}{T}\int_{-T/2}^{T/2} \left[\sum_{k=-\infty}^{\infty} D_k \text{e}^{\text{j}2\pi k f_r t}\right]\left[\sum_{l=-\infty}^{\infty} D_l \text{e}^{\text{j}2\pi l f_r t}\right]^* \text{d}t \\ &= \sum_{k=-\infty}^{\infty}\sum_{l=-\infty}^{\infty} D_k D_l^* \frac{1}{T}\int_{-T/2}^{T/2} \text{e}^{\text{j}2\pi(k-l)f_r t}\text{d}t. \end{aligned} \tag{2.38}$$

Using the orthogonality property of the complex exponential over the period T reduces (2.38) to

$$P_{\text{av}} = \sum_{k=-\infty}^{\infty} |D_k|^2 = |D_0|^2 + 2\sum_{k=1}^{\infty} |D_k|^2. \tag{2.39}$$

The above shows that

$$\frac{1}{T}\int_{-T/2}^{T/2} |s(t)|^2 \text{d}t = \sum_{k=-\infty}^{\infty} |D_k|^2, \tag{2.40}$$

a relationship known as *Parseval's theorem* (in this case for periodic signals).

To judge the quality of a synthesized signal or the required bandwidth needed to pass the signal relatively undistorted one can ask how many terms are needed to "capture" a certain percentage of the signal's average power, i.e., determine N such that:

$$\begin{aligned} \%\text{ power captured} &= \frac{P_{\text{av}}^{[\text{captured}]}}{P_{\text{av}}} \times 100 \\ &= \frac{|D_0|^2 + 2\sum_{k=1}^{N} |D_k|^2}{P_{\text{av}}} \times 100. \end{aligned} \tag{2.41}$$

Analytical expressions are not available for finite sums of the form $\sum_{k=1}^{N} 1/k^n$, but these are readily programmed to obtain plots such as those shown in Figure 2.11 for Examples 2.3–2.6. Table 2.1 lists N, the number of harmonics, needed to capture the specified amount of signal power. From this one can judge the bandwidth

[6] Though $s(t)$ is a real-time signal, its Fourier series representation in terms of D_k is a complex-time signal. Hence the complex conjugate. One also encounters complex-time functions when what is called the equivalent baseband model of a communication system is used.

Table 2.1 Number of harmonics needed to capture the specified amount of signal power

Signal	% captured power			
	90	95	98	99
(a) Square wave	$N = 3$	$N = 9$	$N = 21$	$N = 41$
(b) Sawtooth wave	$N = 1$	$N = 3$	$N = 8$	$N = 15$
(c) Triangular wave	$N = 1$	$N = 1$	$N = 1$	$N = 1$
(d) Rectified sinusoid	$N = 1$	$N = 1$	$N = 1$	$N = 1$

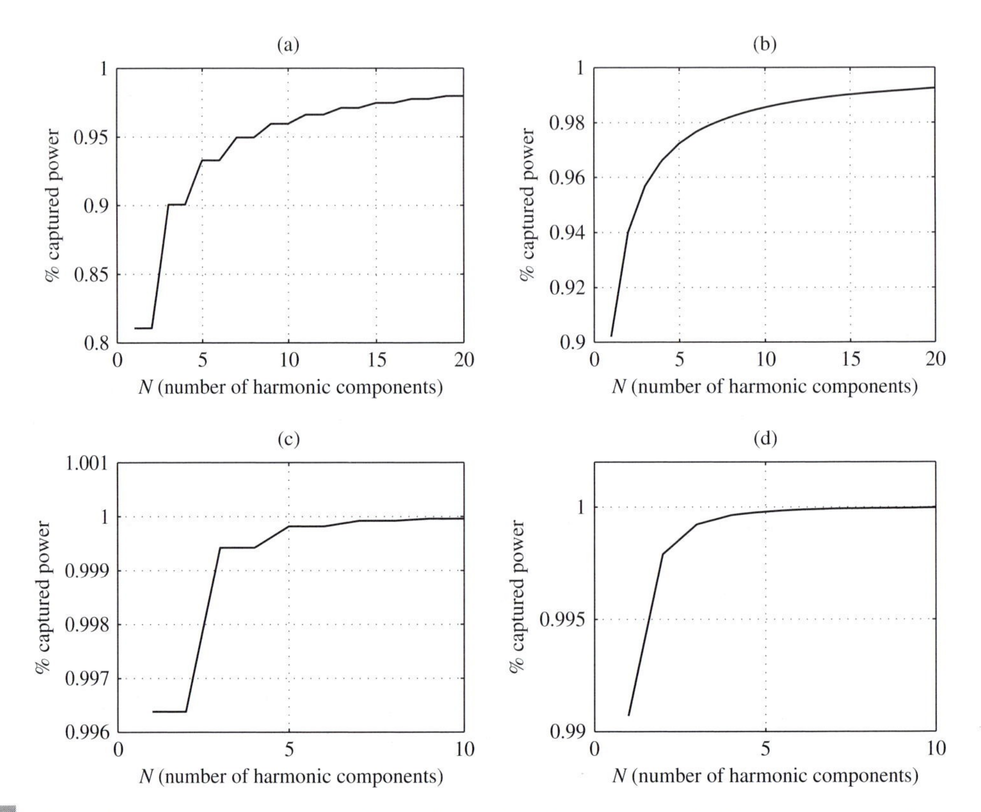

Fig. 2.11 Plots of the percentage power captured for: (a) Example 2.3 (square wave); (b) Example 2.4 (sawtooth wave); (c) Example 2.5 (triangular wave); (d) Example 2.6 (rectified sinusoid).

needed to pass the signal relatively undistorted or the number of terms needed to synthesize the signal. Figures 2.12 and 2.13 compare graphically the synthesized signals using N terms with the actual sawtooth wave and the rectified sinusoid, respectively.

To illustrate the last application of Fourier series, that of filter design, consider the design of a simple power supply circuit shown in Figure 2.14. The full-wave rectifier, under the

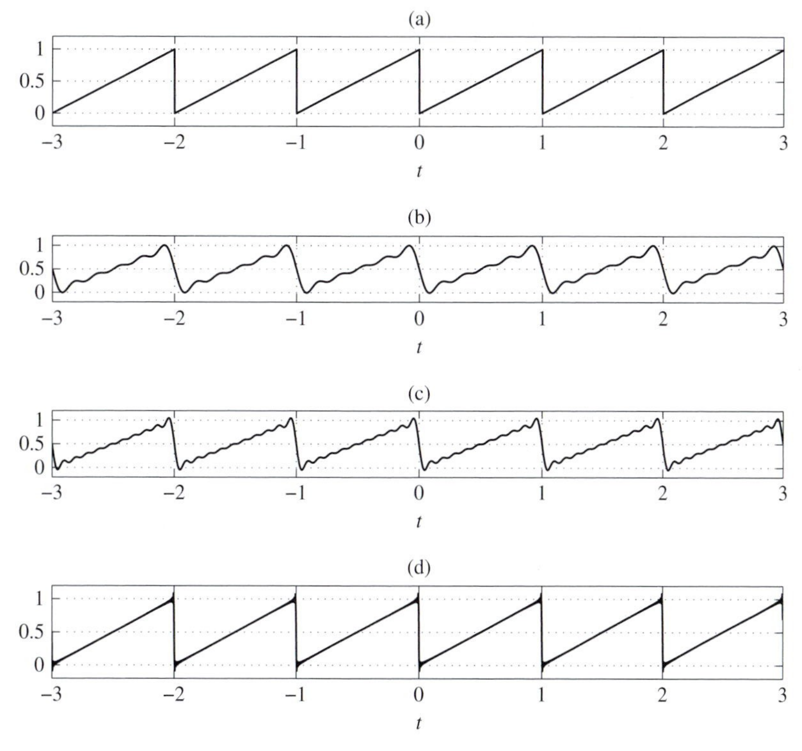

Fig. 2.12 Comparison of synthesized signals and the actual sawtooth wave: (a) actual sawtooth wave, (b) synthesized sawtooth wave $N = 5$; (c) synthesized sawtooth wave $N = 10$; (d) synthesized sawtooth wave $N = 100$.

condition that the current always flows in the output, produces a periodic signal $v_{\text{in}}(t)$ which is that of Example 2.6. The useful component of this signal is the DC value, D_0, while the AC portion of the signal is a ripple which is undesirable. The function of the ripple filter is to filter out this ripple component while passing the DC component undistorted.

In the sinusoidal steady state the output $v_{\text{out}}(t)$, is also a periodic signal with the same period as the input. However, the magnitude and phase of each harmonic is modified by the transfer function, $H(f)$, of the ripple filter/load circuit. From elementary circuit theory one has

$$H(f) = \frac{R_L}{(R_L - 4\pi^2 f^2 R_L LC) + \mathrm{j}2\pi f L} = \frac{1}{(1 - 4\pi^2 f^2 LC) + \mathrm{j}2\pi f L/R_L}. \tag{2.42}$$

The output coefficients are given by:

$$\begin{aligned} D_k^{[v_{\text{out}}(t)]} &= H(f)\bigg|_{f=kf_r} \times D_k^{[v_{\text{in}}(t)]} \\ &= \frac{1}{(1 - 4\pi^2 k^2 f_r^2 LC) + \mathrm{j}2\pi k f_r (L/R_L)} \frac{2V(-1)^k}{\pi(1 - 4k^2)}. \end{aligned} \tag{2.43}$$

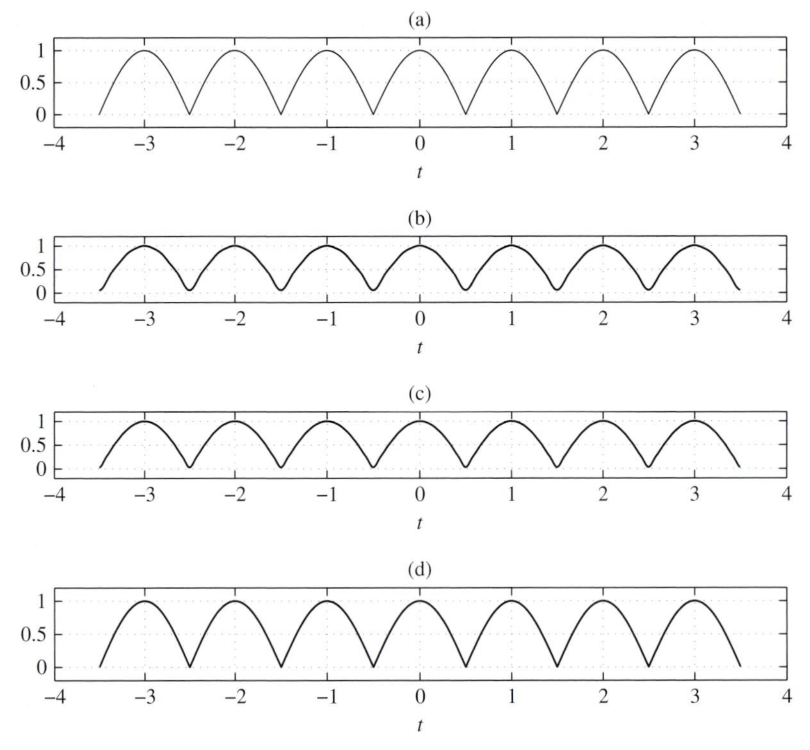

Fig. 2.13 Comparison of synthesized signals and the actual rectified sinusoid: (a) actual rectified sinusoid; (b) synthesized rectified sinusoid $N = 5$; (c) synthesized rectified sinusoid $N = 10$; (d) synthesized rectified sinusoid $N = 100$.

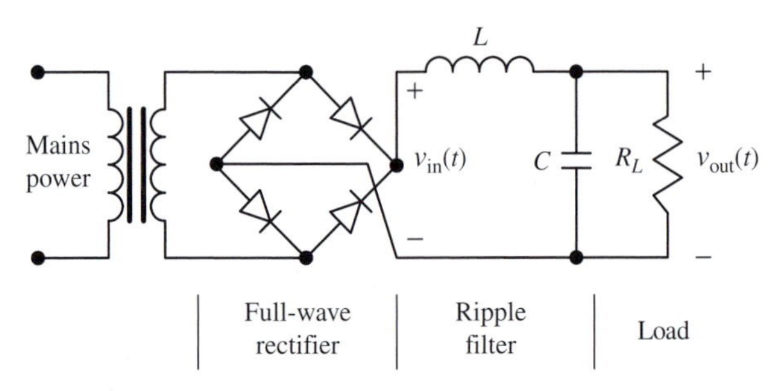

Fig. 2.14 Power supply – ripple filter design.

An important parameter in power supply design is the ripple factor, which is defined as:

$$\text{ripple factor} = \frac{\text{RMS value of the AC signal components}}{\text{DC value of the signal}}$$
$$= \frac{\sqrt{2\sum_{k=1}^{\infty} |D_k|^2}}{D_0}, \tag{2.44}$$

where RMS means root-mean-squared. For the circuit shown the ripple factor is:

$$\begin{aligned}
\text{ripple factor} &= \frac{\sqrt{2\sum_{k=1}^{\infty} |H(kf_r)|^2 \left|D_k^{[v_{\text{in}}(t)]}\right|^2}}{\left|D_0^{[v_{\text{in}}(t)]}\right|} \\
&= \frac{\sqrt{2\sum_{k=1}^{\infty} \frac{1}{(1-4\pi^2k^2f_r^2LC)^2 + (2\pi k f_r(L/R_L))^2} \frac{4V^2}{\pi^2(1-4k^2)^2}}}{2V/\pi} \\
&= \sqrt{2}\sqrt{\sum_{k=1}^{\infty} \frac{1}{(1-4\pi^2k^2f_r^2LC)^2 + (2\pi k f_r L/R_L)^2}} \\
&\quad \times \sqrt{\frac{1}{(1-4k^2)^2}}.
\end{aligned} \tag{2.45}$$

The above is an exact expression for the ripple factor but it does not readily yield any insight for design. Recall, however, that most of the AC power is in the first harmonic ($k = 1$). Therefore only the first term is retained in the sum, which gives

$$\begin{aligned}
\text{ripple factor} &= \sqrt{2}\sqrt{\frac{1}{(1-4\pi^2f_r^2LC)^2 + (2\pi f_r L/R_L)^2}}\sqrt{\frac{1}{9}} \\
&\approx \frac{\sqrt{2}}{3}\frac{1}{\left|1-4\pi^2f_r^2LC\right|}.
\end{aligned} \tag{2.46}$$

Lastly a bit of thought convinces one that if the power supply is to be well designed the ripple factor should be $\ll 1$. This implies that $4\pi^2f_r^2LC$ should be $\gg 1$ and therefore

$$\text{ripple factor} \approx \frac{\sqrt{2}}{3}\frac{1}{4\pi^2f_r^2LC}. \tag{2.47}$$

Using a mains frequency of 60 hertz, $f_r = 120$ hertz, the ripple factor expression is

$$\text{ripple factor} \approx \frac{0.83 \times 10^{-6}}{LC}, \tag{2.48}$$

or

$$\text{ripple factor} \approx \frac{0.83}{LC}, \tag{2.49}$$

where C is in microfaradays and L is in henries.

2.3.6 Fourier series of a product of two signals

To complete the discussion of periodic signals and their Fourier series representation we consider one last property. Namely, what is the Fourier series of a signal that is the product of two periodic signals with the same period.

To this end, let $s(t) = s_1(t)s_2(t)$, where signals $s_1(t)$ and $s_2(t)$ are periodic with the same fundamental period T or $f_r = 1/T$. Substituting in the Fourier series representations of $s_1(t), s_2(t)$, one has

$$s(t) = \sum_{k=-\infty}^{\infty} D_k^{[s_1(t)]} e^{j2\pi k f_r t} \sum_{l=-\infty}^{\infty} D_l^{[s_2(t)]} e^{j2\pi l f_r t}. \tag{2.50}$$

Consider the second sum. By letting the dummy index variable l be $-l$ it can be written as $\sum_{l=-\infty}^{\infty} D_{-l}^{[s_2(t)]} e^{-j2\pi l f_r t}$. The Fourier series representation of $s(t)$ is now

$$s(t) = \sum_{k=-\infty}^{\infty} \left[\sum_{l=-\infty}^{\infty} D_k^{[s_1(t)]} D_{-l}^{[s_2(t)]} e^{j2\pi(k-l) f_r t} \right]. \tag{2.51}$$

Let $m = k - l$, which implies that $l = k - m$. Note that index variable m also ranges from $-\infty$ to ∞. Therefore

$$\begin{aligned} s(t) &= \sum_{k=-\infty}^{\infty} \left[\sum_{m=-\infty}^{\infty} D_k^{[s_1(t)]} D_{m-k}^{[s_2(t)]} e^{j2\pi m f_r t} \right] \\ &= \sum_{m=-\infty}^{\infty} \left[\sum_{k=-\infty}^{\infty} D_k^{[s_1(t)]} D_{m-k}^{[s_2(t)]} \right] e^{j2\pi m f_r t}, \end{aligned} \tag{2.52}$$

where the second equality follows by simply interchanging the order of summation. But the RHS of the last equality is simply the Fourier series representation of $s(t)$ where the coefficient $D_m^{[s(t)]}$ is given in terms of the Fourier series coefficients of $s_1(t)$ and $s_2(t)$ by

$$D_m^{[s(t)]} = \sum_{k=-\infty}^{\infty} D_k^{[s_1(t)]} D_{m-k}^{[s_2(t)]} = \sum_{k=-\infty}^{\infty} D_{m-k}^{[s_1(t)]} D_m^{[s_2(t)]}. \tag{2.53}$$

The above mathematical operation, which in this case arose to determine the coefficients $D_m^{[s(t)]}$, is a very important one in linear system theory and signal processing. It is called *convolution*. It is denoted usually by the symbol $*$ and Equation (2.53) is written as $D_k^{[s(t)]} = D_k^{[s_1(t)]} * D_k^{[s_2(t)]}$. The convolution operation has a graphical interpretation that aids in understanding the mathematical operation. Visualize the coefficients $D_k^{[s_1(t)]}$ and $D_k^{[s_2(t)]}$ plotted on two separate graphs.[7] Flip one set of coefficients, say $D_k^{[s_2(t)]}$, about the vertical axis to produce $D_{-k}^{[s_2(t)]}$. Then slide the coefficients $D_{-k}^{[s_2(t)]}$ to the left or right by the value of index m to get $D_{m-k}^{[s_2(t)]}$. Multiply the set $\{D_k^{[s_1(t)]}\}$ by the set $\{D_{m-k}^{[s_2(t)]}\}$ point by point and sum the resultant products to obtain $D_m^{[s(t)]}$. Repeat for another m. Figure 2.15 illustrates the procedure with a simple toy example.

Since convolution is a mathematical operation that tends to cause some difficulty when first encountered a comment is in order. The graphical interpretation aids one in doing the algebra. The algebra in this case, as shown, is complex number multiplication and addition, no more, no less. Observe that a positive m means a shift of D_{-k+m} to the right

[7] Some imagination is required here since in general the coefficients are complex numbers.

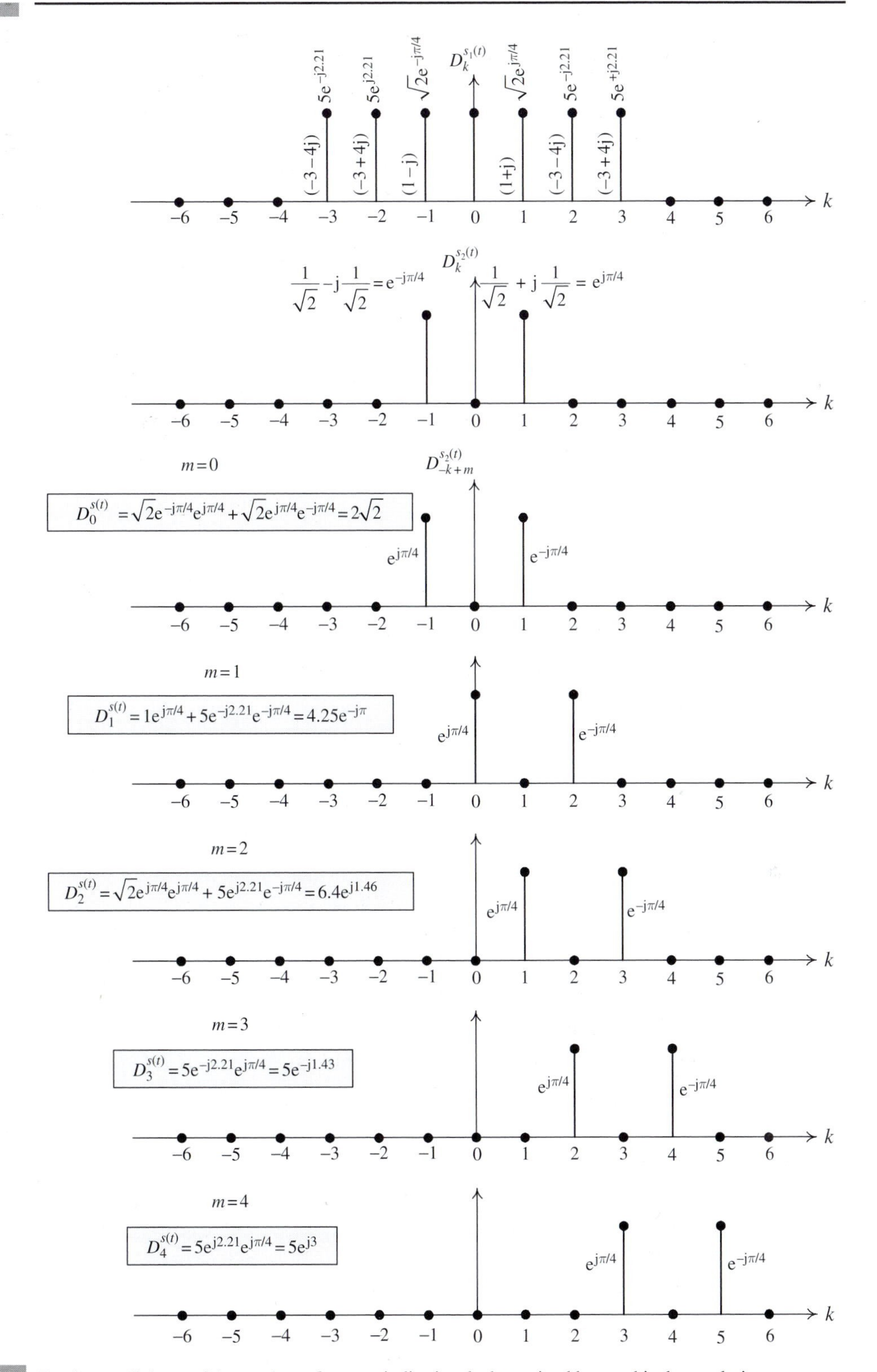

Fig. 2.15 Fourier coefficients of the product of two periodic signals determined by graphical convolution.

by m units. This is a consequence of the fact that we are dealing with D_{-k}, that is the flipped version of D_k.

Lastly an important engineering observation is that $s(t)$ has frequency components not present in either $s_1(t)$ or $s_2(t)$. This is because multiplication is a nonlinear operation and passing signals through nonlinearities such as squarers, rectifiers, saturation, etc., creates frequencies that were not present in the input signal. This is in contrast to a linear operation (or filter), which never creates new frequencies, it only modifies the magnitude/phase of the frequencies in the input signal.

2.4 Nonperiodic signals

2.4.1 Derivation of the Fourier transform representation of a nonperiodic signal

Attention is turned now to nonperiodic signals, specifically signals of finite energy where

$$\int_{-\infty}^{\infty} |s(t)|^2 \mathrm{d}t < \infty. \tag{2.54}$$

As with periodic signals what is desired is a frequency domain representation of the nonperiodic signal. To obtain this representation we start with the periodic signal, $s(t)$, shown in Figure 2.16(a). Keep the shape of the basic pulse, $p(t)$, unchanged and increase the period to $T_1 = \alpha T$, $\alpha > 1$ as shown in Figure 2.16(b). As T_1 or α approaches ∞ the periodic signal becomes nonperiodic, as shown in Figure 2.16(c). Consider now the Fourier series coefficients for the signals of Figure 2.16, note that $f_r^{[s_1(t)]} = 1/\alpha T = f_r/\alpha$:

$$D_k^{[s(t)]} = \frac{1}{T} \int_{-\frac{T}{2}}^{\frac{T}{2}} p(t) \mathrm{e}^{-\mathrm{j}2\pi k f_r t} \mathrm{d}t, \tag{2.55a}$$

$$\begin{aligned} D_k^{[s_1(t)]} &= \frac{1}{\alpha T} \int_{-\frac{\alpha T}{2}}^{\frac{\alpha T}{2}} p(t) \mathrm{e}^{-\mathrm{j}2\pi \frac{k}{\alpha} f_r t} \mathrm{d}t \\ &= \frac{1}{\alpha} \left[\frac{1}{T} \int_{-\frac{T}{2}}^{\frac{T}{2}} p(t) \mathrm{e}^{-\mathrm{j}2\pi \frac{k}{\alpha} f_r t} \mathrm{d}t \right]. \end{aligned} \tag{2.55b}$$

For simplicity visualize α to be integer. Then it is readily seen from (2.55b) that $D_{k=\pm\alpha,\pm 2\alpha,\ldots}^{[s_1(t)]} = (1/\alpha) D_{k=\pm 1,\pm 2,\ldots}^{[s(t)]}$. In other words, $D_{k=\pm\alpha,\pm 2\alpha,\ldots}^{[s_1(t)]}$ is the same as $D_{k=\pm 1,\pm 2,\ldots}^{[s(t)]}$ except for the scaling factor $1/\alpha$. Recall also that the spacing between frequency points, Δf, is equal to the fundamental frequency, f_r or $f_r^{[s_1(t)]} = f_r/\alpha$. Therefore as the period is increased the spacing between harmonics becomes smaller or in a given frequency range the number of harmonics increases. This is illustrated in Figure 2.17 for the case that $p(t)$ is a rectangular of amplitude V over $[-T/3, T/3]$ and $\alpha = 3$. These remarks hold for any $\alpha > 1$. From this discussion one concludes that as $\alpha \to \infty$ the harmonics come closer and closer together, indeed become infinitesimally close, and also that they

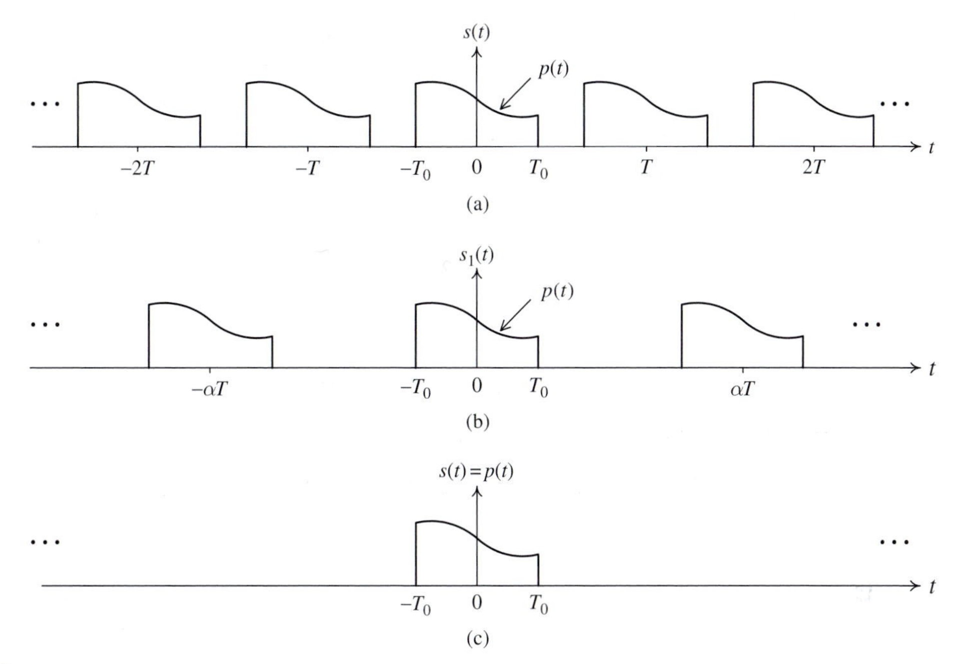

Fig. 2.16 Derivation of the Fourier transform of a nonperiodic signal: fundamental period is (a) T, (b) αT, $\alpha > 1$, (c) approaching ∞.

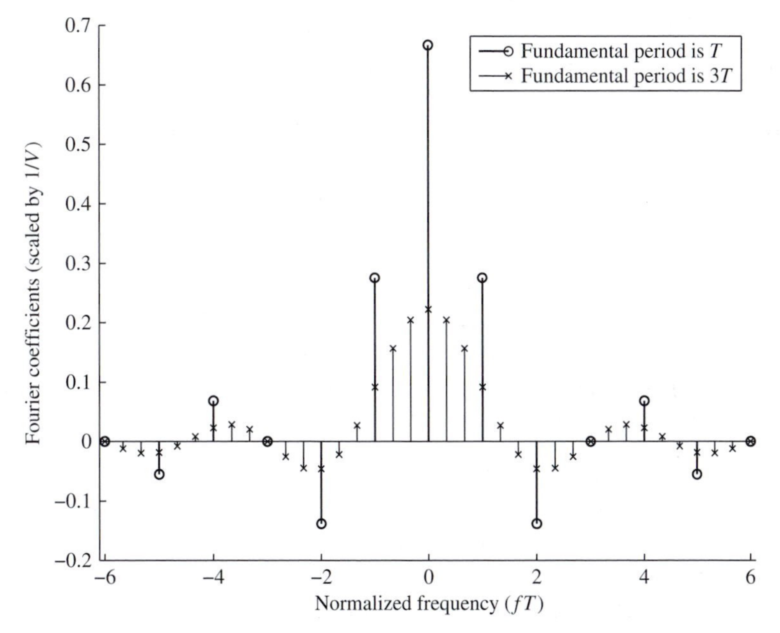

Fig. 2.17 Illustration of the effect of increasing the fundamental period while keeping the basic pulse $p(t)$ unchanged. Here $p(t)$ is a rectangular of amplitude V over $[-T/3, T/3]$.

become infinitesimally small in amplitude. The sum of these infinitesimally small harmonics, spaced infinitesimally apart becomes an integration and results in a useful integral representation of finite energy, nonperiodic signals called the *Fourier transform.*

To derive the Fourier transform representation start as above by assuming $s(t)$ is periodic with fundamental period T. Then the Fourier series representation of $s(t)$ is $s(t) = \sum_{k=-\infty}^{\infty} D_k e^{j2\pi(k/T)t}$ with $D_k = (1/T)\int_{-T/2}^{T/2} s(t)e^{-j2\pi(k/T)t}dt$. Denote the spacing between the harmonics by $\Delta f = 1/T$. Then $D_k^{[s(t)]} = \Delta f \int_{-1/2\Delta f}^{1/2\Delta f} s(t)e^{-j2\pi(k\Delta f)t}dt$. Now let the period T run to infinity which means that Δf goes to zero. Therefore, as long as the integral is finite, then the coefficients $D_k^{[s(t)]} = \Delta f \int_{-1/2\Delta f}^{1/2\Delta f} s(t)e^{-j2\pi(k\Delta f)t}dt \to 0$. However, if the integral is finite, then the ratio $D_k^{[s(t)]}/\Delta f$ is a definite value. The fact that $s(t)$ is a finite energy signal is a sufficient condition for the integral to be finite. Therefore consider the ratio $D_k^{[s(t)]}/\Delta f$ in the limit as $\Delta f \to 0$. As $\Delta f \to 0$ the spacing between the harmonics, Δf, becomes infinitesimal. The frequency, f, of the kth harmonic is $f = k\Delta f$, which suggests that as $\Delta f \to 0$ the value of k must approach infinity to keep the product $k\Delta f$ equal to f. That is, instead of discrete frequencies corresponding to the harmonics, every frequency value is allowed and the discrete index k is replaced by the continuous frequency variable, f. Therefore:

$$\lim_{\Delta f \to 0} \frac{D_k^{[s(t)]}}{\Delta f} = \lim_{\Delta f \to 0} \int_{-1/2\Delta f}^{1/2\Delta f} s(t)e^{-j2\pi(k\Delta f)t}dt = \int_{-\infty}^{\infty} s(t)e^{-j2\pi ft}dt. \tag{2.56}$$

Note that the limit becomes a function of f. It is typically denoted by a capital letter and is called the Fourier transform of the signal $s(t)$,

$$S(f) = \int_{-\infty}^{\infty} s(t)e^{-j2\pi ft}dt. \tag{2.57}$$

The units of $S(f)$ are those of $D_k^{[s(t)]}/\Delta f$ or units of $s(t)$/hertz. For this reason $S(f)$ is called a spectral density: it shows how the amplitude of $s(t)$ is distributed in the frequency domain.

Having represented $s(t)$ by sinusoids with a continuous range of frequencies, albeit of infinitesimal amplitude, we consider how to reconstruct $s(t)$ from $S(f)$. Again the reasoning proceeds from the Fourier series representation,

$$s(t) = \sum_{k=-\infty}^{\infty} D_k e^{j2\pi(k\Delta f)t}. \tag{2.58}$$

But $D_k = S(f)\Delta f$ and therefore $s(t) = \sum_{k\Delta f/\Delta f=-\infty}^{\infty} S(f)\Delta f e^{j2\pi(k\Delta f)t}$. As $\Delta f \to 0$ it becomes infinitesimally small, i.e., $\Delta f \to df$ and $k\Delta f \to f$, and the summation becomes an integration. Therefore

$$s(t) = \int_{-\infty}^{\infty} S(f)e^{j2\pi ft}df; \tag{2.59}$$

this is called the *inverse Fourier transform.* The operations of direct and inverse Fourier transformations will be denoted by $\mathcal{F}$ and $\mathcal{F}^{-1}$, respectively.

Before considering a few examples of the Fourier transform and its properties an interpretation of the transform is given. The Fourier transform was arrived at through mathematical manipulation and, as mentioned, it is a representation of the signal $s(t)$, where $s(t)$ is resolved into sinusoids of infinitesimal amplitude at frequencies that are "spaced" infinitesimally apart. This frequency domain representation results in a spectrum; the spectrum, however, is not discrete, as was the case for periodic signals, but continuous. Furthermore, though the D_k coefficients have the same unit as $s(t)$, the unit of $S(f)$ is the $s(t)$ unit per hertz. For this reason the spectrum is not only a continuous one but it is also a density, i.e., a *continuous amplitude spectral density*. It tells one how the amplitude of $s(t)$ is concentrated with regard to frequency. Higher values of $S(f)$ mean that more amplitude (and eventually more energy) is concentrated around these frequencies. This is directly analogous to the concept of mass density or gas concentration encountered in mechanics. Though an object may have a uniform density throughout its volume (units of mass/volume), no point within the object has finite mass. Finite mass is associated with finite volume only.

2.4.2 Examples of the Fourier transform

The Fourier transform of $s(t)$, $S(f)$, is, in general, a complex function. Therefore, as with D_k the coefficients for periodic signals, $S(f)$ can be resolved into either a magnitude and phase or into real and imaginary components, i.e., $S(f) = |S(f)|e^{j\angle S(f)} = \mathcal{R}\{S(f)\} + j\mathcal{I}\{S(f)\}$. Usually the magnitude/phase representation is chosen since it is the most informative. It is important to note that the Fourier transform pair is unique, important because it assures us that when one goes from the time domain, $s(t)$, to the frequency domain, $S(f)$, one can get to the same $s(t)$ (proof of this is left as an exercise for the reader).

Example 2.8 one-sided decaying exponential The signal under consideration is shown in Figure 2.18(a). The Fourier transform pair is as follows:

$$s(t) = Ve^{-at}u(t), \ a > 0, \tag{2.60a}$$

$$\begin{aligned} S(f) &= \int_{-\infty}^{\infty} Ve^{-at}u(t)e^{-j2\pi ft}dt \\ &= \left.\frac{Ve^{-(a+j2\pi f)t}}{-(a+j2\pi f)}\right|_{t=0}^{\infty} = \frac{V}{a+j2\pi f}, \end{aligned} \tag{2.60b}$$

$$|S(f)| = \frac{V}{\sqrt{a^2+4\pi^2f^2}}, \quad \angle S(f) = -\tan^{-1}\left(\frac{2\pi f}{a}\right), \tag{2.60c}$$

$$\mathcal{R}\{S(f)\} = \frac{Va}{a^2+4\pi^2f^2}, \quad \mathcal{I}\{S(f)\} = -\frac{V2\pi f}{a^2+4\pi^2f^2}. \tag{2.60d}$$

The following observations can be made from (2.60a) and (2.60b). $|S(f)|$ is an even function in f, while $\angle S(f)$ is an odd function. Similarly $\mathcal{R}\{S(f)\}$ is even while $\mathcal{I}\{S(f)\}$ is odd. Furthermore $S(-f) = V/(a - j2\pi f) = S^*(f)$. For real nonperiodic signals these observations are always true with the proof left as an exercise. Plots of $S(f)$ and $\angle S(f)$ are given in Figure 2.18(b). It shows that the signal amplitude is concentrated around $f = 0$. At

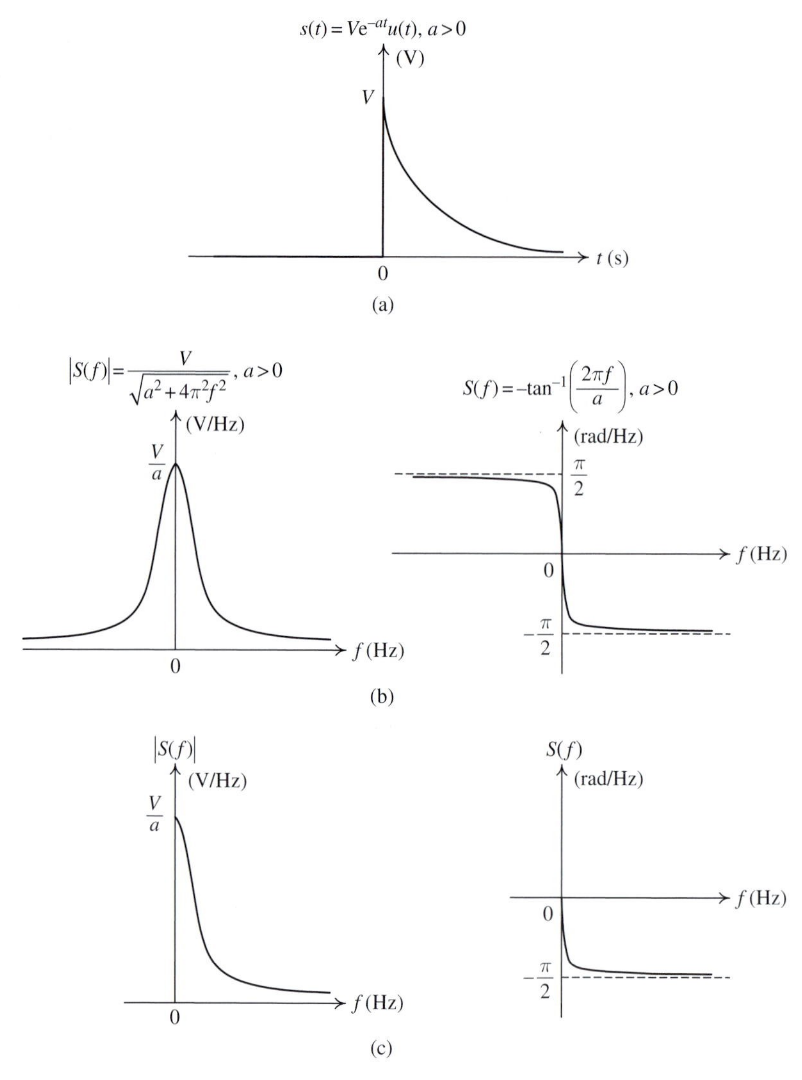

Fig. 2.18 (a) One-sided decaying exponential, (b) two-sided magnitude and phase spectra, and (c) one-sided magnitude and phase spectra.

$f = a/2\pi$, $|S(f)| = S(0)/\sqrt{2}$ and is known as the 3 dB frequency or the half-power point. The plot of Figure 2.18(b) is the two-sided spectral density of amplitude and phase. The same information is easily presented by a one-sided spectral density as shown in Figure 2.18(c). Though the one-sided graph represents the physical phenomena the two-sided spectral density is the one used in analysis.[8] The one-sided spectral density is directly analogous to C_k and θ_k where $|C_k| = 2|D_k|$ and $\theta_k = \angle D_k$. ■

[8] It is a standing challenge of the authors to have someone demonstrate a negative frequency in the lab.

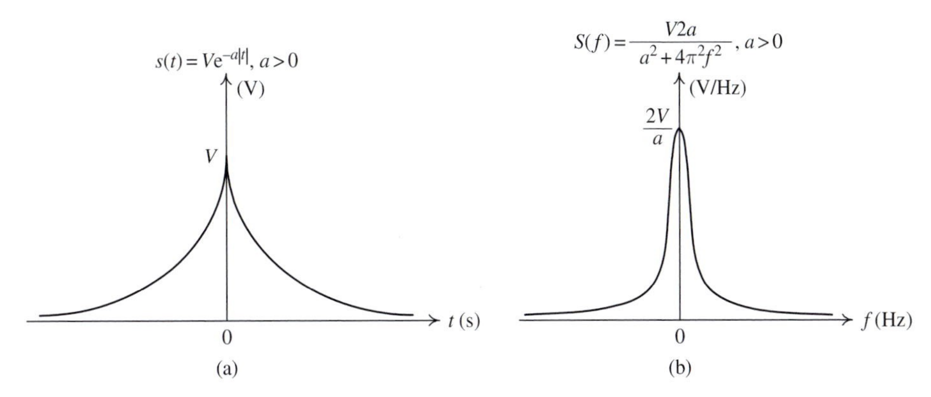

Fig. 2.19 (a) Tent function and (b) its Fourier transform.

Example 2.9 tent function For the tent function shown in Figure 2.19(a), one has

$$s(t) = Ve^{-a|t|}, \ a > 0, \tag{2.61a}$$

$$\begin{aligned} S(f) &= \int_{-\infty}^{\infty} Ve^{-a|t|}e^{-j2\pi ft}dt \\ &= \int_{-\infty}^{0} Ve^{at}e^{-j2\pi ft}dt + \int_{0}^{\infty} Ve^{-at}e^{-j2\pi ft}dt \\ &= \frac{V2a}{a^2 + 4\pi^2 f^2}. \end{aligned} \tag{2.61b}$$

Observe that $S(f)$ is real, i.e., $\mathcal{I}\{S(f)\} = 0$. This is a consequence of $s(t)$ being an even function. The general statement is: if (and only if) $s(t)$ is an even function, then $S(f)$ has only a real component. ■

Example 2.10 For the function plotted in Figure 2.20(a), one has

$$s(t) = Ve^{-a|t|}\text{sgn}(t), \ a > 0, \tag{2.62a}$$

$$\begin{aligned} S(f) &= -\int_{-\infty}^{0} Ve^{at}e^{-j2\pi ft}dt + \int_{0}^{\infty} Ve^{-at}e^{-j2\pi ft}dt \\ &= -j\frac{V4\pi f}{a^2 + 4\pi^2 f^2}, \end{aligned} \tag{2.62b}$$

$$|S(f)| = \frac{V4\pi|f|}{a^2 + 4\pi^2 f^2}, \quad \angle S(f) = -\frac{\pi}{2}\text{sgn}(f). \tag{2.62c}$$

Since $s(t)$ is an odd function, $\mathcal{R}\{S(f)\} = 0$. In general, if (and only if) $s(t)$ is an odd function, then $S(f)$ has only an imaginary component. ■

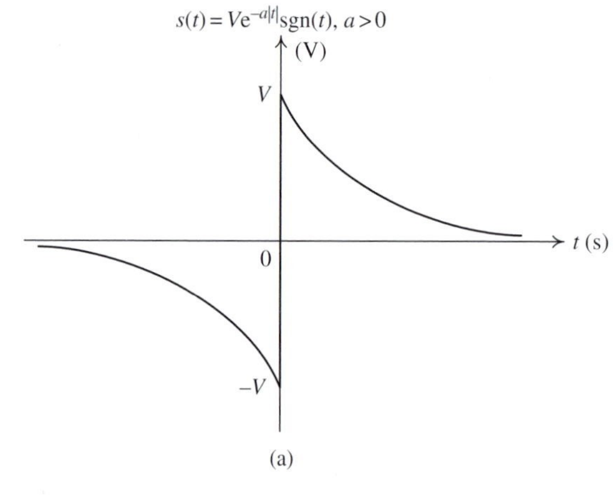

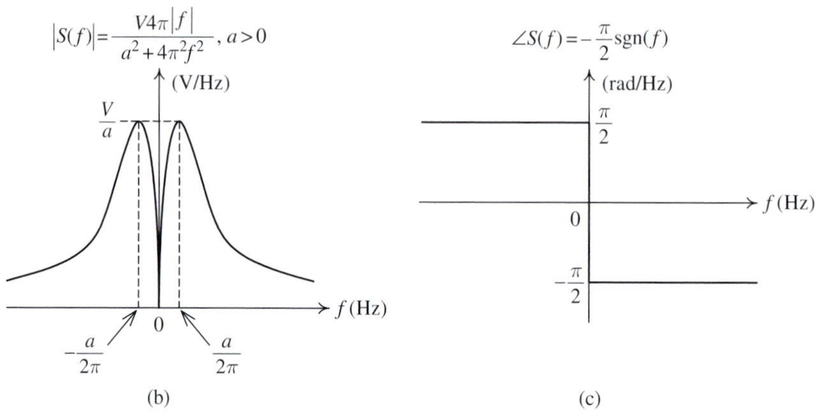

Fig. 2.20 (a) The function considered in Example 2.10, (b) its magnitude spectrum, and (c) its phase spectrum.

Example 2.11 rectangular pulse Figure 2.21(a) shows the rectangular pulse. Its Fourier transform is determined as

$$s(t) = V\left[u\left(t+\frac{T}{2}\right) - u\left(t-\frac{T}{2}\right)\right], \tag{2.63a}$$

$$S(f) = \int_{-\frac{T}{2}}^{\frac{T}{2}} Ve^{-j2\pi ft}dt = VT\frac{\sin(\pi fT)}{(\pi fT)}, \tag{2.63b}$$

$$|S(f)| = VT\left|\frac{\sin(\pi fT)}{(\pi fT)}\right|,$$

$$\angle S(f) = \begin{cases} 0, & \frac{k}{T} < |f| < \frac{k+1}{T},\ k \text{ even and} \geq 0 \\ \pi\,\text{sgn}(f), & \frac{k}{T} < |f| < \frac{k+1}{T},\ k \text{ odd and} > 0 \end{cases}. \tag{2.63c}$$

Since $s(t)$ is an even function, it only has a real component. However, observe that this does not mean that the phase is zero. It can be 0 or $\pm\pi$ depending on whether the real component is positive or negative. There is no difference between a phase of $+\pi$ or $-\pi$

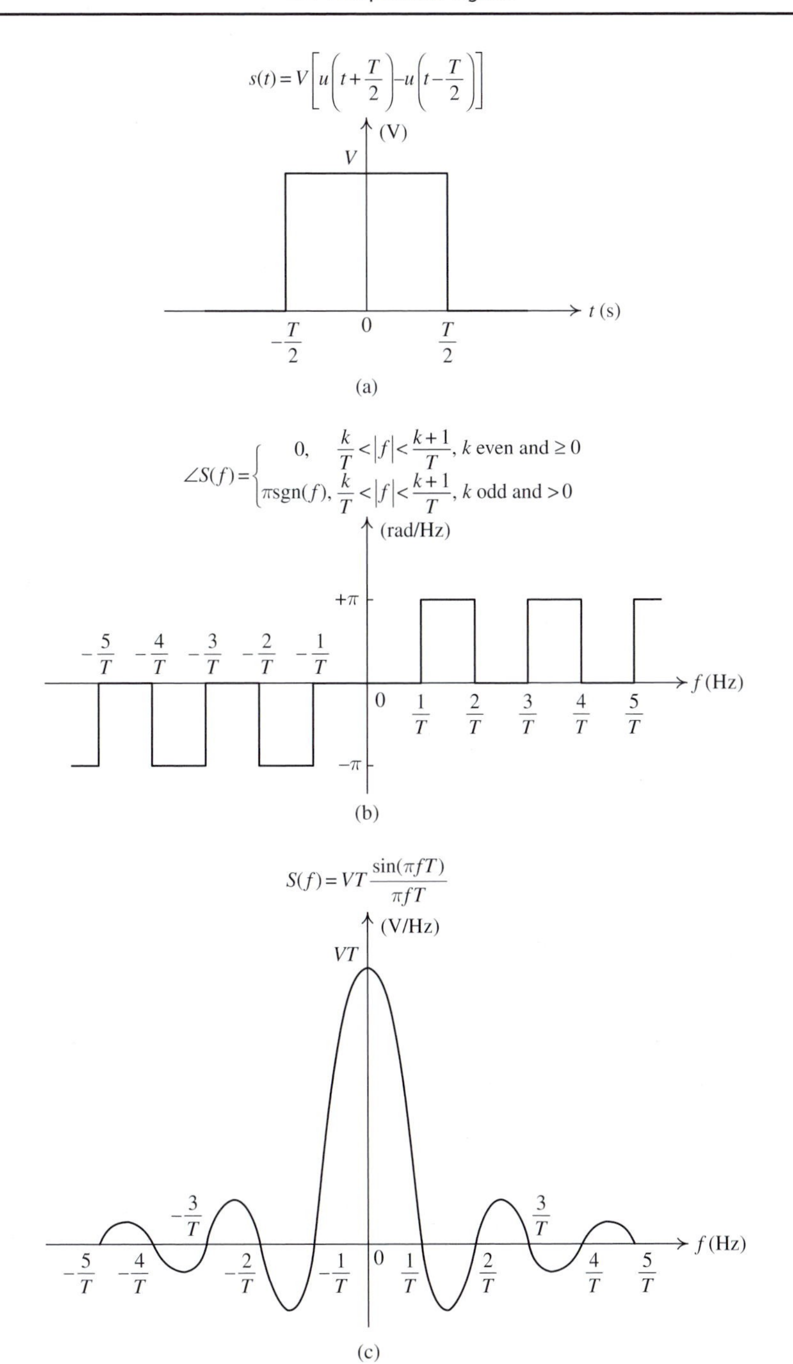

Fig. 2.21 (a) Rectangular pulse, (b) its phase spectrum, and (c) its combined magnitude–phase spectrum.

but it helps to show graphically that the phase function is odd (Figure 2.21(b)). Typically the magnitude/phase function for the rectangular pulse is represented by a single graph as shown in Figure 2.21(c). ■

Example 2.12 impulse function

$$s(t) = V\delta(t), \tag{2.64a}$$

$$S(f) = \int_{-\infty}^{\infty} V\delta(t)\mathrm{e}^{-\mathrm{j}2\pi ft}\mathrm{d}t = V. \tag{2.64b}$$

Though this function is not a finite energy signal, it still has a Fourier transform (recall that finite energy was a sufficient condition, not a necessary one). The signal is equally concentrated at all frequencies, i.e., the spectrum is flat. ■

Example 2.13 DC signal

$$s(t) = V, \tag{2.65a}$$

$$\begin{aligned} S(f) &= \int_{-\infty}^{\infty} V\mathrm{e}^{-\mathrm{j}2\pi ft}\mathrm{d}t = \frac{\mathrm{e}^{\mathrm{j}2\pi f\infty} - \mathrm{e}^{-\mathrm{j}2\pi f\infty}}{\mathrm{j}2\pi f} \\ &= \frac{\sin(\mathrm{j}2\pi f\infty)}{\pi f}. \end{aligned} \tag{2.65b}$$

Unfortunately the above approach fails since $\sin(\infty)$ is undefined. However, recall that $S(f)$ is a spectral density and given that the signal is a DC one it seems reasonable for all of it to be concentrated at $f = 0$. Therefore postulate that $S(f) = V\delta(f)$ and check what time function corresponds to this $S(f)$. Evaluating $\int_{-\infty}^{\infty} V\delta(f)\mathrm{e}^{\mathrm{j}2\pi ft}\mathrm{d}f$ we get V and therefore

$$s(t) = V \longleftrightarrow S(f) = V\delta(f), \tag{2.66a}$$

$$|S(f)| = V\delta(f), \quad \angle S(f) = 0. \tag{2.66b}$$

■

Note that the above implies that $\int_{-\infty}^{\infty} V\mathrm{e}^{-\mathrm{j}2\pi ft}\mathrm{d}t = V\delta(f)$, or that the integral $\int_{-\infty}^{\infty} \mathrm{e}^{-\mathrm{j}2\pi xy}\mathrm{d}y = \delta(x)$. This is a useful relationship that is used in the next example.

Example 2.14 sinusoid

$$s(t) = V\cos(2\pi f_c t + \theta) = \frac{V\left[\mathrm{e}^{\mathrm{j}(2\pi f_c t+\theta)} + \mathrm{e}^{-\mathrm{j}(2\pi f_c t+\theta)}\right]}{2}, \tag{2.67a}$$

$$\begin{aligned} S(f) &= \int_{-\infty}^{\infty} \frac{V}{2}\mathrm{e}^{\mathrm{j}\theta}\mathrm{e}^{-\mathrm{j}2\pi(f-f_c)t}\mathrm{d}t + \int_{-\infty}^{\infty} \frac{V}{2}\mathrm{e}^{-\mathrm{j}\theta}\mathrm{e}^{-\mathrm{j}2\pi(f+f_c)t}\mathrm{d}t \\ &= \frac{V}{2}\mathrm{e}^{\mathrm{j}\theta}\delta(f-f_c) + \frac{V}{2}\mathrm{e}^{-\mathrm{j}\theta}\delta(f+f_c), \end{aligned} \tag{2.67b}$$

$$|S(f)| = \frac{V}{2}\delta(f - f_c) + \frac{V}{2}\delta(f + f_c),\ \angle S(f) = \begin{cases} \theta, & f = f_c \\ -\theta, & f = -f_c \end{cases}. \tag{2.67c}$$

For a pure sinusoid, one lasting from $-\infty$ to ∞, all the amplitude is concentrated at $\pm f_c$, the frequency of the sinusoid. Cosine and sine have the same magnitude density spectrum, the difference between them is in the phase, 0 (rad) for the cosine and $-\pi/2$ (rad) for the sine. But we knew this already. ■

Example 2.15 step function Figure 2.22(a) shows a step function, which is not a finite energy signal. Also as in Examples 2.13 and 2.14, it does not go to 0 as $t \to \infty$. But like the previous two examples it has finite, nonzero average power since

$$\lim_{T\to\infty} \frac{1}{T}\int_{-\frac{T}{2}}^{\frac{T}{2}} s^2(t)\mathrm{d}t = \lim_{T\to\infty} \frac{1}{T}\int_{0}^{\frac{T}{2}} V^2\mathrm{d}t = \frac{V^2}{2}\ \text{(watts)}. \tag{2.68}$$

The direct approach to determine $S(f) = \int_{-\infty}^{\infty} Vu(t)\mathrm{e}^{-\mathrm{j}2\pi ft}\mathrm{d}t = \int_0^{\infty} V\mathrm{e}^{-\mathrm{j}2\pi ft}\mathrm{d}t = V(1 - \mathrm{e}^{\mathrm{j}2\pi f\infty})/2\mathrm{j}\pi f$ fails because $\mathrm{e}^{\mathrm{j}2\pi f\infty}$ is undefined. Therefore decompose $s(t)$ as shown in Figures 2.22(b) and 2.22(c):

$$s(t) = \frac{V}{2} + \frac{V}{2}\mathrm{sgn}(t) \Rightarrow \mathcal{F}\{s(t)\} = \mathcal{F}\left\{\frac{V}{2}\right\} + \mathcal{F}\left\{\frac{V}{2}\mathrm{sgn}(t)\right\}. \tag{2.69}$$

We already know that $\mathcal{F}\{V/2\} = (V/2)\delta(f)$. To determine $\mathcal{F}\{(V/2)\mathrm{sgn}(t)\}$, write $(V/2)\mathrm{sgn}(t) = \lim_{a\to 0}(V/2)\mathrm{e}^{-a|t|}\mathrm{sgn}(t)$. Therefore

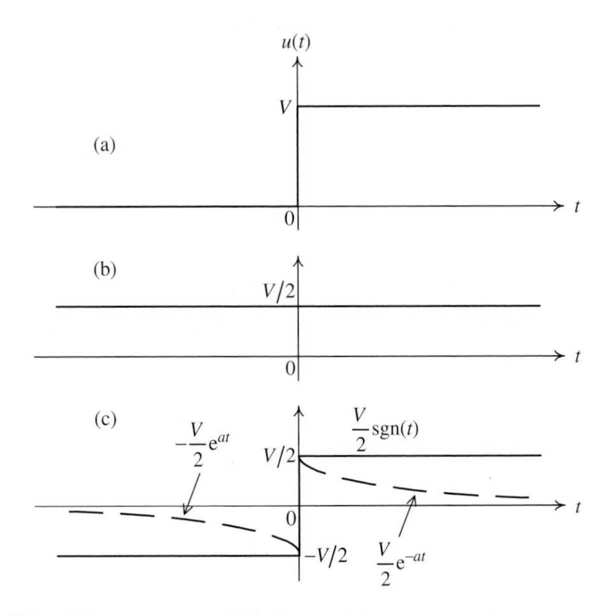

Fig. 2.22 (a) Step function, (b) its DC component, and (c) the remaining component.

$$\mathcal{F}\left\{\frac{V}{2}\text{sgn}(t)\right\} = \mathcal{F}\left\{\lim_{a\to 0}\frac{V}{2}\text{e}^{-a|t|}\text{sgn}(t)\right\} = \lim_{a\to 0}\mathcal{F}\left\{\frac{V}{2}\text{e}^{-a|t|}\text{sgn}(t)\right\}$$
$$= \lim_{a\to 0}\frac{-4V\text{j}\pi f}{a^2+4\pi^2 f^2}\frac{1}{2} = \frac{V}{\text{j}2\pi f}, \tag{2.70}$$

where the result of Example 2.10 has been used to arrive at the last expression. Finally,

$$s(t) = Vu(t) \longleftrightarrow S(f) = \frac{V}{2}\delta(f) + \frac{V}{\text{j}2\pi f}. \tag{2.71}$$

■

The three previous signals had in common three interrelated characteristics: (i) they do not tend to zero as t goes to infinity (either $-\infty$ or $+\infty$ or both); (ii) their energy is infinite but they have finite average power; and (iii) their spectral density contains an impulse(s). In general signals that do not go to zero as $t \to \pm\infty$ and have finite, nonzero, average power have impulses in their amplitude density spectrum.

The next two examples consider signals that illustrate aspects of the frequency bandwidth occupied by a signal. Bandwidth is a precious and expensive resource in communications.

Example 2.16 tone burst Figure 2.1(e) shows a tone burst, which can be expressed as $s(t) = V\cos(2\pi f_c t)\left[u\left(t+T/2\right) - u\left(t-T/2\right)\right]$, where T is such that there are n cycles of the sinusoid in the interval of T seconds. Since $T_c = 1/f_c$ and $nT_c = T$, then $T = n/f_c$. The Fourier transform of $s(t)$ is

$$S(f) = \frac{Vn}{2f_c}\left[\frac{\sin\left(n\pi\left(\frac{f}{f_c}-1\right)\right)}{n\pi\left(\frac{f}{f_c}-1\right)} + \frac{\sin\left(n\pi\left(\frac{f}{f_c}+1\right)\right)}{n\pi\left(\frac{f}{f_c}+1\right)}\right]. \tag{2.72}$$

Figure 2.23 shows a plot of $S(f)/(V/2f_c)$ versus f/f_c for various values of n, the number of cycles. The important observation is that as n increases the spectrum becomes more and more concentrated around $f = f_c$. To increase the number of cycles, n, one must increase T since f_c is fixed. This is precisely what is done in M-ary (or multilevel) digital modulation which is discussed in Chapter 8. As $n \to \infty$ the signal tends to a pure sinusoid and in the limit the spectrum becomes an impulse (see Example 2.14). ■

Example 2.17 Gaussian signal The Gaussian signal is plotted in Figure 2.24(a). Its expression and its Fourier transform are as follows:

$$s(t) = V\text{e}^{-at^2},\ a > 0, \tag{2.73a}$$

$$S(f) = \int_{-\infty}^{\infty} V\text{e}^{-at^2}\text{e}^{-\text{j}2\pi ft}\text{d}t$$
$$= \int_{-\infty}^{\infty} V\text{e}^{-a(t^2+\text{j}2\pi ft/a)}\text{d}t. \tag{2.73b}$$

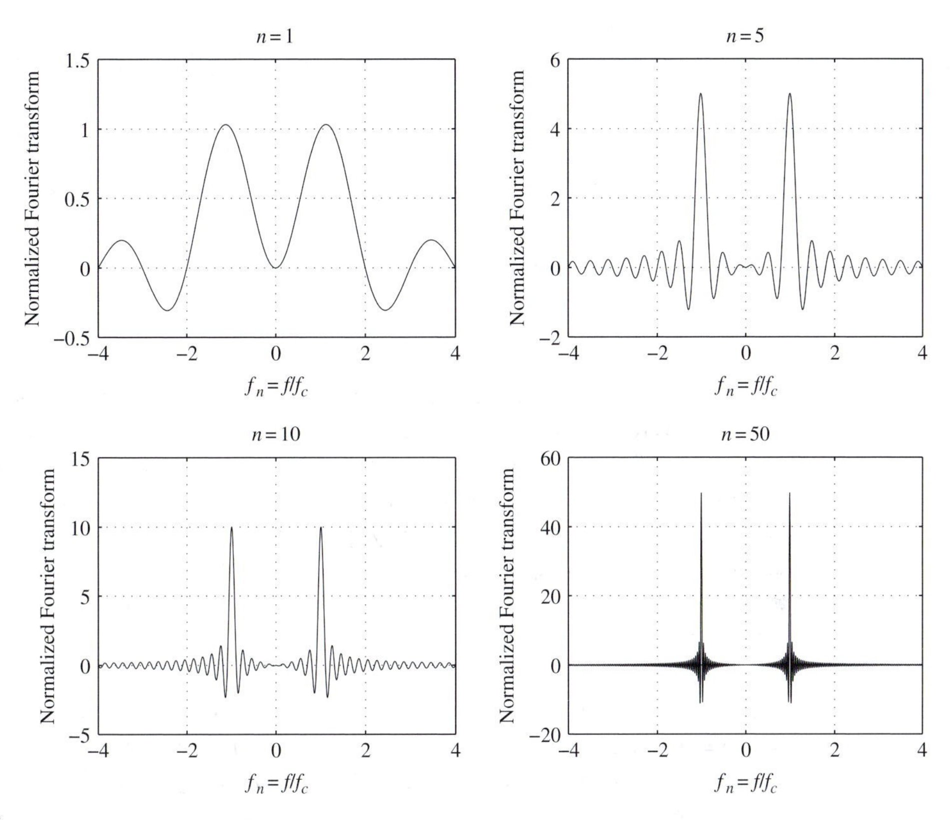

Fig. 2.23 Fourier transforms of tone bursts with different numbers of cycles.

To perform the integration complete the exponent to make it a perfect square as follows:

$$\begin{aligned} t^2 + 2\frac{\mathrm{j}\pi f}{a}t &= t^2 + 2\frac{\mathrm{j}\pi f}{a}t + \left(\frac{\mathrm{j}\pi f}{a}\right)^2 - \left(\frac{\mathrm{j}\pi f}{a}\right)^2 \\ &= \left(t + \frac{\mathrm{j}\pi f}{a}\right)^2 + \frac{\pi^2 f^2}{a^2}. \end{aligned} \tag{2.74}$$

Therefore $S(f) = \mathrm{e}^{-\pi^2 f^2/a}\int_{-\infty}^{\infty} V\mathrm{e}^{-a(t+\mathrm{j}\pi f/a)^2}\mathrm{d}t$. It can be shown that $\left(1/\sqrt{2\pi}\sigma\right)\int_{-\infty}^{\infty}$ $\mathrm{e}^{-(x-\mu)^2/2\sigma^2}\mathrm{d}x = 1$ for any μ (real or complex) and σ (real and positive).[9] Therefore, after some algebraic manipulations, one obtains

$$S(f) = \mathcal{F}\{s(t)\} = V\sqrt{\frac{\pi}{a}}\mathrm{e}^{-\pi^2 f^2/a}. \tag{2.75}$$

The transform has the interesting property that it is of the same form as the time signal. It is sketched in Figure 2.24(b). A more interesting property is that the time–bandwidth product is the minimum possible. ■

[9] The astute reader or one with the appropriate background may recognize that this is the area under a Gaussian probability density function.

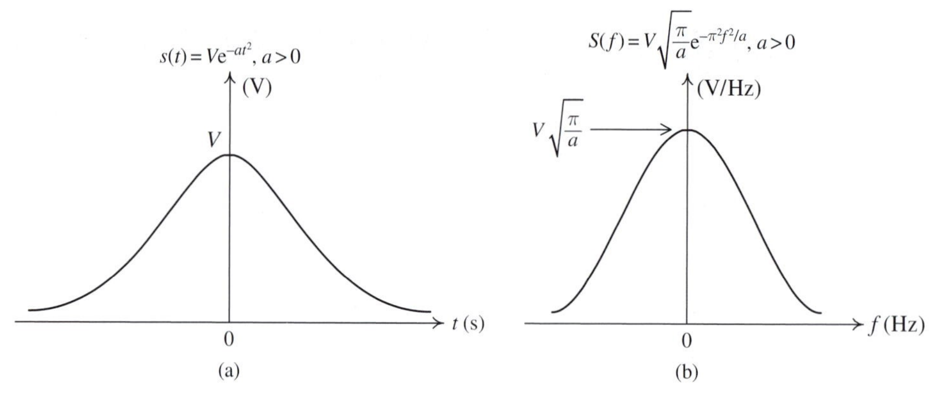

Fig. 2.24 (a) Gaussian signal and (b) its Fourier transform.

Table 2.2 Fourier transform properties

$$s(t) = \int_{-\infty}^{\infty} S(f)e^{j2\pi ft}df \longleftrightarrow S(f) = \int_{-\infty}^{\infty} s(t)e^{-j2\pi ft}dt$$

	Property	Signal type	Remark
1.	$\lvert S(f)\rvert = \lvert S(-f)\rvert$	Real	Magnitude spectrum is even
2.	$\angle S(f) = -\angle S(-f)$	Real	Phase spectrum is odd
3.	$\mathcal{R}\{S(f)\} = \mathcal{R}\{S(-f)\}$	Real	Real part is even
4.	$\mathcal{I}\{S(f)\} = -\mathcal{I}\{S(-f)\}$	Real	Imaginary part is odd
5.	$S(-f) = S^*(f)$	Real	Complex conjugate symmetry
6.	$s(0) = \int_{-\infty}^{\infty} S(f)df$	Complex	Total area under $S(f)$
7.	$S(0) = \int_{-\infty}^{\infty} s(t)dt$	Complex	Total area under $s(t)$

2.4.3 Properties of the Fourier transform

Having presented several common and important Fourier transform pairs we consider next several properties of the Fourier transform. These are presented in tabular form in Tables 2.2 and 2.3, mostly without proof.

Table 2.2 provides several properties pertaining to $S(f)$, the first five of which apply only to real signals. Table 2.3 is concerned with various properties of the Fourier transform operation that are applicable to both real and complex signals.[10] The reader is urged to prove them for her/himself. The transform operation properties are discussed next with proofs given for some along with illustrative examples.

[10] In analysis it is convenient sometimes to deal with complex signals. This occurs with Hilbert transforms for instance.

Table 2.3 Fourier transform relationships for general signals, real or complex

$$s(t) = \int_{-\infty}^{\infty} S(f)e^{j2\pi ft}df \longleftrightarrow S(f) = \int_{-\infty}^{\infty} s(t)e^{-j2\pi ft}dt$$

	Operation	$s(t)$	$S(f)$
1.	Superposition	$\alpha s_1(t) + \beta s_2(t)$	$\alpha S_1(f) + \beta S_2(f)$
2.	Time shifting	$s(t - t_0)$	$S(f)e^{-j2\pi ft_0}$
3.	Time scaling	$s(at)$	$\frac{1}{\lvert a\rvert}S\left(\frac{f}{a}\right)$
	Time reversal	$s(-t)$	$S(-f)$
4.	Frequency-shifting	$s(t)e^{j2\pi f_c t}$	$S(f - f_c)$
	Amplitude modulation	$s(t)\cos(2\pi f_c t)$	$\frac{1}{2}S(f - f_c) + \frac{1}{2}S(f + f_c)$
5.	Time differentiation	$\frac{d^n s(t)}{dt^n}$	$(j2\pi f)^n S(f)$
6.	Time integration	$\int_{-\infty}^{t} s(\lambda)d\lambda$	$\frac{1}{j2\pi f}S(f) + \frac{S(0)}{2}\delta(f)$
7.	Time multiplication	$s_1(t)s_2(t)$	$S_1(f) * S_2(f)$
8.	Time convolution	$s_1(t) * s_2(t)$	$S_1(f)S_2(f)$
9.	Duality	$S(t)$	$s(-f)$

Note: α and β are arbitrary constants,
$S_1(f) * S_2(f) = \int_{-\infty}^{\infty} S_1(\lambda)S_2(f - \lambda)d\lambda = \int_{-\infty}^{\infty} S_2(\lambda)S_1(f - \lambda)d\lambda$,
$s_1(t) * s_2(t) = \int_{-\infty}^{\infty} s_1(\lambda)s_2(t - \lambda)d\lambda = \int_{-\infty}^{\infty} s_2(\lambda)s_1(t - \lambda)d\lambda$.

Superposition (linearity) Fourier transformation is an integral operation on the signal, $s(t)$. Since integration is a linear operation, superposition in an obvious but extremely important property.

Time shifting A shift of the time signal, $s(t)$, leaves its shape unchanged. In essence what is done is that the $t = 0$ point is redefined. In the frequency domain $S(f)$ is multiplied by the factor $e^{-j2\pi ft_0}$. Observe that the magnitude of the transform is unchanged by a time shift. The phase, however, is changed by a linear factor $-2\pi ft_0$ where the slope is determined by the time shift, i.e.,

$$s(t) \longleftrightarrow S(f), \tag{2.76a}$$

$$s_1(t) = s(t - t_0) \longleftrightarrow S_1(f) = S(f)e^{-j2\pi ft_0} = |S(f)|e^{j(\angle S(f) - 2\pi ft_0)}. \tag{2.76b}$$

Time scaling The relationship is easy enough to prove for $a > 0$ since

$$\begin{aligned}\mathcal{F}\{s(at)\} &= \int_{-\infty}^{\infty} s(at)e^{-j2\pi ft}dt \\ &\overset{\lambda=at}{=} \int_{-\infty}^{\infty} s(\lambda)e^{-j2\pi(f/a)\lambda}\frac{d\lambda}{a} = \frac{1}{a}S\left(\frac{f}{a}\right).\end{aligned} \tag{2.77}$$

For $a < 0$, write at as $at = -|a|t$ since $a = -|a|$ for $a < 0$. Now let $\lambda = at = -|a|t$. The Fourier integral becomes

$$\int_{-\infty}^{\infty} s(\lambda)\mathrm{e}^{-\mathrm{j}2\pi(-f/|a|)\lambda}\frac{\mathrm{d}\lambda}{|a|} = \frac{1}{|a|}S\left(\frac{-f}{|a|}\right) = \frac{1}{|a|}S\left(\frac{f}{a}\right). \tag{2.78}$$

The case $a = -1$ is special and corresponds to playing the signal backwards.

Frequency shifting This is extremely important in communications; it is not an exaggeration to call it the cornerstone of communications. Multiplication of the signal $s(t)$ by a sinusoid of frequency f_c shifts the spectrum to $\pm f_c$. Typically f_c is much larger than the bandwidth of $s(t)$ and its spectrum is shifted up to frequencies where antenna design is easier, spectrum sharing is possible, etc.

Time differentiation Since $s(t) = \int_{-\infty}^{\infty} S(f)\mathrm{e}^{\mathrm{j}2\pi ft}\mathrm{d}f$, we have $\mathrm{d}s(t)/\mathrm{d}t = \int_{-\infty}^{\infty}(\mathrm{j}2\pi f) S(f)\mathrm{e}^{\mathrm{j}2\pi ft}\mathrm{d}f$, which implies that $\mathrm{d}s(t)/\mathrm{d}t$ and $(\mathrm{j}2\pi f)S(f)$ are a Fourier transform pair. Typically this relationship is used to determine $S(f)$ where $S(f) = \mathcal{F}\{\mathrm{d}s(t)/\mathrm{d}t\}/\mathrm{j}2\pi f$. Care, however, must be exercised since if $s(t)$ has a DC value it will not show up after differentiation. Consider the step function $s(t) = Vu(t)$. Since $\mathrm{d}s(t)/\mathrm{d}t = V\delta(t)$ and $\mathcal{F}\{\mathrm{d}s(t)/\mathrm{d}t\} = V$, it implies that $S(f) = V/\mathrm{j}2\pi f$. But from Example 2.15 we know that

$$S(f) = \frac{V}{\mathrm{j}2\pi f} + \frac{V}{2}\delta(f),$$

where the second term represents the DC value of $s(t)$ which is lost during the differentiation. A sufficient condition is for $s(t)$ to be absolutely integrable, i.e., $\int_{-\infty}^{\infty}|s(t)|\mathrm{d}t < \infty$. Then it does not have a DC component and the relationship $S(f) = \mathcal{F}\{\mathrm{d}s(t)/\mathrm{d}t\}/\mathrm{j}2\pi f$ holds.

As another example, consider the rectangular signal of Example 2.11, where $s(t) = V[u(t + T/2) - u(t - T/2)]$. By inspection this signal does not have a DC component and

$$\frac{\mathrm{d}s(t)}{\mathrm{d}t} = V\left[\delta\left(t + \frac{T}{2}\right) - \delta\left(t - \frac{T}{2}\right)\right], \tag{2.79}$$

$$\mathcal{F}\left\{\frac{\mathrm{d}s(t)}{\mathrm{d}t}\right\} = V\int_{-\infty}^{\infty}\left[\delta\left(t + \frac{T}{2}\right) - \delta\left(t - \frac{T}{2}\right)\right]\mathrm{e}^{-\mathrm{j}2\pi ft}\mathrm{d}t. \tag{2.80}$$

Using the sifting property of the impulse function one has $\mathcal{F}\{\mathrm{d}s(t)/\mathrm{d}t\} = V\left[\mathrm{e}^{\mathrm{j}\pi fT} - \mathrm{e}^{-\mathrm{j}\pi fT}\right]$. Therefore

$$S(f) = \frac{V\left[\mathrm{e}^{\mathrm{j}\pi fT} - \mathrm{e}^{-\mathrm{j}\pi fT}\right]}{\mathrm{j}2\pi f} = VT\frac{\sin(\pi fT)}{\pi fT}. \tag{2.81}$$

Time integration Consider $s_1(t) = \int_{-\infty}^{t} s(\lambda)\mathrm{d}\lambda$, where $s(t)$ has a Fourier transform $S(f)$. The goal is to find the Fourier transform of $s_1(t)$, i.e., $S_1(f)$, in terms of $S(f)$.

Since $\mathrm{d}s_1(t)/\mathrm{d}t = s(t)$, one may quickly conclude that $S_1(f) = S(f)/\mathrm{j}2\pi f$. However, as pointed out above, any DC component in $s_1(t)$ disappears when differentiated, and therefore is not present in $S_1(f)$. To account for this, rewrite $s_1(t)$ as $s_1(t) = \tilde{s}_1(t) + s_1^{\mathrm{DC}}$, where $\tilde{s}_1(t)$ has zero DC and s_1^{DC} is precisely the DC component of $s_1(t)$. Now $S_1(f) = \tilde{S}_1(f) + s_1^{\mathrm{DC}}\delta(f)$. But $\mathrm{d}s_1(t)/\mathrm{d}t = \mathrm{d}\tilde{s}_1(t)/\mathrm{d}t = s(t)$, and one has $\tilde{S}_1(f) = S(f)/\mathrm{j}2\pi f$. Next, find s_1^{DC} from the basic definition as follows:

$$s_1^{\mathrm{DC}} = \lim_{T\to\infty}\left\{\frac{1}{2T}\int_{t=-T}^{T}\mathrm{d}t\int_{-\infty}^{t}s(\lambda)\mathrm{d}\lambda\right\}. \tag{2.82}$$

Assume $s(t) = 0$ for $t < \tau$, where τ is any finite number. Obviously one also has $s_1(t) = 0$ for $t < \tau$. Then $s_1^{\mathrm{DC}} = \lim_{T\to\infty}\{(1/2T)\int_\tau^T \mathrm{d}t \int_\tau^t s(\lambda)\mathrm{d}\lambda\}$. As $T\to\infty$, the inner integral becomes $\int_{-\infty}^{\infty} s(\lambda)\mathrm{d}\lambda = S(0)$. Therefore $s_1^{\mathrm{DC}} = \lim_{T\to\infty}(S(0)/2T)\int_\tau^T \mathrm{d}t = S(0)\lim_{T\to\infty}$ $\{(T-\tau)/2T\} = S(0)/2$. Finally,

$$S_1(f) = \frac{S(f)}{\mathrm{j}2\pi f} + \frac{S(0)}{2}\delta(f). \tag{2.83}$$

Let us consider the step function, $Vu(t)$, which is the integral of the impulse function $V\delta(t)$. The Fourier transform of $V\delta(t)$ is V, which means that $S(0) = V$. Therefore the Fourier transform of $Vu(t)$, by the integration property is $S(f) = V/\mathrm{j}2\pi f + (V/2)\delta(f)$, in agreement with Example 2.15.

Time multiplication Consider $s(t) = s_1(t)s_2(t)$, where $s_1(t) \longleftrightarrow S_1(f)$ and $s_2(t) \longleftrightarrow S_2(f)$ Therefore

$$\begin{aligned} S(f) &= \int_{-\infty}^{\infty} s_1(t)s_2(t)\mathrm{e}^{-\mathrm{j}2\pi ft}\mathrm{d}t \\ &= \int_{t=-\infty}^{\infty}\left[\int_{\lambda=-\infty}^{\infty} S_1(\lambda)\mathrm{e}^{\mathrm{j}2\pi\lambda t}\mathrm{d}\lambda\right] \\ &\quad\times\left[\int_{u=-\infty}^{\infty} S_2(u)\mathrm{e}^{\mathrm{j}2\pi ut}\mathrm{d}u\right]\mathrm{e}^{-\mathrm{j}2\pi ft}\mathrm{d}t. \end{aligned} \tag{2.84}$$

Interchanging the order of integration, one has

$$S(f) = \int_{\lambda=-\infty}^{\infty}\int_{u=-\infty}^{\infty}\mathrm{d}\lambda\mathrm{d}u S_1(\lambda)S_2(u)\int_{t=-\infty}^{\infty}\mathrm{e}^{\mathrm{j}2\pi(\lambda+u-f)t}\mathrm{d}t. \tag{2.85}$$

The integral with respect to t is an impulse $\delta(\lambda+u-f)$. The impulse sifts out the value of $S_2(u)$ at $u = f-\lambda$ or $S_1(f)$ at $\lambda = f-u$. The result is

$$S(f) = \int_{-\infty}^{\infty} S_1(\lambda)S_2(f-\lambda)\mathrm{d}\lambda = \int_{-\infty}^{\infty} S_2(\lambda)S_1(f-\lambda)\mathrm{d}\lambda. \tag{2.86}$$

The above operation is known as *convolution* and it is directly analogous to that encountered earlier for Fourier series. Indeed, in both cases it arose from multiplication in the time domain. The next property discussed shows that convolution in the time domain results in multiplication in the frequency domain. This we would suspect from the duality property but of course duality has not been discussed yet.

The graphical interpretation of convolution is similar to that of the discrete case. It involves flipping $S_2(\lambda)$ (or $S_1(\lambda)$) about the vertical axis to obtain $S_2(-\lambda)$ (or $S_1(-\lambda)$), shifting it by the value f which gives $S_2(f-\lambda)$ (or $S_1(f-\lambda)$), multiplying by $S_1(\lambda)$ (or

$S_2(\lambda)$), and finding the area under the product. The shift is to the right if f is positive and to the left if f is negative.

As an example consider the tone burst, given as (see Example 2.16) $s(t) = V\cos(2\pi f_c t)\left[u(t+T/2) - u(t-T/2)\right]$. Identify $s_1(t) = \cos(2\pi f_c t)$ and $s_2(t) = V\left[u(t+T/2) - u(t-T/2)\right]$. From Examples 2.11 and 2.14 we know that $S_1(f) = [\delta(f-f_c) + \delta(f+f_c)]/2$ and $S_2(f) = VT\sin(\pi fT)/(\pi fT)$. Therefore

$$\begin{aligned} S(f) &= \int_{-\infty}^{\infty} S_1(\lambda)S_2(f-\lambda)\mathrm{d}\lambda \\ &= \int_{-\infty}^{\infty} \frac{\delta(\lambda - f_c) + \delta(\lambda + f_c)}{2} VT\frac{\sin(\pi(f-\lambda)T)}{\pi(f-\lambda)T}\mathrm{d}\lambda \\ &= \frac{VT}{2}\frac{\sin(\pi(f-f_c)T)}{\pi(f-f_c)T} + \frac{VT}{2}\frac{\sin(\pi(f+f_c)T)}{\pi(f+f_c)T}. \end{aligned} \tag{2.87}$$

If $T = n/f_c$, i.e., T contains n cycles of the sinusoid, we arrive at the same result as in Example 2.16.

Time convolution The convolution referred to here is in the time domain and it leads to multiplication in the frequency domain. The result is of paramount importance in linear system theory, where the output of a linear, time-invariant system is given (in the time domain) by the convolution of the input with the impulse response of the system and by the product of the transforms in the frequency domain. This is discussed more fully in the next section. The proof can be obtained by the same approach as above or by invoking duality.

To illustrate the application of the convolution property, we derive the time integration property in a different way. Recognize that

$$u(t-\lambda) = \begin{cases} 1, & \lambda \le t \\ 0, & \lambda > t \end{cases} \tag{2.88}$$

and write $\int_{-\infty}^{t} s(\lambda)\mathrm{d}\lambda$ in the form of convolution as follows:

$$\int_{-\infty}^{t} s(\lambda)\mathrm{d}\lambda = \int_{-\infty}^{\infty} s(\lambda)u(t-\lambda)\mathrm{d}\lambda = s(t) * u(t). \tag{2.89}$$

Making use of the result $\mathcal{F}\{u(t)\} = \frac{1}{2}\delta(f) + 1/\mathrm{j}2\pi f$ established in Example 2.15 and applying the convolution property gives

$$\begin{aligned} \int_{-\infty}^{t} s(\lambda)\mathrm{d}\lambda = s(t) * u(t) \longleftrightarrow S(f)&\left[\frac{1}{2}\delta(f) + \frac{1}{\mathrm{j}2\pi f}\right] \\ &= \frac{S(0)}{2}\delta(f) + \frac{S(f)}{\mathrm{j}2\pi f}. \end{aligned} \tag{2.90}$$

Duality Consider Examples 2.12 and 2.13. In the first an impulse in the time domain results in a DC in the frequency domain, while in the second the opposite is true, an impulse in the frequency domain corresponds to DC in the time domain. The transform pairs are said to be duals of each other. Duality arises from the fact that aside from a sign difference in the exponent of the exponential the forms of the direct and inverse transforms are the same.

Proof of the duality relationship proceeds as follows. Let $s(t) = \int_{-\infty}^{\infty} S(f)e^{j2\pi ft}df \longleftrightarrow S(f) = \int_{-\infty}^{\infty} s(t)e^{-j2\pi ft}dt$. Now consider the time function $s_1(t) = S(t)$ (be clear on what S means). The Fourier transform of $s_1(t)$ is $S_1(f) = \int_{-\infty}^{\infty} S(t)e^{-j2\pi ft}dt$. For the moment, call the frequency variable λ. Then $S_1(\lambda) = \int_{-\infty}^{\infty} S(t)e^{-j2\pi\lambda t}dt$. But t is a dummy variable and can be changed to a variable f. Therefore $S_1(\lambda) = \int_{-\infty}^{\infty} S(f)e^{j2\pi(-\lambda)f}df$. Compare the RHS of this expression with that for $s(t)$ and realize that it is $s(-\lambda)$. Restoring the usual name for the frequency variable one has $S_1(f) = s(-f)$. That is,

$$S(t) \longleftrightarrow s(-f), \text{ or } S(-t) \longleftrightarrow s(f). \tag{2.91}$$

As an example consider the time function $s(t) = A\sin(at)/at$. To determine the Fourier transform $S(f) = \int_{-\infty}^{\infty} A(\sin(at)/at)e^{-j2\pi ft}dt$ would tax either your integration skills or your integration tables. However, from Example 2.11 we expect, by duality, that the spectrum will be rectangular. This is shown using the duality relationship as follows. Start with the known Fourier transform pair:

$$s(t) = V\left[u\left(t+\frac{T}{2}\right) - u\left(t-\frac{T}{2}\right)\right] \longleftrightarrow S(f) = VT\frac{\sin(\pi fT)}{(\pi fT)}. \tag{2.92}$$

Let $A = VT$ and $a = \pi T$. Then $s(t) = VT\sin(\pi tT)/(\pi tT) = S(-t)$. Therefore $\mathcal{F}\{s(t)\} = s(f) = V\left[u(f+T/2) - u(f-T/2)\right]$. It is customary to let $T/2 = W$ in which case we have the Fourier transform pair:

$$s(t) = 2VW\frac{\sin(2\pi Wt)}{2\pi Wt} \longleftrightarrow S(f) = V\left[u(f+W) - u(f-W)\right]. \tag{2.93}$$

2.4.4 Relationship between Fourier series and the Fourier transform

The Fourier transform was introduced as a limiting case of the Fourier series. We now relate the Fourier series coefficients to the Fourier transform. Given the Fourier transform of the basic pulse, $p(t)$, that defines the periodic signal over $[-T/2, T/2]$, one can readily determine the Fourier series coefficients, D_k. Consider

$$P(f) = \int_{-\infty}^{\infty} p(t)e^{-j2\pi ft}dt = \int_{-T/2}^{T/2} p(t)e^{-j2\pi ft}dt, \tag{2.94}$$

where the second equality follows from the fact that $p(t) = 0$ outside $[-T/2, T/2]$. On the other hand, the Fourier series coefficients of the periodic signal are

$$D_k = \frac{1}{T}\int_{-T/2}^{T/2} p(t)e^{-j2\pi kf_r t}dt. \tag{2.95}$$

Comparing (2.94) and (2.95) gives

$$D_k = \frac{1}{T}P(f)\bigg|_{f=kf_r}. \tag{2.96}$$

One may also determine the Fourier transform of a periodic signal in terms of the Fourier series expansion $s(t) = \sum_{k=-\infty}^{\infty} D_k e^{j2\pi k f_r t}$. It is simply $S(f) = \mathcal{F}\{s(t)\} = \sum_{k=-\infty}^{\infty} D_k \delta(f - kf_r)$, which is a train of impulses located at the harmonics kf_r. Recall that $S(f)$ is a *density* indicating the concentration of the amplitude of $s(t)$ in the frequency domain. Since these amplitudes are concentrated at discrete frequencies this extreme concentration results in an impulse of strength $|D_k|$ and phase $\angle D_k$.

There are two more important topics with regard to the frequency representation of non-periodic signals that remain to be developed: the concentration of energy (or power) in the frequency domain and a lower bound on the time–bandwidth product. Both topics have implications for the bandwidth transmission needed to transmit the signal with reasonable fidelity. But before delving into these topics we digress slightly and discuss linear, time-invariant (LTI) systems, in particular the input/output relationship in both the time and frequency domains.

2.5 Input/output relationship of linear, time-invariant systems

There are two distinct approaches to characterizing the input/output relationship of an LTI system. The first approach, if the system is analog, is to obtain the differential equation relating the output to the input (or the difference equation if the system is discrete). The second approach is to apply a test input as shown in Figure 2.25 and use the response to determine the system's output for any other input. The first approach requires us to know in detail the structure of the system, whereas in the second nothing needs to be known about the internal system structure, it is the so-called black box approach. However, the

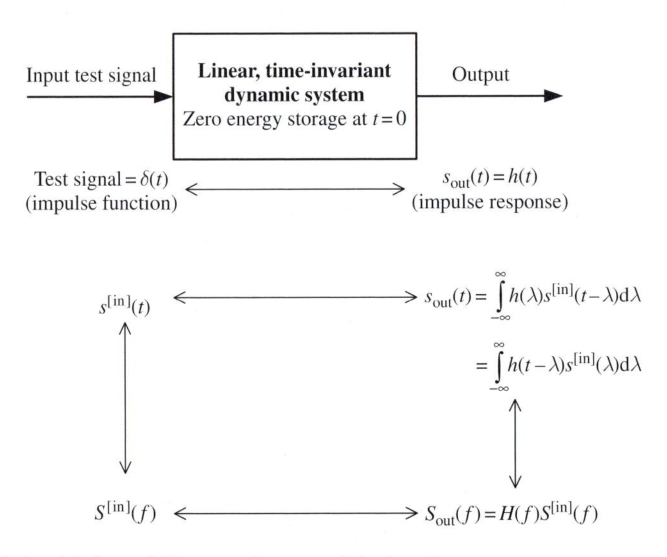

Fig. 2.25 Input/output relationship for an LTI system in terms of the impulse response.

system must be in a relaxed state, i.e., there is no energy stored in it, when the test signal is applied. This assures us that the response we see is due only to the test input.

The black box approach, the one developed here, relies heavily on the superposition property of an LTI system. Superposition means that if the individual responses to two different inputs are known, then the response to an input signal which is a weighted sum of the two inputs is the same weighted sum of the individual responses. Mathematically, if

$$s_1^{[\mathrm{in}]}(t) \longrightarrow s_{\mathrm{out}}^{[s_1]}(t),\ s_2^{[\mathrm{in}]}(t) \longrightarrow s_{\mathrm{out}}^{[s_2]}(t),$$

then

$$s^{[\mathrm{in}]}(t) = \alpha s_1^{[\mathrm{in}]}(t) + \beta s_2^{[\mathrm{in}]}(t) \longrightarrow \alpha s_{\mathrm{out}}^{[s_1]}(t) + \beta s_{\mathrm{out}}^{[s_2]}(t).$$

The approach also depends on the time-invariant property which states that if input $s^{[\mathrm{in}]}(t)$ produces $s_{\mathrm{out}}(t)$, then shifting the input in time shifts the output *accordingly*, i.e., $s^{[\mathrm{in}]}(t-\tau) \longrightarrow s_{\mathrm{out}}(t-\tau)$. Though there is more than one potential candidate that may be chosen for a test signal the one selected here is the *impulse function*. Recall that in the frequency domain the impulse function has equal amplitude at all frequencies. It thus excites the system simultaneously at all frequencies. The system response to a unit impulse applied at $t = 0$ is called the *impulse response* of the system and is denoted typically by $h(t)$.

It is now necessary to determine the output to any input in terms of the impulse response. To this end express the input by the seemingly facetious equation $s^{[\mathrm{in}]}(t) = \int_{\lambda=-\infty}^{\infty} s^{[\mathrm{in}]}(\lambda)\delta(t-\lambda)\mathrm{d}\lambda$ and express the integral as the limiting operation of the sum $s^{[\mathrm{in}]}(t) = \lim_{\Delta\lambda\to 0}\sum_{k=-\infty}^{\infty} s^{[\mathrm{in}]}(k\Delta\lambda)\delta(t-k\Delta\lambda)\Delta\lambda$. Consider each term of the sum to be an individual input to the system. Since it is an impulse of strength $s^{[\mathrm{in}]}(k\Delta\lambda)$ applied at $t = k\Delta\lambda$ the corresponding output, using the time-invariant property, is $s^{[\mathrm{in}]}(k\Delta\lambda)h(t-k\Delta\lambda)$. Now use superposition to determine the total output, $s_{\mathrm{out}}(t) = \lim_{\Delta\lambda\to 0}\sum_{k=-\infty}^{\infty} s^{[\mathrm{in}]}(k\Delta\lambda)h(t-k\Delta\lambda)\Delta\lambda$. In the limit the sum becomes an integration:

$$s_{\mathrm{out}}(t) = \int_{-\infty}^{\infty} s^{[\mathrm{in}]}(\lambda)h(t-\lambda)\mathrm{d}\lambda, \tag{2.97}$$

which is recognized as a convolution.

Therefore for an LTI system, for any input, the output is a convolution of the input with the system's impulse response, $h(t)$. Using *Property 8* of Table 2.3 it is seen that in the frequency domain the input and output are related by a simple product:

$$S_{\mathrm{out}}(f) = S^{[\mathrm{in}]}(f)H(f), \tag{2.98}$$

where $H(f)$ is called the transfer function of the system. It represents a complex gain, i.e., it modifies the gain and phase at any specific frequency where $|S_{\mathrm{out}}(f)| = |S^{[\mathrm{in}]}(f)| \times |H(f)|$ and $\angle S_{\mathrm{out}}(f) = \angle S^{[\mathrm{in}]}(f) + \angle H(f)$. The unit of $H(f)$ (and $|H(f)|$) is the unit of $s_{\mathrm{out}}(t)$ divided by that of $s^{[\mathrm{in}]}(t)$. Note that even when both input and output have the same units, say volts, it is a common practice to refer to the unit of $H(f)$ as volt/volt.

2.5.1 Energy/power relationships for a signal

As in the case of periodic signals it is important for nonperiodic signals to know how the signal's energy or power is distributed in the frequency domain. This knowledge allows one to make judgements regarding the amount of spectrum a signal occupies, the bandwidth needed to transmit the signal, how to design a filter to suppress unwanted signals, etc. Energy signals are dealt with here. The results are readily extended to power signals and this is done in the problem set at the end of the chapter. The energy of a nonperiodic signal is given by

$$E = \int_{-\infty}^{\infty} s^2(t)\mathrm{d}t \quad \text{(joules)}, \tag{2.99}$$

where for our purposes we assume $s(t)$ is a real voltage signal applied to a 1 ohm resistor so that $s^2(t) = s^2(t)/(1 \text{ ohm})$, $\left(\text{volts}^2/\text{ohm} = \text{watts}\right)$ is the instantaneous power. Now (2.99) can be written as

$$\begin{aligned}\int_{-\infty}^{\infty} s(t)s^*(t)\mathrm{d}t &= \int_{t=-\infty}^{\infty} \mathrm{d}t s(t) \int_{f=-\infty}^{\infty} S(f)\mathrm{e}^{\mathrm{j}2\pi ft}\mathrm{d}f \\ &= \int_{f=-\infty}^{\infty} \mathrm{d}f S(f) \int_{t=-\infty}^{\infty} s(t)\mathrm{e}^{\mathrm{j}2\pi ft}\mathrm{d}t.\end{aligned} \tag{2.100}$$

But the inner integral is $S(-f) = S^*(f)$. Therefore

$$E = \int_{-\infty}^{\infty} s^2(t)\mathrm{d}t = \int_{-\infty}^{\infty} S(f)S^*(f)\mathrm{d}f = \int_{-\infty}^{\infty} |S(f)|^2\mathrm{d}f \quad \text{(joules)}, \tag{2.101}$$

which is *Parseval's theorem* for nonperiodic energy signals. Note that the unit of $|S(f)|^2$ is joules/hertz. Therefore $|S(f)|^2$ is an *energy spectral density* and tells us how the energy is distributed in the frequency domain.

Furthermore, $|S(f)|^2$ is a function of f and as such, via the inverse Fourier transform, represents (or corresponds to) a function in the time domain. This function is not $s(t)$ but it is of interest to determine how this function is related to $s(t)$. So let us try to determine the inverse transform of $|S(f)|^2 = S(f)S^*(f)$. This inverse we shall call $R_s(\tau)$, where τ is a time variable and is named so for reasons that shall be given later. The inverse transform is

$$\begin{aligned}R_s(\tau) &= \int_{-\infty}^{\infty} S(f)S^*(f)\mathrm{e}^{\mathrm{j}2\pi f\tau}\mathrm{d}f \\ &= \int_{f=-\infty}^{\infty} \left[\int_{t=-\infty}^{\infty} s(t)\mathrm{e}^{-\mathrm{j}2\pi ft}\mathrm{d}t\right]\left[\int_{\lambda=-\infty}^{\infty} s(\lambda)\mathrm{e}^{\mathrm{j}2\pi f\lambda}\mathrm{d}\lambda\right]\mathrm{e}^{\mathrm{j}2\pi f\tau}\mathrm{d}f.\end{aligned} \tag{2.102}$$

Integrating first with respect to variable f gives

$$R_s(\tau) = \int_{t=-\infty}^{\infty} \mathrm{d}t s(t) \int_{\lambda=-\infty}^{\infty} \mathrm{d}\lambda s(\lambda) \int_{f=-\infty}^{\infty} \mathrm{e}^{\mathrm{j}2\pi(-t+\tau+\lambda)f}\mathrm{d}f. \tag{2.103}$$

But the integral with respect to f we recognize as an impulse function, $\int_{f=-\infty}^{\infty} \mathrm{e}^{\mathrm{j}2\pi(-t+\tau+\lambda)f}$ $\mathrm{d}f = \delta(-t+\tau+\lambda)$, which then sifts out the value of $s(\lambda)$ at $\lambda = t-\tau$, $\int_{\lambda=-\infty}^{\infty} \mathrm{d}\lambda s(\lambda)$ $\delta(-t+\tau+\lambda) = s(t-\tau)$. Finally,

$$R_s(\tau) = \int_{t=-\infty}^{\infty} s(t)s(t-\tau)\mathrm{d}t, \tag{2.104}$$

a time function that is called the *autocorrelation* of $s(t)$. The time variable, τ, is called the *delay* since to compute $R_s(\tau)$, $s(t)$ is delayed by τ to form $s(t-\tau)$, the product is formed and then the area under the product is determined. Here, the subscript s refers to the signal whose autocorrelation we are talking about and R is a commonly used symbol to denote correlation.

Though $R_s(\tau) \longleftrightarrow |S(f)|^2$ is a unique Fourier transform pair, it is important to realize that different signals, $s(t)$, can have the same autocorrelation function. This is due to the fact that phase information is ignored and phase plays as an important role as magnitude in determining the actual time function, $s(t)$. Thus we can take the magnitude function, $|S(f)|$, attach any phase function $\theta(f)$ we wish, the only restriction is that it be an odd function of frequency, and end up with the same correlation function. An example of this is the two time functions $s_1(t) = 1/(1+t^2)$ and $s_2(t) = t/(1+t^2)$. They have the same autocorrelation function and hence their energy is identically distributed in the frequency domain.

Parseval's relationship can be generalized to

$$\int_{-\infty}^{\infty} s_1(t)s_2(t)\mathrm{d}t = \int_{-\infty}^{\infty} S_1(f)S_2^*(f)\mathrm{d}f = \int_{-\infty}^{\infty} S_1^*(f)S_2(f)\mathrm{d}f. \tag{2.105}$$

The two integrands $S_1(f)S_2^*(f)$ and $S_1^*(f)S_2(f)$, though when integrated result in the same value, are different functions and in general represent different time functions. To see what time function $S_1(f)S_2^*(f)$ corresponds to proceed as in the case of the autocorrelation to obtain

$$R_{s_1(t)s_2(t)}(\tau) = \int_{-\infty}^{\infty} s_1(t)s_2(t-\tau)\mathrm{d}t \longleftrightarrow S_1(f)S_2^*(f), \tag{2.106a}$$

$$R_{s_2(t)s_1(t)}(\tau) = \int_{-\infty}^{\infty} s_2(t)s_1(t-\tau)\mathrm{d}t \longleftrightarrow S_2(f)S_1^*(f). \tag{2.106b}$$

The time domain operation is called *crosscorrelation*, i.e., the correlation of signal $s_1(t)$ with signal $s_2(t)$. The variable τ again is interpreted as a delay variable where either $s_1(t)$ is delayed with respect to $s_2(t)$ or vice versa. Note that $R_{s_1(t)s_2(t)}(\tau) \neq R_{s_2(t)s_1(t)}(\tau)$ and that the subscript notation signifies that the second signal is the delayed one, though the notation is not universal. It can be shown that $R_{s_1(t)s_2(t)}(\tau) = R_{s_2(t)s_1(t)}(-\tau)$.

In the frequency domain the crosscorrelation function can be interpreted as a cross energy spectral density while in the time domain it is a measure of the similarity between the two time functions and at what time(s), or value of τ this similarity is most pronounced. It is this latter interpretation that is of interest to us and which we now illustrate with an example.

Example 2.18 a toy radar problem Consider the highly simplified radar detection problem shown in Figure 2.26. A signal, $p(t)$, known to the sender (here we take it to be a square pulse), is transmitted to determine the range of the incoming aircraft. The signal is reflected off the aircraft and the received signal is an attenuated, delayed, and corrupted

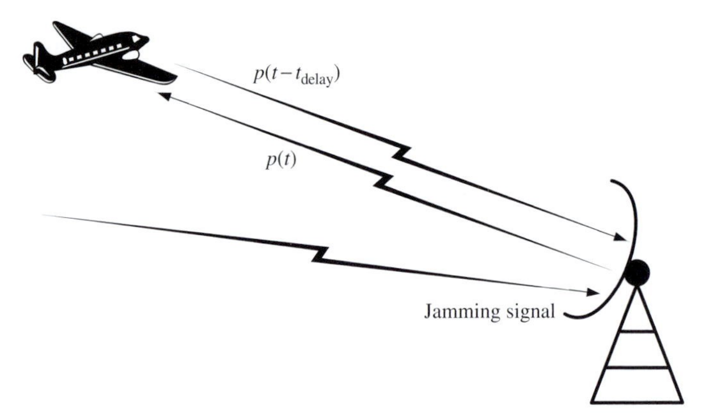

Fig. 2.26 Illustration of the toy radar problem.

version of the transmitted signal, $r(t) = Kp(t - t_{\text{delay}}) + n_{\text{jam}}(t) + n_{\text{noise}}(t)$, where $K \ll 1$ and the transmitted pulse duration $T \ll t_{\text{delay}}$. The term $n_{\text{noise}}(t)$ represents the ever present background thermal noise, while $n_{\text{jam}}(t)$ is an enemy jamming signal. The enemy is not very sophisticated and has limited technology. Thus the jamming signal is the simplest possible, a high-power, high-frequency sinusoid, $n_{\text{jam}}(t) = V_{\text{jam}} \cos(2\pi f_{\text{jam}} t + \theta)$, where $T_{\text{jam}} = 1/f_{\text{jam}} \ll T$. It is desired to estimate t_{delay} which will be used to determine the aircraft range from the relationship, range $= (c \times t_{\text{delay}})/2$, where c is the speed of light.

To obtain this estimate we crosscorrelate $r(t)$ with $p(t)$, the premise being that $p(t)$ is not correlated with either noise component but should be highly correlated with the reflected pulse at delay time t_{delay},

$$\begin{aligned} R_{r(t)p(t)}(\tau) &= \int_{-\infty}^{\infty} r(t)p(t-\tau)\mathrm{d}t \\ &= \int_{-\infty}^{\infty} p(t - t_{\text{delay}})p(t-\tau)\mathrm{d}t + \int_{-\infty}^{\infty} n_{\text{jam}}(t)p(t-\tau)\mathrm{d}t \\ &+ \int_{-\infty}^{\infty} n_{\text{noise}}(t)p(t-\tau)\mathrm{d}t. \end{aligned} \tag{2.107}$$

The crosscorrelation has three components: (i) $\int_{-\infty}^{\infty} p(t - t_{\text{delay}})p(t-\tau)\mathrm{d}t$, which intuitively we expect to be maximum at $\tau = t_{\text{delay}}$. This indeed is the case since in general $R_s(0) \geq R_s(\tau)$ (the proof is left as an exercise). (ii) $\int_{-\infty}^{\infty} n_{\text{jam}}(t)p(t-\tau)\mathrm{d}t = \int_{-\infty}^{\infty} V_{\text{jam}}(t)\cos(2\pi f_{\text{jam}} t + \theta)p(t-\tau)\mathrm{d}t$, which, since $f_{\text{jam}} \gg T$, crosscorrelates out to zero since it is simply the area under the sinusoid over T seconds. (iii) $\int_{-\infty}^{\infty} n_{\text{noise}}(t)p(t-\tau)\mathrm{d}t$ which is also the area under the background noise over T seconds and again intuitively we expect this to be close to zero. Therefore a good estimate of t_{delay} can be obtained from the time at which the crosscorrelation, $R_{r(t)p(t)}(\tau)$, is maximum. This is the delay τ at which $r(t)$ and $p(t)$ are most "similar." Figure 2.27 illustrates the above discussion. ■

The concepts of energy spectral density, and auto- and crosscorrelation are equally applicable for power signals. The approach is similar to that just presented for energy

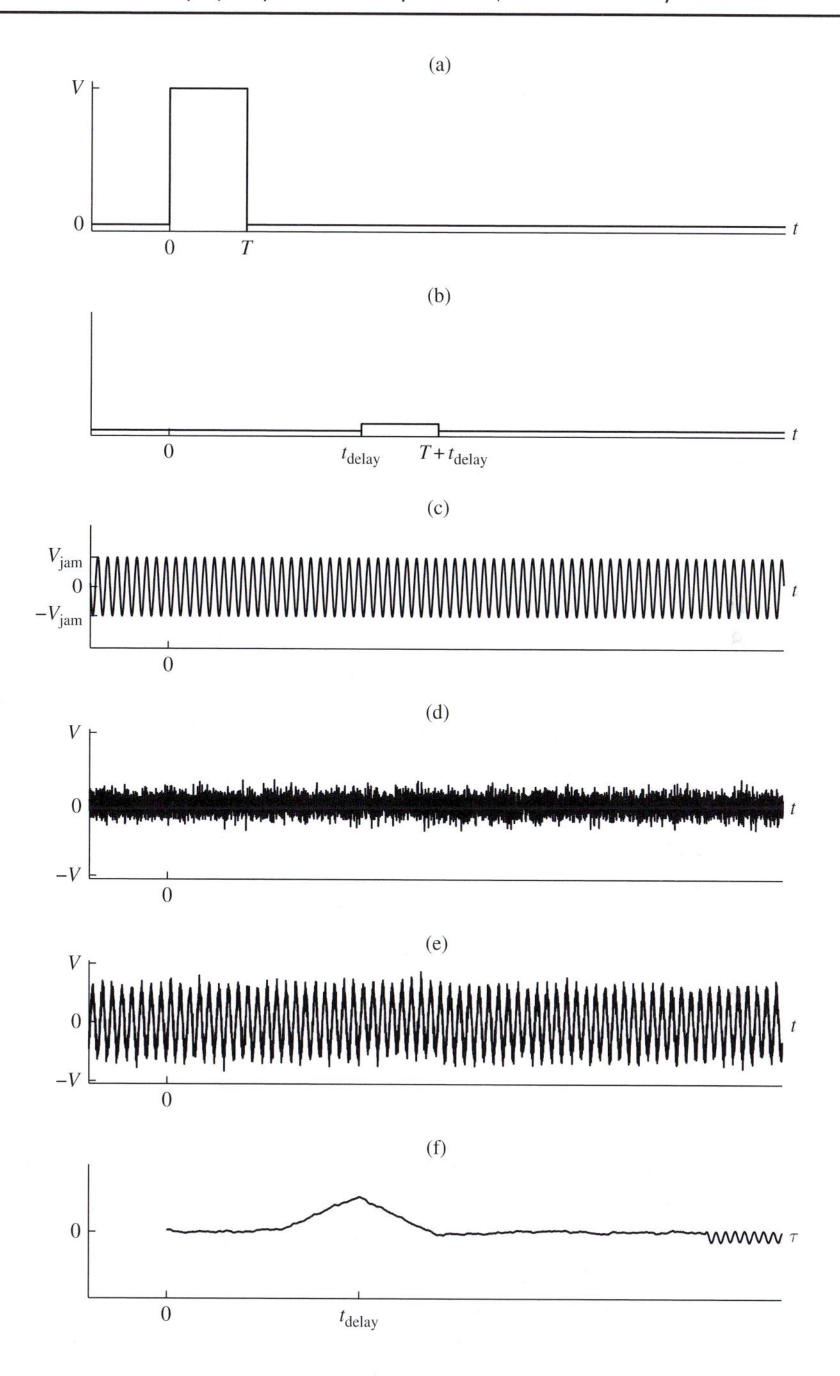

Fig. 2.27 Estimation of t_{delay}: (a) transmitted pulse, $p(t)$; (b) reflected pulse $p(t - t_{delay})$; (c) jamming signal, $n_{jam}(t)$; (d) background noise, $n_{noise}(t)$; (e) received signal $r(t) = Kp(t - t_{delay}) + n_{jam}(t) + n_{noise}(t)$; (f) crosscorrelation between $r(t)$ and $p(t)$.

signals. It, along with various general properties of the correlation functions, is developed through a set of problems at the end of this chapter.

2.6 Time–bandwidth product

A time-limited signal, i.e., one that is nonzero only over a finite time interval cannot be bandlimited in the frequency domain and by *duality* the opposite is true. [11] Thus it appears that the time–bandwidth product is always infinite, at least mathematically. It therefore behooves engineers to come up with time duration, bandwidth duration definitions that are sensible physically and tractable mathematically. The definitions chosen here, though not the only ones possible, are based on energy. They are measures that indicate how a signal's energy is concentrated in the time and frequency domains.

Consider the arbitrary waveform, $s(t)$, shown in Figure 2.28. Then $s^2(t)$, which is also shown in Figure 2.28, is a signal that represents the instantaneous power dissipated by $s(t)$ in a 1 ohm resistor. We now define a measure of the effective width of $s^2(t)$. For any time waveform one can find an axis of symmetry on the time axis such that $\int_{-\infty}^{\infty}(t-t_0)s^2(t)\mathrm{d}t = 0$. A mechanical engineer would call t_0 the center of gravity, a statistician the first moment,

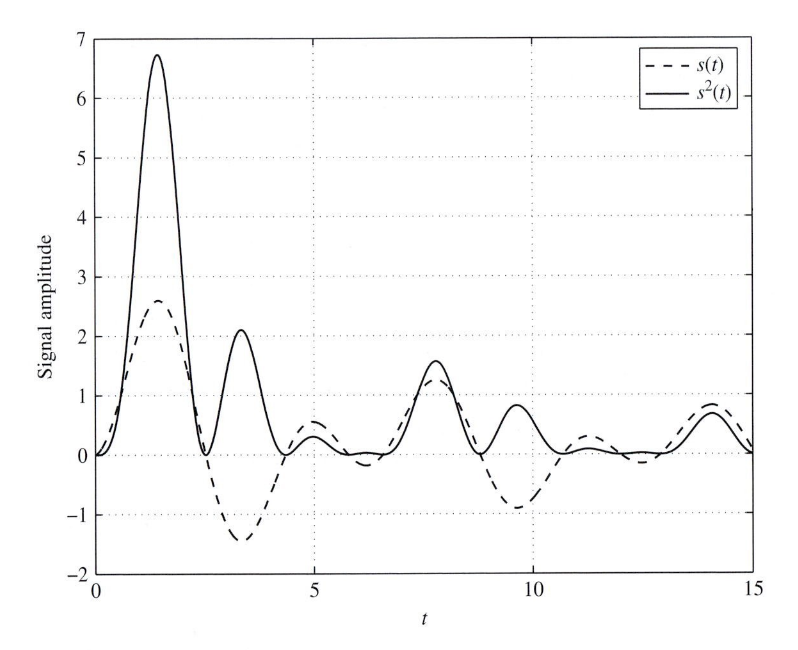

Fig. 2.28 An arbitrary waveform and its squared waveform.

[11] The sampling theorem is used to prove this. See http://en.wikipedia.org/wiki/Bandlimited.

an interior designer or carpenter the picture hanging moment. As a measure of the concentration of the instantaneous power of $s(t)$ around t_0, consider $\int_{-\infty}^{\infty}(t-t_0)^2 s^2(t)\mathrm{d}t$. Again for a mechanical engineer this is the moment of inertia and for a statistician the second central moment; it is not known what an interior designer or carpenter would call it. Now consider all the energy of $s(t)$, $\int_{-\infty}^{\infty} s^2(t)\mathrm{d}t$, to be concentrated at a single point at a distance that gives the same moment of inertia. This distance is known as the *radius of gyration* and we have (radius of gyration)$^2 \int_{-\infty}^{\infty} s^2(t)\mathrm{d}t = \int_{-\infty}^{\infty}(t-t_0)^2 s^2(t)\mathrm{d}t$. We define the radius of gyration to be the effective width or duration of the waveform $s(t)$. Denoting it by W_t we have

$$W_t^2 \equiv \frac{\int_{-\infty}^{\infty}(t-t_0)^2 s^2(t)\mathrm{d}t}{\int_{-\infty}^{\infty} s^2(t)\mathrm{d}t}. \tag{2.108}$$

As examples of this measure take the rectangular pulse of Figure 2.21 and the tent function of Figure 2.19(a). By inspection the axis of symmetry of both signals squared is $t_0 = 0$. For the rectangular pulse,

$$W_t^2 = \frac{\int_{-T/2}^{T/2} t^2 V^2 \mathrm{d}t}{V^2 T} = \frac{T^2}{12} \Rightarrow W_t = \frac{T}{2\sqrt{3}} \text{ (seconds)}. \tag{2.109}$$

As expected the width is proportional to T. For the tent function,

$$W_t^2 = \frac{V^2 \int_{-\infty}^{\infty} t^2 \mathrm{e}^{-2a|t|}\mathrm{d}t}{V^2 \int_{-\infty}^{\infty} \mathrm{e}^{-2a|t|}\mathrm{d}t} = \frac{\int_0^{\infty} t^2 \mathrm{e}^{-2at}\mathrm{d}t}{\int_0^{\infty} \mathrm{e}^{-2at}\mathrm{d}t} = \frac{1}{2a^2} \Rightarrow W_t = \frac{1}{\sqrt{2}a} \text{ (seconds)}. \tag{2.110}$$

The determined width is proportional to the time constant, $1/a$, again a result that is intuitively satisfying.

The width of the signal in the frequency domain is defined in the same way, except now we deal with the energy spectral density, $|S(f)|^2$. Since $|S(f)|^2$ is an even function the axis of symmetry is $f_0 = 0$. The width is defined as

$$W_f^2 \equiv \frac{\int_{-\infty}^{\infty} f^2 |S(f)|^2 \mathrm{d}f}{\int_{-\infty}^{\infty} |S(f)|^2 \mathrm{d}f}. \tag{2.111}$$

We are now in a position to prove a *lower bound* on the time–bandwidth product, $W_t W_f$, but before doing this we recast the numerator of (2.111) in the time domain by application of Parseval's theorem. Rewrite the numerator as $f^2|S(f)|^2 = \left[(\mathrm{j}2\pi f)S(f)(-\mathrm{j}2\pi f)S^*(f)\right](1/4\pi^2)$ and recall that if $s(t) \longleftrightarrow S(f)$ then $\mathrm{d}s(t)/\mathrm{d}t \longleftrightarrow (\mathrm{j}2\pi f)S(f)$. Therefore by Parseval's theorem

$$\int_{-\infty}^{\infty} |(\mathrm{j}2\pi f)S(f)|^2 \,\mathrm{d}f = \int_{-\infty}^{\infty} \left(\frac{\mathrm{d}s(t)}{\mathrm{d}t}\right)^2 \mathrm{d}t. \tag{2.112}$$

Therefore

$$W_f^2 \equiv \frac{\left(1/4\pi^2\right) \int_{-\infty}^{\infty} (\mathrm{d}s(t)/\mathrm{d}t)^2 \,\mathrm{d}t}{\int_{-\infty}^{\infty} s^2(t)\mathrm{d}t}. \tag{2.113}$$

To show that

$$W_t W_f \geq 1/4\pi \tag{2.114}$$

use (2.108) and (2.113) to write the product

$$W_t^2 W_f^2 = \frac{\left[\int_{-\infty}^{\infty}(t-t_0)^2 s^2(t)\mathrm{d}t\right]\left[(1/4\pi^2)\int_{-\infty}^{\infty}(\mathrm{d}s(t)/\mathrm{d}t)^2\,\mathrm{d}t\right]}{\left[\int_{-\infty}^{\infty}s^2(t)\mathrm{d}t\right]^2}. \tag{2.115}$$

Note that t_0 can be chosen to be 0 without loss of generality. Now invoke the Cauchy–Schwartz inequality, which states that $\int_a^b f^2(x)\mathrm{d}x \int_a^b g^2(x)\mathrm{d}x \geq [\int_a^b f(x)g(x)\mathrm{d}x]^2$ and apply it to the numerator. One then has

$$\left[\int_{-\infty}^{\infty} t^2 s^2(t)\mathrm{d}t\right]\left[\frac{1}{4\pi^2}\int_{-\infty}^{\infty}\left(\frac{\mathrm{d}s(t)}{\mathrm{d}t}\right)^2\mathrm{d}t\right] \geq \frac{1}{4\pi^2}\left[\int_{-\infty}^{\infty} ts(t)\frac{\mathrm{d}s(t)}{\mathrm{d}t}\mathrm{d}t\right]^2. \tag{2.116}$$

Integrate $\int_{t=-\infty}^{\infty} ts(t)\mathrm{d}s(t)/\mathrm{d}t = \int_{s=s(-\infty)}^{s=s(+\infty)} ts(t)\mathrm{d}s(t)$ by parts, with $u = t$, $\mathrm{d}v = s(t)\mathrm{d}s(t)$ and $v(t) = s^2(t)/2$. Then

$$\int_{s=s(-\infty)}^{s=s(+\infty)} ts(t)\mathrm{d}s(t) = \left.\frac{ts^2(t)}{2}\right|_{s=s(-\infty)}^{s=s(+\infty)} - \frac{1}{2}\int_{-\infty}^{\infty} s^2(t)\mathrm{d}t, \tag{2.117}$$

which, provided that $\lim_{t\to\pm\infty} ts^2(t) = 0$, is equal to $-\frac{1}{2}\int_{-\infty}^{\infty} s^2(t)\mathrm{d}t$. From this it follows that

$$W_t^2 W_f^2 \geq \frac{(1/4\pi^2)\left[-\frac{1}{2}\int_{-\infty}^{\infty}s^2(t)\mathrm{d}t\right]^2}{\left[\int_{-\infty}^{\infty}s^2(t)\mathrm{d}t\right]^2} = \frac{1}{16\pi^2}, \tag{2.118a}$$

or

$$W_t W_f \geq \frac{1}{4\pi}. \tag{2.118b}$$

As examples, let us determine the time–bandwidth products for the tent function (Example 2.9) and the Gaussian signal (Example 2.17).

Example 2.19 time-bandwidth product of the tent signal To determine W_f use (2.113). The quantity $(\mathrm{d}s(t)/\mathrm{d}t)^2$ is equal to $a^2\mathrm{e}^{-2a|t|}$ and therefore

$$W_f^2 = \frac{(1/4\pi^2)\int_{-\infty}^{\infty} a^2\mathrm{e}^{-2a|t|}\mathrm{d}t}{\int_{-\infty}^{\infty}\mathrm{e}^{-2a|t|}\mathrm{d}t} = \frac{a^2}{4\pi^2} \Rightarrow W_f = \frac{a}{2\pi}. \tag{2.119}$$

It follows that $W_t W_f = (1/\sqrt{2}a)(a/2\pi) = \sqrt{2}(1/4\pi)$ which is almost 50% larger than the lower bound. Note that the product is independent of the time constant, therefore to decrease the spectrum width one must either increase the time constant and hence the effective time duration, or the converse. It is the well-known adage: one cannot have one's cake and eat it too. ■

Example 2.20 time-bandwidth product of the Gaussian signal The Fourier transform pair is repeated from Example 2.17:

$$s(t) = V\mathrm{e}^{-at^2} \longleftrightarrow S(f) = V\sqrt{\frac{\pi}{a}}\mathrm{e}^{-\pi^2 f^2/a}. \tag{2.120}$$

To perform the necessary integration, we state two results, again one that was used in Example 2.17:[12]

$$\frac{1}{\sqrt{2\pi}\sigma}\int_{-\infty}^{\infty} e^{-x^2/2\sigma^2}\,dx = 1, \tag{2.121a}$$

and

$$\frac{1}{\sqrt{2\pi}\sigma}\int_{-\infty}^{\infty} x^2 e^{-x^2/2\sigma^2}\,dx = \sigma^2. \tag{2.121b}$$

Therefore

$$W_t^2 = \frac{\int_{-\infty}^{\infty} t^2 e^{-2at^2}\,dt}{\int_{-\infty}^{\infty} e^{-2at^2}\,dt}.$$

Now write the exponent as $2at^2 = t^2/2\left(1/2\sqrt{a}\right)^2$ and identify $\sigma^2 = \left(1/2\sqrt{a}\right)^2 = 1/4a$. The expression for the width becomes

$$W_t^2 = \frac{\sqrt{2\pi}\sigma\left[\frac{1}{\sqrt{2\pi}\sigma}\int_{-\infty}^{\infty} t^2 e^{-t^2/2\sigma^2}\,dt\right]}{\sqrt{2\pi}\sigma\left[\frac{1}{\sqrt{2\pi}\sigma}\int_{-\infty}^{\infty} e^{-t^2/2\sigma^2}\,dt\right]} = \sigma^2 = \frac{1}{4a}. \tag{2.122}$$

Turning now to the effective bandwidth, using (2.111) one has

$$W_f^2 = \frac{\int_{-\infty}^{\infty} f^2 e^{-2\pi^2 f^2/a}\,df}{\int_{-\infty}^{\infty} e^{-2\pi^2 f^2/a}\,df}.$$

As above, identify $\sigma^2 = a/4\pi^2$ and apply the identities in (2.121a) to arrive at $W_f^2 = a/4\pi^2$, or $W_f = \sqrt{a}/2\pi$. Finally,

$$W_t W_f = \frac{1}{2\sqrt{a}}\frac{\sqrt{a}}{2\pi} = \frac{1}{4\pi}, \tag{2.123}$$

which is the lower bound on the time–bandwidth product. So not only does the Gaussian signal have the interesting property that both its time and frequency descriptions are of the same functional form but it occupies the least possible bandwidth, at least with respect to the bandwidth measure here. ■

2.7 Summary

This chapter has presented the Fourier representation of deterministic signals, both periodic and nonperiodic. This representation leads to a frequency domain characterization of the signals, namely in terms of amplitude and energy/power spectrum densities. The representation is a powerful tool for design purposes, most signal processing/filtering techniques

[12] The proof of these results is left to the next chapter when the Gaussian random variable is discussed.

are based largely on this representation. However, signals can also be represented by other orthogonal sets of basis functions. Some of these sets are explored in the problems at the end of this chapter. These sets have not proven to be as useful for engineering practice with one notable exception. This is the case in which the set of deterministic time functions is finite in number, the situation in digital communications. Then a special orthogonal basis set is developed for detection purposes. This is fully explained in Chapter 5.

The chapter ends with a derivation of a lower bound on the time duration–frequency bandwidth product. It was based on a specific definition, one that makes engineering sense, of a signal's time duration and frequency bandwidth. Other definitions are possible that equally make for good engineering sense, one of which is given in the problem set. The problem set, hopefully, not only gives the reader further insight into the concepts of the chapter but it also develops further results based on these concepts. In particular the Hilbert transform and the baseband representation of passband signals is developed.

The reader is strongly encouraged to at least read through the problems.

2.8 Problems

The first two problems on Fourier series look at the interrelationships between the coefficient sets and properties of the coefficients. Familiarity with these properties can simplify computation of the coefficients and/or help in exposing errors in the computation.

2.1 The Fourier series of a *complex signal* can be expressed in three ways: (i) trigonometric with coefficient set $\{A_k, B_k\}$, (ii) amplitude/phase with coefficient set $\{C_k, \theta_k\}$, and (iii) complex exponential with coefficient set $\{D_k\}$. Given one set of coefficients show how the other two sets would be determined.

2.2 Prove the following properties of the Fourier series coefficients for a *real signal*:
(a) A_k is always an even function of k.
(b) B_k is always an odd function of k.
(c) C_k is always an even function of k.
(d) θ_k is always an odd function of k.
(e) $D_k^* = D_{-k}$ (always).

Note that the relationships and properties of Problems 2.1 and 2.2 have direct analogs for the Fourier transform.

2.3 A continuous-time periodic signal $s(t)$ is real-valued and has a fundamental period $T = 8$. The nonzero Fourier series coefficients for $s(t)$ are specified as

$$D_1 = D_{-1}^* = \mathrm{j}, \qquad D_5 = D_{-5} = 2.$$

Express $s(t)$ in the amplitude-phase and trigonometric forms.

2.4 (*Fourier series by inspection*) Consider, arguably, the simplest periodic signal: $s(t) = V\cos(2\pi f_c t + \alpha)$.

(a) Determine the Fourier series coefficients D_k of $s(t)$ for: (i) α general, (ii) $\alpha = 0$, (iii) $\alpha = \pi$, (iv) $\alpha = -\pi$, (v) $\alpha = \pi/2$, (vi) $\alpha = -\pi/2$.

(b) For each case of α above, plot the magnitude and phase spectra of $s(t)$.

(c) For each case of α, determine the coefficients $\{A_k, B_k\}$, $\{C_k, \theta_k\}$.

2.5 (*Another Fourier series by inspection*) Consider the continuous-time periodic signal: $s(t) = 2\sin(2\pi t - 3) + \sin(6\pi t)$.

(a) Determine the fundamental period and the Fourier series coefficients D_k.

(b) Plot the magnitude and phase spectra of $s(t)$.

2.6 Several symmetries and their influence on the form of the Fourier coefficients are discussed in the chapter. Classify the signals shown in Figure 2.29 as (a) even, (b) odd, (c) half-wave symmetric, (d) quarter-wave even symmetric, (e) quarter-wave odd symmetric, none of the preceding. Use as many symmetries as are applicable.

(f) If a DC value is added to any of the signals in Figure 2.29 how would it change the classification? How would it affect the value of the Fourier series coefficients? Explain.

(g) How would a time shift of τ seconds affect the classification? The coefficients?

2.7 Consider the signal $s(t) = s_1(t) + s_2(t)$, where $s_1(t)$ and $s_2(t)$ are periodic with the same period, T. What can you say about the symmetries or nonsymmetries of $s(t)$ for the following cases:

(a) $s_1(t)$ is even, $s_2(t)$ is even.

(b) $s_1(t)$ is even, $s_2(t)$ is odd.

(c) $s_1(t)$ and $s_2(t)$ are both half-wave symmetric.

(d) $s_1(t)$ is even, $s_2(t)$ is half-wave symmetric.

(e) $s_1(t)$ is odd, $s_2(t)$ is half-wave symmetric.

(f) $s_1(t)$, $s_2(t)$ are both quarter-wave symmetric, either even or odd.

(g) $s_1(t)$ is even, $s_2(t)$ is quarter-wave even symmetric.

2.8 Repeat Problem 2.7 for $s(t) = s_1(t)s_2(t)$.

2.9 (*Properties of Fourier series coefficients*) Figure 2.30 shows the magnitude and phase spectra of three continuous-time periodic signals.

(a) For each signal, determine whether the signal is real or complex. Explain.

(b) Which signal is a real-valued and even function? Which signal is a real-valued and odd function? Explain.

The next three problems introduce one to analog modulation concepts, both amplitude and frequency, where the message signal is the special one of a sinusoid.

2.10 (*Amplitude modulation with suppressed carrier*) Consider the signal:

$$s(t) = V_m \cos(2\pi f_m t) V_c \cos(2\pi f_c t), \tag{P2.1}$$

where $f_m \ll f_c$ and f_c/f_m is a rational number. It is easy enough to find the spectrum of $s(t)$ using the trigonometric relationship $\cos(x)\cos(y) = (\cos(x+y) + \cos(x-y))/2$. However, let $s_1(t) = V_m \cos(2\pi f_m t)$ and $s_2(t) = V_c \cos(2\pi f_c t)$ and use convolution to find the Fourier series of what is $s(t) = s_1(t)s_2(t)$.

2.11 (*Amplitude modulation with carrier*) Change the modulation process of Problem 2.10 slightly to the following:

$$s(t) = \left[V_{DC} + V_m \cos(2\pi f_m t)\right] V_c \cos(2\pi f_c t), \tag{P2.2}$$

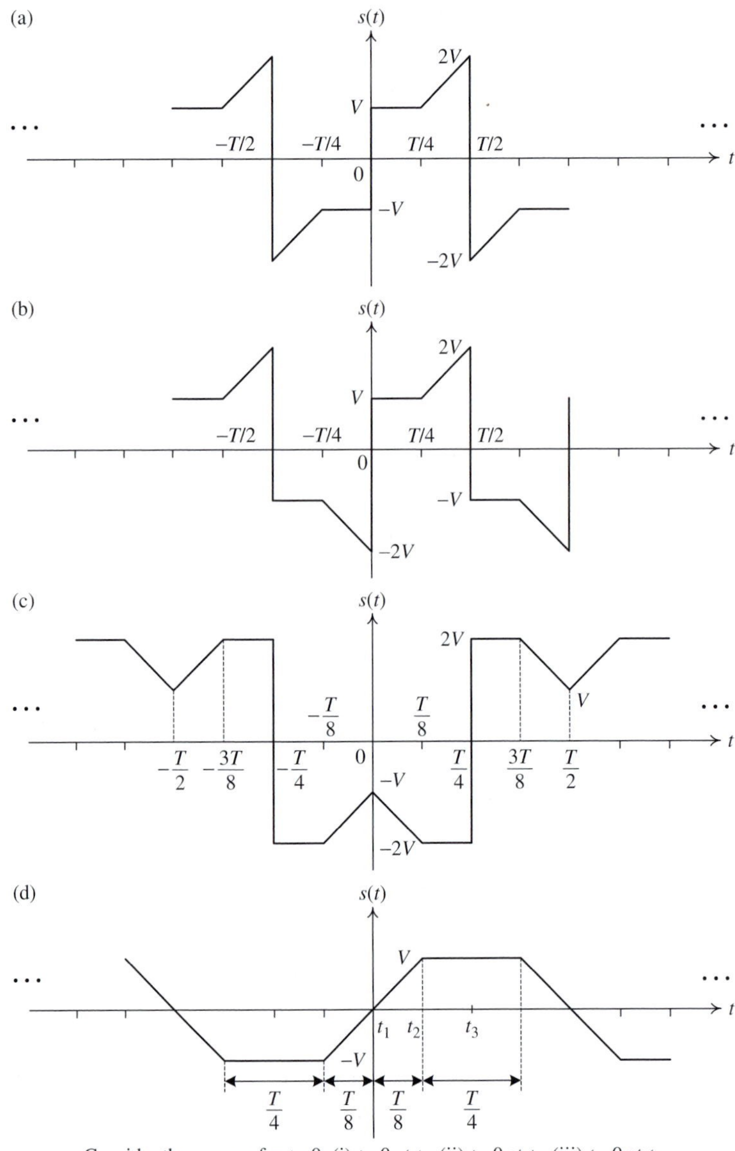

Consider three cases for $t=0$: (i) $t=0$ at t_1, (ii) $t=0$ at t_2, (iii) $t=0$ at t_3.

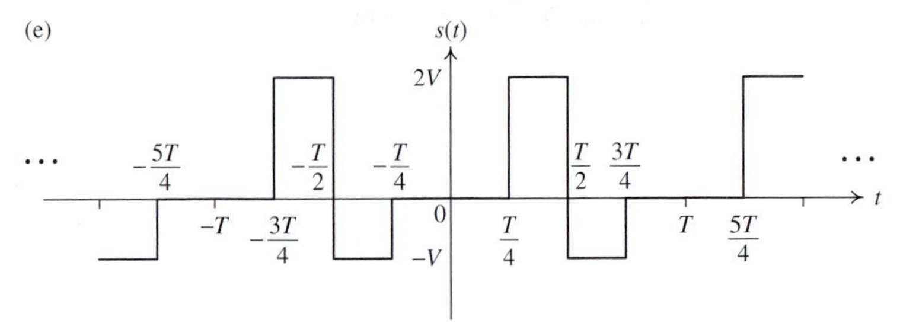

Fig. 2.29 Signals for Problem 2.6.

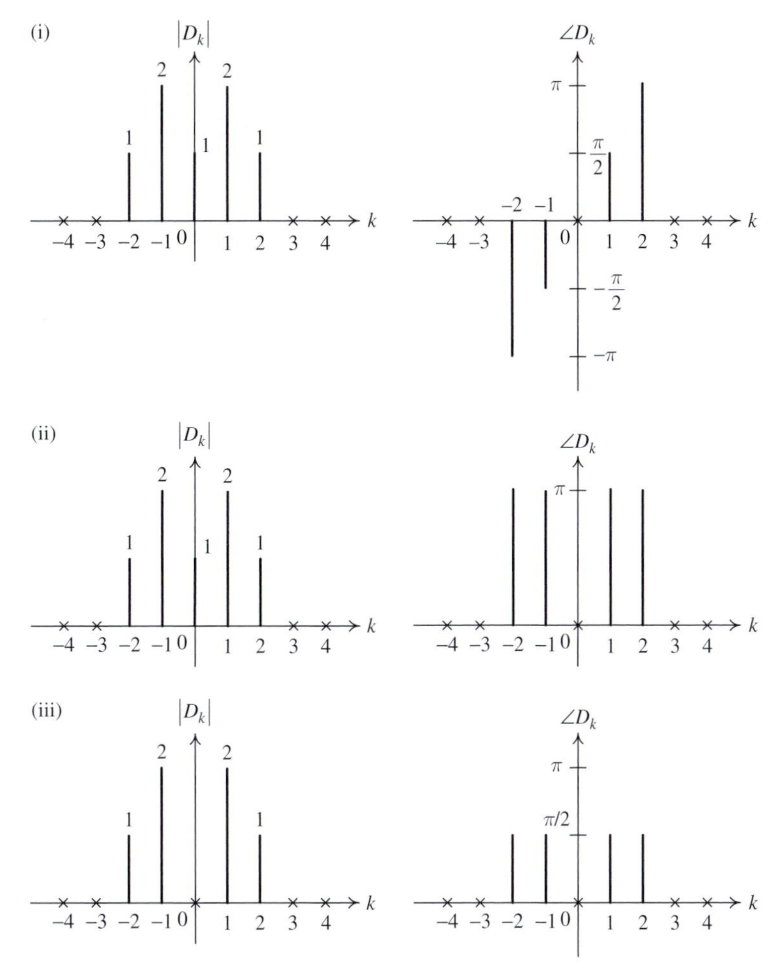

Fig. 2.30 Magnitude and phase spectra of three signals.

where $f_m \ll f_c$ and f_c/f_m is a rational number. Again determine the Fourier series using convolution and plot the two-sided spectrum. Compare the spectrum with that of Problem 2.10.

2.12 (*Frequency modulation*) Consider the signal

$$s(t) = V_c \cos\left(2\pi f_c t - V_m \sin(2\pi f_m t)\right), \tag{P2.3}$$

where $f_c = nf_m$, n an integer $\ggg 1$.

(a) The message $V_m \sin(2\pi f_m t)$, it is more appropriate to call it a test signal, controls the instantaneous frequency of the sinusoid $V_c \cos(2\pi f_c t)$. The instantaneous frequency of a sinusoid is $f_i(t) = (1/2\pi)\mathrm{d}\theta_i(t)/\mathrm{d}t$ (hertz) where $\theta_i(t)$ is the instantaneous phase, which here is $\theta_i(t) = 2\pi f_c t - V_m \sin(2\pi f_m t)$ (radians). Determine $f_i(t)$ and plot it.

(b) Show that $s(t)$ is periodic with period $T = 1/f_m$ (seconds).

(c) Since $s(t)$ is periodic it can be expanded in a Fourier series. To determine coefficients, D_k, write $s(t)$ as

$$s(t) = \mathcal{R}\left\{V_c e^{j[2\pi f_c t - V_m \sin(2\pi f_m t)]}\right\} \tag{P2.4}$$

and observe that $e^{j[2\pi f_c t - V_m \sin(2\pi f_m t)]}$ is a periodic signal, albeit a complex one, with period $T = 1/f_m$. Therefore find the Fourier series of it and then take the real part. D_k is given by, as usual, $D_k = \frac{1}{T}\int_{-T/2}^{T/2} e^{j[2\pi f_c t - V_m \sin(2\pi f_m t)]} e^{-j2\pi k f_m t} dt$. To do the integration, change the integration variable to $\lambda = 2\pi f_m t$ and recognize the resultant integral as a *Bessel function*, namely as $J_{n-k}(V_m)$.

2.13 Consider Figure 2.15. To check the convolution determine the Fourier series directly by the following steps.

(a) Determine the time domain functions $s_1(t)$ and $s_2(t)$ from the given Fourier coefficients. Recall that $\cos x = \frac{1}{2}\left(e^{jx} + e^{-jx}\right)$.

(b) Determine the product, $s(t) = s_1(t)s_2(t)$. Recall that

$$\cos x \cos y = \frac{1}{2}\left[\cos(x+y) + \cos(x-y)\right].$$

(c) Determine the Fourier series coefficients, $D_k^{[s(t)]}$. Do they agree with those of Figure 2.15?

2.14 Consider the output of a half-wave rectifier shown in Figure 2.31. Express the signal as $s(t) = s_1(t)V\cos(2\pi f_c t)$. Determine the Fourier coefficients of $s_1(t)$ and of $V\cos(2\pi f_c t)$, and use convolution to determine the Fourier coefficients of $s(t)$.

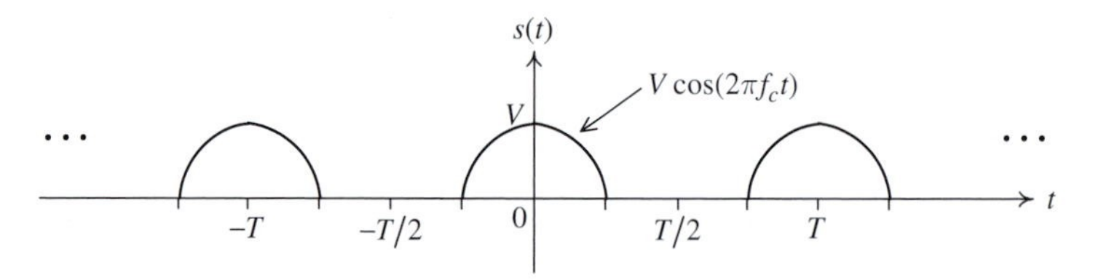

Fig. 2.31 Output of a half-wave rectifier.

2.15 Clipping or saturation is a common phenomenon in amplifiers leading to a distortion of the input signal. To investigate this distortion the frequency spectrum of the output is to be determined when the input is a sinusoid. The block diagram is shown in Figure 2.32.

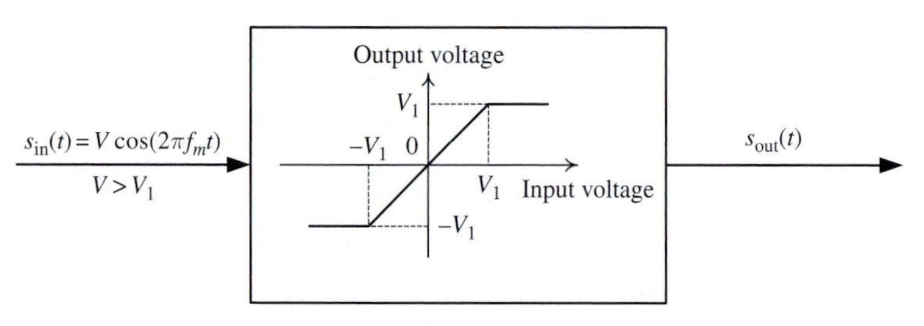

Fig. 2.32 Illustration of signal clipping.

Determine the Fourier series of $s_{out}(t)$. Then sketch the magnitude spectrum when $V = 2V_1$. *Hint* Model the output signal as $s_{out}(t) = s_1(t)s_{in}(t) + s_2(t)$, where $s_1(t)$, $s_2(t)$ are periodic signals chosen to represent the distortion. Use convolution and superposition to determine the Fourier series.

The following problems are concerned with the Fourier transform.

2.16 Consider the signal shown in Figure 2.33.

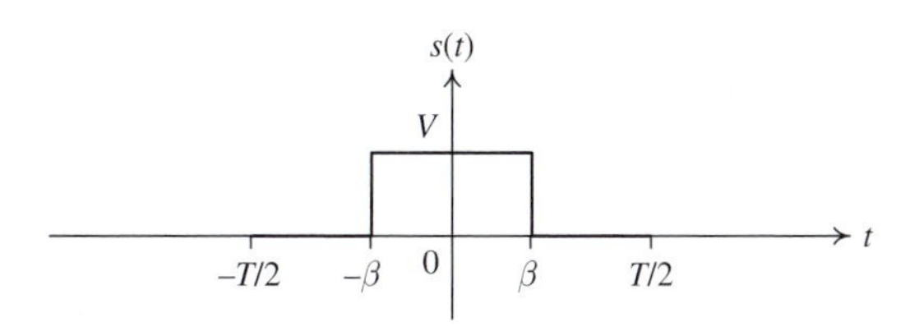

Fig. 2.33 Signal considered in Problem 2.16.

(a) Find the Fourier transform of $s(t)$.

(b) Now make the signal periodic by repeating the signal $s(t)$ every αT seconds, where $\alpha \geq 1$. Determine the Fourier coefficients of the periodic signal by using the relationship $D_k^{[\alpha T]} = (1/T_1)S(f)\big|_{f=kf_r}$, where $T_1 = \alpha T$.

(c) View the coefficients found in (b) as a function of α. Using Matlab plot on the same frequency and amplitude scale the coefficients for several values of α. Values of $V = \beta = 1, T = 4, \alpha = 1,\ 1.5,\ 2,\ 3,\ 5$ are suggested.

(d) Explain qualitatively what happens as α becomes larger and larger. What happens as $\alpha \to \infty$?

2.17 Find the Fourier transforms of the signals in Problem 2.3 and Problem 2.4(a). Sketch the magnitude and phase spectra.

2.18 (a) Determine the Fourier transforms of the following signals: (i) $s(t) = V_1 \cos(2\pi\sqrt{2}t) + V_2 \cos(4\pi t)$, (ii) $s(t) = V_1 \cos(2\pi\sqrt{2}t) + V_2 \cos(4\pi\sqrt{2}t)$.

(b) If appropriate, determine the Fourier series for the two signals.

2.19 (*Amplitude modulation-suppressed carrier*) Consider a more general version of Problem 2.10. Let $s(t) = m(t)\cos(2\pi f_c t)$.

(a) Determine the Fourier transform $S(f)$.

(b) Let $m(t)$ be a bandlimited signal with $M(f) = 0$ for $|f| > W$. Sketch the spectrum of $S(f)$ for the case of $f_c \gg W$.

2.20 (*Frequency modulation*) Generalize Problem 2.12 (but only slightly) by not requiring that f_c and f_m be related harmonically. Write $s(t)$ as $s(t) = \mathcal{R}\left\{V_c e^{j2\pi f_c t} e^{-jV_m \sin(2\pi f_m t)}\right\}$. Find the Fourier series of $e^{-jV_m \sin(2\pi f_m t)}$, then the Fourier transform of the series and apply the shifting property that results from the multiplication by $e^{j2\pi f_c t}$.

2.21 Determine the Fourier transforms of the two signals shown in Figure 2.34.

Remark Though you are welcome to integrate the functions, another approach might be to differentiate them until impulses occur. Then find the Fourier transform of the impulses and use the differentiation property.

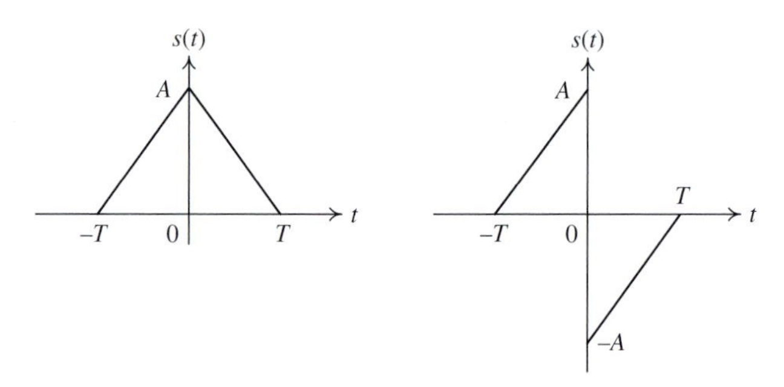

Fig. 2.34 Figure for Problem 2.21.

2.22 Use the Fourier transform of a rectangular pulse (already established in this chapter) and apply the *linear and time-shifting properties* of the Fourier transform to obtain the Fourier transform for the signal $s(t)$ shown in Figure 2.35.

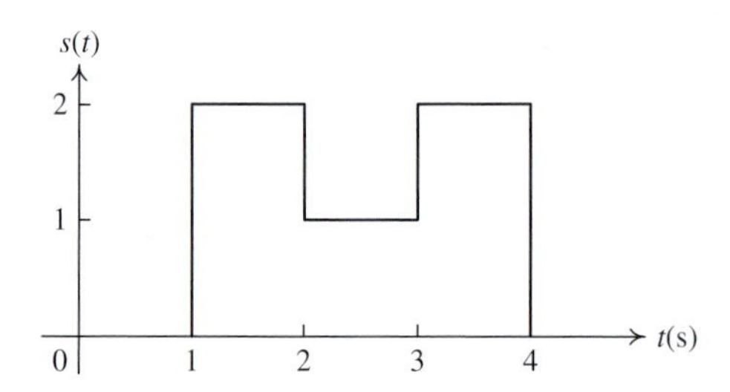

Fig. 2.35 Figure for Problem 2.22.

2.23 (*Inverse Fourier transform*) Determine and plot the continuous-time signal $s(t)$ with the magnitude and phase spectra shown in Figure 2.36.

2.24 (*Fourier transform of a half cosine*) Consider the windowed cosine in Figure 2.37 as a product of a rectangular pulse and a "pure" cosine whose Fourier transforms have already been established and use the convolution property to find the Fourier transform $S(f)$. Simplify the expression as much as you can. Then plot $S(f)$ for $T = 1$ (second) and $A = 1$ (volt).

2.25 (*Properties of Fourier transform*) Consider the signal $s(t)$ shown in Figure 2.38. Here we shall derive the Fourier transform of $s(t)$ using different properties of the Fourier transform.

(a) Use the Fourier transform of a rectangular pulse (already established in this chapter) and apply the *linear and time-shifting properties* to obtain the Fourier transform for $s(t)$.

(b) First find the Fourier transform of $\mathrm{d}s(t)/\mathrm{d}t$ and then apply the *differentiation property* to find the Fourier transform for $s(t)$.

2.26 (*Parseval's theorem for nonperiodic signals*) Consider real, aperiodic signals $x(t)$ and $y(t)$ with Fourier transforms $X(f)$ and $Y(f)$, respectively.

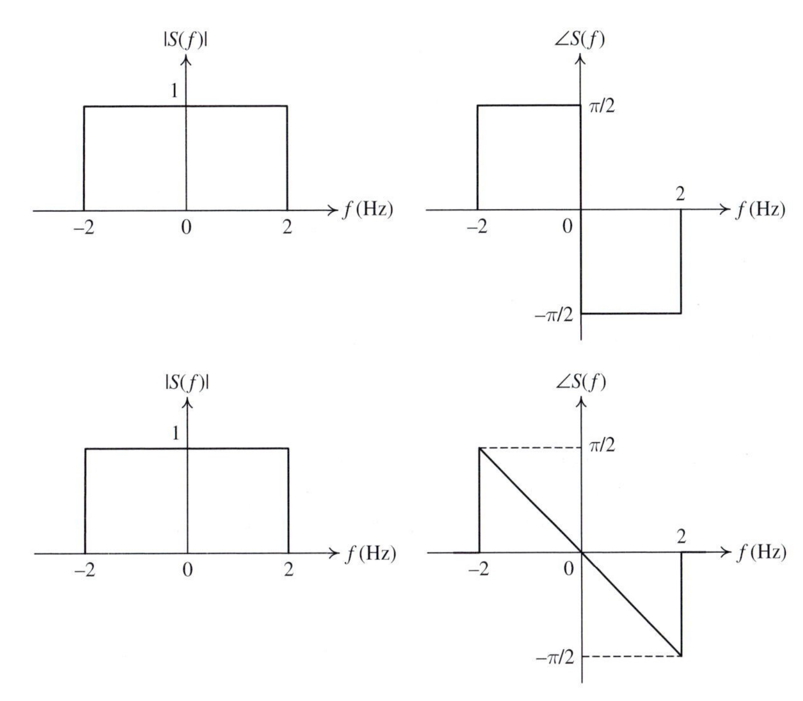

Fig. 2.36 Figures for Problem 2.23.

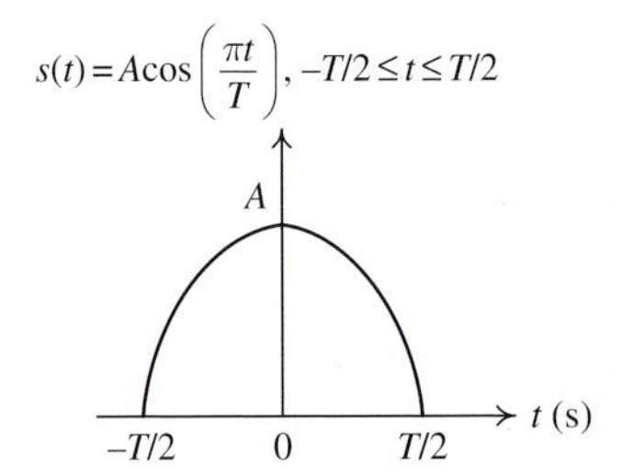

Fig. 2.37 A windowed cosine.

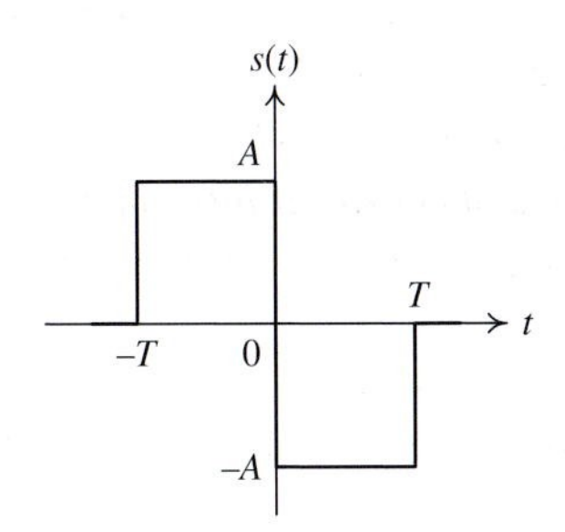

Fig. 2.38 Plot of $s(t)$ for Problem 2.25.

(a) Show that $X(-f) = X^*(f)$.

(b) Prove that

$$\int_{-\infty}^{\infty} x(t)y(t)\mathrm{d}t = \int_{-\infty}^{\infty} X(f)Y^*(f)\mathrm{d}f = \int_{-\infty}^{\infty} X^*(f)Y(f)\mathrm{d}f.$$

Write the above expression for the special case of $x(t) = y(t)$ and interpret it.

2.27 (*An application of Parseval's theorem*) It has been shown that

$$s(t) = \mathrm{e}^{-at}u(t) \overset{\mathcal{F}}{\longleftrightarrow} S(f) = \frac{1}{a + \mathrm{j}2\pi f}, \quad a > 0. \tag{P2.5}$$

(a) Compute the energy E_s of $s(t)$ in the time domain, i.e., by directly performing $\int_{-\infty}^{\infty} s^2(t)\mathrm{d}t$.

(b) Verify the result obtained in (a) using Parseval's theorem.

(c) Determine the frequency W of $s(t)$ so that the energy contributed by all the frequencies *below* W is 95% of the total signal energy E_s. *Hint* $\int \mathrm{d}f/(f^2 + a^2) = (1/a)\tan^{-1}(f/a)$.

2.28 (*Parseval's theorem for power signals*) A power signal is one that has infinite energy but finite average power, i.e., $\int_{-\infty}^{\infty} s^2(t)\mathrm{d}t = \infty$ but $P = \lim_{T\to\infty} \int_{-T/2}^{T/2} s^2(t)\mathrm{d}t = \lim_{T\to\infty} \int_{-\infty}^{\infty} s_T^2(t)\mathrm{d}t$ is finite. Here $s_T(t) = s(t)[u(t + T/2) - u(t - T/2)]$, i.e., $s(t)$ truncated to the interval $[-T/2, T/2]$.

(a) Employing an approach directly analogous to that used in this chapter for energy signals show that:

$$P = \lim_{T\to\infty} \int_{-\infty}^{\infty} s_T^2(t)\mathrm{d}t = \int_{-\infty}^{\infty} \lim_{T\to\infty} \frac{|S_T(f)|^2}{T}\mathrm{d}f. \tag{P2.6}$$

Keep in mind that $\int_{-\infty}^{\infty} \mathrm{e}^{\mathrm{j}2\pi xy}\mathrm{d}y = \delta(x)$.

(b) Do a dimensional analysis to determine the unit of P and $S(f) = \lim_{T\to\infty} |S_T(f)|^2/T$. What name would you assign to $S(f)$?

2.29 (*Autocorrelation for power signals*) Show that in the time domain, $S(f)$ in Problem 2.28 represents the time function

$$R(\tau) = \mathcal{F}^{-1}\{S(f)\} = \lim_{T\to\infty} \frac{1}{T}\int_{-\infty}^{\infty} s_T(t)s_T(t + \tau)\mathrm{d}t. \tag{P2.7}$$

2.30 Determine the autocorrelation, $R(\tau)$, and $S(f)$ for the following power signals:

(a) $s(t) = V$ (a DC signal).

(b) $s(t) = Vu(t)$ (half a DC signal).

(c) $s(t) = V\cos(2\pi f_c t + \theta)$.

2.31 Any periodic signal is a power signal. Show that its autocorrelation is also periodic.

2.32 (a) Differentiation in the time domain results in multiplication in the frequency domain. Use the concept of duality to complete the statement: *Multiplication in the time domain results in* ________________ *in the frequency domain.*

(b) Derive the exact relationship between the Fourier transform $S(f)$ of $s(t)$ and that of $s_1(t) = ts(t)$.

2.33 Given the Fourier transform pair $V\mathrm{e}^{-a|t|} \longleftrightarrow V2a/(a^2 + 4\pi^2 f^2)$, use duality to determine the Fourier transform of $1/(1 + t^2)$.

2.34 Use the results of Problems 2.32 and 2.33 to show that $s(t) = t/(1+t^2)$ and $S(f) = -\mathrm{j}\pi \mathrm{e}^{-2\pi|f|}\,\mathrm{sgn}(f)$ are a Fourier transform pair.

2.35 (*Cauchy–Schwartz inequality*) Consider two energy signals, $s_1(t)$ and $s_2(t)$. To prove the inequality form a quadratic in λ, where λ is an arbitrary real constant, $q(\lambda) = \int_{-\infty}^{\infty}[s_1(t) + \lambda s_2(t)]^2 \mathrm{d}t$, which obviously is nonnegative. Since $q(\lambda) \geq 0$, it cannot have real roots. Use this to prove that

$$\left|\int_{-\infty}^{\infty} s_1(t)s_2(t)\mathrm{d}t\right| \leq \sqrt{\left[\int_{-\infty}^{\infty} s_1^2(t)\mathrm{d}t\right]\left[\int_{-\infty}^{\infty} s_2^2(t)\mathrm{d}t\right]}. \quad \text{(P2.8)}$$

What is the relationship between $s_1(t)$ and $s_2(t)$ for equality to hold?

2.36 Use Parseval's theorem to express the Cauchy–Schwartz inequality in the frequency domain.

2.37 Use the Cauchy–Schwartz inequality to prove the autocorrelation property $|R_s(\tau)| \leq R_s(0)$. *Hint* Let $s_1(t) = s(t)$ and $s_2(t) = s(t+\tau)$.

2.38 Consider the following definitions of time and bandwidth durations based on equal area. Let both $s(t)$ and $|S(f)|$ be maximum at the origin. Define the time duration as the interval T such that $s(0)T = \int_{-\infty}^{\infty}|s(t)|dt$ and the frequency bandwidth W to satisfy $2S(0)W = \int_{-\infty}^{\infty}|S(f)|\mathrm{d}f$. Show that the time–bandwidth product WT is lower bounded by $1/2$.

Remark It is a common practice to take WT to be equal to 1. This will be used in later chapters as a quick means to determine the bandwidth of digital communication systems.

Neither the time-bandwidth product definition given in the chapter nor the one based on equal area defined in Problem 2.38 works for the signals of Examples 2.8 and 2.10. Since both signals have a discontinuity, their Fourier amplitude spectra do not decay rapidly enough with f and therefore the appropriate integrals do not exist. Here another bandwidth definition that is commonly used in communications is introduced.

2.39 Consider the signal of Example 2.8 with $V = 1$ volt.

(a) From either the time signal or the Fourier transform show that the total energy in the signal is $1/2a$ joules.

(b) Now determine the frequency bandwidth needed by an ideal lowpass filter to pass a certain percentage of the signal energy, i.e., find f_κ such that

$$2\int_0^{f_\kappa} |S(f)|^2 \mathrm{d}f = \kappa\left(\frac{1}{2a}\right), \quad \text{(P2.9)}$$

where $0 \leq \kappa \leq 1$. (Note: $\kappa = 0.9$ for 90%, etc.)

Hint Change the variable of integration so that f_κ is normalized and becomes $f_n = 2\pi f_\kappa/a$. Also the integral $\int \mathrm{d}x/(x^2+a^2) = (1/a)\tan^{-1}(x/a)$ is useful.

(c) As a measure of time duration, T, choose the equal-energy definition of Problem 2.38. Show that $T = 1/a$ and that the time duration–frequency bandwidth product $T \times f_\kappa = (1/2\pi)\tan(\pi\kappa/2)$.

(d) Plot $T \times f_\kappa$ versus κ. The suggested range on κ is 0.1–0.99. Interpret and comment on the plot.

2.40 Repeat Problem 2.39 for the signal in Example 2.10.

2.41 (*Asymptotic behavior of the amplitude spectrum*) Consider an energy signal, $s(t)$, that can be differentiated n times before an impulse appears. The nth derivative can therefore be written as $\mathrm{d}^n s(t)/\mathrm{d}t^n = s_1(t) + K\delta(t)$. Practically the Fourier transform of $s_1(t)$ is a ratio of two functions in f such that the limit as f tends to infinity is finite, typically zero. Show that $S(f)$ decays as $1/f^n$ for large f.

The following set of problems considers various versions of analog amplitude modulation. For illustrative purposes let $m(t)$ be a bandlimited signal, bandwidth W hertz with magnitude and phase spectra as shown in Figure 2.39.

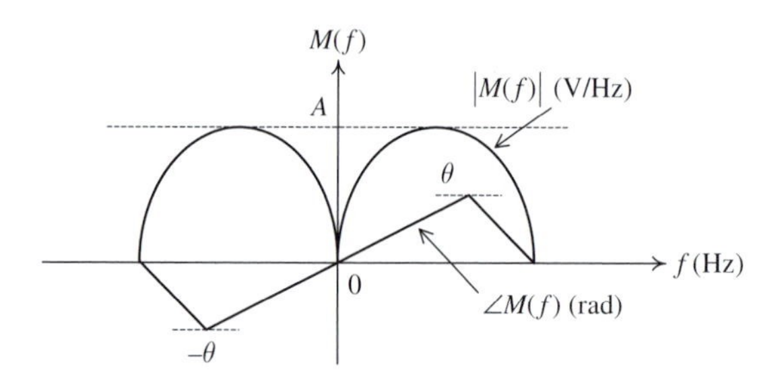

Fig. 2.39 Illustration of magnitude and phase spectra of $m(t)$.

2.42 (*Double sideband suppressed-carrier modulation*) Consider a general version of Problem 2.10. Let $s(t) = m(t)\cos(2\pi f_c t)$.
   (a) Determine the Fourier transform, $S_{\text{DSB-SC}}(f)$, where the mnemonic DSB-SC stands for double sideband suppressed-carrier.
   (b) Sketch the spectrum $S_{\text{DSB-SC}}(f)$ for the case of $f_c \gg W$.
   (c) What is the bandwidth of the transmitted signal, i.e., of $S_{\text{DSB-SC}}(f)$?

2.43 (*Amplitude modulation*) Let $s(t) = [V_c + m(t)]\cos(2\pi f_c t)$. Determine $S_{\text{AM}}(f)$ and sketch its spectrum. As in Problem 2.42 assume $f_c \gg W$.

2.44 (*Coherent demodulation of DSB-SC or AM*) For the signals, $s(t)$, of Problems 2.42 and 2.43, consider $s_1(t) = s(t)\cos(2\pi f_c t)$.
   (a) Determine $S_1(f)$ and sketch the resultant spectrum.
   (b) Let $s_1(t)$ be input to an ideal lowpass filter, bandwidth W hertz. What is the output?
   (c) Repeat (a) and (b) but let $s_1(t) = s(t)\sin(2\pi f_c t)$.

2.45 (*Envelope detection or noncoherent demodulation*) Let the AM signal, $s(t) = [V_c + m(t)]\cos(2\pi f_c t)$, be input to a full-wave rectifier followed by an ideal lowpass filter of bandwidth W (see Figure 2.40).

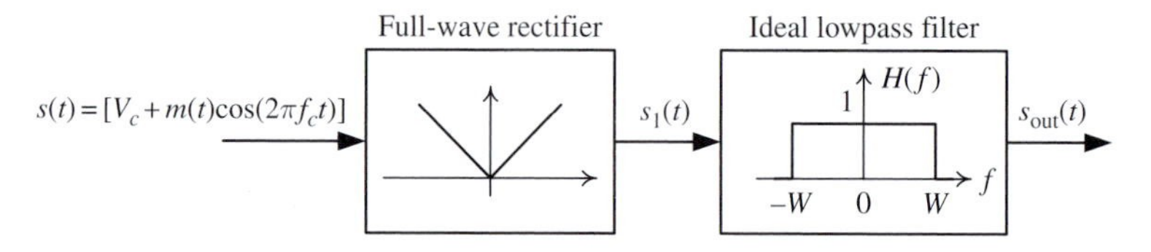

Fig. 2.40 Block diagram of envelope detection of amplitude modulation.

(a) Determine the spectrum of $s_1(t)$ for the case where $V_c \geq \max|m(t)|$. What is $s_{\text{out}}(t)$?

(b) Let $V_c = 0$. What is $s_1(t)$ in this case? Note that in general a sinusoid can be written as $E(t)\cos(\theta(t))$, where $E(t)$ is the instantaneous amplitude of the sinusoid while $\theta(t)$ is the instantaneous phase. The envelope is defined to be $|E(t)|$ and this is what an ideal envelope detector should detect.

The modulation/demodulation above involves shifting the spectrum of the message, $m(t)$, to around f_c for transmission and then back down to baseband at the receiver. The required transmission bandwidth is $2W$ hertz. The next set of problems looks at reducing the bandwidth to W hertz. The technique developed is called single-sideband modulation and leads one to the Hilbert transform.

2.46 Consider the spectrum shown in Figure 2.41.

(a) Show that $S_{\text{USSB}}(f)$ is related to $S_{\text{DSB-SC}}(f)$ of Problem 2.42 by

$$S_{\text{USSB}}(f) = \left[\frac{1+\text{sgn}(f-f_c)}{2} + \frac{1-\text{sgn}(f+f_c)}{2}\right] S_{\text{DSB-SC}}(f). \quad \text{(P2.10)}$$

(b) Let $S_{\text{USSB}}(f) \longleftrightarrow s_{\text{USSB}}(t)$ be a Fourier transform pair. Consider $s(t) = s_{\text{USSB}}(t)2\cos(2\pi f_c t)$. Sketch the spectrum of $s(t)$.

(c) Let $s(t)$ be input to an ideal lowpass filter, bandwidth W hertz. Show that the output is $m(t)$.

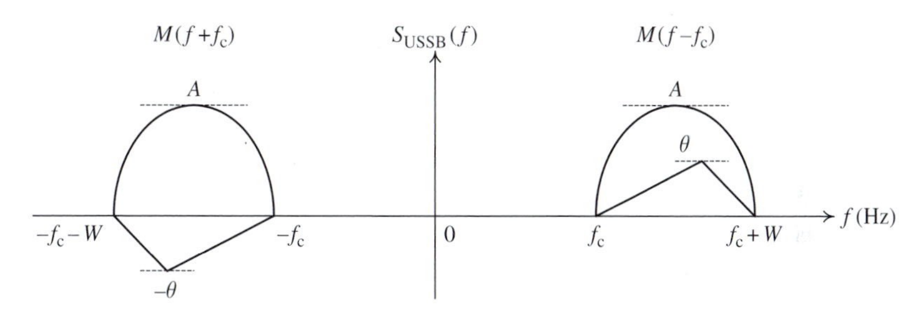

Fig. 2.41 Spectrum of upper single-sideband (USSB) modulation.

2.47 Problem 2.46 shows that $m(t)$ can be recovered from $s_{\text{USSB}}(t)$ but the required transmission bandwidth is only W hertz. The question thus becomes how to modulate $m(t)$ to obtain $s_{\text{USSB}}(t)$. The following steps show how it is done.

(a) Show that $S_{\text{USSB}}(f)$ can be written as

$$S_{\text{USSB}}(f) = \frac{M(f-f_c)+M(f+f_c)}{2} + \frac{M(f-f_c)\text{sgn}(f-f_c) - M(f+f_c)\text{sgn}(f+f_c)}{2}. \quad \text{(P2.11)}$$

(b) What time function does $[M(f-f_c)+M(f+f_c)]/2$ represent.

(c) Consider the term $M(f-f_c)\text{sgn}(f-f_c)$. Argue that it is the time function $[m(t) * h(t)]e^{j2\pi f_c t}$, where $h(t)$ is the inverse Fourier transform of $\text{sgn}(f)$.

(d) Similarly show that $M(f+f_c)\mathrm{sgn}(f+f_c)$ represents the time function $[m(t) * h(t)]\mathrm{e}^{-\mathrm{j}2\pi f_c t}$.

(e) Combine the results of (c) and (d) to show that

$$\frac{M(f-f_c)\mathrm{sgn}(f-f_c) - M(f+f_c)\mathrm{sgn}(f+f_c)}{2} \longleftrightarrow [m(t) * h(t)]\frac{\mathrm{e}^{\mathrm{j}2\pi f_c t} - \mathrm{e}^{-\mathrm{j}2\pi f_c t}}{2} \quad \text{(P2.12)}$$

are a Fourier transform pair.

(f) Is $h(t)$, the inverse Fourier transform of $\mathrm{sgn}(f)$, a real time function? Give the reason(s).

(g) Does $\mathrm{jsgn}(f)$ represent a real time function?

(h) Rewrite $[m(t) * h(t)][\mathrm{e}^{\mathrm{j}2\pi f_c t} - \mathrm{e}^{-\mathrm{j}2\pi f_c t}]/2$ as $[m(t) * \mathrm{j}h(t)][\mathrm{e}^{\mathrm{j}2\pi f_c t} - \mathrm{e}^{-\mathrm{j}2\pi f_c t}]/(2\mathrm{j})$. What is the Fourier transform of $\mathrm{j}h(t)$? Does it represent a real time function?

(i) All of the above leads to the block diagram shown in Figure 2.42. Show that it indeed does produce an USSB spectrum.

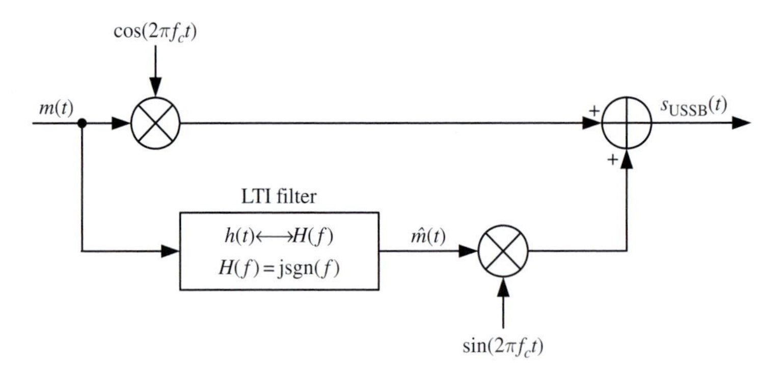

Fig. 2.42 Block diagram of a USSB modulator.

2.48 Modify the reasoning of Problems 2.45 and 2.46 to produce a lower single-sideband signal, $s_{\mathrm{LSSB}}(t)$.

2.49 What, if anything, would change if in Problem 2.46(a) we started with $S_{\mathrm{AM}}(f)$ instead of $S_{\mathrm{DSB\text{-}SC}}(f)$?

The LTI filter of Figure 2.42 in Problem 2.46 produces what is known as a Hilbert transformation. The functions $m(t)$ and $\hat{m}(t)$ are called a Hilbert transform pair. The next set of problems looks at properties of the Hilbert transform and another application.

2.50 Consider the LTI system shown in Figure 2.43. The system has equal gain at all frequencies and a phase shift of $-\pi/2$ for $f \geq 0$ and $\pi/2$ for $f < 0$.

Show that the system's impulse response is $-1/\pi t$. *Hint* Start with $H(f) = \lim_{a\to 0} -\mathrm{j}\mathrm{e}^{-a|f|}\mathrm{sgn}(f)$, find $h_a(t)$ corresponding to $-\mathrm{j}\mathrm{e}^{-a|f|}\mathrm{sgn}(f)$ and then let a go to zero, i.e., $h(t) = \lim_{a\to 0} h_a(t)$.

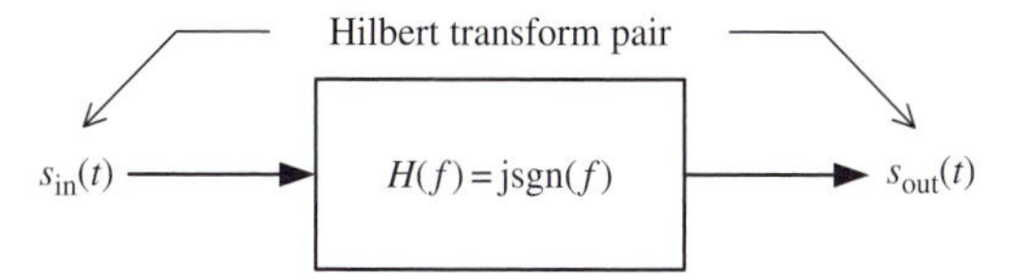

Fig. 2.43 The LTI system considered in Problem 2.50.

2.51 Show that a Hilbert transform pair is orthogonal.

2.52 Show that the autocorrelation of the output in Figure 2.43 is equal to the autocorrelation of the input. *Hint* Work in the frequency domain. Note that this confirms the remark made in the chapter that distinct signals can have the same autocorrelation function.

2.53 Find the Hilbert transform of $V\cos(2\pi f_c t)$.

One application of the Hilbert transform is the representation of a passband signal by a baseband (lowpass) equivalent signal. The analysis/design is done at baseband and then the results are translated to the appropriate passband. The procedure to obtain an equivalent baseband signal is shown in the next problem.

2.54 Consider a passband signal with the magnitude and phase spectra illustrated in Figure 2.44.

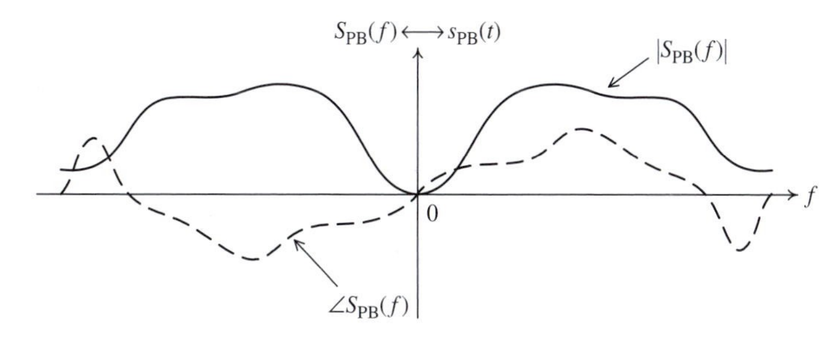

Fig. 2.44 Illustration of the magnitude and phase spectra of a passband signal.

(a) Show that the spectrum, $S(f)$, of the signal $s(t) = s_{PB}(t) - \mathrm{j}\hat{s}_{PB}(t)$ is one-sided and sketch it. Here $\hat{s}_{PB}(t)$ is the Hilbert transform of $s_{PB}(t)$.

(b) Choose a frequency f_s and obtain a baseband signal by shifting $S(f)$ to the left by f_s, i.e., form $s_{BB}(t) = s(t)\mathrm{e}^{-\mathrm{j}2\pi f_s t}$. Sketch the spectrum $S_{BB}(f)$.

(c) Note that, in general, $s_{BB}(t)$ is a complex signal. Why? Express it as $s_{BB}(t) = s_{BB}^{[real]}(t) + \mathrm{j}s_{BB}^{[imag]}(t)$. Show that

$$\begin{aligned} s_{PB}(t) &= \mathcal{R}\left\{s_{BB}(t)\mathrm{e}^{\mathrm{j}2\pi f_s t}\right\} \\ &= s_{BB}^{[real]}(t)\cos(2\pi f_s t) - s_{BB}^{[imag]}(t)\sin(2\pi f_s t). \end{aligned} \tag{P2.13}$$

(d) The choice of the shift frequency, f_s, is somewhat arbitrary. Typically either it is chosen to simplify the problem or there is a natural choice such as the carrier frequency in amplitude modulation. Therefore consider only the positive frequency axis. If there exists an axis of symmetry, f_s, such that $S_{PB}(f)$ is even and $\angle S_{PB}(f)$ is odd about it, then it becomes a natural choice. In this case what can one say about the baseband equivalent signal?

2.55 Determine the baseband equivalent signal for an USSB signal as discussed in Problem 2.46.

3 Probability theory, random variables and random processes

The main objective of a communication system is the transfer of information over a channel. By its nature, the message signal that is to be transmitted is best modeled by a random signal. This is due to the fact that any signal that conveys information must have some uncertainty in it, otherwise its transmission is of no interest. When a signal is transmitted through a communication channel, there are two types of imperfections that cause the received signal to be different from the transmitted signal. One type of imperfection is deterministic in nature, such as linear and nonlinear distortions, intersymbol interference (ISI), etc. The second type is nondeterministic, such as addition of noise, interference, multipath fading, etc. For a quantitative study of these nondeterministic phenomena, a random model is required.

This chapter is concerned with the methods used to describe and characterize a random signal, generally referred to as a *random process* and also commonly called a *stochastic process*.[1] Since a random process is in essence a *random variable* evolving in time we first consider random variables.

3.1 Random variables

3.1.1 Sample space and probability

The fundamental concept in any probabilistic model is the concept of a *random experiment*. In general, an experiment is called random if its outcome, for some reason, cannot be predicted with certainty. Examples of simple random experiments are throwing a die, flipping a coin, and drawing a card from a deck. The common property of all these experiments is that the outcome (or result) of the experiment is uncertain. In throwing a die, the possible outcomes are sides with 1, 2, 3, 4, 5, and 6 dots. In flipping a coin, "head" and "tail" are the possible outcomes. The set of all possible outcomes is called the *sample space* and is denoted by Ω. Outcomes are denoted by ωs and each ω lies in Ω, i.e., $\omega \in \Omega$.

A sample space is *discrete* if the number of its elements is finite or countably infinite, otherwise it is a *nondiscrete* or continuous sample space. All the random experiments mentioned before have discrete sample spaces. If one predicts tomorrow's highest temperature, then the sample space corresponding to this random experiment is nondiscrete.

[1] *Stochastic* is a word of Greek origin and means "pertaining to chance."

Events are subsets of the sample space for which measures of their occurrences, called probabilities, can be defined or determined. For discrete sample spaces, any subset of the sample space is an event, i.e., a probability can be assigned for it. For example, in throwing a die various events such as "the outcome is odd," "the outcome is smaller than 3," and "the outcome divides by 2" can be considered. For a nondiscrete sample space, it might not be possible to assign a probability to every subset of Ω without changing the basic intuitive properties of probability.

For a discrete sample space Ω, define a probability measure P on Ω as a set function that assigns nonnegative values to all events, denoted by E, in Ω such that the following conditions are satisfied

Axiom 1 $0 \le P(E) \le 1$ for all $E \in \Omega$ (on a percentage scale probability ranges from 0 to 100%; despite popular sports lore, it is impossible to have more than 100%).
Axiom 2 $P(\Omega) = 1$ (when an experiment is conducted there has to be an outcome).
Axiom 3 For mutually exclusive events[2] $E_1, E_2, E_3, \ldots$ we have $P\left(\bigcup_{i=1}^{\infty} E_i\right) = \sum_{i=1}^{\infty} P(E_i)$.

From these three axioms several important properties of the probability measure follow. The more important ones are:

(1) $P(E^c) = 1 - P(E)$, where E^c denotes the complement of E. This property implies that $P(E^c) + P(E) = 1$, i.e., something has to happen.
(2) $P(\emptyset) = 0$ (again, something has to happen).
(3) $P(E_1 \cup E_2) = P(E_1) + P(E_2) - P(E_1 \cap E_2)$. Note that if two events E_1 and E_2 are mutually exclusive, then $P(E_1 \cup E_2) = P(E_1) + P(E_2)$, otherwise the nonzero common probability $P(E_1 \cap E_2)$ needs to be subtracted.
(4) If $E_1 \subseteq E_2$, then $P(E_1) \le P(E_2)$. This says that if event E_1 is contained in E_2, then occurrence of E_2 means E_1 has occurred but the converse is not true.

In many situations we observe or are told that event E_1 has occurred but are actually interested in event E_2. Knowledge that E_1 has occurred changes, in general, the probability of E_2 occurring. If it was $P(E_2)$ before, it now becomes $P(E_2|E_1)$, which is read as the probability of E_2 occurring given that event E_1 has occurred. This probability is called *conditional probability* and is given by

$$P(E_2|E_1) = \begin{cases} \dfrac{P(E_2 \cap E_1)}{P(E_1)}, & \text{if } P(E_1) \neq 0 \\ 0, & \text{otherwise} \end{cases}. \tag{3.1}$$

The numerator $P(E_2 \cap E_1)$ is the probability of the two events occurring jointly, or loosely speaking, how much of E_2 is contained within E_1. The denominator reflects that since we know that E_1 has occurred, we are dealing with a new sample space. Note that $P(E|\Omega) = P(E \cap \Omega)/P(\Omega) = P(E)/1 = P(E)$. The probabilities $P(E_2)$ and $P(E_2|E_1)$ are known as the *a priori* (before) and *a posteriori* (after) probabilities.

If knowledge of E_1 does not change the unconditional probability of E_2 occurring, i.e., $P(E_2|E_1) = P(E_2)$, then the two events E_1 and E_2 are said to be *statistically independent*.

[2] The events $E_1, E_2, E_3, \ldots$ are mutually exclusive if $E_i \cap E_j = \emptyset$ for all $i \neq j$, where $\emptyset$ is the null set.

From (3.1) we have $P(E_2|E_1) = P(E_2) = P(E_2 \cap E_1)/P(E_1)$. It follows that[3] $P(E_2 \cap E_1) = P(E_1)P(E_2)$.

Also from (3.1) we have $P(E_2 \cap E_1) = P(E_2|E_1)P(E_1)$. Interchange the roles of E_1 and E_2 and we also have $P(E_1 \cap E_2) = P(E_1|E_2)P(E_2)$. Of course, $P(E_2 \cap E_1) = P(E_1 \cap E_2)$, which means that $P(E_2|E_1)P(E_1) = P(E_1|E_2)P(E_2)$. Written in another way, this relationship is

$$P(E_2|E_1) = \frac{P(E_1|E_2)P(E_2)}{P(E_1)}, \tag{3.2}$$

a result known as *Bayes' rule*.

Example 3.1 In throwing a fair die (i.e., any of the six faces are equally probable to occur), the probability of event E_1 = "the outcome is even" is

$$P(E_1) = P(2) + P(4) + P(6) = \frac{1}{2}. \tag{3.3}$$

The probability of event E_2 = "the outcome is smaller than 4" is

$$P(E_2) = P(1) + P(2) + P(3) = \frac{1}{2}. \tag{3.4}$$

In this case

$$P(E_2|E_1) = \frac{P(E_2 \cap E_1)}{P(E_1)} = \frac{P(2)}{1/2} = \frac{1/6}{1/2} = \frac{1}{3}. \tag{3.5}$$

■

The events $\{E_i\}_{i=1}^n$ partition the sample space Ω if the following conditions are satisfied:

(i)

$$\bigcup_{i=1}^{n} E_i = \Omega \tag{3.6a}$$

(they cover the entire sample space),

(ii)

$$E_i \cap E_j = \oslash \quad \text{for all } 1 \leq i,j \leq n \text{ and } i \neq j \tag{3.6b}$$

(they are mutually exclusive).

Then, if for an event A we have the conditional probabilities $\{P(A|E_i)\}_{i=1}^n$, $P(A)$ can be obtained by applying the *total probability theorem*, which is

$$P(A) = \sum_{i=1}^{n} P(E_i)P(A|E_i). \tag{3.7}$$

[3] This is usually taken as the definition of statistical independence but the definition given here is much preferred.

Furthermore, *Bayes' rule* gives the conditional probabilities $P(E_i|A)$ by the following relation:

$$P(E_i|A) = \frac{P(A|E_i)P(E_i)}{P(A)} = \frac{P(A|E_i)P(E_i)}{\sum_{j=1}^{n} P(A|E_j)P(E_j)}. \tag{3.8}$$

3.1.2 Random variables

A random variable is a *mapping* from the sample space Ω to the set of real numbers. Figure 3.1 shows a schematic diagram describing a random variable. We shall denote random variables (or the mapping) by boldface, i.e., $\mathbf{x}$, $\mathbf{y}$, etc., while individual or specific values of the mapping $\mathbf{x}$ are denoted by $\mathbf{x}(\omega)$. A random variable is discrete if the set of its values is either finite or countably infinite.

A complete description of the random variable is given by the *cumulative distribution function* (cdf), defined as

$$F_{\mathbf{x}}(x) = P(\omega \in \Omega : \mathbf{x}(\omega) \leq x), \tag{3.9}$$

which can be written simply as

$$F_{\mathbf{x}}(x) = P(\mathbf{x} \leq x). \tag{3.10}$$

It is important that the reader is clear as to the difference between $\mathbf{x}$ and x in (3.9) and (3.10). The boldface $\mathbf{x}$ refers to the random variable (i.e., mapping) being considered, while x is simply an indeterminate or dummy variable. The cdf has the following properties:

(1) $0 \leq F_{\mathbf{x}}(x) \leq 1$ (this follows from Axiom 1 of the probability measure).
(2) $F_{\mathbf{x}}(x)$ is nondecreasing: $F_{\mathbf{x}}(x_1) \leq F_{\mathbf{x}}(x_2)$ if $x_1 \leq x_2$ (this is because event $\mathbf{x}(\omega) \leq x_1$ is contained in event $\mathbf{x}(\omega) \leq x_2$).
(3) $F_{\mathbf{x}}(-\infty) = 0$ and $F_{\mathbf{x}}(+\infty) = 1$ ($\mathbf{x}(\omega) \leq -\infty$ is the empty set, hence an impossible event, while $\mathbf{x}(\omega) \leq \infty$ is the whole sample space, i.e., a certain event).
(4) $P(a < \mathbf{x} \leq b) = F_{\mathbf{x}}(b) - F_{\mathbf{x}}(a)$.

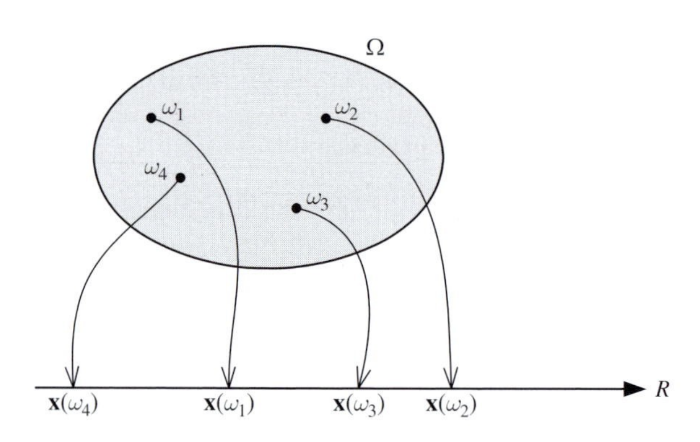

Fig. 3.1 Random variable as a mapping from Ω to R.

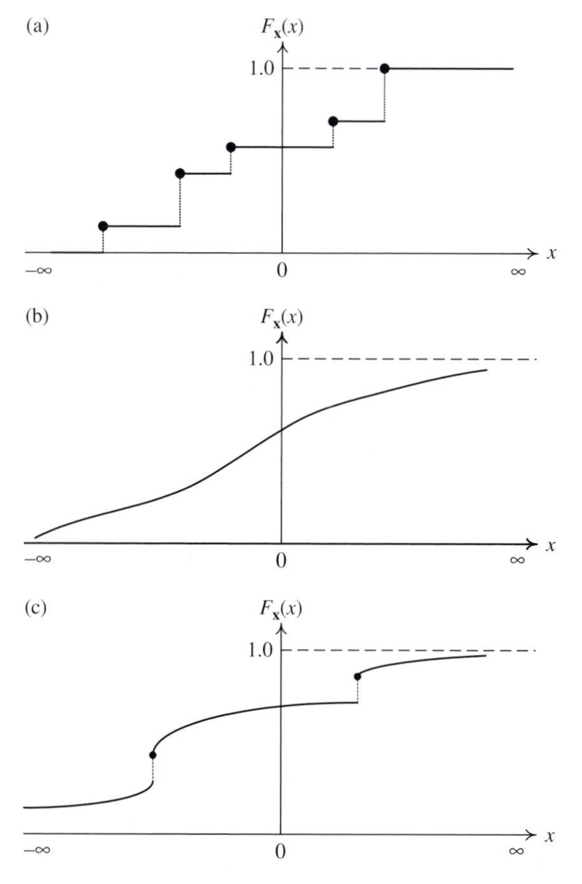

Fig. 3.2 Typical cdfs: (a) a *discrete* random variable, (b) a *continuous* random variable, and (c) a *mixed* random variable.

For a discrete random variable, $F_{\mathbf{x}}(x)$ is a staircase function, whereas a random variable is called *continuous* if $F_{\mathbf{x}}(x)$ is a continuous function. A random variable is called *mixed* if it is neither discrete nor continuous. Typical cdfs for discrete, continuous, and mixed random variables are shown in Figures 3.2(a), 3.2(b), and 3.2(c), respectively.

Rather than dealing with the cdf, it is more common to deal with the *probability density function* (pdf), which is defined as the derivative of $F_{\mathbf{x}}(x)$, i.e.,

$$f_{\mathbf{x}}(x) = \frac{\mathrm{d}F_{\mathbf{x}}(x)}{\mathrm{d}x}. \tag{3.11}$$

From the definition it follows that

$$\begin{aligned} P(x_1 \le \mathbf{x} \le x_2) &= P(\mathbf{x} \le x_2) - P(\mathbf{x} \le x_1) \\ &= F_{\mathbf{x}}(x_2) - F_{\mathbf{x}}(x_1) \\ &= \int_{x_1}^{x_2} f_{\mathbf{x}}(x)\mathrm{d}x. \end{aligned} \tag{3.12}$$

The above says that the probability that the random variable lies in a certain range is the area under the pdf in that range. An important and useful interpretation of this is that $P(x \le \mathbf{x} \le x + \mathrm{d}x) = f_{\mathbf{x}}(x)\mathrm{d}x$.

The pdf has the following basic properties:

(1) $f_{\mathbf{x}}(x) \ge 0$ (because $F_{\mathbf{x}}(x)$ is nondecreasing).
(2) $\int_{-\infty}^{\infty} f_{\mathbf{x}}(x)\mathrm{d}x = 1$ (because it equals $F_{\mathbf{x}}(+\infty) - F_{\mathbf{x}}(-\infty) = 1$).
(3) In general, $P(\mathbf{x} \in \mathcal{A}) = \int_{\mathcal{A}} f_{\mathbf{x}}(x)\mathrm{d}x$.

In the case where the random variable is discrete or mixed, the cdf has discontinuities and therefore the pdf will involve impulses. The pdf is a density and tells us how the probability is concentrated with respect to x and so the impulses signify that a finite amount of probability is concentrated at each discontinuity point. For discrete random variables, it is more common to define the *probability mass function*, or pmf, which is defined as $\{p_i\}$, where $p_i = P(\mathbf{x} = x_i)$. Obviously for all i one has $p_i \ge 0$ and $\sum_i p_i = 1$.

Any function that satisfies conditions (1) and (2) above can serve as a pdf. Whether it is a useful model for the physical situation is another matter. Typically $f_{\mathbf{x}}(x)$ is determined by intuition, logical reasoning, or experimentally. Some of the most commonly encountered random variables in communications and their pdfs are discussed below.

Bernoulli random variable This is a discrete random variable that takes two values, 1 and 0, with probabilities p and $1 - p$. The pdf and cdf of the Bernoulli random variable are shown in Figure 3.3.

A Bernoulli random variable is a good model for a binary data source whose output is bit 1 or 0. Furthermore, when a bit stream is transmitted over a communication channel, some bits might be received in error. An error can be accounted for by a modulo-2 addition of a 1 to the source bit, thus changing a 0 into a 1 and a 1 into a 0. Therefore, a Bernoulli random variable can also be used to model the channel errors.

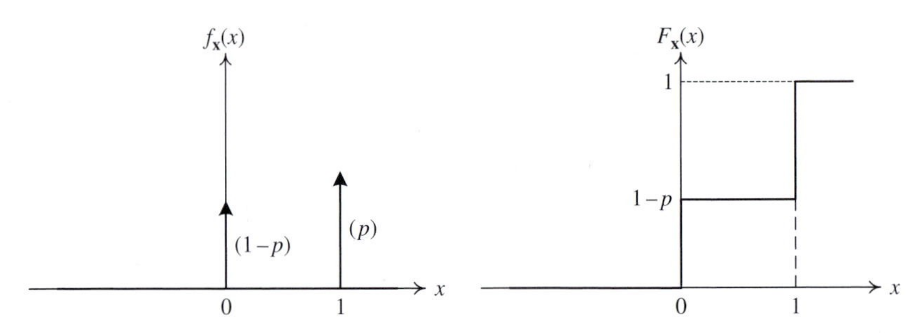

Fig. 3.3 The pdf and cdf for the Bernoulli random variable.

Binomial random variable This is also a discrete random variable that gives the number of 1s in a sequence of n *independent* Bernoulli trials. The pdf is given by

$$f_{\mathbf{x}}(x) = \sum_{k=0}^{n} \binom{n}{k} p^k (1-p)^{n-k} \delta(x-k), \tag{3.13}$$

where the binomial coefficient is defined as

$$\binom{n}{k} = \frac{n!}{k!\,(n-k)!}. \tag{3.14}$$

An example of the pdf for a binomial random variable is shown in Figure 3.4.

If the Bernoulli random variable discussed above is used to model the channel errors of a communication system (with a cross error probability of p), then the binomial random variable can be used to model the total number of bits received in error when a sequence of n bits is transmitted over the channel.

Uniform random variable This is a continuous random variable that takes values between a and b with equal probabilities over intervals of equal length. The pdf is given by

$$f_{\mathbf{x}}(x) = \begin{cases} \dfrac{1}{b-a}, & a < x < b \\ 0, & \text{otherwise} \end{cases}. \tag{3.15}$$

General plots of the pdf and cdf for a uniform random variable are shown in Figure 3.5.

This model is often used for continuous random variables whose range is known, but where nothing else is known about the likelihood of various values that the random variable can take on. For example, in communications the phase of a received sinusoidal carrier is usually modeled as a uniform random variable between 0 and 2π. Quantization error is also typically modeled as uniform.

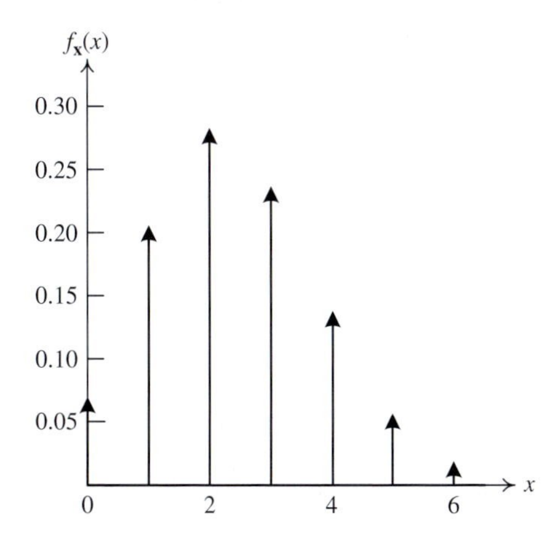

Fig. 3.4 The pdf for the binomial random variable.

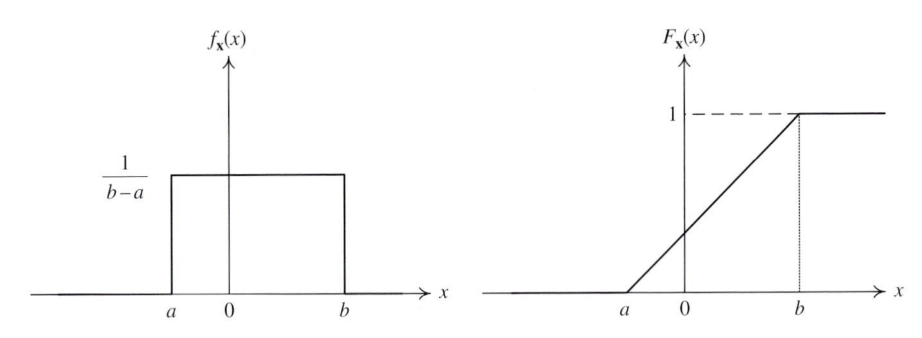

Fig. 3.5 The pdf and cdf for the uniform random variable.

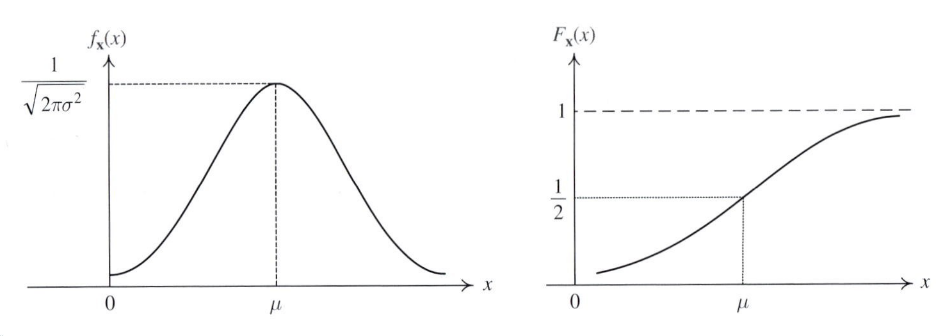

Fig. 3.6 The pdf and cdf of a Gaussian random variable.

Gaussian (or normal) random variable This is a continuous random variable that is described by the following pdf:

$$f_{\mathbf{x}}(x) = \frac{1}{\sqrt{2\pi\sigma^2}} \exp\left\{-\frac{(x-\mu)^2}{2\sigma^2}\right\}, \tag{3.16}$$

where μ and σ^2 are two parameters whose meaning is described later. It is usually denoted as $\mathcal{N}(\mu, \sigma^2)$. Figure 3.6 shows sketches of the pdf and cdf of a Gaussian random variable.

The Gaussian random variable is the most important and frequently encountered random variable in communications. This is because thermal noise, which is the major source of noise in communication systems, has a Gaussian distribution. Gaussian noise and the Gaussian pdf are discussed in more depth at the end of this chapter.

The problems explore other pdf models. Some of these arise when a random variable is passed through a nonlinearity. How to determine the pdf of the random variable in this case is discussed next.

Functions of a random variable A function of a random variable $\mathbf{y} = g(\mathbf{x})$ is itself a random variable. From the definition, the cdf of $\mathbf{y}$ can be written as

$$F_{\mathbf{y}}(y) = P(\omega \in \Omega : g(\mathbf{x}(\omega)) \leq y). \tag{3.17}$$

In practice one usually finds the pdf of $\mathbf{y}$. Assume that for all y, the equation $g(x) = y$ has a countable number of solutions (or roots) and, at each solution point, $\mathrm{d}g(x)/\mathrm{d}x$ exists and is nonzero. Then the pdf of the random variable $\mathbf{y} = g(\mathbf{x})$ is given by

$$f_{\mathbf{y}}(y) = \sum_i \frac{f_{\mathbf{x}}(x_i)}{\left| \left. \frac{\mathrm{d}g(x)}{\mathrm{d}x} \right|_{x=x_i} \right|}, \tag{3.18}$$

where $\{x_i\}$ are the solutions of $g(x) = y$.

Example 3.2 Consider a random variable $\mathbf{x}$ with pdf $f_{\mathbf{x}}(x)$. Find the pdf of the random variable $\mathbf{y} = a\mathbf{x} + b$ in terms of $f_{\mathbf{x}}(x)$. Then examine the special case that $\mathbf{x}$ is a Gaussian random variable with parameters $\mu_{\mathbf{x}}$ and $\sigma_{\mathbf{x}}^2$.

Solution

In this example $g(x) = ax + b$, therefore $\mathrm{d}g(x)/\mathrm{d}x = a$. The equation $ax + b = y$ has only one solution given by $x_1 = (y - b)/a$. Therefore

$$f_{\mathbf{y}}(y) = \frac{f_{\mathbf{x}}\left((y-b)/a\right)}{|a|}. \tag{3.19}$$

The cdf of $\mathbf{y}$ can also be obtained directly as follows. We assume that $a > 0$ (if $a < 0$ the approach is similar). Since the mapping $\mathbf{y} = a\mathbf{x} + b$ is linear and monotonic, then

$$\begin{aligned} F_{\mathbf{y}}(y) &= P(\mathbf{y} \le y) = P(a\mathbf{x} + b \le y) = P\left(\mathbf{x} \le \frac{y-b}{a}\right) \\ &= \int_{-\infty}^{(y-b)/a} f_{\mathbf{x}}(x)\mathrm{d}x = F_{\mathbf{x}}\left(\frac{y-b}{a}\right). \end{aligned} \tag{3.20}$$

Differentiating (3.20) with respect to y gives the following relationship between the pdfs of $\mathbf{y}$ and $\mathbf{x}$:

$$f_{\mathbf{y}}(y) = \frac{f_{\mathbf{x}}\left((y-b)/a\right)}{a}, \tag{3.21}$$

which is the same as before. Though in this case there is not much difference in difficulty between the approaches, typically the approach based on (3.18) is easier to apply.

Finally, for the important case that $\mathbf{x}$ is $\mathcal{N}(\mu_{\mathbf{x}}, \sigma_{\mathbf{x}}^2)$, one has

$$f_{\mathbf{y}}(y) = \frac{1}{\sqrt{2\pi a^2 \sigma_{\mathbf{x}}^2}} \exp\left\{-\frac{(y - a\mu_{\mathbf{x}} - b)^2}{2a^2\sigma_{\mathbf{x}}^2}\right\}. \tag{3.22}$$

Note that $\mathbf{y}$ is a Gaussian random variable with parameters $\mu_{\mathbf{y}} = a\mu_{\mathbf{x}} + b$ and $\sigma_{\mathbf{y}}^2 = a^2\sigma_{\mathbf{x}}^2$. ■

The above result leads to the important conclusion that *a linear function of a Gaussian random variable is itself a Gaussian random variable*.

Example 3.3 Consider a random variable $\mathbf{x}$ with pdf $f_{\mathbf{x}}(x)$. Find the pdf of the random variable $\mathbf{y} = a\mathbf{x}^2$, where $a > 0$, in terms of $f_{\mathbf{x}}(x)$. Then examine the case when $\mathbf{x}$ is $\mathcal{N}(0, \sigma_{\mathbf{x}}^2)$.

Solution

The equation $g(x) = ax^2 = y$ has two real solutions $x_1 = \sqrt{y/a}$ and $x_2 = -\sqrt{y/a}$ for $y \geq 0$, and $\mathrm{d}g(x)/\mathrm{d}x = 2ax$. The pdf $f_{\mathbf{y}}(y)$ consists of two terms:

$$f_{\mathbf{y}}(y) = \frac{f_{\mathbf{x}}\left(x = \sqrt{y/a}\right) + f_{\mathbf{x}}\left(x = -\sqrt{y/a}\right)}{2\sqrt{ay}} u(y). \tag{3.23}$$

Now consider the case $f_{\mathbf{x}}(x) \sim \mathcal{N}(0, \sigma_{\mathbf{x}}^2)$. It is not hard to show that the pdf of $\mathbf{y}$ is given by

$$f_{\mathbf{y}}(y) = \begin{cases} \left(1/\sqrt{2\pi a y \sigma_{\mathbf{x}}^2}\right) \mathrm{e}^{-y/(2a\sigma_{\mathbf{x}}^2)}, & y \geq 0 \\ 0, & y < 0 \end{cases}. \tag{3.24}$$

■

3.1.3 Expectation of random variables

The pdf of a random variable provides a complete description of it. However, in many situations only a partial description is either needed or feasible. This is provided by *statistical averages* which play an important role in the characterization of the random variable defined on the sample space of an experiment. These statistical averages are known as *moments*. Of particular interest are the first and second moments of a single random variable and the joint moments such as the correlation and the covariance, between any pair of random variables in a multidimensional set of random variables. For a Gaussian random variable(s) it turns out that these moments completely characterize it. This section is devoted to the definition of these important statistical averages. In general, the mathematical operation of finding a statistical average is called an *expectation* operation.

The *expected value* (also called the mean value) of the random variable $\mathbf{x}$ is defined as

$$m_{\mathbf{x}} = E\{\mathbf{x}\} \equiv \int_{-\infty}^{\infty} x f_{\mathbf{x}}(x) \mathrm{d}x, \tag{3.25}$$

where E denotes the *statistical expectation operator*. Since $f_{\mathbf{x}}(x)\mathrm{d}x$ is the probability of random variable $\mathbf{x}$ lying in the infinitesimal strip $\mathrm{d}x$, $m_{\mathbf{x}}$ is interpreted as the *weighted average* of $\mathbf{x}$, where each weight is the probability of that specific value of $\mathbf{x}$ occurring. The expected value of a random variable is an average of the values that the random variable takes in a large number of experiments and is called the first moment of a random variable.[4]

[4] The terminology is from mechanics, where $f_{\mathbf{x}}(x)$ is interpreted as a mass density and x is the moment arm. $m_{\mathbf{x}}$ would then be the center of gravity. Colloquially one can call it the picture hanging moment for random variables.

In general, the nth moment of a random variable $\mathbf{x}$ is defined as

$$E\{\mathbf{x}^n\} \equiv \int_{-\infty}^{\infty} x^n f_{\mathbf{x}}(x)\mathrm{d}x. \tag{3.26}$$

The most important moments are the first ($n = 1$) which we have just discussed and the second, $n = 2$, which is also known as the mean-squared value of the random variable. Knowledge of all the moments of a random variable also completely describes the random variable just as the pdf does. This is explored in Problem 3.19.

One can look upon (3.26) as the expected value of a random variable which is the result of passing $\mathbf{x}$ through a nonlinearity, i.e., determining $E\{\mathbf{y}\}$, where $\mathbf{y} = \mathbf{x}^n$. Now consider the more general case $\mathbf{y} = g(\mathbf{x})$, where $g(\mathbf{x})$ is some arbitrary function of the random variable $\mathbf{x}$. The expected value of $\mathbf{y}$ is

$$E\{\mathbf{y}\} = E\{g(\mathbf{x})\} = \int_{-\infty}^{\infty} g(x) f_{\mathbf{x}}(x)\mathrm{d}x. \tag{3.27}$$

In particular, if $\mathbf{y} = (\mathbf{x} - m_{\mathbf{x}})^n$, where $m_{\mathbf{x}}$ is the mean value of $\mathbf{x}$, then

$$E\{\mathbf{y}\} = E\{(\mathbf{x} - m_{\mathbf{x}})^n\} = \int_{-\infty}^{\infty} (x - m_{\mathbf{x}})^n f_{\mathbf{x}}(x)\mathrm{d}x. \tag{3.28}$$

This expected value is called the nth *central moment* of the random variable $\mathbf{x}$, because it is a moment taken relative to the mean. When $n = 2$ the central moment is called the *variance* of the random variable and commonly denoted as $\sigma_{\mathbf{x}}^2$. That is,

$$\sigma_{\mathbf{x}}^2 = \mathrm{var}(\mathbf{x}) = E\{(\mathbf{x} - m_{\mathbf{x}})^2\} = \int_{-\infty}^{\infty} (x - m_{\mathbf{x}})^2 f_{\mathbf{x}}(x)\mathrm{d}x. \tag{3.29}$$

This parameter provides a measure of the dispersion of the random variable $\mathbf{x}$. In some sense it is a measure of the variable's "randomness." By specifying the variance $\sigma_{\mathbf{x}}^2$, one essentially constrains the effective width of the pdf $f_{\mathbf{x}}(x)$ of the random variable $\mathbf{x}$ about its mean value $m_{\mathbf{x}}$. A mathematical statement of this constraint is the *Chebyshev inequality*:

$$P(|\mathbf{x} - m_{\mathbf{x}}|) \geq \epsilon \leq \frac{\sigma_{\mathbf{x}}^2}{\epsilon^2} \tag{3.30}$$

for any positive number ϵ. From the above discussion it can be seen that the mean and variance of a random variable give a *partial description* of its pdf.

By expanding the term $(x - m_{\mathbf{x}})^2$ in the integral of (3.29) and noting that the expected value of a constant is the constant itself, one obtains the following expression that relates the variance to the first and second moments:

$$\sigma_{\mathbf{x}}^2 = E\{\mathbf{x}^2\} - [E\{\mathbf{x}\}]^2 = E\{\mathbf{x}^2\} - m_{\mathbf{x}}^2. \tag{3.31}$$

To put an electrical engineering interpretation on the mean and variance, the mean value would be the DC component of the random variable, while the variance is the AC power. With this interpretation (3.31) states that the AC power equals total power minus DC power. Lastly, the square-root of the variance is known as the *standard deviation* and it is interpreted as the root-mean-squared (RMS) value of the AC component.

3.1.4 Multiple random variables

When dealing with combined experiments or repeated trials of a single experiment, one often encounters multiple random variables and their cdfs and pdfs. Multiple random variables are basically multidimensional functions defined on a sample space of a combined experiment. Consider the two-dimensional case. Let $\mathbf{x}$ and $\mathbf{y}$ be the two random variables defined on the same sample space Ω. For those two random variables, the *joint cdf* $F_{\mathbf{x},\mathbf{y}}(x, y)$ is defined as

$$F_{\mathbf{x},\mathbf{y}}(x, y) = P(\mathbf{x} \leq x, \mathbf{y} \leq y), \tag{3.32}$$

where the notation $(\mathbf{x} \leq x, \mathbf{y} \leq y)$ means $(\mathbf{x} \leq x \cap \mathbf{y} \leq y)$, or $(\mathbf{x} \leq x$ and $\mathbf{y} \leq y)$.

Similarly, the *joint pdf* $f_{\mathbf{x},\mathbf{y}}(x, y)$ of $\mathbf{x}$ and $\mathbf{y}$ is

$$f_{\mathbf{x},\mathbf{y}}(x, y) = \frac{\partial^2 F_{\mathbf{x},\mathbf{y}}(x, y)}{\partial x \partial y}, \tag{3.33}$$

where it is assumed that $F_{\mathbf{x},\mathbf{y}}(x, y)$ is continuous everywhere.

When the joint pdf $f_{\mathbf{x},\mathbf{y}}(x, y)$ is integrated over one of the variables, one obtains the pdf of the other variable. That is,

$$\int_{-\infty}^{\infty} f_{\mathbf{x},\mathbf{y}}(x, y)\mathrm{d}x = f_{\mathbf{y}}(y), \tag{3.34}$$

$$\int_{-\infty}^{\infty} f_{\mathbf{x},\mathbf{y}}(x, y)\mathrm{d}y = f_{\mathbf{x}}(x). \tag{3.35}$$

The above pdfs that are obtained from integrating over one of the variables are called *marginal pdfs*. Furthermore, if $f_{\mathbf{x},\mathbf{y}}(x, y)$ is integrated over both variables one obtains

$$\int_{-\infty}^{\infty}\int_{-\infty}^{\infty} f_{\mathbf{x},\mathbf{y}}(x, y)\mathrm{d}x\mathrm{d}y = F(\infty, \infty) = 1. \tag{3.36}$$

Also note that $F_{\mathbf{x},\mathbf{y}}(-\infty, -\infty) = F_{\mathbf{x},\mathbf{y}}(-\infty, y) = F_{\mathbf{x},\mathbf{y}}(x, -\infty) = 0$.

As with the pdf of a single random variable the joint pdf of two random variables needs to satisfy simple conditions, namely: (i) it is nonnegative for all $\mathbf{x}$ and $\mathbf{y}$, (ii) the volume under it is 1, (3.36) above, (iii) the marginal densities should also be valid pdfs. Again any two-dimensional function that satisfies these conditions is a valid joint pdf. But again only a few are of any use. Indeed, we shall only be concerned with the random variables that are jointly Gaussian. Many might be called but only a few get to serve.

The conditional pdf of the random variable $\mathbf{y}$, given that the value of the random variable $\mathbf{x}$ is equal to x, is denoted by $f_{\mathbf{y}}(y|x)$ and defined as

$$f_{\mathbf{y}}(y|x) = \begin{cases} \dfrac{f_{\mathbf{x},\mathbf{y}}(x, y)}{f_{\mathbf{x}}(x)}, & f_{\mathbf{x}}(x) \neq 0 \\ 0, & \text{otherwise} \end{cases}. \tag{3.37}$$

We have already defined statistical independence of two or more events of a sample space Ω. The concept of statistical independence can be extended to random variables defined on a sample space generated by a combined experiment or by repeated trials of a single experiment. In particular, two random variables $\mathbf{x}$ and $\mathbf{y}$ are statistically independent if and only if

$$f_{\mathbf{y}}(y|x) = f_{\mathbf{y}}(y) \quad \text{(statistical independence)} \tag{3.38}$$

or equivalently

$$f_{\mathbf{x},\mathbf{y}}(x, y) = f_{\mathbf{x}}(x) f_{\mathbf{y}}(y). \tag{3.39}$$

The joint pdf provides a complete description of the two random variables but as in the single variable case a partial description is provided by the moments of the random variables. In the case of two random variables, $\mathbf{x}$ and $\mathbf{y}$, we define the *joint moment* as

$$E\{\mathbf{x}^j \mathbf{y}^k\} = \int_{-\infty}^{\infty} \int_{-\infty}^{\infty} x^j y^k f_{\mathbf{x},\mathbf{y}}(x, y) \mathrm{d}x \mathrm{d}y \tag{3.40}$$

and the *joint central moment* as

$$E\{(\mathbf{x} - m_{\mathbf{x}})^j (\mathbf{y} - m_{\mathbf{y}})^k\} = \int_{-\infty}^{\infty} \int_{-\infty}^{\infty} (x - m_{\mathbf{x}})^j (y - m_{\mathbf{y}})^k f_{\mathbf{x},\mathbf{y}}(x, y) \mathrm{d}x \mathrm{d}y, \tag{3.41}$$

where $m_{\mathbf{x}} = E\{\mathbf{x}\}$ and $m_{\mathbf{y}} = E\{\mathbf{y}\}$.

Of particular importance are the joint moment and the joint central moment corresponding to $j = k = 1$:

$$E\{\mathbf{xy}\} \equiv \int_{-\infty}^{\infty} \int_{-\infty}^{\infty} xy f_{\mathbf{x},\mathbf{y}}(x, y) \mathrm{d}x \mathrm{d}y \quad \text{(correlation)} \tag{3.42}$$

and

$$\begin{aligned} \operatorname{cov}\{\mathbf{x}, \mathbf{y}\} &\equiv E\{(\mathbf{x} - m_{\mathbf{x}})(\mathbf{y} - m_{\mathbf{y}})\} \\ &= E\{\mathbf{xy}\} - m_{\mathbf{x}} m_{\mathbf{y}} \quad \text{(covariance)}. \end{aligned} \tag{3.43}$$

These joint moments are called the *correlation* and the *covariance* of the random variables $\mathbf{x}$ and $\mathbf{y}$, respectively.[5]

Let $\sigma_{\mathbf{x}}^2$ and $\sigma_{\mathbf{y}}^2$ be the variance of $\mathbf{x}$ and $\mathbf{y}$ respectively. The covariance normalized with respect to $\sigma_{\mathbf{x}}\sigma_{\mathbf{y}}$ is called the *correlation coefficient*, denoted by $\rho_{\mathbf{x},\mathbf{y}}$; i.e.,

$$\rho_{\mathbf{x},\mathbf{y}} = \frac{\operatorname{cov}\{\mathbf{x}, \mathbf{y}\}}{\sigma_{\mathbf{x}}\sigma_{\mathbf{y}}}. \tag{3.44}$$

Using the Cauchy–Schwartz inequality, it can be shown that $|\rho_{\mathbf{x},\mathbf{y}}| \leq 1$.

The two random variables $\mathbf{x}$ and $\mathbf{y}$ are said to be *uncorrelated* if and only if their correlation coefficient (or covariance) is zero. So again, uncorrelatedness means $E\{\mathbf{xy}\} = E\{\mathbf{x}\}E\{\mathbf{y}\}$. It is easy to verify that if $\mathbf{x}$ and $\mathbf{y}$ are independent, then $\operatorname{cov}\{\mathbf{x}, \mathbf{y}\} = \rho_{\mathbf{x},\mathbf{y}} = 0$. That is, independence implies lack of correlation ($\rho_{\mathbf{x},\mathbf{y}} = 0$). It should be noted that lack of correlation does not in general imply statistical independence. That is, $\rho_{\mathbf{x},\mathbf{y}}$ might be zero but the random variables $\mathbf{x}$ and $\mathbf{y}$ may still be statistically dependent.

[5] The "co" in covariance has the same connotation as "co" in co-ed.

3.2 Random processes

Many random phenomena that occur in nature are functions of time. For example, meteorological phenomena such as the random fluctuations in air temperature and air pressure are functions of time. The thermal noise voltage generated in a resistor of an electronic device, say a radio receiver, is also a random function of time. Similarly, the signal at the output of a source that generates information is characterized as a random signal that varies with time. An audio signal that is transmitted over a telephone channel is an example of such a signal. All these are examples of random (stochastic) processes. In our study of digital communications, we encounter random processes in the characterization and modeling of signals generated by information sources, in the characterization of communication channel used to transmit the information, and in the design of the optimum receiver for processing the received random signal.

As with random variables a random process can be visualized as a mapping from a sample space that is the set of experimental outcomes to a set of time functions as shown in Figure 3.7. In practice this mapping is seldom known and indeed is not needed to be known. What is important is the set of possible time functions that one sees, i.e., the *ensemble*. Denote this set by $\mathbf{x}(t)$, where the time functions $x_1(t,\omega_1)$, $x_2(t,\omega_2)$, $x_3(t,\omega_3)$,... are

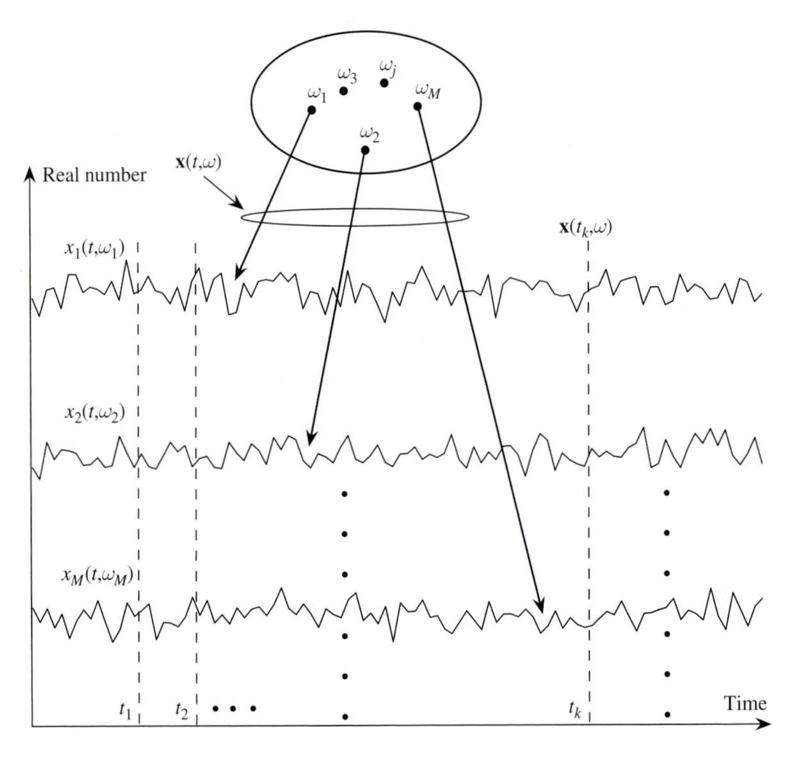

Fig. 3.7 An illustration of a random process.

specific members of the ensemble. At any time instant, say $t = t_k$, we have a random variable $\mathbf{x}(t_k)$, a mapping from the sample space to $x_1(t_k, \omega_1)$, $x_2(t_k, \omega_2)$, $x_3(t_k, \omega_3)$, ... Therefore the random process can be looked upon as a random variable that is a function of time.

One can obtain a partial description of the random variable by determining the pdf of $\mathbf{x}(t)$ at time t, $f_{\mathbf{x}(t)}(x; t)$, where the variable t is included in the argument of the pdf to signify that the pdf may change with time. It should be pointed out that at any two time instants, say t_1 and t_2, we have two different random variables $\mathbf{x}(t_1)$ and $\mathbf{x}(t_2)$. Any relationship between them is described by the joint pdf $f_{\mathbf{x}(t_1),\mathbf{x}(t_2)}(x_1, x_2; t_1, t_2)$ where again the times t_1 and t_2 are included in the argument since in general the joint pdf may depend on the actual times t_1 and t_2.

A complete description of the random process is accomplished by taking *any set* of time instants $t_1 < t_2 < \cdots < t_N$, where N is any positive integer, $N = 1, 2, \ldots$, and either determining or postulating the joint pdf of the N random variables, $\mathbf{x}(t_1), \mathbf{x}(t_2), \ldots, \mathbf{x}(t_N)$, i.e., the Nth-order pdf $f_{\mathbf{x}(t_1),\mathbf{x}(t_2),\ldots,\mathbf{x}(t_N)}(x_1, x_2, \ldots, x_N; t_1, t_2, \ldots, t_N)$, where x_j in the argument is the (dummy) variable for random variable $\mathbf{x}(t_j)$. The variables t_j again signify that the joint pdf may depend on where the time instants are taken to be. For our purposes the most important joint pdfs are the first-order pdf $f_{\mathbf{x}(t)}(x; t)$ and the second-order pdf $f_{\mathbf{x}(t_1),\mathbf{x}(t_2)}(x_1, x_2; t_1, t_2)$.

In general the random variable may be discrete or continuous, depending on the characteristics of the source that generates the random process. The parameter t is considered here to be continuous, though it may be also discrete. Figure 3.8 shows four examples of commonly encountered random processes.

The sample functions of the first process are what ensemble members of thermal noise, discussed later in the chapter, look like on an oscilloscope. The pdf of the amplitude is Gaussian. The second process is encountered in communication systems where it is not feasible to establish timing at the receiver. The mapping is $\mathbf{x}(t) = V\cos(2\pi f_c t + \boldsymbol{\Theta})$, where $\boldsymbol{\Theta}$ is often taken to be uniform over $[0, 2\pi)$. This is a mapping from the random variable $\boldsymbol{\Theta}$ to the time function.

The process illustrated in Figure 3.8(c) is one commonly encountered in wireless channels where the communication system experiences fading. The mapping is given by $\mathbf{x}(t) = \mathbf{V}\cos(2\pi f_c t + \boldsymbol{\Theta})$, where $\mathbf{V} = \sqrt{\mathbf{x}_1^2 + \mathbf{x}_2^2}$ and $\boldsymbol{\Theta} = \tan^{-1}(\mathbf{x}_2/\mathbf{x}_1)$. Furthermore, $\mathbf{x}_1$ and $\mathbf{x}_2$ are usually modeled as being zero mean, of equal variance σ^2, statistically independent Gaussian random variables. In this case $\mathbf{V}$ has a Rayleigh pdf, whereas $\boldsymbol{\Theta}$ is uniform over $[0, 2\pi)$. It is commonly referred to as a Rayleigh process. The Rayleigh pdf is given by

$$f_{\mathbf{V}}(v) = \frac{v}{\sigma^2} e^{-v^2/(2\sigma^2)} u(v). \tag{3.45}$$

The last process is the binary random process where transmitted bits 0 and 1 are mapped to $+V$ and $-V$ (volts). The mapping is assumed to be statistically independent from one bit interval to another. Further, it is also usual to assume that the starting time, $\boldsymbol{\epsilon}$, of transmission is random, uniform over the bit interval. A reasonable assumption from the channel's point of view since it does not know when the transmitter starts to transmit. An analytical expression for the process is

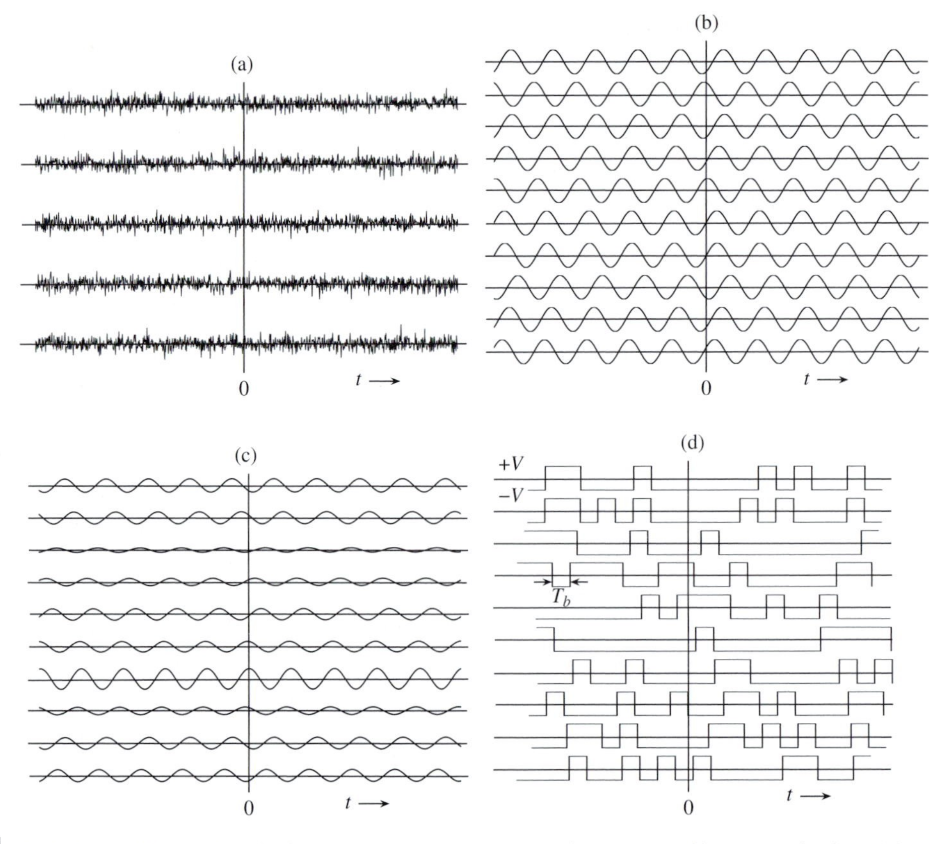

Fig. 3.8 Typical ensemble members for four random processes commonly encountered in communications: (a) thermal noise, (b) uniform phase, (c) Rayleigh fading process, and (d) binary random data process.

$$\mathbf{x}(t) = \sum_{k=-\infty}^{\infty} V(2\mathbf{a}_k - 1)[u(t - kT_b + \boldsymbol{\epsilon}) - u(t - (k+1)T_b + \boldsymbol{\epsilon})], \qquad (3.46)$$

where $\mathbf{a}_k = 1$ with probability p, 0 with probability $(1 - p)$ (usually $p = 1/2$), T_b is the bit duration, and $\boldsymbol{\epsilon}$ is uniform over $[0, T_b)$.

Observe that two of these ensembles have member functions that look very deterministic, one is "quasi" deterministic but for the last one even individual time functions look random. But the point is not whether any one member function looks deterministic or not; the issue when dealing with random processes is that we do not know for sure which member function we shall have to deal with.

3.2.1 Classification of random processes

The basic classification of random processes arises from whether its statistics change with time or not, i.e., whether the process is *nonstationary* or *stationary*. Even if stationary there are different levels of stationarity. A strictly stationary process is one where the joint pdf of any order is independent of a shift in time. Consider again the ensemble of Figure 3.7, the time instants $t_1, \ldots, t_N$, associated random variables $\mathbf{x}(t_1), \ldots, \mathbf{x}(t_N)$ with joint pdf $f_{\mathbf{x}(t_1),\ldots,\mathbf{x}(t_N)}(x_1, \ldots, x_N; t_1, \ldots, t_N)$. If we consider a new set of N random variables, obtained by an arbitrary time shift of t, then the process is said to be Nth-order stationary if the joint pdf does not change, i.e.,

$$\begin{aligned} &f_{\mathbf{x}(t_1),\mathbf{x}(t_2),\ldots,\mathbf{x}(t_N)}(x_1, x_2, \ldots, x_N; t_1, t_2, \ldots, t_N) \\ &= f_{\mathbf{x}(t_1+t),\mathbf{x}(t_2+t),\ldots,\mathbf{x}(t_N+t)}(x_1, x_2, \ldots, x_N; t_1 + t, t_2 + t, \ldots, t_N + t). \end{aligned} \tag{3.47}$$

If (3.47) is true for all t and all N, then the process is strictly stationary. Observe that though Nth-order stationarity does not depend on the time shift, it does depend on the spacing between the time instants. Taking t_1 as a reference point, the pdf depends on $\tau_1 = t_2 - t_1$, $\tau_2 = t_3 - t_1, \ldots, \tau_{N-1} = t_N - t_1$. Put simply, we can throw N lines arbitrarily across the ensemble and shift these lines as we wish to see different sets of random variables. But as long as we maintain the spacings between the lines the joint pdf of any set of encountered random variables is the same. Provided, of course, that the process is Nth-order stationary.

Strict stationarity is a very strong condition that only a few physical processes may satisfy. And indeed for our purposes it is not necessary since we shall be interested exclusively in the first- and second-order stationarity, i.e., that

$$f_{\mathbf{x}(t_1)}(x, t_1) = f_{\mathbf{x}(t_1+t)}(x; t_1 + t) = f_{\mathbf{x}(t)}(x) \tag{3.48}$$

and

$$\begin{aligned} f_{\mathbf{x}(t_1),\mathbf{x}(t_2)}(x_1, x_2; t_1, t_2) &= f_{\mathbf{x}(t_1+t),\mathbf{x}(t_2+t)}(x_1, x_2; t_1 + t, t_2 + t) \\ &= f_{\mathbf{x}(t_1),\mathbf{x}(t_2)}(x_1, x_2; \tau), \quad \tau = t_2 - t_1. \end{aligned} \tag{3.49}$$

3.2.2 Statistical averages or joint moments

As in the case of random variables, statistical averages or expected values can provide a partial but useful characterization for a random process. Consider N random variables $\mathbf{x}(t_1), \mathbf{x}(t_2), \ldots, \mathbf{x}(t_N)$. The most general form for the joint moments of these random variables is

$$\begin{aligned} E\{\mathbf{x}^{k_1}(t_1), \mathbf{x}^{k_2}(t_2), \ldots, \mathbf{x}^{k_N}(t_N)\} = &\int_{x_1=-\infty}^{\infty} \cdots \int_{x_N=-\infty}^{\infty} x_1^{k_1} x_2^{k_2} \cdots x_N^{k_N} \\ &\times f_{\mathbf{x}(t_1),\mathbf{x}(t_2),\ldots,\mathbf{x}(t_N)}(x_1, x_2, \ldots, x_N; t_1, t_2, \ldots, t_N) \mathrm{d}x_1 \mathrm{d}x_2 \ldots \mathrm{d}x_N, \end{aligned} \tag{3.50}$$

for all integers $k_j \geq 1$ and $N \geq 1$.

However, we shall be satisfied by considering only the first- and second-order moments, i.e., $E\{\mathbf{x}(t)\}$, $E\{\mathbf{x}^2(t)\}$, and $E\{\mathbf{x}(t_1)\mathbf{x}(t_2)\}$. They are the *mean value*, *mean-squared value*, and *(auto)correlation*, respectively, of the process.

Mean value or the first moment The mean value of the process at time t is

$$m_{\mathbf{x}}(t) = E\{\mathbf{x}(t)\} = \int_{-\infty}^{\infty} x f_{\mathbf{x}(t)}(x;t)\mathrm{d}x. \tag{3.51}$$

Note that the average is *across the ensemble* and if the pdf varies with time, then the mean value is a (deterministic) function of time. If, however, the process is stationary, then the mean is independent of t or a constant, i.e.,

$$m_{\mathbf{x}} = E\{\mathbf{x}(t)\} = \int_{-\infty}^{\infty} x f_{\mathbf{x}}(x)\mathrm{d}x. \tag{3.52}$$

Mean-squared value or the second moment This is defined, as per the discussion for random variables as

$$\mathrm{MSV}_{\mathbf{x}}(t) = E\{\mathbf{x}^2(t)\} = \int_{-\infty}^{\infty} x^2 f_{\mathbf{x}(t)}(x;t)\mathrm{d}x \tag{3.53}$$

or for a stationary process

$$\mathrm{MSV}_{\mathbf{x}} = E\{\mathbf{x}^2(t)\} = \int_{-\infty}^{\infty} x^2 f_{\mathbf{x}}(x)\mathrm{d}x. \tag{3.54}$$

As with random variables we have a second central moment, which is the variance:

$$\sigma_{\mathbf{x}}^2(t) = E\left\{[\mathbf{x}(t) - m_{\mathbf{x}}(t)]^2\right\} = \mathrm{MSV}_{\mathbf{x}}(t) - m_{\mathbf{x}}^2(t) \text{ (nonstationary)}, \tag{3.55}$$

$$\sigma_{\mathbf{x}}^2 = E\left\{[\mathbf{x}(t) - m_{\mathbf{x}}]^2\right\} = \mathrm{MSV}_{\mathbf{x}} - m_{\mathbf{x}}^2 \text{ (stationary)}. \tag{3.56}$$

The physical interpretations of these moments are DC value ($m_{\mathbf{x}}$), total power ($\mathrm{MSV}_{\mathbf{x}}$), AC power ($\sigma_{\mathbf{x}}^2$), and DC power ($m_{\mathbf{x}}^2$).

Correlation This is another statistical average that plays a very important role in our study of random processes. It is a second-order moment which is called the autocorrelation of the random process. Its importance arises from the fact that it completely describes the power spectral density (PSD) of the random process, i.e., it tells how the power is distributed in frequency. It is therefore analogous to the correlation functions encountered in Chapter 2 for deterministic signals. However, in general, the averaging is done here *across the ensemble* and not in time. The autocorrelation function for random process $\mathbf{x}(t)$ is the correlation between the two random variables $\mathbf{x}_1 = \mathbf{x}(t_1)$ and $\mathbf{x}_2 = \mathbf{x}(t_2)$ and is defined as

$$\begin{aligned} R_{\mathbf{x}}(t_1, t_2) &= E\{\mathbf{x}(t_1)\mathbf{x}(t_2)\} \\ &= \int_{x_1=-\infty}^{\infty} \int_{x_2=-\infty}^{\infty} x_1 x_2 f_{\mathbf{x}_1,\mathbf{x}_2}(x_1, x_2; t_1, t_2)\mathrm{d}x_1\mathrm{d}x_2. \end{aligned} \tag{3.57}$$

For a stationary process, since the joint pdf depends only on the time difference (also known as the lag time) $\tau = t_2 - t_1$ $(t_2 > t_1)$ the above becomes

$$R_{\mathbf{x}}(\tau) = E\{\mathbf{x}(t)\mathbf{x}(t+\tau)\} = \int_{x_1=-\infty}^{\infty}\int_{x_2=-\infty}^{\infty} x_1 x_2 f_{\mathbf{x}_1,\mathbf{x}_2}(x_1, x_2; \tau)\mathrm{d}x_1\mathrm{d}x_2. \tag{3.58}$$

An even less restrictive condition than demanding first- and second-order stationarity of a process is what is termed *wide-sense stationarity* (WSS). A process is said to be WSS if the following two conditions are satisfied:

(1) The mean value is independent of time, i.e., $E\{\mathbf{x}(t)\} = m_{\mathbf{x}}$ for any t.
(2) The autocorrelation depends only on the time difference $\tau = t_2 - t_1$, i.e., $R_{\mathbf{x}}(t_1, t_2) = R_{\mathbf{x}}(\tau)$.

Note that the first- and second-order stationarity implies WSS but that the converse is not true in general. It is fortunate that most information signals and noise sources encountered in communication systems are well modeled as WSS random processes. Hereafter, the term stationary with no adjective will mean WSS and unless explicitly stated the random processes will be considered to be WSS.

The autocorrelation function $R_{\mathbf{x}}(\tau)$ has several properties:

(1) $R_{\mathbf{x}}(\tau) = R_{\mathbf{x}}(-\tau)$. It is an even function of τ because the same set of product values is averaged across the ensemble, regardless of the direction of translation.
(2) $|R_{\mathbf{x}}(\tau)| \leq R_{\mathbf{x}}(0)$. The maximum always occurs at $\tau = 0$, though there may be other values of τ for which it is as big. Further $R_{\mathbf{x}}(0)$ is the mean-squared value of the random process.
(3) If for some τ_0 we have $R_{\mathbf{x}}(\tau_0) = R_{\mathbf{x}}(0)$, then for all integers k, $R_{\mathbf{x}}(k\tau_0) = R_{\mathbf{x}}(0)$.
(4) If $m_{\mathbf{x}} \neq 0$, then $R_{\mathbf{x}}(\tau)$ will have a constant component equal to $m_{\mathbf{x}}^2$.
(5) Autocorrelation functions cannot have an arbitrary shape. The restriction on the shape arises from the fact that the Fourier transform of an autocorrelation function must be greater than or equal to zero, i.e., $\mathcal{F}\{R_{\mathbf{x}}(\tau)\} \geq 0$.

The Fourier transform of the autocorrelation function is the power spectral density of the process, as discussed next. The power of a physical process is always nonnegative (≥ 0), otherwise we could design a perpetual motion machine. Among other things, the restriction means that $R_{\mathbf{x}}(\tau)$ cannot have any discontinuities or flat tops.

A final point is that there may be many different random processes that have the same autocorrelation function. Knowledge of $R_{\mathbf{x}}(\tau)$ is not sufficient to specify the joint pdf, not even $f_{\mathbf{x}}(x)$. The autocorrelation function is the Fourier transform pair of the PSD, where all phase information is disregarded. It is directly analogous to the situation for deterministic signals discussed in Chapter 2. Furthermore, as shown later, the effect of linear systems on the autocorrelation function of the input can be determined without any knowledge of the pdfs. Therefore the correlation function represents a considerably smaller amount of information.

3.2.3 Power spectral density (PSD) of a random process

In seeking a frequency domain representation of a random process it must be realized at the outset that the ensemble members of a stationary process are not energy signals but ones that have finite average power. This is simply due to the fact that $\text{MSV}_\mathbf{x} = E\{\mathbf{x}^2(t)\}$ represents average power and is the same for all time for a stationary process. Therefore the straightforward approach of taking the Fourier transform of the random process, i.e., determining $\mathbf{X}(f) = \int_{-\infty}^{\infty} \mathbf{x}(t)\text{e}^{-\text{j}2\pi ft}\text{d}t$ runs into mathematical difficulties due to convergence problems. As an aside, note that what is meant by the above operation is that each time function of the ensemble, $\mathbf{x}(t)$, is Fourier transformed into the frequency domain, see Figure 3.9.

Even if one ignores the difficulties associated with taking the Fourier transform of a power signal all one winds up with is another random process. Perhaps a way out is to define an average member for the ensemble, $E\{\mathbf{X}(f)\}$, and take it as being representative of the entire process. But this leads nowhere since[6]

$$E\{\mathbf{X}(f)\} = E\left\{\int_{-\infty}^{\infty} \mathbf{x}(t)\text{e}^{-\text{j}2\pi ft}\text{d}t\right\} = \int_{-\infty}^{\infty} E\{\mathbf{x}(t)\}\text{e}^{-\text{j}2\pi ft}\text{d}t$$
$$= \int_{-\infty}^{\infty} m_\mathbf{x}\text{e}^{-\text{j}2\pi ft}\text{d}t = m_\mathbf{x}\delta(f). \tag{3.59}$$

Surely not every random process has all its amplitude concentrated at $f = 0$.

The approach taken to obtain a frequency-domain characterization of the process is to determine how the average power of the process is distributed in frequency. To this end, define a truncated process as follows:

$$\mathbf{x}_T(t) = \begin{cases} \mathbf{x}(t), & -T \le t \le T \\ 0, & \text{otherwise} \end{cases}. \tag{3.60}$$

The truncation ensures that, provided the process $\mathbf{x}(t)$ has a finite mean-squared value, the truncated process will have finite energy, i.e., $\int_{-\infty}^{\infty} |\mathbf{x}_T(t)|^2\text{d}t < \infty$. Now consider the Fourier transform of this truncated process:

$$\mathbf{X}_T(f) = \int_{-\infty}^{\infty} \mathbf{x}_T(t)\text{e}^{-\text{j}2\pi ft}\text{d}t. \tag{3.61}$$

Again the truncation ensures that this is well defined. By Parseval's theorem it follows that

$$\int_{-\infty}^{\infty} \mathbf{x}_T^2(t)\text{d}t = \int_{-\infty}^{\infty} |\mathbf{X}_T(f)|^2\,\text{d}f \quad \text{(joules)}. \tag{3.62}$$

Since we are seeking to determine how the average power is distributed in frequency, we average the energy over the total time, $2T$, i.e.,

$$\mathbf{P} = \frac{1}{2T}\int_{-T}^{T} \mathbf{x}_T^2(t)\text{d}t = \frac{1}{2T}\int_{-\infty}^{\infty} |\mathbf{X}_T(f)|^2\,\text{d}f \quad \text{(watts)}. \tag{3.63}$$

[6] Note that expectation is a linear operation and that it only acts on random quantities, i.e., as far as it is concerned, $\text{e}^{-\text{j}2\pi ft}$ is simply a constant.

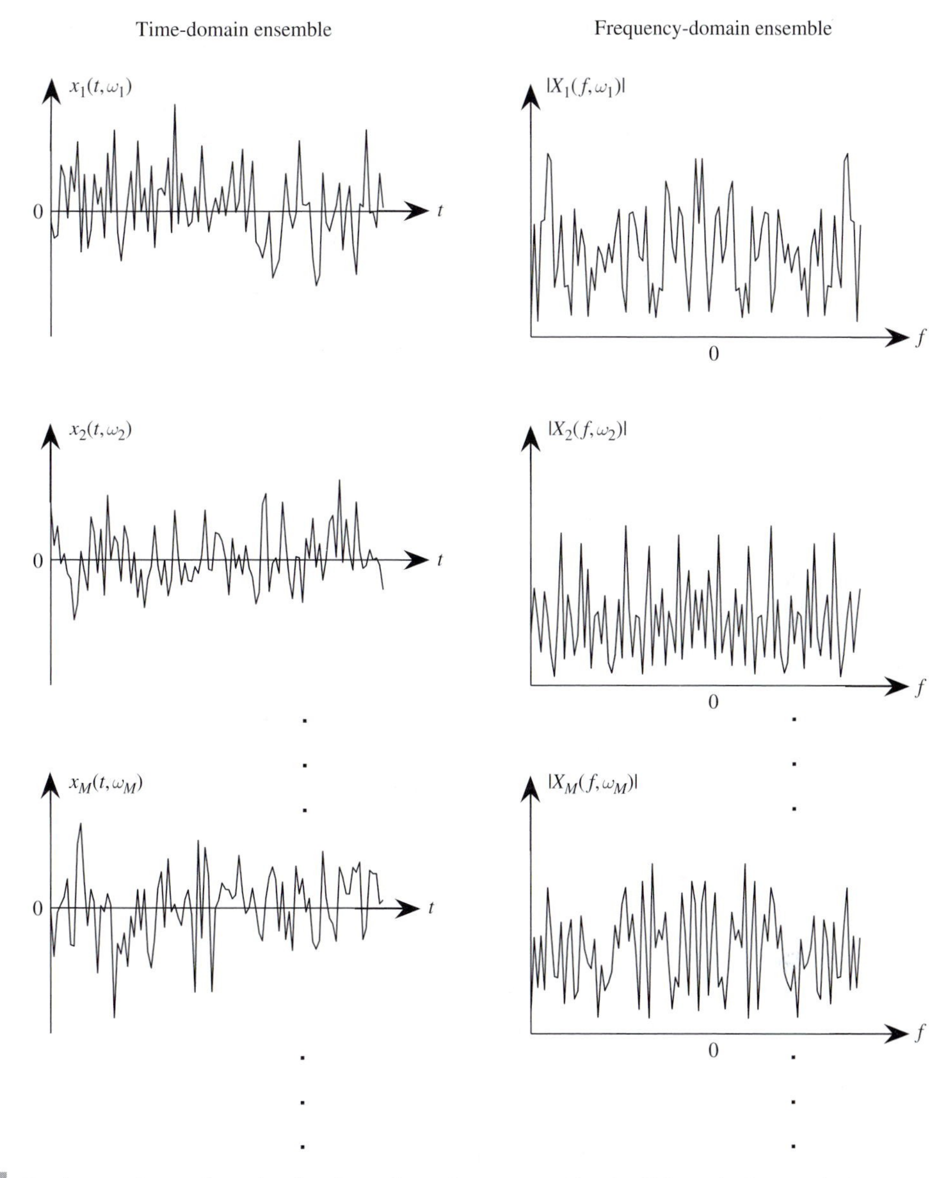

Fig. 3.9 Fourier transforms of member functions of a random process. For simplicity, only the magnitude spectra are shown.

What we have managed to accomplish thus far is to create the random variable, **P**, which in some sense represents the power in the process. Now we find the average value of **P**, i.e.,

$$E\{\mathbf{P}\} = E\left\{\frac{1}{2T}\int_{-T}^{T}\mathbf{x}_T^2(t)\mathrm{d}t\right\} = E\left\{\frac{1}{2T}\int_{-\infty}^{\infty}|\mathbf{X}_T(f)|^2\,\mathrm{d}f\right\}. \tag{3.64}$$

To return to the original process, we take the limit as $T \to \infty$:

$$\lim_{T\to\infty} \frac{1}{2T} \int_{-T}^{T} E\left\{\mathbf{x}_T^2(t)\right\} \mathrm{d}t = \lim_{T\to\infty} \frac{1}{2T} \int_{-\infty}^{\infty} E\left\{|\mathbf{X}_T(f)|^2\right\} \mathrm{d}f, \tag{3.65}$$

where the expectation and integration operations have been interchanged. Note that for a stationary process, within the interval $[-T, T]$, $E\{\mathbf{x}_T^2(t)\}$ and $E\{\mathbf{x}^2(t)\}$ are the same as the mean-squared value ($\mathrm{MSV}_\mathbf{x}$) of the process $\mathbf{x}(t)$. Therefore $\lim_{T\to\infty}(1/2T)\int_{-T}^{T} E\left\{\mathbf{x}_T^2(t)\right\}$ $\mathrm{d}t = \mathrm{MSV}_\mathbf{x} \lim_{T\to\infty}(1/2T)\int_{-T}^{T}\mathrm{d}t = \mathrm{MSV}_\mathbf{x}$. It follows that

$$\mathrm{MSV}_\mathbf{x} = \int_{-\infty}^{\infty} \lim_{T\to\infty} \frac{E\left\{|\mathbf{X}_T(f)|^2\right\}}{2T} \mathrm{d}f \quad \text{(watts)}. \tag{3.66}$$

Since $\mathrm{MSV}_\mathbf{x}$ has the unit of watts, the quantity $\lim_{T\to\infty} E\left\{|\mathbf{X}_T(f)|^2\right\}/2T$ must have the unit of watts/hertz. Call the limit $S_\mathbf{x}(f)$ and thus

$$S_\mathbf{x}(f) = \lim_{T\to\infty} \frac{E\left\{|\mathbf{X}_T(f)|^2\right\}}{2T} \quad \text{(watts/hertz)}, \tag{3.67}$$

which is the PSD of the process.

As in the case of deterministic signals we now relate the PSD, $S_\mathbf{x}(f)$, to the corresponding time function. It turns out to be the autocorrelation function defined in (3.58). We start with $\mathbf{X}_T(f) = \int_{-T}^{T} \mathbf{x}_T(t)\mathrm{e}^{-\mathrm{j}2\pi ft}\mathrm{d}t$, which means that

$$|\mathbf{X}_T(f)|^2 = \mathbf{X}_T(f)\mathbf{X}_T^*(f) = \int_{-T}^{T} \mathbf{x}_T(t)\mathrm{e}^{-\mathrm{j}2\pi ft}\mathrm{d}t \int_{-T}^{T} \mathbf{x}_T(\lambda)\mathrm{e}^{\mathrm{j}2\pi f\lambda}\mathrm{d}\lambda. \tag{3.68}$$

Now consider

$$\begin{aligned} S_\mathbf{x}(f) &= \lim_{T\to\infty} \frac{E\left\{|\mathbf{X}_T(f)|^2\right\}}{2T} \\ &= \lim_{T\to\infty} \left[\frac{1}{2T} E\left\{\int_{t=-T}^{T}\int_{\lambda=-T}^{T} \mathbf{x}_T(t)\mathbf{x}_T(\lambda)\mathrm{e}^{-\mathrm{j}2\pi f(t-\lambda)}\mathrm{d}\lambda\mathrm{d}t\right\}\right] \\ &= \lim_{T\to\infty} \left[\frac{1}{2T} \int_{t=-T}^{T}\int_{\lambda=-\infty}^{\infty} E\left\{\mathbf{x}_T(t)\mathbf{x}_T(\lambda)\right\}\mathrm{e}^{-\mathrm{j}2\pi f(t-\lambda)}\mathrm{d}\lambda\mathrm{d}t\right], \end{aligned} \tag{3.69}$$

where we let λ range from $-\infty$ to ∞ since $\mathbf{x}_T(\lambda)$ is zero outside the range $[-T, T]$. Now[7] $E\{\mathbf{x}_T(t)\mathbf{x}_T(\lambda)\} = R_\mathbf{x}(t-\lambda)$. Therefore

$$S_\mathbf{x}(f) = \lim_{T\to\infty} \left[\frac{1}{2T} \left\{\int_{t=-T}^{T} \mathrm{d}t \int_{\lambda=-\infty}^{\infty} R_\mathbf{x}(t-\lambda)\mathrm{e}^{-\mathrm{j}2\pi f(t-\lambda)}\mathrm{d}\lambda\right\}\right]. \tag{3.70}$$

[7] The astute reader might observe that $\mathbf{x}_T(t)$ must be a nonstationary process and therefore the correlation function should be a function of both t and λ, not just the difference. But within the interval $[-T, T]$, $\mathbf{x}_T(t)$ and $\mathbf{x}(t)$ are identical and $\mathbf{x}(t)$ is stationary. So for t and λ in this range the equation is valid and eventually we let the range go to $[-\infty, \infty]$.

Now change variables in the second integral to $\tau = t - \lambda$, it becomes $\int_{\tau=-\infty}^{\infty} R_{\mathbf{x}}(\tau) \mathrm{e}^{-\mathrm{j}2\pi f\tau}\mathrm{d}\tau = \mathcal{F}\{R_{\mathbf{x}}(\tau)\}$. Therefore,

$$
\begin{aligned}
S_{\mathbf{x}}(f) &= \lim_{T\to\infty}\left[\frac{1}{2T}\left\{\int_{t=-T}^{T}\mathcal{F}\{R_{\mathbf{x}}(\tau)\}\mathrm{d}t\right\}\right] \\
&= \mathcal{F}\{R_{\mathbf{x}}(\tau)\}\lim_{T\to\infty}\left[\frac{1}{2T}\left\{\int_{t=-T}^{T}\mathrm{d}t\right\}\right] \\
&= \mathcal{F}\{R_{\mathbf{x}}(\tau)\}. \qquad (3.71)
\end{aligned}
$$

To conclude, *the PSD and the autocorrelation function are a Fourier transform pair*, i.e.,

$$R_{\mathbf{x}}(\tau) \longleftrightarrow S_{\mathbf{x}}(f). \qquad (3.72)$$

It is important to recognize that there is a fundamental difference regarding this relationship between the autocorrelation function and the corresponding PSD, and that presented for deterministic signals. The autocorrelation function here is *defined on the ensemble*, while for deterministic signals it is defined for a single time function. For the random process it is an *average across the ensemble* while for the deterministic signal it is a *time average*. Would it not be fortuitous if one could determine the autocorrelation function and indeed any of the other parameters, such as the mean value, mean-squared value, $f_{\mathbf{x}}(x)$, from any one member function of the ensemble. It certainly would be efficient from the engineering perspective since one would not have to conduct the experiment many times to create the ensemble and then average across it to develop the model. One would only do the experiment once and use the resultant time function to estimate the mean, autocorrelation, $f_{\mathbf{x}}(x)$, etc.

A process where any one ensemble member function can be used as a representative of the ensemble is called *ergodic*. Ergodicity is discussed next.

3.2.4 Time averaging and ergodicity

An ergodic random process is one where any member of the ensemble exhibits the same statistical behavior as that of the whole ensemble.[8] In particular all time averages on a single ensemble member are equal to the corresponding ensemble average. Thus for an ergodic process

$$E\{\mathbf{x}^n(t)\} = \int_{-\infty}^{\infty} x^n f_{\mathbf{x}}(x)\mathrm{d}x = \lim_{T\to\infty}\frac{1}{2T}\int_{-T}^{T}[\mathbf{x}_k(t,\omega_k)]^n\mathrm{d}t,\ \forall\, n,\, k. \qquad (3.73)$$

A natural consequence of ergodicity is that in measuring various statistical averages (mean and autocorrelation, for example), it is sufficient to look at only one realization of the process and find the corresponding time average, rather than consider a large number of realizations and averaging over them. Since time averages equal ensemble averages for

[8] More precisely, almost every member of the ensemble since it may be possible to have a set of sample functions with a total probability of zero that do not have the appropriate behavior. Note that zero probability does not mean that the sample function cannot occur.

ergodic processes, fundamental electrical engineering parameters, such as DC value, RMS value, and average power can be related to the moments of an ergodic random process, as indeed we have been doing.

Other consequences of ergodicity are that the autocorrelation, $R_{\mathbf{x}}(\tau)$, and the pdfs, $f_{\mathbf{x}}(x)$ and $f_{\mathbf{x}_1,\mathbf{x}_2}(x_1, x_2; \tau)$, can be determined from a single sample function. It is important to note that *for a process to be ergodic it must be stationary*. However, not all stationary processes are necessarily ergodic. To prove ergodicity is in general difficult, if not impossible. However, it is customary to assume ergodicity unless there are compelling physical reasons for not doing so, such as an obvious nonstationarity. Three examples to illustrate this discussion are presented next.

Example 3.4 A random process is defined by $\mathbf{x}(t) = A\cos(2\pi f_0 t + \boldsymbol{\Theta})$, where $\boldsymbol{\Theta}$ is a random variable uniformly distributed on $[0, 2\pi]$. Note that in this case, we have an analytic description of the random process. The mean of this random process can be obtained by noting that

$$f_{\boldsymbol{\Theta}}(\Theta) = \begin{cases} \dfrac{1}{2\pi}, & 0 \le \Theta < 2\pi \\ 0, & \text{otherwise} \end{cases}. \tag{3.74}$$

Hence

$$m_{\mathbf{x}}(t) = E\{\mathbf{x}(t)\} = \int_0^{2\pi} A\cos(2\pi f_0 t + \Theta)\frac{1}{2\pi}\mathrm{d}\Theta = 0. \tag{3.75}$$

Observe that $m_{\mathbf{x}}(t)$ is independent of t. The autocorrelation function is

$$\begin{aligned} R_{\mathbf{x}}(t_1, t_2) &= E\{A\cos(2\pi f_0 t_1 + \boldsymbol{\Theta})A\cos(2\pi f_0 t_2 + \boldsymbol{\Theta})\} \\ &= A^2 E\left\{\frac{1}{2}\cos\left(2\pi f_0(t_1 - t_2)\right) + \frac{1}{2}\cos(2\pi f_0(t_1 + t_2) + 2\boldsymbol{\Theta})\right\} \\ &= \frac{A^2}{2}\cos\left(2\pi f_0(t_1 - t_2)\right), \end{aligned}$$

where we have used the fact that

$$E\{\cos(2\pi f_0(t_1 + t_2) + 2\boldsymbol{\Theta})\} = \int_0^{2\pi} \cos[2\pi f_0(t_1 + t_2) + 2\Theta]\frac{1}{2\pi}\mathrm{d}\Theta = 0. \tag{3.76}$$

Since $m_{\mathbf{x}}(t) = 0$ and $R_{\mathbf{x}}(t_1, t_2) = (A^2/2)\cos\left(2\pi f_0(t_1 - t_2)\right)$, the random process is WSS.

Furthermore, for any value of $0 \le \Theta < 2\pi$ (i.e., for any realization of the process), consider the following time average of a sample function:

$$\begin{aligned}
\lim_{T\to\infty} \frac{1}{2T} & \int_{-T}^{T} [A\cos(2\pi f_0 t + \Theta)]^n \mathrm{d}t \\
&= \lim_{N\to\infty} \frac{1}{2NT_0} \int_{-NT_0}^{NT_0} [A\cos(2\pi f_0 t + \Theta)]^n \mathrm{d}t \quad (\text{where } T_0 = 1/f_0) \\
&= \frac{1}{T_0} \int_0^{T_0} [A\cos(2\pi f_0 t + \Theta)]^n \mathrm{d}t \\
&= \frac{1}{T_0} \int_{\Theta}^{2\pi+\Theta} [A\cos u]^n \frac{\mathrm{d}u}{2\pi f_0} \quad (\text{where } u = 2\pi f_0 t + \Theta) \\
&= \frac{1}{2\pi} \int_0^{2\pi} [A\cos u]^n \mathrm{d}u. \qquad (3.77)
\end{aligned}$$

On the other hand,

$$\begin{aligned}
E\left\{\mathbf{x}^n(t)\right\} &= \int_0^{2\pi} [A\cos(2\pi f_0 t + \Theta)]^n \frac{1}{2\pi} \mathrm{d}\Theta \\
&= \frac{1}{2\pi} \int_{2\pi f_0 t}^{2\pi f_0 t + 2\pi} [A\cos(u)]^n \mathrm{d}u \quad (\text{where } u = 2\pi f_0 t + \Theta) \\
&= \frac{1}{2\pi} \int_0^{2\pi} [A\cos u]^n \mathrm{d}u. \qquad (3.78)
\end{aligned}$$

The above shows that the time averages are the same as the ensemble averages. Therefore the process is also ergodic.

Finally, since the process is both stationary and ergodic, we have

$$P_{\mathbf{x}} = R_{\mathbf{x}}(0) = \frac{A^2}{2}\cos(2\pi f_0 \tau)\bigg|_{\tau=0} = \frac{A^2}{2}. \qquad (3.79)$$

This is, in fact, the power content of each sample function in the process since each realization is a sinusoidal waveform of constant amplitude A. ■

Example 3.5 The process $\mathbf{x}(t)$ is defined by $\mathbf{x}(t) = \mathbf{x}$, where $\mathbf{x}$ is a random variable uniformly distributed on $[-A, A]$, where $A > 0$. In this case, again an analytic description of the random process is given. For this random process, each sample is a constant signal. The mean value is $E\{\mathbf{x}\} = 0$. The autocorrelation function of this random process is

$$R_{\mathbf{x}}(t_1, t_2) = E\{\mathbf{x}^2\} = \int_{-A}^{A} \frac{1}{2A} x^2 \mathrm{d}x = \frac{A^2}{3}, \qquad (3.80)$$

which is a constant. The process is therefore WSS, indeed it is strictly stationary.

However, the process is not ergodic. Given a specific member function of the ensemble function, say x where $-A \le x \le A$, the mean and mean-squared values obtained by a time average are x and x^2, respectively, not $m_{\mathbf{x}} = 0$ and $P_{\mathbf{x}} = R_{\mathbf{x}}(0) = A^2/3$ as obtained by an ensemble average. ■

Example 3.6 Consider the random process of Example 3.4 but let the amplitude of the sinusoid be random as well. The process is then $\mathbf{x}(t) = \mathbf{A}\cos(2\pi f_0 t + \mathbf{\Theta})$ where $\mathbf{A}$ is a zero-mean random variable with variance, $\sigma_{\mathbf{A}}^2$, and $\mathbf{\Theta}$ is uniform in $[0, 2\pi]$. Furthermore, $\mathbf{A}$ and $\mathbf{\Theta}$ are statistically independent.

Let us now compute the first and second moments of $\mathbf{x}(t)$ by an ensemble average and a time average.

For ensemble average, the first and second moments are:

$$E\{\mathbf{x}(t)\} = E\{\mathbf{A}\cos(2\pi f_0 t + \mathbf{\Theta})\}$$
$$= E\{\mathbf{A}\}E\{\cos(2\pi f_0 t + \mathbf{\Theta})\} = 0, \tag{3.81}$$
$$E\{\mathbf{x}^2(t)\} = E\{[\mathbf{A}\cos(2\pi f_0 t + \mathbf{\Theta})]^2\}$$
$$= E\{\mathbf{A}^2\}E\{[\cos(2\pi f_0 t + \mathbf{\Theta})]^2\} = \frac{\sigma_{\mathbf{A}}^2}{2}. \tag{3.82}$$

For the time average, one has:

$$\lim_{T\to\infty}\frac{1}{2T}\int_{-T}^{T} A\cos(2\pi f_0 t + \Theta)\mathrm{d}t$$
$$= A\lim_{T\to\infty}\frac{1}{2T}\int_{-T}^{T}\cos(2\pi f_0 t + \Theta)\mathrm{d}t = 0 \quad \text{(first moment)}, \tag{3.83}$$
$$\lim_{T\to\infty}\frac{1}{2T}\int_{-T}^{T}[A\cos(2\pi f_0 t + \Theta)]^2\mathrm{d}t$$
$$= A^2\lim_{T\to\infty}\frac{1}{2T}\int_{-T}^{T}[\cos(2\pi f_0 t + \Theta)]^2\mathrm{d}t = \frac{A^2}{2} \quad \text{(second moment)}. \tag{3.84}$$

This example illustrates several points. The second moments obtained by ensemble and time averages do not agree, therefore, the process is not ergodic. However, the first moment is the same for both averages. Thus the process can be said to be ergodic in the mean.[9] Furthermore, it can be shown that $\mathbf{x}(t)$ is strictly stationary, which illustrates that stationarity is a necessary not sufficient condition for ergodicity. ■

3.3 Random processes and LTI systems

Since random processes are invariably subjected to filtering we now consider the effect of a linear, time-invariant (LTI) system on the statistics of the input process. Consider an LTI system with impulse response, $h(t)$ (see Figure 3.10). Though one may desire a complete or

[9] As for stationarity, there are different levels of ergodicity which leads to the definitions of: strictly ergodic, ergodic in the mean as above, wide sense ergodicity which is ergodicity in the mean and autocorrelation. Wide-sense ergodicity is the ergodicity of most interest to us.

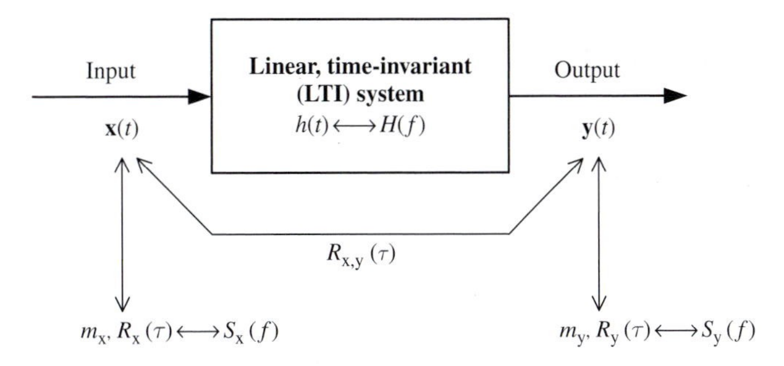

Fig. 3.10 Illustration of the response of an LTI system to an input random process.

at least a partial characterization of the output process in terms of say the first- and second-order pdfs, it turns out, somewhat ironically, that this is impossible to determine, except in the important Gaussian case. Ironically, because when the random process is input to a nonlinearity we have a well-established method to determine the output pdf (see Equation (3.18)). Therefore we shall have to be content with determining only the output mean, autocorrelation function, PSD, and crosscorrelation between input and output. The PSD is of particular importance since it informs us as to the filtering (shaping) capability of the system. Finally if the input is a Gaussian process, then so is the output because the output is a weighted linear combination of the input (shown in a following section). In this very important case we can obtain a complete characterization of the output random process.

For the LTI system of Figure 3.10 the output is given by the convolution integral $\mathbf{y}(t) = \int_{-\infty}^{\infty} h(\lambda)\mathbf{x}(t-\lambda)\mathrm{d}\lambda = \int_{-\infty}^{\infty} h(t-\lambda)\mathbf{x}(\lambda)\mathrm{d}\lambda$ in the time domain or $\mathbf{Y}(f) = H(f)\mathbf{X}(f)$ in the frequency domain. We assume that the input is at least WSS and because the system is time-invariant the output will also be WSS. The mean $m_{\mathbf{y}}$, autocorrelation $R_{\mathbf{y}}(\tau)$, and PSD $S_{\mathbf{y}}(f)$ are then determined as follows.

For the mean value, one has

$$\begin{aligned} m_{\mathbf{y}} &= E\{\mathbf{y}(t)\} = E\left\{\int_{-\infty}^{\infty} h(\lambda)\mathbf{x}(t-\lambda)\mathrm{d}\lambda\right\} \\ &= \int_{-\infty}^{\infty} h(\lambda)E\{\mathbf{x}(t-\lambda)\}\,\mathrm{d}\lambda = m_{\mathbf{x}}\int_{-\infty}^{\infty} h(\lambda)\mathrm{d}\lambda. \end{aligned} \tag{3.85}$$

But $\int_{-\infty}^{\infty} h(\lambda)\mathrm{d}\lambda = H(0)$, which is the DC gain of the LTI system. Therefore we have the following very sensible relationship:

$$m_{\mathbf{y}} = m_{\mathbf{x}}H(0). \tag{3.86}$$

Regarding the relationship between $S_{\mathbf{y}}(f)$ and $S_{\mathbf{x}}(f)$, it turns out to be somewhat simpler to compute the output power spectral density from the basic definition. As in the case when the basic definition for the power spectral density was given, we start with the truncated version of $\mathbf{x}_T(t)$ and the corresponding output $\mathbf{y}_{(T)}(t)$, which certainly should be well behaved. The subscript (T) for the output indicates that it corresponds to a truncated input. The process $\mathbf{y}_{(T)}(t)$ itself is not necessarily truncated but this is not the issue here. Now we have that $\mathbf{Y}_{(T)}(f) = H(f)\mathbf{X}_T(f)$ and $\left|\mathbf{Y}_{(T)}(f)\right|^2 = |H(f)|^2\,|\mathbf{X}_T(f)|^2$. Therefore

$$S_{\mathbf{y}}(f) = \lim_{T\to\infty} E\left\{\frac{\left|\mathbf{Y}_{(T)}(f)\right|^2}{2T}\right\} = \lim_{T\to\infty} E\left\{\frac{|H(f)|^2\,|\mathbf{X}_T(f)|^2}{2T}\right\}$$
$$= |H(f)|^2 \lim_{T\to\infty} E\left\{\frac{|\mathbf{X}_T(f)|^2}{2T}\right\} = |H(f)|^2\,S_{\mathbf{x}}(f). \tag{3.87}$$

Due to the above relationship, it is obvious to call $|H(f)|^2$ the *power transfer function.*

Next, the output autocorrelation function is obtained readily from (3.87) by writing $|H(f)|^2$ as $H(f)H^*(f)$. Since $S_{\mathbf{y}}(f) = H(f)H^*(f)S_{\mathbf{x}}(f)$ is the product of three terms in the frequency domain, the inverse Fourier transform of $S_{\mathbf{y}}(f)$ is the convolution of the corresponding time functions in the time domain, i.e.,

$$R_{\mathbf{y}}(\tau) = h(\tau) * h(-\tau) * R_{\mathbf{x}}(\tau), \tag{3.88}$$

where we have used the fact that $H^*(f) \longleftrightarrow h(-\tau)$.

Finally consider the crosscorrelation, $R_{\mathbf{x},\mathbf{y}}(\tau)$, between the input and output. The function $R_{\mathbf{x},\mathbf{y}}(\tau)$ is not that important in communications but it does have an interesting application in system theory, where it can be used for system identification. This is explored in a problem at the end of the chapter. Left as an exercise to the reader, it can be shown that

$$R_{\mathbf{x},\mathbf{y}}(\tau) = h(\tau) * R_{\mathbf{x}}(\tau). \tag{3.89}$$

3.4 Noise in communication systems

Noise refers to unwanted electrical signals that are always present in electronic circuits; it limits the receiver's ability to detect the desired signal and thereby limits the rate of information transmission. Noise can be human made or occur naturally. A natural noise source is thermal noise, which is caused by the omnipresent random motion of free electrons in conducting material. If one were to put a sensitive enough oscilloscope across a resistor one would see a voltage across the resistor terminals due to this motion (assuming the oscilloscope circuitry did not itself produce significant noise). Further, if one were to observe the waveforms of N resistors all having the same resistance, at the same temperature, and built from the same material, one would see N different waveforms, i.e., *an ensemble*. Therefore it is appropriate to model the noise as a random process.

Because the terminal voltage is produced by the random motion of a great many free electrons,[10] due to the central limit theorem (discussed in Section 10.6) the amplitude statistics are well modeled to be Gaussian. The Gaussian process is determined by the mean and autocorrelation. The mean value of thermal noise can be shown, both analytically and

[10] It is of interest to determine how many free electrons there are in our ubiquitous 1 ohm resistor.

experimentally, to be zero. Intuitively one would not expect the electrons to huddle at one terminal or the other for any length of time.

A statistical analysis of the random motion of electrons shows that the autocorrelation of thermal noise $\mathbf{w}(t)$ is well modeled as

$$R_{\mathbf{w}}(\tau) = k\theta G \frac{e^{-|\tau|/t_0}}{t_0} \quad \text{(watts)}, \tag{3.90}$$

where k is Boltzmann's constant ($k = 1.38 \times 10^{-23}$ joule/degree Kelvin), G is the conductance of the resistor (mhos), θ is temperature in degrees Kelvin, and t_0 is the statistical average of time intervals between collisions of free electrons in the resistor, which is on the order of 10^{-12} seconds. The corresponding PSD is

$$S_{\mathbf{w}}(f) = \frac{2k\theta G}{1 + (2\pi f t_0)^2} \quad \text{(watts/hertz)}. \tag{3.91}$$

Observe that the noise PSD is approximately flat over the frequency range 0–10 gigahertz. As far as a typical communication system is concerned we might as well let the spectrum be flat from 0 to ∞, i.e.,

$$S_{\mathbf{w}}(f) = \frac{N_0}{2} \quad \text{(watts/hertz)}, \tag{3.92}$$

where N_0 is a constant; in this case $N_0 = 4k\theta G$.

The factor 2 in the denominator is included to indicate that $S_{\mathbf{w}}(f)$ is a two-sided spectrum. Noise that has a uniform spectrum over the entire frequency range is referred to as *white noise*. The adjective "white" comes from white light, which contains equal amounts of all frequencies within the visible band of electromagnetic radiation.

The autocorrelation of white noise $\mathbf{w}(t)$ is given by

$$R_{\mathbf{w}}(\tau) = \frac{N_0}{2}\delta(\tau) \quad \text{(watts)}. \tag{3.93}$$

Thus the autocorrelation of white noise is a delta function of strength $N_0/2$ occurring at $\tau = 0$. Since $R_{\mathbf{w}}(\tau) = 0$ for $\tau \neq 0$, any two different samples of white noise, no matter how close in time they are taken, are *uncorrelated*. The models of PSD and autocorrelation for white noise are illustrated in Figure 3.11 together with the counterparts of thermal noise as given in (3.91) and (3.90). In particular Equations (3.91) and (3.90) are plotted on Figure 3.11 by assuming that $G = 1/10$ (mhos), $\theta = 298.15$ K and $t_0 = 3 \times 10^{-12}$ seconds. Note also that the frequency range 0–15 GHz on Figure 3.11(a) covers the spectrum of almost all commercial wired and wireless communication systems.

The average power of white noise, which equals the area under the power spectral density, is obviously infinite. White noise is therefore an abstraction since no physical noise process can truly be white. Nonetheless, it is a useful abstraction. The noise encountered in many real systems can be assumed to be approximately white. This is because we can only observe such noise after it has passed through a real system, which will have a finite bandwidth. Thus, as long as the bandwidth of the noise is significantly larger than that of the system, the noise can be considered to have an infinite bandwidth. As a rule of thumb, noise is well modeled as white when its PSD is flat over a frequency band that is 3–5 times that of the communication system under consideration.

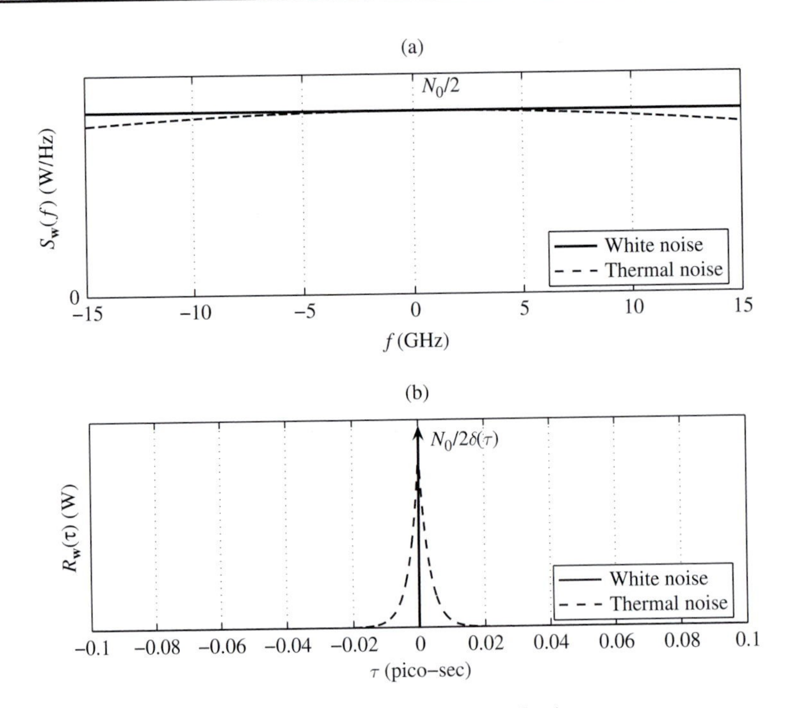

Fig. 3.11 (a) The PSD ($S_w(f)$), and (b) the autocorrelation ($R_w(\tau)$) of thermal noise.

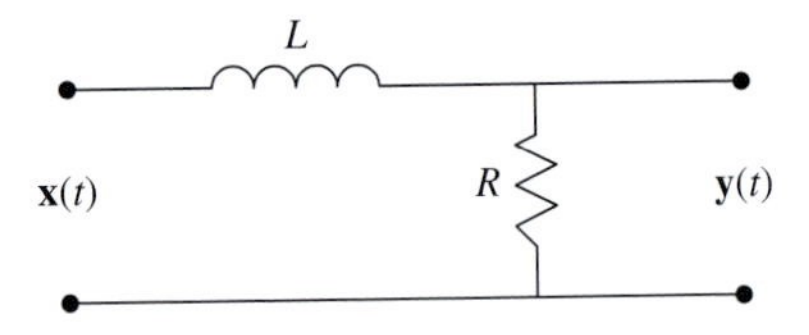

Fig. 3.12 A lowpass filter.

Finally, since the noise samples of white noise are uncorrelated, if the noise is both white and Gaussian (for example, thermal noise) then the noise samples are also *independent*.

Example 3.7 Consider the lowpass filter given in Figure 3.12. Suppose that a (WSS) white noise process, $\mathbf{x}(t)$, of zero-mean and PSD $N_0/2$ is applied to the input of the filter.

(a) Find and sketch the PSD and autocorrelation function of the random process $\mathbf{y}(t)$ at the output of the filter.
(b) What are the mean and variance of the output process $\mathbf{y}(t)$?

Solution

(a) Since $\mathbf{x}(t)$ is WSS white noise of zero-mean and PSD $N_0/2$, $S_{\mathbf{x}}(f) = N_0/2$, for all f. The transfer function of the lowpass filter is:

$$H(f) = \frac{R}{R + \mathrm{j}2\pi fL} = \frac{1}{1 + \mathrm{j}2\pi fL/R}. \tag{3.94}$$

Therefore

$$|H(f)|^2 = \frac{1}{1+(2\pi L/R)^2 f^2}. \quad (3.95)$$

One obtains

$$S_{\mathbf{y}}(f) = \frac{N_0}{2}\frac{1}{1+(2\pi L/R)^2 f^2}. \quad (3.96)$$

The autocorrelation function of $\mathbf{y}(t)$ is the inverse Fourier transform of $S_{\mathbf{y}}(f)$:

$$R_{\mathbf{y}}(\tau) = \mathcal{F}^{-1}\{S_{\mathbf{y}}(f)\} = \frac{N_0}{2}\mathcal{F}^{-1}\left\{\frac{1}{1+(2\pi L/R)^2 f^2}\right\}. \quad (3.97)$$

We have established in Chapter 2 that

$$\mathcal{F}^{-1}\left\{\frac{2a}{a^2+(2\pi f)^2}\right\} = \mathrm{e}^{-a|t|},\ a > 0. \quad (3.98)$$

Therefore

$$\begin{aligned} R_{\mathbf{y}}(\tau) &= \frac{N_0}{2}\mathcal{F}^{-1}\left\{\frac{(R/L)^2}{(R/L)^2+(2\pi f)^2}\right\} \\ &= \frac{N_0}{4}\frac{R}{L}\mathcal{F}^{-1}\left\{\frac{2(R/L)}{(R/L)^2+(2\pi f)^2}\right\} \\ &= \frac{N_0 R}{4L}\mathrm{e}^{-(R/L)|\tau|}. \end{aligned} \quad (3.99)$$

Plots of $S_{\mathbf{y}}(f)$ and $R_{\mathbf{y}}(\tau)$ are provided in Figure 3.13.

(b) The mean of $\mathbf{y}(t)$ is

$$m_{\mathbf{y}} = m_{\mathbf{x}} H(0) = 0 \times H(0) = 0 \quad (3.100)$$

and the variance of $\mathbf{y}(t)$ is

$$E\{[\mathbf{y}(t) - m_{\mathbf{y}}]^2\} = E\{\mathbf{y}^2(t)\} = R_{\mathbf{y}}(0) = \frac{N_0 R}{4L}. \quad (3.101)$$

■

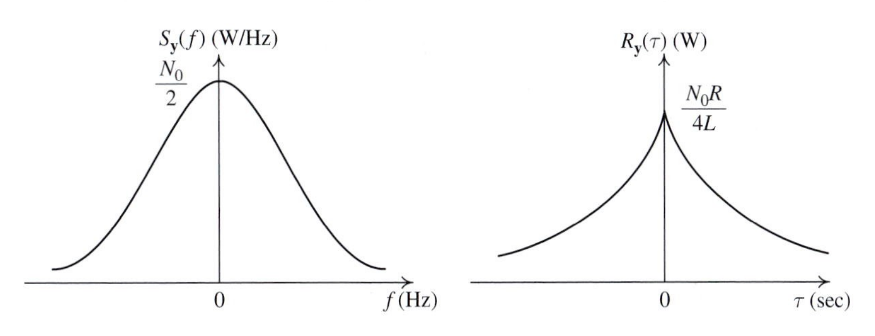

Fig. 3.13 PSD and autocorrelation function of the output process.

3.5 The Gaussian random variable and process

Unquestionably the Gaussian pdf model is the one most frequently encountered in nature. This, as mentioned earlier, is because most random phenomena are due to the action of many different factors. By the central limit theorem, the resultant random variable tends to be Gaussian regardless of the underlying probabilities of the individual actors. In this section the Gaussian pdf is discussed in some detail. We look at its form and its properties and also try to gain graphical insight into its behavior.

To motivate the discussion, consider the electrical activity of a skeletal muscle. A sample function is shown in Figure 3.14(a), along with the experimentally determined histogram in Figure 3.14(b).[11] The histogram is an estimate of the amplitude's pdf and we attempt to determine a reasonable analytical model for the pdf. To this end two functional forms are proposed for the pdf. Both are decaying exponentials: one has an exponent that is linear in x, of the form $K\mathrm{e}^{-a|x|}$, while the exponent in the other is quadratic in x, $K\mathrm{e}^{-(ax^2+bx+c)}$. The latter is known as a Gaussian pdf but is also frequently referred to as a normal pdf. Simply put, *a Gaussian random variable is one whose pdf is of the form of an exponential where the exponent is a quadratic function in the variable.*

We now proceed to write the Gaussian pdf in the standard form in which one always sees it written. Of course $f_{\mathbf{x}}(x)$ should be a valid pdf. This means that it needs to satisfy the two basic conditions: (i) $f_{\mathbf{x}}(x) \geq 0$ for all x and (ii) $\int_{-\infty}^{\infty} f_{\mathbf{x}}(x)\mathrm{d}x = 1$. The constants K, a, b, and c therefore are not completely arbitrary. The first condition means that $K > 0$, the second that $a > 0$. Now by a series of fairly straightforward algebraic steps we write the pdf in the standard form as follows:

$$\begin{aligned}
f_{\mathbf{x}}(x) &= K\mathrm{e}^{-a\left(x^2+(b/a)x+(c/a)\right)} \\
&= K\mathrm{e}^{-a\left(x^2+2b_1x+c_1\right)}, \text{ where } b_1 = b/2a \text{ and } c_1 = c/a \\
&= K\mathrm{e}^{-a\left(x^2+2b_1x+b_1^2-b_1^2+c_1\right)} = K\mathrm{e}^{-a(c_1-b_1^2)}\mathrm{e}^{-a(x+b_1)^2} \\
&= K_1\mathrm{e}^{-a(x+b_1)^2}, \text{ where } K_1 = K\mathrm{e}^{-a(c_1-b_1^2)}.
\end{aligned} \tag{3.102}$$

Note that regardless of K_1, the center of gravity of $f_{\mathbf{x}}(x)$ is $x = -b_1$. This then is the first moment, $E\{\mathbf{x}\}$, call it $m_{\mathbf{x}}$, i.e., $b_1 = -m_{\mathbf{x}}$. Therefore $f_{\mathbf{x}}(x) = K_1\mathrm{e}^{-a(x-m_{\mathbf{x}})^2}$. The constant K_1 simply ensures that the area under $f_{\mathbf{x}}(x)$ is 1. However, we shall use this constraint to determine the relationship between K_1 and a. Realizing that the area is independent of $m_{\mathbf{x}}$, set the mean to zero and proceed as follows:

[11] The histogram shows how many samples fall within a certain amplitude interval normalized by the total number of samples. In essence, it is an estimate of $P[x < \mathbf{x} \leq x + \Delta x]$, where Δx is the width of the amplitude interval.

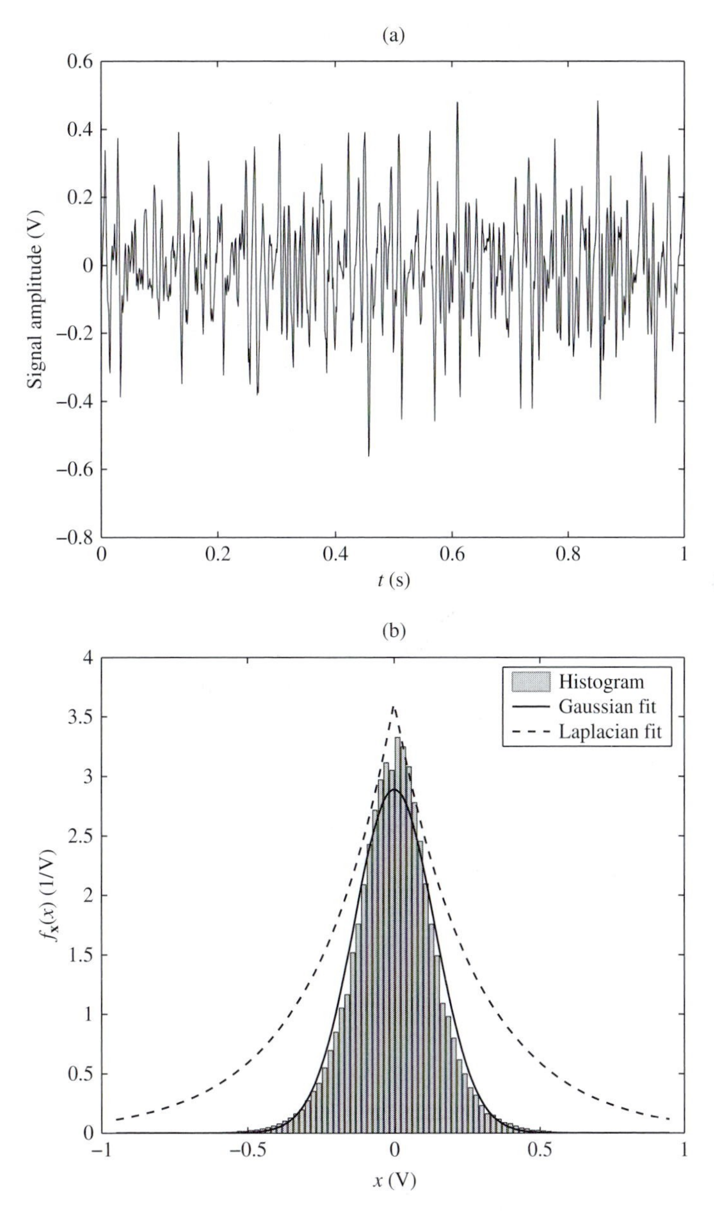

Fig. 3.14 (a) A sample skeletal muscle (emg) signal, and (b) its histogram and pdf fits.

$$
\begin{aligned}
1 = \left[\int_{-\infty}^{\infty} f_{\mathbf{x}}(x)\mathrm{d}x\right]^2 &= \left[\int_{-\infty}^{\infty} K_1 \mathrm{e}^{-ax^2}\mathrm{d}x\right]^2 \\
&= K_1^2 \int_{x=-\infty}^{\infty} \mathrm{e}^{-ax^2}\mathrm{d}x \int_{y=-\infty}^{\infty} \mathrm{e}^{-ay^2}\mathrm{d}y \\
&= K_1^2 \int_{x=-\infty}^{\infty}\int_{y=-\infty}^{\infty} \mathrm{e}^{-a(x^2+y^2)}\mathrm{d}x\mathrm{d}y. \qquad (3.103)
\end{aligned}
$$

Now find the volume under $\mathrm{e}^{-a(x^2+y^2)}$ in polar coordinates with $\rho = \sqrt{x^2 + y^2}$ as

$$\begin{aligned} 1 &= K_1^2 \int_{\alpha=0}^{2\pi} \int_{\rho=0}^{\infty} \mathrm{e}^{-a\rho^2} \rho \mathrm{d}\rho \mathrm{d}\alpha = 2\pi K_1^2 \int_{\rho=0}^{\infty} \rho \mathrm{e}^{-a\rho^2} \mathrm{d}\rho \\ &\stackrel{\lambda=\rho^2}{=} \pi K_1^2 \int_0^{\infty} \mathrm{e}^{-a\lambda} \mathrm{d}\lambda = \frac{\pi K_1^2}{a}. \end{aligned} \tag{3.104}$$

The above gives $K_1 = \sqrt{a/\pi}$ and the pdf becomes $f_{\mathbf{x}}(x) = \sqrt{a/\pi}\mathrm{e}^{-a(x-m_{\mathbf{x}})^2}$.

To write the equation in the standard form consider the second central moment or variance:

$$\begin{aligned} \sigma_{\mathbf{x}}^2 &= E\left\{(\mathbf{x} - m_{\mathbf{x}})^2\right\} = \sqrt{\frac{a}{\pi}} \int_{-\infty}^{\infty} (x - m_{\mathbf{x}})^2 \mathrm{e}^{-a(x-m_{\mathbf{x}})^2} \mathrm{d}x \\ &\stackrel{\lambda=x-m_{\mathbf{x}}}{=} \sqrt{\frac{a}{\pi}} \int_{-\infty}^{\infty} \lambda^2 \mathrm{e}^{-a\lambda^2} \mathrm{d}\lambda. \end{aligned} \tag{3.105}$$

Integrate by parts with $u = \lambda$ and $\mathrm{d}v = \lambda \mathrm{e}^{-a\lambda^2} \mathrm{d}\lambda$, i.e., $v = -(1/2a)\mathrm{e}^{-a\lambda^2}$, to obtain:

$$\sigma_{\mathbf{x}}^2 = \underbrace{-\frac{1}{2a}\sqrt{\frac{a}{\pi}} \lambda \mathrm{e}^{-a\lambda^2}\Big|_{-\infty}^{\infty}}_{=0} + \frac{1}{2a}\left[\sqrt{\frac{a}{\pi}} \int_{-\infty}^{\infty} \mathrm{e}^{-a\lambda^2} \mathrm{d}\lambda\right] = \frac{1}{2a}. \tag{3.106}$$

Therefore $a = 1/2\sigma_{\mathbf{x}}^2$ and the final form for the Gaussian pdf is

$$f_{\mathbf{x}}(x) = \frac{1}{\sqrt{2\pi\sigma_{\mathbf{x}}^2}} \mathrm{e}^{-(x-m_{\mathbf{x}})^2/2\sigma_{\mathbf{x}}^2}, \tag{3.107}$$

where, to repeat, the parameters $m_{\mathbf{x}}$ and $\sigma_{\mathbf{x}}^2$ are the mean and variance of the random variable.

The other proposed pdf model, using the unit area constraint, can be written as $f_{\mathbf{x}}(x) = (a/2)\mathrm{e}^{-a|x|}$. It is known as a *Laplacian* density, and in this case it has zero mean. Figure 3.14(b) also shows curve fits of both pdfs to the histogram and, as can be seen, the Gaussian pdf provides a better fit. This is to be expected since the signal under consideration is the result of the underlying activity of many (> 100) active motor neurons that activate the skeletal muscle.

Before going on to the discussion of jointly Gaussian random variables, consider the influence of $\sigma_{\mathbf{x}}^2$ on the shape of the pdf (see Figure 3.15). As mentioned, the variance, more precisely the standard deviation, $\sigma_{\mathbf{x}}$, of a random variable is a measure of the dispersion of the random variable about the mean. Therefore consider how much of the probability of the Gaussian random variable lies within different ranges of $\sigma_{\mathbf{x}}$. To this end, we determine $P(m_{\mathbf{x}} - k\sigma_{\mathbf{x}} < \mathbf{x} \leq m_{\mathbf{x}} + k\sigma_{\mathbf{x}})$, $k = 1, 2, 3 \ldots$. Table 3.1 shows that a range of $\pm 3\sigma_{\mathbf{x}}$ captures most of the probability.[12] However, one should not conclude from this that the tails

[12] The erf(·) function available in Matlab finds the area under a zero-mean Gaussian pdf with $\sigma_{\mathbf{x}}^2$ set to 1/2.

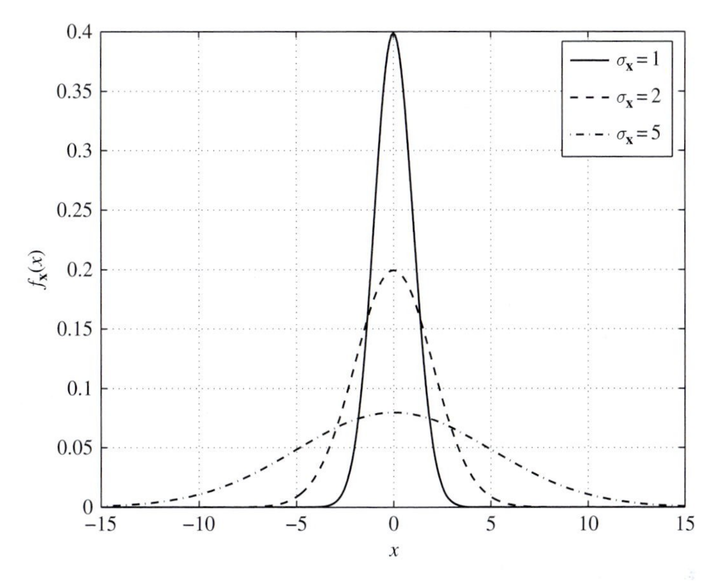

Fig. 3.15 Plots of the zero-mean Gaussian pdf for different values of standard deviation, $\sigma_{\mathbf{x}}$.

Table 3.1 Influence of $\sigma_{\mathbf{x}}$ on different quantities

Range ($\pm k\sigma_{\mathbf{x}}$)	$k=1$	$k=2$	$k=3$	$k=4$
$P(m_{\mathbf{x}} - k\sigma_{\mathbf{x}} < \mathbf{x} \leq m_{\mathbf{x}} + k\sigma_{\mathbf{x}})$	0.683	0.955	0.997	0.999
Error probability	10^{-3}	10^{-4}	10^{-6}	10^{-8}
Distance from the mean	3.09	3.72	4.75	5.61

of the pdf are ignorable. Indeed when communication systems are considered later it is the presence of these tails that results in bit errors. The probabilities are on the order of 10^{-3}–10^{-12}, very small, but still significant in terms of system performance. It is of interest to see how far, in terms of $\sigma_{\mathbf{x}}$, one must be from the mean value to have the different levels of error probabilities. As shall be seen in later chapters this translates to the required SNR to achieve a specified bit error probability. This is also shown in Table 3.1.

Having considered the single (or *univariate*) Gaussian random variable, we turn our attention to the case of two jointly Gaussian random variables (or the *bivariate* case). Again they are described by their joint pdf which, in general, is an exponential whose exponent is a quadratic in the two variables, i.e., $f_{\mathbf{x},\mathbf{y}}(x,y) = K\mathrm{e}^{(ax^2+bx+cxy+dy+ey^2+f)}$, where the constants K, a, b, c, d, e, and f are chosen to satisfy the basic properties of a valid joint pdf, namely being always nonnegative (≥ 0), having unit volume, and also that the marginal pdfs, $f_{\mathbf{x}}(x) = \int_{-\infty}^{\infty} f_{\mathbf{x},\mathbf{y}}(x,y)\mathrm{d}y$ and $f_{\mathbf{y}}(y) = \int_{-\infty}^{\infty} f_{\mathbf{x},\mathbf{y}}(x,y)\mathrm{d}x$, are valid. Written in standard form the joint pdf is

$$f_{\mathbf{x},\mathbf{y}}(x,y) = \frac{1}{2\pi\sigma_{\mathbf{x}}\sigma_{\mathbf{y}}\sqrt{1-\rho_{\mathbf{x},\mathbf{y}}^2}} \exp\left\{ -\frac{1}{2(1-\rho_{\mathbf{x},\mathbf{y}}^2)} \times \left[\frac{(x-m_{\mathbf{x}})^2}{\sigma_{\mathbf{x}}^2} - \frac{2\rho_{\mathbf{x},\mathbf{y}}(x-m_{\mathbf{x}})(y-m_{\mathbf{y}})}{\sigma_{\mathbf{x}}\sigma_{\mathbf{y}}} + \frac{(y-m_{\mathbf{y}})^2}{\sigma_{\mathbf{y}}^2} \right] \right\}, \quad (3.108)$$

where the parameters $m_{\mathbf{x}}$, $m_{\mathbf{y}}$, $\sigma_{\mathbf{x}}$, $\sigma_{\mathbf{y}}$ have the usual significance of being the mean and variance of the respective random variables, $\mathbf{x}$ and $\mathbf{y}$. The parameter $\rho_{\mathbf{x},\mathbf{y}}$ is the correlation coefficient defined earlier, $\rho_{\mathbf{x},\mathbf{y}} = E\{(\mathbf{x}-m_{\mathbf{x}})(\mathbf{y}-m_{\mathbf{y}})\}/(\sigma_{\mathbf{x}}\sigma_{\mathbf{y}})$. One very important property of jointly Gaussian random variables is that any weighted sum of them is also Gaussian. That is, if $\mathbf{z} = a\mathbf{x} + b\mathbf{y}$ and both $\mathbf{x}$ and $\mathbf{y}$ are jointly Gaussian, then $\mathbf{z}$ is also Gaussian. Before proving this, we show that (a) the marginal density of jointly Gaussian random variables is also Gaussian, and (b) the correlation coefficient is indeed the correlation coefficient.

(a) The marginal density of jointly Gaussian random variables is Gaussian

Applying the basic equation to determine the marginal pdf, $f_{\mathbf{x}}(x) = \int_{-\infty}^{\infty} f_{\mathbf{x},\mathbf{y}}(x,y)\mathrm{d}y$, we have

$$f_{\mathbf{x}}(x) = \frac{1}{2\pi\sigma_{\mathbf{x}}\sigma_{\mathbf{y}}\sqrt{1-\rho_{\mathbf{x},\mathbf{y}}^2}} \int_{y=-\infty}^{\infty} \exp\left\{ -\frac{1}{2\sigma_{\mathbf{y}}^2(1-\rho_{\mathbf{x},\mathbf{y}}^2)} \times \left[(y-m_{\mathbf{y}})^2 - \frac{2\rho_{\mathbf{x},\mathbf{y}}\sigma_{\mathbf{y}}(x-m_{\mathbf{x}})(y-m_{\mathbf{y}})}{\sigma_{\mathbf{x}}} + \frac{\rho_{\mathbf{x},\mathbf{y}}^2\sigma_{\mathbf{y}}^2(x-m_{\mathbf{x}})^2}{\sigma_{\mathbf{x}}^2} - \frac{\rho_{\mathbf{x},\mathbf{y}}^2\sigma_{\mathbf{y}}^2(x-m_{\mathbf{x}})^2}{\sigma_{\mathbf{x}}^2} + \frac{\sigma_{\mathbf{y}}^2(x-m_{\mathbf{x}})^2}{\sigma_{\mathbf{x}}^2} \right] \mathrm{d}y \right\}, \quad (3.109)$$

where the square in the exponent is completed in terms of y. After some straightforward algebra, one obtains

$$f_{\mathbf{x}}(x) = \frac{1}{\sqrt{2\pi}\sigma_{\mathbf{x}}} \exp\left\{ -\frac{(x-m_{\mathbf{x}})^2}{2\sigma_{\mathbf{x}}^2} \right\} \left[\frac{1}{\sqrt{2\pi}\sigma_{\mathbf{y}}\sqrt{1-\rho_{\mathbf{x},\mathbf{y}}^2}} \times \int_{y=-\infty}^{\infty} \exp\left\{ -\frac{1}{2(1-\rho_{\mathbf{x},\mathbf{y}}^2)\sigma_{\mathbf{y}}^2} \left[(y-m_{\mathbf{y}}) - \frac{\rho_{\mathbf{x},\mathbf{y}}\sigma_{\mathbf{y}}}{\sigma_{\mathbf{x}}}(x-m_{\mathbf{x}}) \right]^2 \right\} \mathrm{d}y \right] \quad (3.110)$$

$$= \frac{1}{\sqrt{2\pi}\sigma_{\mathbf{x}}} \exp\left\{ -\frac{(x-m_{\mathbf{x}})^2}{2\sigma_{\mathbf{x}}^2} \right\}, \quad (3.111)$$

where the last equality follows by recognizing that the integration in the square brackets is simply the total area under a valid Gaussian pdf and hence evaluates to 1. By symmetry, one also has

$$f_{\mathbf{y}}(y) = \frac{1}{\sqrt{2\pi}\sigma_{\mathbf{y}}} \exp\left\{ -\frac{(y-m_{\mathbf{y}})^2}{2\sigma_{\mathbf{y}}^2} \right\}. \quad (3.112)$$

(b) The parameter $\rho_{\mathbf{x},\mathbf{y}}$ is indeed the correlation coefficient Since the mean values do not affect $\rho_{\mathbf{x},\mathbf{y}}$, we set them to zero and show that $E\{\mathbf{xy}\} = \rho_{\mathbf{x},\mathbf{y}}\sigma_{\mathbf{x}}\sigma_{\mathbf{y}}$. Again showing this fact is a matter of completing the square and doing the algebra as follows:

$$E\{\mathbf{xy}\} = \frac{1}{\sqrt{2\pi}\sigma_{\mathbf{x}}}\int_{-\infty}^{\infty}\int_{-\infty}^{\infty}\frac{xy}{\sqrt{2\pi}\sigma_{\mathbf{y}}\sqrt{1-\rho_{\mathbf{x},\mathbf{y}}^2}} \times \exp\left\{-\frac{1}{2\sigma_{\mathbf{y}}^2\left(1-\rho_{\mathbf{x},\mathbf{y}}^2\right)}\left[y^2 - 2\rho_{\mathbf{x},\mathbf{y}}\frac{\sigma_{\mathbf{y}}}{\sigma_{\mathbf{x}}}xy + \frac{\sigma_{\mathbf{y}}^2}{\sigma_{\mathbf{x}}^2}x^2\right]\right\}\mathrm{d}x\mathrm{d}y. \tag{3.113}$$

Integrating first with respect to y yields

$$\begin{aligned}
&\int_{y=-\infty}^{\infty}\frac{y}{\sqrt{2\pi}\sigma_{\mathbf{y}}\sqrt{1-\rho_{\mathbf{x},\mathbf{y}}^2}}\exp\left\{-\frac{1}{2\sigma_{\mathbf{y}}^2\left(1-\rho_{\mathbf{x},\mathbf{y}}^2\right)}\right. \\
&\quad\times\left.\left[y^2 - 2\rho_{\mathbf{x},\mathbf{y}}\frac{\sigma_{\mathbf{y}}}{\sigma_{\mathbf{x}}}xy + \frac{\rho_{\mathbf{x},\mathbf{y}}^2\sigma_{\mathbf{y}}^2}{\sigma_{\mathbf{x}}^2}x^2 - \frac{\rho_{\mathbf{x},\mathbf{y}}^2\sigma_{\mathbf{y}}^2}{\sigma_{\mathbf{x}}^2}x^2 + \frac{\sigma_{\mathbf{y}}^2}{\sigma_{\mathbf{x}}^2}x^2\right]\right\}\mathrm{d}y \\
&= \exp\left\{-\frac{1}{2\sigma_{\mathbf{y}}^2\left(1-\rho_{\mathbf{x},\mathbf{y}}^2\right)}\left[-\frac{\rho_{\mathbf{x},\mathbf{y}}^2\sigma_{\mathbf{y}}^2}{\sigma_{\mathbf{x}}^2}x^2 + \frac{\sigma_{\mathbf{y}}^2}{\sigma_{\mathbf{x}}^2}x^2\right]\right\} \\
&\quad\times\underbrace{\left[\frac{1}{\sqrt{2\pi}\sigma_{\mathbf{y}}\sqrt{1-\rho_{\mathbf{x},\mathbf{y}}^2}}\int_{y=-\infty}^{\infty} y\exp\left\{-\frac{\left(y-\frac{\rho_{\mathbf{x},\mathbf{y}}\sigma_{\mathbf{y}}}{\sigma_{\mathbf{x}}}x\right)^2}{2\sigma_{\mathbf{y}}^2\left(1-\rho_{\mathbf{x},\mathbf{y}}^2\right)}\right\}\mathrm{d}y\right]}_{=\frac{\rho_{\mathbf{x},\mathbf{y}}\sigma_{\mathbf{y}}}{\sigma_{\mathbf{x}}}x} \\
&= \frac{\rho_{\mathbf{x},\mathbf{y}}\sigma_{\mathbf{y}}}{\sigma_{\mathbf{x}}}x\exp\left\{-\frac{x^2}{2\sigma_{\mathbf{x}}^2}\right\}.
\end{aligned} \tag{3.114}$$

Now integrating with respect to x gives

$$E\{\mathbf{xy}\} = \frac{\rho_{\mathbf{x},\mathbf{y}}\sigma_{\mathbf{y}}}{\sigma_{\mathbf{x}}}\left[\frac{1}{\sqrt{2\pi}\sigma_{\mathbf{x}}}\int_{-\infty}^{\infty}x^2\exp\left\{-\frac{x^2}{2\sigma_{\mathbf{x}}^2}\right\}\right] = \rho_{\mathbf{x},\mathbf{y}}\sigma_{\mathbf{y}}\sigma_{\mathbf{x}} \tag{3.115}$$

as required.

The correlation coefficient, $\rho_{\mathbf{x},\mathbf{y}}$, is an indication of the amount of "interdependence" between the random variables $\mathbf{x}$ and $\mathbf{y}$. Note that when $\rho_{\mathbf{x},\mathbf{y}} = 0$ the joint density becomes $f_{\mathbf{x},\mathbf{y}}(x,y) = f_{\mathbf{x}}(x)f_{\mathbf{y}}(y)$, which means that the random variables are statistically independent. Thus *uncorrelatedness means that jointly Gaussian random variables are statistically independent*, a result that is used extensively in later chapters. However, it should be stressed that just because two random variables are uncorrelated does not mean that they are

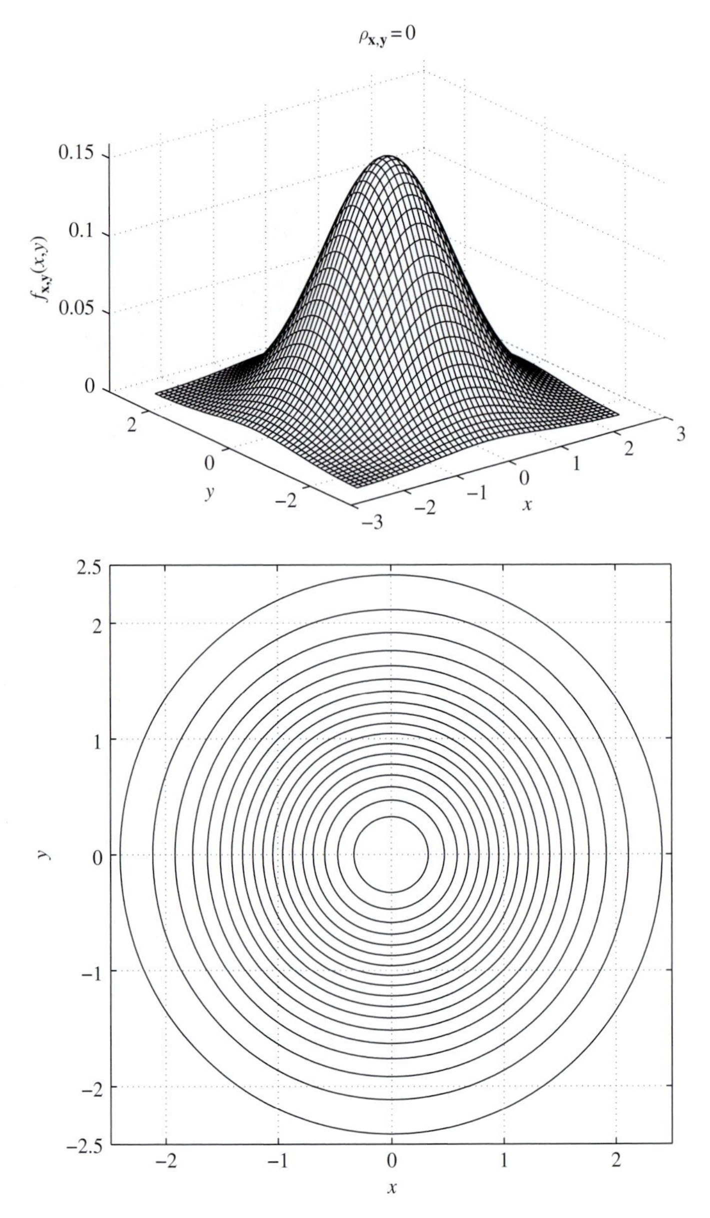

Fig. 3.16 Plots of a joint Gaussian pdf and its contours: $\sigma_{\mathbf{x}} = \sigma_{\mathbf{y}} = 1$ and $\rho_{\mathbf{x},\mathbf{y}} = 0$.

statistically independent. Indeed this is only true for random variables that are jointly Gaussian.

Figures 3.16–3.19 show surface and contour plots of $f_{\mathbf{x},\mathbf{y}}(x, y)$ for various values of $\rho_{\mathbf{x},\mathbf{y}}$. Note that as $\rho_{\mathbf{x},\mathbf{y}}$ increases to 1 the probability becomes more and more concentrated

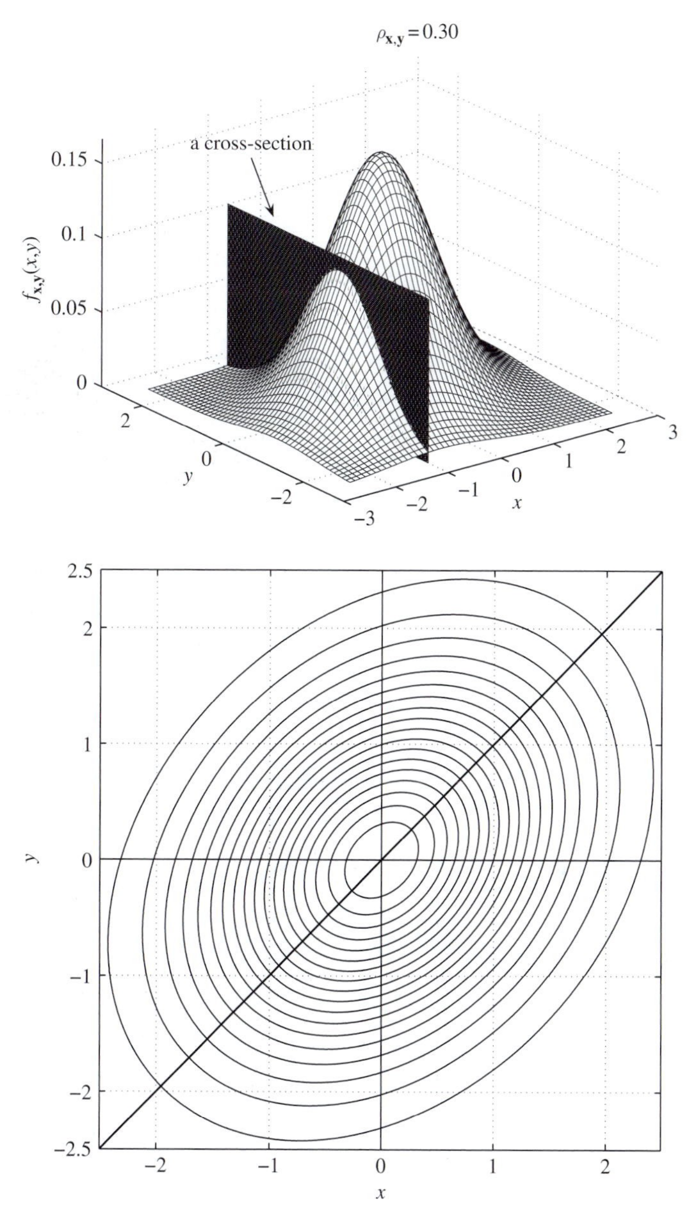

Fig. 3.17 Plots of a joint Gaussian pdf and its contours: $\sigma_{\mathbf{x}} = \sigma_{\mathbf{y}} = 1$ and $\rho_{\mathbf{x},\mathbf{y}} = 0.3$.

around the $x = y$ line, implying there is a stronger linear relationship between x and y. Any cross-section is a univariate Gaussian pdf and in the case of $\rho_{\mathbf{x},\mathbf{y}} = 0$ the joint pdf exhibits circular symmetry. Hence for $\rho_{\mathbf{x},\mathbf{y}} = 0$ all cross-sections through the origin, more generally through the point $(m_{\mathbf{x}}, m_{\mathbf{y}})$, are identical. The plots are done for zero means since all nonzero means do is shift the surface to the $(m_{\mathbf{x}}, m_{\mathbf{y}})$ point. The variances are set to be

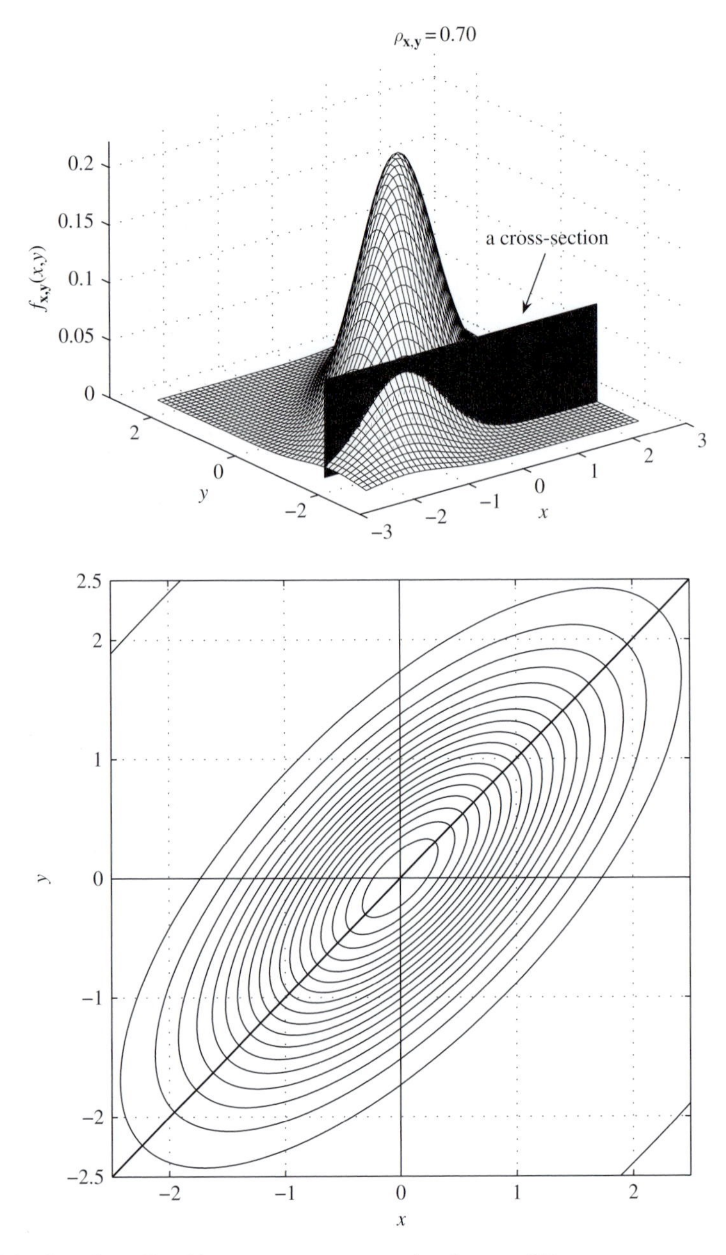

Fig. 3.18 Plots of a joint Gaussian pdf and its contours: $\sigma_{\mathbf{x}} = \sigma_{\mathbf{y}} = 1$ and $\rho_{\mathbf{x},\mathbf{y}} = 0.7$.

equal but similar observations can be made for unequal variances, except that in the case of $\rho_{\mathbf{x},\mathbf{y}} = 0$ we would not have circular symmetry.

We now go on to show that the weighted sum of two jointly Gaussian random variables is also Gaussian. To do this we need the following from probability theory.

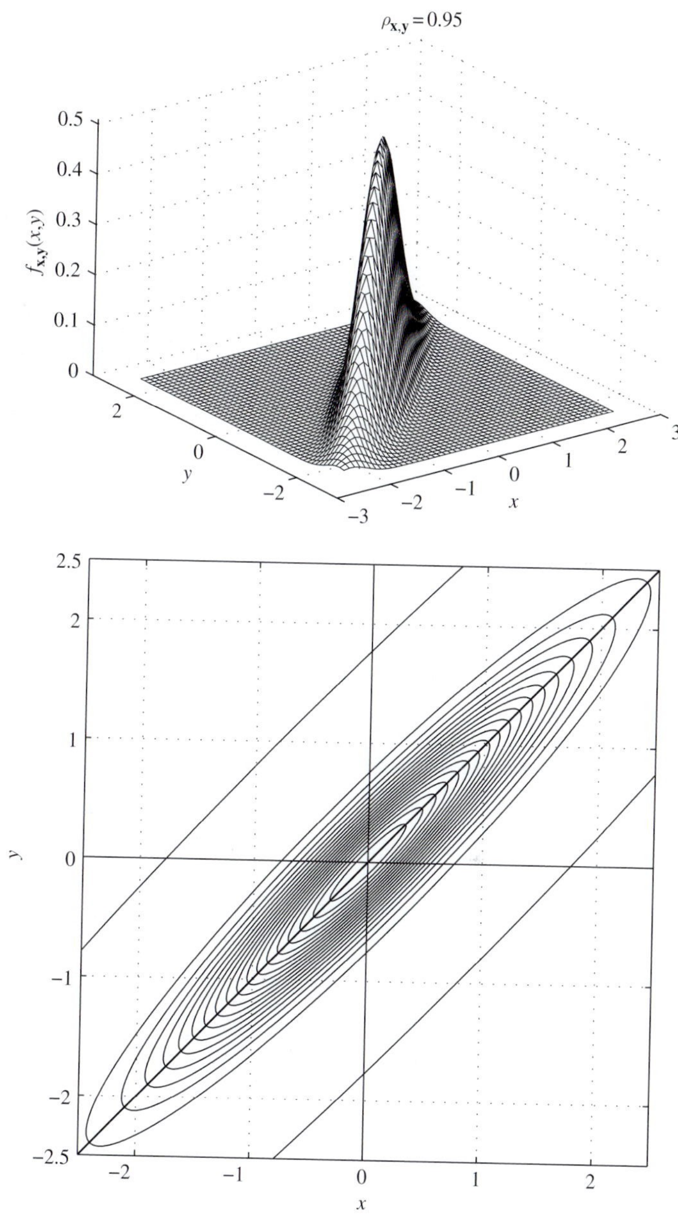

Fig. 3.19 Plots of a joint Gaussian pdf and its contours: $\sigma_{\mathbf{x}} = \sigma_{\mathbf{y}} = 1$ and $\rho_{\mathbf{x},\mathbf{y}} = 0.95$.

Joint density function of two functions of two random variables Consider two deterministic functions, $g(x, y)$, $h(x, y)$, two random variables, $\mathbf{x}$, $\mathbf{y}$, with joint pdf $f_{\mathbf{x},\mathbf{y}}(x, y)$. Form two random variables $\mathbf{z} = g(\mathbf{x}, \mathbf{y})$ and $\mathbf{w} = h(\mathbf{x}, \mathbf{y})$. The joint density function, $f_{\mathbf{z},\mathbf{w}}(z, w)$ is given by

$$f_{\mathbf{z},\mathbf{w}}(z, w) = \sum_{i=1}^{n} \frac{f_{\mathbf{x},\mathbf{y}}(x_i, y_i)}{|J(x_i, y_i)|}, \tag{3.116}$$

where the pair (x_i, y_i) is one of the n real solutions to the equation set $g(x, y) = z; h(x, y) = w$ and $J(x, y)$ is the *Jacobian* of the above transformation, given as

$$J(x, y) = \det \begin{bmatrix} \frac{\partial g(x,y)}{\partial x} & \frac{\partial g(x,y)}{\partial y} \\ \frac{\partial h(x,y)}{\partial x} & \frac{\partial h(x,y)}{\partial y} \end{bmatrix}. \tag{3.117}$$

The result is a direct analog of (3.18) for the single variable case.

The (linear) transformation discussed here is

$$\mathbf{z} = a\mathbf{x} + b\mathbf{y}, \tag{3.118}$$

$$\mathbf{w} = c\mathbf{x} + d\mathbf{y}. \tag{3.119}$$

The random variable $\mathbf{w}$ is introduced as an auxiliary variable. We determine the joint pdf $f_{\mathbf{z},\mathbf{w}}(z, w)$ and use it to find the marginal pdf $f_{\mathbf{z}}(z) = \int_{-\infty}^{\infty} f_{\mathbf{z},\mathbf{w}}(z, w)\mathrm{d}w$. There is only one solution pair to the transformation as shown below:

$$\begin{bmatrix} a & b \\ c & d \end{bmatrix} \begin{bmatrix} x \\ y \end{bmatrix} = \begin{bmatrix} z \\ w \end{bmatrix} \tag{3.120}$$

$$\Rightarrow \begin{bmatrix} x \\ y \end{bmatrix} = \frac{1}{ad - bc} \begin{bmatrix} d & -b \\ -c & a \end{bmatrix} \begin{bmatrix} z \\ w \end{bmatrix} = \begin{bmatrix} a_1 z + b_1 w \\ c_1 z + d_1 w \end{bmatrix}, \tag{3.121}$$

where it has been assumed that $(ad - bc) \neq 0$, i.e., the equations are linearly independent (which is always possible to arrange). The Jacobian is

$$J(x, y) = \det \begin{bmatrix} a & b \\ c & d \end{bmatrix} = ad - bc. \tag{3.122}$$

Therefore

$$f_{\mathbf{z},\mathbf{w}}(z, w) = \frac{f_{\mathbf{x},\mathbf{y}}(a_1 z + b_1 w, c_1 z + d_1 w)}{|ad - bc|}. \tag{3.123}$$

The above is a general result. However, we are considering the case where $f_{\mathbf{x},\mathbf{y}}(x, y)$ is a jointly Gaussian pdf and from (3.123) we conclude that $f_{\mathbf{z},\mathbf{w}}(z, w)$ is of the following form: an exponential whose exponent is *a quadratic in z and w*. But this is exactly what a joint Gaussian pdf is!

We have already shown that if the joint pdf is Gaussian, then the marginal pdfs are also Gaussian. Therefore $f_{\mathbf{z}}(z)$ is Gaussian. To complete the discussion we determine the mean and variance of $f_{\mathbf{z}}(z)$ since they completely specify the pdf. They are simply computed as follows:

$$\begin{aligned} m_{\mathbf{z}} &= E\{\mathbf{z}\} = E\{a\mathbf{x} + b\mathbf{y}\} \\ &= aE\{\mathbf{x}\} + bE\{\mathbf{y}\} = am_{\mathbf{x}} + bm_{\mathbf{y}}, \end{aligned} \tag{3.124}$$

$$\begin{aligned} \operatorname{var}(\mathbf{z}) &= E\left\{(\mathbf{z} - m_{\mathbf{z}})^2\right\} = E\left\{\left[(a\mathbf{x} + b\mathbf{y}) - \left(am_{\mathbf{x}} + bm_{\mathbf{y}}\right)\right]^2\right\} \\ &= E\left\{\left[a\left(\mathbf{x} - m_{\mathbf{x}}\right) + b\left(\mathbf{y} - m_{\mathbf{y}}\right)\right]^2\right\} \\ &= a^2E\left\{(\mathbf{x} - m_{\mathbf{x}})^2\right\} + 2abE\left\{(\mathbf{x} - m_{\mathbf{x}})\left(\mathbf{y} - m_{\mathbf{y}}\right)\right\} \\ &\quad + b^2E\left\{\left(\mathbf{y} - m_{\mathbf{y}}\right)^2\right\}, \end{aligned} \tag{3.125}$$

$$\sigma_{\mathbf{z}}^2 = a^2\sigma_{\mathbf{x}}^2 + 2ab\rho_{\mathbf{x},\mathbf{y}}\sigma_{\mathbf{x}}\sigma_{\mathbf{y}} + b^2\sigma_{\mathbf{y}}^2. \tag{3.126}$$

The importance of the result that we have just proven can be appreciated, for example, by considering the case of a random process applied to a linear system. The output is a weighted linear sum (more precisely an integral where the weight is the impulse response, $h(t)$) of the input. Therefore if the input is a Gaussian random process, then so is the output. All we need to determine is the mean and autocorrelation of the output and we know the pdf of the output process.

To conclude the section consider the *multivariate* Gaussian pdf. The pdf for two jointly Gaussian random variables can be extended to n jointly Gaussian random variables $\mathbf{x}_1, \mathbf{x}_2, \ldots, \mathbf{x}_n$. If we define the random vector $\vec{\mathbf{x}} = [\mathbf{x}_1, \mathbf{x}_2, \ldots, \mathbf{x}_n]$, a vector of the means $\vec{m} = [m_1, m_2, \ldots, m_n]$, and the $n \times n$ covariance matrix C such that $C_{i,j} = \operatorname{cov}(\mathbf{x}_i, \mathbf{x}_j) = E\{(\mathbf{x}_i - m_i)(\mathbf{x}_j - m_j)\}$, then the random variables $\{\mathbf{x}_i\}_{i=1}^n$ are jointly Gaussian if the n-dimensional pdf is

$$\begin{aligned} f_{\mathbf{x}_1,\mathbf{x}_2,\ldots,\mathbf{x}_n}(x_1, x_2, \ldots, x_n) &= \frac{1}{\sqrt{(2\pi)^n \det(C)}} \\ &\quad \times \exp\left\{-\frac{1}{2}(\vec{x} - \vec{m})C^{-1}(\vec{x} - \vec{m})^{\top}\right\}. \end{aligned} \tag{3.127}$$

Once again, the joint pdf is in the form of an exponential with an exponent that is a quadratic in the variables $x_1, x_2, \ldots, x_n$. Furthermore, in the special case that the covariance matrix C is *diagonal* (i.e., the random variables $\{\mathbf{x}_i\}_{i=1}^n$ are all uncorrelated), there are no cross-terms in the quadratic exponent and the joint pdf in (3.127) is a product of the marginal pdfs. So again, uncorrelatedness implies statistical independence for multiple Gaussian random variables.

Finally, it should be observed that for jointly Gaussian random variables all that need be known to specify the joint pdf is the first and second moments. Again, even though these moments only provide, in general, a partial description of the random variables, for the Gaussian random variables they provide a complete description. This is of importance when, for example, we are trying to determine the pdf experimentally. If we have good reason to believe that the random variables are Gaussian, then we need only measure the first two moments.

3.6 Summary

Philosophically randomness is the spice of life and living in a world where everything is predetermined does not strike one as being particularly fulfilling. Therefore random signals (or processes) and ways of characterizing them have been described in this chapter. The principal characterization is through the amplitude pdf but of equal importance is the partial characterization through moments. Specifically the first and second moments, i.e., the DC value and the variance respectively.

As will be seen in later chapters, at the receiver of a communication system the first moment is determined by the transmitted signal, whereas the variance is due to noise added by the channel. The second moment also includes the autocorrelation or equivalently the PSD of a random signal. In contrast to deterministic signals the frequency bandwidth requirements for transmitting random signals are based largely on the bandwidth occupied by a certain percentage of the random signal's power. This is obtained from the PSD.

3.7 Problems

3.1 Consider the random experiment of simultaneously tossing two fair coins.
(a) Clearly illustrate the sample space of the above random experiment.
(b) Let A denote the event that "at least one head shows" and B denote the event that "there is a match of two coins." Find $P(A)$ and $P(B)$.
(c) Compute $P(A|B)$ and $P(B|A)$.
(d) Determine whether the two events A and B are statistically independent.

3.2 A person has a test for a nasty disease. Use the random variable $\mathbf{x}$ to describe the person's health condition:

$$\begin{aligned}\mathbf{x}=1: &\quad \text{the person has the disease,}\\ \mathbf{x}=0: &\quad \text{the person does not have the disease.}\end{aligned}$$

Similarly, we use the random variable $\mathbf{y}$ to describe the test result. The result of the test is either "positive" ($\mathbf{y}=1$) or "negative" ($\mathbf{y}=0$). The test is 95% reliable, i.e., in 95% of cases of people who really have the disease, a positive result is returned, and in 95% of cases of people who do not have the disease, a negative result is obtained. A final piece of information is that 1% of those of the person's age and background have the disease.

The person has the test, and the result is positive. What is the probability that the person has the disease? *Hint* Use the conditional probability.

3.3 An information source produces 0 and 1 with probabilities 0.6 and 0.4, respectively. The output of the source is transmitted over a channel that has a probability of error (turning a 0 into a 1 or a 1 into a 0) equal to 0.1.
(a) What is the probability that at the output of the channel a 1 is observed?

(b) What is the probability that a 1 is the output of the source if at the output of the channel a 1 is observed?

(c) Relate the question in (b) to that of Problem 3.2.

3.4 (*This problem generalizes Problem 3.3*) Consider a binary symmetric channel (BSC), depicted in Figure 3.20. The channel's binary input is represented by the random variable $\mathbf{x}$ and the channel's binary output is represented by the random variable $\mathbf{y}$. Let ϵ be the crossover probability of the channel such that $P(\mathbf{y} = 1|\mathbf{x} = 0) = P(\mathbf{y} = 0|\mathbf{x} = 1) = \epsilon$. The probability of 0 being the channel's input (i.e., 0 is transmitted) is p.

(a) Given that a 1 is received, what is the probability that a 1 was transmitted?

(b) Consider the decision rule which guesses that the transmitted symbol is the same as the received symbol. What is the probability of making an error?

(c) If a sequence of n independent symbols is transmitted in succession, what is the probability that k 1s are received?

(d) Suppose we want to transmit two equally likely messages. Represent one message by a sequence of n 1s and the other by a sequence of n 0s. The sequence is passed through the channel. Assume $\epsilon < 1/2$. Give a natural decision rule for deciding which message is transmitted based on the received sequence (of length n). Find the error probability.

(e) The representation in (d) is a *repetition code*. What happens to the error probability as $n \to \infty$? What is the problem with using this code when n is large?

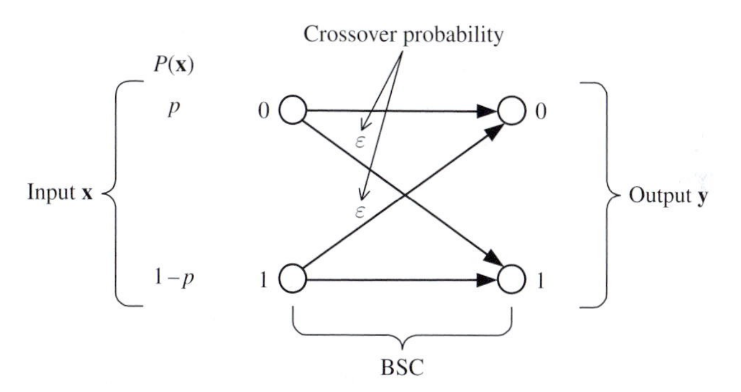

Fig. 3.20 Binary symmetric channel (BSC) considered in Problem 3.4.

3.5 A random variable $\mathbf{x}$ is defined by the following cdf:

$$F_{\mathbf{x}}(x) = \begin{cases} 0, & x \leq 0 \\ Ax^3, & 0 \leq x \leq 10 \\ B, & 10 < x \end{cases}. \tag{P3.1}$$

(a) Find the proper values for A and B.

(b) Obtain and plot the pdf $f_{\mathbf{x}}(x)$.

(c) Find the mean and variance of $\mathbf{x}$.

(d) What is the probability that $3 \leq \mathbf{x} \leq 7$?

3.6 $\mathbf{x}$ is a Gaussian random variable with a pdf $\mathcal{N}(0, \sigma^2)$. This random variable is passed through a limiter whose input–output relationship is given by

$$y = g(x) = \begin{cases} -b, & x \le -b \\ b, & x \ge b \\ x, & x < |b| \end{cases}. \quad \text{(P3.2)}$$

Find the pdf of the output random variable $\mathbf{y}$. *Hint* First find the cdf of $\mathbf{y}$ based on the definition of the cdf.

3.7 Find the mean and variance of the random variable $\mathbf{x}$ for the following cases:

(a) $\mathbf{x}$ is a uniformly distributed random variable, whose pdf is

$$p_{\mathbf{x}}(x) = \begin{cases} \frac{1}{b-a}, & a \le x \le b \\ 0, & \text{otherwise} \end{cases}. \quad \text{(P3.3)}$$

Also consider the special case when $a = -b$.

(b) $\mathbf{x}$ is a Rayleigh distributed random variable, whose pdf is

$$f_{\mathbf{x}}(x) = \begin{cases} \frac{x}{\sigma^2} e^{-x^2/2\sigma^2}, & x > 0 \\ 0, & \text{otherwise} \end{cases}. \quad \text{(P3.4)}$$

(c) $\mathbf{x}$ is a Laplacian distributed random variable, whose pdf is

$$f_{\mathbf{x}}(x) = \frac{c}{2} e^{-c|x|}. \quad \text{(P3.5)}$$

(d) $\mathbf{y}$ is a discrete random variable, given by $\mathbf{y} = \sum_{i=1}^{n} \mathbf{x}_i$, where the random variables $\mathbf{x}_i$, $i = 1, 2, \ldots, n$, are statistically independent and identically distributed (i.i.d.) random variables with the pmf: $P(\mathbf{x}_i = 1) = p$ and $P(\mathbf{x}_i = 0) = 1 - p$. *Remark* $\mathbf{y}$ is, in fact, a binomial distributed random variable. However, you do not need the binomial distribution to calculate the mean and variance of $\mathbf{y}$. All you need are the linear property of the expectation operation $E\{\cdot\}$ and the fact that if $\mathbf{U}$ and $\mathbf{V}$ are statistically independent random variables, then $E\{\mathbf{UV}\} = E\{\mathbf{U}\} \cdot E\{\mathbf{V}\}$.

Given a sample space Ω and events such as A, B, C, set operations on these events of union, $(A \cup B)$, intersection $(A \cap C)$, complement $(\bar{B})$ can be conveniently visualized geometrically by a drawing called the *Venn diagram*, shown in Figure 3.21. To use the Venn diagram, the areas of the different events represent the probabilities of these events. Thus the area of Ω, the whole sample space, is equal to 1.

Use Venn diagrams to solve the next set of problems.

3.8 Prove the following relationships:

(a) $P(A \cup B) = P(A) + P(B) - P(A \cap B)$.

(b) $P(A \cup B \cup C) = P(A) + P(B) + P(C) - P(A \cap B) - P(A \cap C) - P(B \cap C) + P(A \cap B \cap C)$.

Remark One can do this directly from the Venn diagram or by making use of the result in (a).

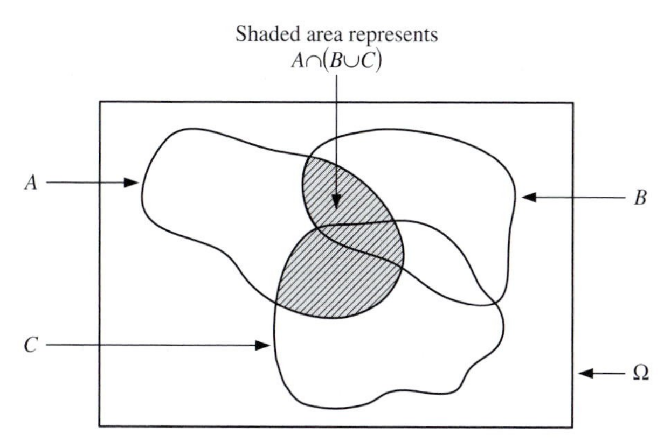

Fig. 3.21 Visualization of set operations with the Venn diagram.

3.9 (*Bayes' rule*)
(a) Show that $P(A|B) = P(A \cap B)/P(B)$.
(b) Interpret statistical independence in terms of the areas of events A and B.
(c) Are mutually exclusive events statistically independent? Explain.

3.10 Conditional probabilities restrict consideration to a subspace, say B, of the sample space Ω. In effect B becomes the new sample space, call it Ω_B. In terms of the Venn diagram, the picture could look as shown in Figure 3.22, where events A_i, $i = 1, \ldots, n$, are mutually exclusive and partition the sample space Ω, i.e., $A_i \cap A_j = \varnothing$ for $i \neq j$, and $\bigcup_{i=1}^{n} A_i = \Omega$. Show that the probability system created by knowledge of event B satisfies the appropriate axioms.

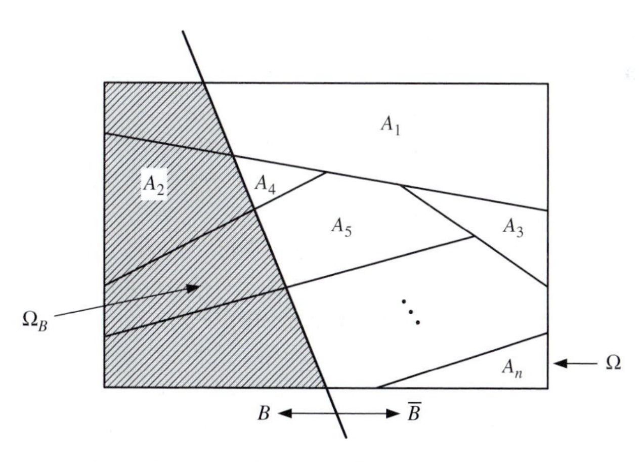

Fig. 3.22 Visualization of conditional probabilities with the Venn diagram.

3.11 Statistical independence was defined in terms of two events in this chapter. For n events $\{A_i\}$, $i = 1, \ldots, n$, statistical independence means that the probability of the intersection of any group of k or fewer events equals the product of the probabilities of that group of events. Stated mathematically, for all possible combinations of k sets, $k = 2, 3, \ldots, n$, one has $P\left(\bigcap_{i=q_1}^{q_k} A_i\right) = \prod_{i=q_1}^{q_k} P(A_i)$, where $q_1, \ldots, q_k$ are indexes of the events belonging to a given set. Consider three events A, B, and C.

(a) State the four conditions that the event probabilities must satisfy for them to be statistically independent.

(b) Consider the Venn diagrams shown in Figure 3.23. Note that the areas are not drawn to scale. For each Venn diagram determine whether they are or are not statistically independent.

(c) Is it possible for events to be pairwise statistically independent but not statistically independent?

(d) Given n events, what is the expression for the number of conditions that must be satisfied for them to be statistically independent?

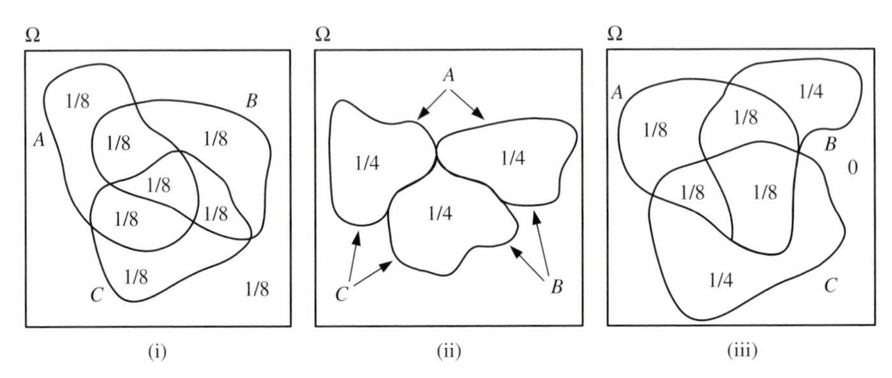

Fig. 3.23 Three Venn diagrams considered in Problem 3.11.

3.12 Consider the Venn diagram in Figure 3.24. Show that events A, B, C are statistically dependent but that events A, B are conditionally independent, i.e., $P(A \cap B|C) = P(A|C)P(B|C)$.

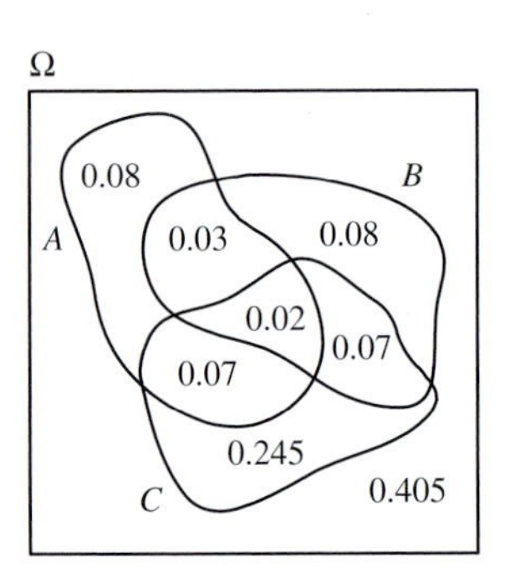

Fig. 3.24 The Venn diagram considered in Problem 3.12.

3.13 Consider the "toy" channel model shown in Figure 3.25. The channel is toy only in the sense of the transition probability values; the structure of it represents what is known as an "*error and erasure channel*," where output y_2 represents the erasure symbol.

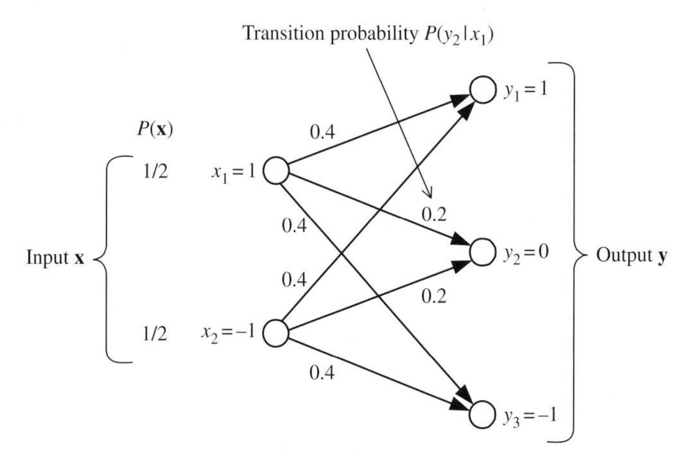

Fig. 3.25 The toy channel model considered in Problem 3.13.

(a) Determine $P(\mathbf{y}=1)$, $P(\mathbf{y}=1|\mathbf{x}=1)$, $P(\mathbf{y}=1|\mathbf{x}=-1)$. Are the random variables $\mathbf{x}$, $\mathbf{y}$ statistically independent?

(b) Determine the crosscorrelation $E\{\mathbf{xy}\}$. Comment on the result if a comment is appropriate.

3.14 Consider the toy channel model of Problem 3.13 again but with the different set of transition probabilities indicated in Figure 3.26. Again determine whether random variables $\mathbf{x}$, $\mathbf{y}$ are statistically independent and/or uncorrelated.

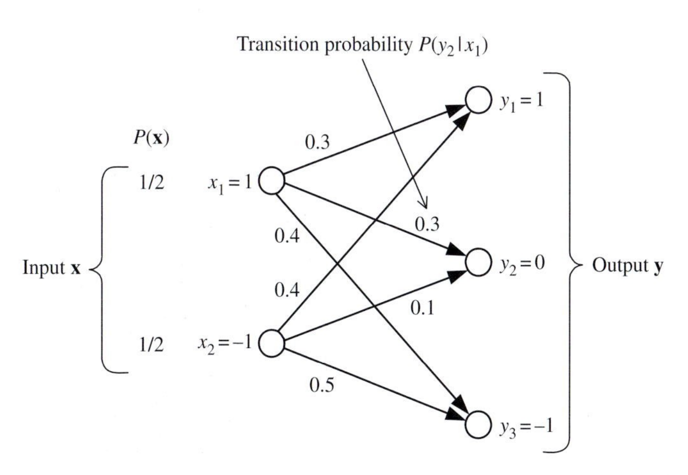

Fig. 3.26 The toy channel model considered in Problem 3.14.

3.15 The channel model of Problem 3.13 is repeated in Figure 3.27 in a more general form. Realistically the transition probabilities p and ϵ are $\lll 1$, on the order of 10^{-3} or smaller. Under these conditions determine whether $\mathbf{x}$, $\mathbf{y}$ are statistically independent and/or correlated.

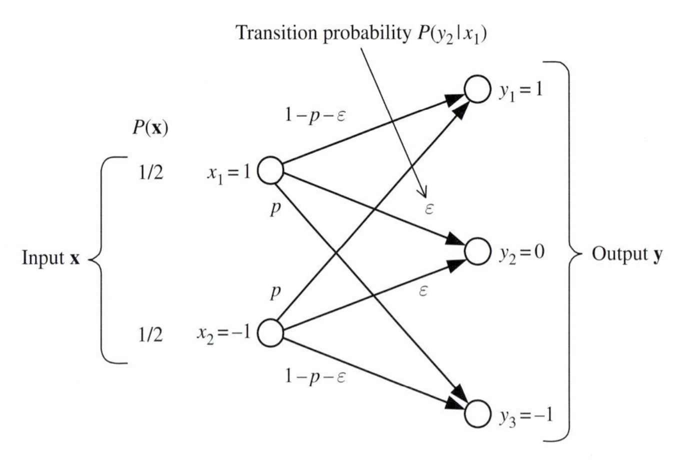

Fig. 3.27 The channel model considered in Problem 3.15.

3.16 In developing the optimal receiver for a binary communication system in the next chapter we shall use the concept of "coordinate rotation" to simplify the receiver. Here we shall look at the effect of rotation on the correlation of two random variables.

Consider two random variables $\mathbf{x}$ and $\mathbf{y}$ with means $m_{\mathbf{x}}$, $m_{\mathbf{y}}$ and variances $\sigma_{\mathbf{x}}^2$, $\sigma_{\mathbf{y}}^2$ respectively. Further, let the two random variables be *uncorrelated*. Now "rotate" $\mathbf{x}$ and $\mathbf{y}$ to obtain new random variables $\mathbf{x}_R$ and $\mathbf{y}_R$ as follows:

$$\begin{bmatrix} \mathbf{x}_R \\ \mathbf{y}_R \end{bmatrix} = \begin{bmatrix} \cos\theta & \sin\theta \\ -\sin\theta & \cos\theta \end{bmatrix} \begin{bmatrix} \mathbf{x} \\ \mathbf{y} \end{bmatrix}, \quad \text{(P3.6)}$$

where θ is some arbitrary angle.

Under what condition on $\mathbf{x}$ and $\mathbf{y}$, are the two random variables $\mathbf{x}_R$ and $\mathbf{y}_R$ also uncorrelated?

3.17 (*Theorem of expectation*) Consider random variable $\mathbf{y}$, defined by the transformation of random variable $\mathbf{x}$, $\mathbf{y} = g(\mathbf{x})$, where $g(\cdot)$ maps every value of $\mathbf{x}$ into a real number. As usual the expected value of $\mathbf{y}$ is given by $E\{\mathbf{y}\} = \int_{-\infty}^{\infty} y f_{\mathbf{y}}(y)\mathrm{d}y$, which means that $f_{\mathbf{y}}(y)$ needs to be determined. This can be avoided by the following reasoning based on Figure 3.28.

(a) The probability that $\mathbf{x}$ lies in the infinitesimal $\mathrm{d}x$, $P(x < \mathbf{x} \leq x + \mathrm{d}x)$ is given by ________________.

(b) The probability that $\mathbf{y}$ lies in the infinitesimal $\mathrm{d}y$, $P(y < \mathbf{y} \leq y + \mathrm{d}y)$ is given by ________________.

(c) Are the two probabilities determined in (a) and (b) equal or unequal?

(d) Combine the answers of (a), (b), and (c) to show that $E\{\mathbf{y}\} = \int_{-\infty}^{\infty} y f_{\mathbf{y}}(y)\mathrm{d}y = \int_{-\infty}^{\infty} g(x) f_{\mathbf{x}}(x)\mathrm{d}x$, which is (3.27).

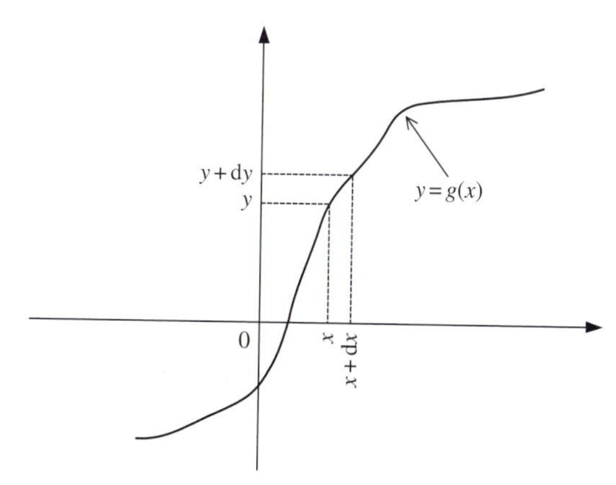

Fig. 3.28 Figure for Problem 3.17.

An interesting and important transformation on $\mathbf{x}$ is $\mathbf{y} = e^{j2\pi f\mathbf{x}}$. The expected value of $\mathbf{y}$, $E\{\mathbf{y}\} = \int_{-\infty}^{\infty} f_{\mathbf{x}}(x)e^{j2\pi fx}dx$ is known as the *characteristic function* or *moment generating function* of the random variable $\mathbf{x}$. It is a function of f and is typically denoted as $\Phi_{\mathbf{x}}(f)$. Note that, in principle, $\Phi_{\mathbf{x}}(f)$ is the Fourier transform of $f_{\mathbf{x}}(x)$. The only difference with the Fourier transform encountered in Chapter 2 is in the sign of the exponent, but mathematically this was an arbitrary choice. The implication is that the inverse transform is given by $f_{\mathbf{x}}(x) = \int_{-\infty}^{\infty} \Phi_{\mathbf{x}}(f)e^{-j2\pi xf}df$. With this in mind all the usual properties and results for Fourier transforms and transform pairs hold. The function $\Phi_{\mathbf{x}}(f)$ is useful to prove a number of basic results. This is done in the next set of problems.

3.18 Show that the nth moment of $\mathbf{x}$, $E\{\mathbf{x}^n\}$, is equal to

$$\frac{1}{(2\pi j)^n} \frac{d^n \Phi_{\mathbf{x}}(f)}{df^n}\bigg|_{f=0} : \tag{P3.7}$$

hence the name, moment generating function, for $\Phi_{\mathbf{x}}(f)$.

3.19 Expand $e^{j2\pi xf}$ in a McLaurin series to show that $\Phi_{\mathbf{x}}(f)$ is completely determined by the moments of $\mathbf{x}$. Since $\Phi_{\mathbf{x}}(f)$ determines $f_{\mathbf{x}}(x)$, knowledge of all the moments provides the same description of a random variable as does the pdf.

3.20 Determine the characteristic function of the following zero-mean random variables:

(a) Gaussian,

(b) Laplacian,

(c) uniform of width $2A$.

Remark You may determine $\Phi_{\mathbf{x}}(f)$ directly from the basic definition or use the fact that $\Phi_{\mathbf{x}}(f)$ is the Fourier transform of $f_{\mathbf{x}}(x)$ and use results from Chapter 2.

3.21 Expand the characteristic function of the zero-mean Gaussian random variable in a McLaurin series in f. Derive the relationship between the nth moment and the variance $\sigma_{\mathbf{x}}^2$.

3.22 Consider a random variable $\mathbf{x}$ with nonzero mean, $m_{\mathbf{x}}$. Let $\mathbf{y} = \mathbf{x} - m_{\mathbf{x}}$. Obviously $E\{\mathbf{y}\} = 0$.

(a) What is the relationship between the two characteristic functions, $\Phi_{\mathbf{x}}(f)$ and $\Phi_{\mathbf{y}}(f)$?

(b) What then is the characteristic function of a Gaussian random variable of mean $m_{\mathbf{x}}$, variance $\sigma_{\mathbf{x}}^2$?

Remark This shows that the pdf of a Gaussian random variable is completely specified by the first and second moments.

3.23 The expectation theorem is readily extended to the mapping $\mathbf{y} = g(\mathbf{x})$, where $\mathbf{x}$ is a k-dimensional random variable $\{\mathbf{x}_1, \ldots, \mathbf{x}_k\}$ with joint pdf $f_{\mathbf{x}}(x_1, \ldots, x_k)$. Now

$$E\{\mathbf{y}\} = \int_{-\infty}^{\infty} \cdots \int_{-\infty}^{\infty} g(x_1, \ldots, x_k) f_{\mathbf{x}}(x_1, \ldots, x_k) \mathrm{d}x_1 \cdots \mathrm{d}x_k. \qquad \text{(P3.8)}$$

Consider $\mathbf{y} = \mathbf{x}_1 + \mathbf{x}_2$, where $\mathbf{x}_1$, $\mathbf{x}_2$ are statistically independent random variables.

(a) Show that the characteristic function of $\mathbf{y}$ is

$$\Phi_{\mathbf{y}}(f) = \Phi_{\mathbf{x}_1}(f)\Phi_{\mathbf{x}_2}(f). \qquad \text{(P3.9)}$$

(b) Multiplication in the frequency (transform) domain means ____________ in the original domain. Use this to find the pdf of $\mathbf{y}$ when $\mathbf{x}_1$, $\mathbf{x}_2$ are statistically independent uniform random variables, each zero mean and of width $2A$.

(c) Generalize the result of (a) to $\mathbf{y} = \sum_{k=1}^{n} \mathbf{x}_k$, where $\mathbf{x}_k$s are statistically independent Gaussian random variables.

3.24 Though to find the expected value of $\mathbf{y} = g(\mathbf{x})$ one does not need to determine $f_{\mathbf{y}}(y)$ as shown in Problem 3.17, there are situations where it is desirable to do so. This is accomplished by (3.18) and here you are guided through a derivation of this relationship. Consider the graph plotted on Figure 3.29, where x_1, x_2, x_3 are the real roots of the equation $g(x) = y$ (for the shown y).

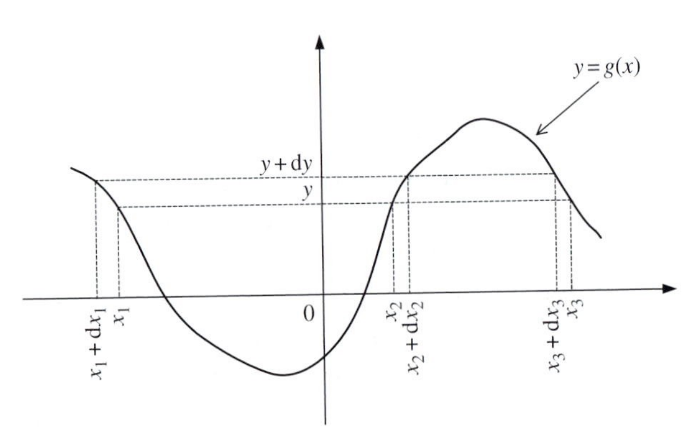

Fig. 3.29 Figure for Problem 3.24.

(a) Reason that the probability of the event $(y < \mathbf{y} \leq y + \mathrm{d}y)$ is given by

$$f_{\mathbf{y}}(y)\mathrm{d}y = f_{\mathbf{x}}(x_1)|\mathrm{d}x_1| + f_{\mathbf{x}}(x_2)\mathrm{d}x_2 + f_{\mathbf{x}}(x_3)|\mathrm{d}x_3|. \qquad \text{(P3.10)}$$

(b) Why are there magnitude signs on $\mathrm{d}x_1$ and $\mathrm{d}x_3$?

(c) In terms of infinitesimals is there any difference among $|\mathrm{d}x_1|$, $|\mathrm{d}x_2|$, $|\mathrm{d}x_3|$, or can they be called $|\mathrm{d}x|$?

(d) Combine (a), (b), (c) and generalize to obtain the desired relationship

$$f_{\mathbf{y}}(y) = \sum_i \frac{f_{\mathbf{x}}(x_i)}{\left| \left.\frac{\mathrm{d}g(x)}{\mathrm{d}x}\right|_{x=x_i} \right|},$$

where $\{x_i\}$ are the real roots of $g(x) = y$.

Note that as y changes the number of real roots may change. Also there may be intervals for which there are no real roots, in which case $f_{\mathbf{y}}(y) = 0$ for that interval and finally there may be specific values of y where there is an infinite number of roots. In this case $f_{\mathbf{y}}(y)$ has an impulse at this y of strength equal to the probability of the values of $\mathbf{x}$ producing this y. All these cases are illustrated by the very artificial nonlinearity of the next problem.

3.25 Let $y = g(x)$ be defined as:

$$y = \begin{cases} -1, & -\infty < x \leq -1 \\ x, & -1 < x \leq 0 \\ 1-(1-x)^2, & 0 \leq x \end{cases}. \tag{P3.11}$$

Determine $f_{\mathbf{y}}(y)$ when $f_{\mathbf{x}}(x) = \frac{1}{2}\mathrm{e}^{-|x|}$. *Hint* Consider the following ranges of y, $-\infty < y < -1$, $y = -1$, $-1 < y \leq 0$, $0 < y \leq 1$, $1 \leq y$.

3.26 A more realistic nonlinearity is $\mathbf{y} = \cos(\alpha + \boldsymbol{\theta})$, where α is a constant while $\boldsymbol{\theta}$ is a uniform random variable over the range $[0, 2\pi)$.

(a) Determine $f_{\mathbf{y}}(y)$. Does it depend on α?

(b) Determine $f_{\mathbf{y}}(y)$ where $\boldsymbol{\theta}$ is still uniform but over the range $[0, 20°)$.

3.27 (*Cauchy–Schwartz inequality for random variables*) The inequality is proven in a manner directly analogous to that used in Chapter 2 for signals.

(a) The expected or average value $E\{(\mathbf{x} + \lambda\mathbf{y})^2\}$ where λ is an arbitrary real number is obviously $\geq$ ________________.

(b) Consider the expression in (a) as a quadratic in λ. Reason that it cannot have any real nonzero roots and therefore that $E\{\mathbf{xy}\} \leq \sqrt{E\{\mathbf{x}^2\}E\{\mathbf{y}^2\}}$.

(c) When does the equality hold?

(d) Use the results of (b) to prove that the correlation coefficient $\rho_{\mathbf{x},\mathbf{y}} = E\{(\mathbf{x} - m_{\mathbf{x}})(\mathbf{y} - m_{\mathbf{y}})\}/\sigma_{\mathbf{x}}\sigma_{\mathbf{y}}$ lies in the range $[-1, 1]$.

(e) Let the random variables $\mathbf{x}$ and $\mathbf{y}$ be $\mathbf{x}(t)$ and $\mathbf{x}(t + \tau)$, respectively, where $\mathbf{x}(t)$ is a WSS process. Use the result of (b) to show that $|R_{\mathbf{x}}(\tau)| \leq R_{\mathbf{x}}(0)$.

3.28 Equation (3.37) gives the relationship for the conditional pdf $f_{\mathbf{y}}(y|x)$ as

$$f_{\mathbf{y}}(y|x) = \begin{cases} \dfrac{f_{\mathbf{x},\mathbf{y}}(x,y)}{f_{\mathbf{x}}(x)}, & f_{\mathbf{x}}(x) \neq 0 \\ 0, & \text{otherwise} \end{cases}. \tag{P3.12}$$

To derive this relationship consider the events $(y < \mathbf{y} < y + \Delta y)$ and $(x < \mathbf{x} < x + \Delta x)$.

(a) Use Bayes' rule to determine the probability $P(y < \mathbf{y} < y + \Delta y | x < \mathbf{x} < x + \Delta x)$.
(b) As $\Delta x \to 0$ what value does the random variable become?
(c) Multiply and divide both sides of the expression obtained in (a) appropriately by Δx, Δy and then let $\Delta x, \Delta y \to 0$ to obtain (P3.12).

3.29 Show that if $\mathbf{x}$, $\mathbf{y}$ are jointly Gaussian, then the conditional densities $f_{\mathbf{y}}(y|x)$, $f_{\mathbf{x}}(x|y)$ are also Gaussian.

3.30 The relation in Problem 3.28 involves two random variables. This problem considers a "mixed" form of the relation where one variable is a random variable, $\mathbf{r}$, and the other is a random event, A. Using an approach similar to that of Problem 3.28 show that

$$P(A|r) = \frac{f(r|A)P(A)}{f_{\mathbf{r}}(r)}. \tag{P3.13}$$

Remark Typically in communications A is the event that a 0 or 1 was transmitted while $\mathbf{r}$ is the observation at the output of the demodulator.

3.31 Two first-order pdfs are proposed for the random process, $\mathbf{x}(t)$, as follows:

$$\text{(i)}\, f_{\mathbf{x}}(x;t) = \frac{1}{\sqrt{2\pi}\sigma} e^{-(x-\text{sgn}(t))^2/2\sigma^2}; \tag{P3.14}$$

$$\text{(ii)}\, f_{\mathbf{x}}(x;t) = \frac{1}{\sqrt{2\pi}\sigma} e^{-x^2/2\sigma^2} \text{sgn}(t), \tag{P3.15}$$

where $\text{sgn}(t) = 1$ for $t \geq 0$ and $= -1$ for $t < 0$.
(a) Which density function, if either, is valid?
(b) If a density function is valid, find the mean and variance of $\mathbf{x}(t)$.

3.32 A random process is generated as follows: $\mathbf{x}(t) = e^{-\mathbf{a}|t|}$, where $\mathbf{a}$ is a random variable with pdf $f_{\mathbf{a}}(a) = u(a) - u(a-1)$ (1/seconds).
(a) Sketch several members of the ensemble.
(b) For a specific time, t, over what values of amplitude does the random variable $\mathbf{x}(t)$ range?
(c) Find the mean and mean-squared value of $\mathbf{x}(t)$.
(d) Determine the first-order pdf of $\mathbf{x}(t)$.

3.33 Figure 3.30 shows a simple lowpass filter with impulse response $h(t) = (1/RC)e^{-t/RC}u(t)$.
(a) Design the filter for a 3 decibel bandwidth of 10 kilohertz. Use a 1.5 nanofaraday capacitor value and the closest available resistor value. In practice the values of

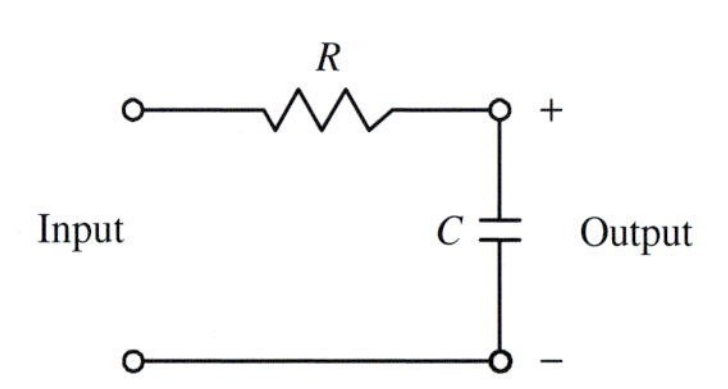

Fig. 3.30 The lowpass filter considered in Problem 3.33.

the resistor and capacitor vary around their nominal specifications, typically by $\pm 5\%$ and $\pm 10\%$, respectively, i.e., $R = R_{\text{nom}} \pm \Delta R$, $C = C_{\text{nom}} \pm \Delta C$. The time constant becomes $RC = R_{\text{nom}}C_{\text{nom}} \pm R_{\text{nom}}\Delta C \pm C_{\text{nom}}\Delta R \pm \Delta R \Delta C$. The term $\Delta R \Delta C$ is quite small, compared to the others and can be ignored. Furthermore, the variation in the component values is modeled as random and uniform over their ranges. Lastly the resistor value is statistically independent of the capacitor value.

(b) For the values of R_{nom} and C_{nom} determined in (a), using the discussion above and any other reasonable assumptions determine the first-order pdf of the product RC.

(c) The impulse response is a random process, denoted by $\mathbf{h}(t)$. Find the average impulse response and the worst-case impulse response.

(d) Determine the first-order pdf of $\mathbf{h}(t)$.

(e) Find the mean and variance of the 3 decibel filter bandwidth.

3.34 The first-order pdf of $\mathbf{x}(t)$ is proposed to be $f_{\mathbf{x}}(x;t) = (|t|/2)\mathrm{e}^{-|x|/|t|}$.

(a) Is the pdf a valid one?

(b) If valid, find the mean and variance of $\mathbf{x}(t)$.

(c) Discuss the case of $t = 0$.

3.35 You take a number of measurements of the electrical activity produced by your bicep muscle as you contract it from rest to a constant force level. As the output power of the muscle increases so does the power or variance of the electrical activity. Based on the ensemble you propose the following first-order pdf for the random signal: $f_{\mathbf{x}}(x;t) = (1/\sqrt{2\pi}\sigma(t))\mathrm{e}^{-x^2/2\sigma^2(t)}$, where $\sigma(t) = [1 - \mathrm{e}^{-10t}]u(t)$.

(a) Is the proposed pdf valid? Explain.

(b) If it is valid, what is the mean and variance of the signal?

(c) It appears that the random process is nonstationary. At what time (in milliseconds) would you feel comfortable to judge the process to be at least first-order stationary?

3.36 Let $g(t)$ be a deterministic signal of duration $[0, T_b]$ whose energy is E_g, i.e., $\int_0^{T_b} g^2(t)\mathrm{d}t = E_g$. Let $\mathbf{w}(t)$ be a zero-mean, white noise with two-sided PSD $N_0/2$. In the development of the optimal receiver, you will often see that the noise $\mathbf{w}(t)$ is correlated with the function $g(t)$ to form the random variable $\mathbf{x}$ as $\mathbf{x} = \int_0^{T_b} \mathbf{w}(t)g(t)\mathrm{d}t$. Find the mean and variance of $\mathbf{x}$.

3.37 The input noise $\mathbf{x}(t)$ applied to the filter in Figure 3.31 is modeled as a WSS, white, Gaussian random process with a zero mean and two-sided PSD $N_0/2$. Let $\mathbf{y}(t)$ denote the random process at the output of the filter.

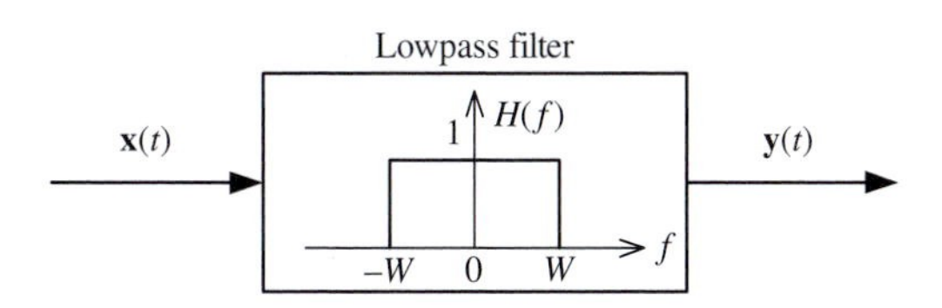

Fig. 3.31 The lowpass filter considered in Problem 3.37.

(a) Find and sketch the power spectral density of $\mathbf{y}(t)$.
(b) Find and sketch the autocorrelation function of $\mathbf{y}(t)$.
(c) What are the average DC level and the average power of $\mathbf{y}(t)$?
(d) Suppose that the output noise is sampled every T_s seconds to obtain the noise samples $\mathbf{y}(kT_s)$, $k = 0, 1, 2, \ldots$. Find the smallest values of T_s so that the noise samples are *statistically independent*. Explain.

3.38 Repeat Problem 3.37 with the system shown in Figure 3.32.

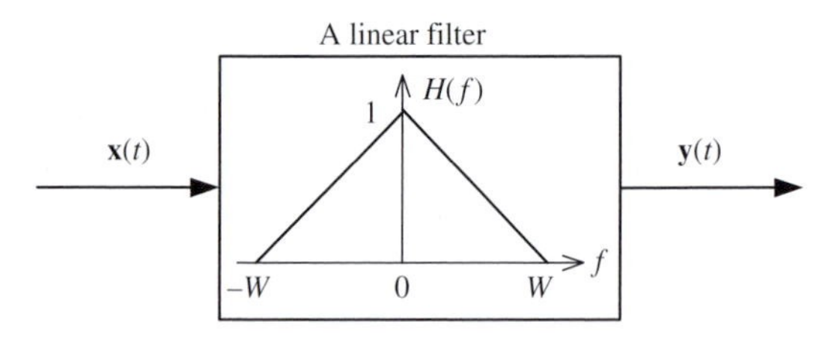

Fig. 3.32 System under consideration in Problem 3.38.

3.39 Let the random process $\mathbf{x}(t)$ in Problem 3.37 be a linear sum of two signals, i.e., $\mathbf{x}(t) = \mathbf{s}(t) + \mathbf{w}(t)$, where $\mathbf{s}(t)$ is the desired signal with autocorrelation $R_{\mathbf{s}}(\tau) = Ke^{-a|\tau|}$ and $\mathbf{w}(t)$ is white noise of spectral strength $N_0/2$.
(a) Derive the expression for the power SNR at the filter's output in terms of the bandwidth W.
(b) Obtain the differential equation needed to maximize the SNR and solve for W.

3.40 The noise $\mathbf{x}(t)$ applied to the filter in Figure 3.33 is modeled as a WSS random process with PSD $S_{\mathbf{x}}(f)$. Let $\mathbf{y}(t)$ denote the random noise process at the output of the filter.

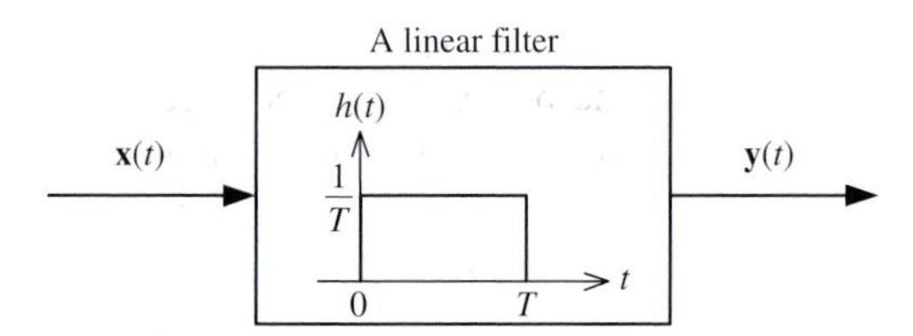

Fig. 3.33 System under consideration in Problem 3.40.

(a) Is $\mathbf{y}(t)$ a WSS noise process? Why?
(b) Find the frequency response, $H(f)$, of the filter.
(c) If $\mathbf{x}(t)$ is a *white* noise process with PSD $N_0/2$, find the PSD of the noise process $\mathbf{y}(t)$.
(d) What frequency components cannot be present in the output process? Explain.

3.41 Repeat Problem 3.40 with the system shown in Figure 3.34.

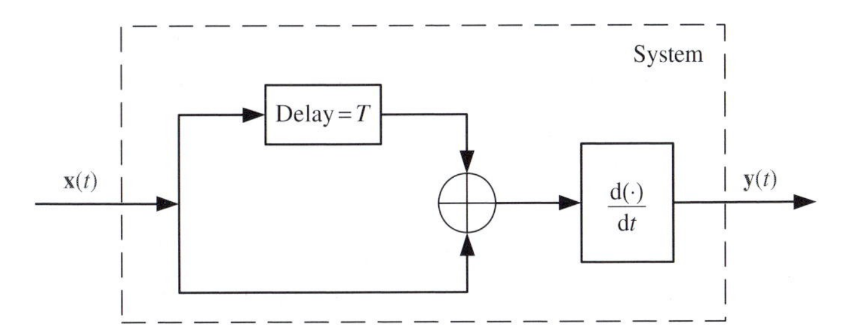

Fig. 3.34 System under consideration in Problem 3.41.

3.42 In digital communications, both message and noise are modeled as WSS random processes. Consider a message $\mathbf{m}(t)$, whose autocorrelation function is $R_{\mathbf{m}}(\tau) = Ae^{-|\tau|}$ (watts). The message $\mathbf{m}(t)$ is corrupted by zero-mean additive white Gaussian noise (AWGN) $\mathbf{n}(t)$ of spectral strength $N_0/2$ (watts/hertz) and the received signal is $\mathbf{r}(t) = \mathbf{m}(t) + \mathbf{n}(t)$. You decide to filter the noise by passing $\mathbf{r}(t)$ through an ideal lowpass filter with bandwidth W. The procedure is depicted in block diagram form in Figure 3.35.

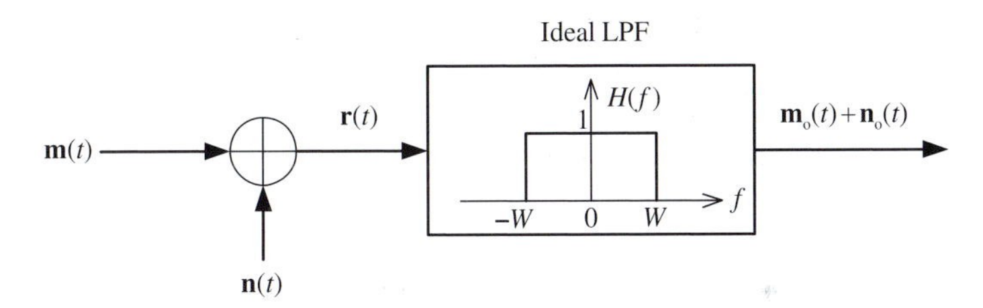

Fig. 3.35 System under consideration in Problem 3.42.

(a) Show that the PSD of the message is given by $S_{\mathbf{m}}(f) = 2A/(1 + 4\pi^2 f^2)$, $-\infty \leq f \leq \infty$. Then sketch $S_{\mathbf{m}}(f)$.
(b) Determine the power of the message at the output of the filter. What are the percentages of the input power passed through the filter when $W = 10$ hertz, $W = 50$ hertz and $W = 100$ hertz? *Hint* $\int(1/(a^2 + x^2))\mathrm{d}x = (1/a)\tan^{-1}(x/a)$.
(c) Determine the power of the noise at the output of the filter.
(d) Let $A = 4 \times 10^{-3}$ watts, $W = 4$ kilohertz and $N_0 = 10^{-8}$ watts/hertz. Determine the SNR in decibels at the filter output.

3.43 (*System identification*) Identification of an LTI system can be accomplished by finding the gain/phase response of the system at all frequencies, the sinusoidal steady-state approach, or by applying an impulse at the input and measuring $h(t)$, the resultant impulse response. Both approaches require the system to be taken "off-line." The frequency response method takes considerable time while the impulse method, if not actually causing damage to the system due to the extremely large input amplitude, tends to drive the system into nonlinear operation. Here another approach based on a white noise input is explored.

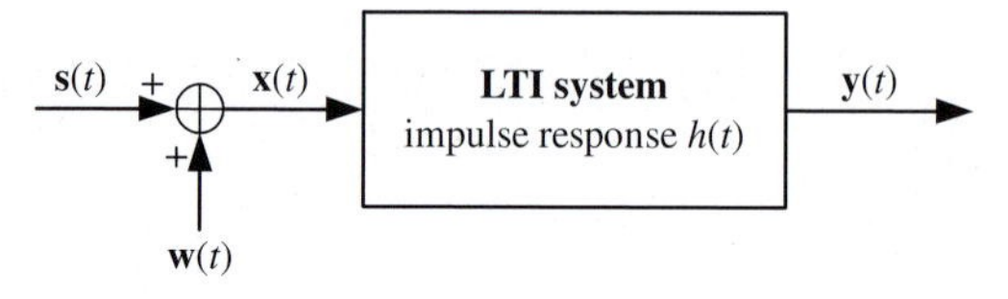

Fig. 3.36 System setup for Problem 3.43.

Consider the setup shown in Figure 3.36, where $\mathbf{s}(t)$ is the operating input and $\mathbf{w}(t)$ is a white process. Assume $\mathbf{s}(t)$ and $\mathbf{w}(t)$ are uncorrelated. Determine the crosscorrelation between the output $\mathbf{y}(t)$ and $\mathbf{w}(t)$. Comment on the result.

4 Sampling and quantization

The previous two chapters have reviewed and discussed the various characteristics of signals along with methods of describing and representing them. This chapter begins the discussion of transmitting the signals or messages using a digital communication system. Though one can easily visualize messages produced by sources that are inherently digital in nature, witness text messaging via the keyboard or keypad, two of the most common message sources, audio and video, are analog, i.e., they produce continuous time signals. To make them amenable for digital transmission it is first required to transform the analog information source into digital symbols which are compatible with digital processing and transmission.

The first step in this transformation process is to discretize the time axis, which involves sampling the continuous time signal at discrete values of time. The sampling process, primarily how many samples per second are needed to exactly represent the signal, practical sampling schemes, and how to reconstruct, at the receiver, the analog message from the samples is considered first. This is followed by a brief discussion of three pulse modulation techniques, a sort of half-way house between the analog modulation methods of AM and FM and the various digital modulation–demodulation methods which are the focus of the rest of the text.

Though time has been discretized by the sampling process the sample values are still analog, i.e., they are continuous variables. To represent the sample value by a digital symbol chosen from a finite set necessitates the choice of a discrete set of amplitudes to represent the continuous range of possible amplitudes. This process is known as quantization and unlike discretization of the time axis, it results in a distortion of the original signal since it is a many-to-one mapping. The measure of this distortion is commonly expressed by the signal power to quantization noise power ratio, $\mathrm{SNR_q}$. Various approaches to quantization and the resultant $\mathrm{SNR_q}$ are the major focus of this chapter.

The final step is to map (or encode) the quantized signal sample into a string of digital, typically binary, symbols, commonly known as pulse-code modulation (PCM). The complete process of analog-to-digital (A/D) conversion is a special, but important, case of source coding.[1]

[1] A more general source coding process not only involves A/D conversion but also some form of data compression to remove the redundancy of the information source.

4.1 Sampling of continuous-time signals

The first operation in A/D conversion is the sampling process. Both the theoretical and practical implementations of this process are studied in this section.

4.1.1 Ideal (or impulse) sampling

The sampling process converts an analog waveform into a sequence of discrete samples that are usually spaced *uniformly* in time. This process can be mathematically described as in Figure 4.1(a). Here the analog waveform $m(t)$ is multiplied by a periodic train of unit *impulse* functions $s(t)$ to produce the sampled waveform $m_s(t)$. The expression for $s(t)$ is as follows:

$$s(t) = \sum_{n=-\infty}^{\infty} \delta(t - nT_s). \tag{4.1}$$

Thus the sampled waveform $m_s(t)$ can be expressed as

$$\begin{aligned} m_s(t) &= m(t)s(t) \\ &= \sum_{n=-\infty}^{\infty} m(t)\delta(t - nT_s) = \sum_{n=-\infty}^{\infty} m(nT_s)\delta(t - nT_s). \end{aligned} \tag{4.2}$$

The parameter T_s in (4.1) and (4.2) is the period of the impulse train, also referred to as the *sampling period*. The inverse of the sampling period, $f_s = 1/T_s$, is called the sampling frequency or *sampling rate*. Figures 4.1(b)–(d) graphically illustrate the ideal sampling process. It is intuitive that the higher the sampling rate is, the more accurate the representation of $m(t)$ by $m_s(t)$ is. However, to achieve a high efficiency, it is desired to use as low a sampling rate as possible. Thus an important question is: what is the minimum sampling rate for the sampled version $m_s(t)$ to exactly represent the original analog signal $m(t)$? For the family of bandlimited signals, this question is answered by the *sampling theorem*, which is derived next.

Consider the Fourier transform of the sampled waveform $m_s(t)$. Since $m_s(t)$ is the product of $m(t)$ and $s(t)$, the Fourier transform of $m_s(t)$ is the *convolution* of the Fourier transforms of $m(t)$ and $s(t)$. Recall that the Fourier transform of an impulse train is another impulse train, where the values of the periods of the two trains are reciprocally related to one another. The Fourier transform of $s(t)$ is given by

$$S(f) = \frac{1}{T_s} \sum_{n=-\infty}^{\infty} \delta(f - nf_s). \tag{4.3}$$

Also note that convolution with an impulse function simply shifts the original function as follows:

$$X(f) * \delta(f - f_0) = X(f - f_0). \tag{4.4}$$

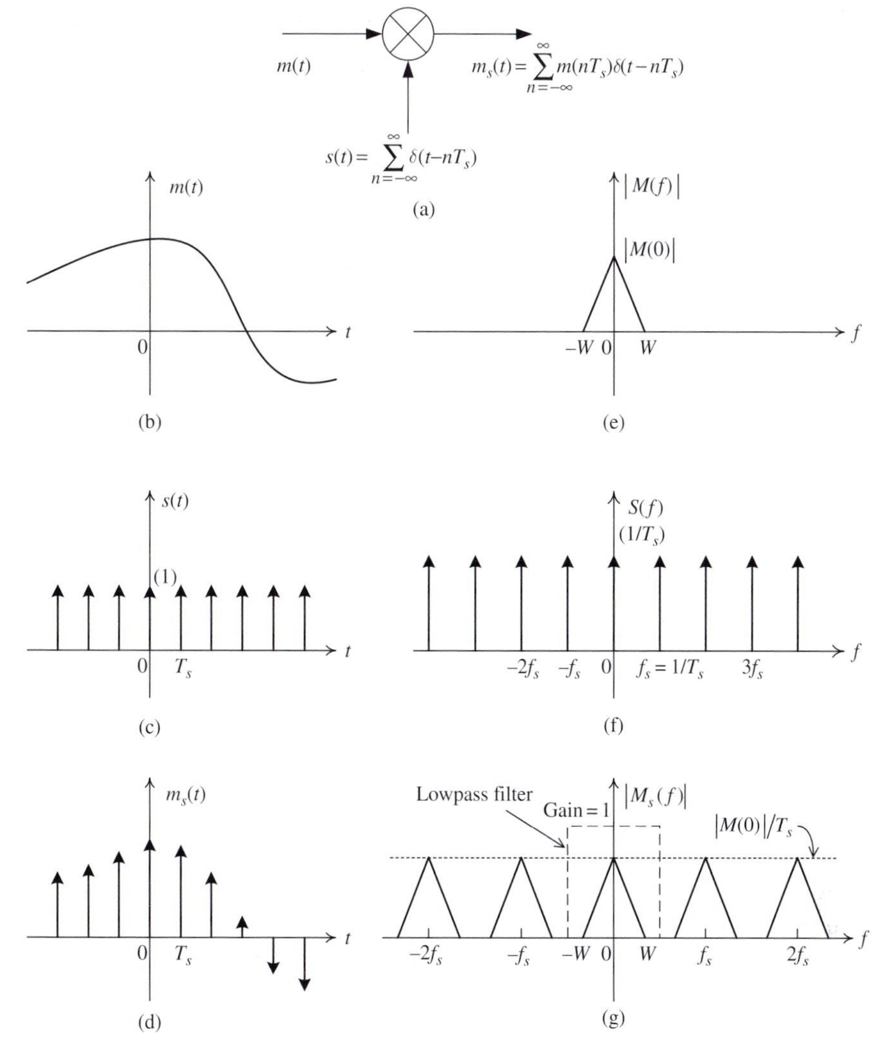

Fig. 4.1 Ideal sampling process: (a) mathematical model, (b) analog signal, (c) impulse train, (d) sampled version of the analog signal, (e) spectrum of bandlimited signal, (f) spectrum of the impulse train, (g) spectrum of the sampled waveform.

From the above equations, the transform of the sampled waveform can be written as

$$M_s(f) = M(f) * S(f) = M(f) * \left[\frac{1}{T_s}\sum_{n=-\infty}^{\infty} \delta(f - nf_s)\right]$$

$$= \frac{1}{T_s}\sum_{n=-\infty}^{\infty} M(f - nf_s). \tag{4.5}$$

Equation (4.5) shows that the spectrum of the sampled waveform consists of an infinite number of scaled and shifted copies of the spectrum of the original signal $m(t)$. More precisely, the spectrum $M(f)$ is scaled by $1/T_s$ and periodically repeated every f_s. It should be noted that the relation in (4.5) holds for any continuous-time signal $m(t)$, even if it is not bandlimited, or of finite energy.

However, for the bandlimited waveform $m(t)$ with bandwidth limited to W hertz, a generic Fourier transform of the sampled signal $m_s(t)$ is illustrated in Figure 4.1(e). The triangular shape chosen for the magnitude spectrum of $m(t)$ is only for ease of illustration. In general, $M(f)$ can be of arbitrary shape, as long as it is confined to $[-W, W]$. Since within the original bandwidth (around zero frequency) the spectrum of the sampled waveform is the same as that of the original signal (except for a scaling factor $1/T_s$), it suggests that the original waveform $m(t)$ can be completely recovered from $m_s(t)$ by an ideal lowpass filter (LPF) of bandwidth W as shown in Figure 4.1(g). However, a closer investigation of Figure 4.1(g) reveals that this is only possible if the sampling rate f_s is high enough that there is no overlap among the copies of $M(f)$ in the spectrum of $m_s(t)$. It is easy to see that the condition for no overlapping of the copies of $M(f)$ is $f_s \geq 2W$, therefore the minimum sampling rate is $f_s = 2W$. When the sampling rate $f_s < 2W$ (undersampling), then the copies of $M(f)$ overlap in the frequency domain and it is not possible to recover the original signal $m(t)$ by filtering. The distortion of the recovered signal due to undersampling is referred to as *aliasing*.

It has been shown using the frequency domain that the original continuous signal $m(t)$ can be completely recovered from the sampled signal $m_s(t)$. Next we wish to show how to reconstruct the continuous signal $m(t)$ from its sampled values $m(nT_s)$, $n = 0, \pm 1, \pm 2, \ldots$. To this end, write the Fourier transform of $m_s(t)$ as follows:

$$\begin{aligned} M_s(f) = \mathcal{F}\{m_s(t)\} &= \sum_{n=-\infty}^{\infty} m(nT_s)\mathcal{F}\{\delta(t - nT_s)\} \\ &= \sum_{n=-\infty}^{\infty} m(nT_s)\mathrm{e}^{-\mathrm{j}2\pi nfT_s}. \end{aligned} \tag{4.6}$$

Since $M(f) = M_s(f)/f_s$, for $-W \leq f \leq W$, one can write

$$M(f) = \frac{1}{f_s}\sum_{n=-\infty}^{\infty} m(nT_s)\mathrm{e}^{-\mathrm{j}2\pi nfT_s}, \quad -W \leq f \leq W. \tag{4.7}$$

The signal $m(t)$ is the inverse Fourier transform of $M(f)$ and it can be found as follows:

$$\begin{aligned} m(t) = \mathcal{F}^{-1}\{M(f)\} &= \int_{-\infty}^{\infty} M(f)\mathrm{e}^{\mathrm{j}2\pi ft}\mathrm{d}f \\ &= \int_{-W}^{W} \frac{1}{f_s}\sum_{n=-\infty}^{\infty} m(nT_s)\mathrm{e}^{-\mathrm{j}2\pi nfT_s}\mathrm{e}^{\mathrm{j}2\pi ft}\mathrm{d}f \\ &= \frac{1}{f_s}\sum_{n=-\infty}^{\infty} m(nT_s)\int_{-W}^{W} \mathrm{e}^{\mathrm{j}2\pi f(t-nT_s)}\mathrm{d}f \\ &= \sum_{n=-\infty}^{\infty} m(nT_s)\frac{\sin[2\pi W(t - nT_s)]}{\pi f_s(t - nT_s)} \end{aligned}$$

$$= \sum_{n=-\infty}^{\infty} m\left(\frac{n}{2W}\right) \frac{\sin(2\pi Wt - n\pi)}{(2\pi Wt - n\pi)} \tag{4.8}$$

$$= \sum_{n=-\infty}^{\infty} m\left(\frac{n}{2W}\right) \operatorname{sinc}(2Wt - n), \tag{4.9}$$

where, to arrive at the last two equations, the minimum sampling rate $f_s = 2W$ has been used and $\operatorname{sinc}(x) \equiv \sin(\pi x)/(\pi x)$.

Equation (4.9) provides an *interpolation formula* for the construction of the original signal $m(t)$ from its sampled values $m(n/2W)$. The sinc function $\operatorname{sinc}(2Wt)$ plays the role of an *interpolating function* (also known as the sampling function). In essence, each sample is multiplied by a delayed version of the interpolating function and all the resulting waveforms are added up to obtain the original signal.

Now the sampling theorem can be stated as follows.

Theorem 4.1 (sampling theorem) *A signal having no frequency components above W hertz is completely described by specifying the values of the signal at periodic time instants that are separated by at most* $1/2W$ *seconds.*

The theorem stated in terms of the sampling rate, $f_s \geq 2W$, is known as the *Nyquist criterion*. The sampling rate $f_s = 2W$ is called the *Nyquist rate* with the reciprocal called the *Nyquist interval*.

The sampling process considered so far is known as *ideal sampling* because it involves ideal impulse functions. Obviously, ideal sampling is not practical. In the next two sections two practical methods of implementing sampling of continuous-time signals are introduced.

4.1.2 Natural sampling

Figure 4.2(a) shows the mathematical model of natural sampling. Here, the analog signal $m(t)$ is multiplied by the pulse train, or gating waveform $p(t)$ shown in Figure 4.2(c). Let $h(t) = 1$ for $0 \leq t \leq \tau$ and $h(t) = 0$ otherwise. Then the pulse train $p(t)$ can be written as,

$$p(t) = \sum_{n=-\infty}^{\infty} h(t - nT_s). \tag{4.10}$$

Natural sampling is therefore very simple to implement since it requires only an on/off gate. Figure 4.2(d) shows the resultant sampled waveform, which is simply $m_s(t) = m(t)p(t)$. As in ideal sampling, here the sampling rate f_s also equals the inverse of the period T_s of the pulse train, i.e., $f_s = 1/T_s$. Next it is shown that with natural sampling, a bandlimited waveform can also be reconstructed from its sampled version as long as the sampling rate satisfies the Nyquist criterion. As before, the analysis is carried out in the frequency domain.

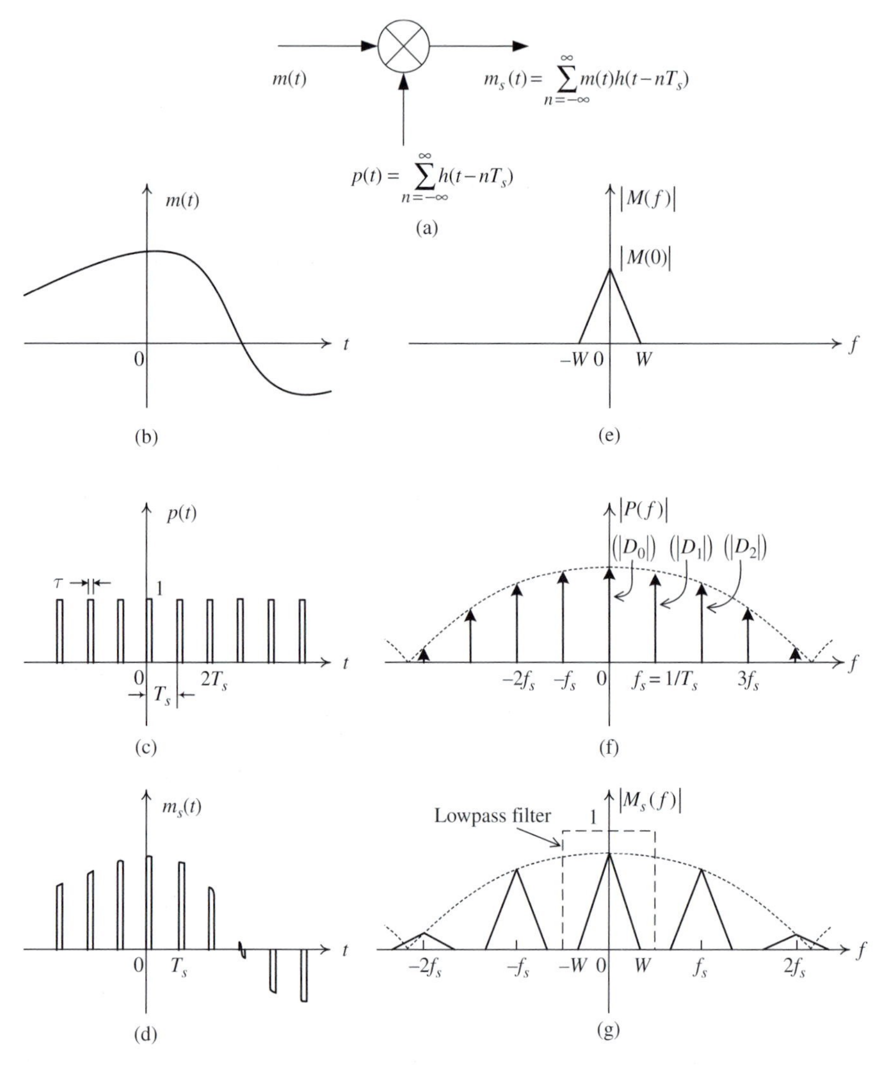

Fig. 4.2 The natural sampling process: (a) mathematical model, (b) analog signal, (c) rectangular pulse train, (d) sampled version of the analog signal, (e) spectrum of bandlimited signal, (f) spectrum of the pulse train, (g) spectrum of the sampled waveform.

Recall that the periodic pulse train $p(t)$ can be expressed in a Fourier series as follows:

$$p(t) = \sum_{n=-\infty}^{\infty} D_n e^{j2\pi n f_s t}, \tag{4.11}$$

where D_n is the Fourier coefficient, given by

$$D_n = \frac{\tau}{T_s} \operatorname{sinc}\left(\frac{n\tau}{T_s}\right) e^{-j\pi n\tau/T_s}. \tag{4.12}$$

Thus the sampled waveform is

$$m_s(t) = m(t) \sum_{n=-\infty}^{\infty} D_n e^{j2\pi n f_s t}. \tag{4.13}$$

The Fourier transform of $m_s(t)$ can be found as follows:

$$\begin{aligned} M_s(f) = \mathcal{F}\{m_s(t)\} &= \sum_{n=-\infty}^{\infty} D_n \mathcal{F}\left\{m(t)e^{j2\pi n f_s t}\right\} \\ &= \sum_{n=-\infty}^{\infty} D_n M(f - nf_s). \end{aligned} \tag{4.14}$$

Similarly to the ideal sampling case, (4.14) shows that $M_s(f)$ consists of an infinite number of copies of $M(f)$, which are periodically shifted in frequency every f_s hertz. However, here the copies of $M(f)$ are not uniformly weighted (scaled) as in the ideal sampling case, but rather they are weighted by the Fourier series coefficients of the pulse train. The spectrum of the sampled waveform $m_s(t)$ is shown in Figure 4.2(g), where $m(t)$ is again a bandlimited waveform. It can be seen from Figure 4.2(g) that, despite the above difference, the original signal $m(t)$ can be equally well reconstructed using an LPF as long as the Nyquist criterion is satisfied. Finally, it should be noted that natural sampling can be considered to be a practical approximation of ideal sampling, where an ideal impulse is approximated by a narrow rectangular pulse. With this perspective, it is not surprising that when the width τ of the pulse train approaches zero, the spectrum in (4.14) converges to (4.5).

4.1.3 Flat-top sampling

Flat-top sampling is the most popular sampling method. This sampling process involves two simple operations:

(i) Instantaneous sampling of the analog signal $m(t)$ every T_s seconds. As in the ideal and natural sampling cases, it will be shown that to reconstruct the original signal $m(t)$ from its sampled version, the sampling rate $f_s = 1/T_s$ must satisfy the Nyquist criterion.
(ii) Maintaining the value of each sample for a duration of τ seconds.

In circuit technology, these two operations are referred to as *sample and hold*. The flat-top sampled waveform is illustrated in Figure 4.3.

It is straightforward to verify that the flat-top sampling described above can be mathematically modeled as shown in Figure 4.4(a). Note that Figure 4.4(a) is an extension of Figure 4.1(a), where a filter with impulse response $h(t)$ is added at the end.

Again, we are interested in the spectrum of the sampled signal $m_s(t)$, which is related to the original signal $m(t)$ through the following expression:

$$m_s(t) = \left[m(t) \sum_{n=-\infty}^{\infty} \delta(t - nT_s) \right] * h(t). \tag{4.15}$$

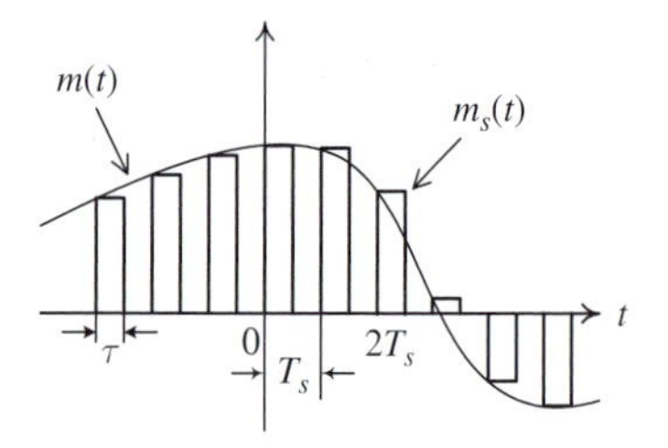

Fig. 4.3 Flat-top sampling: $m(t)$ is the original analog signal and $m_s(t)$ is the sampled signal.

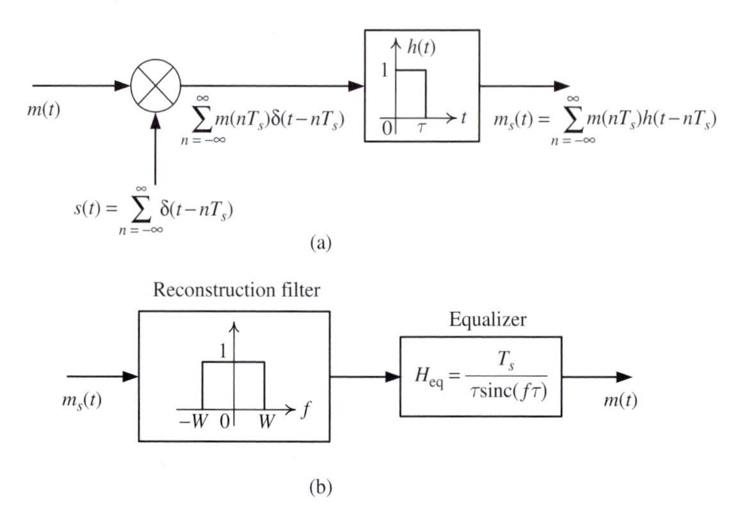

Fig. 4.4 The flat-top sampling process: (a) sampling, (b) reconstruction.

The Fourier transform of $m_s(t)$ is determined as follows:

$$\begin{aligned} M_s(f) &= \mathcal{F}\left\{ m(t) \sum_{n=-\infty}^{\infty} \delta(t - nT_s) \right\} \mathcal{F}\{h(t)\} \\ &= \frac{1}{T_s} H(f) \sum_{n=-\infty}^{\infty} M(f - nf_s), \end{aligned} \tag{4.16}$$

where $H(f)$ is the Fourier transform of the rectangular pulse $h(t)$, given by $H(f) = \tau \text{sinc}(f\tau)\text{e}^{-\text{j}\pi f\tau}$.

Equations (4.16) and (4.5) imply that the spectrum of the signal produced by flat-top sampling is essentially the spectrum of the signal produced by ideal sampling shaped by $H(f)$. Since $H(f)$ has the form of a sinc function, each spectral component of the ideal sampled signal is weighted differently, hence causing amplitude distortion. As a consequence of this distortion, it is not possible to reconstruct the original signal using an LPF, even when the Nyquist criterion is satisfied. This is illustrated in Figure 4.5.

In fact, if the Nyquist criterion is satisfied, then passing the flat-top sampled signal through an LPF (with a bandwidth of W) produces the signal whose Fourier transform is

$(1/T_s)M(f)H(f)$. Thus the distortion due to $H(f)$ can be corrected by connecting an *equalizer* in cascade with the lowpass reconstruction filter, as shown in Figure 4.4(b). Ideally, the amplitude response of the equalizer is given by

$$|H_{\text{eq}}| = \frac{T_s}{|H(f)|} = \frac{T_s}{\tau\,\text{sinc}(f\tau)} \tag{4.17}$$

Finally, it should be noted that for a duty cycle $\tau/T_s \leq 0.1$, the amplitude distortion is less than 0.5%. In this case, equalization may not be necessary in practical applications.

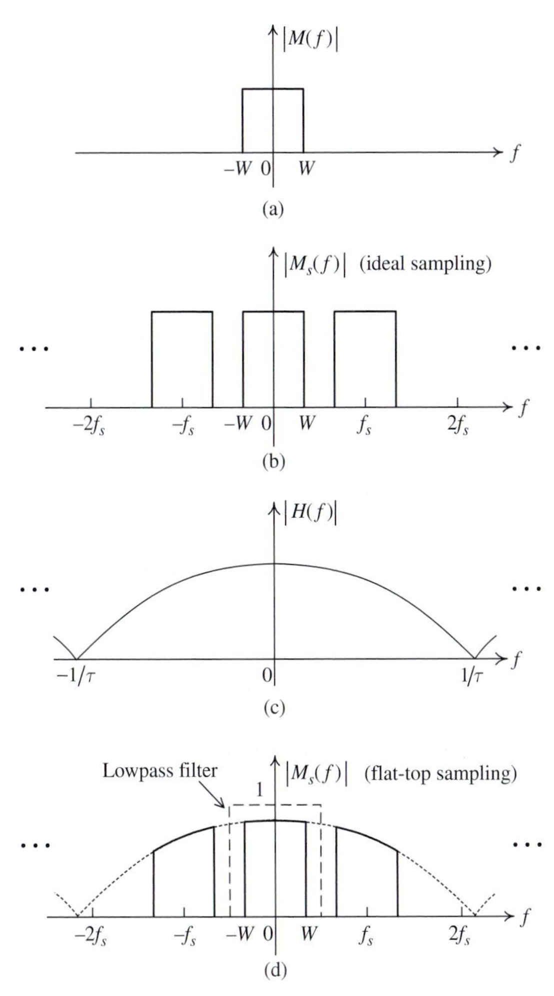

Fig. 4.5 Various spectra in the flat-top sampling process: (a) magnitude spectrum of the original message, (b) spectrum of the ideal sampled waveform, (c) $|H(f)| = |\tau\,\text{sinc}(f\tau)|$, (d) spectrum of the flat-top sampled waveform.

4.2 Pulse modulation

Recall that in analog (continuous-wave) modulation, some parameter of a sinusoidal carrier $A_c \cos(2\pi f_c t + \theta)$, such as the amplitude A_c, the frequency f_c, or the phase θ, is varied *continuously* in accordance with the message signal $m(t)$. Similarly, in pulse modulation, some parameter of a *pulse train* is varied in accordance with the sample values of a message signal.

Pulse-amplitude modulation (PAM) is the simplest and most basic form of analog pulse modulation. In PAM, the *amplitudes* of regularly spaced pulses are varied in proportion to the corresponding sample values of a continuous message signal. In general, the pulses can be of some appropriate shape. In the simplest case, when the pulse is rectangular, then the PAM signal is identical to the signal produced by flat-top sampling described in Section 4.1.3.

It should be noted that PAM transmission does not improve the noise performance over baseband modulation (which is the transmission of the original continuous signal). The main (perhaps the only) advantage of PAM is that it allows multiplexing, i.e., the sharing of the same transmission media by different sources (or users). This is because a PAM signal only occurs in slots of time, leaving the idle time for the transmission of other PAM signals. However, this advantage comes at the expense of a larger transmission bandwidth, as can be seen from Figures 4.5(a) and 4.5(d).

It is well known that in analog FM, bandwidth can be traded for noise performance. As mentioned before, PAM signals require a larger transmission bandwidth without any improvement in noise performance. This suggests that there should be better pulse modulations than PAM in terms of noise performance. Two such forms of pulse modulation are:

- Pulse-width modulation (PWM): in PWM, the samples of the message signal are used to vary the *width* of the individual pulses in the pulse train.
- Pulse-position modulation (PPM): in PPM, the *position* of a pulse relative to its original time of occurrence is varied in accordance with the sample values of the message.

Examples of PWM and PPM waveforms are shown in Figure 4.6 for a sinusoidal message.

Note that in PWM, long pulses (corresponding to large sample values) expend considerable power, while bearing no additional information. In fact, if only time transitions are preserved, then PWM becomes PPM. Accordingly, PPM is a more power-efficient form of pulse modulation than PWM.

Regarding the noise performance of PWM and PPM systems, since the transmitted information (the sample values) is contained in the relative positions of the modulated pulses, the additive noise, which mainly introduces amplitude distortion, has much less effect. As a consequence, both PWM and PPM systems have better noise performance than PAM.

Pulse modulation techniques, however, are still analog modulation. For digital communications of an analog source, one needs to proceed to the next step, i.e., quantization.

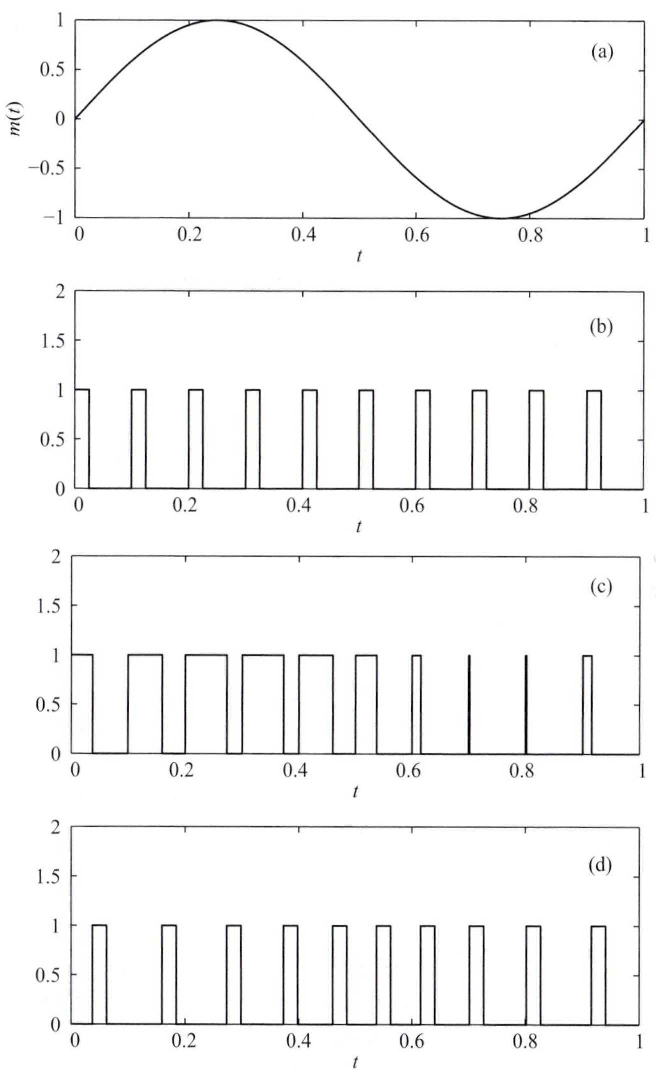

Fig. 4.6 Examples of PWM and PPM: (a) a sinusoidal message, (b) pulse carrier, (c) PWM waveform, (d) PPM waveform.

4.3 Quantization

In all the sampling processes described in the previous section, the sampled signals are discrete in time but still continuous in amplitude. To obtain a fully digital representation of a continuous signal, two further operations are needed: *quantization* of the amplitude of the sampled signal and *encoding* of the quantized values, as illustrated in Figure 4.7. This section discusses quantization.

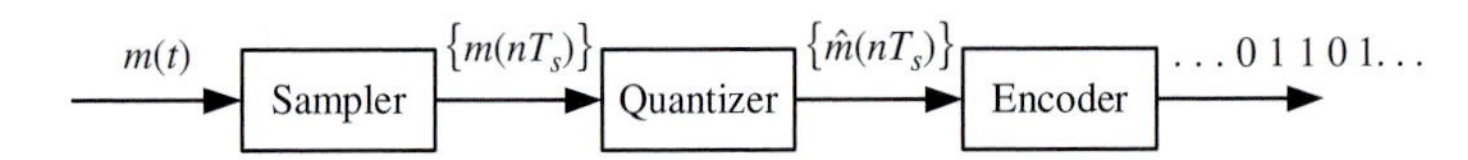

Fig. 4.7 Quantization and encoding operations.

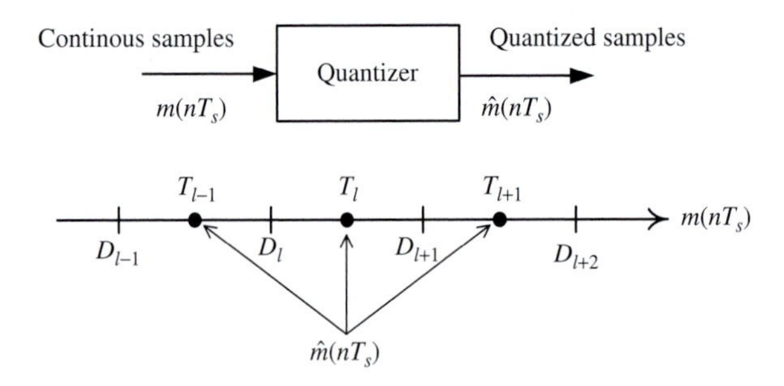

Fig. 4.8 Description of a memoryless quantizer.

By definition, amplitude quantization is the process of transforming the sample amplitude $m(nT_s)$ of a message signal $m(t)$ at time $t = nT_s$ into a discrete amplitude $\hat{m}(nT_s)$ taken from a *finite* set of possible amplitudes. Clearly, if the finite set of amplitudes is chosen such that the spacing between two adjacent amplitude levels is sufficiently small, then the approximated (or quantized) signal, $\hat{m}(nT_s)$, can be made practically indistinguishable from the continuous sampled signal, $m(nT_s)$. Nevertheless, unlike the sampling process, there is always a loss of information associated with the quantization process, no matter how finely one may choose the finite set of the amplitudes for quantization. This implies that it is not possible to *completely* recover the sampled signal from the quantized signal.

In this section, we shall assume that the quantization process is *memoryless* and *instantaneous*, meaning that the quantization of sample value at time $t = nT_s$ is independent of earlier or later samples. With this assumption, the quantization process can be described as in Figure 4.8. Let the amplitude range of the continuous signal be partitioned into L intervals, where the lth interval, denoted by $\mathcal{I}_l$, is determined by the *decision levels* (also called the *threshold levels*) D_l and D_{l+1}:

$$\mathcal{I}_l : \{D_l < m \leq D_{l+1}\}, \quad l = 1, \ldots, L. \tag{4.18}$$

Then the quantizer represents all the signal amplitudes in the interval $\mathcal{I}_l$ by some amplitude $T_l \in \mathcal{I}_l$ referred to as the *target level* (also known as the *representation level* or *reconstruction level*). The spacing between two adjacent decision levels is called the *step-size*. If the step-size is the same for each interval, then the quantizer is called a *uniform quantizer*, otherwise the quantizer is nonuniform. The uniform quantizer is the simplest and most practical one. Besides having equal decision intervals, the target level is chosen to lie in the middle of the interval.

4.3.1 Uniform quantizer

From the description of the quantizer, it follows that the input–output characteristic of the quantizer (or *quantizer characteristic*) is a staircase function. Figures 4.9(a) and 4.9(b) display two uniform quantizer characteristics, called *midtread* and *midrise*. As can be seen from these figures, the classification whether a characteristic is midtread or midrise depends on whether the origin lies in the middle of a tread, or a rise of the staircase characteristic. For both characteristics, the decision levels are equally spaced and the lth target level is the midpoint of the lth interval, i.e.,

$$T_l = \frac{D_l + D_{l+1}}{2}. \tag{4.19}$$

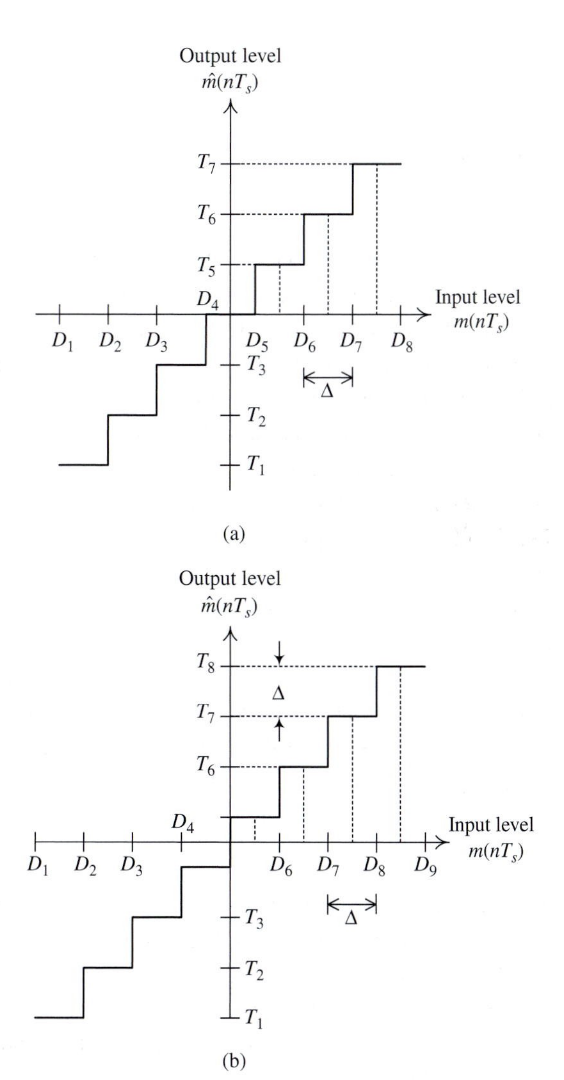

Fig. 4.9 Two types of uniform quantization: (a) midtread and (b) midrise.

As an example, Figure 4.10 plots the input and output waveforms of a midrise uniform quantizer.

As mentioned before, the quantization process always introduces an error. The performance of a quantizer is usually evaluated in terms of its SNR. In what follows, this parameter is derived for the uniform quantizer.

Since we concentrate only on memoryless quantization, we can ignore the time index and simply write m and $\hat{m}$ instead of $m(nT_s)$ and $\hat{m}(nT_s)$ for the input and output of the quantizer respectively. Typically, the input of the quantizer can be modeled as a zero-mean random variable $\mathbf{m}$ with some pdf $f_{\mathbf{m}}(m)$. Furthermore, assume that the amplitude range of $\mathbf{m}$ is $-m_{\max} \leq \mathbf{m} \leq m_{\max}$, that the uniform quantizer is of midrise type, and that the number of quantization levels is L. Then the quantization step-size is given by

$$\Delta = \frac{2m_{\max}}{L}. \tag{4.20}$$

Let $\mathbf{q} = \mathbf{m} - \hat{\mathbf{m}}$ be the error introduced by the quantizer, then $-\Delta/2 \leq \mathbf{q} \leq \Delta/2$. If the step-size Δ is sufficiently small (i.e., the number of quantization intervals L is sufficiently large), then it is reasonable to assume that the quantization error $\mathbf{q}$ is a *uniform* random variable over the range $[-\Delta/2, \Delta/2]$. The pdf of the random variable $\mathbf{q}$ is therefore given by

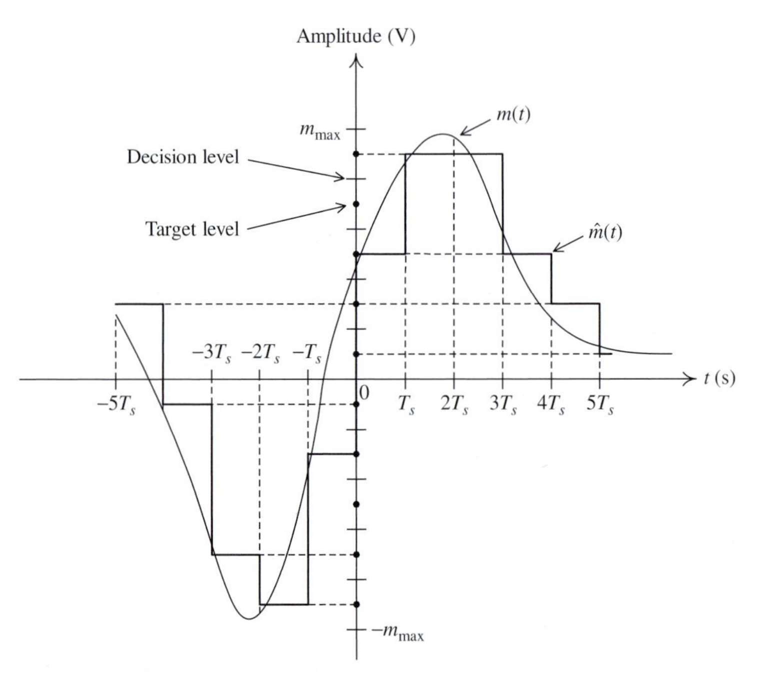

Fig. 4.10 An example of the input and output of a midrise uniform quantizer.

$$f_{\mathbf{q}}(q) = \begin{cases} 1/\Delta, & -\Delta/2 < q \leq \Delta/2 \\ 0, & \text{otherwise} \end{cases}. \tag{4.21}$$

Note that with this assumption, the mean of the quantization error is zero, while its variance can be calculated as follows:

$$\begin{aligned} \sigma_{\mathbf{q}}^2 &= \int_{-\Delta/2}^{\Delta/2} q^2 f_{\mathbf{q}}(q)\mathrm{d}q = \int_{-\Delta/2}^{\Delta/2} q^2 \left(\frac{1}{\Delta}\right) \mathrm{d}q \\ &= \frac{\Delta^2}{12} = \frac{m_{\max}^2}{3L^2}, \end{aligned} \tag{4.22}$$

where the last equality follows from (4.20).

Each target level at the output of a quantizer is typically encoded (or represented) in binary form, i.e., a binary string. For convenience, the number of quantization levels is usually chosen to be a power of 2, i.e., $L = 2^R$, where R is the number of bits needed to represent each target level. Substituting $L = 2^R$ into (4.22) one obtains the following expression for the variance of the quantization error:

$$\sigma_{\mathbf{q}}^2 = \frac{m_{\max}^2}{3 \times 2^{2R}}. \tag{4.23}$$

Since the message sample $\mathbf{m}$ is a zero-mean random variable whose pdf is $f_{\mathbf{m}}(m)$, the average power of the message is equal to the variance of $\mathbf{m}$, i.e., $\sigma_{\mathbf{m}}^2 = \int_{-m_{\max}}^{m_{\max}} m^2 f_{\mathbf{m}}(m)\mathrm{d}m$. Therefore, the $\mathrm{SNR_q}$ can be expressed as,

$$\mathrm{SNR_q} = \left(\frac{3\sigma_{\mathbf{m}}^2}{m_{\max}^2}\right) 2^{2R} \tag{4.24}$$

$$= \frac{3 \times 2^{2R}}{F^2}. \tag{4.25}$$

The parameter F in (4.25) is called the *crest factor* of the message, defined as

$$F = \frac{\text{peak value of the signal}}{\text{RMS value of the signal}} = \frac{m_{\max}}{\sigma_{\mathbf{m}}}. \tag{4.26}$$

Equation (4.25) shows that the $\mathrm{SNR_q}$ of a uniform quantizer increases *exponentially* with the number of bits per sample R and decreases with the square of the message's crest factor. The message's crest factor is an inherent property of the signal source, while R is a technical specification, i.e., under an engineer's control.

Expressed in decibels, the $\mathrm{SNR_q}$ is given by

$$10\log_{10}\mathrm{SNR_q} = 6.02R + 10\log_{10}\left(\frac{\sigma_{\mathbf{m}}^2}{m_{\max}^2}\right) + 4.77 \tag{4.27}$$

$$= 6.02R - 20\log_{10}F + 4.77. \tag{4.28}$$

The above equation, called the *6 decibel rule*, points out a significant performance characteristic of a uniform quantizer: *an additional 6 decibel improvement in* $\mathrm{SNR_q}$ *is obtained for each bit added to represent the continuous signal sample.*

4.3.2 Optimal quantizer

In the previous section the uniform quantizer was discussed where all the quantization regions are of equal size and the target (quantized) levels are at the midpoint of the quantization regions. Though simple, uniform quantizers are not optimal in terms of minimizing the SNR_q. In this section the optimal quantizer that maximizes the SNR_q is studied.

Consider a message signal $m(t)$ drawn from some stationary process. Let$[-m_{\max}, m_{\max}]$ be the amplitude range of the message, which is partitioned into L quantization regions as in (4.18). Instead of being equally spaced, the decision levels are constrained to satisfy only the following three conditions:

$$\begin{aligned} D_1 &= -m_{\max}, \\ D_{L+1} &= m_{\max}, \\ D_l &\le D_{l+1}, \quad \text{for } l = 1, 2, \ldots, L. \end{aligned} \tag{4.29}$$

A target level may lie anywhere within its quantization region and as before, is denoted by T_l, $l = 1, \ldots, L$. Then the average quantization noise power is given by

$$N_\text{q} = \sum_{l=1}^{L} \int_{D_l}^{D_{l+1}} (m - T_l)^2 f_\mathbf{m}(m) \text{d}m. \tag{4.30}$$

We need to find the set of $2L - 1$ variables $\{D_2, D_3, \ldots, D_L, T_1, T_2, \ldots, T_L\}$ to maximize the SNR_q, or equivalently to minimize the average power of the quantization noise N_q. Differentiating N_q with respect to a specific threshold, say D_j (using Leibniz's rule)[2] and setting the result to 0 yields

$$\frac{\partial N_\text{q}}{\partial D_j} = f_\mathbf{m}(D_j)\left[(D_j - T_{j-1})^2 - (D_j - T_j)^2\right] = 0, \quad j = 2, 3, \ldots, L. \tag{4.31}$$

The above gives $L - 1$ equations with solutions

$$D_l^{\text{opt}} = \frac{T_{l-1} + T_l}{2}, \quad l = 2, 3, \ldots, L. \tag{4.32}$$

This result simply means that, in an optimal quantizer, the decision levels are the midpoints of the target values (note that the target values of the optimal quantizer are not yet known).

To determine the L target values T_l, differentiate N_q with respect to a specific target level T_j and set the result to zero:

$$\frac{\partial N_\text{q}}{\partial T_j} = -2 \int_{D_j}^{D_{j+1}} (m - T_j) f_\mathbf{m}(m) \text{d}m = 0, \quad j = 1, 2, \ldots, L. \tag{4.33}$$

The above equation gives

$$T_l^{\text{opt}} = \frac{\int_{D_l}^{D_{l+1}} m f_\mathbf{m}(m) \text{d}m}{\int_{D_l}^{D_{l+1}} f_\mathbf{m}(m) \text{d}m}, \quad l = 1, 2, \ldots, L. \tag{4.34}$$

[2] Leibniz's rule states that if $f(p) = \int_{l(p)}^{u(p)} g(x;p) \text{d}x$, where p is a parameter, then $\partial f(p)/\partial p = \int_{l(p)}^{u(p)} (\partial g(x;p)/\partial p) \text{d}x + g(x = u(p); p) \partial u(p)/\partial p - g(x = l(p); p) \partial l(p)/\partial p$.

Equation (4.34) states that in an optimal quantizer the target value for a quantization region should be chosen to be the *centroid* (conditional expected value) of that region.

In summary, (4.32) and (4.34) give the necessary and sufficient conditions for a memoryless quantizer to be optimal and are known as the *Lloyd–Max conditions*. Although the conditions are very simple, an analytical solution to the optimal quantizer design is not possible except in some exceptional cases. Instead, the optimal quantizer is designed in an iterative manner as follows: start by specifying an arbitrary set of decision levels (for example the set that results in equal-length regions) and find the target values using (4.34). Then determine the new decision levels using (4.32). The two steps are iterated until the parameters do not change significantly from one step to the next.

Though optimal, a major disadvantage of the optimal quantizer is that it requires knowledge of the statistical properties of the message source, namely the pdf $f_{\mathbf{m}}(m)$ of the message amplitude. In practice, the quantizer in use may have to deal with a variety of sources. Another disadvantage is that the quantizer is designed for a specific $m_{\max}$, while typically the signal level varies, resulting in poor performance. These disadvantages prevent the use of the optimal quantizer in practical applications. In the next section, a different quantization method that overcomes these disadvantages and which is used is practice is examined. The method is quite robust to both the source statistics and changes in the signal's power level.

4.3.3 Robust quantizers

It can be easily verified from (4.32) and (4.34) that for the special case where the message signal is uniformly distributed, the optimal quantizer is a uniform quantizer. Thus, as long as the distribution of the message signal is close to uniform, the uniform quantizer works well. However, for certain signals such as voice, the input distribution is far from uniform. For a voice signal, in particular, there exists a higher probability for smaller amplitudes (corresponding to silent periods and soft speech) and a lower probability for larger amplitudes (corresponding to loud speech). Therefore it is more efficient to design a quantizer with more quantization regions at lower amplitudes and less quantization regions at larger amplitudes to overcome the variations in power levels that the quantizer sees at its input. The resulting quantizer would be, in essence, a *nonuniform* quantizer having quantization regions of various sizes.

The usual and robust method for performing nonuniform quantization is to first pass the continuous samples through a monotonic nonlinearity called a *compressor* that compresses the large amplitudes (which essentially reduces the dynamic range of the signal). One view of the compressor is that it acts like a variable-gain amplifier: it has high gain at low amplitudes and less gain at high amplitudes. The compressed signal is applied to a uniform quantizer. At the receiving end, the inverse of compression is carried out by the *expander* to obtain the sampled values. The combination of a *com*pressor and an ex*pander* is called a *compander*. Figure 4.11 shows the block diagram for this technique, where $g(m)$ and $g^{-1}(m)$ are the compressing and expanding functions, respectively.

Fig. 4.11 Block diagram of the compander technique.

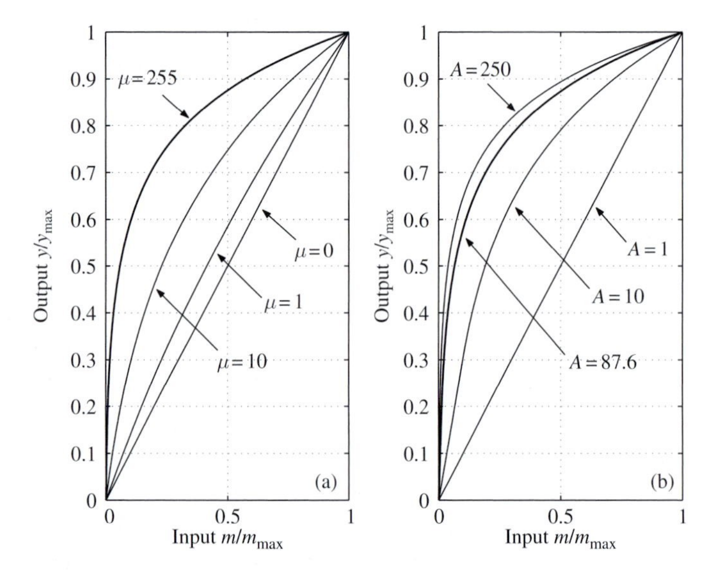

Fig. 4.12 Compression characteristic: (a) μ-law and (b) A-law. Note that both characteristics are *odd* functions and only the positive half of the input signal is illustrated.

There are two types of companders that are widely used for voice signals in practical telecommunication systems. The μ-law compander used in the USA, Canada, and Japan employs the following logarithmic compressing function:

$$y = y_{\max} \frac{\ln\left[1 + \mu\left(|m|/m_{\max}\right)\right]}{\ln(1+\mu)} \operatorname{sgn}(m), \tag{4.35}$$

where

$$\operatorname{sgn}(x) = \begin{cases} +1, & \text{for } x \geq 0 \\ -1, & \text{for } x < 0 \end{cases}. \tag{4.36}$$

In (4.35) $m_{\max}$ and $y_{\max}$ are the maximum positive levels of the input and output voltages respectively. The compression characteristic is shown in Figure 4.12(a) for several values of μ. Note that $\mu = 0$ corresponds to a linear amplification, i.e., there is no compression. In the USA and Canada, the parameter μ was originally set to 100 for use with a seven-bit PCM encoder. It was later changed to 255 for use with an eight-bit encoder.

Another compression characteristic, used mainly in Europe, Africa, and the rest of Asia, is the A law, defined as

$$y = \begin{cases} y_{\max} \dfrac{A\left(|m|/m_{\max}\right)}{1+\ln A} \operatorname{sgn}(m), & 0 < \dfrac{|m|}{m_{\max}} \leq \dfrac{1}{A} \\ y_{\max} \dfrac{1+\ln\left[A\left(|m|/m_{\max}\right)\right]}{1+\ln A} \operatorname{sgn}(m), & \dfrac{1}{A} < \dfrac{|m|}{m_{\max}} < 1 \end{cases}, \tag{4.37}$$

where A is a positive constant. The A-law compression characteristic is shown in Figure 4.12(b) for several values of A. Note that a standard value for A is 87.6.

4.3.4 SNR$_\mathrm{q}$ of nonuniform quantizers

Consider a general compression characteristic as shown in Figure 4.13, where $g(m)$ maps the interval $[-m_{\max}, m_{\max}]$ into the interval $[-y_{\max}, y_{\max}]$. Note that the output of the compressor is uniformly quantized. Let y_l and Δ denote, respectively, the target level and the (equal) step-size of the lth quantization region for the compressed signal y. Recall that for an L-level midrise quantizer one has $\Delta = 2y_{\max}/L$. The corresponding target level and step-size of the lth region for the original signal m are m_l and Δ_l respectively.

Assume that the number of quantization intervals L is very large ($L \gg 1$) and the density function of the message m is smooth enough. Then both Δ and Δ_l are small. Thus one can approximate $f_{\mathbf{m}}(m)$ to be a constant $f_{\mathbf{m}}(m_l)$ over Δ_l and, as a consequence, the target level

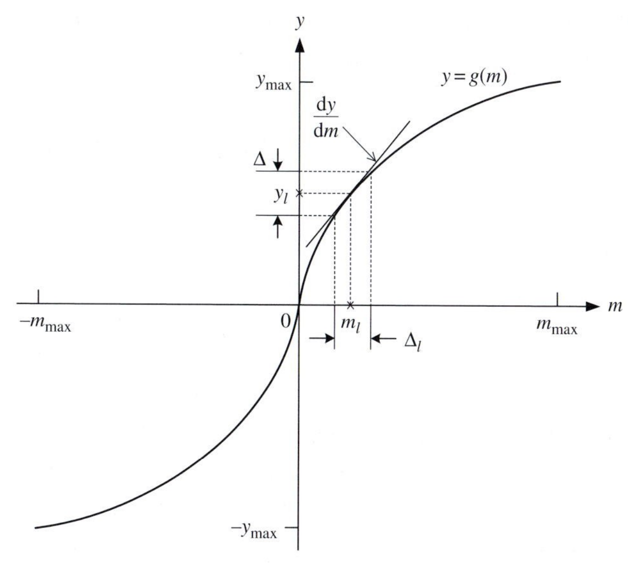

Fig. 4.13 Compression characteristic $g(m)$.

m_l is basically at the midpoint of the lth quantization region. With this approximation, N_q can be evaluated as follows:

$$N_q = \sum_{l=1}^{L} \int_{m_l-\Delta_l/2}^{m_l+\Delta_l/2} (m - m_l)^2 f_{\mathbf{m}}(m) \mathrm{d}m$$
$$\cong \sum_{l=1}^{L} f_{\mathbf{m}}(m_l) \int_{m_l-\Delta_l/2}^{m_l+\Delta_l/2} (m - m_l)^2 \mathrm{d}m$$
$$= \sum_{l=1}^{L} \frac{\Delta_l^3}{12} f_{\mathbf{m}}(m_l). \tag{4.38}$$

Furthermore, it can be seen from Figure 4.13 that Δ is related to Δ_l through the slope of $g(m)$ as follows:

$$\frac{\Delta}{\Delta_l} = \left.\frac{\mathrm{d}g(m)}{\mathrm{d}m}\right|_{m=m_l}. \tag{4.39}$$

Substituting Δ_l found from (4.39) into (4.38) produces

$$N_q = \frac{\Delta^2}{12} \sum_{l=1}^{L} \frac{f_{\mathbf{m}}(m_l)}{\left(\left.\frac{\mathrm{d}g(m)}{\mathrm{d}m}\right|_{m=m_l}\right)^2} \Delta_l. \tag{4.40}$$

Finally, since $L \gg 1$ one can approximate the summation by an integral to obtain

$$N_q = \frac{\Delta^2}{12} \int_{-m_{\max}}^{m_{\max}} \frac{f_{\mathbf{m}}(m)}{(\mathrm{d}g(m)/\mathrm{d}m)^2} \mathrm{d}m = \frac{y_{\max}^2}{3L^2} \int_{-m_{\max}}^{m_{\max}} \frac{f_{\mathbf{m}}(m)}{(\mathrm{d}g(m)/\mathrm{d}m)^2} \mathrm{d}m. \tag{4.41}$$

Example 4.1 (SNR_q of the μ-law compander) As an example, let us determine SNR_q for the μ-law compander. To this end, evaluate the derivative of $g(m)$ in (4.35) to get

$$\frac{\mathrm{d}g(m)}{\mathrm{d}m} = \frac{y_{\max}}{\ln(1+\mu)} \frac{\mu(1/m_{\max})}{1 + \mu(|m|/m_{\max})}. \tag{4.42}$$

Note that the derivative of $g(m)$ is an even function. This is expected from the symmetry of the μ-law compression characteristic. Now substituting (4.42) into (4.41) gives

$$N_q = \frac{y_{\max}^2}{3L^2} \frac{\ln^2(1+\mu)}{y_{\max}^2 (\mu/m_{\max})^2} \int_{-m_{\max}}^{m_{\max}} \left[1 + \mu\left(\frac{|m|}{m_{\max}}\right)\right]^2 f_{\mathbf{m}}(m) \mathrm{d}m$$
$$= \frac{m_{\max}^2}{3L^2} \frac{\ln^2(1+\mu)}{\mu^2}$$
$$\times \int_{-m_{\max}}^{m_{\max}} \left[1 + 2\mu\left(\frac{|m|}{m_{\max}}\right) + \mu^2\left(\frac{|m|}{m_{\max}}\right)^2\right] f_{\mathbf{m}}(m) \mathrm{d}m. \tag{4.43}$$

Then since $\int_{-m_{\max}}^{m_{\max}} f_{\mathbf{m}}(m)\mathrm{d}m = 1$, $\int_{-m_{\max}}^{m_{\max}} m^2 f_{\mathbf{m}}(m)\mathrm{d}m = \sigma_{\mathbf{m}}^2$, and $\int_{-m_{\max}}^{m_{\max}} |m| f_{\mathbf{m}}(m)\mathrm{d}m = E\{|\mathbf{m}|\}$, the quantization noise power can be rewritten as,

$$N_{\mathrm{q}} = \frac{m_{\max}^2}{3L^2}\frac{\ln^2(1+\mu)}{\mu^2}\left[1 + 2\mu\frac{E\{|\mathbf{m}|\}}{m_{\max}} + \mu^2\frac{\sigma_{\mathbf{m}}^2}{m_{\max}^2}\right]. \tag{4.44}$$

Finally, by noting that the average power of the message signal is $\sigma_{\mathbf{m}}^2$, the $\mathrm{SNR}_{\mathrm{q}}$ of the μ-law compander is given by

$$\mathrm{SNR}_{\mathrm{q}} = \frac{\sigma_{\mathbf{m}}^2}{N_{\mathrm{q}}} = \frac{3L^2\mu^2}{\ln^2(1+\mu)}\frac{(\sigma_{\mathbf{m}}^2/m_{\max}^2)}{1 + 2\mu(E\{|\mathbf{m}|\}/m_{\max}) + \mu^2(\sigma_{\mathbf{m}}^2/m_{\max}^2)}. \tag{4.45}$$

To express $\mathrm{SNR}_{\mathrm{q}}$ as a function of the normalized power level $\sigma_n^2 = \sigma_{\mathbf{m}}^2/m_{\max}^2$, rewrite the term $E\{|\mathbf{m}|\}/m_{\max}$ in the denominator as

$$\frac{E\{|\mathbf{m}|\}}{\sigma_{\mathbf{m}}}\frac{\sigma_{\mathbf{m}}}{m_{\max}} = \frac{E\{|\mathbf{m}|\}}{\sigma_{\mathbf{m}}}\sigma_n.$$

Therefore,

$$\mathrm{SNR}_{\mathrm{q}}(\sigma_n^2) = \frac{3L^2\mu^2}{\ln^2(1+\mu)}\frac{\sigma_n^2}{1 + 2\mu\sigma_n E\{|\mathbf{m}|\}/\sigma_{\mathbf{m}} + \mu^2\sigma_n^2}. \tag{4.46}$$

Equation (4.46) shows that the $\mathrm{SNR}_{\mathrm{q}}$ of the μ-law compander depends on the statistics of the message through $E\{|\mathbf{m}|\}/\sigma_{\mathbf{m}}$. For example, for a message with Gaussian density $E\{|\mathbf{m}|\}/\sigma_{\mathbf{m}} = \sqrt{2/\pi} = 0.798$, and $E\{|\mathbf{m}|\}/\sigma_{\mathbf{m}} = 1/\sqrt{2} = 0.707$ for a Laplacian-distributed message.

Furthermore, if $\mu \gg 1$ then the dependence of $\mathrm{SNR}_{\mathrm{q}}$ on the message's characteristics is very small and $\mathrm{SNR}_{\mathrm{q}}$ can be approximated as

$$\mathrm{SNR}_{\mathrm{q}} = \frac{3L^2}{\ln^2(1+\mu)}. \tag{4.47}$$

For practical values of $\mu = 255$ and $L = 256$, one has $\mathrm{SNR}_{\mathrm{q}} = 38.1$ decibels. ■

To compare the μ-law quantizer with the uniform quantizer, Figure 4.14 plots the $\mathrm{SNR}_{\mathrm{q}}$ of the μ-law quantizer and the uniform quantizer for the Gaussian-distributed message over the same range of normalized input power. As can be seen from this figure, the μ-law quantizer can maintain a fairly constant $\mathrm{SNR}_{\mathrm{q}}$ over a wide range of input power levels (it is also true for different input pdfs). In contrast, the $\mathrm{SNR}_{\mathrm{q}}$ of the uniform quantizer decreases linearly as the input power level drops. Note that at $\sigma_{\mathbf{m}}^2/m_{\max}^2 = 0$ decibels, the $\mathrm{SNR}_{\mathrm{q}}$s for the μ-law and uniform quantizers are 38.1 decibels and 52.9 decibels respectively. Thus with the μ-law quantizer, one sacrifices performance for larger input power levels to obtain a performance that remains robust over a wide range of input levels.

4.3.5 Differential quantizers

In the quantizers looked at thus far each signal sample is quantized independently of all the others. However, most message signals (such as voice or video signals) sampled at the

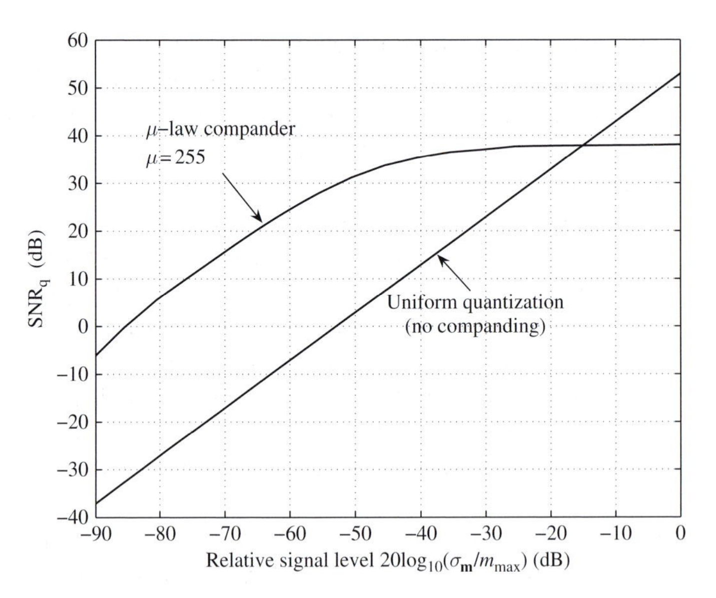

Fig. 4.14 $\mathrm{SNR_q}$ of an eight-bit quantizer with and without companding for the Gaussian-distributed message.

Nyquist rate or faster exhibit a high degree of correlation between successive samples. This essentially means that the signal does not change rapidly from one sample to the next. This redundancy can be exploited to obtain a better $\mathrm{SNR_q}$ for a given number of levels, L, or conversely for a specified $\mathrm{SNR_q}$ the number of levels, L, can be reduced. The number of bits, R, needed to represent L levels is $R = \log_2 L$. Reducing L means that the number of bits needed to represent a sample is reduced and hence the bit rate is reduced. A lower bit rate eventually implies that less bandwidth is needed by the communication system.

To motivate the discussion further two examples are considered: one is a toy example which illustrates how redundancy helps in reducing bit rate, the other leads one into the major topic of this section, *differential quantizers*. First let us consider, as a toy example, a message source whose output is the random process $\mathbf{m}(t) = \mathbf{A}\mathrm{e}^{-\mathbf{c}t}u(t)$, where $\mathbf{A}$ and $\mathbf{c}$ are random variables with arbitrary pdfs. In any transmission the output of the message source is a specific member of the ensemble, i.e., the signal $m(t) = A\mathrm{e}^{-ct}u(t)$, where A and c are now specific, but unknown, values. One could, of course, sample this signal at a chosen rate T_s, quantize these samples, and transmit the resultant sequence of samples. However, after some thought one could equally just quantize two samples, say at $t = 0$ and $t = t_1 > 0$, and transmit the quantized values $m(0) = A$, $m(t_1) = A\mathrm{e}^{ct_1}$. At the receiver the two sample values are used to determine the values of A and c, which are then used to reconstruct the entire waveform.

Of course, the above is an extreme example of redundancy in the message waveform. Indeed once one has the first two samples, then one can predict exactly the sample values at all other sampling times. In the more practical case one does not have good analytical expressions for how the waveform is generated to enable this perfect prediction. The typical information one has about the redundancy in the samples, as mentioned above, is the correlation between them. The approach, as shown in Figure 4.15 where $\mathbf{m}[n]$ is employed instead of $\mathbf{m}(nT_s)$ to refer to the sampled values, is to use the previous sample values to predict the next sample value and then transmit the difference. Recall from (4.23) that the quantization noise, $\sigma_{\mathbf{q}}^2$, depends directly on the message range, i.e., $m_{\max}$. The message being quantized and transmitted now is the prediction error, $\mathbf{e}[n] = \mathbf{m}[n] - \tilde{\mathbf{m}}[n] = \mathbf{m}[n] - k\mathbf{m}[n]$. Therefore if $|e_{\max}| = |m_{\max} - km_{\max}| = |1-k|m_{\max}$ is less than $m_{\max}$, then the quantization noise power is reduced. Therefore the predictor should be such that $0 < k < 2$, ideally $k = 1$, hopefully $k \approx 1$. The approach just described is called differential quantization. The main design issue for this quantizer is how to predict the message sample value. Ideally one would base the design on a criterion that minimizes $e_{\max}$ but this turns out to be intractable. Thus the predictor design is based on the minimization of the error variance, $\sigma_{\mathbf{e}}^2$. Further, a linear structure is imposed on the predictor since it is straightforward to design, quite practical to realize, and performs well.

Linear predictor A linear predictor forecasts the current sample based on a weighted sum of the previous p samples. It is called a pth-order linear predictor. As can be seen in Figure 4.16, the predicted sample, $\tilde{\mathbf{m}}[n]$, is given by

$$\tilde{\mathbf{m}}[n] = \sum_{i=1}^{p} w_i \mathbf{m}[n-i]. \tag{4.48}$$

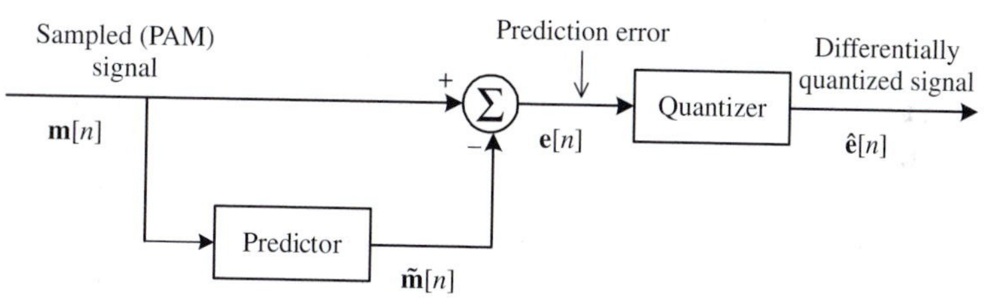

Fig. 4.15 Illustration of a differential quantizer.

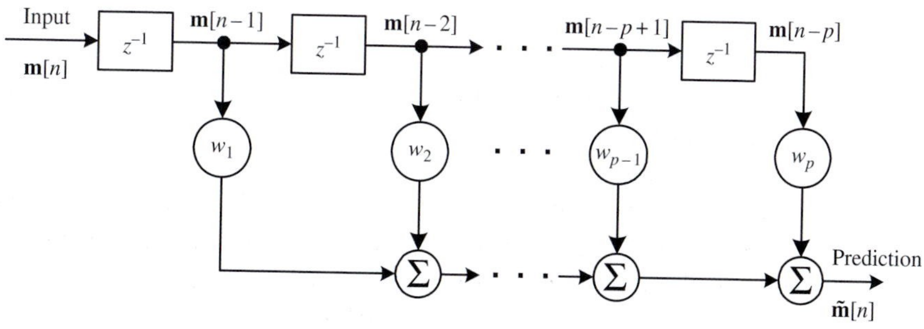

Fig. 4.16 Block diagram of a pth-order linear predictor.

The blocks labeled with z^{-1} in Figure 4.16 signify unit-delay elements (z-transform notation).

In general, the coefficients $\{w_i\}$ are selected to *minimize* some function of the error between $\mathbf{m}[n]$ and $\tilde{\mathbf{m}}[n]$. However, as mentioned it is the average power of the prediction error that is of interest. The coefficient set $\{w_i\}$ is selected to minimize

$$\begin{aligned}\sigma_{\mathbf{e}}^2 = E\{\mathbf{e}^2[n]\} &= E\left\{\left(\mathbf{m}[n] - \sum_{i=1}^{p} w_i\mathbf{m}[n-i]\right)^2\right\} \\ &= E\{\mathbf{m}^2[n]\} - 2\sum_{i=1}^{p} w_i E\{\mathbf{m}[n]\mathbf{m}[n-i]\} \\ &\quad + \sum_{i=1}^{p}\sum_{j=1}^{p} w_i w_j E\{\mathbf{m}[n-i]\mathbf{m}[n-j]\}.\end{aligned} \tag{4.49}$$

Since the quantity $E\{\mathbf{m}[n]\mathbf{m}[n+k]\}$ is the *autocorrelation* function, $R_{\mathbf{m}}(k)$, of the sampled signal sequence $\{\mathbf{m}[n]\}$, (4.49) can be expressed as follows:

$$\sigma_{\mathbf{e}}^2 = R_{\mathbf{m}}(0) - 2\sum_{i=1}^{p} w_i R_{\mathbf{m}}(i) + \sum_{i=1}^{p}\sum_{j=1}^{p} w_i w_j R_{\mathbf{m}}(i-j). \tag{4.50}$$

Now take the partial derivative of $\sigma_{\mathbf{e}}^2$ with respect to each coefficient w_i and set the result to zero to yield a set of linear equations:

$$\sum_{i=1}^{p} w_i R_{\mathbf{m}}(i-j) = R_{\mathbf{m}}(j), \quad j = 1, 2, \ldots, p. \tag{4.51}$$

The above collection of equations can be arranged in matrix form as

$$\begin{bmatrix} R_{\mathbf{m}}(0) & R_{\mathbf{m}}(1) & R_{\mathbf{m}}(2) & \cdots & R_{\mathbf{m}}(p-1) \\ R_{\mathbf{m}}(-1) & R_{\mathbf{m}}(0) & R_{\mathbf{m}}(1) & \cdots & R_{\mathbf{m}}(p-2) \\ R_{\mathbf{m}}(-2) & R_{\mathbf{m}}(-1) & R_{\mathbf{m}}(0) & \cdots & R_{\mathbf{m}}(p-3) \\ \vdots & \vdots & \vdots & \ddots & \vdots \\ R_{\mathbf{m}}(-p+1) & R_{\mathbf{m}}(-p+2) & R_{\mathbf{m}}(-p+3) & \cdots & R_{\mathbf{m}}(0) \end{bmatrix} \begin{bmatrix} w_1 \\ w_2 \\ w_3 \\ \vdots \\ w_p \end{bmatrix} = \begin{bmatrix} R_{\mathbf{m}}(1) \\ R_{\mathbf{m}}(2) \\ R_{\mathbf{m}}(3) \\ \vdots \\ R_{\mathbf{m}}(p) \end{bmatrix}. \tag{4.52}$$

This equation set is also known as the *normal equations* or the *Yule–Walker equations*.

The remaining design issue is the choice of p, the predictor's order. In practice, for voice signals it has been found that the greatest improvement happens when one goes from no prediction to a first-order prediction, i.e., $p = 1$, but for other sources one should be prepared to experiment.

Reconstruction of m[n] from the differential samples Though at the receiver one sees the quantized samples, $\hat{\mathbf{e}}[n]$, for discussion purposes we assume that the quantization noise or error is negligible. The distortion introduced by the quantizer is ignored and therefore we look at the reconstruction of $\mathbf{m}[n]$ from the differential samples $\mathbf{e}[n]$. For the pth-order linear predictor the differential sample $\mathbf{e}[n]$ is related to $\mathbf{m}[n]$, using (4.48), by

$$\mathbf{e}[n] = \mathbf{m}[n] - \sum_{i=1}^{p} w_i \mathbf{m}[n-i]. \tag{4.53}$$

Using the z-transform, the above relationship becomes

$$\begin{aligned} \mathbf{e}(z^{-1}) &= \mathbf{m}(z^{-1}) - \sum_{i=1}^{p} w_i z^{-i} \mathbf{m}(z^{-1}) \\ &= \mathbf{m}(z^{-1}) - \mathbf{m}(z^{-1}) \sum_{i=1}^{p} w_i z^{-i} \\ &= \mathbf{m}(z^{-1}) - \mathbf{m}(z^{-1}) H(z^{-1}), \end{aligned} \tag{4.54}$$

where $H(z^{-1}) = \sum_{i=1}^{p} w_i z^{-i}$ is the transfer function of the linear predictor.

Equation (4.54) states that $\mathbf{e}(z^{-1}) = \mathbf{m}(z^{-1}) \left[1 - H(z^{-1})\right]$, or

$$\mathbf{m}(z^{-1}) = \frac{1}{1 - H(z^{-1})} \mathbf{e}(z^{-1}). \tag{4.55}$$

Recall from feedback control theory that the input–output relationship for negative feedback is $G/(1 + GH)$, where G is the forward loop gain and H is the feedback loop gain. Comparing this with (4.55) one concludes that $G = 1$ and $H = -H(z^{-1})$. Therefore $\mathbf{m}(z^{-1})$ is reconstructed by the block diagram of Figure 4.17. Note that even if one has forgotten or has never learned feedback control theory one should be able to convince oneself that the block diagram on the right in Figure 4.17 reconstructs $\mathbf{m}[n]$.

However, the input to the reconstruction filter is not $\mathbf{e}[n]$ but $\hat{\mathbf{e}}[n]$, which is equal to $\mathbf{e}[n] + \mathbf{q}[n]$, where $\mathbf{q}[n]$ is the quantization noise or error. The difficulty this poses with the proposed reconstruction filter is that not only does the noise in the present differential sample affect the reconstructed signal but so do all the previous quantization noise samples, though with diminishing effect. In essence this is because the reconstruction filter has infinite memory. To add to this effect one should also be aware that the channel will also induce noise in $\mathbf{e}[n]$.

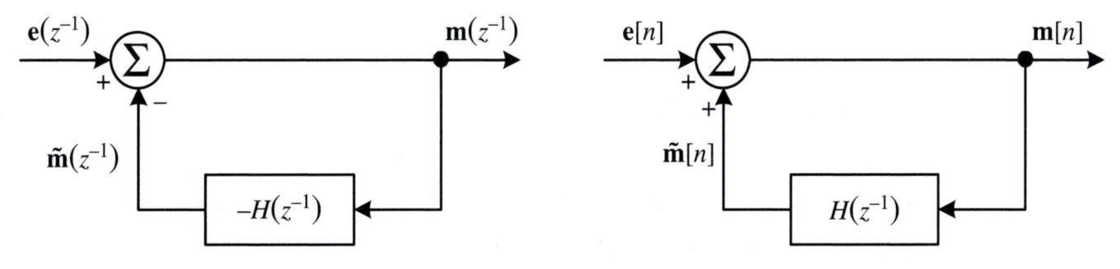

Fig. 4.17 Reconstruction of $\mathbf{m}[n]$ from $\mathbf{e}[n]$.

A solution that would eliminate this effect is quite desirable. Since engineers are supposed to be, if anything, ingenious, this is precisely what is accomplished by the system shown in Figure 4.18, known as differential pulse-code modulation (DPCM). The pulse-code aspect is discussed in the next section but here we show that indeed $\hat{\mathbf{m}}[n]$ is only affected by the quantization noise, $\mathbf{q}[n]$. The analysis shows that this is true regardless of the predictor used, it can even be nonlinear, and the conclusion is also independent of the form of the quantizer. This is truly ingenious.

Referring to Figure 4.18(a), the input signal to the quantizer is given by

$$\mathbf{e}[n] = \mathbf{m}[n] - \tilde{\mathbf{m}}[n], \tag{4.56}$$

which is the difference between the unquantized sample $\mathbf{m}[n]$ and its prediction, denoted by $\tilde{\mathbf{m}}[n]$. The predicted value can be obtained using a linear prediction filter, whose input is $\hat{\mathbf{m}}[n]$. The output of the quantizer is then encoded to produce the DPCM signal. The quantizer output can be expressed as

$$\hat{\mathbf{e}}[n] = \mathbf{e}[n] - \mathbf{q}[n], \tag{4.57}$$

where $\mathbf{q}[n]$ is the quantization error. According to Figure 4.18(a), the input to the predictor can be written as follows:

$$\begin{aligned}\hat{\mathbf{m}}[n] &= \tilde{\mathbf{m}}[n] + \hat{\mathbf{e}}[n] = \tilde{\mathbf{m}}[n] + (\mathbf{e}[n] - \mathbf{q}[n]) \\ &= (\tilde{\mathbf{m}}[n] + \mathbf{e}[n]) - \mathbf{q}[n] = \mathbf{m}[n] - \mathbf{q}[n].\end{aligned} \tag{4.58}$$

Since $\mathbf{q}[n]$ is the quantization error, (4.58) implies that $\hat{\mathbf{m}}[n]$ is just the quantized version of the input sample $\mathbf{m}[n]$.

The receiver for a DPCM system is shown in Figure 4.18(b). In the absence of channel noise, the output of the decoder is identical to the input of the encoder (at the transmitter), which is $\hat{\mathbf{e}}[n]$. If the predictor at the receiver is the same as the one in the transmitter,

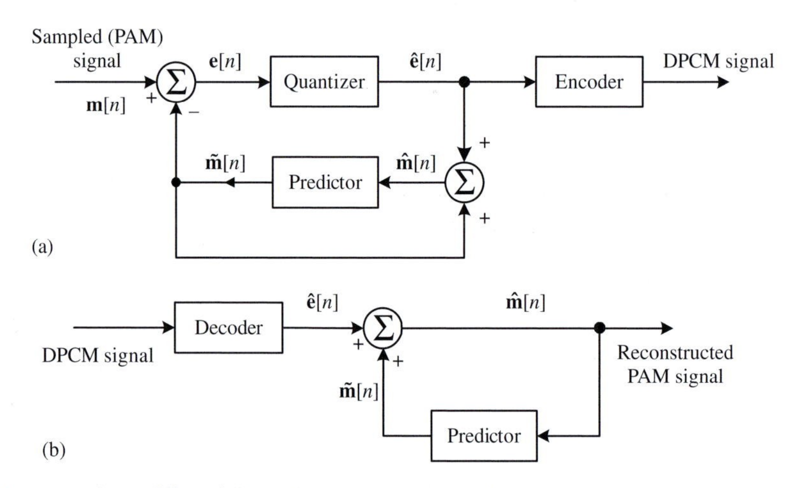

Fig. 4.18 DPCM system using a differential quantizer: (a) transmitter and (b) receiver.

then its output should be equal to $\widetilde{\mathbf{m}}[n]$. It follows that the final output of the receiver is $\hat{\mathbf{e}}[n] + \widetilde{\mathbf{m}}[n]$, which is equal to $\hat{\mathbf{m}}[n]$. Thus, according to (4.58), the output of the receiver differs from the original input $\mathbf{m}[n]$ only by the quantization error $\mathbf{q}[n]$ incurred as a result of quantizing the prediction error. Note that the quantization errors do not accumulate in the receiver.

4.4 Pulse-code modulation (PCM)

The last block in Figure 4.7 to be discussed is the *encoder*. A PCM signal is obtained from the quantized PAM signal by encoding each quantized sample to a *digital codeword*. If the PAM signals are quantized using L target levels, then in binary PCM each quantized sample is digitally encoded into an R-bit binary codeword, where $R = \lceil \log_2 L \rceil + 1$. The quantizing and encoding operations are usually performed in the same circuit known as an A/D converter. The advantage of having a PCM signal over a quantized PAM signal is that the binary digits of a PCM signal can be transmitted using many efficient modulation schemes compared to the transmission of a PAM signal. The topics of baseband and passband modulation of binary digits are covered in Chapters 6 and 7.

There are several ways to establish a one-to-one correspondence between target levels and the codeword. A convenient method, known as natural binary coding (NBC), is to express the ordinal number of the target level as a binary number as described in Figure 4.19. Another popular mapping method is called *Gray* mapping. Gray mapping is important in the demodulation of the signal because the most likely errors caused by noise involve the erroneous selection of a target level that is adjacent to the transmitted target level. Still another mapping, called foldover binary coding (FBC), is also sometimes encountered.

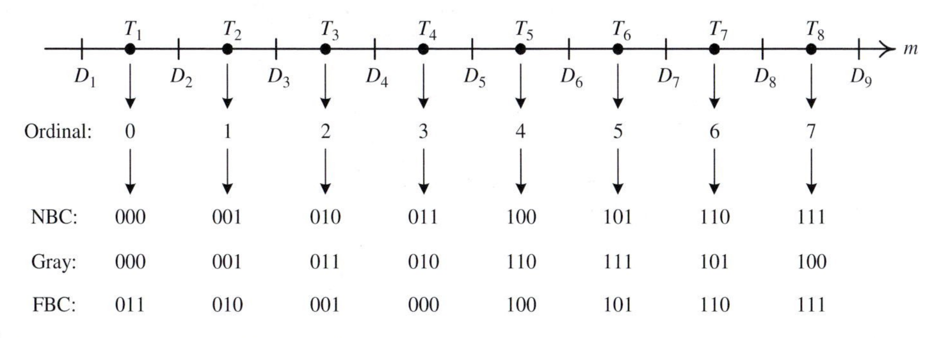

Fig. 4.19 Encoding of quantized levels into PCM codewords, $L = 8$.

4.5 Summary

The basic concepts behind converting analog information to digital have been presented in this chapter. The first step in this process is to sample the continuous-time output of the analog source. The sampling theorem shows that if the sampling rate is at least twice the highest frequency present in the analog signal, then the original signal can be recovered exactly.

However, not only does the time axis need to be discretized, the analog amplitude values of the samples also must be quantized. Since this step is a many-to-one mapping, it inherently produces a distortion or a loss. Various quantizers to minimize this distortion, with the performance measure of the SNR_q (signal power to quantization noise power ratio), have been analyzed and discussed.

The final step, encoding, is to map quantized sample values to a set of discrete symbols, typically a sequence of binary digits called variously a binary string, binary word, or simply codeword. The codeword, however, is an abstract quantity, i.e., a nonphysical entity. To transmit the information that a codeword represents one needs to map each codeword to an electrical signal or waveform. As has been mentioned several times, this mapping is called modulation and it, along with demodulation, is the subject of succeeding chapters. To conclude we are finally in a position to discuss digital communications.

However, a few further closing remarks are in order. The quantization and encoding discussed in this chapter go under the general topic of source coding. But some sources, e.g., the keyboard, are already discrete in nature. The sampling/quantization steps are not necessary for them and only the encoding step, that of converting the discrete source symbols to a codeword, is required. Traditional source coding is concerned primarily with this step. Some aspects of this are presented in the problems at the end of the chapter but the interested reader is referred to the literature for a more comprehensive treatment.

4.6 Problems

4.1 Consider signal $s(t) = 5\cos(1000\pi t) + 2\cos(3600\pi t)$ (volts) which is ideally sampled at a frequency of $f_s = 2000$ hertz. The sampled signal is then passed through an ideal LPF. What is the output signal if the filter's bandwidth is (a) 1 kilohertz, and (b) 2 kilohertz.

4.2 The sampling theorem states that a bandlimited signal, $m(t)$, can be expressed as

$$m(t) = T_s \sum_{n=-\infty}^{\infty} m(nT_s) \frac{\sin[2\pi W(t - nT_s)]}{\pi(t - nT_s)}, \tag{P4.1}$$

where W is the bandwidth and $2T_s = 1/W$. Show that

$$\left\{\sqrt{T_s}\frac{\sin[2\pi W(t-nT_s)]}{\pi(t-nT_s)}, n = 0, \pm 1, \pm 2, \ldots\right\}$$

is an orthonormal set.

Hint Use Parseval's theorem which states that $\int_{-\infty}^{\infty} s_1(t)s_2^*(t)\mathrm{d}t = \int_{-\infty}^{\infty} S_1(f) S_2^*(f)\mathrm{d}f$.

4.3 Consider a white Gaussian process, $\mathbf{w}(t)$, of spectral strength $N_0/2$ (watts/hertz). A discrete random process $\{\mathbf{w}_{\text{out}}(kT_s)\}$, where $k \in$ integer, is generated by passing $\mathbf{w}(t)$ through an ideal LPF of bandwidth W hertz to obtain the signal $\mathbf{w}_{\text{out}}(t)$ and sampling $\mathbf{w}_{\text{out}}(t)$ at the Nyquist sampling rate, $T_s = 1/(2W)$.

(a) The output of the LPF is a correlated process. Determine the correlation function of $\mathbf{w}_{\text{out}}(t)$.

(b) Does the correlation function depend on the fact that $\mathbf{w}(t)$ is Gaussian? However, when the input is Gaussian, is $\mathbf{w}_{\text{out}}(t)$ then Gaussian? Why?

(c) Show that the sampled process $\{\mathbf{w}_{\text{out}}(kT_s)\}$ is a set of statistically independent random variables. Is this true when the input is still white but no longer Gaussian?

(d) If the ideal LPF is unchanged but the sampling rate is increased or decreased, are the samples still statistically independent? Uncorrelated?

4.4 Common input voltage ranges for commercial A/D converters (i.e., quantizers) are ± 1 volt, ± 5 volts, ± 10 volts. Take an A/D converter with a ± 1 volt range and compute the number of levels and the step size if it is (a) 8-bit, (b) 12-bit, and (c) 16-bit.

The next set of problems principally deals with the optimum quantizer. Recall that the equations to be solved for the target levels and thresholds of the optimum quantizer are:

$$D_l = \frac{T_{l-1} + T_l}{2}, \quad l = 2, \ldots, L, \tag{P4.2}$$

$$T_l = \frac{\int_{D_l}^{D_{l+1}} m f_{\mathbf{m}}(m)\mathrm{d}m}{\int_{D_l}^{D_{l+1}} f_{\mathbf{m}}(m)\mathrm{d}m}, \quad l = 1, \ldots, L. \tag{P4.3}$$

4.5 Consider a source, $\mathbf{m}(t)$, whose amplitude statistics are as follows:

$$f_{\mathbf{m}}(m) = \begin{cases} \frac{1}{4}, & -1 \le m \le +1 \\ \frac{1}{12}, & -4 \le m \le -1 \\ \frac{1}{12}, & +1 \le m \le +4 \\ 0, & \text{otherwise} \end{cases}. \tag{P4.4}$$

(a) Design an optimum three-bit quantizer.

(b) Determine the SNR_q. *Hint* Try to solve the problem graphically and use (P4.2) and (P4.3) to confirm your solution.

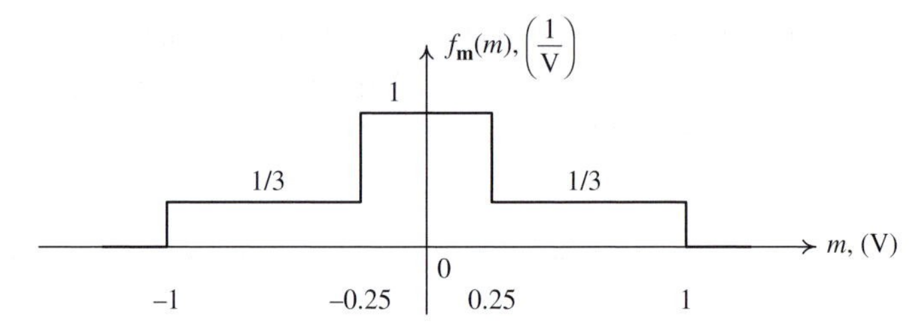

Fig. 4.20 Amplitude pdf considered in Problem 4.6.

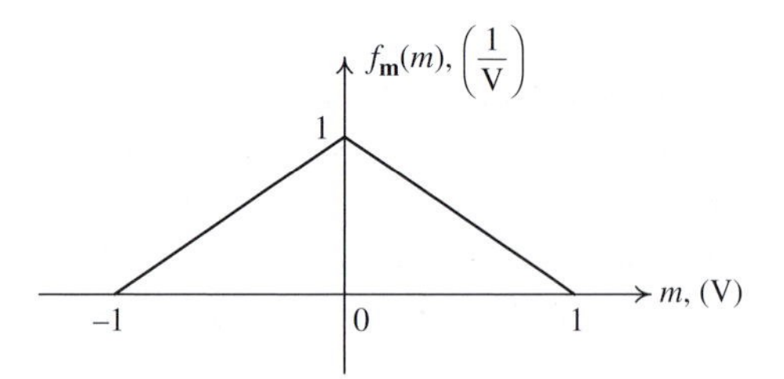

Fig. 4.21 Amplitude pdf considered in Problem 4.7.

4.6 Consider a message with the amplitude pdf shown in Figure 4.20. Derive the set of equations that the optimum two-bit quantizer should satisfy and solve them for the threshold and target values. *Hint* Exploit the symmetry and assume one of the thresholds to be greater than 0.25.

4.7 Consider a signal with the amplitude pdf shown in Figure 4.21. Determine the target and decision levels and SNR_q for the following quantizers:
(a) one-bit ($L = 2$ levels) *uniform* quantizer.
(b) one-bit *optimum* quantizer.
(c) Repeat (a) and (b) for a two-bit quantizer. Derive the equations to be solved for the optimum target and decision levels and check the following answers: $T_1 = 0.176$, $T_2 = 0.588$ (two positive target levels) and $D_1 = 0.382$ (a positive decision level).
(d) Repeat (a) for a three-bit quantizer. For the optimum quantizer simply derive the set of equations to be solved.
Remark Use symmetry to reduce the number of equations.

4.8 Consider a message source that with confidence you feel is well modeled by a first-order Laplacian density function, i.e., $f_{\mathbf{m}}(m) = \frac{1}{2}\text{c}e^{-c|m|}$.
(a) Derive the general equation for the target level T_l for an optimum quantizer. For simplicity let $c = 1$. However, once you obtain a solution for the thresholds and target levels with $c = 1$, you should be able to obtain a solution for any c. How? Also because of the symmetry of $f_{\mathbf{m}}(m)$, you only need to determine the equation for $m > 0$.
(b) Find the target levels, thresholds, and SNR_q for a one-bit *uniform* quantizer.
(c) Find the target levels, thresholds, and SNR_q for a one-bit *optimum* quantizer.

(d) Repeat (b) and (c) for two-bit uniform and optimum quantizers. Note that, for the optimum quantizer, all that need be determined are T_3, D_3 and T_4 (see Figure 4.22). Obtain an equation for D_3 and solve it.

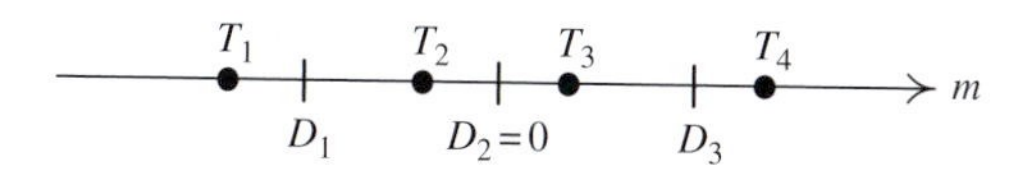

Fig. 4.22 Thresholds and target levels of the two-bit optimum quantizer considered in Problem 4.8.

4.9 For a very coarse quantization of one and two bits obtaining the solutions for the target values and thresholds can be done explicitly. However, if $L = 256$ or larger solving the set of coupled simultaneous nonlinear algebraic equations poses a challenge. The following attempts to set up a recursive algorithm to obtain the solution. It is illustrated by the message model of Problem 4.8 and for $L = 4$. It is as follows.

(i) Guess D_3 and determine T_3 so that it is the centroid of $f_{\mathbf{m}}(m)$ over the interval 0 to D_3.

(ii) Compute T_4 using the optimum equation $D_3 = (T_3 + T_4)/2$. Call this solution $T_4^{(*)}$.

(iii) Compute T_4 to be the centroid of $f_{\mathbf{m}}(m)$ over the interval D_3 to infinity. Call this solution $T_4^{(**)}$.

(iv) Compare $T_4^{(*)}$ (obtained after the first iteration) and $T_4^{(**)}$ (obtained from the centroid condition). There are three possibilities:

(a) $T_4^{(*)} = T_4^{(**)}$. In this case heave a sigh of relief, say eureka and pat yourself on the back for either your tremendous insight or lucky guess. Why?

(b) $T_4^{(*)} < T_4^{(**)}$.

(c) $T_4^{(*)} > T_4^{(**)}$.

If (b) or (c) occurs you obviously have not been that lucky and will have to try again. Develop some method of correcting the initial guess T_3 and repeat steps (ii)–(iv). Continue until you are comfortable that you have a good engineering solution which means that you will have to come up with a stopping criterion.

Remark An initial guess of $D_3 = 1.5$ is suggested.

4.10 Generalize the recursive algorithm outlined in Problem 4.9 to any number of levels, L. Write a Matlab program that implements it for the message model of Problem 4.8 and $L = 256$.

4.11 Usually the number of levels, L, in a quantizer is quite large, which implies that $\Delta = 2m_{\max}/L$ is "small." Thus you feel justified in approximating $f_{\mathbf{m}}(m)$ by a stepwise approximation which graphically looks as in Figure 4.23. Based on $f_{\mathbf{m}}^{(a)}(m)$ what is the optimum quantizer?

4.12 Since a uniform quantizer is optimum for a message that has a uniform amplitude pdf and since commercial A/D converters use uniform quantizers, as a hotshot newly

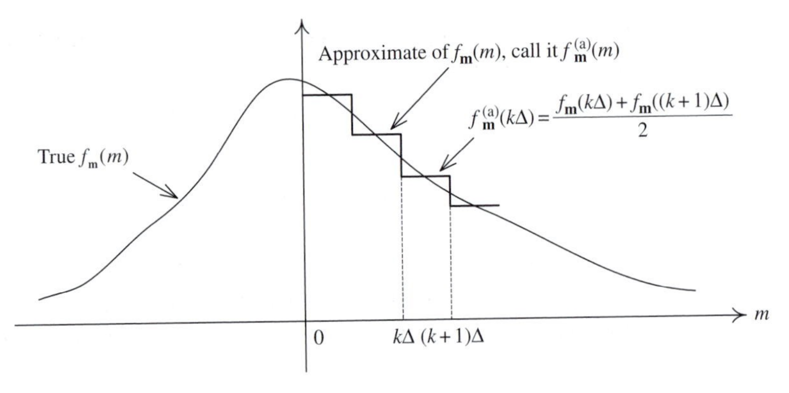

Fig. 4.23 Stepwise approximation of $f_{\mathbf{m}}(m)$ considered in Problem 4.11.

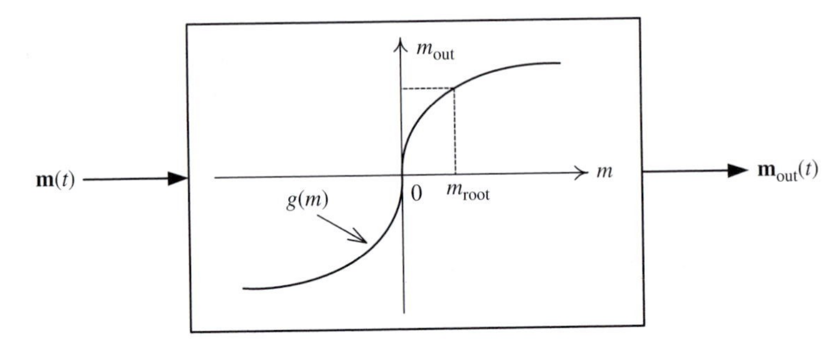

Fig. 4.24 Nonlinearity considered in Problem 4.12.

graduated engineer you propose the following: pass the message $\mathbf{m}(t)$ through a nonlinearity, $g(m)$, which produces a signal $\mathbf{m}_{\text{out}}(t)$ that is uniform over the interval $[0, 1]$. Since you eventually wish to recover $\mathbf{m}(t)$ the nonlinearity of necessity must be invertible, i.e., monotonically increasing. In block diagram form, it looks as in Figure 4.24.

From probability theory you know that

$$f_{\mathbf{m}_{\text{out}}}(m_{\text{out}}) = \frac{f_{\mathbf{m}}(m)\big|_{m=m_{\text{root}}}}{|\mathrm{d}g(m)/\mathrm{d}m|_{m=m_{\text{root}}}}. \tag{P4.5}$$

If $f_{\mathbf{m}_{\text{out}}}(m_{\text{out}})$ is to be uniform over the interval $[0, 1]$ what must be the RHS of this equation over this interval? From this observation conclude what the form of the nonlinearity $g(\cdot)$ is. However, it is usually desired to have a uniform density over the interval $[-1, 1]$. To conclude the design, how should $\mathbf{m}_{\text{out}}(t)$ be processed to achieve this.

Excited by the above you present the idea and design to your supervisor. She, however, refuses to okay the project. Why?

4.13 (*Crest factor*) A signal parameter of interest in designing a quantizer is the crest factor F, defined as

$$F = \frac{\text{peak value of the signal}}{\text{RMS value of the signal}}. \tag{P4.6}$$

Determine the crest factors for the following signals:

(a) A sinusoid with peak amplitude of $m_{\max}$.

(b) A square wave of period T and amplitude range $[-m_{\max}, m_{\max}]$.

The above are deterministic signals and most of you are quite familiar with their crest factors. Now consider random signals with the following pdfs and determine their crest factors.

(c) Uniform over the amplitude range $[-m_{\max}, m_{\max}]$.

(d) Zero-mean Gaussian:

$$f_{\mathbf{m}}(m) = \frac{1}{\sqrt{2\pi\sigma_{\mathbf{m}}^2}} \exp\left(-\frac{m^2}{2\sigma_{\mathbf{m}}^2}\right).$$

(e) Zero-mean Laplacian:

$$f_{\mathbf{m}}(m) = \frac{c}{2}\exp(-c|m|).$$

(f) Zero-mean Gamma:

$$f_{\mathbf{m}}(m) = \sqrt{\frac{k}{4\pi|m|}}\exp(-k|m|)$$

The Gamma pdf is sometimes used to model the voice signal. The following should be of use: $\int_0^\infty x^{\nu-1}\mathrm{e}^{-\lambda x}\mathrm{d}x = \Gamma(\nu)/\lambda^\nu$, where $\Gamma(\nu)$ is the Gamma function with the property $\Gamma(\nu) = (\nu-1)\Gamma(\nu-1)$ and $\Gamma(1/2) = \sqrt{\pi}$.

Remark For (d), (e), and (f), the peak value appears to be infinity. Thus take an "engineering approach" and define the peak value $m_{\max}$ such that the probability that the signal amplitude falls into $[-m_{\max}, m_{\max}]$ is 99%. Furthermore, evaluation of $m_{\max}$ for (d) and (f) involves the *error function*, defined as

$$\operatorname{erf}(x) = \frac{2}{\sqrt{\pi}}\int_0^x \mathrm{e}^{-t^2}\mathrm{d}t. \tag{P4.7}$$

The error function is available in Matlab under the name `erf`.

4.14 The $\mathrm{SNR_q}$ for a compander involves the following signal parameter:

$$\frac{\text{average value of the absolute value of the signal}}{\text{RMS value of the signal}} = \frac{E\{|\mathbf{m}|\}}{\sigma_{\mathbf{m}}}. \tag{P4.8}$$

Determine the values of this parameter for the signals of Problem 4.13.

4.15 Determine and on a single graph plot $\mathrm{SNR_q}(\sigma_n^2)$ for the μ-law compander with $L = 256$ and $\mu = 255$. Assume the following models for the message statistics: (a) Gaussian, (b) Laplacian, and (c) Gamma.

Remark Let σ_n^2 range from -100 decibels to 0 decibels. Also note that

$$E\left\{\frac{|\mathbf{m}|}{m_{\max}}\right\} = E\left\{\frac{\sigma_{\mathbf{m}}|\mathbf{m}|}{\sigma_{\mathbf{m}} m_{\max}}\right\} = \frac{\sigma_{\mathbf{m}}}{m_{\max}} E\left\{\frac{|\mathbf{m}|}{\sigma_{\mathbf{m}}}\right\}$$
$$= \sqrt{\sigma_n^2} E\left\{\frac{|\mathbf{m}|}{\sigma_{\mathbf{m}}}\right\}. \tag{P4.9}$$

4.16 Derive SNR_q for the A-law compander. As compared with the μ-law compander how dependent is the A-law compander on the assumed signal model?

4.17 Intuitively one may feel that the output of the compressor block of a compander, since it is quantized with a uniform quantizer, has a pdf that is uniform or close to it. How true this intuition is, is explored in this problem. Let the input to the compressor be Gaussian, zero-mean, unit variance. Determine and plot the pdf of the output where the compressor is μ-law with $\mu = 255$. Draw conclusions.

4.18 Consider the following proposed S-law compressor (S stands for student):

$$y(x) = (1+S)\frac{|x|}{1+S|x|}\text{sgn}(x), \quad |x| < 1. \tag{P4.10}$$

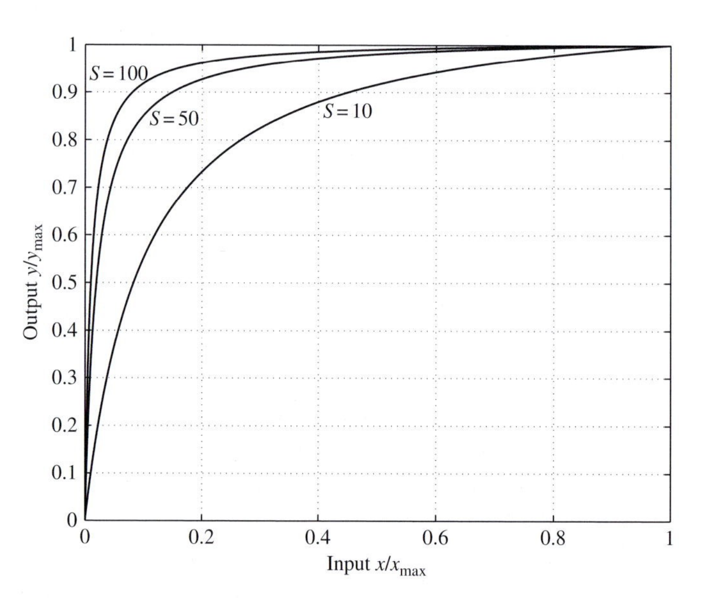

Fig. 4.25 Characteristics of the S-law compressor.

Figure 4.25 plots, as examples, the characteristics of the above compressor for $S = 10, 50, 100$. Derive $\text{SNR}_\text{q}(\sigma_n^2)$ for it and plot for a Gaussian input. Choose S to be 10 and 100.

The next set of problems is on the topic of differential quantization.

4.19 Given an R-bit quantizer (i.e., the number of target levels is $L = 2^R$), how many mappings are there from the target levels to the R-bit sequences?

4.20 For the mappings presented in the chapter, what is SNR_q when the channel is such that only one bit in the R-bit sequence is in error and that with probability $P[\text{error}]$. Since $P[\text{error}]$ is typically 10^{-4} or smaller, how much do the channel errors affect SNR_q?

4.21 (*Delta modulation*) Assume that the sampling period is normalized to be $T_s = 1$ second and the amplitude of the signal $\mathbf{m}(kT_s)$ is zero mean and uniformly distributed in $(-V_{\max}, V_{\max})$ (first-order statistics). Also the correlation function (second-order statistics) is given by

$$R_m(\tau) = \mathrm{e}^{-|\tau|/100}.$$

(a) Assume that the number of quantization levels is 256. Calculate the mean-square quantization error for an optimum quantizer designed for $\mathbf{m}(kT_s)$.

(b) Consider the differential quantization system and assume that a first-order linear predictor has been used as illustrated in Figure 4.26. Choose the coefficient of the linear predictor under the minimum mean-square error (MMSE) criterion.

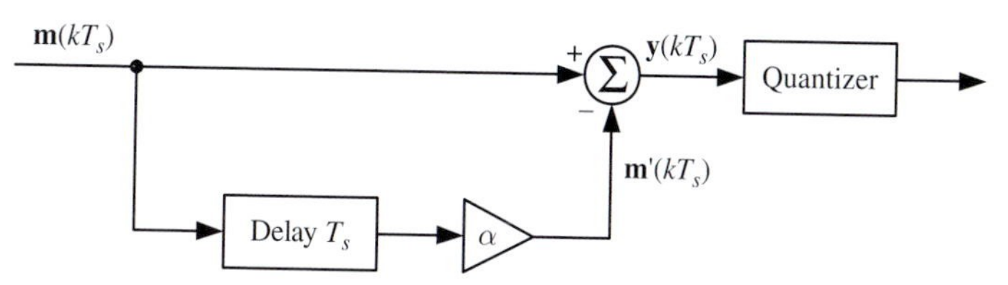

Fig. 4.26 Differential quantization with a first-order linear predictor.

(c) In (b), since $\mathbf{y}(kT_s) = \mathbf{m}(kT_s) - \alpha\mathbf{m}((k-1)T_s)$, in general the amplitude of $\mathbf{y}(kT_s)$ is no longer uniformly distributed. However, this signal can be approximated by another signal $\mathbf{y}'(kT_s)$. The amplitude distribution of $\mathbf{y}'(kT_s)$ is uniform and has the same mean and variance as $\mathbf{y}(kT_s)$. Find the pdf of the amplitude of $\mathbf{y}'(kT_s)$. Suppose an optimum quantizer with 256 quantization levels is designed based on $\mathbf{y}'(kT_s)$, calculate the mean-square quantization error and compare with the result in (a).

4.22 (*PCM representation*) Your USB drive has a capacity of 10^9 bytes (1 gigabyte). You wish to store a digital representation of an information source on the drive. Using a straightforward PCM representation, determine the maximum recording duration for each of the following sources:

(a) 4 kilohertz speech signal, with eight bits per sample.

(b) 22 kilohertz stereo audio signal, with 16 bits per sample in each stereo channel.

(c) 5 megahertz video signal, with 12 bits per sample, and combined with the audio signal in (b).

(d) Digital surveillance video signal, with 1024×768 pixels per frame, eight bits per pixel, and one frame per second.

4.23 (*A different definition for* $\mathrm{SNR_q}$) For an analog signal with peak magnitude $m_{\max}$, let us define the *peak signal-to-quantization noise ratio* as

$$\mathrm{PSNR_q} = 20\log_{10}\left(\frac{m_{\max}}{q_{\max}}\right), \tag{P4.11}$$

where $q_{\max} = \max\{|m - \hat{m}|\}$ is the peak magnitude of the quantization noise. Assuming an R-bit uniform quantizer, what is the smallest value of R that ensures the $\mathrm{PSNR_q}$ is at least D decibels?

5 Optimum receiver for binary data transmission

In Chapter 4 it was shown how continuous waveforms are transformed into binary digits (or bits) via the processes of sampling, quantization, and encoding (commonly referred to as pulse-code modulation, or PCM). However, it should be pointed out that bits are just abstractions: there is nothing *physical* about bits. Thus, for the transmission of the information, we need something physical to represent or "carry" the bits.

Here, we represent the binary digit $\mathbf{b}_k$ (0 or 1) by one of two electrical waveforms $s_1(t)$ and $s_2(t)$. Such a representation is the function of the *modulator* as shown in Figure 5.1. These waveforms are transmitted through the channel and perturbed by the *noise*. At the receiving side, the *receiver* needs to make a decision, $\hat{\mathbf{b}}_k$, on the transmitted bit based on the received signal $\mathbf{r}(t)$.

In this chapter we study the *optimum* receiver for a *binary* digital communication system as illustrated in Figure 5.1. The performance of the receiver will be measured in terms of error probability. A channel model of *infinite* bandwidth with AWGN is assumed. Though somewhat idealized, this channel model does provide us with a starting point to develop the approaches necessary to design and analyze receivers. Besides, many communication channels such as satellite and deep space communications are well represented by this model.

Other assumptions and the notation which pertain to the discussion of the communication system in Figure 5.1 are as follows.

- The bit duration of $\mathbf{b}_k$ is T_b seconds, or the bit rate is $r_b = 1/T_b$ (bits/second).
- Bits in two different time slots are *statistically independent.*
- The *a priori* probabilities (i.e., before we observe the received signal $\mathbf{r}(t)$) of $\mathbf{b}_k$ are as follows:

$$P[\mathbf{b}_k = 0] = P_1, \tag{5.1}$$

$$P[\mathbf{b}_k = 1] = P_2, \tag{5.2}$$

where $P_1 + P_2 = 1$. Most often, we will assume that the bits are equally likely, i.e., $P_1 = P_2 = 1/2$.

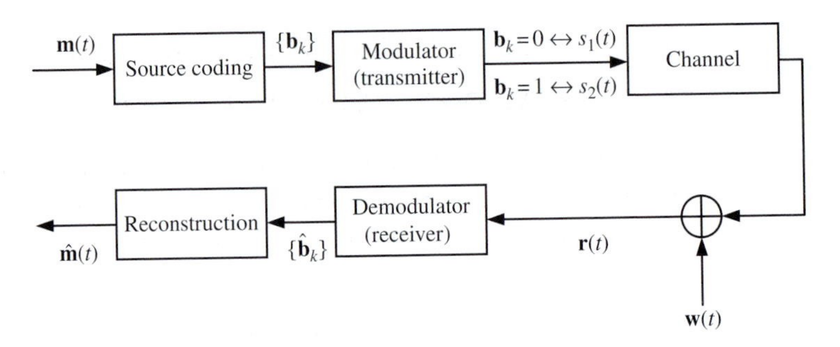

Fig. 5.1 Block diagram of a binary digital communication system.

- The bit $\mathbf{b}_k$ is mapped by the modulator into one of the two signals $s_1(t)$ and $s_2(t)$. Each signal is of duration T_b seconds and has a finite energy:

$$E_1 = \int_0^{T_b} s_1^2(t)\mathrm{d}t \text{ (joules)}, \quad E_1 < \infty, \tag{5.3}$$

$$E_2 = \int_0^{T_b} s_2^2(t)\mathrm{d}t \text{ (joules)}, \quad E_2 < \infty. \tag{5.4}$$

  For now we shall consider the signal waveforms to be *arbitrary*, but known to the receiver. Later, in Chapters 6 and 7, by employing certain waveforms for $s_1(t)$ and $s_2(t)$, many popular baseband/passband modulation schemes are analyzed.
- The channel is sufficently wideband that the signals $s_1(t)$ and $s_2(t)$ pass through without any distortion. Essentially, this means that there is no intersymbol interference (ISI) between successive bits.
- The noise $\mathbf{w}(t)$ is stationary *Gaussian*, zero-mean, *white* noise with a two-sided PSD of $N_0/2$ (watts/hertz). That it is Gaussian means that the pdf of the noise amplitude at any time instant t is Gaussian. Also if the noise is passed through any linear operation (or filter) the output will be Gaussian. That the noise is white means that

$$E\{\mathbf{w}(t)\} = 0, \quad E\{\mathbf{w}(t)\mathbf{w}(t+\tau)\} = \frac{N_0}{2}\delta(\tau). \tag{5.5}$$

Given the above, the received signal over the time interval $[(k-1)T_b, kT_b]$, i.e., the interval in which we send the bit $\mathbf{b}_k$, can be written as

$$\mathbf{r}(t) = s_i(t-(k-1)T_b) + \mathbf{w}(t), \quad (k-1)T_b \le t \le kT_b. \tag{5.6}$$

The objective is to design a receiver (or demodulator) such that by observing the signal $\mathbf{r}(t)$, the probability of making an error is minimized. The development proceeds by reducing the problem from the observation of a time waveform to that of observing a set of numbers (which are random variables).[1]

The approach is basically to represent the signal $s_1(t)$, $s_2(t)$ and the noise $\mathbf{w}(t)$ by a specifically chosen series. The coefficients of this series then constitute the set of numbers on which our decision is based. This approach leads to a geometrical interpretation which readily lends insight and interpretation to the relationship between the signal energy, the distance between signals, and the error probability.

[1] A common name for this set is sufficient statistics, another is observables.

5.1 Geometric representation of signals $s_1(t)$ and $s_2(t)$

Consider two arbitrary signals $s_1(t)$ and $s_2(t)$. We wish to represent them as linear combinations of two *orthonormal* basis functions $\phi_1(t)$ and $\phi_2(t)$. Recall that two functions $\phi_1(t)$ and $\phi_2(t)$ are said to be orthonormal if the following two conditions are satisfied:

$$\int_0^{T_b} \phi_1(t)\phi_2(t)\mathrm{d}t = 0 \ (\textit{ortho}\text{gonality}), \tag{5.7}$$

$$\int_0^{T_b} \phi_1^2(t)\mathrm{d}t = \int_0^{T_b} \phi_2^2(t)\mathrm{d}t = 1 \ (\textit{normal}\text{ized to have unit energy}). \tag{5.8}$$

The signals $s_1(t)$ and $s_2(t)$ should have the following expansions in terms of $\phi_1(t)$ and $\phi_2(t)$:

$$s_1(t) = s_{11}\phi_1(t) + s_{12}\phi_2(t), \tag{5.9}$$

$$s_2(t) = s_{21}\phi_1(t) + s_{22}\phi_2(t). \tag{5.10}$$

By direct substitution and integral evaluation, it is simple to verify that if one can represent $s_1(t)$ and $s_2(t)$ as in (5.9) and (5.10), then the coefficients s_{ij}, $i, j \in \{1, 2\}$ must be given as follows:

$$s_{ij} = \int_0^{T_b} s_i(t)\phi_j(t)\mathrm{d}t, \quad i, j \in \{1, 2\}, \tag{5.11}$$

where we can view the mathematical operation $\int_0^{T_b} s_i(t)\phi_j(t)\mathrm{d}t$ as the projection of signal $s_i(t)$ onto the jth axis, $\phi_j(t)$. Equations (5.9) and (5.10) can be geometrically interpreted as shown in Figure 5.2.

The question now is how do we choose the time functions $\phi_1(t)$ and $\phi_2(t)$, so that (5.7) is satisfied and also so that $s_1(t)$ and $s_2(t)$ can be exactly represented by them. More fundamentally, we could ask if such a choice is possible. If possible, is the choice *unique*?

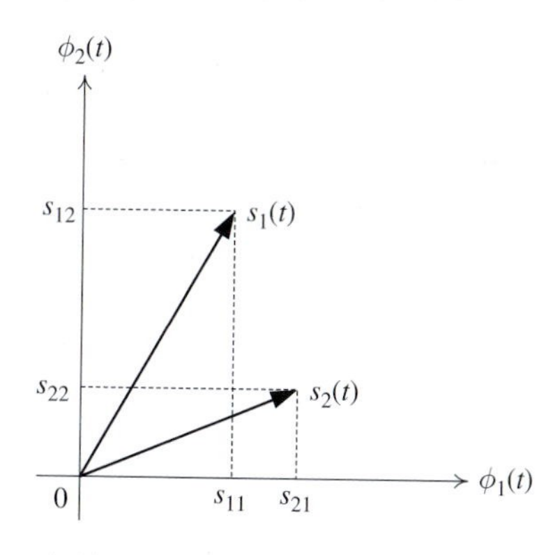

Fig. 5.2 Projection of $s_1(t)$ and $s_2(t)$ onto $\phi_1(t)$ and $\phi_2(t)$.

As we shall see in the following, the choice is possible, indeed there are many possible choices!

The most straightforward way to find $\phi_1(t)$ and $\phi_2(t)$ is as follows:

(i) Let $\phi_1(t) \equiv s_1(t)/\sqrt{E_1}$. This means that the first orthonormal function is chosen to be identical in shape to the first signal, but is normalized by the the square root of the signal energy. Note that from (5.9), one has $s_{11} = \sqrt{E_1}$ and $s_{12} = 0$.

(ii) To find $\phi_2(t)$, we first project $s_2^{'}(t) = s_2(t)/\sqrt{E_2}$ onto $\phi_1(t)$ and call this quantity ρ, the *correlation coefficient*:

$$\rho = \int_0^{T_b} \frac{s_2(t)}{\sqrt{E_2}} \phi_1(t) \mathrm{d}t = \frac{1}{\sqrt{E_1 E_2}} \int_0^{T_b} s_1(t) s_2(t) \mathrm{d}t. \tag{5.12}$$

(iii) Now subtract the projection $\rho\phi_1(t)$ from $s_2^{'}(t) = s_2(t)/\sqrt{E_2}$ to obtain

$$\phi_2^{'}(t) = \frac{s_2(t)}{\sqrt{E_2}} - \rho\phi_1(t). \tag{5.13}$$

Intuitively we would expect $\phi_2^{'}(t)$ to be orthogonal to $\phi_1(t)$. To verify that this is true, consider

$$\begin{aligned} \int_0^{T_b} \phi_2^{'}(t)\phi_1(t)\mathrm{d}t &= \int_0^{T_b} \frac{s_2(t)}{\sqrt{E_2}}\phi_1(t)\mathrm{d}t - \rho \int_0^{T_b} \phi_1^2(t)\mathrm{d}t \\ &= \rho - \rho = 0. \end{aligned} \tag{5.14}$$

(iv) The only thing left to do is to normalize $\phi_2^{'}(t)$ to make sure that $\phi_2(t)$ has unit energy (see Problem 5.1):

$$\begin{aligned} \phi_2(t) &= \frac{\phi_2^{'}(t)}{\sqrt{\int_0^{T_b} \left[\phi_2^{'}(t)\right]^2 \mathrm{d}t}} = \frac{\phi_2^{'}(t)}{\sqrt{1-\rho^2}} \\ &= \frac{1}{\sqrt{1-\rho^2}} \left[\frac{s_2(t)}{\sqrt{E_2}} - \frac{\rho s_1(t)}{\sqrt{E_1}} \right]. \end{aligned} \tag{5.15}$$

To summarize, the two orthonormal functions are given by,

$$\phi_1(t) = \frac{s_1(t)}{\sqrt{E_1}}, \tag{5.16}$$

$$\phi_2(t) = \frac{1}{\sqrt{1-\rho^2}} \left[\frac{s_2(t)}{\sqrt{E_2}} - \frac{\rho s_1(t)}{\sqrt{E_1}} \right]. \tag{5.17}$$

Furthermore, the coefficients s_{21} and s_{22} can be determined as follows:

$$s_{21} = \int_0^{T_b} s_2(t)\phi_1(t)\mathrm{d}t = \rho\sqrt{E_2}, \tag{5.18}$$

$$\begin{aligned} s_{22} &= \int_0^{T_b} s_2(t)\phi_2(t)\mathrm{d}t = \int_0^{T_b} s_2(t) \frac{1}{\sqrt{1-\rho^2}} \left[\frac{s_2(t)}{\sqrt{E_2}} - \frac{\rho s_1(t)}{\sqrt{E_1}} \right] \mathrm{d}t \\ &= \frac{1}{\sqrt{1-\rho^2}} \left(\sqrt{E_2} - \rho^2\sqrt{E_2} \right) = \left(\sqrt{1-\rho^2} \right) \sqrt{E_2}. \end{aligned} \tag{5.19}$$

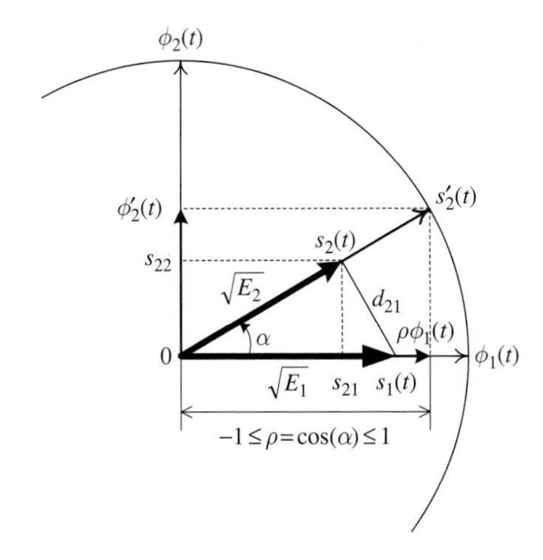

Fig. 5.3 Obtaining $\phi_1(t)$ and $\phi_2(t)$ from $s_1(t)$ and $s_2(t)$.

Geometrically, the above procedure is illustrated in Figure 5.3.

To carry the geometric ideas further, define the distance between the signals $s_2(t)$ and $s_1(t)$ as

$$d_{21} = \sqrt{\int_0^{T_b} [s_2(t) - s_1(t)]^2 \mathrm{d}t}. \tag{5.20}$$

From the geometry of the signal representation we see that

$$\begin{aligned} d_{21}^2 &= s_{22}^2 + \left(\sqrt{E_1} - s_{21}\right)^2 \\ &= \left(1 - \rho^2\right) E_2 + \left(E_1 - 2\rho\sqrt{E_1 E_2} + \rho^2 E_2\right) \\ &= E_1 - 2\rho\sqrt{E_1 E_2} + E_2. \end{aligned} \tag{5.21}$$

It can be verified (try it yourself) that evaluating (5.20) yields the same result of (5.21) for the distance d_{21}.

An important geometric interpretation follows from Figure 5.3. The distance of either signal point to the origin is the square root of the energy of that signal. For example, in the case of $s_2(t)$, the distance is $\sqrt{s_{21}^2 + s_{22}^2}$, which from (5.18), (5.19) is $\sqrt{E_2}$. The general statement is that if a signal is expressed as a linear combination of orthonormal functions, then the distance of the signal from the origin is always equal to the square root of its energy.

The above procedure has shown that two arbitrary deterministic waveforms can be represented *exactly* by at most two properly chosen orthonormal functions. This procedure can be generalized to a set with an arbitrary number of finite-energy signals (which are not necessarily limited to $[0, T_b]$ as in the case of two signals just considered) and it is known as the *Gram–Schmidt* orthogonalization procedure. The Gram–Schmidt procedure is outlined below for the set of M waveforms.

Gram–Schmidt procedure Suppose that we have a set of M finite-energy waveforms $\{s_i(t), i = 1, 2, \ldots, M\}$ and we wish to construct a set of orthonormal waveforms that can represent $\{s_i(t), i = 1, 2, \ldots, M\}$ exactly. The first orthonormal function is simply constructed as

$$\phi_1(t) = \frac{s_1(t)}{\sqrt{\int_{-\infty}^{\infty} s_1^2(t)\mathrm{d}t}}. \tag{5.22}$$

The subsequent orthonormal functions are found as follows:

$$\phi_i(t) = \frac{\phi_i'(t)}{\sqrt{\int_{-\infty}^{\infty} \left[\phi_i'(t)\right]^2 \mathrm{d}t}}, \quad i = 2, 3, \ldots, N, \tag{5.23}$$

where

$$\phi_i'(t) = \frac{s_i(t)}{\sqrt{E_i}} - \sum_{j=1}^{i-1} \rho_{ij}\phi_j(t), \tag{5.24}$$

$$\rho_{ij} = \int_{-\infty}^{\infty} \frac{s_i(t)}{\sqrt{E_i}}\phi_j(t)\mathrm{d}t, \quad j = 1, 2, \ldots, i-1. \tag{5.25}$$

In general, the number of orthonormal functions, N, is less than or equal to the number of given waveforms, M, depending on one of the two possibilities:

(i) If the waveforms $\{s_i(t), i = 1, 2, \ldots, M\}$ form a *linearly independent set*, then $N = M$.
(ii) If the waveforms $\{s_i(t), i = 1, 2, \ldots, M\}$ are not linearly independent, then $N < M$.

To illustrate the Gram–Schmidt procedure, let us consider a few examples.

Example 5.1 Consider the signal set shown in Figure 5.4(a). This signal set is in a sense a degenerate case since only one basis function $\phi_1(t)$ is needed to represent it. The signal energies are given by

$$E_1 = \int_0^{T_b} s_1^2(t)\mathrm{d}t = V^2 T_b = E_2 \equiv E \text{ (joules)}. \tag{5.26}$$

The first orthonormal function is given by

$$\phi_1(t) = \frac{s_1(t)}{\sqrt{E}} = \frac{s_1(t)}{\sqrt{V^2 T_b}}. \tag{5.27}$$

The correlation coefficient, ρ, is $\rho = \int_0^{T_b} (s_2(t)s_1(t)/E)\,\mathrm{d}t = -1$, reflecting the fact that $s_2(t) = -s_1(t)$. Therefore the unnormalized basis function $\phi_2'(t)$ is

$$\phi_2'(t) = \frac{s_2(t)}{\sqrt{E}} - \rho\phi_1(t) = \frac{s_2(t) + s_1(t)}{\sqrt{E}} = 0. \tag{5.28}$$

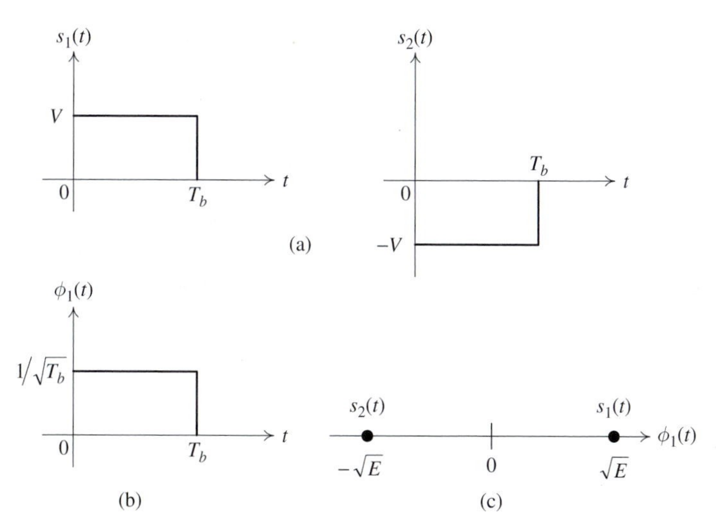

Fig. 5.4 (a) Signal set for Example 5.1, (b) orthonormal function, (c) signal space representation.

The basis function $\phi_1(t)$ is plotted in Figure 5.4(b). In terms of $\phi_1(t)$ the two signals can be expressed as

$$s_1(t) = \sqrt{E}\phi_1(t), \tag{5.29}$$

$$s_2(t) = -\sqrt{E}\phi_1(t). \tag{5.30}$$

The geometrical representation of the two signals $s_1(t)$ and $s_2(t)$ is presented in Figure 5.4(c). Note that the distance between the two signals is $d_{21} = 2\sqrt{E}$. ■

Example 5.2 The signal set considered in this example is given in Figure 5.5(a). Again this is a special case because the two signals are *orthogonal*. The energy in each signal is equal to $V^2T_b \equiv E$ (joules). The first orthonormal basis function is

$$\phi_1(t) = \frac{s_1(t)}{\sqrt{E}}. \tag{5.31}$$

The correlation coefficient, ρ, is

$$\rho = \int_0^{T_b} \frac{s_1(t)s_2(t)}{E} \mathrm{d}t = 0, \tag{5.32}$$

which shows that the two signals are orthogonal. Therefore $\phi_2'(t) = s_2(t)/\sqrt{E} = \phi_2(t)$. Thus the signals $s_1(t)$ and $s_2(t)$ are expressed as

$$s_1(t) = \sqrt{E}\phi_1(t), \tag{5.33}$$

$$s_2(t) = \sqrt{E}\phi_2(t). \tag{5.34}$$

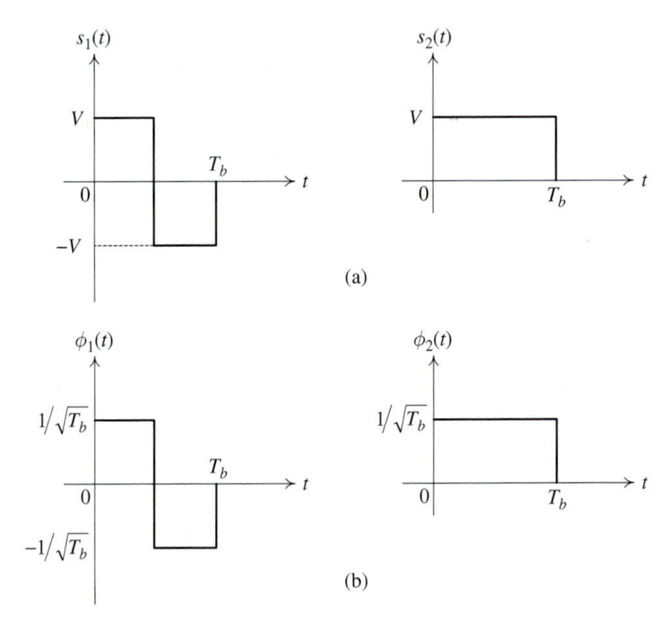

Fig. 5.5 (a) Signal set for Example 5.2, (b) orthonormal functions.

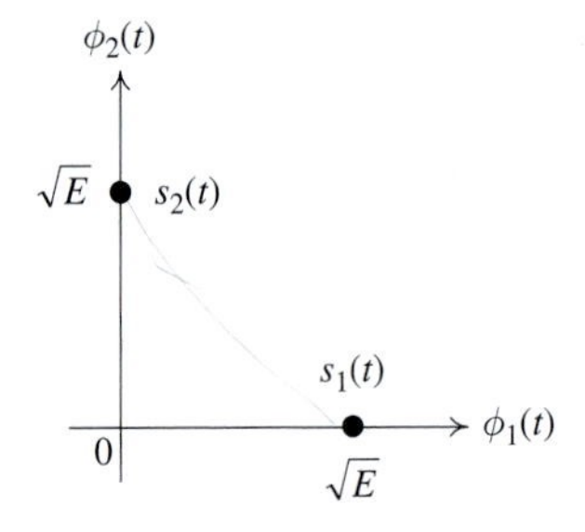

Fig. 5.6 Signal space representation for Example 5.2.

Graphically, the orthonormal basis functions $\phi_1(t)$ and $\phi_2(t)$ look as in Figure 5.5(b) and the signal space is plotted in Figure 5.6. The distance between the two signals can be easily computed as follows:

$$d_{21} = \sqrt{E+E} = \sqrt{2E} = \sqrt{2}\sqrt{E}. \tag{5.35}$$

■

In comparing Examples 5.1 and 5.2 we observe that the *energy per bit* at the transmitter or sending end is the same in each example. The signals in Example 5.2, however, are closer together and therefore at the receiving end, in the presence of noise, we would expect more difficulty in distinguishing which signal was sent. We shall see presently that this is the case and quantitatively express this increased difficulty.

Example 5.3 This is a generalization of Examples 5.1 and 5.2. It is included principally to illustrate the geometrical representation of two signals. The signal set is shown in Figure 5.7, where each signal has energy equal to $E = V^2 T_b$. The first basis function is $\phi_1(t) = s_1(t)/\sqrt{E}$. The correlation coefficient ρ depends on parameter α and is given by

$$\rho = \frac{1}{E}\int_0^{T_b} s_2(t)s_1(t)\mathrm{d}t = \frac{1}{V^2 T_b}\left[V^2\alpha - V^2(T_b - \alpha)\right] = \frac{2\alpha}{T_b} - 1. \tag{5.36}$$

As a check, for $\alpha = 0$, $\rho = -1$ and for $\alpha = \frac{1}{2}T_b$, $\rho = 0$, as expected. The second basis function is

$$\phi_2(t) = \frac{1}{\sqrt{E(1-\rho^2)}}[s_2(t) - \rho s_1(t)]. \tag{5.37}$$

To obtain the geometrical picture, consider the case when $\alpha = \frac{1}{4}T_b$. For this value of α one has $\rho = -\frac{1}{2}$. As before $\phi_1(t) = s_1(t)/\sqrt{E}$, whereas the second orthonormal function is given by

$$\phi_2(t) = \frac{2}{\sqrt{3}V\sqrt{T_b}}\left[s_2(t) + \frac{1}{2}s_1(t)\right]. \tag{5.38}$$

The two orthonormal basis functions are plotted in Figure 5.8. The geometrical representation of $s_1(t)$ and $s_2(t)$ is given in Figure 5.9. Note that the coefficients for the representation of $s_2(t)$ are $s_{21} = -\frac{1}{2}\sqrt{E}$ and $s_{22} = \frac{\sqrt{3}}{2}\sqrt{E}$. Since $s_{21}^2 + s_{22}^2 = E$, the signal $s_2(t)$ is at a distance of $\sqrt{E}$ from the origin. In general as α varies from 0 to T_b, the function $\phi_2(t)$ changes. However, for each specific $\phi_2(t)$ the signal $s_2(t)$ is always at distance $\sqrt{E}$ from the origin. The locus of $s_2(t)$ is also plotted in Figure 5.9. Note that as α increases, ρ increases and the distance between the two signals decreases. ■

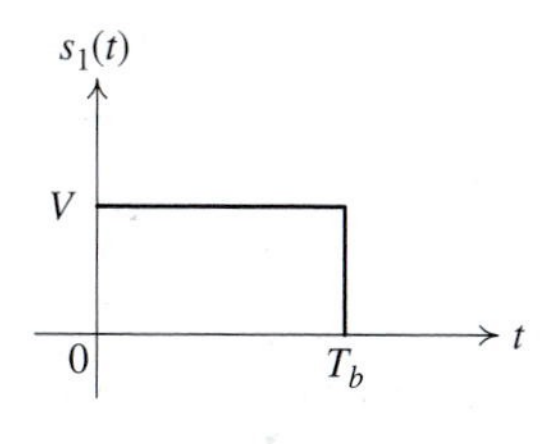

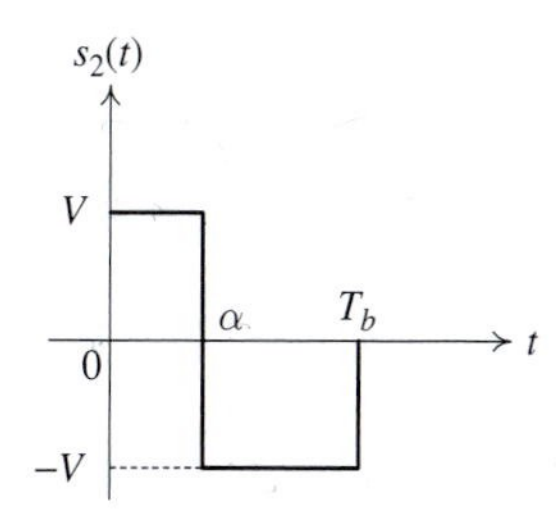

Fig. 5.7 Signal set for Example 5.3.

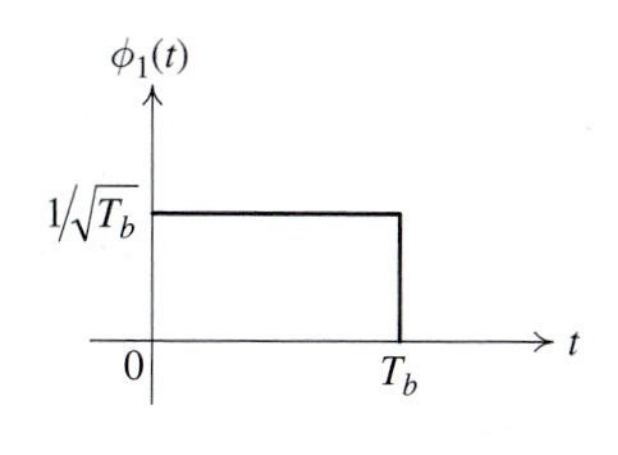

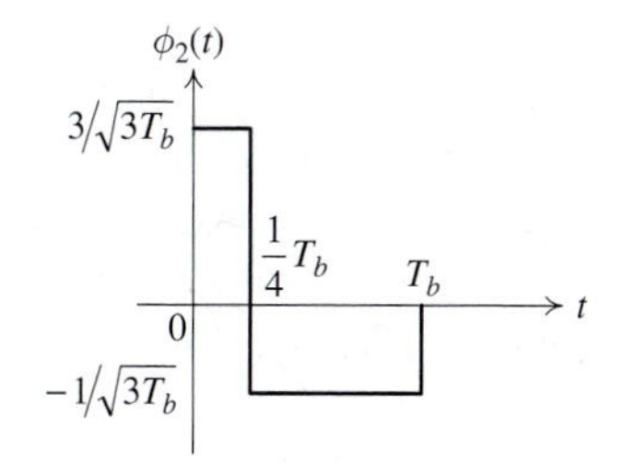

Fig. 5.8 Orthonormal functions for Example 5.3.

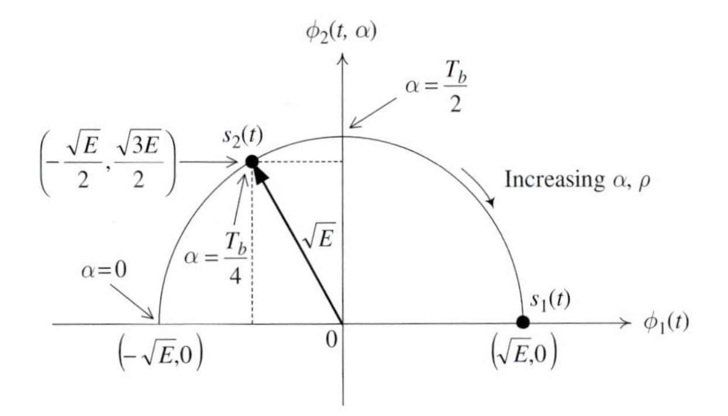

Fig. 5.9 Signal space representation for the signal set in Example 5.3.

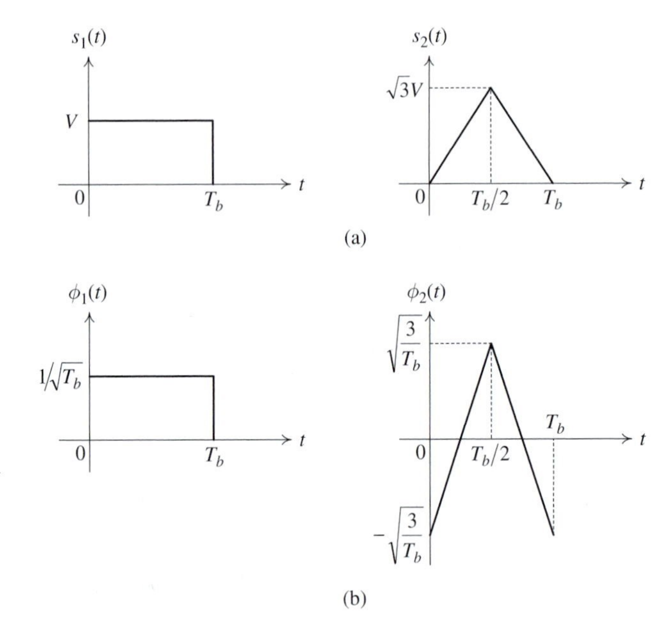

Fig. 5.10 (a) Signal set for Example 5.4, (b) orthonormal functions.

Example 5.4 Consider the signal set shown in Figure 5.10(a). Again the energy in each signal is $E = V^2 T_b$ joules. We have the following:

$$\phi_1(t) = \frac{s_1(t)}{\sqrt{E}}, \tag{5.39}$$

$$\rho = \frac{1}{E}\int_0^{T_b} s_2(t)s_1(t)\mathrm{d}t = \frac{2}{E}\int_0^{T_b/2}\left(\frac{2\sqrt{3}}{T_b}Vt\right)V\mathrm{d}t = \frac{\sqrt{3}}{2}, \tag{5.40}$$

$$\phi_2(t) = \frac{1}{(1-\frac{3}{4})^{\frac{1}{2}}}\left[\frac{s_2(t)}{\sqrt{E}} - \rho\frac{s_1(t)}{\sqrt{E}}\right] = \frac{2}{\sqrt{E}}\left[s_2(t) - \frac{\sqrt{3}}{2}s_1(t)\right], \tag{5.41}$$

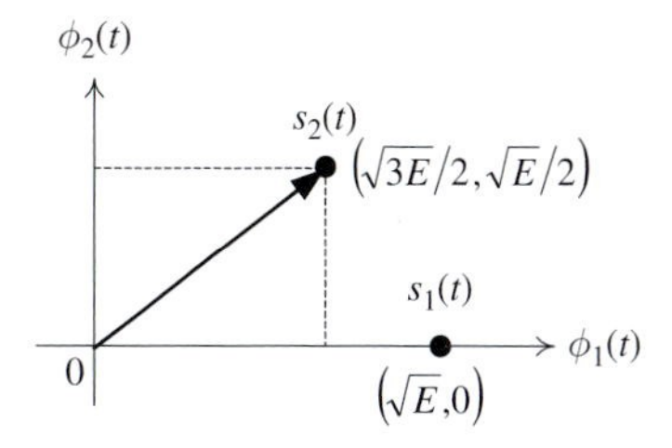

Fig. 5.11 Signal space representation for signal set in Example 5.4.

$$s_{21} = \frac{\sqrt{3}}{2}\sqrt{E}, \quad s_{22} = \frac{1}{2}\sqrt{E}. \tag{5.42}$$

The distance between the two signals is

$$\begin{aligned} d_{21} &= \left[\int_0^{T_b} [s_2(t) - s_1(t)]^2 \mathrm{d}t\right]^{\frac{1}{2}} = \left\{\left[\sqrt{E}\left(1 - \frac{\sqrt{3}}{2}\right)\right]^2 + \left(\frac{\sqrt{E}}{2}\right)^2\right\}^{\frac{1}{2}} \\ &= E\left(2 - \sqrt{3}\right)^{\frac{1}{2}} = \sqrt{\left(2 - \sqrt{3}\right)E}. \end{aligned} \tag{5.43}$$

The orthonormal functions are plotted in Figure 5.10(b), whereas the signal space representation is illustrated in Figure 5.11. ■

Example 5.5 As a final example to illustrate the Gram–Schmidt procedure, consider two sinusoidal signals of the same frequency but different phase:

$$s_1(t) = \sqrt{E}\sqrt{\frac{2}{T_b}}\cos(2\pi f_c t), \tag{5.44}$$

$$s_2(t) = \sqrt{E}\sqrt{\frac{2}{T_b}}\cos(2\pi f_c t + \theta). \tag{5.45}$$

Choose $f_c = k/2T_b$, k an integer. This choice means that $\cos(2\pi f_c t)$ and $\sin(2\pi f_c t)$ are orthogonal over a duration of T_b seconds. The energy in each signal is

$$E_1 = E\int_0^{T_b} \frac{2}{T_b}\cos^2(2\pi f_c t)\mathrm{d}t = E, \tag{5.46}$$

$$E_2 = E\int_0^{T_b} \frac{2}{T_b}\cos^2(2\pi f_c t + \theta)\mathrm{d}t = E. \tag{5.47}$$

The first orthonormal function is

$$\phi_1(t) = \frac{s_1(t)}{\sqrt{E}} = \sqrt{\frac{2}{T_b}}\cos(2\pi f_c t). \tag{5.48}$$

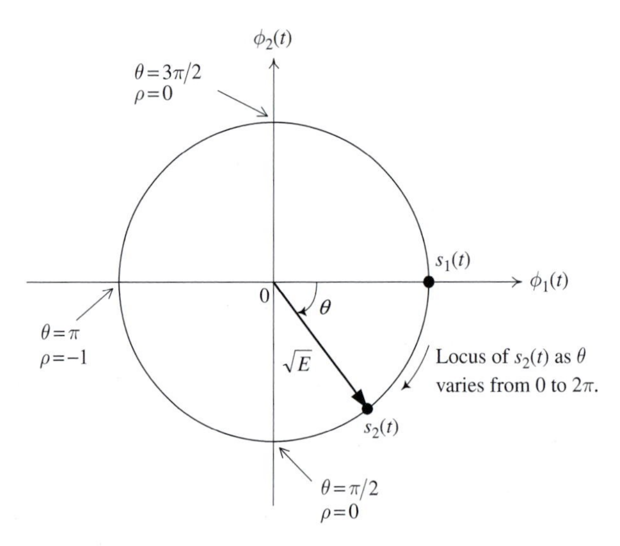

Fig. 5.12 Signal space representation for signal set in Example 5.5.

Now, using the basic trigonometric identity $\cos(x+y) = \cos x \cos y - \sin x \sin y$, write $s_2(t)$ as

$$s_2(t) = \left(\sqrt{E}\cos\theta\right)\left[\sqrt{\frac{2}{T_b}}\cos(2\pi f_c t)\right] + \left(-\sqrt{E}\sin\theta\right)\left[\sqrt{\frac{2}{T_b}}\sin(2\pi f_c t)\right]. \quad (5.49)$$

Since we have said that $\sin(2\pi f_c t)$ is orthogonal to $\cos(2\pi f_c t)$ (you might want to show this for yourself) over the interval $(0, T_b)$, our second basis function is chosen to be

$$\phi_2(t) = \sqrt{\frac{2}{T_b}}\sin(2\pi f_c t). \quad (5.50)$$

Because $s_2(t) = s_{21}\phi_1(t) + s_{22}\phi_2(t)$, by inspection we have

$$s_{21} = \sqrt{E}\cos\theta, \quad s_{22} = -\sqrt{E}\sin\theta. \quad (5.51)$$

Finally, the signal space plot looks as shown in Figure 5.12. ■

5.2 Representation of the noise

As seen in the previous section, the signal set $\{s_1(t), s_2(t)\}$ used to transmit the binary data needs at most only two orthonormal functions to be represented exactly. In contrast, to represent the random noise signal, $\mathbf{w}(t)$, in the time interval $[(k-1)T_b, kT_b]$ (the interval of bit $\mathbf{b}_k$) we need to use a *complete* orthonormal set of known deterministic functions. The series representation of the noise is given by

$$\mathbf{w}(t) = \sum_{i=1}^{\infty} \mathbf{w}_i \phi_i(t), \tag{5.52}$$

where the coefficients $\mathbf{w}_i$ are given as follows:

$$\mathbf{w}_i = \int_0^{T_b} \mathbf{w}(t)\phi_i(t)\mathrm{d}t. \tag{5.53}$$

Again the coefficient $\mathbf{w}_i$ can be viewed as the projection of the noise process onto the $\phi_i(t)$ axis. Note also that the above representation can be used for any random signal.

The coefficients $\mathbf{w}_i$ are *random variables* and understanding their statistical properties is imperative in developing the optimum receiver. When the noise process $\mathbf{w}(t)$ is zero-mean and white, one has the following important properties of these coefficients.[2]

(i) The average (also known as the mean, or DC, or expected) value of $\mathbf{w}_i$ is zero. This is because $\mathbf{w}(t)$ is a zero-mean random process. Mathematically,

$$\begin{aligned} E\{\mathbf{w}_i\} &= E\left\{\int_0^{T_b} \mathbf{w}(t)\phi_i(t)\mathrm{d}t\right\} \\ &= \int_0^{T_b} E\{\mathbf{w}(t)\}\phi_i(t)\mathrm{d}t = 0. \end{aligned} \tag{5.54}$$

(ii) The correlation between $\mathbf{w}_i$ and $\mathbf{w}_j$ is given by

$$\begin{aligned} E\left\{\mathbf{w}_i\mathbf{w}_j\right\} &= E\left\{\int_0^{T_b} \mathrm{d}\lambda \mathbf{w}(\lambda)\phi_i(\lambda) \int_0^{T_b} \mathrm{d}\tau \mathbf{w}(\tau)\phi_j(\tau)\right\} \\ &= \int_0^{T_b} \phi_i(\lambda)\left[\int_0^{T_b} \phi_j(\tau)E\{\mathbf{w}(\lambda)\mathbf{w}(\tau)\}\mathrm{d}\tau\right]\mathrm{d}\lambda. \end{aligned} \tag{5.55}$$

But $\mathbf{w}(t)$ is *white* noise, therefore $E\{\mathbf{w}(\lambda)\mathbf{w}(\tau)\} = (N_0/2)\delta(\lambda - \tau)$. Equation (5.55) can now be written as

$$\begin{aligned} E\{\mathbf{w}_i\mathbf{w}_j\} &= \int_0^{T_b} \phi_i(\lambda)\left[\int_0^{T_b} \phi_j(\tau)\frac{N_0}{2}\delta(\lambda-\tau)\mathrm{d}\tau\right]\mathrm{d}\lambda \\ &= \frac{N_0}{2}\int_0^{T_b} \phi_i(\lambda)\phi_j(\lambda)\mathrm{d}\lambda \\ &= \begin{cases} \frac{N_0}{2}, & i = j \\ 0, & i \neq j \end{cases}. \end{aligned} \tag{5.56}$$

The above result means that $\mathbf{w}_i$ and $\mathbf{w}_j$ are *uncorrelated* if $i \neq j$.

[2] It is important to stress the simple fact that if the noise $\mathbf{w}(t)$ is zero-mean and white, it is not necessarily Gaussian.

In summary, the coefficients $\{\mathbf{w}_1, \mathbf{w}_2, \ldots\}$ are zero-mean and uncorrelated random variables.

Now, if the noise process $\mathbf{w}(t)$ is not only zero-mean and white, but also Gaussian, then each $\mathbf{w}_i$ is a Gaussian random variable. This is because $\mathbf{w}_i$ is obtained by passing the Gaussian process $\mathbf{w}(t)$ through a linear transformation (system). Because the random variables $\{\mathbf{w}_1, \mathbf{w}_2, \ldots\}$ are Gaussian and uncorrelated, we deduce the important property that they are *statistically independent*. Furthermore, we note that the above properties do not depend on the set $\{\phi_i(t), i = 1, 2, \ldots\}$ chosen, i.e., on the orthonormal basis functions. The set of orthonormal basis functions that we shall choose will have as the first two functions the functions $\phi_1(t)$ and $\phi_2(t)$ used to represent the two signals $s_1(t)$ and $s_2(t)$ exactly. The remaining functions, i.e., $\phi_3(t), \phi_4(t), \ldots$, are chosen simply to complete the set. However, as will be seen shortly, in practice, we do not need to find these functions.

5.3 Optimum receiver

In any bit interval we receive the noise-corrupted signal. Let us concentrate, without any loss of generality, on the first bit interval. The received signal is

$$\begin{aligned}\mathbf{r}(t) &= s_i(t) + \mathbf{w}(t), \quad 0 \le t \le T_b \\ &= \begin{cases} s_1(t) + \mathbf{w}(t), & \text{if a "0" is transmitted} \\ s_2(t) + \mathbf{w}(t), & \text{if a "1" is transmitted} \end{cases}.\end{aligned} \tag{5.57}$$

Besides $\mathbf{r}(t)$, the following are known to us. We know exactly when the time interval begins, i.e., the receiver is synchronized with the transmitter. The receiver knows precisely the two signals $s_1(t)$ and $s_2(t)$. Finally the *a priori* probability of a 0 or 1 being transmitted is also known.

To obtain the optimum receiver, we first expand $\mathbf{r}(t)$ into a series using the orthonormal functions $\{\phi_1(t), \phi_2(t), \phi_3(t), \ldots\}$. As mentioned before, the first two functions are chosen so that they represent the signals $s_1(t)$ and $s_2(t)$ exactly (in effect the signals $s_1(t)$ and $s_2(t)$ determine them, for example, by applying the Gram–Schmidt procedure). The rest are chosen to complete the orthonormal set. Over the time interval $[0, T_b]$ the signal $\mathbf{r}(t)$ can therefore be expressed as

$$\begin{aligned}\mathbf{r}(t) &= s_i(t) + \mathbf{w}(t), \quad 0 \le t \le T_b \\ &= \underbrace{[s_{i1}\phi_1(t) + s_{i2}\phi_2(t)]}_{s_i(t)} + \underbrace{[\mathbf{w}_1\phi_1(t) + \mathbf{w}_2\phi_2(t) + \mathbf{w}_3\phi_3(t) + \mathbf{w}_4\phi_4(t) + \cdots]}_{\mathbf{w}(t)} \\ &= (s_{i1} + \mathbf{w}_1)\phi_1(t) + (s_{i2} + \mathbf{w}_2)\phi_2(t) + \mathbf{w}_3\phi_3(t) + \mathbf{w}_4\phi_4(t) + \cdots \\ &= \mathbf{r}_1\phi_1(t) + \mathbf{r}_2\phi_2(t) + \mathbf{r}_3\phi_3(t) + \mathbf{r}_4\phi_4(t) + \cdots,\end{aligned} \tag{5.58}$$

where $\mathbf{r}_j = \int_0^{T_b} \mathbf{r}(t)\phi_j(t)\mathrm{d}t$, and

$$\begin{aligned} \mathbf{r}_1 &= s_{i1} + \mathbf{w}_1, \\ \mathbf{r}_2 &= s_{i2} + \mathbf{w}_2, \\ \mathbf{r}_3 &= \mathbf{w}_3, \\ \mathbf{r}_4 &= \mathbf{w}_4, \\ &\vdots \end{aligned} \tag{5.59}$$

It is important to note that $\mathbf{r}_j$, for $j = 3, 4, 5, \ldots$, does not depend on which signal ($s_1(t)$ or $s_2(t)$) was transmitted.

The decision as to which signal has been transmitted can now be based on the observations[3] $r_1, r_2, r_3, r_4, \ldots$. Before proceeding we need a criterion which the decision should optimize. The criterion that we choose is the perfectly natural one, namely *to minimize the bit error probability*.

To derive the receiver that minimizes the error probability, consider the set of observations $\vec{r} = \{r_1, r_2, r_3, \ldots\}$. If we consider only the first n terms in the set, then they form an n-dimensional observation space which we have to partition into *decision regions* in order to minimize the error probability. This is illustrated in Figure 5.13.

The probability of making an error can be expressed as

$$\begin{aligned} P[\text{error}] = &P[(\text{"0" decided and "1" transmitted) or} \\ &(\text{"1" decided and "0" transmitted})]. \end{aligned} \tag{5.60}$$

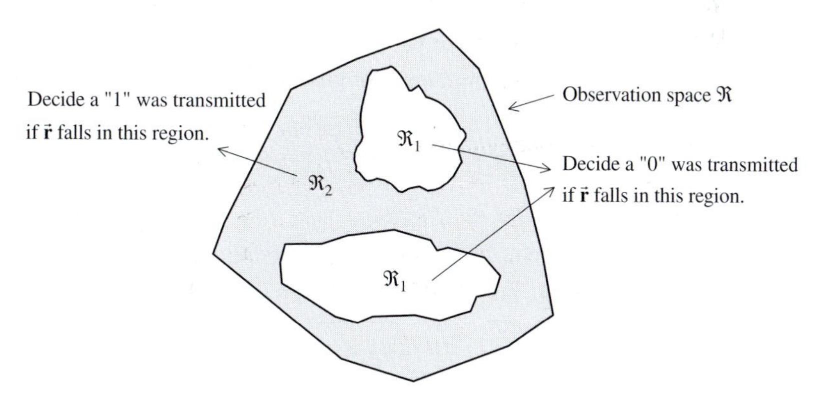

Fig. 5.13 Observation space and decision regions.

[3] Though, in general, the coefficients of the expansion in (5.58) are random quantities, in the discussion that follows we take the viewpoint that the experiment, i.e., the transmission and reception of the signal, has been conducted and that we are observing its outcome. Namely, the set of projections, of the received signal onto the basis functions. Therefore the observations are treated as being a specific set of values.

Using obvious notation and the fact that the two events that constitute the error event are *mutually exclusive*, the above can be written as

$$\begin{aligned} P[\text{error}] &= P[0_D, 1_T] + P[1_D, 0_T] \\ &= P[0_D|1_T]P[1_T] + P[1_D|0_T]P[0_T], \end{aligned} \tag{5.61}$$

where the second equality follows from Bayes' rule. Consider now the quantity $P[0_D|1_T]$. A zero is chosen when the observation $\vec{r}$ falls into region $\Re_1$. Therefore the probability that a "0" is chosen given that a "1" was transmitted is the same as the probability that $\vec{\mathbf{r}}$ falls in region $\Re_1$ when a "1" is transmitted. But this latter probability is given by the volume under the conditional density[4] $f(\vec{r}|1_T)$ in region $\Re_1$, i.e.,

$$P[0_D|1_T] = \int_{\Re_1} f(\vec{r}|1_T)\mathrm{d}\vec{r} \tag{5.62}$$

(note that the above is a *multiple* integral). Since $P[0_T] = P_1$ and $P[1_T] = P_2$, one has

$$\begin{aligned} P[\text{error}] &= P_2 \int_{\Re_1} f(\vec{r}|1_T)\mathrm{d}\vec{r} + P_1 \int_{\Re_2} f(\vec{r}|0_T)\mathrm{d}\vec{r} \\ &= P_2 \int_{\Re-\Re_2} f(\vec{r}|1_T)\mathrm{d}\vec{r} + P_1 \int_{\Re_2} f(\vec{r}|0_T)\mathrm{d}\vec{r} \\ &= P_2 \int_{\Re} f(\vec{r}|1_T)\mathrm{d}\vec{r} + \int_{\Re_2} [P_1 f(\vec{r}|0_T) - P_2 f(\vec{r}|1_T)]\mathrm{d}\vec{r} \\ &= P_2 + \int_{\Re_2} \left[P_1 f(\vec{r}|0_T) - P_2 f(\vec{r}|1_T)\right] \mathrm{d}\vec{r}. \end{aligned} \tag{5.63}$$

The error, of course, depends on how the observation space is partitioned. Looking at the above expression for the error probability, we observe that if each $\vec{r}$ that makes the integrand $[P_1 f(\vec{r}|0_T) - P_2 f(\vec{r}|1_T)]$ negative is assigned to $\Re_2$, then the error probability is minimized. Note also that, for an observation $\vec{r}$ that gives $[P_1 f(\vec{r}|0_T) - P_2 f(\vec{r}|1_T)] = 0$, it does not matter whether $\vec{r}$ is put into $\Re_1$ or into $\Re_2$, i.e., the receiver can arbitrarily decide "1" or "0." Thus the minimum error probability decision rule can be expressed as

$$\begin{cases} P_1 f(\vec{r}|0_T) - P_2 f(\vec{r}|1_T) \geq 0 & \Rightarrow \text{ decide "0" } (0_D) \\ P_1 f(\vec{r}|0_T) - P_2 f(\vec{r}|1_T) < 0 & \Rightarrow \text{ decide "1" } (1_D) \end{cases}. \tag{5.64}$$

Equivalently,

$$\frac{f(\vec{r}|1_T)}{f(\vec{r}|0_T)} \underset{0_D}{\overset{1_D}{\gtrless}} \frac{P_1}{P_2}. \tag{5.65}$$

The expression $f(\vec{r}|1_T)/f(\vec{r}|0_T)$ is commonly called the *likelihood ratio*. The receiver consists of computing this ratio and comparing the result to a threshold determined by the *a priori* probabilities. Furthermore, observe that the decision rule in (5.65) was derived without specifying any statistical properties of the noise process $\mathbf{w}(t)$. In other words, it is true for any set of observations $\vec{r}$.

[4] A more consistent notation for the conditional pdf of $\vec{\mathbf{r}}$ would be $f_{\vec{\mathbf{r}}}(\vec{r}|1_T)$. However, for simplicity of presentation, the subscript $\vec{\mathbf{r}}$ is dropped.

The decision rule in (5.65) can, however, be greatly simplified when the noise $\mathbf{w}(t)$ is zero-mean, white, and Gaussian. In this case, the observations $\mathbf{r}_1, \mathbf{r}_2, \mathbf{r}_3, \mathbf{r}_4, \ldots$, are statistically independent Gaussian random variables. The variance of each coefficient is equal to the variance of the noise coefficient, $\mathbf{w}_j$, which is $N_0/2$ (watts). The average or expected values of the first two coefficients $\mathbf{r}_1$ and $\mathbf{r}_2$ are (s_{11}, s_{12}) and (s_{21}, s_{22}), respectively. The other coefficients, $\mathbf{r}_j, j \geq 3$, have zero mean. The *conditional densities* $f(\vec{r}|1_T)$ and $f(\vec{r}|0_T)$ can simply be written as products of the individual pdfs as follows:

$$f(\vec{r}|1_T) = f(r_1|1_T)f(r_2|1_T)f(r_3|1_T)\cdots f(r_j|1_T)\cdots, \tag{5.66}$$

$$f(\vec{r}|0_T) = f(r_1|0_T)f(r_2|0_T)f(r_3|0_T)\cdots f(r_j|0_T)\cdots \tag{5.67}$$

or

$$f(\vec{r}|1_T) = \frac{1}{\sqrt{\pi N_0}} \exp\left[-\frac{(r_1 - s_{21})^2}{N_0}\right] \frac{1}{\sqrt{\pi N_0}} \exp\left[-\frac{(r_2 - s_{22})^2}{N_0}\right] \times f(r_3|1_T)\cdots \tag{5.68}$$

$$f(\vec{r}|0_T) = \frac{1}{\sqrt{\pi N_0}} \exp\left[-\frac{(r_1 - s_{11})^2}{N_0}\right] \frac{1}{\sqrt{\pi N_0}} \exp\left[-\frac{(r_2 - s_{12})^2}{N_0}\right] \times f(r_3|0_T)\cdots \tag{5.69}$$

Note that the terms $f(r_j|1_T) = f(r_j|0_T) = (1/\sqrt{\pi N_0}) \exp\left(-r_j^2/N_0\right)$ for $j \geq 3$, therefore they will cancel in the likelihood ratio in (5.65). Thus the decision rule becomes

$$\frac{\exp\left[-(r_1 - s_{21})^2/N_0\right] \exp\left[-(r_2 - s_{22})^2/N_0\right]}{\exp\left[-(r_1 - s_{11})^2/N_0\right] \exp\left[-(r_2 - s_{12})^2/N_0\right]} \mathop{\gtrless}_{0_D}^{1_D} \frac{P_1}{P_2}. \tag{5.70}$$

To further simplify the expression, take the natural logarithm of both sides of (5.70). Since the natural logarithm is a *monotonic* function, the inequality is preserved for each (r_1, r_2) pair. The resulting decision rule becomes

$$(r_1 - s_{11})^2 + (r_2 - s_{12})^2 \mathop{\gtrless}_{0_D}^{1_D} (r_1 - s_{21})^2 + (r_2 - s_{22})^2 + N_0 \ln\left(\frac{P_1}{P_2}\right). \tag{5.71}$$

The above rule has an interesting geometrical interpretation. The quantity $(r_1 - s_{11})^2 + (r_2 - s_{12})^2$ is the *distance squared* from the projection (r_1, r_2) of the received signal $r(t)$ to the transmitted signal $s_1(t)$. Similarly, $(r_1 - s_{21})^2 + (r_2 - s_{22})^2$ is the distance squared from (r_1, r_2) to $s_2(t)$. For the special case of $P_1 = P_2$, i.e., each signal is equally likely, the decision rule becomes

$$(r_1 - s_{11})^2 + (r_2 - s_{12})^2 \mathop{\gtrless}_{0_D}^{1_D} (r_1 - s_{21})^2 + (r_2 - s_{22})^2. \tag{5.72}$$

In essence, the above implies that the optimum receiver needs to determine the distance from $r(t)$ to both $s_1(t)$ and $s_2(t)$ and then to choose the signal to which $r(t)$ is closest.

For the more general case of $P_1 \neq P_2$ the decision regions are determined by a line *perpendicular* to the line joining the two signals $\{s_1(t), s_2(t)\}$, except that it now shifts toward $s_2(t)$ (if $P_1 > P_2$) or toward $s_1(t)$ (if $P_1 < P_2$). This is illustrated in Figure 5.14.

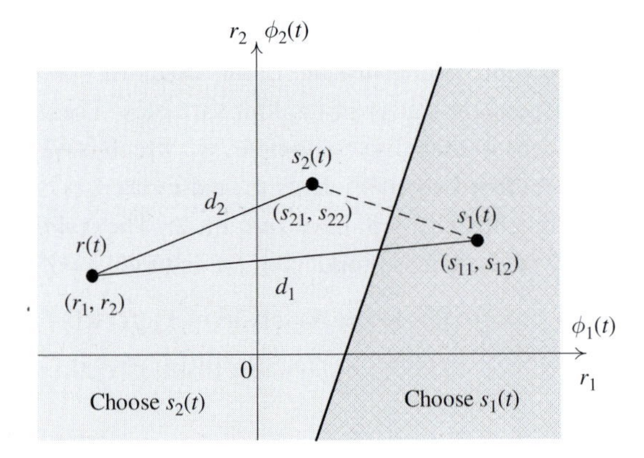

Fig. 5.14 Decision regions are determined by a line perpendicular to the line joining $s_1(t)$ to $s_2(t)$.

5.4 Receiver implementation

To determine, in a minimum error probability sense, whether a "0" or "1" was transmitted we need to determine (r_1, r_2) and then use (5.71) to make a decision. The receiver would therefore look as shown in Figure 5.15.

The process of multiplying $r(t)$ by $\phi_1(t)$ and integrating over the bit duration is that of *correlation* and therefore the above is called a correlation receiver configuration. Since at the end of each bit duration the integrator must be reset to zero initial condition, the above is also commonly called an integrate-and-dump receiver. Moreover, the decision block can be simplified somewhat by expanding the terms in (5.71) to arrive at

$$\begin{matrix} (r_1 - s_{11})^2 + (r_2 - s_{12})^2 \\ -N_0 \ln P_1 \end{matrix} \underset{0_D}{\overset{1_D}{\gtrless}} \begin{matrix} (r_1 - s_{21})^2 + (r_2 - s_{22})^2 \\ -N_0 \ln P_2 \end{matrix} \tag{5.73}$$

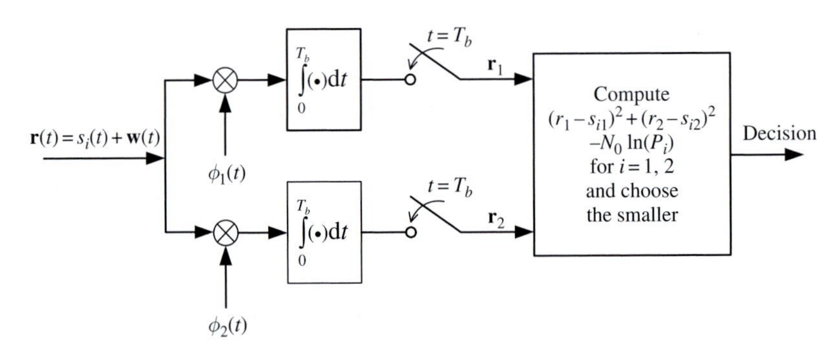

Fig. 5.15 Correlation receiver implementation.

or

$$\begin{matrix} r_1^2 - 2r_1s_{11} + s_{11}^2 + r_2^2 \\ -2r_2s_{12} + s_{12}^2 - N_0 \ln P_1 \end{matrix} \underset{0_D}{\overset{1_D}{\gtrless}} \begin{matrix} r_1^2 - 2r_1s_{21} + s_{21}^2 + r_2^2 \\ -2r_2s_{22} + s_{22}^2 - N_0 \ln P_2. \end{matrix} \tag{5.74}$$

Canceling the common terms, recognizing that $E_1 = s_{11}^2 + s_{12}^2$, $E_2 = s_{21}^2 + s_{22}^2$ and rearranging yields

$$r_1s_{21} + r_2s_{22} - \frac{E_2}{2} + \frac{N_0}{2} \ln P_2 \underset{0_D}{\overset{1_D}{\gtrless}} r_1s_{11} + r_2s_{12} - \frac{E_1}{2} + \frac{N_0}{2} \ln P_1. \tag{5.75}$$

The term $r_1s_{i1} + r_2s_{i2} = (r_1, r_2)\begin{pmatrix} s_{i2} \\ s_{i1} \end{pmatrix}$ can be interpreted as the dot product between the vector $\vec{r} = (r_1, r_2)$ representing the received signal $r(t)$ and the vector $\vec{s}_i = (s_{i1}, s_{i2})$ representing the signal $s_i(t)$. Therefore the receiver implementation can be redrawn as in Figure 5.16.

The correlation part of the optimum receiver involves a multiplier and integrator. Multipliers are difficult to realize physically. As an alternative, the correlator section can be implemented by a finite impulse response filter as shown in Figure 5.17, where

$$h_1(t) = \phi_1(T_b - t), \tag{5.76}$$

$$h_2(t) = \phi_2(T_b - t). \tag{5.77}$$

The above filters are generally referred to as *matched filters*, in this case they are "matched" to the basis functions.

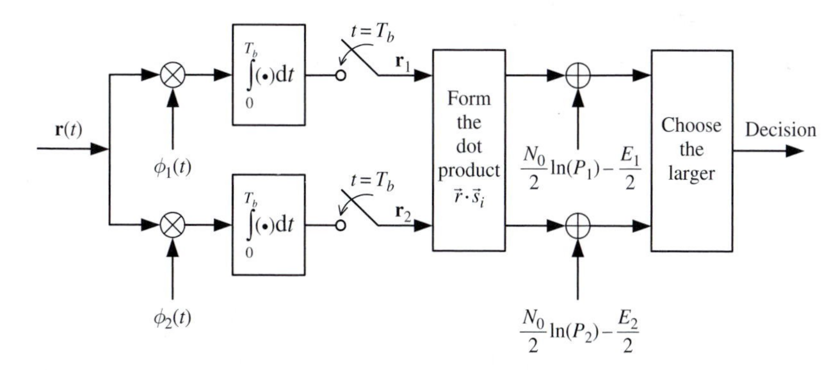

Fig. 5.16 A different implementation of a correlation receiver.

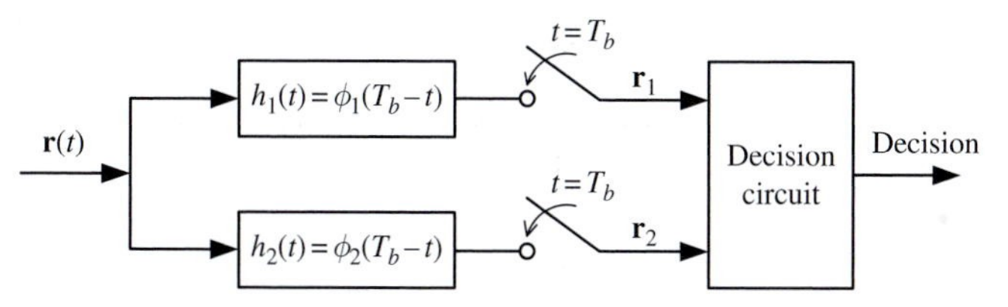

Fig. 5.17 Receiver implementation using matched filters.

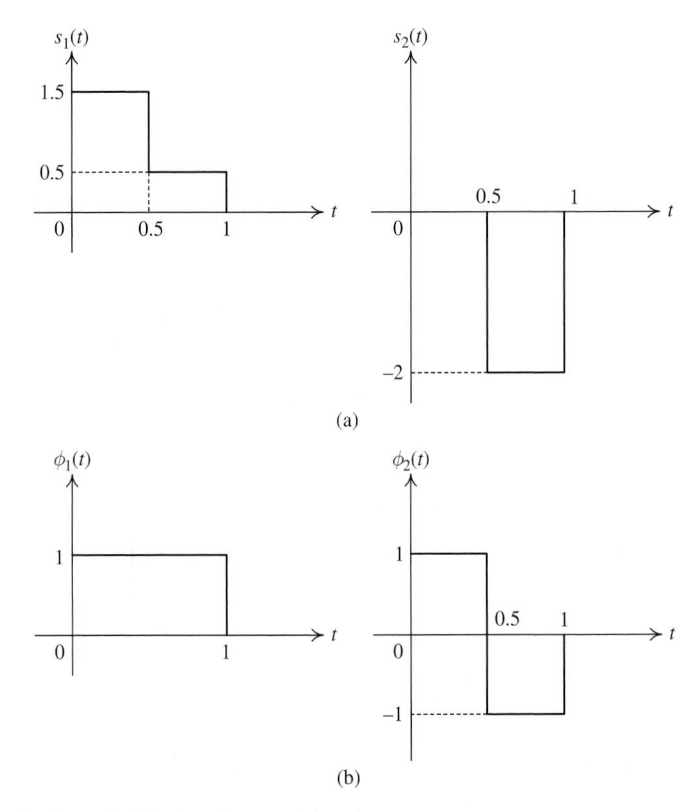

Fig. 5.18 (a) Signal set for Example 5.6, (b) orthonormal functions.

Example 5.6 Consider the signal set $\{s_1(t), s_2(t)\}$ shown in Figure 5.18(a). Simple inspection shows that this signal set can be represented by the orthonormal functions $\{\phi_1(t), \phi_2(t)\}$ in Figure 5.18(b) as follows:

$$s_1(t) = \phi_1(t) + \tfrac{1}{2}\phi_2(t),$$
$$s_2(t) = -\phi_1(t) + \phi_2(t).$$

The energy in each of the signals is given as

$$E_1 = \int_0^{T_b} s_1^2(t)\mathrm{d}t = 1.25 \text{ (joules)}, \quad E_2 = \int_0^{T_b} s_2^2(t)\mathrm{d}t = 2 \text{ (joules)}. \tag{5.78}$$

Unlike the examples considered up to now, here the two signals have unequal energy. The signal space plot looks as shown in Figure 5.19. Consider now the detection of the transmitted signal over a bit interval where the PSD of the white Gaussian noise is $N_0/2 = 0.5$ (watts/hertz). The optimum receiver to minimize error probability consists of projecting

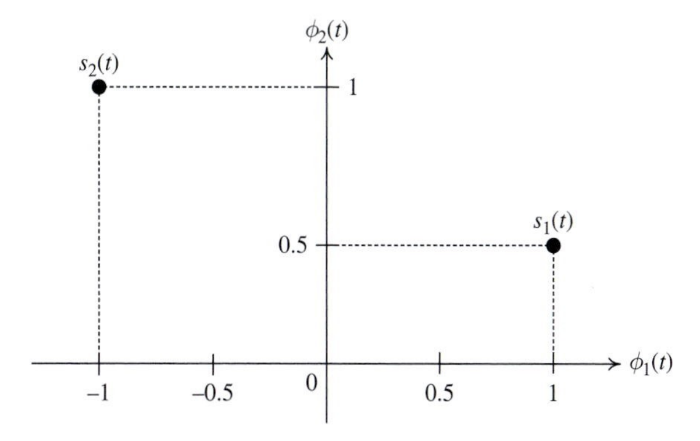

Fig. 5.19 Signal space representation, for Example 5.6.

the received signal $r(t)$ onto $\phi_1(t), \phi_2(t)$ and then applying the decision rule given by (5.71), i.e.,

$$(r_1 - 1)^2 + (r_2 - \tfrac{1}{2})^2 \underset{0_D}{\overset{1_D}{\gtrless}} (r_1 + 1)^2 + (r_2 - 1)^2 + \ln\left(\tfrac{P_1}{P_2}\right). \tag{5.79}$$

Upon expanding, the decision rule becomes that of (5.75) which can be written as

$$-4r_1 + r_2 - \left(\tfrac{3}{4} + \ln\tfrac{P_1}{P_2}\right) \underset{0_D}{\overset{1_D}{\gtrless}} 0. \tag{5.80}$$

The boundary between the two decision regions is given by

$$4r_1 - r_2 + \left(\frac{3}{4} + \ln\frac{P_1}{P_2}\right) = 0, \tag{5.81}$$

which is an equation of a straight line of slope 4 and intercept $\left(\frac{3}{4} + \ln(P_1/P_2)\right)$. In terms of (r_2, r_1) the equation of the straight line joining the $s_2(t)$ and $s_1(t)$ points is given by

$$\frac{r_2 - s_{12}}{s_{22} - s_{12}} = \frac{r_1 - s_{11}}{s_{21} - s_{11}}, \quad \text{or} \quad r_2 = -\frac{1}{4}r_1 + \frac{3}{4}. \tag{5.82}$$

It can be seen that the straight lines defined by (5.81) and (5.82) are perpendicular to each other. The decision regions therefore look as in Figure 5.20 for three different sets of *a priori* probabilities. ■

Example 5.7 For the second example, consider the two orthonormal functions shown in Figure 5.21. Let the signal set be as follows:

$$s_2(t) = \phi_1(t) + \phi_2(t), \tag{5.83}$$

$$s_1(t) = \phi_1(t) - \phi_2(t). \tag{5.84}$$

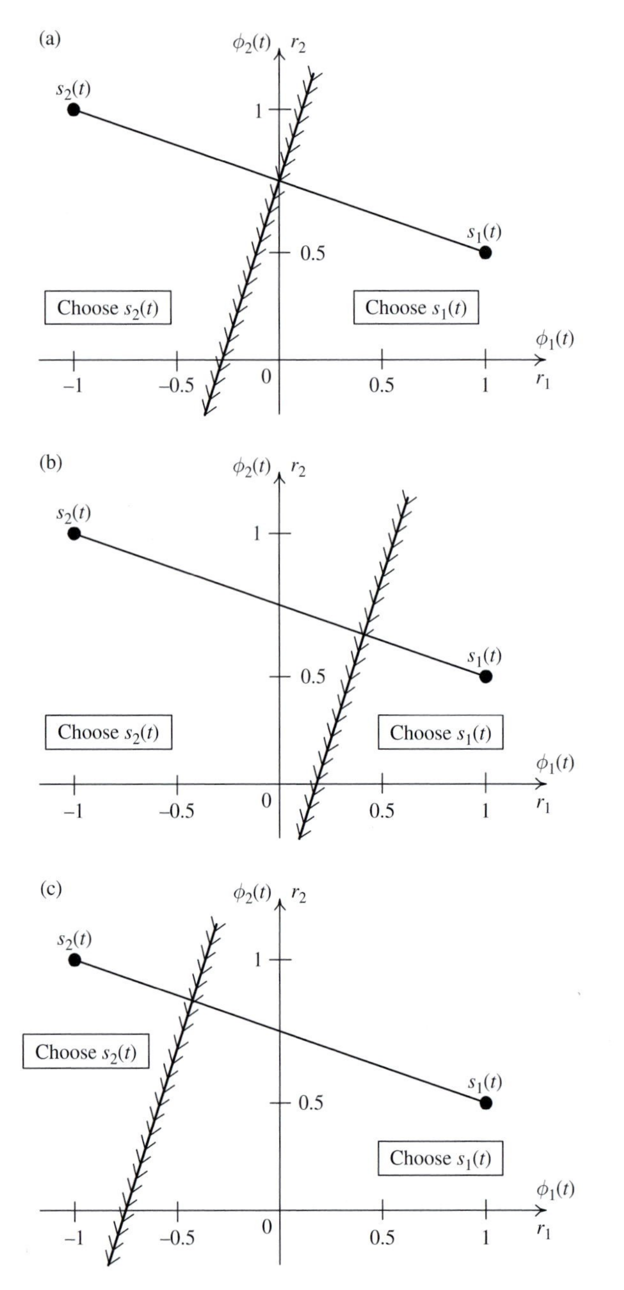

Fig. 5.20 Decision regions for Example 5.6: (a) $P_1 = P_2 = 0.5$. (b) $P_1 = 0.25$, $P_2 = 0.75$. (c) $P_1 = 0.75$, $P_2 = 0.25$.

The two signals have equal energy of $E_1 = E_2 = 2$ (joules). Note that $s_1(t)$ and $s_2(t)$ have a common component $\phi_1(t)$. Thus, intuitively we would expect that only the component along $\phi_2(t)$ will help us to distinguish between the possible transmitted signals in the presence of noise. This is verified formally by applying (5.71) as follows:

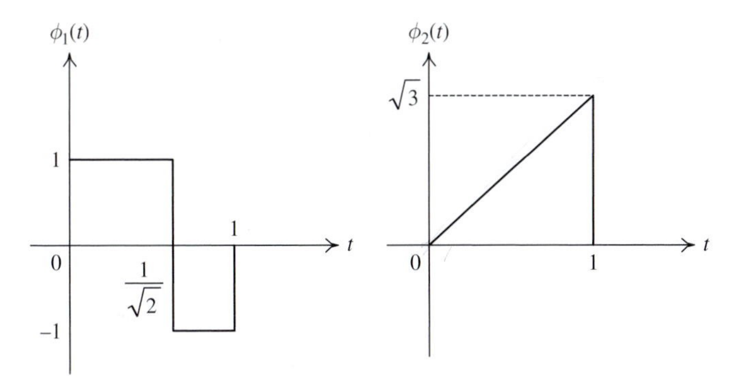

Fig. 5.21 Orthonormal functions for Example 5.7.

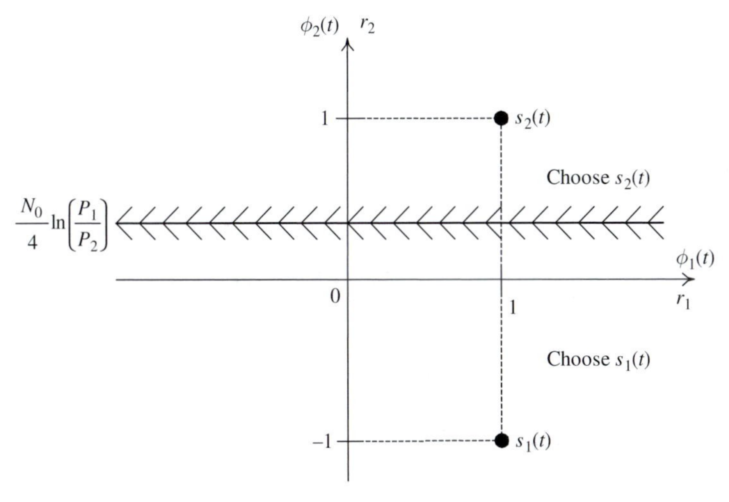

Fig. 5.22 Signal space representation and the decision regions for Example 5.7.

$$(r_1 - 1)^2 + (r_2 + 1)^2 - N_0 \ln P_1 \underset{0_D}{\overset{1_D}{\gtrless}} (r_1 - 1)^2 + (r_2 - 1)^2 - N_0 \ln P_2, \tag{5.85}$$

which simplifies to

$$r_2 \underset{0_D}{\overset{1_D}{\gtrless}} \frac{N_0}{4} \ln\left(\frac{P_1}{P_2}\right). \tag{5.86}$$

The signal set and the decision regions are shown in Figure 5.22. The receiver can be implemented using a correlator or a matched filter, as shown in Figures 5.23(a) and 5.23(b), respectively. ■

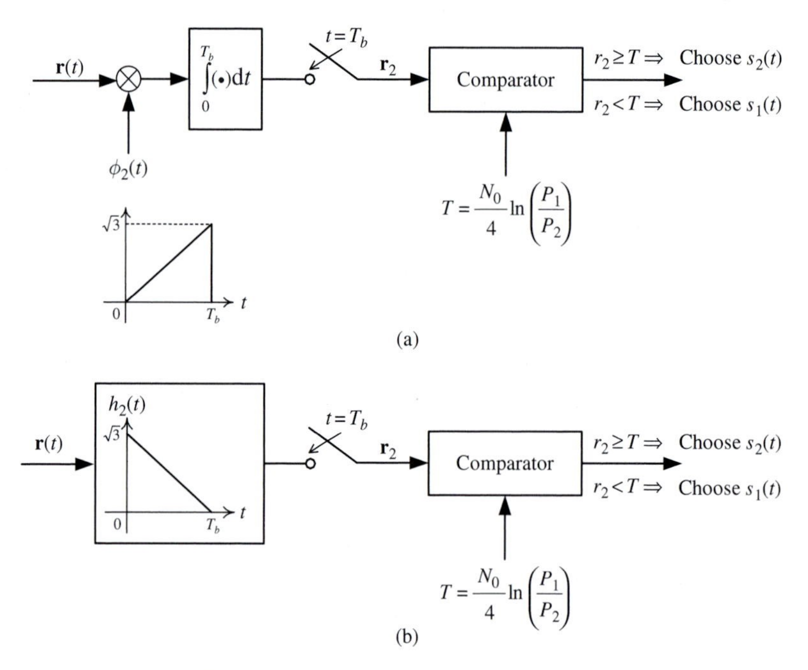

Fig. 5.23 Receiver implementation for Example 5.7: (a) correlation receiver, and (b) matched-filter receiver.

5.5 Receiver implementation with one correlator (or matched filter)

In general, for two arbitrary signals $s_1(t)$ and $s_2(t)$ we need two orthonormal basis functions to represent them exactly. Furthermore, the optimum receiver requires the projection of $\mathbf{r}(t)$ onto the two basis functions. This in turn means that in the receiver implementation either two correlators or two matched filters are required. In Example 5.7, however, only one correlator or matched filter was required. For binary data transmission, where the transmitted symbol is represented by one of two signals, this simplification is *always possible* by a judicious choice of the orthonormal basis.

Consider that $s_1(t)$ and $s_2(t)$ are represented by the orthonormal basis $\phi_1(t)$ and $\phi_2(t)$ as shown in Figure 5.24(a). To simplify the receiver so that only one correlator or matched filter is needed what we need is to find orthonormal basis functions $\hat{\phi}_1(t)$ and $\hat{\phi}_2(t)$ such that along one axis, say $\hat{\phi}_1(t)$, the signals $s_1(t)$ and $s_2(t)$ have an identical component. To determine this basis set we rotate $\phi_1(t)$ and $\phi_2(t)$ through an angle θ until one of the axes is perpendicular to the line joining $s_1(t)$ to $s_2(t)$ (see Problem 5.2). This rotation can be expressed as

$$\begin{bmatrix} \hat{\phi}_1(t) \\ \hat{\phi}_2(t) \end{bmatrix} = \begin{bmatrix} \cos\theta & \sin\theta \\ -\sin\theta & \cos\theta \end{bmatrix} \begin{bmatrix} \phi_1(t) \\ \phi_2(t) \end{bmatrix}. \tag{5.87}$$

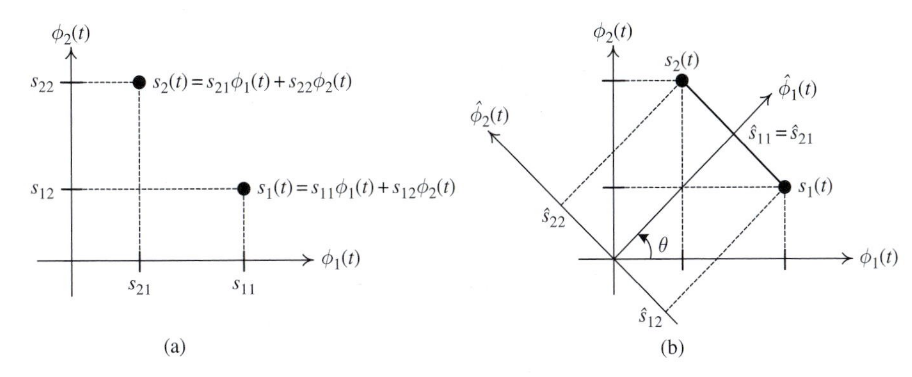

Fig. 5.24 Signal space representation: (a) by $\phi_1(t)$ and $\phi_2(t)$ and (b) by $\hat{\phi}_1(t)$ and $\hat{\phi}_2(t)$.

Now we project the received signal $\mathbf{r}(t) = s_i(t) + \mathbf{w}(t)$ onto the orthonormal basis functions $\hat{\phi}_1(t)$ and $\hat{\phi}_2(t)$. The components of $s_1(t)$ and $s_2(t)$ along $\hat{\phi}_1(t)$, namely $\hat{s}_{11}$ and $\hat{s}_{21}$, are identical. The noise projections $\hat{\mathbf{w}}_1$ and $\hat{\mathbf{w}}_2$ are again statistically independent zero-mean Gaussian random variables with variance $N_0/2$.

As before, the other orthonormal basis functions $\hat{\phi}_3(t), \hat{\phi}_4(t), \ldots,$ are simply chosen to complete the set. Again, given the fact that $\mathbf{w}(t)$ is white and Gaussian, the likelihood ratio of the projections of $r(t)$ onto $\hat{\phi}_1(t), \hat{\phi}_2(t), \hat{\phi}_3(t), \ldots$ is

$$\frac{f(\hat{r}_1, \hat{r}_2, \hat{r}_3, \ldots, |1_T)}{f(\hat{r}_1, \hat{r}_2, \hat{r}_3, \ldots, |0_T)} = \frac{f(\hat{s}_{21} + \hat{w}_1)f(\hat{s}_{22} + \hat{w}_2)f(\hat{w}_3)\cdots}{f(\hat{s}_{11} + \hat{w}_1)f(\hat{s}_{12} + \hat{w}_2)f(\hat{w}_3)\cdots} \underset{0_D}{\overset{1_D}{\gtrless}} \frac{P_1}{P_2}, \tag{5.88}$$

which upon canceling the common terms, remember $\hat{s}_{21} = \hat{s}_{11}$, reduces to

$$\frac{f(\hat{r}_2|1_T)}{f(\hat{r}_2|0_T)} = \frac{f(\hat{s}_{22} + \hat{w}_2)}{f(\hat{s}_{12} + \hat{w}_2)} \underset{0_D}{\overset{1_D}{\gtrless}} \frac{P_1}{P_2}. \tag{5.89}$$

Substituting in the exact expressions of the density functions, the decision rule becomes

$$\frac{(\pi N_0)^{-1/2} \exp[-(\hat{r}_2 - \hat{s}_{22})^2/N_0]}{(\pi N_0)^{-1/2} \exp[-(\hat{r}_2 - \hat{s}_{12})^2/N_0]} \underset{0_D}{\overset{1_D}{\gtrless}} \frac{P_1}{P_2}. \tag{5.90}$$

Taking the natural logarithm and simplifying, the final decision rule is

$$\hat{r}_2 \underset{0_D}{\overset{1_D}{\gtrless}} \frac{\hat{s}_{22} + \hat{s}_{12}}{2} + \left(\frac{N_0/2}{\hat{s}_{22} - \hat{s}_{12}}\right) \ln\left(\frac{P_1}{P_2}\right). \tag{5.91}$$

Note that, to arrive at the above decision rule, it was assumed that $(\hat{s}_{22} - \hat{s}_{12}) > 0$. This is indeed the case since, as will be shown in Section 5.6, $(\hat{s}_{22} - \hat{s}_{12})$ measures the distance between the two signals.

Thus the optimum receiver consists of finding $\hat{r}_2$ by projecting $\mathbf{r}(t)$ onto $\hat{\phi}_2(t)$, i.e., $\hat{\mathbf{r}}_2 = \int_0^{T_b} \mathbf{r}(t)\hat{\phi}_2(t)\mathrm{d}t$, and comparing $\hat{\mathbf{r}}_2 = \hat{r}_2$ to a threshold:

$$T \equiv \frac{\hat{s}_{22} + \hat{s}_{12}}{2} + \left(\frac{N_0/2}{\hat{s}_{22} - \hat{s}_{12}}\right) \ln\left(\frac{P_1}{P_2}\right). \tag{5.92}$$

Two receiver implementations are shown in Figures 5.25(a) and 5.25(b), corresponding to the correlation receiver and the matched filter, respectively.

As evidenced from Figure 5.25, one needs to know $\hat{\phi}_2(t)$ in order to implement the optimum receiver. This basis function can be determined as follows. Observe first that $\hat{\phi}_2(t)$ points along the line joining $s_1(t)$ to $s_2(t)$ which, as a vector, is $s_2(t) - s_1(t)$. Thus normalizing this vector so that the resultant time function has unit energy gives $\hat{\phi}_2(t)$. That is,

$$\hat{\phi}_2(t) = \frac{s_2(t) - s_1(t)}{\left\{\int_0^{T_b} [s_2(t) - s_1(t)]^2 \mathrm{d}t\right\}^{\frac{1}{2}}} = \frac{s_2(t) - s_1(t)}{(E_2 - 2\rho\sqrt{E_1 E_2} + E_1)^{\frac{1}{2}}}. \tag{5.93}$$

Observe that the above expression of $\hat{\phi}_2(t)$ illustrates that only the difference between the two signals is important in making the optimum decision at the receiver.

Example 5.8 Consider the signal set of Example 5.2. The signal space diagram is shown in Figure 5.26. By inspection, a rotation of 45° results in the desired orthonormal basis functions, where $\hat{\phi}_1(t)$ and $\hat{\phi}_2(t)$ can be written as

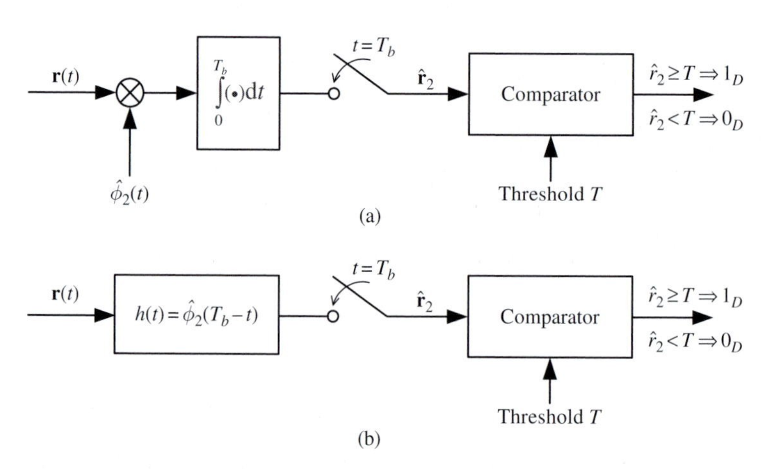

Fig. 5.25 Simplified optimum receiver: (a) using one correlator; (b) using one matched filter.

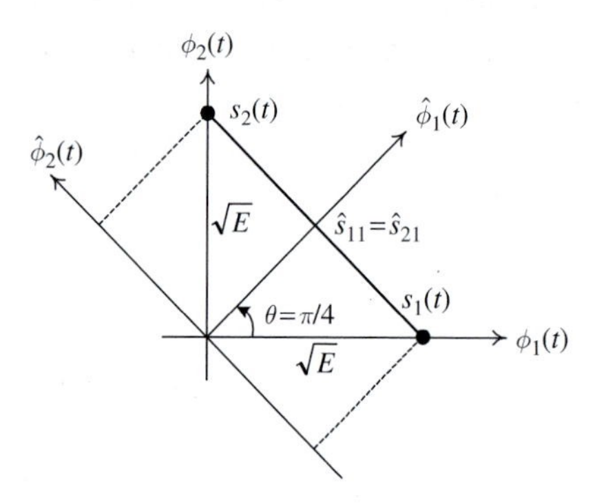

Fig. 5.26 Signal space representation for Example 5.8.

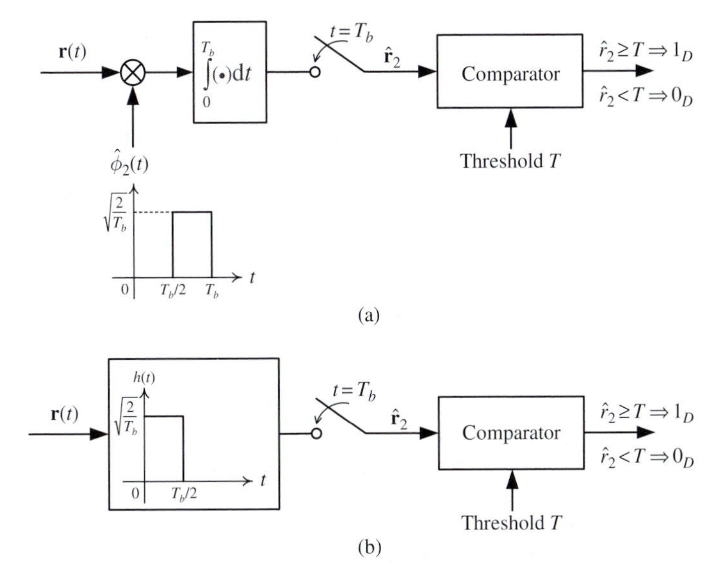

Fig. 5.27 Receiver implementations of Example 5.8: (a) correlation receiver; (b) matched-filter receiver.

$$\hat{\phi}_1(t) = \frac{1}{\sqrt{2}}[\phi_1(t) + \phi_2(t)],$$

$$\hat{\phi}_2(t) = \frac{1}{\sqrt{2}}[-\phi_1(t) + \phi_2(t)].$$

The receiver implementations are illustrated in Figures 5.27(a) and 5.27(b). ■

5.6 Receiver performance

Consider now the determination of the performance of the optimum receiver, which is measured in terms of its bit error probability. Recall from the preceding section that the detection of the information bit $\mathbf{b}_k$ transmitted over the kth interval $[(k-1)T_b, kT_b]$ consists of computing $\hat{\mathbf{r}}_2 = \int_{(k-1)T_b}^{kT_b} \mathbf{r}(t)\hat{\phi}_2(t)\mathrm{d}t$ and comparing $\hat{\mathbf{r}}_2$ to the threshold

$$T = \frac{\hat{s}_{12} + \hat{s}_{22}}{2} + \frac{N_0}{2(\hat{s}_{22} - \hat{s}_{12})} \ln\left(\frac{P_1}{P_2}\right). \tag{5.94}$$

When a "0" is transmitted, $\hat{\mathbf{r}}_2$ is a Gaussian random variable of mean $\hat{s}_{12}$ and variance $N_0/2$ (watts). When a "1" is transmitted $\hat{\mathbf{r}}_2$ is a Gaussian random variable of mean $\hat{s}_{22}$ and variance $N_0/2$ (watts). Graphically the conditional density functions $f(\hat{r}_2|0_T)$ and $f(\hat{r}_2|1_T)$ are illustrated in Figure 5.28.

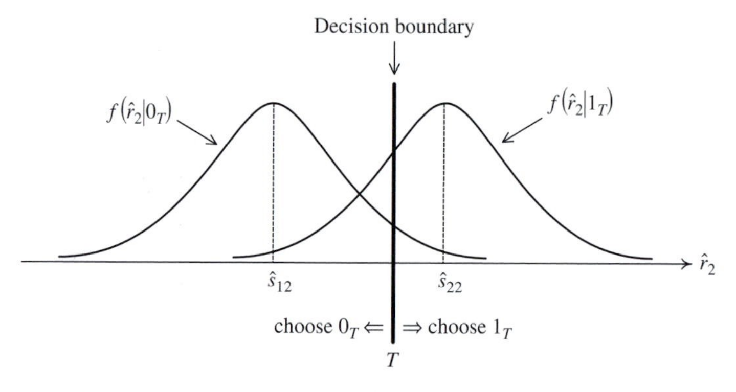

Fig. 5.28 Conditional density functions.

The probability of making an error is given by

$$\begin{aligned} P[\text{error}] &= P[(0 \text{ transmitted and } 1 \text{ decided}) \text{ or } (1 \text{ transmitted and } 0 \text{ decided})] \\ &= P[(0_T, 1_D) \text{ or } (1_T, 0_D)]. \end{aligned} \tag{5.95}$$

Since the two events are mutually exclusive, one has

$$\begin{aligned} P[\text{error}] &= P[0_T, 1_D] + P[1_T, 0_D] \\ &= P[1_D|0_T]P[0_T] + P[0_D|1_T]P[1_T] \\ &= P_1 \underbrace{\int_T^{\infty} f(\hat{r}_2|0_T)\mathrm{d}\hat{r}_2}_{\text{area } \mathbf{B}} + P_2 \underbrace{\int_{-\infty}^{T} f(\hat{r}_2|1_T)\mathrm{d}\hat{r}_2}_{\text{area } \mathbf{A}}. \end{aligned} \tag{5.96}$$

Note that the two integrals in (5.96) are equal to area **B** and area **A** in Figure 5.29, respectively. Thus the error probability can be calculated as follows:

$$\begin{aligned} P[\text{error}] = P_1 &\int_T^{\infty} \frac{1}{\sqrt{\pi N_0}} \exp\left\{-\frac{(\hat{r}_2 - \hat{s}_{12})^2}{N_0}\right\} \mathrm{d}\hat{r}_2 \\ + P_2 &\int_{-\infty}^{T} \frac{1}{\sqrt{\pi N_0}} \exp\left\{-\frac{(\hat{r}_2 - \hat{s}_{22})^2}{N_0}\right\} \mathrm{d}\hat{r}_2. \end{aligned} \tag{5.97}$$

Consider the first integral in (5.97). Change the variable $\lambda \equiv (\hat{r}_2 - \hat{s}_{12})/\sqrt{N_0/2}$, then $\mathrm{d}\lambda = \mathrm{d}\hat{r}_2/\sqrt{N_0/2}$ and the lower limit becomes $(T - \hat{s}_{12})/\sqrt{N_0/2}$. The integral therefore can be rewritten as

$$\begin{aligned} \int_T^{\infty} \frac{1}{\sqrt{\pi N_0}} \exp\left\{-\frac{(\hat{r}_2 - \hat{s}_{12})^2}{N_0}\right\} \mathrm{d}\hat{r}_2 &= \frac{1}{\sqrt{2\pi}} \int_{\frac{T-\hat{s}_{12}}{\sqrt{N_0/2}}}^{\infty} \exp\left(-\frac{\lambda^2}{2}\right) \mathrm{d}\lambda \\ &= Q\left(\frac{T - \hat{s}_{12}}{\sqrt{N_0/2}}\right), \end{aligned} \tag{5.98}$$

where $Q(x)$ is called the *Q-function*. This function is defined as the area under a zero-mean, unit-variance Gaussian curve from x to ∞ (see Figure 5.30 for a graphical interpretation). Mathematically,

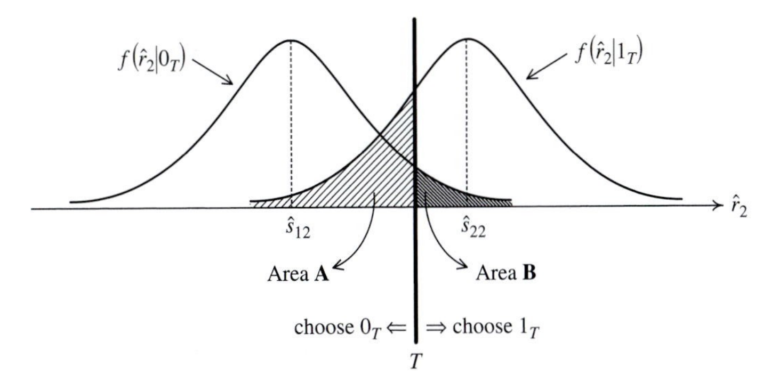

Fig. 5.29 Evaluation of the integrals by areas A and B.

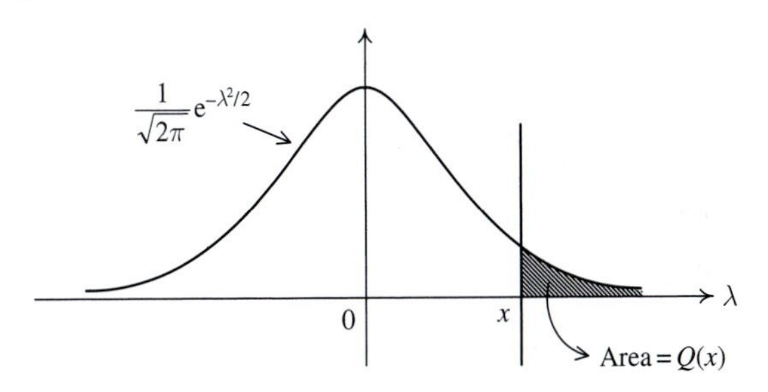

Fig. 5.30 Illustration of the Q-function.

$$Q(x) \equiv \frac{1}{\sqrt{2\pi}} \int_x^{\infty} \exp\left(-\frac{\lambda^2}{2}\right) d\lambda, \quad -\infty \leq x \leq \infty. \tag{5.99}$$

A plot of $Q(x)$ is shown in Figure 5.31. It is also simple to verify the following useful property of the Q-function:

$$Q(x) = 1 - Q(-x), \quad -\infty \leq x \leq \infty. \tag{5.100}$$

Similarly, the second integral in (5.97) can be evaluated as follows:

$$\begin{aligned}
&\int_{-\infty}^{T} \frac{1}{\sqrt{\pi N_0}} \exp\left\{-\frac{(\hat{r}_2 - \hat{s}_{22})^2}{N_0}\right\} d\hat{r}_2 \\
&= 1 - \int_T^{\infty} \frac{1}{\sqrt{\pi N_0}} \exp\left\{-\frac{(\hat{r}_2 - \hat{s}_{22})^2}{N_0}\right\} d\hat{r}_2 \\
&= 1 - \frac{1}{\sqrt{2\pi}} \int_{\frac{T-\hat{s}_{22}}{\sqrt{N_0/2}}}^{\infty} \exp\left(-\frac{\lambda^2}{2}\right) d\lambda \\
&= 1 - Q\left(\frac{T - \hat{s}_{22}}{\sqrt{N_0/2}}\right),
\end{aligned} \tag{5.101}$$

where the second equality in the above evaluation follows by changing $\lambda = (\hat{r}_2 - \hat{s}_{22})/\sqrt{N_0/2}$.

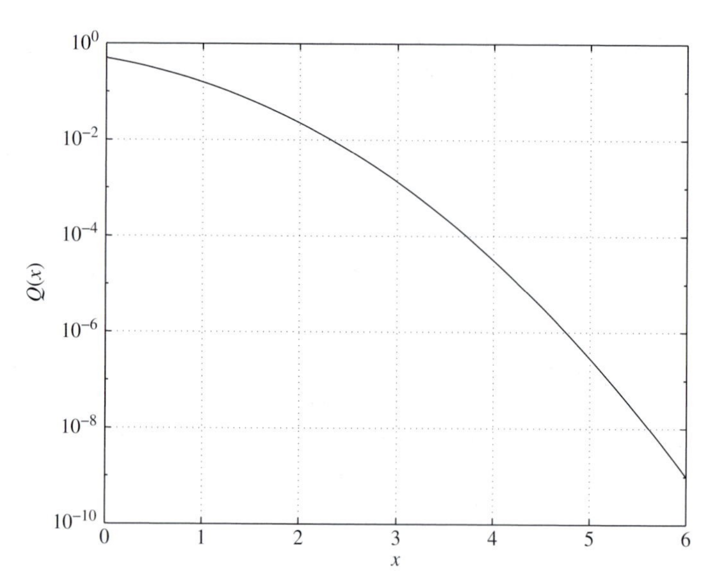

Fig. 5.31 A plot of $Q(x)$.

Therefore, in terms of the Q-function, the probability of error can be compactly expressed as

$$P[\text{error}] = P_1 Q\left(\frac{T - \hat{s}_{12}}{\sqrt{N_0/2}}\right) + P_2\left[1 - Q\left(\frac{T - \hat{s}_{22}}{\sqrt{N_0/2}}\right)\right]. \tag{5.102}$$

Consider now the important case where the *a priori* probabilities are equal, i.e., $P_1 = P_2$. The threshold T becomes $T = (\hat{s}_{12} + \hat{s}_{22})/2$. Substituting this threshold into (5.102) and applying the property $Q(x) = 1 - Q(-x)$ of the Q-function simplifies the expression of the error probability to

$$P[\text{error}] = Q\left(\frac{\hat{s}_{22} - \hat{s}_{12}}{2\sqrt{N_0/2}}\right). \tag{5.103}$$

The term $(\hat{s}_{22} - \hat{s}_{12})$ can be further evaluated as follows:

$$\begin{aligned}
\hat{s}_{22} - \hat{s}_{12} &= \int_0^{T_b} s_2(t)\hat{\phi}_2(t)\mathrm{d}t - \int_0^{T_b} s_1(t)\hat{\phi}_2(t)\mathrm{d}t \\
&= \int_0^{T_b} [s_2(t) - s_1(t)]\hat{\phi}_2(t)\mathrm{d}t \\
&= \int_0^{T_b} [s_2(t) - s_1(t)]\frac{s_2(t) - s_1(t)}{(E_2 - 2\rho\sqrt{E_1E_2} + E_1)^{1/2}}\mathrm{d}t \\
&= \frac{\int_0^{T_b} [s_2(t) - s_1(t)]^2\mathrm{d}t}{(E_2 - 2\rho\sqrt{E_1E_2} + E_1)^{1/2}} = \frac{d_{21}^2}{d_{21}} = d_{21}.
\end{aligned} \tag{5.104}$$

The above result shows that $\hat{s}_{22} - \hat{s}_{12}$ simply measures the distance between the two signals $s_1(t)$ and $s_2(t)$ (see (5.21)), which is consistent with what is observed graphically in

Figure 5.24(b). The above evaluation also shows that, as long as $s_1(t) \neq s_2(t)$, the quantity $\hat{s}_{22} - \hat{s}_{12}$ is indeed positive. This is the assumption we used to arrive at the decision rule in (5.91).

Since the term $\sqrt{N_0/2}$ is interpreted as the RMS value of the noise at the output of the correlator or matched filter after sampling, one can write the expression for the error performance as

$$P[\text{error}] = Q\left(\frac{\text{distance between the signals}}{2 \times \text{noise RMS value}}\right). \tag{5.105}$$

$Q(\cdot)$ is a *monotonically decreasing* function of its argument, the probability of error decreases as the ratio (distance between the signals/$2 \times$ noise RMS value) increases, i.e., as either the two signals become more dissimilar (increasing the distance between them) or the noise power becomes less. Both factors of course make it easier to distinguish between the two possible transmitted signals at the receiver.

Typically the channel noise power is fixed and thus the only way to reduce error is by maximizing the distance between the two signals. One way of doing this is, of course, by increasing the signal energy.

However, the transmitter also has an energy constraint, say $\sqrt{E}$. Recall that in the signal space representation the distance from a signal point to the origin of the $\{\phi_1(t), \phi_2(t)\}$ plane is simply the square root of the signal energy. Thus the signals must lie on or inside a circle of radius $\sqrt{E}$. Therefore, to maximize the distance between the two signals one chooses them so that they are placed 180° from each other. This implies that $s_2(t) = -s_1(t)$ (you might want to show this). This very popular signal set is commonly known as *antipodal signaling*.

A last observation is that the error probability does *not* depend on the signal shapes but only on the distance between them. Vastly different signal sets (in terms of time waveforms) will lead to the same error performance. This is due to the fact that our noise is considered to be white[5] and Gaussian.

Relationship between Q(x) and erfc(x) Besides the Q-function, another function widely used in error probability calculation is the *complementary error function*, denoted by erfc(·). The erfc function is defined as follows:

$$\text{erfc}(x) = \frac{2}{\sqrt{\pi}}\int_x^{\infty} e^{-\lambda^2}\,d\lambda \tag{5.106}$$

$$= 1 - \text{erf}(x). \tag{5.107}$$

By change of variables, it is not hard to show that the erfc function and the Q function are related by

$$Q(x) = \frac{1}{2}\text{erfc}\left(\frac{x}{\sqrt{2}}\right), \tag{5.108}$$

[5] "You can run but you cannot hide." Because white noise has equal power at each frequency one can say that the signals can run but cannot hide. The challenge to boxing fans is: What famous person uttered this phrase and in what circumstance?

or conversely

$$\text{erfc}(x) = 2Q(\sqrt{2}x). \tag{5.109}$$

Moreover, let $Q^{-1}(x)$ and $\text{erfc}^{-1}(x)$ be the inverses of $Q(x)$ and $\text{erfc}(x)$, respectively. Then these functions are related by

$$Q^{-1}(x) = \sqrt{2}\text{erfc}^{-1}(2x). \tag{5.110}$$

It should be pointed out that $\text{erfc}(x)$ and $\text{erfc}^{-1}(x)$ are available in Matlab, not $Q(x)$ nor $Q^{-1}(x)$. However, the above relationships allow one to easily write new Matlab functions for $Q(x)$ and $Q^{-1}(x)$.

Finally, the next example is provided to illustrate some of the most important concepts introduced in this chapter.

Example 5.9 Consider the signal space diagram shown in Figure 5.32.

(a) Determine and sketch the two signals $s_1(t)$ and $s_2(t)$.
(b) The two signals $s_1(t)$ and $s_2(t)$ are used for the transmission of equally likely bits 0 and 1, respectively, over an additive white Gaussian noise (AWGN) channel. Clearly draw the decision boundary and the decision regions of the optimum receiver. Write the expression for the optimum decision rule.
(c) Find and sketch the two orthonormal basis functions $\hat{\phi}_1(t)$ and $\hat{\phi}_2(t)$ such that the optimum receiver can be implemented using only the projection $\hat{\mathbf{r}}_2$ of the received signal $\mathbf{r}(t)$ onto the basis function $\hat{\phi}_2(t)$. Draw the block diagram of such a receiver that uses a matched filter.

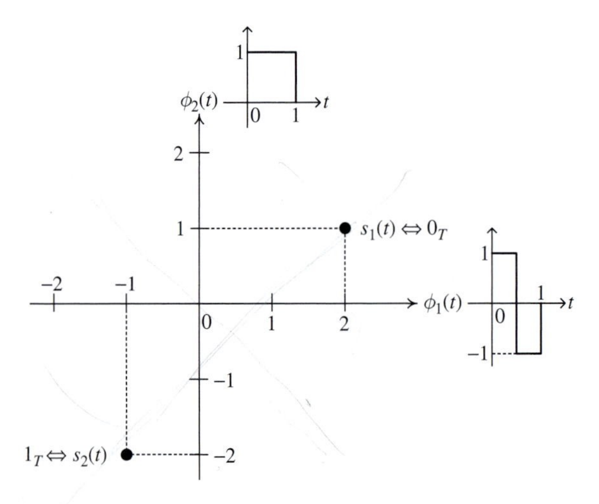

Fig. 5.32 Signal set for Example 5.9.

(d) Consider now the following argument put forth by your classmate. She reasons that since the component of the signals along $\hat{\phi}_1(t)$ is not useful at the receiver in determining which bit was transmitted, one should not even transmit this component of the signal. Thus she modifies the transmitted signal as follows:

$$s_1^{(\mathrm{M})}(t) = s_1(t) - \left(\text{component of } s_1(t) \text{ along } \hat{\phi}_1(t)\right) \tag{5.111}$$

$$s_2^{(\mathrm{M})}(t) = s_2(t) - \left(\text{component of } s_2(t) \text{ along } \hat{\phi}_1(t)\right) \tag{5.112}$$

Clearly identify the locations of $s_1^{(\mathrm{M})}(t)$ and $s_2^{(\mathrm{M})}(t)$ in the signal space diagram. What is the average energy of this signal set? Compare it to the average energy of the original set. Comment.

Solution

(a) The two signals $s_1(t)$ and $s_2(t)$ can be determined simply from their coordinates as

$$s_1(t) = s_{11}\phi_1(t) + s_{12}\phi_2(t) = 2\phi_1(t) + \phi_2(t), \tag{5.113}$$

$$s_2(t) = s_{21}\phi_1(t) + s_{22}\phi_2(t) = -\phi_1(t) - 2\phi_2(t). \tag{5.114}$$

The two signals are plotted in Figure 5.33.

(b) Since the two binary bits are equally likely, the decision boundary of the optimum receiver is the bisector of the line joining the two signals. The optimum decision

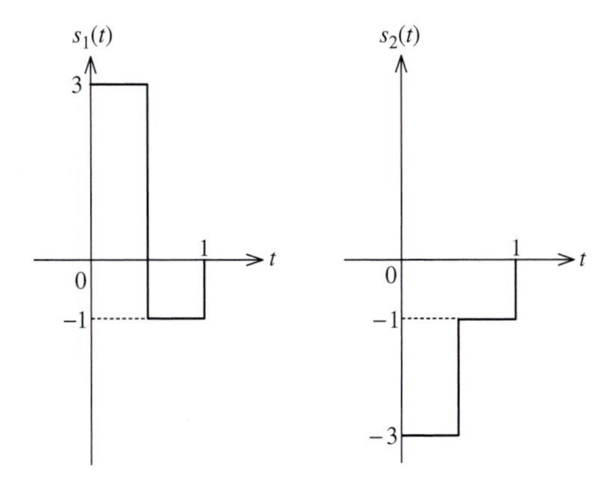

Fig. 5.33 Plots of $s_1(t)$ and $s_2(t)$.

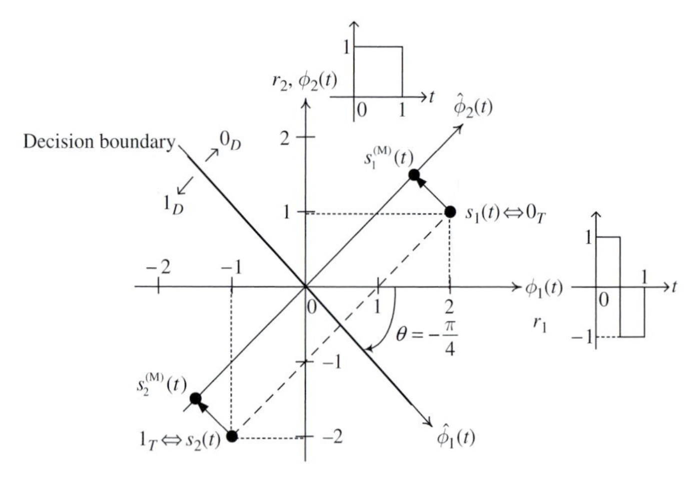

Fig. 5.34 Optimum decision boundary and decision regions.

boundary and the decision regions are shown in Figure 5.34. A simple inspection of the decision boundary in Figure 5.34 gives the following optimum decision rule:

$$r_1 \underset{1_D}{\overset{0_D}{\gtrless}} -r_2. \tag{5.115}$$

Of course, the above expression for the optimum decision rule can also be reached by substituting all the signal coordinates into the following fundamental minimum distance rule:

$$(r_1 - s_{21})^2 + (r_2 - s_{22})^2 \underset{1_D}{\overset{0_D}{\gtrless}} (r_1 - s_{11})^2 + (r_2 - s_{12})^2. \tag{5.116}$$

(c) From the signal space diagram in Figure 5.34, it is clear that the orthonormal basis functions $\hat{\phi}_1(t)$ and $\hat{\phi}_2(t)$ are obtained by rotating $\phi_1(t)$ and $\phi_2(t)$ by 45° clockwise (i.e., $\theta = -\pi/4$), or 135° counterclockwise (i.e., $\theta = 3\pi/4$). This rotation is to ensure that $\hat{\phi}_1(t)$ is perpendicular to the line joining the two signals. Choosing $\theta = -\pi/4$ yields:

$$\begin{aligned}\begin{bmatrix}\hat{\phi}_1(t)\\ \hat{\phi}_2(t)\end{bmatrix} &= \begin{bmatrix}\cos(-\pi/4) & \sin(-\pi/4)\\ -\sin(-\pi/4) & \cos(-\pi/4)\end{bmatrix}\begin{bmatrix}\phi_1(t)\\ \phi_2(t)\end{bmatrix}\\ &= \begin{bmatrix}\frac{1}{\sqrt{2}} & -\frac{1}{\sqrt{2}}\\ \frac{1}{\sqrt{2}} & \frac{1}{\sqrt{2}}\end{bmatrix}\begin{bmatrix}\phi_1(t)\\ \phi_2(t)\end{bmatrix}.\end{aligned} \tag{5.117}$$

It follows that

$$\hat{\phi}_1(t) = \frac{1}{\sqrt{2}}[\phi_1(t) - \phi_2(t)], \tag{5.118}$$

$$\hat{\phi}_2(t) = \frac{1}{\sqrt{2}}[\phi_1(t) + \phi_2(t)]. \tag{5.119}$$

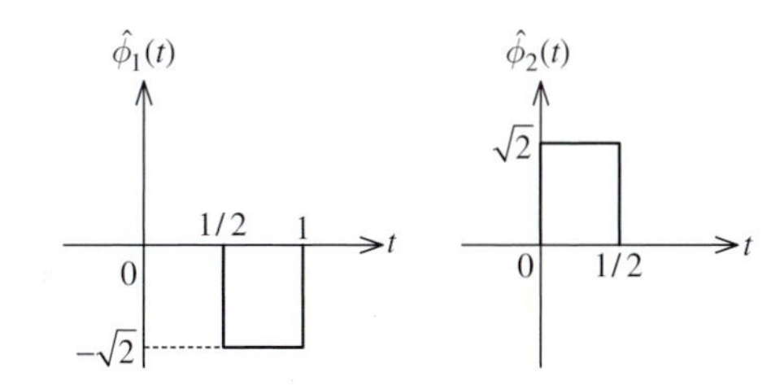

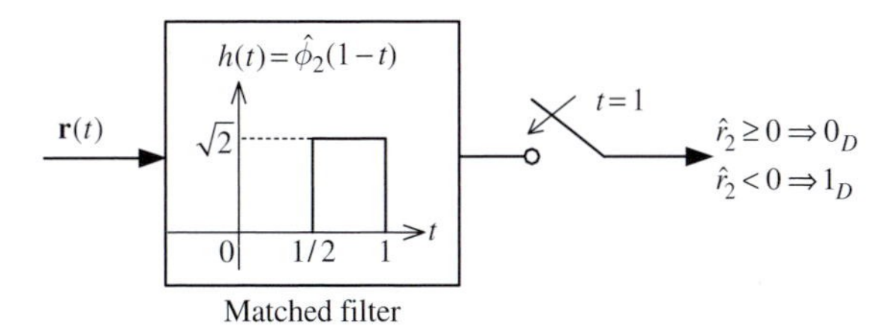

Fig. 5.35 Plots of $\hat{\phi}_1(t)$ and $\hat{\phi}_2(t)$ and the simplified receiver.

These two functions are plotted in Figure 5.35, together with the block diagram of a receiver that uses one matched filter.

(d) The locations of $s_1^{(\mathrm{M})}(t)$ and $s_2^{(\mathrm{M})}(t)$ are shown on the signal space diagram in Figure 5.34. The average energy of this signal set is simply

$$\begin{aligned} E^{(\mathrm{M})} &= \frac{1}{2}(\hat{s}_{12}^2 + \hat{s}_{22}^2) = \frac{1}{2}[(E_1 - \hat{s}_{11}^2) + (E_2 - \hat{s}_{21}^2)] \\ &= \frac{1}{2}(E_1 + E_2) - \frac{1}{2}(\hat{s}_{11}^2 + \hat{s}_{21}^2) = 5 - \left(\frac{1}{\sqrt{2}}\right)^2 \\ &= 4.5 \text{ (joules)}. \end{aligned} \tag{5.120}$$

The average energy of the modified signal set is clearly smaller than the average energy of the original set, which is $(E_1 + E_2)/2 = 5$ joules. Since the distance between the modified signals is the same as that of the original signals, both sets perform identically in terms of the bit error probability performance. The modified set is therefore preferred due to its better energy (or power) efficiency. It should be pointed out, however, that the other set might have better timing or spectral properties. Besides error performance, the issues of timing and bandwidth are also very important considerations and will be explored further in the next chapter. The next two sections derive general expressions of PSD for some families of digital modulation. ■

5.7 Power spectral density of digital modulation

In this section we derive an expression for the PSD of digital signals that, in essence, are produced by amplitude modulation. A general block diagram of the modulation process is shown in Figure 5.36.

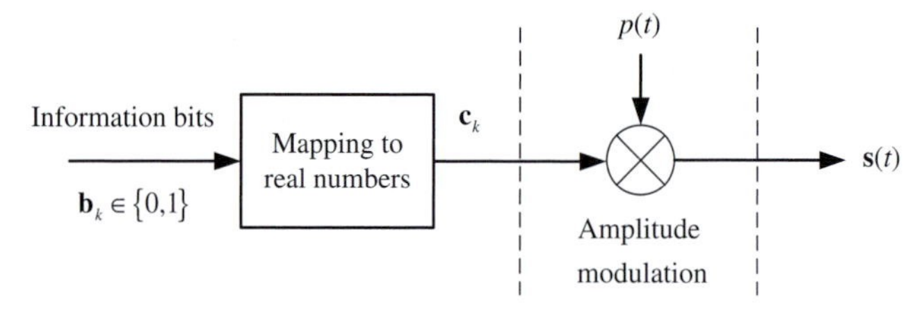

Fig. 5.36 General block diagram of digital amplitude modulation.

Typically $\mathbf{c}_k$ takes on the values from the set $\{-1, +1\}$ (antipodal signaling), $\{0, 1\}$ (on–off keying), $\{-1, 0, +1\}$ (pseudoternary line coding) or $\{\pm 1, \pm 3, \cdots, \pm(M-1)\}$ (M-ary amplitude-shift keying (ASK)). These modulations are discussed in the next three chapters. Here the important thing about $\mathbf{c}_k$ is that it is drawn from a finite set of real numbers with a probability that is known or can be determined. The function $p(t)$ is a pulse waveform of duration T_b, i.e., it is zero outside the interval $[0, T_b]$. The transmitted signal is

$$\mathbf{s}(t) = \sum_{k=-\infty}^{\infty} \mathbf{c}_k p(t - kT_b). \tag{5.121}$$

To find the PSD of $\mathbf{s}(t)$ we follow the procedure of Chapter 3 where the basic definition of PSD was established. First, truncate the random process to a time interval of $-T = -NT_b$ to $T = NT_b$, i.e., $2N + 1$ bit durations. This gives

$$\mathbf{s}_T(t) = \sum_{k=-N}^{N} \mathbf{c}_k p(t - kT_b). \tag{5.122}$$

Take the Fourier transform of the truncated process:

$$\mathbf{S}_T(f) = \sum_{k=-\infty}^{\infty} \mathbf{c}_k \mathcal{F}\{p(t - kT_b)\} = P(f) \sum_{k=-\infty}^{\infty} \mathbf{c}_k \mathrm{e}^{-\mathrm{j}2\pi f k T_b}. \tag{5.123}$$

Now apply the basic definition of (3.67):

$$S(f) = \lim_{T\to\infty} \frac{E\left\{|\mathbf{S}_T(f)|^2\right\}}{2T} = \lim_{N\to\infty} \frac{|P(f)|^2}{(2N+1)T_b} E\left\{\left|\sum_{k=-N}^{N} \mathbf{c}_k \mathrm{e}^{-\mathrm{j}2\pi f k T_b}\right|^2\right\}. \tag{5.124}$$

The expectation in (5.124) is computed as follows:

$$\begin{aligned} E&\left\{\left|\sum_{k=-N}^{N} \mathbf{c}_k \mathrm{e}^{-\mathrm{j}2\pi f k T_b}\right|^2\right\} \\ &= E\left\{\left[\sum_{k=-N}^{N} \mathbf{c}_k \mathrm{e}^{-\mathrm{j}2\pi f k T_b}\right]\left[\sum_{l=-N}^{N} \mathbf{c}_l \mathrm{e}^{-\mathrm{j}2\pi f l T_b}\right]^*\right\} \end{aligned}$$

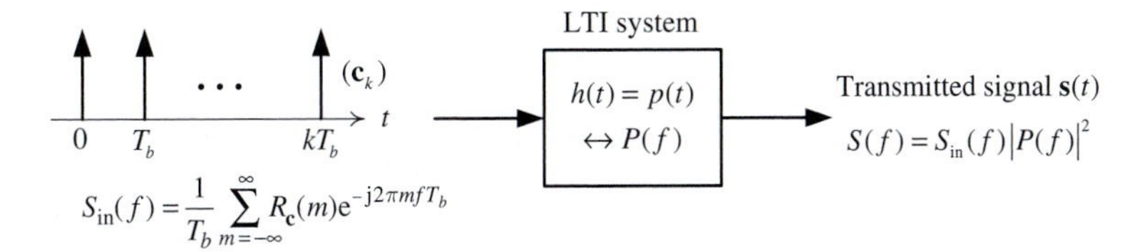

Fig. 5.37 Impulse modulator model.

$$= \sum_{k=-N}^{N}\sum_{l=-N}^{N} E\{\mathbf{c}_k\mathbf{c}_l\}\, e^{-j2\pi f(k-l)T_b}$$

$$\stackrel{m=k-l}{=} \sum_{k=-N}^{N}\sum_{m=k+N}^{k-N} R_{\mathbf{c}}(m)e^{-j2\pi mfT_b}, \tag{5.125}$$

where $R_{\mathbf{c}}(m) = E\{\mathbf{c}_k\mathbf{c}_{k-m}\}$ is the (discrete) autocorrelation of the sequence $\{\mathbf{c}_k\}$, with $R_{\mathbf{c}}(m) = R_{\mathbf{c}}(-m)$.

The important observation is that in (5.125) the inner summation becomes the same, regardless of the index m, as $N \to \infty$. The outer summation then gives $(2N+1)$ inner summations, i.e., for large N, (5.125) becomes

$$E\left\{\left|\sum_{k=-N}^{N} \mathbf{c}_k e^{-j2\pi fkT_b}\right|^2\right\} = (2N+1)\sum_{m=k+N}^{k-N} R_{\mathbf{c}}(m)e^{-j2\pi mfT_b}. \tag{5.126}$$

Therefore,

$$S(f) = \frac{|P(f)|^2}{T_b}\lim_{N\to\infty}\frac{2N+1}{2N+1}\sum_{m=k+N}^{k-N} R_{\mathbf{c}}(m)e^{-j2\pi mfT_b}$$

$$= \frac{|P(f)|^2}{T_b}\sum_{m=-\infty}^{\infty} R_{\mathbf{c}}(m)e^{-j2\pi mfT_b}. \tag{5.127}$$

Basically (5.127) tells us that the output PSD is the input PSD, $(1/T_b)\sum_{m=-\infty}^{\infty} R_{\mathbf{c}}(m)e^{-j2\pi mfT_b}$, multiplied by $|P(f)|^2$, a transfer function. The production of $\mathbf{s}(t)$ can be modeled by what is called *impulse modulator model*. This model is shown in Figure 5.37.

5.8 A PSD derivation for an arbitrary binary modulation

The previous PSD derivation applies for antipodal signaling, in fact to any digital amplitude modulation. However, a modulation scheme such as binary frequency-shift keying (BFSK, to be discussed in Chapter 7) does not fall into the above model. This section presents an approach that is applicable to *any* binary modulation with *arbitrary a priori* probabilities of the information bits. However, it is restricted to the input symbols being *statistically independent*.

Consider binary modulation in which the two signals $s_1(t)$ and $s_2(t)$ are used to transmit information bits "0" and "1," respectively. The bits are statistically independent (from bit to bit) and their *a priori* probabilities are $P[0] = P_1$ and $P[1] = P_2$. Modulations such as NRZ-L, RZ-L, Biϕ, BASK, BPSK, and BFSK, which are discussed in the next two chapters, fall in this category, but for now, no specific forms are assumed for $s_1(t)$ and $s_2(t)$.

An example of the transmitted signal is shown in Figure 5.38. In general, the transmitted signal, viewed as a random process, can be analytically expressed as

$$\mathbf{s}_T(t) = \sum_{k=-\infty}^{\infty} \mathbf{g}_k(t), \tag{5.128}$$

where

$$\mathbf{g}_k(t) = \begin{cases} s_1(t - kT_b), & \text{with probability } P_1 \\ s_2(t - kT_b), & \text{with probability } P_2 \end{cases}. \tag{5.129}$$

The process $\mathbf{s}_T(t)$ can be decomposed into the sum of a DC and an AC component as follows:

$$\mathbf{s}_T(t) = \underbrace{E\{\mathbf{s}_T(t)\}}_{\text{DC}} + \underbrace{\mathbf{s}_T(t) - E\{\mathbf{s}_T(t)\}}_{\text{AC}} = v(t) + \mathbf{q}(t). \tag{5.130}$$

Let $S_{\mathbf{s}_T}(f)$, $S_v(f)$, and $S_{\mathbf{q}}(f)$ denote the PSD of $\mathbf{s}_T(t)$, $v(t)$, and $\mathbf{q}(t)$, respectively. Obviously, $S_{\mathbf{s}_T}(f) = S_v(f) + S_{\mathbf{q}}(f)$. Since $v(t)$ is a deterministic signal (why?), the calculation of $S_v(f)$ is relatively easy.

Write $v(t)$ as follows:

$$\begin{aligned} v(t) = E\{\mathbf{s}_T(t)\} &= \sum_{k=-\infty}^{\infty} E\{\mathbf{g}_k(t)\} \\ &= \sum_{k=-\infty}^{\infty} [P_1 s_1(t - kT_b) + P_2 s_2(t - kT_b)]. \end{aligned} \tag{5.131}$$

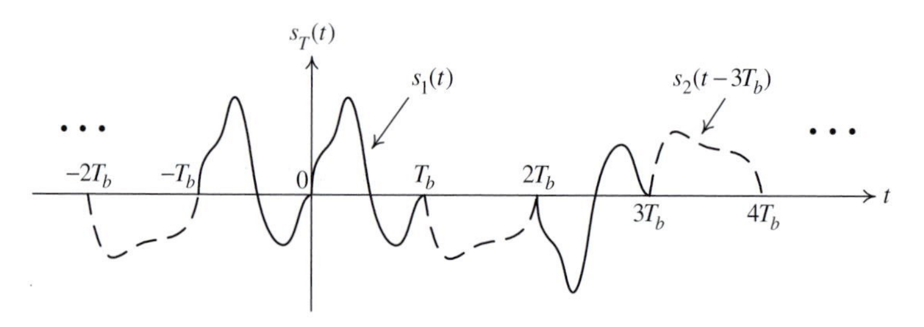

Fig. 5.38 Transmitted signal in binary modulation.

Recognize that $v(t+T_b)=v(t)$ (see Problem 5.26), i.e., $v(t)$ is a periodic signal. Therefore,

$$S_v(f)=\sum_{n=-\infty}^{\infty}|D_n|^2\delta\left(f-\frac{n}{T_b}\right), \tag{5.132}$$

where the D_n are the coefficients of the Fourier series expansion of $v(t)$. They are determined as follows:

$$\begin{aligned}
D_n &= \frac{1}{T_b}\int_{-T_b/2}^{T_b/2} v(t)\mathrm{e}^{-\mathrm{j}2\pi nt/T_b}\mathrm{d}t \\
&= \frac{1}{T_b}\int_{-T_b/2}^{T_b/2} \mathrm{e}^{-\mathrm{j}2\pi nt/T_b}\sum_{k=-\infty}^{\infty}[P_1s_1(t-kT_b)+P_2s_2(t-kT_b)]\mathrm{d}t \\
&\stackrel{\lambda=t-kT_b}{=} \frac{1}{T_b}\sum_{k=-\infty}^{\infty}\int_{(-T_b/2)-kT_b}^{(T_b/2)-kT_b}[P_1s_1(\lambda)+P_2s_2(\lambda)]\mathrm{e}^{-\mathrm{j}2\pi n(\lambda+kT_b)/T_b}\mathrm{d}\lambda \\
&= \frac{1}{T_b}\int_{-\infty}^{\infty}[P_1s_1(\lambda)+P_2s_2(\lambda)]\mathrm{e}^{-\mathrm{j}2\pi n\lambda/T_b}\mathrm{d}\lambda \\
&= \frac{1}{T_b}\left[P_1S_1\left(\frac{n}{T_b}\right)+P_2S_2\left(\frac{n}{T_b}\right)\right],
\end{aligned} \tag{5.133}$$

where $S_1(f)$ and $S_2(f)$ are the Fourier transforms of $s_1(t)$ and $s_2(t)$, respectively. Using (5.133) in (5.132) the PSD of $v(t)$ is

$$S_v(f)=\sum_{n=-\infty}^{\infty}\left|\frac{P_1S_1(n/T_b)+P_2S_2(n/T_b)}{T_b}\right|^2\delta\left(f-\frac{n}{T_b}\right). \tag{5.134}$$

Next, we calculate $S_{\mathbf{q}}(f)$, the PSD of $\mathbf{q}(t)$. To this end, apply the basic definition of power spectral density, i.e., let $\mathbf{q}_T(t)$ denote the *truncated* version of $\mathbf{q}(t)$ to the length T so that

$$S_{\mathbf{q}}(f)=\lim_{T\to\infty}\frac{E\{|\mathbf{Q}_T(f)|^2\}}{T}\quad \text{(watts/hertz)} \tag{5.135}$$

where $\mathbf{Q}_T(f)$ is the Fourier transform of $\mathbf{q}_T(t)$ and T can be chosen as $T=(2N+1)T_b$, with N a very large integer.

Now,

$$\begin{aligned}
\mathbf{q}_T(t) &= [\mathbf{s}_T(t)-E\{\mathbf{s}_T(t)\}]_{\text{truncated}} \\
&= \sum_{k=-N}^{N}\{\mathbf{g}_k(t)-[P_1s_1(t-kT_b)+P_2s_2(t-kT_b)]\} \\
&= \sum_{k=-N}^{N}\mathbf{q}_{T,k}(t),
\end{aligned} \tag{5.136}$$

where

$$\mathbf{q}_{T,k}(t)=\begin{cases} P_2[s_1(t-kT_b)-s_2(t-kT_b)], & \text{with probability } P_1 \\ -P_1[s_1(t-kT_b)-s_2(t-kT_b)], & \text{with probability } P_2 \end{cases}, \tag{5.137}$$

i.e., within each bit interval, $\mathbf{q}_{T,k}(t)$ can be only one of two possible signals. Another way to express $\mathbf{q}_{T,k}(t)$ is to write

$$\mathbf{q}_{T,k}(t) = \mathbf{a}_k[s_1(t - kT_b) - s_2(t - kT_b)], \tag{5.138}$$

where $\mathbf{a}_k$ is a *random variable* that can take only one of two values:[6]

$$\mathbf{a}_k = \begin{cases} P_2, & \text{with probability } P_1 \\ -P_1, & \text{with probability } P_2 \end{cases}. \tag{5.139}$$

Using (5.138) in (5.136), the Fourier transform of $\mathbf{q}_T(t)$ is

$$\begin{aligned}
\mathbf{Q}_T(f) &= \int_{-\infty}^{\infty} \mathbf{q}_T(t) \mathrm{e}^{-\mathrm{j}2\pi ft} \mathrm{d}t \\
&= \int_{-\infty}^{\infty} \sum_{k=-N}^{N} \mathbf{a}_k[s_1(t - kT_b) - s_2(t - kT_b)] \mathrm{e}^{-\mathrm{j}2\pi ft} \mathrm{d}t \\
&= \sum_{k=-N}^{N} \mathbf{a}_k \int_{-\infty}^{\infty} [s_1(t - kT_b) - s_2(t - kT_b)] \mathrm{e}^{-\mathrm{j}2\pi ft} \mathrm{d}t \\
&= \sum_{k=-N}^{N} \mathbf{a}_k [S_1(f) - S_2(f)] \mathrm{e}^{-\mathrm{j}2\pi kfT_b}.
\end{aligned} \tag{5.140}$$

It follows that

$$\begin{aligned}
|\mathbf{Q}_T(f)|^2 &= \mathbf{Q}_T(f)[\mathbf{Q}_T(f)]^* \\
&= \left\{ \sum_{k=-N}^{N} \mathbf{a}_k [S_1(f) - S_2(f)] \mathrm{e}^{-\mathrm{j}2\pi kfT_b} \right\} \\
&\quad \times \left\{ \sum_{m=-N}^{N} \mathbf{a}_m [S_1(f) - S_2(f)]^* \mathrm{e}^{\mathrm{j}2\pi mfT_b} \right\} \\
&= \sum_{k=-N}^{N} \sum_{m=-N}^{N} \mathbf{a}_k \mathbf{a}_m |S_1(f) - S_2(f)|^2 \mathrm{e}^{\mathrm{j}2\pi(m-k)fT_b}.
\end{aligned} \tag{5.141}$$

The expected value of (5.141) is given by

$$E\left\{|\mathbf{Q}_T(f)|^2\right\} = \sum_{k=-N}^{N} \sum_{m=-N}^{N} E\{\mathbf{a}_k \mathbf{a}_m\} |S_1(f) - S_2(f)|^2 \mathrm{e}^{\mathrm{j}2\pi(m-k)fT_b}. \tag{5.142}$$

[6] Be careful and clear about the two different meanings of P_1 and P_2 in (5.139).

The expectation $E\{\mathbf{a}_k\mathbf{a}_m\}$ in (5.142) is computed as follows.

(i) When $k = m$, one has

$$\mathbf{a}_k\mathbf{a}_m = \mathbf{a}_k^2 = \begin{cases} P_2^2, & \text{with probability } P_1 \\ P_1^2, & \text{with probability } P_2 \end{cases}. \tag{5.143}$$

Therefore $E\{\mathbf{a}_k^2\} = P_2^2P_1 + P_1^2P_2 = P_1P_2(P_2 + P_1) = P_1P_2$.

(ii) When $k \neq m$, then

$$\mathbf{a}_k\mathbf{a}_m = \begin{cases} P_2^2, & \text{with probability } P_1^2 \\ P_1^2, & \text{with probability } P_2^2 \\ -P_1P_2, & \text{with probability } 2P_1P_2 \end{cases}. \tag{5.144}$$

Therefore $E\{\mathbf{a}_k\mathbf{a}_m\} = P_2^2P_1^2 + P_1^2P_2^2 - (P_1P_2)(2P_1P_2) = 0$.

Using the above results in (5.142) we obtain:

$$\begin{aligned} E\left\{|\mathbf{Q}_T(f)|^2\right\} &= \sum_{k=-N}^{N} P_1P_2|S_1(f) - S_2(f)|^2 \\ &= P_1P_2|S_1(f) - S_2(f)|^2(2N+1). \end{aligned} \tag{5.145}$$

Substituting (5.145) into (5.135) yields the following PSD of $\mathbf{g}(t)$:

$$\begin{aligned} S_{\mathbf{q}}(f) &= \lim_{T\to\infty} \frac{E\{|\mathbf{Q}_T(f)|^2\}}{T} \\ &= \lim_{N\to\infty} \frac{P_1P_2|S_1(f) - S_2(f)|^2(2N+1)}{(2N+1)T_b} \\ &= \frac{P_1P_2}{T_b}|S_1(f) - S_2(f)|^2. \end{aligned} \tag{5.146}$$

Finally, combining (5.146), (5.134), and (5.130) gives the following expression for the PSD of $\mathbf{s}_T(t)$:

$$\begin{aligned} S_{\mathbf{s}_T}(f) = &\frac{P_1P_2}{T_b}|S_1(f) - S_2(f)|^2 \\ &+ \sum_{n=-\infty}^{\infty} \left|\frac{P_1S_1(n/T_b) + P_2S_2(n/T_b)}{T_b}\right|^2 \delta\left(f - \frac{n}{T_b}\right). \end{aligned} \tag{5.147}$$

Equation (5.147) clearly shows that the PSD depends not only on the Fourier transforms of the two signals chosen to represent bits "0" and "1," but also on the *a priori* probabilities of the data from the source.

5.9 Summary

The foundation for the analysis, evaluation, and design of a wide spectrum of digital communication systems has been developed in this chapter. Since the approach is so fundamental, it is worthwhile to summarize the important features of it.

One starts with the modulator where a signal set comprising two members, $s_1(t)$ and $s_2(t)$, is chosen to represent the binary digits "0" and "1" respectively. Then an orthonormal set, $\{\phi_1(t), \phi_2(t)\}$, is selected to represent the two signals exactly. This representation is expressed geometrically in a signal space plot. The received signal is corrupted by AWGN, $\mathbf{w}(t)$, which is also represented by a series expansion using the orthonormal basis set, $\phi_1(t)$, $\phi_2(t)$, as dictated by $s_1(t)$, $s_2(t)$, and the set $\phi_3(t)$, $\phi_4(t)$, $\ldots$, which are used to complete the set. However, as shown the projections of the noise $\mathbf{w}(t)$ onto the basis functions $\phi_3(t)$, $\phi_4(t)$, $\ldots$, do not provide any information as to which signal was transmitted, they are irrelevant statistics. Therefore these projections can be discarded or ignored which means that in practice the basis functions $\phi_3(t)$, $\phi_4(t)$, $\ldots$, do not need to be determined.

Having developed a representation of the transmitted signal, the additive noise, and hence the received signal, $\mathbf{r}(t)$, attention was turned to the development of the demodulator. To proceed the criterion of minimizing bit error probability was chosen. This resulted in the likelihood ratio test of (5.65). One should observe that in the development of the likelihood ratio test no assumptions regarding the statistics of the received samples, $\mathbf{r}_1$, $\mathbf{r}_2$, $\mathbf{r}_3$, $\ldots$, were made. The developed test is therefore quite general; what one needs to do is to determine the two conditional densities, $f(\vec{r}|1_T)$ and $f(\vec{r}|0_T)$.

Determination of these conditional densities is greatly simplified when the important channel model of AWGN is invoked. This is because the noise projections are Gaussian and uncorrelated and because they are *Gaussian and uncorrelated they are statistically independent*. Therefore, as mentioned above, the projections of the received signal onto $\phi_3(t)$, $\phi_4(t)$, $\ldots$, can be ignored. The receiver, in this situation, has a very intuitive interpretation both algebraically and graphically using the signal space plot. Mainly it can be interpreted either as a minimum-distance receiver, i.e., choose the transmitter's signal to which the received signal is closest, or as a maximum correlation receiver, i.e., choose the transmitter's signal with which the received signal is most correlated. The minimum distance interpretation holds exactly when the transmitter's two signals are equally probable, which is the typical situation. If not, then the distance or correlation needs to reflect this *a priori* knowledge.

The next two chapters apply the concepts developed in this chapter to important baseband and passband modulation schemes for binary digital communication systems. Though a formal approach known as the Gram–Schmidt procedure was presented in this chapter to determine the orthonormal basis $\{\phi_1(t), \phi_2(t)\}$ it will be used sparingly, if at all, in the next two chapters. Indeed, the basis set will be determined by inspection. But to do this one needs to have a clear idea of what orthogonality is. Two signals, $s_1(t)$ and $s_2(t)$, are orthogonal over the interval of T seconds if and only if:

$$\int_T s_1(t)s_2(t)\mathrm{d}t = 0.$$

5.10 Problems

5.1 Verify the final expression of $\phi_2(t)$ in (5.15).

5.2 (*Coordinate rotation*) Assume that $s_1(t)$ and $s_2(t)$ are represented by an orthonormal basis set $\{\phi_1(t), \phi_2(t)\}$ as shown in Figure 5.39.

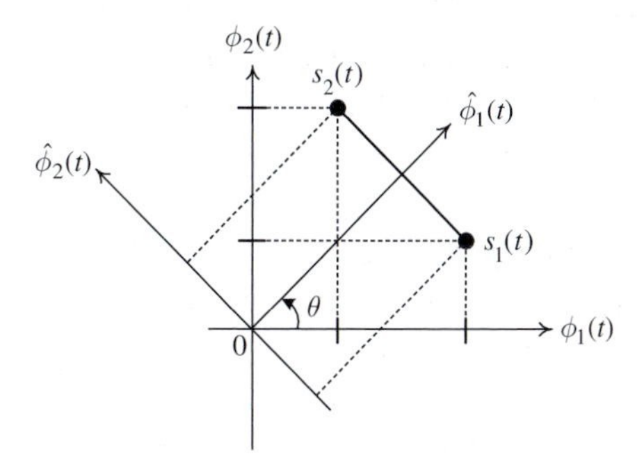

Fig. 5.39 Rotation of $\phi_1(t)$ and $\phi_2(t)$ by θ.

Now consider the rotation of $\phi_1(t)$ and $\phi_2(t)$ by an angle θ to obtain a new set of functions $\{\hat{\phi}_1(t), \hat{\phi}_2(t)\}$. This rotation can be expressed as

$$\begin{bmatrix} \hat{\phi}_1(t) \\ \hat{\phi}_2(t) \end{bmatrix} = \begin{bmatrix} \cos\theta & \sin\theta \\ -\sin\theta & \cos\theta \end{bmatrix} \begin{bmatrix} \phi_1(t) \\ \phi_2(t) \end{bmatrix}. \tag{P5.1}$$

(a) Show that, regardless of the angle θ, the set $\{\hat{\phi}_1(t), \hat{\phi}_2(t)\}$ is also an orthonormal basis set.

Remark Note that the above result means that the projection of white noise of spectral strength $N_0/2$ (watts/hertz) onto the basis set $\{\hat{\phi}_1(t), \hat{\phi}_2(t)\}$ still results in zero-mean uncorrelated random variables with variance $N_0/2$ (watts).

(b) What are the values of θ that make $\hat{\phi}_1(t)$ perpendicular to the line joining $s_1(t)$ to $s_2(t)$? For these values of θ, mathematically show that the components of $s_1(t)$ and $s_2(t)$ along $\hat{\phi}_1(t)$, namely $\hat{s}_{11}$ and $\hat{s}_{21}$, are identical.

5.3 The Q-function is defined as the area under a zero-mean, unit-variance, Gaussian curve from x to ∞. In determining the error performance of the receiver in digital communications, one often needs to compute the area from a threshold T to infinity under a general Gaussian density function with mean μ and variance σ^2 (see Figure 5.40). Show that such an area can be expressed as $Q((T-\mu)/\sigma)$, $-\infty \le T \le \infty$.

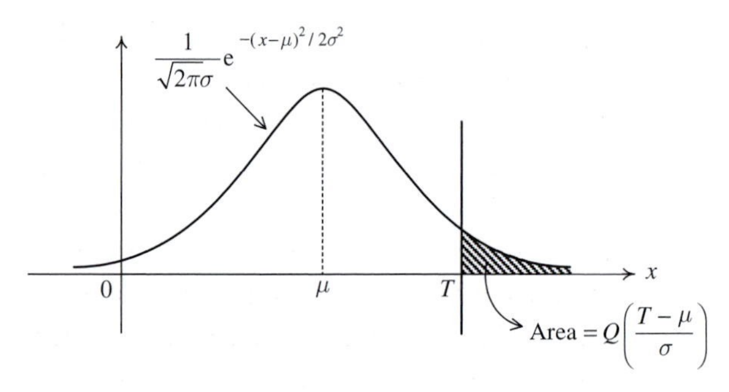

Fig. 5.40 The area under the right tail of a Gaussian pdf.

The next eight problems deal with the representation of signals using the signal space diagram. It is almost impossible to overestimate the importance of this concept in digital communications. The reader is strongly urged to attempt all the problems.

5.4 (a) Consider two arbitrary signals $s_1(t)$ and $s_2(t)$ whose energies are E_1 and E_2, respectively. Both signals are time-limited over $0 \le t \le T_b$. It is known that two *orthonormal* functions $\phi_1(t)$ and $\phi_2(t)$ can be used to represent $s_1(t)$ and $s_2(t)$ exactly as follows:

$$\begin{cases} s_1(t) = s_{11}\phi_1(t) + s_{12}\phi_2(t) \\ s_2(t) = s_{21}\phi_1(t) + s_{22}\phi_2(t) \end{cases}. \tag{P5.2}$$

Show that $E_i = s_{i1}^2 + s_{i2}^2$ by directly evaluating $\int_0^{T_b} s_i^2(t)\mathrm{d}t$, $i = 1, 2$.

(b) Let

$$\phi_1(t) = \begin{cases} \sqrt{2/T_b}\cos(2\pi f_c t), & 0 \le t \le T_b \\ 0, & \text{otherwise} \end{cases}, \tag{P5.3}$$

and

$$\phi_2(t) = \begin{cases} \sqrt{2/T_b}\sin(2\pi f_c t), & 0 \le t \le T_b \\ 0, & \text{otherwise} \end{cases}. \tag{P5.4}$$

Find the minimum value of frequency f_c that makes $\phi_1(t)$ and $\phi_2(t)$ orthogonal.

Remark The signal set considered in (b) is an important one in passband communication systems, not only binary but also M-ary.

5.5 Consider the following two signals that are time-limited to $[0, T_b]$:

$$s_1(t) = V\cos(2\pi f_c t), \tag{P5.5}$$

$$s_2(t) = V\cos(2\pi f_c t + \theta), \tag{P5.6}$$

where $f_c = k/2T_b$ and k is an integer.

(a) Find the energies of both signals. Then determine the value of V for which both signals have an unit energy.

(b) Determine the correlation coefficient ρ of the two signals. Recall that

$$\rho = \frac{1}{\sqrt{E_1 E_2}}\int_0^{T_b} s_1(t)s_2(t)\mathrm{d}t. \tag{P5.7}$$

(c) Plot ρ as a function of θ over the range $0 \le \theta \le 2\pi$. What is the value of θ that makes the two signals orthogonal?

(d) Verify that the distance between the two signals is $d = \sqrt{2E}\sqrt{1-\rho}$. What is the value of θ that maximizes the distance between the two signals?

Remark The above is another important signal set. Relate the value of θ found in (c) to Problem 5.4(b). Sketch the answers in (c) and (d) in the signal space to get geometrical insight. Finally to determine the energies in (a): you may do it directly or reason as follows. For a sinusoid of amplitude V with an integer number of cycles in the time interval T_b, its RMS (root mean-squared value) should be known by you from an elementary signals course. The average dissipated power across a 1 ohm resistor is then $V_{\text{RMS}}^2/1$ (watts). Multiplying by the time interval gives the average energy of the signal $V_{\text{RMS}}^2 T_b$ (watts $\times$ seconds = joules).

5.6 *(In contrast to Problem 5.5 where the phase of a sinusoid is used to distinguish between the two signals, here the frequency is used. Though the frequency can be chosen to maximize the distance between the signals, in practice the frequency separation is chosen so that the two signals are orthogonal. This greatly simplifies receiver design and also makes synchronization easier, as will be seen in later chapters.)*

Consider the following signal set over $0 \le t \le T_b$:

$$s_1(t) = V_1 \cos\left[2\pi\left(f_c - \frac{\Delta f}{2}\right)t\right], \qquad \text{(P5.8)}$$

$$s_2(t) = V_2 \cos\left[2\pi\left(f_c + \frac{\Delta f}{2}\right)t\right], \qquad \text{(P5.9)}$$

where $f_c = k/2T_b$ and k is an integer. The amplitudes V_1 and V_2 are adjusted so that regardless of Δf the energies E_1, E_2 are always the same and equal to $E = V^2 T_b/2$ joules.

(a) Determine and plot the correlation coefficient, ρ, as a function of Δf.

(b) Given that the distance between two equal-energy signals is $d = \sqrt{2E}\sqrt{1-\rho}$, show that the distance between the two above signals is maximum when $\Delta f = 0.715/T_b$. Compute the distance between $s_1(t)$ and $s_2(t)$ for this Δf. How much has the distance increased as compared to the case $\rho = 0$.

5.7 *(This problem results in an orthonormal set that is a subset of the well-known Legendre polynomials. As such they are not particularly important in communications. The problem is included to strengthen your understanding of the Gram–Schmidt procedure.)*

Using the Gram–Schmidt procedure, construct an orthonormal basis for the space of quadratic polynomials $\{a_2t^2 + a_1t + a_0;\ a_0, a_1, a_2 \in \mathbb{R}\}$ over the interval $-1 \le t \le 1$. *Hint* The equivalent problem is to find an orthonormal basis for three signals $\{1, t, t^2\}$ over the interval $-1 \le t \le 1$.

5.8 (*Simplex signal set*) Consider a set of M orthogonal signal waveforms $s_m(t)$, $1 \leq m \leq M$, $0 \leq t \leq T$, all of which have the same energy E. Define a new set of M waveforms as

$$\hat{s}_m(t) = s_m(t) - \frac{1}{M}\sum_{k=1}^{M} s_k(t), \quad 1 \leq m \leq M, \quad 0 \leq t \leq T. \tag{P5.10}$$

Show that the M signal waveforms $\{\hat{s}_m(t)\}$ have equal energy, given by

$$\hat{E} = (M-1)E/M \tag{P5.11}$$

and are equally correlated, with correlation coefficient

$$\rho_{mn} = \frac{1}{\hat{E}}\int_0^T \hat{s}_m(t)\hat{s}_n(t)\mathrm{d}t = -\frac{1}{M-1}. \tag{P5.12}$$

Remark The signal set obtained here is the well-known *simplex set*. Signal space plots of them for $M = 2, 3, 4$ are quite informative.

5.9 (*A generalization of the Fourier approach to approximate an energy signal*) Suppose that $s(t)$ is a deterministic, real-valued signal with finite energy $E_s = \int_{-\infty}^{\infty} s^2(t)\mathrm{d}t$. Furthermore, suppose that there exists a set of orthonormal basis functions $\{\phi_n(t), n = 1, 2, \ldots, N\}$, i.e.,

$$\int_{-\infty}^{\infty} \phi_n(t)\phi_m(t)\mathrm{d}t = \begin{cases} 0, & m \neq n \\ 1, & m = n \end{cases}. \tag{P5.13}$$

We want to approximate the signal $s(t)$ by a weighted linear combination of these basis functions, i.e.,

$$\hat{s}(t) = \sum_{k=1}^{N} s_k\phi_k(t), \tag{P5.14}$$

where $\{s_k\}$, $k = 1, 2, \ldots, N$, are the coefficients in the approximation of $s(t)$. The approximation error incurred is

$$e(t) = s(t) - \hat{s}(t). \tag{P5.15}$$

(a) Find the coefficients $\{s_k\}$ that minimize the energy of the approximation error.
(b) What is the minimum mean square approximation error, i.e., $\int_{-\infty}^{\infty} e^2(t)\mathrm{d}t$?

5.10 (*This problem emphasizes the geometrical approach to signal representation*) Consider two signals $s_1(t)$ and $s_2(t)$ as plotted in Figure 5.41(b). The two orthonormal basis functions $\phi_1(t)$ and $\phi_2(t)$ in Figure 5.41(a) are chosen to represent the two signals $s_1(t)$ and $s_2(t)$, i.e.,

$$\begin{bmatrix} s_1(t) \\ s_2(t) \end{bmatrix} = \begin{bmatrix} s_{11} & s_{12} \\ s_{21} & s_{22} \end{bmatrix} \begin{bmatrix} \phi_1(t) \\ \phi_2(t) \end{bmatrix}. \tag{P5.16}$$

(a) Determine the coefficients s_{ij}, $i, j \in \{1, 2\}$.

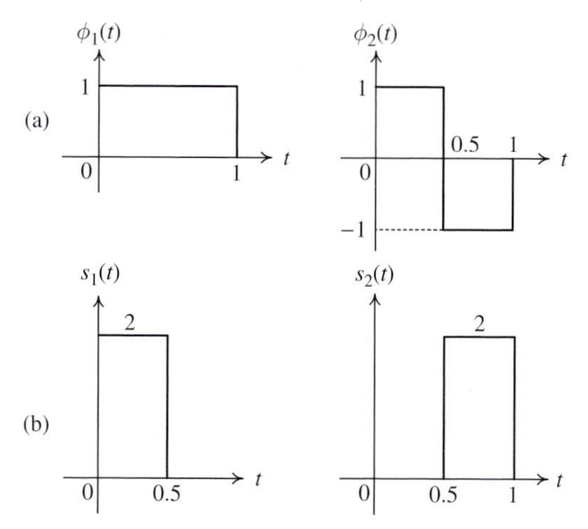

Fig. 5.41 (a) Orthonormal functions, and (b) signal set.

(b) Consider a new set of orthonormal functions $\phi_1^R(t)$ and $\phi_2^R(t)$, which are $\phi_1(t)$ and $\phi_2(t)$ axes rotated by an angle of θ degrees, i.e.,

$$\begin{bmatrix} \phi_1^R(t) \\ \phi_2^R(t) \end{bmatrix} = \begin{bmatrix} \cos\theta & \sin\theta \\ -\sin\theta & \cos\theta \end{bmatrix} \begin{bmatrix} \phi_1(t) \\ \phi_2(t) \end{bmatrix}. \tag{P5.17}$$

Determine and draw the *time waveforms* of the new orthonormal functions for $\theta = 60°$.

(c) Determine the new set of coefficients s_{ij}^R, $i, j \in \{1, 2\}$ in the representation:

$$\begin{bmatrix} s_1(t) \\ s_2(t) \end{bmatrix} = \begin{bmatrix} s_{11}^R & s_{12}^R \\ s_{21}^R & s_{22}^R \end{bmatrix} \begin{bmatrix} \phi_1^R(t) \\ \phi_2^R(t) \end{bmatrix}. \tag{P5.18}$$

(d) Provide the geometrical representation of the signal set using both basis sets on the same figure.

(e) Determine the distance d between the two signals $s_1(t)$ and $s_2(t)$ in two ways:
  (i) algebraically: $d = \sqrt{\int_0^1 [s_1(t) - s_2(t)]^2 \, dt}$;
  (ii) geometrically: from the signal space plot of (d) above.

(f) Though $\phi_1(t)$ and $\phi_2(t)$ can represent the two given signals, they are by no means a complete basis set because they cannot represent an arbitrary, finite-energy signal defined on the time interval $[0, 1]$. Start completing the basis, by plotting the next two possible orthonormal functions $\phi_3(t)$ and $\phi_4(t)$ of the basis set.

5.11 (*Again more practice in determining an orthonormal basis set to represent the signal set exactly but in the M-ary, M = 4, case*) Consider the set of four time-limited waveforms shown in Figure 5.42.

(a) Using the Gram–Schmidt procedure, construct a set of orthonormal basis functions for these waveforms.

(b) By inspection, show that the set of orthonormal functions in Figure 5.43 can also be used to exactly represent the four signals in Figure 5.42.

(c) Plot the geometrical representation of the set of four signals $\{s_1(t), s_2(t), s_3(t), s_4(t)\}$ in the three-dimensional signal space spanned by $\{v_1(t), v_2(t), v_3(t)\}$.

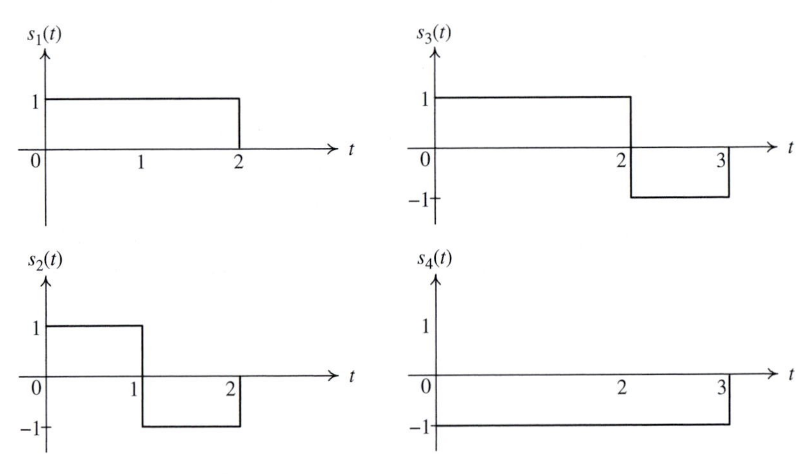

Fig. 5.42 A set of four time-limited waveforms.

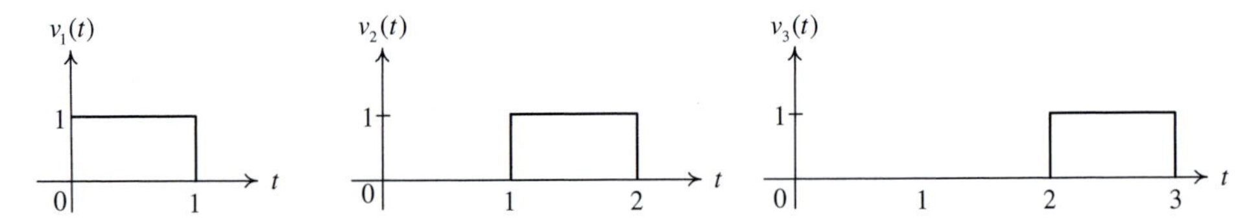

Fig. 5.43 A set of three orthonormal functions.

The next three problems consider aspects of the design of a minimum error probability communication system.

5.12 (*An orthogonal binary signal set*) Consider the following signal set for binary data transmission over a channel disturbed by AWGN:

$$s_1(t) = \begin{cases} \sqrt{3}A\cos(2\pi t/T_b), & 0 \le t \le T_b \\ 0, & \text{otherwise} \end{cases}, \tag{P5.19}$$

$$s_2(t) = \begin{cases} A\sin(2\pi t/T_b), & 0 \le t \le T_b \\ 0, & \text{otherwise} \end{cases}. \tag{P5.20}$$

The noise is zero-mean and has two-sided PSD $N_0/2$. As usual, $s_1(t)$ is used for the transmission of bit "0" and $s_2(t)$ for the transmission of bit "1." Furthermore, the two bits are equally likely.

(a) Show that $s_1(t)$ is *orthogonal* to $s_2(t)$. Then find and draw an orthonormal basis $\{\phi_1(t),\phi_2(t)\}$ for the signal set.

(b) Draw the signal space diagram and the optimum decision regions. Write the expression for the optimum decision rule.

(c) Let $A = 1$ volt and assume that $N_0 = 10^{-8}$ watts/hertz. What is the maximum bit rate that can be sent with a probability of error $P[\text{error}] \leq 10^{-6}$.

(d) Draw the block diagram of an optimum receiver that uses only one matched filter. Give the precise expression for the impulse response of the matched filter.

(e) Assume that, as long as the average energy of the signal set $\{s_1(t), s_2(t)\}$ stays the same, you can freely change (or move) both $s_1(t)$ and $s_2(t)$ in the same signal space. Modify them so that the probability of error is as small as possible. Explain your answer.

5.13 Consider the signal set in Figure 5.44 for binary data transmission over an AWGN channel. The noise is zero-mean and has two-sided PSD $N_0/2$. As usual, $s_1(t)$ and $s_2(t)$ are used for the transmission of equally likely bits "0" and "1," respectively.

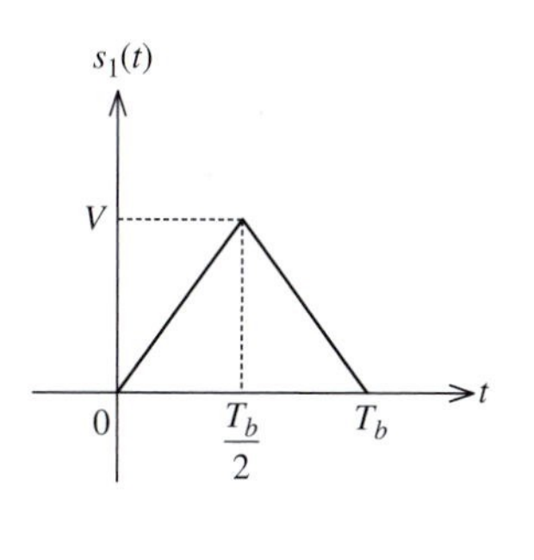

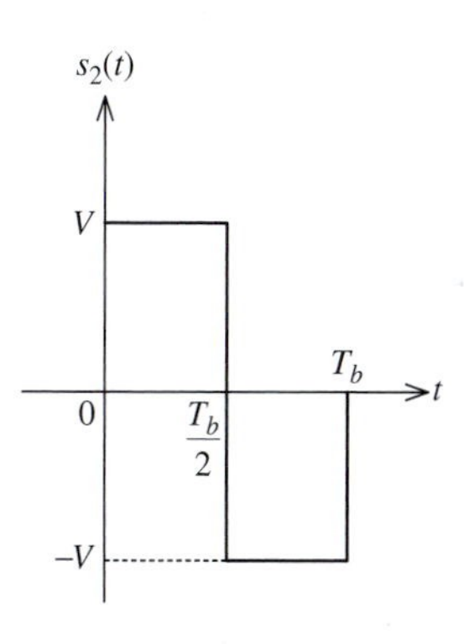

Fig. 5.44 A binary signal set.

(a) Show that $s_1(t)$ is *orthogonal* to $s_2(t)$. Then find and draw an orthonormal basis set $\{\phi_1(t), \phi_2(t)\}$ for the signal set.

(b) Draw the signal space diagram and the optimum decision regions. Write the expression for the optimum decision rule.

(c) Let $V = 1$ volt and assume that $N_0 = 10^{-8}$ watts/hertz. What is the maximum bit rate that can be sent with a probability of error $P[\text{error}] \leq 10^{-6}$.

(d) Draw the block diagram of an optimum receiver that uses only one matched filter and sketch the impulse response of the matched filter.

(e) Assume that the signal $s_1(t)$ is fixed. However, you can change the shape, but not the energy, of $s_2(t)$. Modify $s_2(t)$ so that the probability of error is as small as possible. Explain your answer.

5.14 Consider the signal set in Figure 5.45 for binary data transmission over a channel disturbed by AWGN. The noise is zero-mean and has two-sided PSD $N_0/2$. As usual, $s_1(t)$ is used for the transmission of bit "0" and $s_2(t)$ is for the transmission of bit "1." Furthermore, the two bits are equiprobable.

(a) Find and draw an orthonormal basis $\{\phi_1(t),\phi_2(t)\}$ for the signal set.

(b) Draw the signal space diagram and the optimum decision regions. Write the expression for the optimum decision rule.

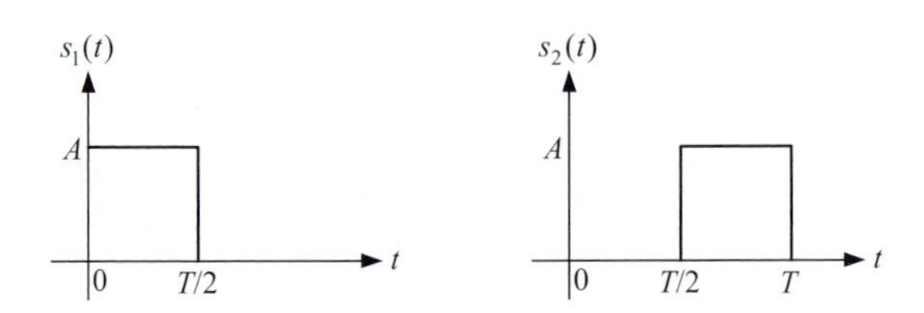

Fig. 5.45 A binary signal set.

(c) Draw the block diagram of the receiver that implements the optimum decision rule in (b).

(d) Let $A = 1$ volt and assume that $N_0 = 10^{-8}$ watts/hertz. What is the maximum bit rate that can be sent with a probability of error $P[\text{error}] \leq 10^{-6}$.

(e) Draw the block diagram of an optimum receiver that uses only one correlator or one matched filter.

(f) Assume that signal $s_1(t)$ is fixed. However, you can change $s_2(t)$. Modify it so that the average energy is maintained at $A^2T/2$ but the probability of error is as small as possible. Explain.

5.15 (*An example of colored noise*) For a binary communication system with AWGN, the error performance of the optimum receiver does not depend on the specific signal shapes but simply on the distance between the two signals. This problem considers noise that is Gaussian but *not* white (i.e., it is *colored* noise).

The communication system under consideration is shown in Figure 5.46.

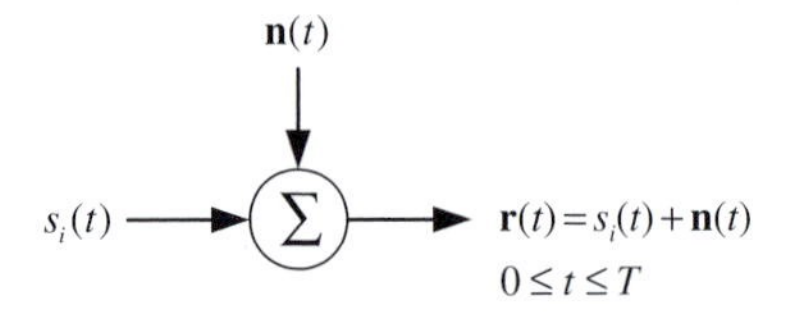

Fig. 5.46 A binary communication system with additive noise.

The noise, $\mathbf{n}(t) = \mathbf{n}$, is simply a DC level but the amplitude of the level is random and is Gaussian distributed with a probability density function given by

$$f_{\mathbf{n}}(n) = \frac{1}{\sqrt{2\pi N_0}} \mathrm{e}^{-n^2/2N_0}. \tag{P5.21}$$

(a) Determine the autocorrelation and the PSD of the noise. Are the successive noise samples correlated?

(b) Consider the signal set and the receiver shown in Figure 5.47. As usual, $s_1(t)$ is used for the transmission of bit "0" and $s_2(t)$ is for the transmission of bit "1." What is the error probability of this receiver?

(c) Next consider the signal set in Figure 5.48. Find a receiver that will have an error probability of 0.

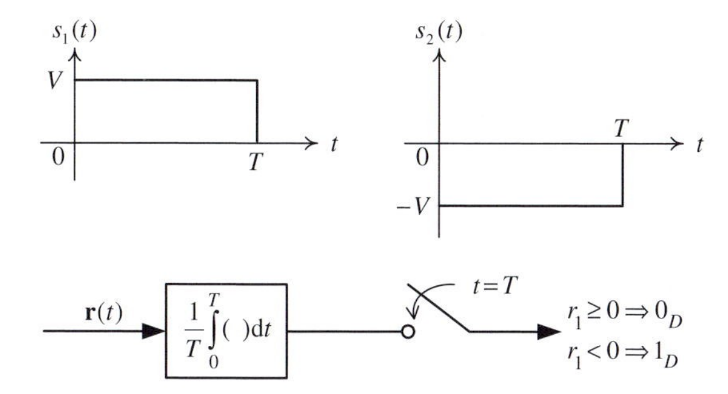

Fig. 5.47 The signal set and receiver considered in Problem 5.15(b).

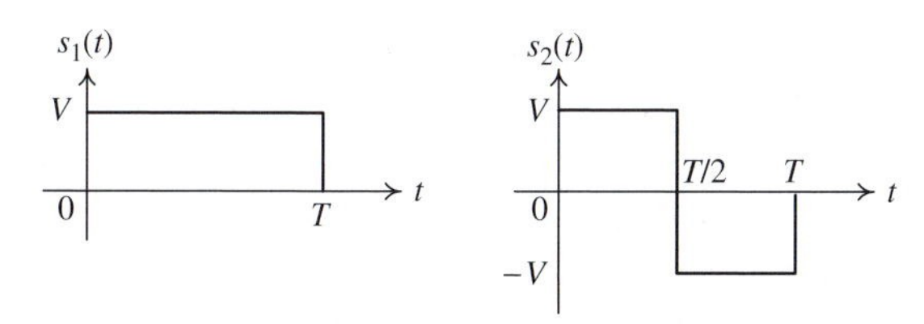

Fig. 5.48 The signal set considered in Problem 5.15(c).

5.16 (*In the development of the optimum demodulator the concept of the matched filter arose: either matched to the orthonormal basis set that is used to represent the signal set or directly to the signal set. The optimality criterion was minimum error probability. This problem shows that the matched filter has another desirable property, namely, it maximizes the signal-to-noise ratio at the output.*)

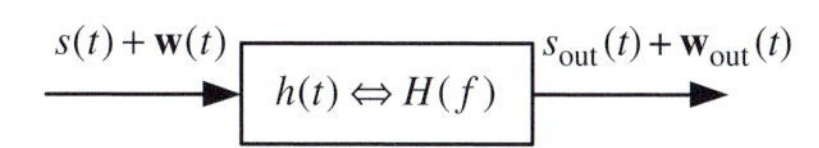

Fig. 5.49 A block diagram of the matched filter, in Problem 5.16.

Consider the block diagram shown in Figure 5.49. The signal $s(t)$ is known, i.e., deterministic, and duration-limited to $[0, T_b]$ seconds. The input noise $\mathbf{w}(t)$ is *white* with spectral strength of $N_0/2$. It is desired to determine the impulse response $h(t)$, or equivalently the transfer function $H(f)$, that *maximizes* the SNR at the output at time $t = t_0$. The problem is therefore to choose $h(t)$ so that

$$\text{SNR}_{\text{out}} \equiv \frac{s^2_{\text{out}}(t_0)}{E\{\mathbf{w}^2_{\text{out}}(t)\}} = \frac{s^2_{\text{out}}(t_0)}{P_{\mathbf{w}_{\text{out}}}} \tag{P5.22}$$

is maximized. To find the solution, proceed as follows:

(a) Write the expression for $s_{\text{out}}(t)$ in terms of $H(f)$ and $S(f)$. From this get the expression for $s^2_{\text{out}}(t_0)$.

(b) Write the expression for the output noise power $P_{\mathbf{w}_{\text{out}}}$ (watts) in terms of $H(f)$ and the input noise power spectral density. Based on the results in (a) and (b) write the expression for SNR_{out}. To proceed further we need the Cauchy–Schwartz inequality, which states that for any two *real* functions $a(t)$ and $b(t)$ one has

$$\left[\int_{-\infty}^{\infty} a(t)b(t)\text{d}t\right]^2 \leq \int_{-\infty}^{\infty} a^2(t)\text{d}t \int_{-\infty}^{\infty} b^2(t)\text{d}t. \tag{P5.23}$$

(c) Prove this inequality. The starting point is to consider $\int_{-\infty}^{\infty}[a(t)+\lambda b(t)]^2\text{d}t \geq 0$ (always). Expand the square and observe that the quadratic in λ cannot have any real roots to arrive at the inequality. When does the equality hold?

(d) Recall that $\int_{-\infty}^{\infty} a(t)b(t)\text{d}t = \int_{-\infty}^{\infty} A(f)B^*(f)\text{d}f$. (Why?) Rewrite the Cauchy–Schwartz inequality in the frequency domain, i.e., in terms of $A(f)$ and $B(f)$.

(e) Identify $A(f) = H(f)$ and $B^*(f) = S(f)\text{e}^{\text{j}2\pi f t_0}$. Argue that $\text{SNR}_{\text{out}} \leq \int_{-\infty}^{\infty}[|S(f)|^2/(N_0/2)]\text{d}f$.

(f) Choose $H(f)$ to achieve equality and to complete the problem, find the corresponding $h(t)$.

5.17 A communication magazine carries an advertisement for bargain basement pieces on a filter that has the impulse response shown in Figure 5.50.

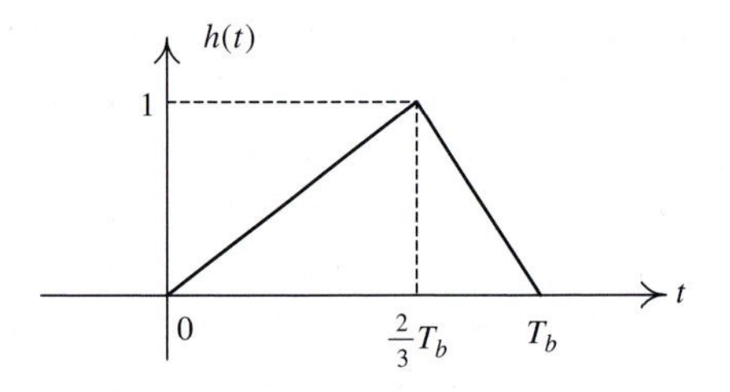

Fig. 5.50 An impulse response.

The manager of your division notices this and asks you to design an antipodal binary modulator for an AWGN channel, where the noise has a two-sided PSD of $N_0/2$ (watts/hertz). The goal of the design is to take advantage of this filter in the receiver for an optimum demodulation.

(a) Design the modulator, basically the signal set to take advantage of this filter. Explain.

(b) Assume that $N_0 = 10^{-9}$ watts/hertz. What is the maximum bit rate that can be sent with a probability of error $P[\text{error}] \leq 10^{-3}$.

Hint One can match the filter to the signal or match the signal to the filter. To paraphrase the biblical statement: one can either bring the mountain to Moses or Moses to the mountain.

5.18 (*Antipodal signaling*) The received signal in a binary communication system that employs antipodal signals is

$$\mathbf{r}(t) = s_i(t) + \mathbf{w}(t) = \begin{cases} s(t) + \mathbf{w}(t), & \text{if "0" is transmitted} \\ -s(t) + \mathbf{w}(t), & \text{if "1" is transmitted} \end{cases}, \tag{P5.24}$$

where $s(t)$ is shown in Figure 5.51 and $\mathbf{w}(t)$ is AWGN with PSD $N_0/2$ (watts/hertz).

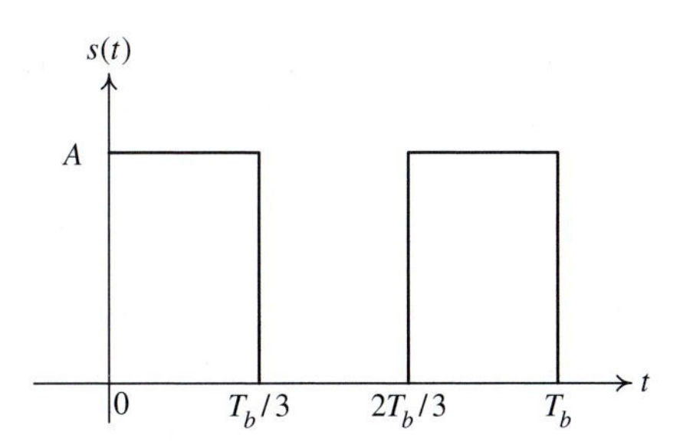

Fig. 5.51 Plot of $s(t)$.

(a) Sketch the impulse response of the filter that is *matched* to $s(t)$.
(b) Precisely sketch the output of the filter to the input $s(t)$ (i.e., "0" was transmitted and noise is ignored).
(c) Precisely sketch the output of the filter to the input $-s(t)$ (i.e., "1" was transmitted and noise is ignored).
(d) Compute the SNR at the output of the filter at $t = T_b$.
(e) Determine the probability of error as a function of SNR.
(f) Plot the probability of error (on a log scale) as a function of SNR (in decibels). What is the minimum SNR to achieve a probability of error of 10^{-6}?

5.19 (*Bandwidth*) Consider an antipodal binary communication system, where the two signals $s(t)$ and $-s(t)$ are used to transmit bits 0 and 1 respectively every T_b seconds. You have learned in this chapter that the error performance of the optimum receiver for such a system depends on the energy, *not* the specific shape of the signal $s(t)$. Despite this fact, the choice of $s(t)$ is important in any practical design because it determines the transmission bandwidth of the system.

Since the signal $s(t)$ is time-limited to T_b seconds, it cannot be bandlimited and a bandwidth definition is required. Here we define W to be the bandwidth of the signal $s(t)$ if $\epsilon\%$ of the total energy of $s(t)$ is contained inside the band $[-W, W]$. Mathematically, this means that

$$\frac{\int_{-W}^{W} |S(f)|^2 \mathrm{d}f}{\int_{-\infty}^{\infty} |S(f)|^2 \mathrm{d}f} = \frac{2\int_{0}^{W} |S(f)|^2 \mathrm{d}f}{E} = \frac{\epsilon}{100}, \tag{P5.25}$$

where $S(f)$ is the Fourier transform of $s(t)$. Consider the following three signals:
(i) rectangular pulse: $s(t) = \sqrt{1/T_b}$, $0 \le t \le T_b$;
(ii) half-sine: $s(t) = \sqrt{2/T_b}\sin(\pi t/T_b)$, $0 \le t \le T_b$;
(iii) raised-cosine: $s(t) = \sqrt{2/3T_b}\left[1 - \cos(2\pi t/T_b)\right]$, $0 \le t \le T_b$.

Note that all the three signals have been normalized to have unit energy, i.e., $E = 1$ (joule).

(a) Derive (P5.25) for the above three signals. Try to put the final expressions in a form such that WT_b appears only in the limit of the integrals.

(b) Based on the expressions obtained in (a), evaluate WT_b for the three signals for each of the following values of ϵ: $\epsilon = 90$, $\epsilon = 95$, and $\epsilon = 99$.

Remark For this part, you need to do the integration numerically. In Matlab, the routine `quadl` is useful for numerical integration. Type `help quadl` to see how to use this routine.

5.20 In antipodal signaling, two signals $s(t)$ and $-s(t)$ are used to transmit equally likely bits 0 and 1, respectively. Consider two communication systems, called system (i) and system (ii), that use two different time-limited waveforms as shown in Figure 5.52.

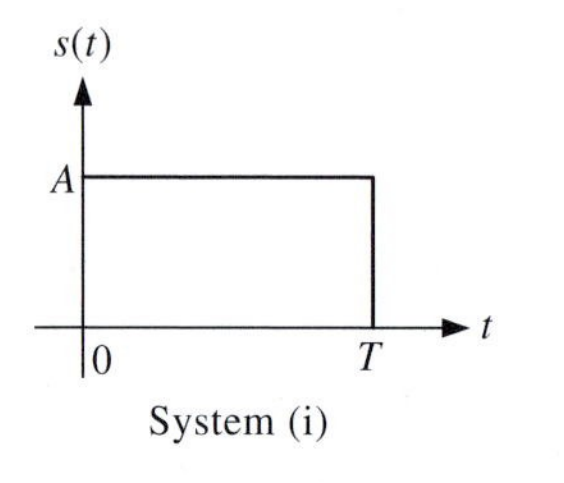

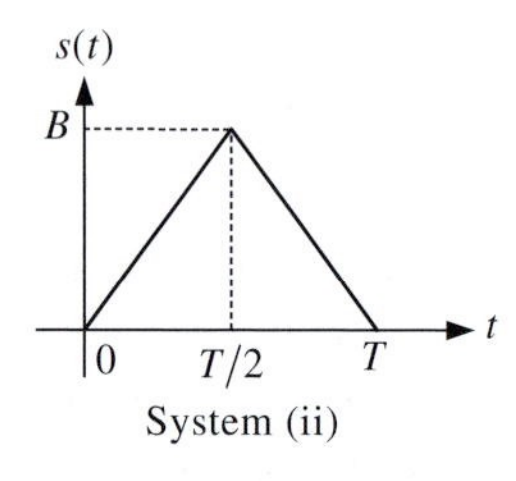

Fig. 5.52 Two possible waveforms for antipodal signaling.

(a) What is the relationship between the parameters A and B of the two waveforms if the two communication systems have the same error performance? Explain your answer.

(b) Since the two signals in Figure 5.52 are time-limited, they cannot be bandlimited and a bandwidth definition is required. Here we define W to be the bandwidth of the signal $s(t)$ if 95% of the total energy of $s(t)$ is contained inside the band $[-W, W]$. Assume that both systems have the same bit rate of 2 Mbps. What are the required bandwidths of the two systems? Which system is preferred and why?

(c) Consider the system (i) in Figure 5.52. How large does the voltage level A need to be to achieve an error probability of 10^{-6} if the bit rate is 2 Mbps and $N_0/2 = 10^{-8}$ (watts/hertz)?

5.21 (*A diversity system*) Consider the antipodal signaling system in Figure 5.53, where the signals $s(t)$ and $-s(t)$ are used to transmit the information bits "1" and "0" respectively. The bit duration is T_b and $s(t)$ is assumed to have unit energy. The signal is sent via two different channels, denoted "A" and "B," to the same destination. Each channel is described by a gain factor (V_A or V_B) and AWGN ($\mathbf{w}_A(t)$ or $\mathbf{w}_B(t)$). The noises $\mathbf{w}_A(t)$ and $\mathbf{w}_B(t)$ both have zero means and PSDs of σ_A^2 and σ_B^2 respectively. Furthermore, they are independent noise sources.

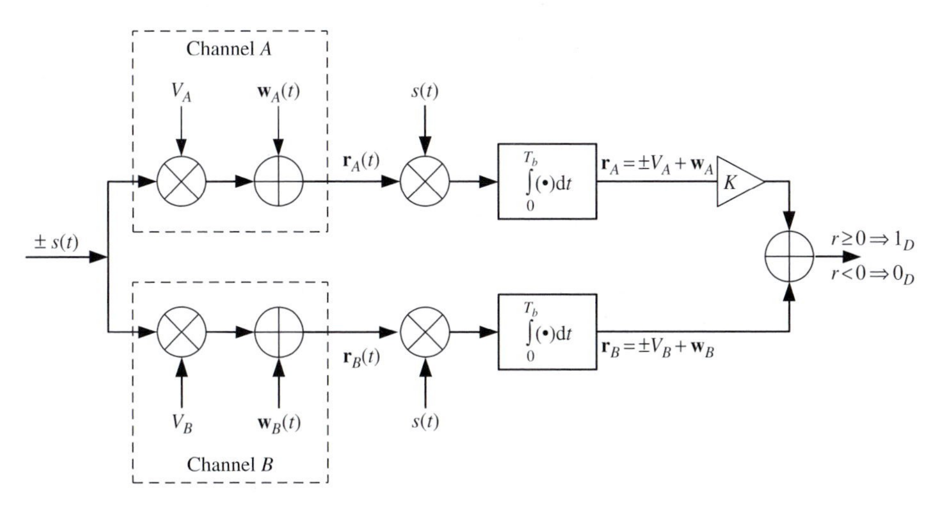

Fig. 5.53 A diversity system.

The receiver for such a system consists of two correlators, one for each channel, as shown in Figure 5.53. The output of one correlator, say the one corresponding to channel A, is passed through an amplifier with an adjustable voltage gain K. The signals are then added before being compared with a threshold of zero to make the decision.

(a) For a fixed K, find the probability of error of this system. *Hint* The noise samples $\mathbf{w}_A$ and $\mathbf{w}_B$ are independent, zero-mean, Gaussian random variables with variances σ_A^2 and σ_B^2 respectively. Furthermore, the sum of two independent Gaussian random variables is a Gaussian random variable whose variance equals the sum of the individual variances.

(b) Find the value of K that minimizes the probability of error. *Hint* Minimizing $Q(x)$ is equivalent to maximizing x.

(c) What is the probability of error when the optimum value of K is used?

(d) What is the probability of error when K is simply set to 1? Comment.

Remark Using diversity is an effective technique that is used to combat fading, a channel degradation commonly experienced in wireless communications.

The next two problems look at channels that are not modeled as AWGN. Though the general results obtained for the AWGN channel do not apply directly one can still use the concepts developed in this chapter. In particular, as mentioned in the summary section, the likelihood ratio test of (5.65) holds.

5.22 (*Detection in Laplacian noise*) Consider the communication system model in Figure 5.54, where $P[s_1(t)] = P_1$ and $P[s_2(t)] = P_2$. The noise is modeled to be Laplacian. At the receiver, you sample $\mathbf{r}(t)$ uniformly m times within the time period $[0, T_b]$. The samples are taken far enough apart so that you feel reasonably confident that the noisy samples $\mathbf{r}_j = \mathbf{r}(jT_b/m), j = 1, \ldots, m$, are statistically independent.

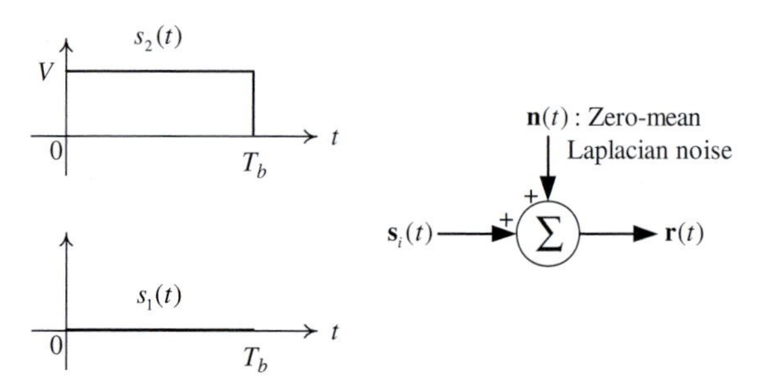

Fig. 5.54 A binary communication system with additive noise.

(a) Determine the two conditional density functions for a single sample, i.e., $f(r_j|1_T)$ and $f(r_j|0_T)$. From this what are the conditional density functions $f(r_1, r_2, \ldots, r_m|1_T)$ and $f(r_1, r_2, \ldots, r_m|0_T)$?

(b) Find the natural logarithm of the likelihood ratio:

$$\frac{f(r_1, r_2, \ldots, r_m|1_T)}{f(r_1, r_2, \ldots, r_m|0_T)}. \tag{P5.26}$$

Simplify the sum as much as possible by considering the samples that fall in three different regions: $r_j < 0$ with m_1 samples; $0 < r_j < V$ with m_2 samples; $V < r_j$ with m_3 samples, where $m_1 + m_2 + m_3 = m$. Considering the three different regions should allow you to eliminate the magnitude operation.

(c) Derive the decision rule that minimizes the error probability.

5.23 (*Optical communications*) As one goes higher and higher in frequency, the wave-like nature of electromagnetic radiation recedes into the background and quantum effects become more pronounced. In optical communications, one turns on the source (say a semiconductor laser) for a fixed interval of time and lets it radiate energy (photons). This represents the binary digit "1." To transmit a "0" the laser is switched off but because of background radiation, there are photons arriving at the receiver even when the laser is switched off. A common model for the number of photons emitted per unit time is that it is random and behaves like a Poisson point process. Thus

$$P[k \text{ photons emitted in a unit interval}] = \frac{\lambda^k e^{-\lambda}}{k!}, \quad k = 0, 1, \ldots, \tag{P5.27}$$

where λ is the mean arrival rate with units of photons per unit time. Let the signaling interval be T_b seconds. The receiver counts the number of photons received in the interval $[0, T_b]$ and based on this count makes a decision as to whether a "0" or "1" was transmitted. Assume that the probability of transmitting a "1" is the same as transmitting "0." Note also that $\lambda = \lambda_s + \lambda_n$ or $\lambda = \lambda_n$, where λ_s is due to the transmitter laser being turned on and λ_n is due to the background radiation.

(a) Design the receiver that minimizes the probability of error.

(b) After designing the receiver, derive an expression for the error performance. Note that the average transmitted energy is hfT_b, where hf is the energy of photon with frequency f, h is Planck's constant ($h = 6.6 \times 10^{-34}$) and $\lambda_s T_b$ is the average number of signal photons per bit interval. Is the error performance dependent on "some" SNR?

5.24 (*Signal design for nonuniform sources*) Consider a binary communication system where $s_1(t)$ and $s_2(t)$ are used for the transmission of bits "0" and "1," respectively, over an AWGN channel. The bit duration is T_b and both $s_1(t)$ and $s_2(t)$ are time-limited to $[0, T_b]$. Furthermore, the *a priori* probabilities of the information bits are

$$P[0_T] = P[s_1(t)] = p \leq 0.5, \; P[1_T] = P[s_2(t)] = 1 - p \geq 0.5. \tag{P5.28}$$

The two signals $s_1(t)$ and $s_2(t)$ have energies E_1 and E_2. The optimum receiver for such a system implements the following decision rule:

$$\int_0^{T_b} r(t)[s_2(t) - s_1(t)]\mathrm{d}t \underset{0_D}{\overset{1_D}{\gtrless}} \frac{E_2 - E_1}{2} + \frac{N_0}{2}\ln\left(\frac{p}{1-p}\right), \tag{P5.29}$$

where $r(t)$ is the received signal and $N_0/2$ is the PSD of the AWGN.

It was determined in this chapter that the error performance of the above optimum receiver is

$$P[\text{error}] = pQ\left(\frac{T - \hat{s}_{12}}{\sqrt{N_0/2}}\right) + (1-p)\left[1 - Q\left(\frac{T - \hat{s}_{22}}{\sqrt{N_0/2}}\right)\right], \tag{P5.30}$$

where

$$T = \frac{\hat{s}_{12} + \hat{s}_{22}}{2} + \frac{N_0}{2(\hat{s}_{22} - \hat{s}_{12})}\ln\left(\frac{p}{1-p}\right) \tag{P5.31}$$

and $\hat{s}_{12}$, $\hat{s}_{22}$ are the projections of $s_1(t)$ and $s_2(t)$ onto $\hat{\phi}_2(t)$, respectively. The basis function $\hat{\phi}_2(t)$ is given by

$$\hat{\phi}_2(t) = \frac{s_2(t) - s_1(t)}{(E_2 - 2\rho\sqrt{E_1E_2} + E_1)^{1/2}}, \tag{P5.32}$$

with

$$\rho = \frac{1}{\sqrt{E_1E_2}}\int_0^{T_b} s_1(t)s_2(t)\mathrm{d}t. \tag{P5.33}$$

The above results apply for an arbitrary source distribution (i.e., arbitrary value of $p \leq 0.5$) and arbitrary signal set $\{s_1(t), s_2(t)\}$. Now consider the case that p is fixed but one can freely design $s_1(t)$ and $s_2(t)$ to minimize $P[\text{error}]$ in (P5.30). Of course, the design is subject to a constraint on the average transmitted energy per bit:

$$\bar{E}_b = E_1 p + E_2(1-p). \tag{P5.34}$$

(a) Show that the error probability in (P5.30) can be written as

$$P[\text{error}] = pQ\left(\sqrt{A} - \frac{B}{\sqrt{A}}\right) + (1-p)Q\left(\sqrt{A} + \frac{B}{\sqrt{A}}\right), \tag{P5.35}$$

where $A = (\hat{s}_{22} - \hat{s}_{12})^2/2N_0 = d_{21}^2/2N_0 = (E_2 - 2\rho\sqrt{E_1E_2} + E_1)/2N_0$ and $B = 0.5\ln((1-p)/p)$. It can be shown from (P5.35) that $P[\text{error}]$ is minimized when A, or equivalently d_{21}, is *maximized*.

(b) Consider the special case of orthogonal signaling, i.e., $\rho = 0$. Design the two signals $s_1(t)$ and $s_2(t)$ to minimize P[error].

(c) Assume that $p = 0.1$. Use Matlab to plot on the same figure the P[error] of the system that uses your signal design obtained in (b) and also the P[error] of the system that uses the "conventional" design of $E_1 = E_2 = \bar{E}_b$. Show your plots over the ranges [0 : 1 : 20] decibels for $\bar{E}_b/N_0$ and $[10^{-7} \to 10^0]$ for P[error]. What is the gain in $\bar{E}_b/N_0$ of your design over the conventional design at $P[\text{error}] = 10^{-5}$?

(d) Now let ρ be arbitrary. Design the two signals $s_1(t)$ and $s_2(t)$ to minimize P[error]. Repeat (c) for the signal set obtained in this part and antipodal signaling (also assume that $p = 0.1$).

Hint Argue that the optimal signal set must correspond to $\rho < 0$.

5.25 (*A ternary communication system*) Three messages m_1, m_2, and m_3 can be broadcasted by transmitting one of three signals, $s(t)$, 0, or $-s(t)$ every T seconds, respectively. The received signal is

$$\mathbf{r}(t) = \begin{cases} s(t) + \mathbf{w}(t), & \text{if } m_1 \text{ is broadcast} \\ \mathbf{w}(t), & \text{if } m_2 \text{ is broadcast} \\ -s(t) + \mathbf{w}(t), & \text{if } m_3 \text{ is broadcast} \end{cases}, \tag{P5.36}$$

where $\mathbf{w}(t)$ is white Gaussian noise with zero mean and PSD of $N_0/2$. The optimum receiver computes the correlation metric

$$\boldsymbol{\ell} = \int_0^T \mathbf{r}(t)s(t)\mathrm{d}t \tag{P5.37}$$

and compares each specific quantity ℓ with a threshold A and a threshold $-A$. If $\ell > A$, the decision is made that m_1 was broadcast. If $\ell < -A$ the decision is made in favor of m_3. If $-A \le \ell \le A$, the decision is made in favor of m_2.

(a) Determine the three conditional probabilities of error: $P[\text{error}|m_1]$, $P[\text{error}|m_2]$, and $P[\text{error}|m_3]$.

(b) Determine the average probability of error P[error] as a function of the threshold A, where the *a priori* probabilities of the three messages are $P[m_1] = P[m_3] = \frac{1}{4}$ and $P[m_2] = \frac{1}{2}$.

(c) Determine the value of A that minimizes the average probability of error when the *a priori* probabilities of three signals are given as in (b). How does the value of A change if three signals are equally probable? *Hint* $\partial Q(x)/\partial x = -(1/\sqrt{2\pi})\mathrm{e}^{-x^2/2}$.

Remark This problem illustrates that the concepts developed in this chapter can also be applied to a communication system that employs more than two waveforms.

5.26 Show that the DC signal given in (5.131) is a periodic signal with fundamental period T_b.

5.27 (*Regenerative repeaters or why go digital*) In analog communication systems, amplifiers (called repeaters) are used to periodically boost the signal level. However, each amplifier also boosts the noise in the system. In contrast, digital communication systems allow one to detect and *regenerate* a clean (noise-free) signal to send over the transmission channel. Such devices, called *regenerative repeaters*, are commonly used in wireline and fiber optic communication channels.

Because a noise-free signal is regenerated at each repeater, the additive noise does not *accumulate*. However, when errors occur in the detection process of a repeater, the errors *propagate* forward to the following repeaters in the channel. To evaluate the effect of errors on the performance of the overall system, suppose that antipodal signaling is used. The bit error probability for one hop (transmission from one repeater to the next) is $Q\left(\sqrt{2E_b/N_0}\right)$, where E_b is the energy per bit, $N_0/2$ is the two-sided PSD of AWGN and E_b/N_0 is the SNR.

Typically errors occur with a low probability and one may ignore the probability that any given bit will be detected incorrectly more than once in transmission through a channel with K repeaters. Under this reasonable assumption, the number of errors increases linearly with the number of regenerative repeaters used in the channel. Therefore, the overall bit error probability can be approximated as

$$P[\text{error}]_{\text{digital}} \approx KQ\left(\sqrt{\frac{2E_b}{N_0}}\right). \tag{P5.38}$$

For analog communication systems, the use of K analog repeaters in the channel reduces the received SNR by K, and hence, the bit error probability is

$$P[\text{error}]_{\text{analog}} \approx Q\left(\sqrt{\frac{2E_b}{KN_0}}\right). \tag{P5.39}$$

It is obvious from the above two expressions that for the same error performance, the use of regenerative repeaters yields a significant saving in the transmitted power as compared to the use of analog repeaters. Because of this regenerative repeaters are preferable in digital communication systems. Nevertheless, in channels that are used to transmit both analog and digital signals (such as the wireline telephone channels), analog repeaters are generally employed.

Consider a binary communication system that transmits data over a wireline channel of length 2000 kilometers. Repeaters are used every 20 kilometers to offset the effect of channel attenuation. Determine the required SNR, E_b/N_0, to achieve a bit error probability of 10^{-6} if:

(a) analog repeaters are used;

(b) regenerative repeaters are used.

5.28 (*Simulation of a binary communication system using antipodal signaling*) The probability of bit error, or bit error rate (BER), is an important performance parameter for any digital communication system. However, obtaining such a performance parameter in a closed-form expression is sometimes very difficult, if not impossible. This is especially true for complex systems that employ error control coding, multiple access techniques, wireless transmission, etc. It is common in the study and design of digital communication systems that the BER is evaluated through computer simulation. In this problem you will use Matlab to write a simple simulation program to test the BER performance of a binary communication system using antipodal signaling. The specific steps in your program are as follows:

- *Information source* Generate a *random* vector `b` that contains L information bits "0" and "1." The two bits should be equally likely. For this step, the functions `rand` and `round` in Matlab might be useful.
- *Modulator* The binary information bits contained in vector `b` are transmitted using antipodal signaling, where a voltage V is used for bit "1" and $-V$ for bit "0." Thus the transmitted signal is simply `y=V*(2*b-1);`
- *Channel* The channel noise is AWGN with two-sided power spectral density $N_0/2 = 1$ (watts/hertz). The effect of this AWGN can be simulated in Matlab by adding a noise vector `w` to the transmitted signal `y`. The vector of independent Gaussian noise samples with variance of 1 can be generated in Matlab as follows: `w=randn(1,length(y));` The received signal `r` is simply `r=y+w;`

  Remark In essence, the above implements a discrete (and equivalent) model of an antipodal signaling system. In particular, the simulated vector `r` is the output of the correlator or matched filter. Also for simplicity, it is assumed that $T_b = 1$ second.
- *Demodulator (or receiver)* With antipodal signaling, the demodulator is very simple. It simply compares the received signal with zero to make the decision.

Determine the minimum values of V to achieve the bit error probability levels of $10^{-1}, 10^{-2}, 10^{-3}, 10^{-4}$, and 10^{-5}. Use each value of V you found to run your Matlab program and record the actual BER. Plot (on a logarithmic scale) both the theoretical and experimental BER versus V^2 (decibels) on the same graph and compare.

Remark If you expect a BER of 10^{-K} for $K = 1, \ldots, 5$, then the length L of the information bit vector `b` should be at least $L = 100 \times 10^K$. This is to ensure that at least about 100 erroneous bits are recorded in each simulation run and the experimental BER value is reasonably reliable.

6 Baseband data transmission

6.1 Introduction

As pointed out in the previous chapter, binary digits (or bits) "0" and "1" are used simply to represent the information content. They are abstract (intangible) quantities and need to be converted into electrical waveforms for effective transmission or storage. How to perform such a conversion is generally governed by many factors, of which the most important one is the available transmission bandwidth of the communication channel or the storage media.

In baseband[1] data transmission, the bits are mapped into two voltage levels for direct transmission without any frequency translation. Such a baseband data transmission is applicable to cable systems (both metallic and fiber optics) since the transmission bandwidth of most cable systems is in the baseband. Various baseband signaling techniques, also known as line codes, have been developed to satisfy a number of criteria. Typical criteria are:

(i) *Signal interference and noise immunity* Depending on the signal sets, certain signaling schemes exhibit superior performance in the presence of noise as reflected by the probability of bit error.

(ii) *Signal spectrum* Typically one would like the transmitted signal to occupy as small a frequency band as possible. For baseband signaling, this implies a lack of high-frequency components. However, it is sometimes also important to have no DC component. Having a signaling scheme which does not have a DC component implies that AC coupling via a transformer may be used in the transmission channel. This provides electrical isolation which tends to reduce interference. Moreover, it is also possible in certain signaling schemes to match the transmitted signal to the special characteristics of a transmission channel.

(iii) *Signal synchronization capability* In implementing the receiver, it is necessary to establish the beginning and the end of each bit transmission period. This typically requires a separate clock to synchronize the transmitter and the receiver. Self-synchronization is, however, also possible if there are adequate transitions in the transmitted baseband signal. Several self-synchronizing baseband schemes have also been developed.

[1] Baseband modulation can be defined, somewhat imprecisely, as a modulation whose PSD is huddled around $f = 0$ (hertz).

(iv) *Error detection capability* Some signaling schemes have an inherent error detection capability. This is made possible by introducing constraints on allowable transitions among the signal levels and exploiting those constraints at the receiver.

(v) *Cost and complexity of transmitter and receiver implementations* This is still a factor which should not be ignored even though the price of digital logic continues to decrease.

This chapter discusses four baseband signaling schemes (also known as line codes) from the aspects of the first three criteria. The four signaling schemes are commonly known as nonreturn-to-zero-level (NRZ-L), return-to-zero (RZ), bi-phase-level (Biϕ-L) or Manchester, and delay modulation or Miller.

6.2 Baseband signaling schemes

Nonreturn-to-zero (NRZ) code The NRZ code can be regarded as the most basic baseband signaling scheme, since it appears "naturally" in synchronous digital circuits. In NRZ code, the signal alternates between the two voltage levels only when the current bit differs from the previous one.

Figure 6.1(a) represents an example of NRZ waveform, where T_b is the bit duration. Note that there is only one polarity in the waveform, hence it is also known as unipolar NRZ waveform. This is the simplest version of NRZ and can easily be generated. However, the DC component of a long random sequence of ones and zeros is nonzero. More precisely, the DC component is $VP[1_T] + 0P[0_T] = VP_2 + 0P_1 = VP_2$. For the common case of equally likely bits, the DC component is one-half of the positive voltage, i.e., $0.5V$ (volts). Therefore it is common to pass the NRZ waveform through a level shifter. The resultant waveform then alternates between $+V$ and $-V$ as shown in Figure 6.1(b). It is called the polar NRZ or NRZ-L waveform, whose DC component is $VP[1_T] + (-V)P[0_T] = V(P_2 - P_1)$. Obviously, the DC component of the NRZ-L waveform is zero if the two bits are equally likely.

Observe that the NRZ-L code produces a transition whenever the current bit in the input sequence differs from the previous one. These transitions can be used for synchronization purposes at the receiver. However, if the transmitted data contain long strings of similar bits, then the timing information is sparse, and regeneration of the clock signal at the receiver can be very difficult.

Return-to-zero (RZ) code The RZ code is similar to the NRZ code except that the information is contained in the first half of the bit interval, while the second half is always at level "zero." An example of an RZ waveform is shown in Figure 6.1(c). Once again the code has a DC component, which is $(0.5V)P[1_T] + 0P[0_T] = (0.5V)P_2$. If $P_1 = P_2 = 0.5$, then the DC component is one-fourth of the positive voltage, i.e., $0.25V$ (volts). Figure 6.2 shows that the RZ code is generated by gating the basic NRZ signal with the transmitter clock.

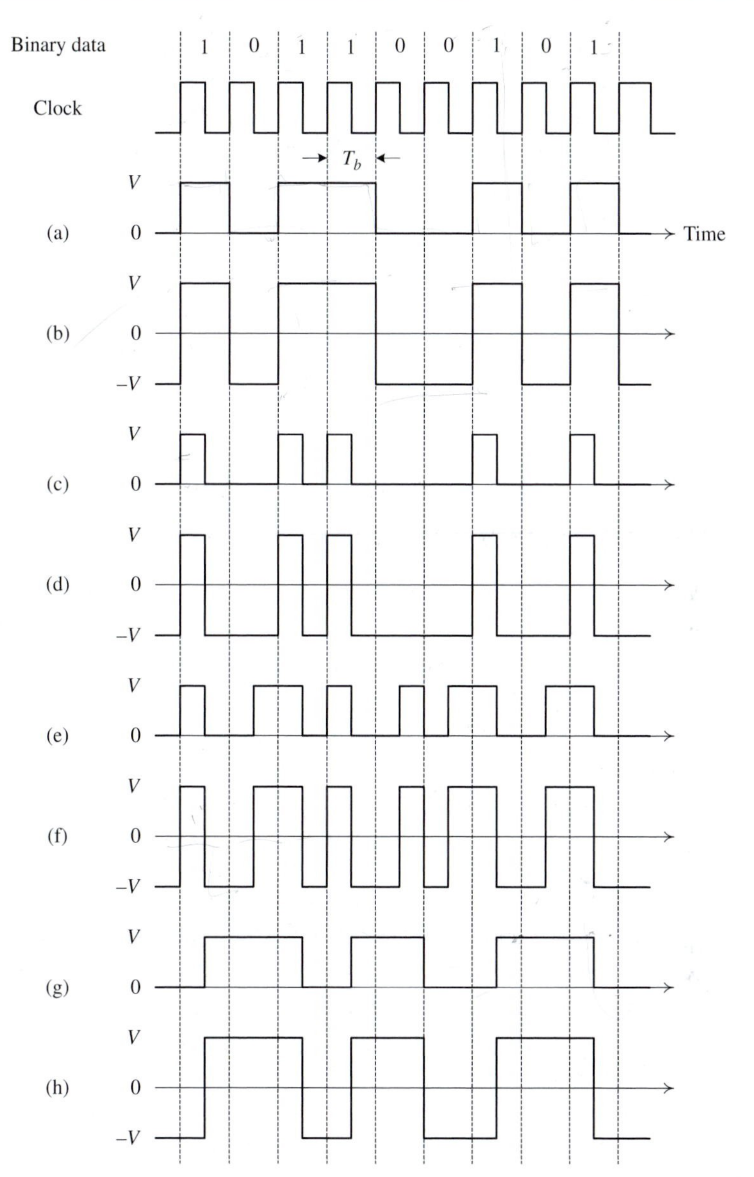

Fig. 6.1 Various binary signaling formats: (a) NRZ; (b) NRZ-L; (c) RZ; (d) RZ-L; (e) Biϕ, (f) Biϕ-L; (g) Miller; (h) Miller-L.

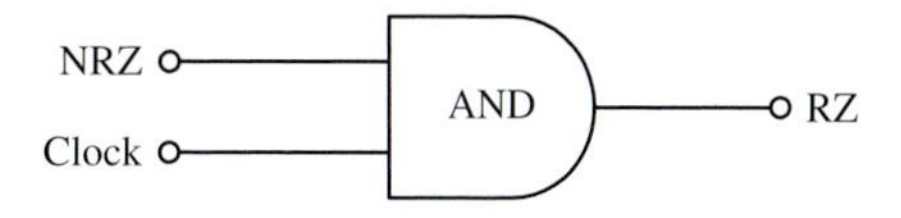

Fig. 6.2 RZ encoder.

In order to compare RZ fairly with the NRZ-L code in terms of error probability, we shall consider the RZ-L (or bipolar RZ) code where the two levels are $+V$ and $-V$ rather than V and 0. The corresponding waveform of RZ-L code is shown in Figure 6.1(d). The DC component of RZ-L waveform is $-VP_1$, which is $-0.5V$ (volts) if $P_1 = P_2 = 0.5$. Thus the code still has a nonzero DC component.

Regarding timing information, note that as opposed to NRZ, with RZ a long string of "1" bits results in transitions from which a clock at the receiver can be regenerated. A string of "0" bits, however, does not have any transitions just as with the NRZ code. For this reason and the fact that it has poor spectral properties and inferior error performance, RZ coding is not used except in some very elementary transmitting and recording equipment.

Biphase (Biϕ) or Manchester code To overcome the poor synchronization capability of NRZ and RZ codes, biphase (Biϕ) coding has been developed. It encodes information in terms of level transitions in the middle of a bit interval. Note that this conversion of bits to electrical waveforms is in a sharp contrast to what done in NRZ and RZ codes, where the information bits are converted to voltage levels. The biphase conversion (or mapping, or encoding) rules are as follows:

- Bit "1" is encoded as a transition from a high level to a low level occurring in the middle of the bit interval.
- Bit "0" is encoded as a transition from a low level to a high level occurring in the middle of the bit interval.
- An additional "idle" transition may have to be added at the beginning of each bit interval to establish the proper starting level for the information carrying transition.

Examples of Biϕ and Biϕ-L waveforms are shown in Figures 6.1(e) and 6.1(f), respectively. The DC component of the Biϕ signal is evaluated as $(0.5V)P[1_T] + (0.5V)P[0_T] = 0.5V[P_2 + P_1] = 0.5V$, while the DC component of the Biϕ-L signal is obviously 0. Note that the above results for the DC component hold regardless of the *a priori* probabilities of the bits. Therefore the Biϕ-L code does not have a DC component. Figure 6.3 shows that the bi-phase code can be generated with an XOR logic whose inputs are the basic NRZ signal and the transmitter clock.

The Biϕ signal, however, occupies a wider frequency band than the NRZ signal. This is due to the fact that for alternating bits there is one transition per bit interval while for two identical consecutive bits, two transitions occur per bit interval. On the other hand, because there is a predictable transition during every bit interval, the receiver can synchronize on

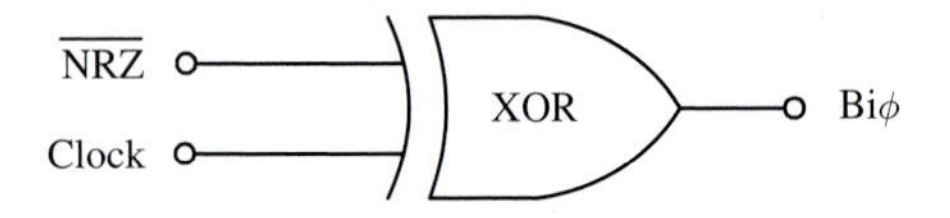

Fig. 6.3 Biϕ encoder.

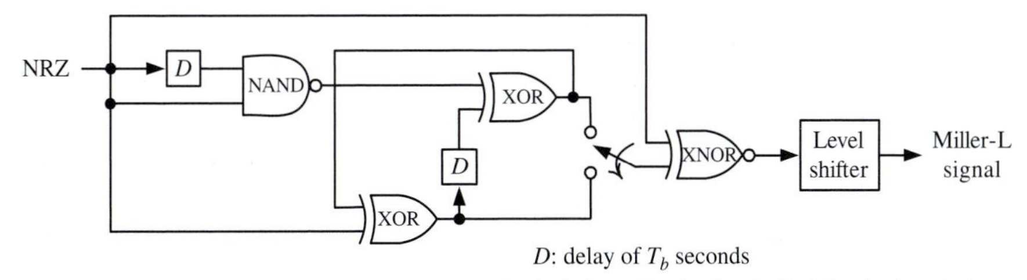

Fig. 6.4 Miller-L encoder.

that transition. The Biϕ code is thus known as a self-synchronizing code. It is commonly used in local area networks (LANs), such as the Ethernet.

Miller code This code is an alternative to the Biϕ code. It has at least one transition every two-bit interval and there are never more than two transitions every two-bit interval. It thus provides good synchronization capabilities, while requiring less bandwidth than the Biϕ signal. The encoding rules are:

- Bit "1" is encoded by a transition in the middle of the bit interval. Depending on the previous bit this transition may be either upward or downward.
- Bit "0" is encoded by a transition at the beginning of the bit interval if the previous bit is "0". If the previous bit is "1," then there is no transition.

The waveforms for Miller and Miller-level (Miller-L) codes are illustrated in Figures 6.1(g) and 6.1(h), respectively. The Miller-L signal can be generated from the NRZ signal by the circuit shown in Figure 6.4.

6.3 Error performance

To determine the probability of bit error for each of the line codes we shall consider that the transmitted signals are corrupted by zero-mean AWGN noise of spectral strength $N_0/2$ (watts/hertz) and that the two bits, "0" and "1," are *equally likely*. As shown in the previous chapter, the error probability of each line code is readily determined by identifying the elementary signals used for bits "0" and "1" and representing them in the signal space diagram. In all cases, the orthonormal basis set for the signal space can be determined simply by inspection. For each signaling scheme, a voltage swing from $-V$ to V volts is also assumed.

NRZ-L code The elementary signals are shown in Figure 6.5(a), where each signal has energy $E_{\text{NRZ-L}} = V^2T_b$ (joules). The single basis function and signal space plot

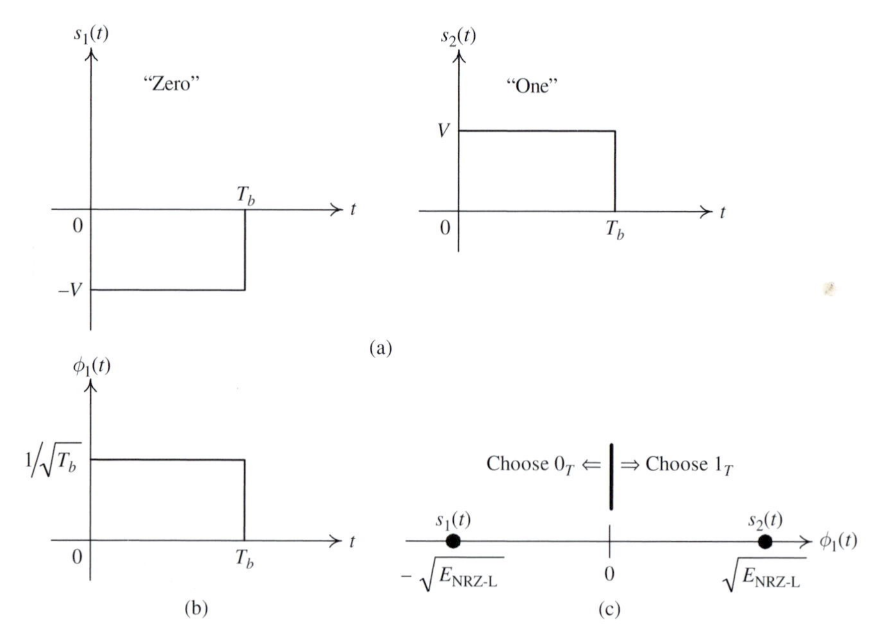

Fig. 6.5 NRZ-L code: (a) elementary signals; (b) basis function; (c) signal space and decision regions.

are given in Figures 6.5(b) and 6.5(c). Applying (5.105), the probability of bit error is given by

$$P[\text{error}]_{\text{NRZ-L}} = Q\left(\sqrt{2E_{\text{NRZ-L}}/N_0}\right). \tag{6.1}$$

RZ-L code Figure 6.6 shows the elementary signals, the basis functions, the signal space together with the optimum decision regions of RZ-L signaling. Each signal has energy $E_{\text{RZ-L}} = V^2T_b = E_{\text{NRZ-L}}$ (joules) and the error probability is given as

$$P[\text{error}]_{\text{RZ-L}} = Q\left(\sqrt{E_{\text{RZ-L}}/N_0}\right). \tag{6.2}$$

Biphase-level (Biϕ-L) code Similar to the NRZ-L code, Biϕ-L code is also an antipodal signaling (see Figure 6.7). Its elementary signals are, however, different in shape compared to that of NRZ-L code (so as to have the self-synchronizing capability discussed before). Here each signal has energy $E_{\text{Bi}\phi\text{-L}} = V^2T_b = E_{\text{NRZ-L}}$ joules and the error probability is expressed as

$$P[\text{error}]_{\text{Bi}\phi\text{-L}} = Q\left(\sqrt{2E_{\text{Bi}\phi\text{-L}}/N_0}\right). \tag{6.3}$$

Miller-level (Miller-L) code For this code there are four elementary signals, two of which represent "1" bits with the other two representing "0" bits. The four elementary

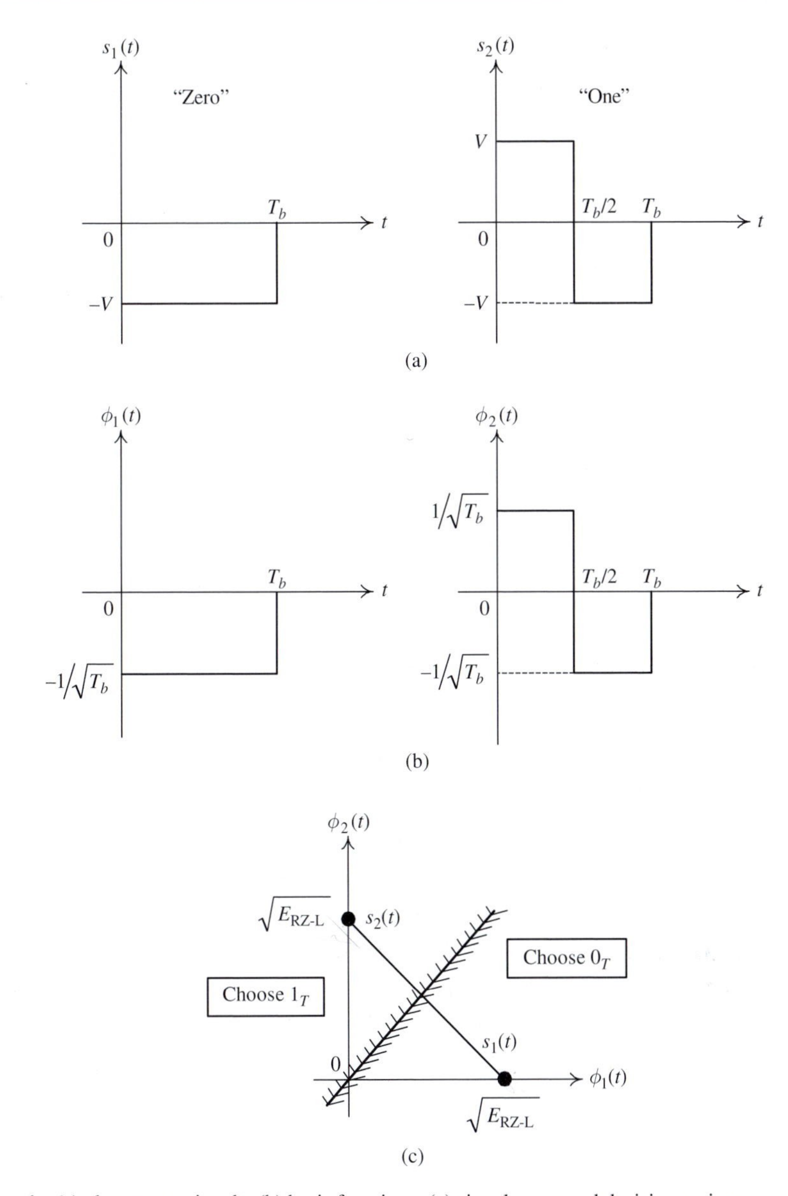

Fig. 6.6 RZ-L code: (a) elementary signals; (b) basis functions; (c) signal space and decision regions.

signals are shown in Figure 6.8(a), whereas the two orthonormal basis functions needed to represent these four signals are plotted in Figure 6.8(b). The energy in each signal is $E_{\text{M-L}} = V^2 T_b = E_{\text{NRZ-L}}$ joules.

The minimum-distance receiver consists of projecting the received signal, $r(t)$, onto $\phi_1(t)$ and $\phi_2(t)$, which generates the statistics (r_1, r_2). Since each signal is equally likely and has an *a priori* probability of $1/4$, the decision rule is to choose the signal to which the point (r_1, r_2) is closest. The decision space is shown in Figure 6.9.

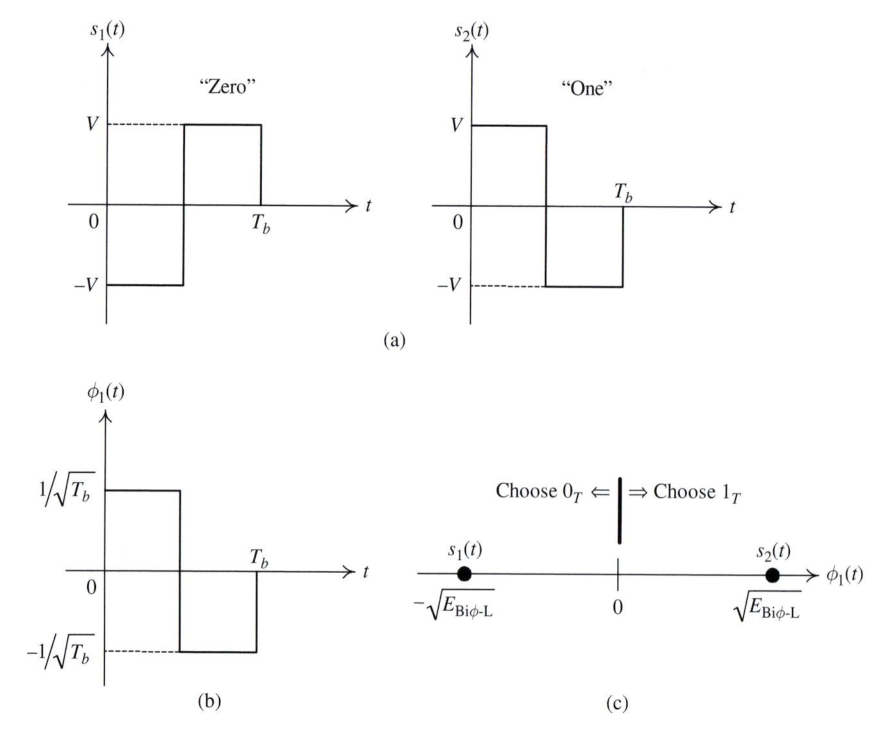

Fig. 6.7 Biϕ-L code: (a) elementary signals; (b) basis function; (c) signal space and decision regions.

To determine the bit error probability, note that

$$P[\text{error}] = 1 - P[\text{correct}], \tag{6.4}$$

where

$$P[\text{correct}] = \sum_{i=1}^{4} P[s_i(t)]P[\text{correct}|s_i(t)]. \tag{6.5}$$

Also $P[s_i(t)] = 1/4$, $i = 1, 2, 3, 4$. Because of the symmetry, $P[\text{correct}|s_i(t)]$ is the same regardless of which signal is considered. Therefore

$$P[\text{correct}] = P[\text{correct}|s_i(t)]. \tag{6.6}$$

The above probability can be found by evaluating the volume under the pdf $f(r_1, r_2|s_1(t))$. Consider the volume under $f(r_1, r_2|s_1(t))$ over region 1 shown as the shaded area in Figure 6.10. The random variables $\mathbf{r}_1$ and $\mathbf{r}_2$ are statistically independent Gaussian random variables, with means of $\sqrt{E_{\text{M-L}}}$ and 0 (volts), respectively, and with the same variance of $N_0/2$ (watts).

Rather than evaluating the integral with the joint pdf of $\mathbf{r}_1$ and $\mathbf{r}_2$, we change variables so that the integral is expressed in terms of the joint pdf of $\hat{\mathbf{r}}_1$ and $\hat{\mathbf{r}}_2$, which are also statistically independent Gaussian random variables with the same mean $\sqrt{E_{\text{M-L}}/2}$ and

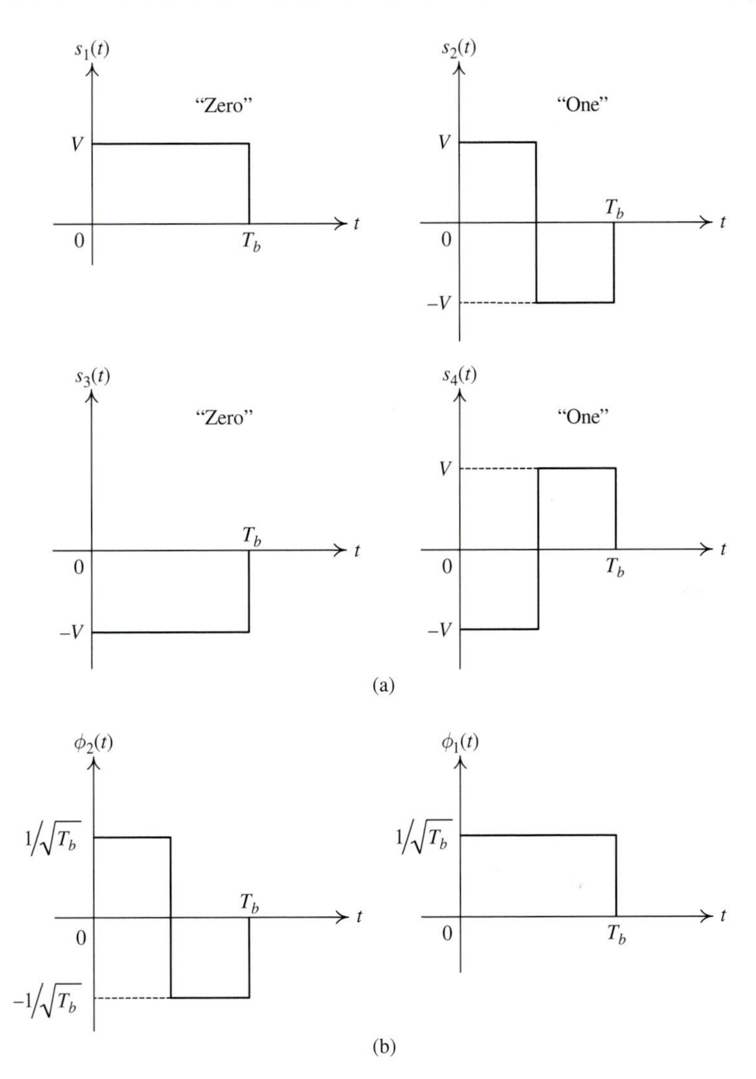

Fig. 6.8 Miller-L code: (a) elementary signals; (b) basis functions.

variance $N_0/2$. This is much more convenient since the region of the double integrals can be expressed as two independent regions of variables $\hat{r}_1, \hat{r}_2$. In particular,

$$\begin{aligned}
P[\text{correct}|s_i(t)] &= \int_0^\infty \int_0^\infty f\left(\hat{r}_1, \hat{r}_2 | s_i(t)\right) \mathrm{d}\hat{r}_1 \mathrm{d}\hat{r}_2 \\
&= \left[\int_0^\infty f\left(\hat{r}_1 | s_i(t)\right) \mathrm{d}\hat{r}_1\right]\left[\int_0^\infty f\left(\hat{r}_2 | s_i(t)\right) \mathrm{d}\hat{r}_2\right] \\
&= \left[\int_0^\infty \frac{1}{\sqrt{\pi N_0}} \exp\left\{\frac{\left(\hat{r} - \sqrt{E_{\text{M-L}}/2}\right)^2}{N_0}\right\} \mathrm{d}\hat{r}\right]^2 = \left[1 - Q\left(\sqrt{E_{\text{M-L}}/N_0}\right)\right]^2.
\end{aligned} \tag{6.7}$$

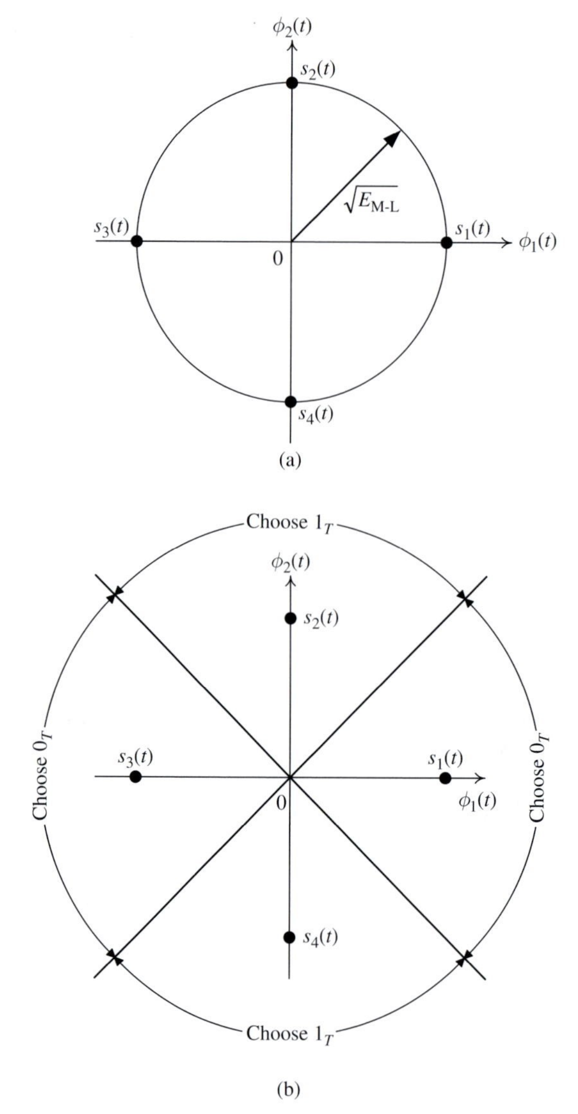

Fig. 6.9 Miller-L code: (a) signal space plot; (b) decision regions.

The bit error probability for the Miller code is therefore given by

$$P[\text{error}]_{\text{M-L}} = 1 - \left[1 - Q\left(\sqrt{E_{\text{M-L}}/N_0}\right)\right]^2$$
$$= 2Q\left(\sqrt{E_{\text{M-L}}/N_0}\right) - \left[Q\left(\sqrt{E_{\text{M-L}}/N_0}\right)\right]^2. \tag{6.8}$$

Comparing the four signaling schemes investigated in this chapter we see that they have the same energy per bit, i.e.,

$$E_{\text{NRZ-L}} = E_{\text{RZ-L}} = E_{\text{Bi}\phi\text{-L}} = E_{\text{M-L}} = V^2 T_b \equiv E_b \text{ (joules/bit)}. \tag{6.9}$$

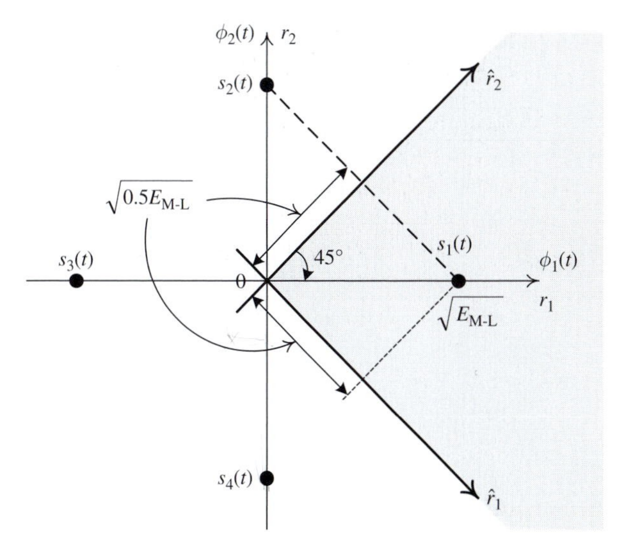

Fig. 6.10 Evaluating the error probability for Miller-L code.

Therefore, in terms of energy per bit, E_b, the error probabilities for the signaling schemes are:

$$P[\text{error}]_{\text{NRZ-L}} = P[\text{error}]_{\text{Bi}\phi\text{-L}} = Q\left(\sqrt{\frac{2E_b}{N_0}}\right), \tag{6.10}$$

$$P[\text{error}]_{\text{RZ-L}} = Q\left(\sqrt{\frac{E_b}{N_0}}\right), \tag{6.11}$$

and

$$P[\text{error}]_{\text{M-L}} = 2Q\left(\sqrt{\frac{E_b}{N_0}}\right) - \left[Q\left(\sqrt{\frac{E_b}{N_0}}\right)\right]^2 \tag{6.12}$$

$$\approx 2Q\left(\sqrt{\frac{E_b}{N_0}}\right). \tag{6.13}$$

The approximation in (6.13) follows from the fact that, because the typical value for the error probability is 10^{-4} or less, the second term in (6.12) is insignificant compared to the first term.

These expressions are plotted in Figure 6.11 as functions of E_b/N_0. What the above expressions say is that the SNR, E_b/N_0, would have to be double (i.e., it requires 3 decibels more transmitted power) for RZ-L (or a little more than double for Miller-L) coding to achieve the same error probability as NRZ-L or Biϕ-L coding. This fact can also be verified from Figure 6.11 at high SNR region.

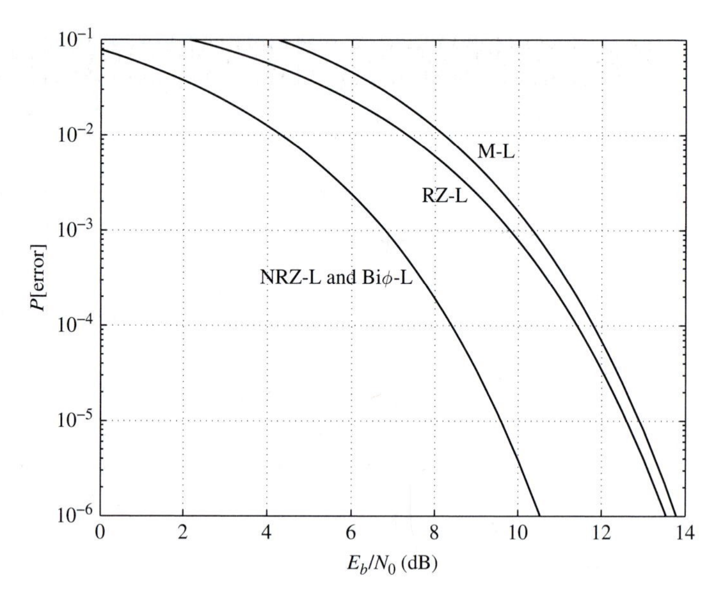

Fig. 6.11 Probabilities of bit error for various baseband signaling schemes.

6.4 Optimum sequence demodulation for Miller signaling

To this point the demodulation developed in Chapter 5 and applied in this chapter to the four line codes is a *symbol-by-symbol* demodulator. This type of demodulator assumes that there is no information about the transmitted sequence other than that in the present bit interval. Though this is true for the first three line codes it does not hold for Miller signaling. Miller modulation has memory, since the transmitted signal in the present bit interval depends on both the present bit and also the previous bit. In demodulation of the received signal knowledge of this memory can, and should be, exploited to make a decision. This leads to sequence demodulation, rather than symbol-by-symbol demodulation.

Before discussing sequence demodulation, a simple example is given to clearly illustrate the shortcoming of the symbol-by-symbol receiver when a modulation has memory.

Example 6.1 Assume that the four Miller signals $s_i(t)$, $i = 1, 2, 3, 4$ have unit energy. The projections of the received signals onto $\phi_1(t)$ and $\phi_2(t)$, denoted by $r_1^{(k)}, r_2^{(k)}$ where k signifies the bit interval, are given as

$$\left\{r_1^{(1)} = -0.2, r_2^{(1)} = -0.4\right\}, \left\{r_1^{(2)} = +0.2, r_2^{(2)} = -0.8\right\},$$
$$\left\{r_1^{(3)} = -0.61, r_2^{(3)} = +0.5\right\}, \left\{r_1^{(4)} = -1.1, r_2^{(4)} = +0.1\right\}.$$

Table 6.1 Distances squared from the received signals to four possible transmitted signals

Transmitted signal	Distance squared			
	$0 \to T_b$	$T_b \to 2T_b$	$2T_b \to 3T_b$	$3T_b \to 4T_b$
$s_1(t)$	1.6	1.28	2.8421	4.42
$s_2(t)$	2.0	3.28	0.6221	2.02
$s_3(t)$	0.8	2.08	**0.4021**	**0.02**
$s_4(t)$	**0.4**	**0.08**	2.6221	2.42

The distances squared from the received signal to all four possible transmitted signals in each bit interval are tabulated in Table 6.1. Thus the symbol-by-symbol minimum-distance receiver decides $\{s_4(t), s_4(t), s_3(t), s_3(t)\}$ as the sequence of transmitted signals, which corresponds to the bit sequence {1100}. However, according to the encoding rule of Miller modulation, the sequence $\{s_4(t), s_4(t), s_3(t), s_3(t)\}$ is not a *valid* transmitted sequence. This implies that there must be an error in the above symbol-by-symbol decisions. ■

The above example suggests that a better decision rule could be achieved by exploiting the memory of the Miller code. One possible way to demodulate the received signal to an allowable transmitted sequence is described next. Assume that a total of n bits is transmitted. Each n-bit pattern results in a transmitted signal over $0 \le t \le nT_b$. Obviously, the total number of different bit patterns (or signals) is $M = 2^n$, i.e., it grows exponentially with n. Typically n is a large number and 2^n can be a very very big number.[2] Denote the entire transmitted signal over the time interval $[0, nT_b]$ as $S_i(t)$, $i = 1, 2, \ldots, M = 2^n$. The signal $S_i(t)$ can also be written as $S_i(t) = \sum_{j=1}^{n} S_{ij}(t)$, where $S_{ij}(t)$ is one of the four possible signals used in Miller code in the bit interval $[(j-1)T_b, jT_b]$ and zero elsewhere. Note the meaning of the subscript notation, i refers to the specific transmitted signal under consideration and j to the Miller signal in the jth bit interval. At the receiver, the received signal over the time interval $[0, nT_b]$ is $r(t)$ and it can also be written as $r(t) = \sum_{j=1}^{n} r_j(t)$, where $r_j(t) = r(t)$ in the interval $[(j-1)T_b, jT_b]$ and zero elsewhere.

To decide which of the M possible signals was transmitted one can compute the distance from each signal $S_i(t)$ to the received signal $r(t)$, which is simply

$$d_i = \sqrt{\int_0^{nT_b} [r(t) - S_i(t)]^2 \mathrm{d}t}. \tag{6.14}$$

After such distances are computed for all $S_i(t)$, $i = 1, 2, \ldots, M = 2^n$, the decision rule is to choose the transmitted signal to be the one that is closest to $r(t)$.

The computation of the distance in the above decision rule can be simplified by projecting the continuous-time transmitted and received signals onto the signal space of Miller

[2] Unless you are a loan shark or a banker (which may amount to the same thing) exponential growth is *bad*, while linear growth is *good*.

code. To this end, proceed as follows. First, note that if the distance is minimum then the square of it will also be minimum. Then compute d_i^2 by splitting the integral up as follows

$$d_i^2 = \sum_{j=1}^{n} \int_{(j-1)T_b}^{jT_b} [r_j(t) - S_{ij}(t)]^2 \mathrm{d}t. \tag{6.15}$$

Now the term of summation $d_{ij}^2 = \int_{(j-1)T_b}^{jT_b} [r_j(t) - S_{ij}(t)]^2 \mathrm{d}t$ is simply the distance squared of $r_j(t)$ to $S_{ij}(t)$. Let $[r_1^{(j)}, r_2^{(j)}]$ be the outputs of the two correlators (or matched filters) in the $[(j-1)T_b, jT_b]$ time interval and $[S_{i1}^{(j)}, S_{i2}^{(j)}]$ the coefficients in the representation of $S_{ij}(t)$. We know that $d_{ij}^2 = (r_1^{(j)} - S_{i1}^{(j)})^2 + (r_2^{(j)} - S_{i2}^{(j)})^2$. Therefore, the distance computation in (6.14) can be rewritten as

$$d_i^2 = \sum_{j=1}^{n} \left[\left(r_1^{(j)} - S_{i1}^{(j)} \right)^2 + \left(r_2^{(j)} - S_{i2}^{(j)} \right)^2 \right]. \tag{6.16}$$

If the transmitted bits are equally likely and if the channel is AWGN, it can be shown that (see Chapter 7) the above decision rule, which is based on the minimum sequence distance, constitutes an optimum receiver for Miller signaling. It is optimum in the sense that the probability of making a sequence error is minimized.

Viterbi algorithm Though the optimum decision rule is rather simple in its interpretation, it requires an extensive and simply impossible amount of computation if direct evaluation of the $M = 2^n$ distances in (6.16) is to be carried out. Fortunately, a much more efficient algorithm, due to A. J. Viterbi and known as the Viterbi algorithm [1, 2], exists to implement the optimum decision rule. At the heart of the Viterbi algorithm are the concepts of the state and trellis diagrams that are used to elegantly represent all the possible transmitted sequences.

In general, the *state* of a system can be looked upon very simply (and perhaps somewhat loosely) as what information from the past do we need at the present time, which together with the present input allows us to determine the system's output for any future input. Here consider the system to be the modulator that produces the Miller encoded signal. Recall that in Miller signaling, the transmitted signal depends on the bit to be transmitted in the present interval and the signal, or bit, transmitted in the previous interval. There are four possible previous signals and therefore there are four states. The *state diagram* that represents the Miller encoding rule is shown in Figure 6.12.

Although the state diagram in Figure 6.12 clearly and concisely describes the encoding rule, it only describes the rule in a single bit interval. To illustrate the modulator's output for any possible input sequence one can follow the path dictated by the input bits and produce the output signal. However, a more informative approach is to use a *trellis diagram*. In essence, a trellis diagram is simply an unfolded state diagram. Figure 6.13 shows the trellis diagram for Miller modulation over the interval $[0, 4T_b]$.

Some important remarks regarding Figure 6.13 are as follows:

- It is assumed that the initial (or starting) state is $s_1(t)$. In practice, this can always be guaranteed by an agreed protocol.
- A transmitted signal in a given bit interval is represented by a *branch* connecting two states. The *solid* line corresponds to bit "0," whereas the *dashed* line corresponds to bit "1."
- In the fourth bit interval ($[3T_b, 4T_b]$) the trellis is fully expanded. From this bit interval on, the trellis pattern is the same.
- Each possible output sequence is represented by a *path* through the trellis. Conversely, each path in the trellis represents a valid (or allowable) Miller signal.

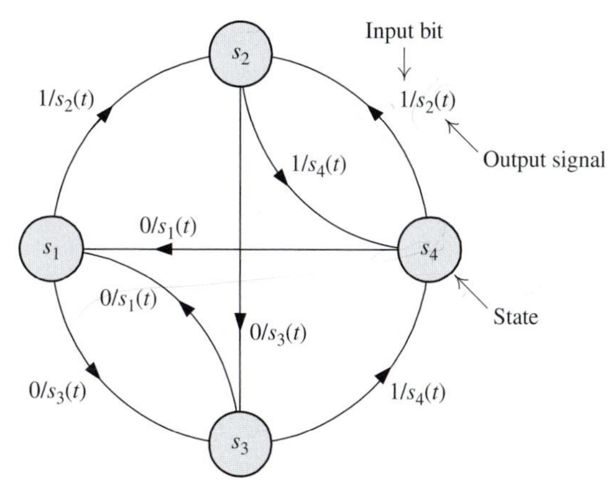

Fig. 6.12 State diagram of the Miller code. The state is defined as the signal transmitted in the previous bit interval.

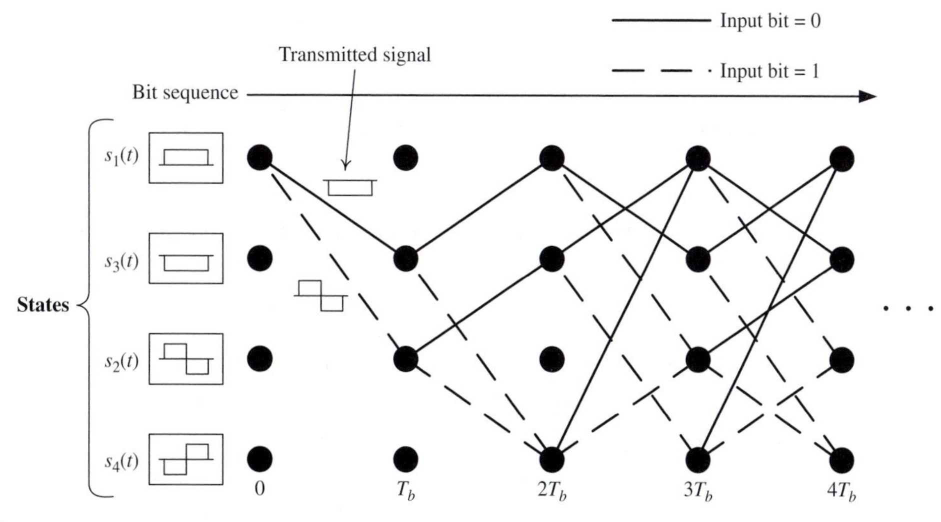

Fig. 6.13 Trellis diagram of Miller code.

Working with the trellis diagram, the main steps in the Viterbi algorithm to find the sequence (i.e., the path through the trellis) that is closest to the received signal are as follows.

Step 1: Start from the initial state ($s_1(t)$ in our case).

Step 2: In each bit interval, calculate the *branch metric*, which is the distance squared between the received signal in that interval with the signal corresponding to each possible branch. Add this branch metric to the previous metrics to get the *partial path metric* for each partial path up to that bit interval.

Step 3: If there are two partial paths entering the same state, discard the one that has a larger partial path metric and call the remaining path the *survivor*.

Step 4: Extend only the survivor paths to the next interval. Repeat steps 2–4 till the end of the sequence.

In step 3 above, the procedure where one of the competing partial paths is discarded at each state requires some justification. Namely, nothing that is received in the future will give one any information about what happened in the past. This is because future noise samples are statistically independent of present ones (recall that the noise is white and Gaussian) and also the bits are assumed to be statistically independent.

Example 6.2 To illustrate the Viterbi algorithm, let us revisit Example 6.1 and apply the Viterbi algorithm to demodulate the received sequence.

For the first interval there are only two possibilities. The distances squared of the received signal to these two possibilities are computed and are shown in Figure 6.14.

At this point no decision is made. Rather we continue to the next interval and compute the distance squared of the received signal (in the interval $[T_b, 2T_b]$) to the signal along a particular branch. The distances squared of each possible transmitted sequence to the received sequence are (see Figure 6.15):

$$00 \to 2.08;\ 10 \to 4.08;\ 01 \to 0.88;\ 11 \to 2.08.$$

If we were forced (or inclined) to make a decision at this stage we would choose sequence 01 for a rather "obvious" reason. Note that symbol-by-symbol demodulation, which ignores the memory, results in the sequence 11 being chosen. Again we do not make

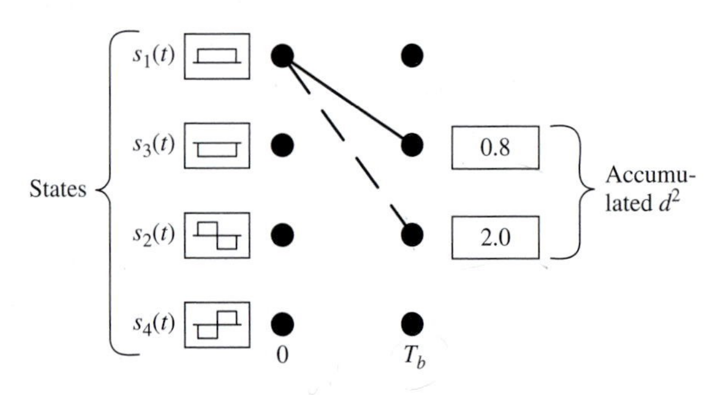

Fig. 6.14 Accumulated distances squared in the first bit interval.

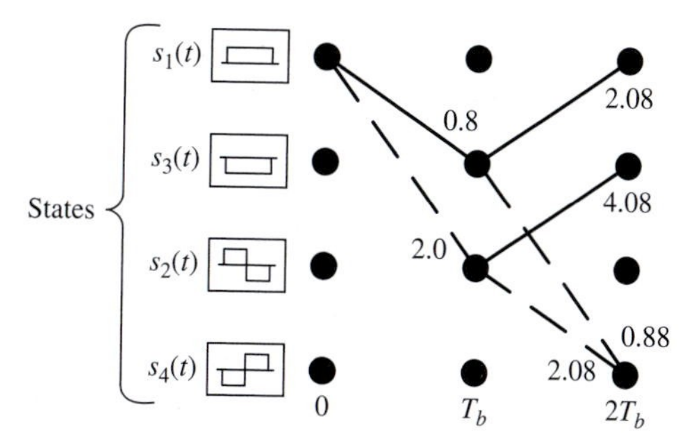

Fig. 6.15 Accumulated squared distances after the second bit interval.

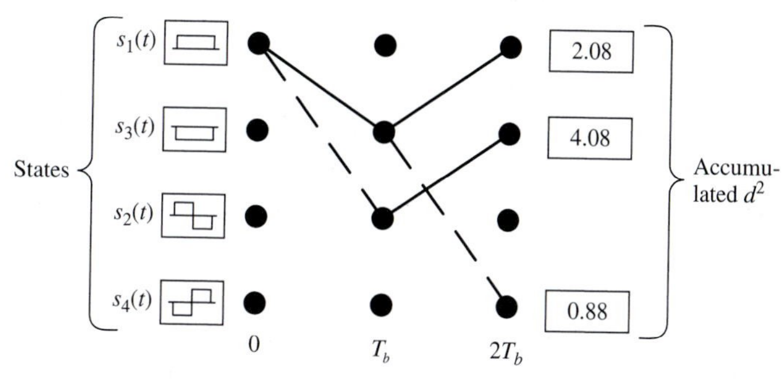

Fig. 6.16 Distances squared of the survivor paths after the second bit interval.

any firm decision as to which sequence is the most likely, but proceed to the next interval. However, we make one decision, namely, which sequence(s) cannot be most likely (step 3 of the procedure). For the two sequences that end in state $s_4(t)$ we discard one, namely, sequence 11. The picture now looks as in Figure 6.16.

In the third bit interval, the distance squared of the received signal to each possible signal during this bit interval (called the *branch metric*) is now computed (see Table 6.1), added to the surviving distances squared and new surviving sequences are determined. After the third bit interval, the surviving sequences are shown in Figure 6.17.

For the fourth bit interval, the branch distances squared are calculated as in Table 6.1 (see column 5). These distances are added appropriately to the survivors after the third bit interval. For the two sequences that converge in a given state, choosing the sequence that is closest to the received signal results in the survivors shown in Figure 6.18.

So, the surviving sequences are (from top to bottom):

$$s_3(t), s_1(t), s_3(t), s_1(t) \Leftrightarrow 0000,$$
$$s_3(t), s_4(t), s_2(t), s_3(t) \Leftrightarrow 0110,$$
$$s_3(t), s_4(t), s_1(t), s_2(t) \Leftrightarrow 0101,$$
$$s_3(t), s_4(t), s_2(t), s_4(t) \Leftrightarrow 0111.$$

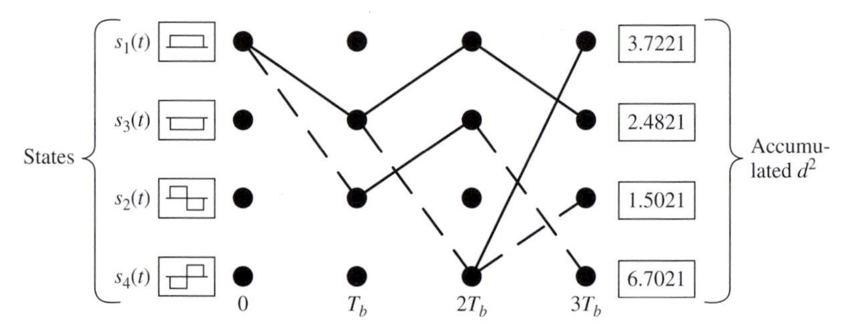

Fig. 6.17 Survivor paths and accumulated distances squared after the third bit interval.

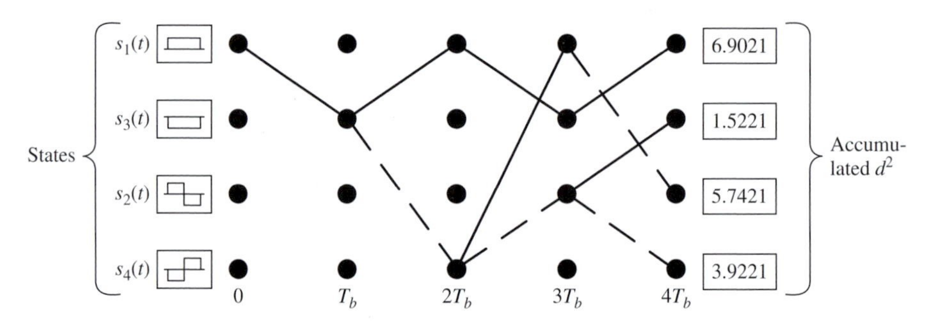

Fig. 6.18 Accumulated distances squared after the fourth bit interval.

We could continue *ad nauseum*, i.e., until we are ready to make a decision on the transmitted sequence or we know (by prior agreement or by protocol) that the sequence transmission has ended. Note that at this time ($4T_b$) one can make a very *firm decision* about the transmitted sequence, i.e., a decision that will not change due to future received signals. This decision is that of the first transmitted bit being "0," which corresponds to the common stem over $[0, T_b]$ of all the four survivor paths.

From the above surviving sequences, it is quite obvious that if one needs to make a decision at $4T_b$, one will decide that the sequence $s_3(t), s_4(t), s_2(t), s_3(t)$, which corresponds to 0110, was transmitted. ■

Finally, it is of interest to compare the bit error performance of the symbol-by-symbol demodulation to that of the sequence demodulation of Miller signaling. The bit error probability of the symbol-by-symbol demodulation was derived earlier and is given in (6.12). Derivation of the bit error probability for the sequence demodulation is, however, quite complicated and is beyond the scope of the current discussion. Instead, a simulation result can be obtained (see Problem 6.23), which is plotted in Figure 6.19 together with the expression in (6.12). Observe that the sequence demodulation consistently outperforms the symbol-by-symbol demodulation. The power gain provided by the sequence demodulation is most significant in the low to medium SNR region, where a gain of 2 decibels can be observed at the error probability of 10^{-2}. The gain becomes smaller in the high SNR region and it approaches about 0.5 decibels at the error probability of 10^{-6}.

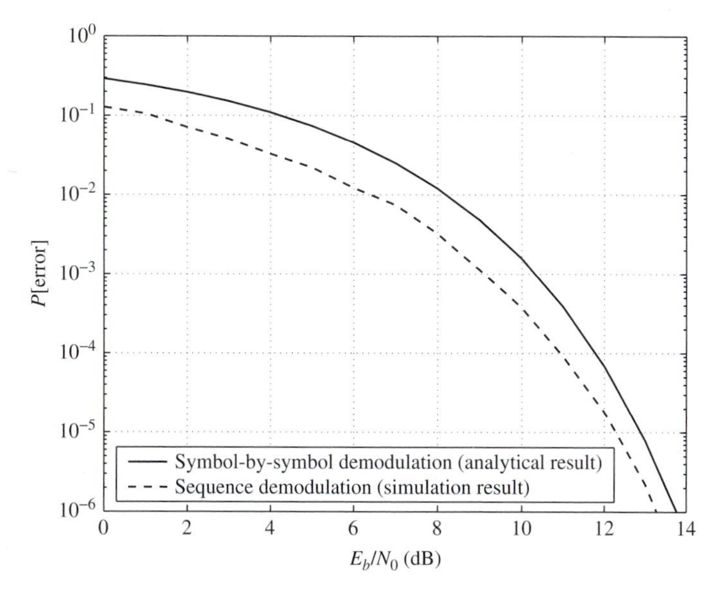

Fig. 6.19 Bit error probability comparison of symbol-by-symbol demodulation and sequence demodulation (Viterbi algorithm) of Miller signaling scheme.

6.5 Spectrum

The transmitted power required by a modulation scheme to achieve a certain error performance is important but equally important is the bandwidth requirement of the modulation. Therefore it is necessary to obtain the power spectral density.

For NRZ-L, RZ-L and Biϕ codes, the PSD expression derived in Chapter 5, (5.147), can be applied, since, as usual, we assume that the bits in different intervals are statistically independent. For each signaling scheme, let the energy in an elementary waveform be $E \triangleq V^2 T_b$ joules. Also let the *a priori* probability of bit "1" be $P_2 = P$ and that of bit "0" be $P_1 = 1 - P$. The normalized (with respect to E) PSDs are as follows.

For NRZ-L code, the spectrum is

$$\frac{S_{\text{NRZ-L}}(f)}{E} = \frac{1}{T_b}(1-2P)^2\delta(f) + 4P(1-P)\frac{\sin^2(\pi f T_b)}{(\pi f T_b)^2}. \tag{6.17}$$

Note that when $P = 0.5$ the impulse at $f = 0$ disappears as expected.

For Biϕ code, the spectrum is given by

$$\frac{S_{\text{Bi}\phi}(f)}{E} = \frac{1}{T_b}(1-2P)^2 \sum_{\substack{n=-\infty \\ n \text{ odd}}}^{\infty} \left(\frac{2}{n\pi}\right)^2 \delta\left(f - \frac{n}{T_b}\right) + 4P(1-P)\frac{\sin^4(\pi f T_b/2)}{(\pi f T_b/2)^2}. \tag{6.18}$$

There is no DC component as expected. Further, if $P = 0$ or $P = 1$, the continuous part of the spectrum disappears and there is only a line spectrum, reflecting the fact that the signal is periodic. When $P = 0.5$ the line spectrum disappears and only the continuous spectrum is present.

The PSD for Miller code is much more involved to derive. Here we simply give the PSD expression. A procedure to derive the PSD can be found in [3, 4]. The expression for the case of $P_1 = P_2 = 0.5$ is

$$\frac{S_{\text{M-L}}(f)}{E} = \frac{1}{2\theta^2(17 + 8\cos 8\theta)}(23 - 2\cos\theta - 22\cos 2\theta - 12\cos 3\theta + 5\cos 4\theta + 12\cos 5\theta + 2\cos 6\theta - 8\cos 7\theta + 2\cos 8\theta), \tag{6.19}$$

where $\theta = \pi f T_b$.

For comparison, the plots of the three different PSDs are shown in Figure 6.20. Observe that the Miller code has several advantages in terms of spectral properties over the NRZ-L and Biϕ codes. These are:

- The majority of the signaling energy lies at frequencies less than one-half the bit rate, $r_b = 1/T_b$.
- The spectrum is minimal in the vicinity of $f = 0$. This is important for channels which have poor DC response (for example, in magnetic recording).
- Bandwidth requirement is approximately 1/2 of that needed by biphase coding.

Miller modulation has another very interesting and important property. Consider any transmitted signal out of the modulator. Suppose that at some point in time there is a polarity reversal, i.e., multiplication of the signal by -1, in one of the various stages of the transmitter or receiver, due, perhaps, to the insertion of an inverting amplifier stage. Though, after the polarity reversal occurs, the entire received signal is the complete opposite of the transmitted signal, it is easy to see that the polarity reversal does not change

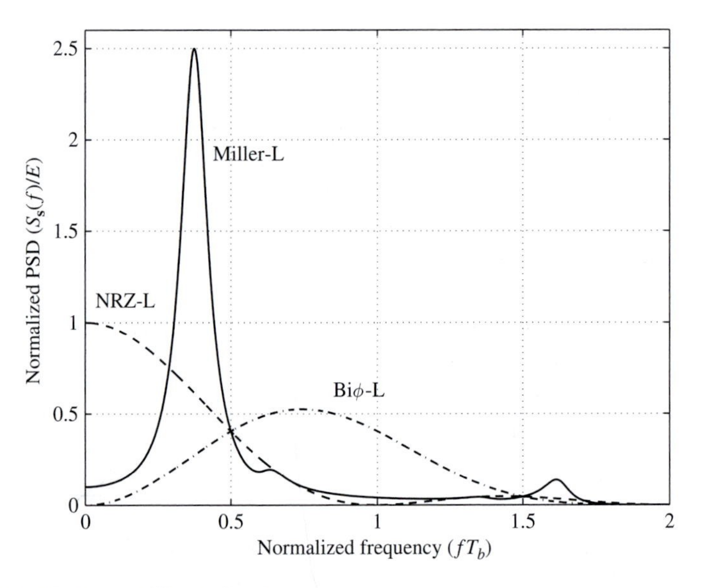

Fig. 6.20 PSDs of NRZ-L, Biϕ-L and Miller-L signals.

the decisions made by optimum symbol-by-symbol demodulation. Thus when symbol-by-symbol demodulation is used, Miller modulation is completely immune to a polarity reversal.

Less obvious and more subtle is that when the Viterbi algorithm is used for sequence demodulation of a Miller signal, it is also partially transparent to a polarity reversal. Because the Viterbi algorithm is constrained to follow a path in the trellis an error(s) will occur at the time of polarity reversal. However, the Viterbi algorithm will recover the most likely path based on the received sequence.

For the other three baseband modulations, a polarity reversal means that the demodulated bits are complemented, i.e., in the absence of AWGN, all bits are in error until another polarity reversal occurs. Can this be avoided? The answer is yes, by using a technique called *differential coding*. Since differential coding is used in later chapters on phase modulation and for channel models that exhibit phase uncertainty, we discuss it here in the context of baseband modulation.

6.6 Differential modulation

The basic concept behind differential modulation or coding[3] is that the signal transmitted in one bit interval is relative to the one transmitted in the previous interval. The actual transmitted signal thus depends on the present information bit and the previously transmitted bit. As a concrete example, take NRZ-L. Rather than mapping "1" to level $+V$ and "0" to level $-V$ irrespective of the previous signal, we change the modulation rule to:

- If the present bit is a "1," then transmit a level that is opposite to that of the previous interval.
- If the present bit is a "0," then stay at the same level.

Put simply, a "1" means a level change, a "0" no change. The transmitted signal in any interval is relative to that of the previous interval. More importantly, if there is a polarity reversal this relativity is still preserved, i.e., two consecutive signals at the same level imply a "0" and a change of level implies a "1."

Though described at the modulator output level, the modulation rule can be implemented by first differentially encoding the information bits, followed by an NRZ-L modulation. This is shown in Figure 6.21.

If $b_k = 1$ then $d_k = \overline{d}_{k-1}$, implying a level change, and if $b_k = 0$, then $d_k = d_{k-1}$, which means no level change. Note that because of the memory one must initialize the shift register circuit, say with a logic 0.

To demodulate the received signal one first determines d_k with minimum error probability. Call this estimate $\hat{d}_k$. To recover an estimate of b_k, which is the bit of interest,

[3] Whether one calls it differential coding or differential modulation depends on whether one considers the process to be a mapping from information bits to differential bits or from information bits to signals. The end result is the same.

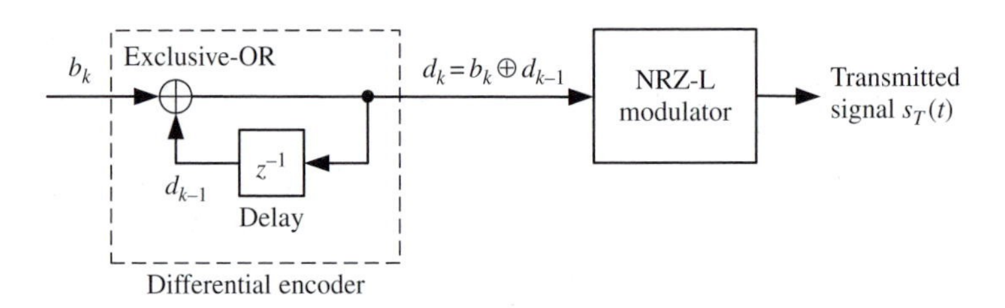

Fig. 6.21 Differential NRZ-L modulation.

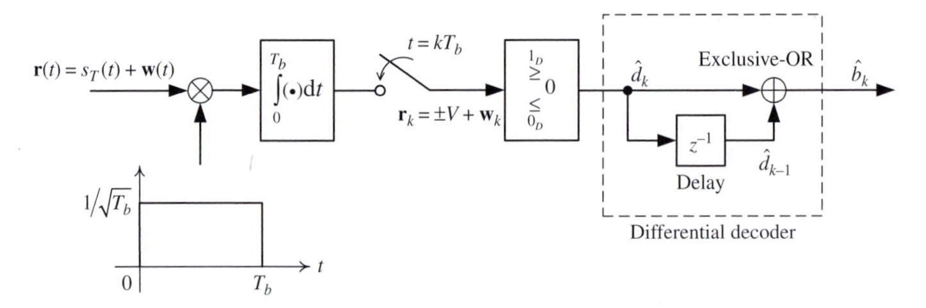

Fig. 6.22 Demodulation of differential NRZ-L modulation.

note that $d_k = b_k \oplus d_{k-1}$ and adding modulo-2 d_{k-1} to both sides results in $d_k \oplus d_{k-1} = b_k \oplus d_{k-1} \oplus d_{k-1} = b_k$, since $d_{k-1} \oplus d_{k-1} = 0$ and $b_k \oplus 0 = b_k$. This is what is done at the receiver, except we do not have d_{k-1} but the next best thing, $\hat{d}_{k-1}$. A block diagram of the demodulation is shown in Figure 6.22.

Once again, an initial condition must be assumed for the delay element, typically 0. Observe that if $\hat{d}_k$ is in error, then there will be two errors in the sequence $\{\hat{b}_k\}$ since $\hat{d}_k$ appears in the computation of $\hat{b}_k$ and $\hat{b}_{k+1}$. Thus the modulation is not totally transparent to polarity reversals. The differential encoding/decoding process can be viewed, somewhat imprecisely mathematically, as follows. The transfer function of the encoder is $1/(1 + z^{-1})$ while that of the decoder is $1 + z^{-1}$. Thus the end-to-end result is a transfer function of $\left(1/(1 + z^{-1})\right)\left(1 + z^{-1}\right) = 1$. Imprecisely, because we are dealing with modulo-2 algebra and not the real (or complex) number system.

Finally the described modulation is known as non-return-to-zero-inverse (NRZI). The problems explore the concepts of this section further.

6.7 Summary

The four baseband modulation methods dealt with in this chapter though basic are still found, and indeed should continue to be found, in digital communication systems. This holds, in particular, for NRZ-L, Biϕ, and Miller modulation. The next chapter considers passband modulation techniques. Though the approach taken there is a direct one, passband modulation can also be considered to be the frequency translation of an equivalent baseband modulation. This is also touched upon in the next chapter.

Miller modulation has memory and this property led to the important concepts of state, trellis, sequence demodulation, and Viterbi algorithm. The next three chapters continue the study of memoryless modulation techniques since they are not only important in their own right, but are found in modulation paradigms that have memory. State, trellis, and sequence modulation/demodulation are returned to in the chapter on bandlimited channels where intersymbol interference (ISI) is present and the one on trellis-coded modulation (TCM).

6.8 Problems

Many different baseband signaling schemes or line codes have been developed (see [3]). The following problems look at some of them. We start with NRZI, which is used primarily in magnetic recording. Because a long string of zeros results in no transitions, its timing characteristics are poor and therefore in telecommunications it is limited to short-haul applications.

6.1 Consider the following information bit sequence $\{b_k\} = \{1011000111\}$.

(a) Draw the NRZI waveform by applying the rule that $b_k = 1$ means there is a level transition while $b_k = 0$ means no transition. Assume that the initial condition is $-V$ volts.

(b) Determine the output of the differential encoder, $\{d_k\}$ and again draw the output waveform of the NRZ-L modulator. Compare with (a). Assume the initial condition $d_{-1} = 0$.

(c) Suppose a polarity reversal occurred after the fourth bit, i.e., at $t = 4T_b$ where transmission starts at 0. Draw the received waveform. Assume that there is no AWGN during transmission. What would the detected differential bits $\hat{d}_k$ be? Keep in mind d_k bits are NRZ-L modulated. What is the detected information sequence, $\{\hat{b}_k\}$? As at the transmitter, the receiver assumes an initial condition of $\hat{d}_{-1} = 0$.

(d) Suppose that the receiver misunderstands the transmitter and that it assumes an initial condition of $\hat{d}_{-1} = 1$. What is the detected information bit sequence now?

(e) Finally assume that the technologist realized at $t = 7T_b$ that there was a polarity reversal and corrected it. Draw the received waveform now and determine the detected differential bit sequence, $\{\hat{d}_k\}$ and the decoded information bit sequence, $\{\hat{b}_k\}$. Assume $\hat{d}_{-1} = 0$ (same as the transmitter).

(f) Draw conclusions as to the influence of polarity reversals on the bit error probability.

6.2 We now consider the influence of AWGN and polarity reversals on the bit error probability for NRZI. Consider the information bit sequence of Problem 6.1, assume that they are equally likely and that $V^2T_b = 1$ joule.

(a) Draw the signal space diagram for the two signals used in NRZI.
(b) The sampled output of the matched filter $\mathbf{r}_k = s_k + \mathbf{w}_k$, where $s_k = \pm\sqrt{E_b}$, in general is due to the signal transmitted and $\mathbf{w}_k$ is AWGN. What is the output sequence $\{s_k\}$ for the given information bit sequence?
(c) Now assume that the noise sample sequence, $\{\mathbf{w}_k\}$, is:

$$\{-0.4, -1.2, 0.2, 0.2, -0.4, -0.2, -0.8, 1.2, 0.2, 0.0\}.$$

Determine the complete sampled output sequence $\{\mathbf{r}_k\}$. Based on these $\{\mathbf{r}_k\}$, what are the detected bit estimates $\{\hat{d}_k\}$ and the corresponding $\{\hat{b}_k\}$? Note, particularly, the errors in $\{\hat{d}_k\}$ and $\{\hat{b}_k\}$. Based on this can you make a general statement about the bit error probability?
(d) Now assume that a polarity reversal has occurred at $t = 4T_b$. What is the $\{s_k\}$ sequence now? Given the noise samples of (c) what is the $\{\mathbf{r}_k\}$ sequence and then what are the $\{\hat{d}_k\}$, $\{\hat{b}_k\}$ sequences?

6.3 Here we derive an expression for bit error probability when NRZI is used. The information bits are equally likely.
(a) Show that the differential bits, $\{d_k\}$, are also equally likely.
(b) Let the voltage level be set so that each signal has E_b joules. Draw the signal space diagram and determine the bit error probability of the detected bits, $\hat{d}_k$, in AWGN, strength $N_0/2$ watts/hertz.
(c) But it is the error probability of $\{\hat{b}_k\}$ that is of interest. To obtain this error probability answer the following questions:
(i) Is $\hat{b}_k$ in error if both $\hat{d}_k$ and $\hat{d}_{k-1}$ are correct? Yes or No.
(ii) Is $\hat{b}_k$ in error if both $\hat{d}_k$ and $\hat{d}_{k-1}$ are incorrect? Yes or No.
(iii) Is $\hat{b}_k$ in error if $\hat{d}_k$ is correct and $\hat{d}_{k-1}$ is incorrect? Yes or No.
(iv) Is $\hat{b}_k$ in error if $\hat{d}_k$ is incorrect and $\hat{d}_{k-1}$ is correct? Yes or No.
Based on the above answers write $P[\hat{b}_k \text{ is in error}]$ as the probability of two mutually exclusive events and show that

$$P[\hat{b}_k \text{ is in error}] = 2Q\left(\sqrt{\frac{2E_b}{N_0}}\right)\left[1 - Q\left(\sqrt{\frac{2E_b}{N_0}}\right)\right]. \tag{P6.1}$$

Compare with NRZ-L and comment. How would you explain on an intuitive basis the factor 2 in the above expression?

6.4 NRZI has memory and therefore we expect that it can be represented by a state diagram and trellis. The memory can be looked upon in two ways: it occurs in the mapping from b_k to d_k or from b_k to the modulator output $\pm V$.
(a) Define a state set for both ways of looking at the memory and draw a state diagram. Label the transitions with the input bit, b_k, and the corresponding output quantity, either d_k or voltage level $\pm V$.
(b) Now draw a trellis corresponding to the above state diagrams. Start at $t = 0$ and assume that before $t = 0$ either $b_k = 0$ or the voltage level is $-V$ volts (the initial condition). Note you do not really need to draw two trellises, one trellis with clear labeling should suffice.

(c) Now assume that the sampled output sequence, $\{\mathbf{r}_k\}$, is that of Problem 6.2(d). Assume that the source bits are equally likely and that $V^2 T_b = E_b = 1$ joule. Use the signal space diagram of the NRZ-L modulator and the trellis of (b) to sequence demodulate using the Viterbi algorithm. Note that we can demodulate in two ways: either first demodulate to the differential bits and then pass them through the differential decoder to obtain the sequence $\{\hat{b}_k\}$, or demodulate directly to the sequence $\{\hat{b}_k\}$.

(d) Now assume that the sampled output sequence, $\{\mathbf{r}_k\}$, is that of Problem 6.2(d) where a polarity reversal takes place at $t = 4T_b$. Repeat (c).

6.5 The trellis determined in Problem 6.4 leads to the following interesting observation. Consider any two consecutive bit intervals.

(a) What transmitted signals do the following b_k bit patterns result in:

b_{k-1}	b_k
0	0
1	0
0	1
1	1

(P6.2)

(b) Based on these waveforms draw a signal space diagram which shows the signal points that represent the b_k bit, i.e., the present transmitted bit. Compare with the Miller modulation signal space.

(c) Based on the signal space of (b) propose a demodulator for bit b_k.

(d) What is the bit error probability, i.e., $P[b_k \text{ is in error}]$, for this demodulator? Compare with the answer of Problem 6.3(c) where differential decoding was used.

6.6 Of interest is the PSD of NRZI. To determine it show that the correlation between the differential bits d_k and d_{k-1} is zero. Note that d_k is an abstract quantity, logic 0 or logic 1. To find the crosscorrelation map d_k to the real number ± 1, (i.e., $d_k = 1 \rightarrow +1$ and $d_k = 0 \rightarrow -1$) and find the crosscorrelation between these quantities. Based on this reason that the PSD of NRZI is the same as that of NRZ-L. Finally, though uncorrelated, would you consider d_k and d_{k-1} to be statistically independent?

6.7 What is called *conditioned* Biϕ or Manchester code is in essence a combination of a differential encoder whose output modulates a Biϕ or Manchester modulator. It looks as shown in Figure 6.23:

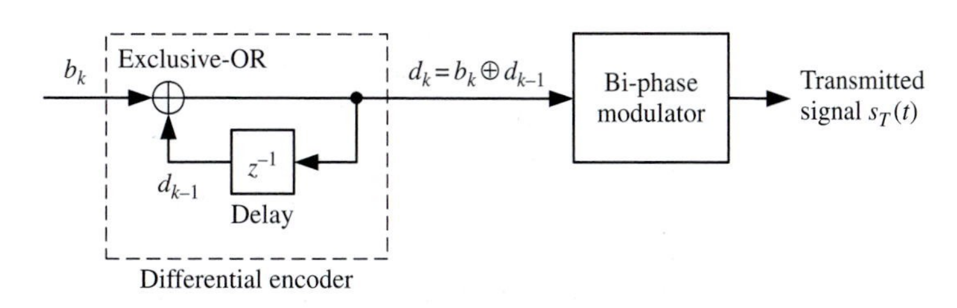

Fig. 6.23 Block diagram of conditioned Biϕ modulator.

(a) Take the information bit sequence of Problem 6.1 and draw the output signal.

(b) Is the modulation reasonably transparent to polarity reversals?

(c) If you have assiduously worked through Problems 6.1–6.5 how would the solutions change if the modulation is conditioned Biϕ instead of NRZI?

6.8 Consider the following statement. "*Any antipodal binary signal set can be made immune to polarity reversals by differential encoding and decoding*." Please discuss the statement from the point of view of error performance, PSD, demodulator implementation. Assume the information bits are equally probable and statistically independent and that the only impairment is AWGN.

Of interest is that Biϕ formats are used in magnetic recording, optical communications, and some satellite telemetry links. It is specified in the IEEE 802.3 Ethernet standard for baseband coaxial cable using carrier sense multiple access and collision detection (CSMA/CD). Differential Manchester coding is specified in the IEEE 802.5 standard for Token ring using either coaxial cable or twisted-pair wire.

6.9 Another so called Biϕ modulation has the following mapping rule(s). There is always a transition at the beginning of a bit interval. If $b_k = 1$ there is a transition in the middle of the bit interval, while if $b_k = 0$ there is no transition. As always the transitions are between $\pm V$.

(a) Based on this, draw the transmitted signal for the information bit sequence of Problem 6.1.

(b) Is the modulation immune to polarity reversals?

(c) Draw the signal space diagram for this modulation. Based on this signal set, how would you demodulate the received signal set with minimum error probability. Assume a symbol-by-symbol demodulator and equally probable bits.

(d) What is the error probability of this modulation?

Remark The PSD of this is identical to that of Manchester coding. However, the modulation process has memory. It can be modeled as a first-order Markov process but the PSD derivation involves concepts beyond the scope of the text.

6.10 Try to extend differential modulation as follows. Any orthogonal binary signal set can be made immune to polarity reversals by modifying the demodulator. What is this modification and what is the price you pay for achieving this immunity?

The next set of problems, Problems 6.11–6.16, considers baseband modulation that can be classed as pseudoternary. Ternary because it uses three levels, $\pm V$ volts and 0.

6.11 (*AMI-NRZ coding*) Alternate-mark-inverse NRZ is a binary coding scheme. It belongs to the family of pseudoternary codes where three levels $\pm V$ and 0 are used. The output signal is determined from the source bit stream as follows:

- If the bit to be transmitted is a "0," then the signal is 0 volt over the period of T_b seconds.
- If the bit to be transmitted is a "1," then the signal is either $+V$ volts if $-V$ volts was previously used to represent bit 1, or vice versa. Hence the name and mnemonic for the modulation.

(a) Draw the transmitted signal when the input bit sequence is that of Problem 6.1. Is the modulation immune to polarity reversals?

(b) Obtain a signal space representation of this modulation in terms of $E = V^2 T_b$. Based on the signal space what is the sufficient statistic on which to base a decision?

(c) Using this sufficient statistic find the conditional pdfs, $f(r|1_T)$ and $f(r|0_T)$. To find $f(r|1_T)$ average the two conditional pdfs $f(r|1_T, -V), f(r|1_T, +V)$. Note that $f(r|1_T)$ is non-Gaussian. Sketch the two pdfs.

(d) Based on $f(r|1_T)$ and $f(r|0_T)$ from (c) show that the decision rule is

$$\cosh\left(\frac{2\sqrt{E}}{N_0}r\right) \underset{0_D}{\overset{1_D}{\gtrless}} e^{E/N_0}. \qquad \text{(P6.3)}$$

(e) You wish to find the error performance versus the SNR, E_b/N_0. To do this first let $\ell = (2\sqrt{E}/N_0)r$, which makes the decision rule become

$$\ell \underset{0_D}{\overset{1_D}{\gtrless}} \cosh^{-1}\left(e^{2E/N_0}\right) \equiv T_h. \qquad \text{(P6.4)}$$

Note that $\cosh^{-1}(\cdot)$ has two solutions, one at $+T_h$ and one at $-T_h$. Sketch the decision regions and the individual pdfs, $f(\ell|0_T), f(\ell| - V), f(\ell| + V)$ on the ℓ axis. Note that $\boldsymbol{\ell}$ in the conditional pdfs is Gaussian, indicate the mean and the variance in each case. Based on this sketch show that

$$P[\text{error}|0_T] = 2Q\left(\frac{T_h}{\sqrt{\frac{4E_b}{N_0}}}\right), \qquad \text{(P6.5)}$$

$$P[\text{error}|1_T] = Q\left(\frac{(4E_b/N_0) - T_h}{\sqrt{4E_b/N_0}}\right) - Q\left(\frac{(4E_b/N_0) + T_h}{\sqrt{4E_b/N_0}}\right). \qquad \text{(P6.6)}$$

6.12 Your friend decides to obtain the decision rule for demodulation as follows. First she demodulates the received signal to the signal level, i.e., detects whether a $+V$ or a $-V$ or a 0 was transmitted and then maps the detected signal to the corresponding bit. She does this on symbol-by-symbol basis and since there are now three symbols (signals) she needs to extend the theory. She thus reads ahead to Chapter 8 on M-ary modulation and finds out that she needs to determine three likelihoods, one for each signal. That is, the decision rule is

$$\begin{gathered}\text{compute} \\ P_1 f(r| - V),\ P_2 f(r|0),\ P_3 f(r| + V) \\ \text{and choose the } \textit{largest}, \end{gathered} \qquad \text{(P6.7)}$$

where $P_1 = P[-V \text{ is transmitted}] = \frac{1}{4}$, $P_3 = P[+V \text{ is transmitted}] = \frac{1}{4}$, $P_2 = P[0 \text{ is transmitted}] = \frac{1}{2}$, and r is the same sufficient statistic as in Problem 6.11.

(a) Based on this show that the decision rule can be expressed as:

$$\begin{cases} -\dfrac{N_0}{2\sqrt{E}}\ln 2 + \dfrac{\sqrt{E}}{2} \le r \le \dfrac{N_0}{2\sqrt{E}}\ln 2 + \dfrac{\sqrt{E}}{2}, & \text{choose } 0 \\ \text{otherwise,} & \text{choose } 1 \end{cases}. \qquad \text{(P6.8)}$$

(b) Let $\ell = (2\sqrt{E}/N_0)r$ and express the decision rule in terms of ℓ. Determine $f(\ell| - V), f(\ell|0), f(\ell| + V)$.

(c) Using ℓ as the decision variable show that:

$$P[\text{error}|0_T] = 2Q\left(\frac{\ln 2 + (2E_b/N_0)}{\sqrt{4E_b/N_0}}\right), \tag{P6.9}$$

$$P[\text{error}|1_T] = 2\left[Q\left(\frac{(2E_b/N_0) - \ln 2}{\sqrt{4E_b/N_0}}\right) - Q\left((6E_b/N_0) + \ln 2\sqrt{4E_b/N_0}\right)\right], \tag{P6.10}$$

where $E_b = E/2$ (joules/bit). Then find P[bit error].

6.13 Another friend neither wants to read ahead nor do the algebra of the two previous problems. He simply sketches the three pdfs $f(r|-V)$, $f(r|0)$, and $f(r|+V)$, notes which is the largest at each point on the axis, and chooses the signal and bit corresponding to that density.

(a) What is the decision rule now? Express it as simply as possible.

(b) Determine the P[bit error] in terms of E_b/N_0 for this approach.

6.14 The above three approaches resulted in three different decision rules and three different expressions for error probability. The three approaches can be termed as: (i) demodulate to bit, (ii) demodulate to signal and then to bit taking *a priori* probability into account, and (iii) demodulate to signal or bit ignoring the *a priori* probability.

(a) Do you think there is any difference in error probability between the three approaches? If so, then which approach should have the best performance?

(b) To confirm or disprove your intuition in (a), use Matlab to plot the error performance for the three approaches for E_b/N_0 in the range 0–10 dB. Look also at sections of the plot, say E_b/N_0 from 0.6–0.7 dB and E_b/N_0 from 8–8.1 dB. Discuss the plot.

6.15 One could, of course, consider another demodulator for AMI-NRZ. Since the signal transmitted in any bit period depends on what happened previously, the modulation has memory. Therefore there is a state diagram that describes the modulation and a corresponding trellis.

(a) Decide on what you need to know from the past, define the states, and draw the state diagram. Label the transitions between the states with the input bit and the corresponding transmitted signal.

(b) Now draw the trellis corresponding to the above state diagram. As usual you will have to decide on the initial state, i.e., the starting point. The decision is not crucial as long as the transmitter and the receiver agree on it.

Remark To sequence demodulate the transmitted signal using the trellis and Viterbi algorithm as discussed in the chapter one needs to derive a branch metric for this modulation. The sufficient statistic is the same but $f(r|1_T)$ is *not* Gaussian and therefore the minimum distance metric no longer applies.

6.16 (*AMI-RZ*) Instead of NRZ, one can use RZ signals. Modify the AMI-NRZ mapping rule to reflect this and redo Problems 6.11 and 6.12.

Remark AMI-RZ is used in the T1 carrier system by AT&T.

Optical fiber communication systems use baseband modulation. However, though pseudoternary line codes have been widely used in coaxial or twisted-pair cable systems, they cannot be used readily in fiber because of the three levels, $-V$, 0, $+V$.

Though one could map $-V$ to the laser diode being off, 0 to $1/2$ intensity and $+V$ to full intensity, the laser diode has nonlinearity. This has led to the line coding schemes called coded mode inversion (CMI) and differential mode inversion (DMI).

6.17 (*CMI*) The modulation rule is as follows:

- Bit "1" is represented by either $+I$ (laser diode full on) or 0 (laser diode off) with the levels alternating between successive ones.
- A "0" is represented by level $+I$ for the first half of the bit interval and by level 0 for the second half (or vice versa).

(a) Draw the output waveform for the bit sequence of Problem 6.1.

(b) Determine a basis set for the modulator and then plot the signal space. Note that $+I$ represents light intensity, it does not have the unit of volts. So plot the signal points in terms of E, the energy transmitted over a full period.

(c) The AWGN model does not apply readily for fiber optics. One demodulation method is to compare the intensity in the first half with that in the second half. Based on this decide on an intuitive basis what the decision rule is and what the decision space looks like. Is the decision insensitive to polarity inversions?

Remark CMI has been chosen for the 139.246 Mbps multiplex within the European digital hierarchy [5].

6.18 Consider using the CMI scheme for optical communications, where the receiver measures the light intensity by counting the number of received photons. As in Problem 5.23, the number of photons emitted per unit time is modeled with a Poisson point process:

$$P[k \text{ photons emitted in a unit interval}] = \frac{\lambda^k e^{-\lambda}}{k!}, \quad k = 0, 1, \ldots, \qquad \text{(P6.11)}$$

where λ is the mean arrival rate with units of photons per unit time. Here $\lambda = \lambda_s + \lambda_n$ or $\lambda = \lambda_n$, where λ_s is due to the transmitter laser being turned on and λ_n is due to the background radiation. Assume that the probability of transmitting a "1" is the same as transmitting a "0."

(a) Design the receiver that minimizes the probability of error.

(b) Derive an expression for the error performance of the receiver obtained in (a). *Remark* The average transmitted energy is hfT_b, where hf is the energy of photon with frequency f, $h = 6.6 \times 10^{-34}$ is Planck's constant and $\lambda_s T_b$ is the average number of signal photons per bit interval.

6.19 (*DMI*) The mapping rule for bit "1" is the same as that for CMI. However, it is changed for bit zero to prevent any pulse widths that are wider than T_b, the bit interval, i.e., bit "0" is mapped to either $(I, 0)$ or $(0, I)$, chosen so as to prevent a pulse width wider than T_b.

(a) Plot the transmitted signal for the bit sequence of Problem 6.1.

(b) Is the modulation immune to polarity reversals?

(c) Obtain a signal space plot for the modulator.

(d) Obtain a state diagram for the modulator.

6.20 To determine the PSD of AMI-NRZ one needs to determine the autocorrelation of the input information sequence after the bits have been mapped to real numbers, c_k, as follows:

$$\mathbf{c}_k = \begin{cases} 1 \text{ or } -1, & \text{if } \mathbf{b}_k = 1 \\ 0, & \text{if } \mathbf{b}_k = 0 \end{cases}. \tag{P6.12}$$

(a) Assuming that the $\mathbf{b}_k$ are equally probable, what are the following probabilities: $P[\mathbf{c}_k = 1]$; $P[\mathbf{c}_k = -1]$; $P[\mathbf{c}_k = 0]$?

(b) Based on this determine the autocorrelation of sequence $\{\mathbf{c}_k\}$, i.e., find $E\{\mathbf{c}_k^2\}$; $E\{\mathbf{c}_k\mathbf{c}_{k+1}\}$ and $E\{\mathbf{c}_k\mathbf{c}_{k+n}\}$, $n > 1$. *Hint* Determine the probabilities by considering all possible products and the probability of each product.

(c) Once the autocorrelation has been found the PSD is determined from (5.127). Show that it is

$$S(f) = V^2 T_b \left(\frac{\sin \pi f T_b}{\pi f T_b} \right)^2 \sin^2(\pi f T_b). \tag{P6.13}$$

(d) What is the null-to-null bandwidth?

6.21 (*Miller code*) Consider the optimum decision rule of Miller code that is based on the minimum sequence distance.

(a) Show that finding the sequence with the minimum distance as computed in (6.14) is equivalent to finding the sequence with the maximum correlation metric, computed as follows:

$$\int_0^{nT_b} r(t)S_i(t)\mathrm{d}t = \sum_{j=1}^{n} \left[r_1^{(j)}, r_2^{(j)} \right] \left[\begin{array}{c} S_{i1}^{(j)} \\ S_{i2}^{(j)} \end{array} \right]. \tag{P6.14}$$

(b) Clearly describe all the steps of the Viterbi algorithm to find the transmitted sequence based on the maximization of the above metric.

(c) Redo Example 6.2 using the above metric.

6.22 (*Class2K modulation*) In deriving the error probability of the symbol-by-symbol demodulation of the Miller signaling, the factor of 2 arises from the fact that each bit has two nearest neighbors crowding it. To overcome this the scheme shown in Figure 6.24, which we call *Class2K* (class of 2000) modulation, is proposed.

Several questions about the mapping in Figure 6.24 arise.

(a) The first question is what is the error probability of the symbol-by-symbol demodulation? Determine this and compare it to the Miller scheme.

(b) More importantly, what is its PSD? Like the Miller scheme, this modulation has memory, as described by the state diagram in Figure 6.25. The PSDs of signals with memory are very difficult to determine analytically. However, there is more than "one way to skin a cat" and here we shall skin the cat as follows.

The PSD of *Class2K* modulation can be found by simulation in Matlab. An outline as how to accomplish this is given next.

(i) Generate a sequence of equally likely bits using `rand` and `round` commands (a sequence of 100 bits is suggested).

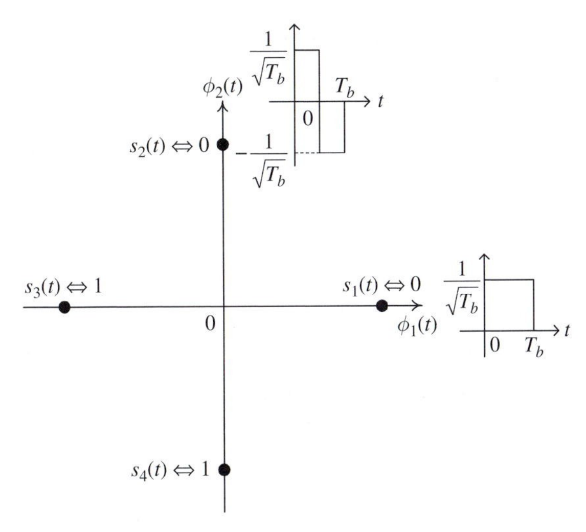

Fig. 6.24 Signal space diagram of *Class2K* modulation.

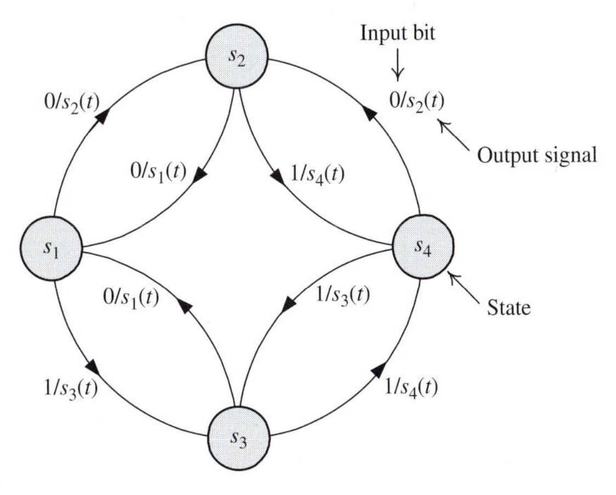

Fig. 6.25 State diagram of *Class2K* modulation.

(ii) Generate the corresponding time signal (using 100 points per bit per interval is suggested). Note that you will need to follow the state diagram in Figure 6.25.

(iii) Find the `fft` of the time signal, `fftshift` it, and then magnitude square the Fourier transform. Plot the result. It will in all likelihood look quite "jagged." To overcome this one needs to perform the last step in the basic definition of the PSD of a random process: *ensemble average*.

(iv) Therefore repeat the above another nine times and average the resultant spectra.

Compare the determined spectrum with that of Miller modulation and comment on which you would choose.

6.23 (*Sequence demodulation of Miller signaling*) Write a simulation program in Matlab to perform Miller-L modulation and its *sequence* demodulation (using the Viterbi algorithm). Verify the simulation result shown in Figure 6.19.

References

[1] A. J. Viterbi, "Error bounds for convolutional codes and asymptotically optimum decoding algorithm," *IEEE Transactions on Information Theory*, vol. IT-13, pp. 260–269, Apr. 1967.

[2] G. D. Forney, Jr., "The Viterbi algorithm," *Proc. IEEE*, vol. 61, pp. 268–278, Mar. 1973.

[3] F. Xiong, *Digital Modulation Techniques*. Artech House, second edn, 2006.

[4] M. Hecht and A. Guida, "Delay modulation," *Proc. IEEE*, vol. 57, pp. 1314–1316, July 1969.

[5] CCITT Yellow Book, *Digital Networks – Transmission Systems and Multiplexing Equipment*. ITU, Geneva, 1981.

7 Basic digital passband modulation

7.1 Introduction

In baseband transmission the transmitted signal power lies at low frequencies, typically around zero. It is desirable in many digital communication systems, for the same reasons as in analog communication systems, for the transmitted signal to lie in a frequency band toward the high end of the spectrum. As an example satellite communication is normally conducted in the 6–8 gigahertz band, while mobile phones systems are implemented in the 800 megahertz–2.0 gigahertz band.

The digital information is encoded as a variation of the parameters of a sinusoidal signal, called the *carrier* signal. Typically, as for analog modulation systems, the carrier frequency is much higher than the highest frequency of the modulating signals (or messages). Digital passband modulation is based on variation of the *amplitude*, *phase*, or *frequency* of the sinusoidal carrier, or some combination of these parameters.

Amplitude-shift keying (ASK) was probably the first type of digital modulation to be practically applied. In its simplest form it has been used for radio telegraphy transmission in Morse code. Another name for ASK is "on–off keying" (OOK), since a binary "1" corresponds to the sinusoid being transmitted while a binary "0" suppresses the carrier. *Phase-shift keying* (PSK) is an efficient, in terms of signal power, digital modulation method. It is widely used in modern digital communication systems, such as satellite links, wideband microwave radio relay systems, etc. The digital information is encoded in the phase function of a *constant-amplitude* carrier signal. *Frequency-shift keying* (FSK) is also a constant-amplitude modulation technique. Its main applications are in narrowband digital radio equipment, such as portable radio sets. It has also been successfully used in relatively wideband digital microwave radio relay equipment, primarily as modification of existing analog frequency-division multiplexing (FDM) and frequency modulation (FM) systems. The main advantages of FSK are simplicity, low implementation cost, and good performance level, especially under signal fading conditions. As the name suggests the digital information is encoded in the frequency of the sinusoidal carrier.

The discussion on digital modulation in this chapter first considers *binary* modulation of the carrier. As with baseband transmission, binary ASK (BASK), binary PSK (BPSK), and binary FSK (BFSK) modulations are discussed from the viewpoint of error performance. Optimum receivers are developed for the AWGN channels. The spectra of the three modulation schemes are also considered. After discussing these basic modulation techniques, extensions of them, including quadrature phase-shift keying (QPSK), offset QPSK (OQPSK) and minimum shift keying (MSK), are presented.

It is assumed throughout the chapter that there is perfect synchronization at the receiver and that the phase of the reference carrier is known. Optimum reception, when the channel causes phase distortion and/or fading, is developed in Chapter 10, while the general M-ary modulation techniques are covered in Chapter 8.

7.2 Binary amplitude-shift keying (BASK)

In BASK a sinusoidal carrier is simply gated on and off by the bit sequence to be transmitted. A typical transmitted waveform would be that shown in Figure 7.1(c), corresponding to the information sequence of Figure 7.1(a). The transmitted signal can be written as

$$s(t) = m(t)c(t), \tag{7.1}$$

where $m(t)$ is the modulating signal (the baseband signal, an NRZ signal) and $c(t) = V\cos(2\pi f_c t)$ is the sinusoidal carrier. The logic "1" and "0" are represented during any bit interval, T_b, by the following signal set:

$$\begin{cases} s_1(t) = 0, & \text{if "}0_T\text{"} \\ s_2(t) = V\cos(2\pi f_c t), & \text{if "}1_T\text{"} \end{cases}, \quad 0 < t \le T_b, \tag{7.2}$$

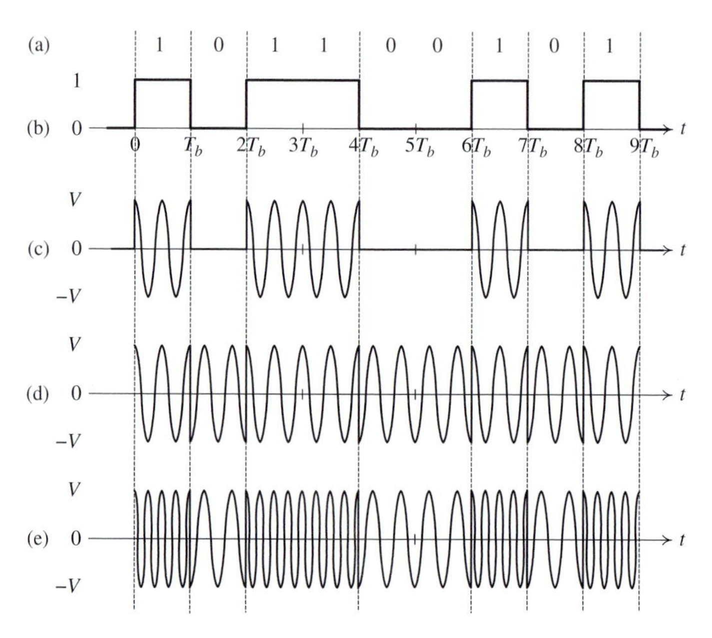

Fig. 7.1 Binary passband modulation techniques (a) binary data; (b) modulating signal $m(t)$; (c) BASK signal; (d) BPSK signal; (e) BFSK signal.

where the carrier frequency is usually chosen such that there is an integer number of cycles over the bit duration T_b, i.e., $f_c = n/T_b$, n an integer. The energy in $s_2(t)$ is $E_{\text{BASK}} = V^2 T_b/2$ joules.

The received signal is $\mathbf{r}(t) = s_i(t) + \mathbf{w}(t)$, where $i = 1$ or 2 depending on the transmitted signal, and $\mathbf{w}(t)$ is a zero-mean Gaussian noise process with two-sided PSD $N_0/2$. Only one orthonormal basis function, $\phi_1(t) = s_2(t)/\sqrt{E_{\text{BASK}}}$, is needed to represent the signal set. The signal space plot is shown in Figure 7.2(a). The optimum receiver, i.e., the one with the minimum error probability, is shown in Figure 7.2(b) in the form of a correlation receiver. The threshold, T_h, in Figure 7.2(b) is given by

$$T_h = \frac{N_0}{2\sqrt{E_{\text{BASK}}}} \ln\left(\frac{P_1}{P_2}\right) + \frac{\sqrt{E_{\text{BASK}}}}{2}. \tag{7.3}$$

For $P_1 = P_2$, the decision regions are depicted in Figure 7.2(c). The error probability for this case is

$$P[\text{error}]_{\text{BASK}} = Q\left(\sqrt{\frac{E_{\text{BASK}}}{2N_0}}\right). \tag{7.4}$$

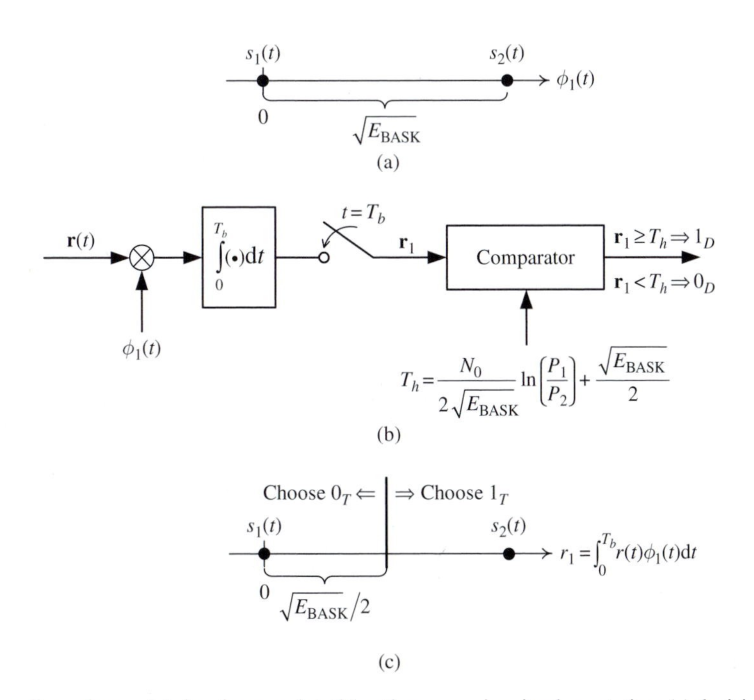

Fig. 7.2 BASK signaling scheme: (a) signal space plot; (b) optimum receiver implementation; (c) decision regions.

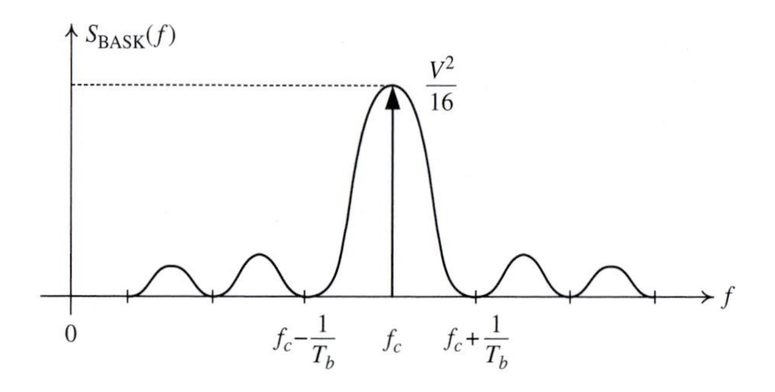

Fig. 7.3 PSD of BASK.

To determine the bandwidth requirement of BASK signaling use (5.147) to obtain the PSD, which is[1]

$$S_{\text{BASK}}(f) = \frac{V^2}{16}\left[\delta(f - f_c) + \delta(f + f_c) + \frac{\sin^2[\pi T_b(f + f_c)]}{\pi^2 T_b(f + f_c)^2} + \frac{\sin^2[\pi T_b(f - f_c)]}{\pi^2 T_b(f - f_c)^2}\right]. \tag{7.5}$$

A sketch of the PSD for positive frequencies is shown in Figure 7.3. It can be shown that approximately 95% of the total transmitted power lies in a band of $3/T_b$ (hertz), centered at f_c.

7.3 Binary phase-shift keying (BPSK)

A BPSK signal is generated by amplitude modulating the sinusoidal carrier with a NRZ-L signal of amplitude ± 1. The transmitted signal is $s(t) = m(t)c(t)$ (where $m(t)$ is a NRZ-L signal) with a resultant phase that is either 0 or π radians. The waveforms are plotted in Figure 7.1(d). The signal set is given by

$$\begin{cases} s_1(t) = -V\cos(2\pi f_c t), & \text{if "}0_T\text{"} \\ s_2(t) = +V\cos(2\pi f_c t), & \text{if "}1_T\text{"} \end{cases}, \quad 0 < t \leq T_b, \tag{7.6}$$

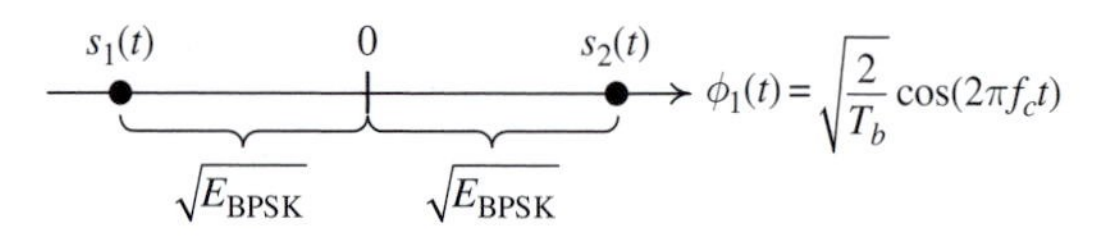

Fig. 7.4 Signal space plot of BPSK.

[1] Strictly speaking, there is a cross product term

$$\frac{V^2}{8}\frac{\sin[\pi(f - f_c)T_b]}{\pi(f - f_c)T_b} \cdot \frac{\sin[\pi(f + f_c)T_b]}{\pi(f + f_c)T_b}.$$

But since f_c is typically very large, it is ignored.

where $f_c = n/T_b$ for some integer n. Each signal has energy $E_{\text{BPSK}} = V^2T_b/2$. The signal space plot is shown in Figure 7.4, where $\phi_1(t) = s_2(t)/\sqrt{E_{\text{BPSK}}} = \sqrt{2/T_b}\cos(2\pi f_c t)$.

The minimum-error-probability (optimum) receiver projects the received signal $\mathbf{r}(t)$ onto $\phi_1(t-(k-1)T_b)$, samples the output of the projection at $t = kT_b$ and compares it to the threshold, which in the special case of $P_1 = P_2$ is equal to zero. The error probability for $P_1 = P_2$ is given by

$$P[\text{error}]_{\text{BPSK}} = Q\left(\sqrt{\frac{2E_{\text{BPSK}}}{N_0}}\right). \tag{7.7}$$

BPSK is an antipodal modulation and using the result in (5.147), the PSD for the BPSK signal is given by (as in BASK, the cross product is negligible)

$$S_{\text{BPSK}}(f) = \frac{V^2}{4}\left[\frac{\sin^2[\pi(f-f_c)T_b]}{\pi^2(f-f_c)^2T_b} + \frac{\sin^2[\pi(f+f_c)T_b]}{\pi^2(f+f_c)^2T_b}\right]. \tag{7.8}$$

The PSD of BPSK is similar to that of BASK except that there are no impulse functions at $\pm f_c$, reflecting the fact that there is no power at the carrier. This is reasonable since BPSK is really a "double" sideband suppressed carrier modulation.

7.4 Binary frequency-shift keying (BFSK)

The most basic method of generating BFSK is to gate two oscillators with the modulating signal, as illustrated in Figure 7.5(a).

The elementary signals are

$$\begin{cases} s_1(t) = V\cos(2\pi f_1 t + \theta_1), & \text{if "}0_T\text{"} \\ s_2(t) = V\cos(2\pi f_2 t + \theta_2), & \text{if "}1_T\text{"} \end{cases}, \quad 0 < t \le T_b. \tag{7.9}$$

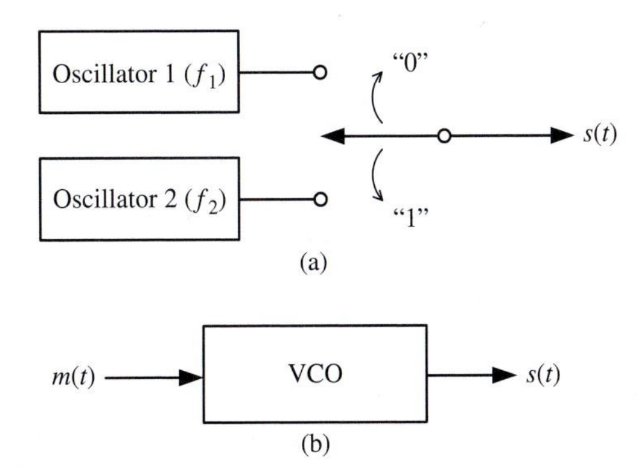

Fig. 7.5 Simple BFSK modulators: (a) by gating two oscillators; (b) using a voltage controlled oscillator (VCO).

The two carrier frequencies are chosen to be integer multiples of $1/T_b$, while the two phases θ_1 and θ_2 need not be the same. Furthermore, the frequencies f_1 and f_2 are chosen so that $s_1(t)$ and $s_2(t)$ are *orthogonal* over the interval $[0, T_b]$, i.e.,

$$\int_0^{T_b} s_1(t)s_2(t)\mathrm{d}t = 0. \tag{7.10}$$

To see what this implies about the two frequencies assume, without loss of generality, that $f_2 > f_1$. Substitute for $s_1(t)$, $s_2(t)$ in (7.10) and integrate to obtain

$$\frac{\sin[2\pi(f_2+f_1)T_b + (\theta_2+\theta_1)] - \sin(\theta_2+\theta_1)}{(f_2+f_1)} + \frac{\sin[2\pi(f_2-f_1)T_b + (\theta_2-\theta_1)] - \sin(\theta_2-\theta_1)}{(f_2-f_1)} = 0. \tag{7.11}$$

Equation (7.11) gives the following conditions on f_1 and f_2:

(i) If the two phases are the same, i.e., $\theta_1 = \theta_2$, then

$$f_2 - f_1 = \frac{m}{2T_b}, \quad m = 1, 2, \ldots. \tag{7.12}$$

The minimum frequency separation $(f_2 - f_1)$ for orthogonality occurs when $m = 1$ and is given by

$$(\Delta f)_{\min}^{[\text{coherent}]} = \frac{1}{2T_b}. \tag{7.13}$$

In this case the two sinusoidal carriers are said to be *coherently* orthogonal (coherent because the two phases are the same).

(ii) If the two phases are different, i.e., $\theta_1 \neq \theta_2$, then

$$f_2 - f_1 = \frac{m}{T_b}, \quad m = 1, 2, \ldots. \tag{7.14}$$

The minimum frequency separation for this case, called *noncoherent* orthogonality (noncoherent because there is no relationship between the two phases), is

$$(\Delta f)_{\min}^{[\text{noncoherent}]} = \frac{1}{T_b}. \tag{7.15}$$

The above shows that relaxing phase synchronization of the two carriers requires a doubling of their *minimum spacing* in order to maintain the orthogonality of the two carriers.

It is also possible to express the signal set in a different way:

$$\begin{cases} s_1(t) = V\cos 2\pi(f_c - f_d)t \\ s_2(t) = V\cos 2\pi(f_c + f_d)t \end{cases}, \quad 0 < t \leq T_b, \tag{7.16}$$

where f_c is the carrier frequency and f_d is the *frequency deviation*. From this viewpoint the transmitted signal is generated by frequency modulating a voltage-controlled oscillator

(VCO) with the random binary sequence, $m(t)$. This is illustrated in Figure 7.5(b). It can be verified that the orthogonal condition requires that

$$f_c = n/4T_b, \tag{7.17}$$

$$f_d = \begin{cases} m/4T_b & \text{(coherent orthogonality)} \\ m/2T_b & \text{(noncoherent orthogonality)} \end{cases}, \tag{7.18}$$

where n and m are positive integers, and $n \gg m$.

With either coherent or noncoherent orthogonality, the energy in each signal of BFSK is given by $E_{\text{BFSK}} = V^2 T_b/2$ (joules). Two orthogonal basis functions are required to represent the signal set, namely

$$\phi_1(t) = \frac{s_1(t)}{\sqrt{E_{\text{BFSK}}}}, \quad \phi_2(t) = \frac{s_2(t)}{\sqrt{E_{\text{BFSK}}}}. \tag{7.19}$$

The signal space plot is shown in Figure 7.6. The optimum receiver projects the received signal along the $[\phi_2(t) - \phi_1(t)]/\sqrt{2}$ axis and compares the projection to a threshold. For the case of $P_1 = P_2$, this threshold is 0 and the decision region is geometrically shown in Figure 7.6. The error probability is given by

$$P[\text{error}]_{\text{BFSK}} = Q\left(\sqrt{\frac{E_{\text{BFSK}}}{N_0}}\right). \tag{7.20}$$

BFSK is a frequency modulation, i.e., a nonlinear modulation, and one would expect that its PSD, as in the analog FM situation, is more difficult to determine. However, (5.147) applies to BFSK under the provision that the signals are statistically independent from bit interval to bit interval. The PSD of BFSK is therefore given by (again, ignoring the cross product term)

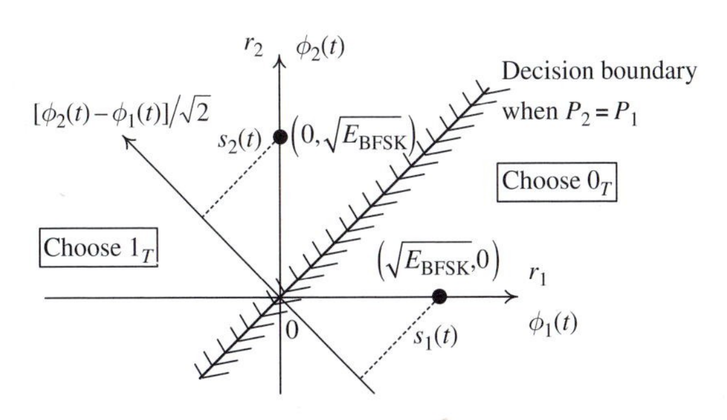

Fig. 7.6 Signal space plot and decision regions of BFSK.

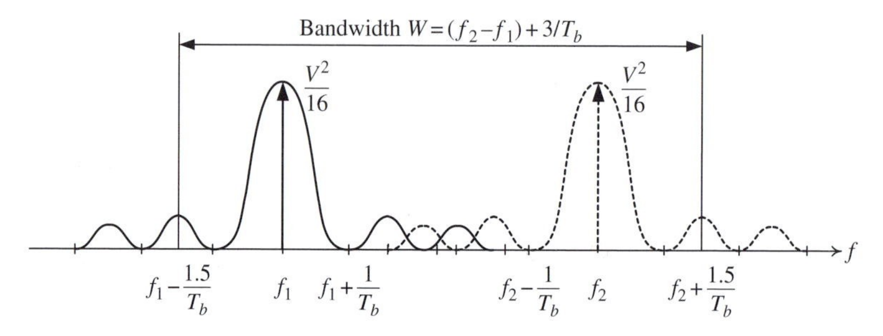

Fig. 7.7 Bandwidth approximation for BFSK.

$$\begin{aligned} S_{\text{BFSK}}(f) = \frac{V^2}{16}\Bigg[& \delta(f-f_2)+\delta(f+f_2) \\ & + \frac{\sin^2[\pi T_b(f+f_2)]}{\pi^2 T_b(f+f_2)^2} + \frac{\sin^2[\pi T_b(f-f_2)]}{\pi^2 T_b(f-f_2)^2}\Bigg] \\ + \frac{V^2}{16}\Bigg[& \delta(f-f_1)+\delta(f+f_1) \\ & + \frac{\sin^2[\pi T_b(f+f_1)]}{\pi^2 T_b(f+f_1)^2} + \frac{\sin^2[\pi T_b(f-f_1)]}{\pi^2 T_b(f-f_1)^2}\Bigg]. \end{aligned} \tag{7.21}$$

The above result shows that the PSD of BFSK is that of two interleaved BASK signals. Thus 95% of the power in each signal lies in a band of $3/T_b$ hertz centered respectively on f_2 and f_1. Based on this the total 95% bandwidth can be taken to be $(f_2 + 1.5/T_b) - (f_1 - 1.5/T_b) = (f_2 - f_1) + 3/T_b = \Delta f + 3/T_b$. This is conceptually illustrated in Figure 7.7.

It is clear that the bandwidth of BFSK is kept to a minimum by using the minimum frequency separation between the two orthogonal carriers. As discussed before, such a minimum frequency separation is $1/2T_b$ for the case of coherently orthogonal carriers (which have the same phase), while it is $1/T_b$ for noncoherently orthogonal carriers (which have different phases).

7.5 Performance comparison of BASK, BPSK, and BFSK

To compare the error performance of the three signaling schemes, it is necessary to express the error probabilities in terms of the average *energy per bit*, or E_b. With equally likely information bits "0" and "1" one has the following relationships for different signaling schemes: $E_b = E_{\text{BPSK}}$, $E_b = E_{\text{BASK}}/2$, and $E_b = E_{\text{BFSK}}$. Thus the error performances of different modulation schemes are as follows:

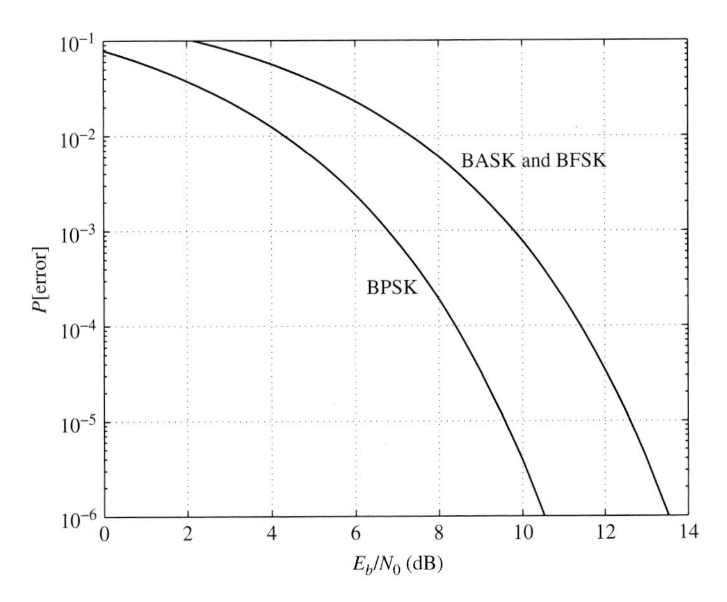

Fig. 7.8 Error performance of binary passband modulation techniques.

$$P[\text{error}]_{\text{BPSK}} = Q\left(\sqrt{\frac{2E_b}{N_0}}\right), \tag{7.22}$$

$$P[\text{error}]_{\text{BASK}} = P[\text{error}]_{\text{BFSK}} = Q\left(\sqrt{\frac{E_b}{N_0}}\right). \tag{7.23}$$

The above shows that BPSK is 3 dB more efficient than BFSK, which has the same performance as BASK. This is shown graphically in Figure 7.8.

In terms of bandwidth, BFSK occupies a larger bandwidth than BPSK and BASK (recall that BPSK and BASK occupy the same bandwidth). Each of the three modulation techniques has a spectrum that decays as $1/f^2$ for frequencies away from the carrier, reflecting the fact that for each modulation the transmitted signal has discontinuities. In the next section, other modulation techniques are introduced which are more spectrally efficient.

7.6 Digital modulation techniques for spectral efficiency

To increase the bit rate without increasing the bandwidth, various modulation techniques have evolved. Straightforward extensions of the techniques considered in the previous section are QPSK, OQPSK, and MSK. They are a form of M-ary ($M = 4$) modulation which is considered in a greater detail in Chapter 8.

Table 7.1 QPSK signals and a mapping to the messages

Bit pattern	Message	Signal transmitted	
00	m_1	$s_1(t) = V\cos(2\pi f_c t)$,	$0 \le t \le T_s = 2T_b$
01	m_2	$s_2(t) = V\sin(2\pi f_c t)$,	$0 \le t \le T_s = 2T_b$
11	m_3	$s_3(t) = -V\cos(2\pi f_c t)$,	$0 \le t \le T_s = 2T_b$
10	m_4	$s_4(t) = -V\sin(2\pi f_c t)$,	$0 \le t \le T_s = 2T_b$

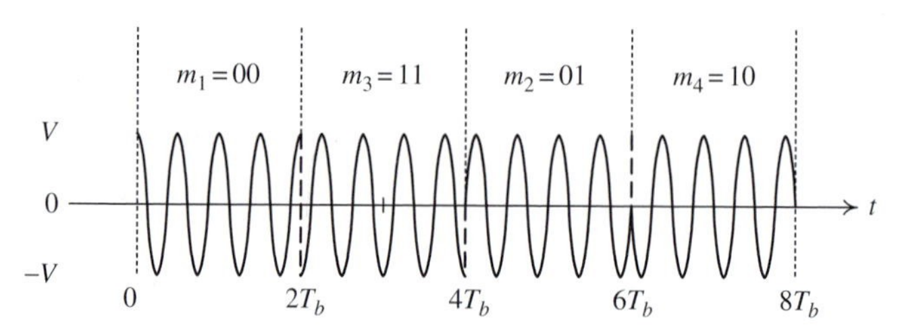

Fig. 7.9 An example of a QPSK signal.

7.6.1 Quadrature phase-shift keying (QPSK)

The basic idea behind QPSK exploits the fact that $\cos(2\pi f_c t)$ and $\sin(2\pi f_c t)$ are orthogonal over the interval $[0, T_b]$ when $f_c = k/T_b$, k integer. Just as in analog modulation, this can be used to transmit two different messages over the same frequency band. To accomplish this the bit stream is taken two bits at a time and mapped into signals as shown in Table 7.1. An example QPSK signal is shown in Figure 7.9. Since each bit occupies T_b seconds, the signals corresponding to the "digits" (or symbols), 00, 01, 11, 10, last for a *symbol duration* of $T_s = 2T_b$ seconds. The symbol signaling rate or what is commonly called the *baud rate* is therefore $r_s = 1/T_s = 1/(2T_b) = r_b/2$ (symbols/second), i.e., *halved*. Since the bandwidth requirement (the PSD will be derived later) is proportional to r_s, it can also be reduced by half for a given bit rate r_b. Conversely, for a fixed bandwidth the bit rate r_b can be doubled.

Though the bit rate has been increased without a corresponding increase in bandwidth it is also necessary to look at what happens to the bit error probability. To accomplish this the signals $s_1(t)$, $s_2(t)$, $s_3(t)$, and $s_4(t)$ are represented, as usual, by an orthonormal basis set. As mentioned, the signals satisfy the following:

$$\int_0^{T_s} s_i^2(t)\mathrm{d}t = \frac{V^2}{2}T_s = V^2 T_b = E_s, \tag{7.24}$$

$$\int_0^{T_s} V\sin(2\pi f_c t)V\cos(2\pi f_c t)\mathrm{d}t = 0. \tag{7.25}$$

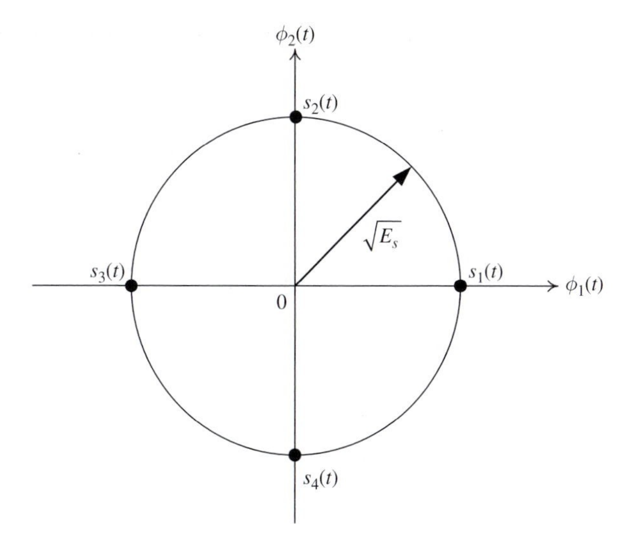

Fig. 7.10 Signal space plot of QPSK modulation.

Therefore only two orthonormal functions are needed to represent the four signals, namely,

$$\phi_1(t) = \frac{s_1(t)}{\sqrt{E_s}}, \quad \phi_2(t) = \frac{s_2(t)}{\sqrt{E_s}}. \tag{7.26}$$

The signal space plot is shown in Figure 7.10.

To derive a minimum-error-probability receiver the results of the binary (two messages) case have to be extended slightly. Rather than bit error, the criterion will be to find a receiver that minimizes the *symbol* (message) *error probability*. In the present situation there are four messages corresponding to $s_1(t)$, $s_2(t)$, $s_3(t)$, and $s_4(t)$. The optimum receiver is derived by expanding the received signal $\mathbf{r}(t) = s_i(t) + \mathbf{w}(t)$ over the interval of T_s seconds into a series as follows:

$$\mathbf{r}(t) = \mathbf{r}_1\phi_1(t) + \mathbf{r}_2\phi_2(t) + \mathbf{r}_3\phi_3(t) + \cdots . \tag{7.27}$$

Once again $\phi_1(t)$ and $\phi_2(t)$ are determined by the signal set, while $\phi_i(t),\ i > 2$ are chosen to simply complete the orthonormal set. Regardless of which signal is sent, the coefficients $\mathbf{r}_i$, $i > 2$, are due only to the noise, $\mathbf{w}(t)$. They are uncorrelated Gaussian random variables with zero mean and variance $N_0/2$.

Based on the set of $\mathbf{r}_1, \mathbf{r}_2, \mathbf{r}_3, \ldots$ it is desired to make a decision as to the actual signal transmitted at the modulator output. Consider for the moment only the first m projections, $\mathbf{r}_1, \mathbf{r}_2, \ldots, \mathbf{r}_m$. The receiver is required to partition the m-dimensional space into four regions in a manner which achieves the minimum error probability. Geometrically this can be visualized as shown in Figure 7.11.

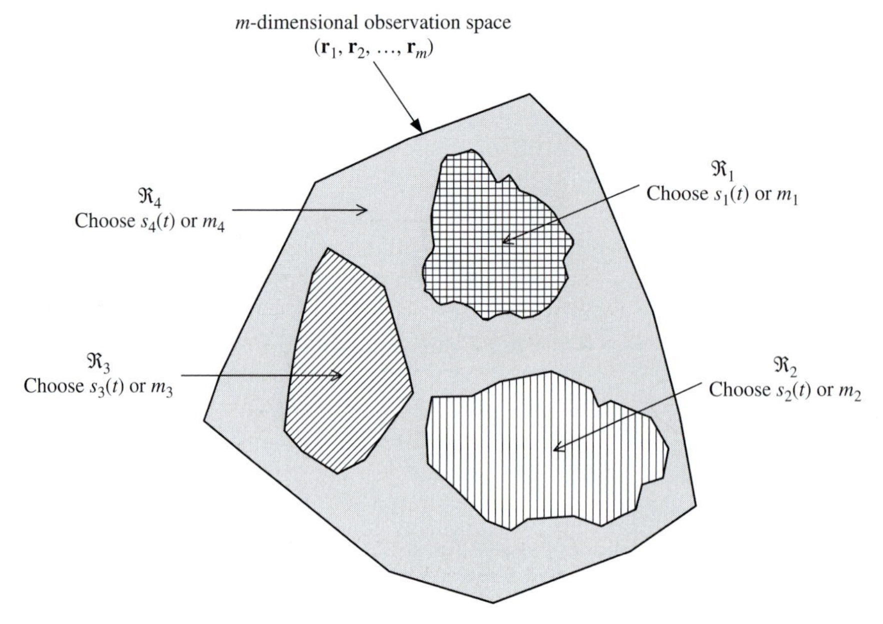

Fig. 7.11 Decision regions of QPSK modulation.

Rather than minimizing the error, consider instead the equivalent criterion, that of maximizing the probability of a *correct* decision. The expression for this can be written as

$$\begin{aligned} P[\text{correct}] &= P[\vec{\mathbf{r}} \in \Re_1 | s_1(t)]P[s_1(t)] + P[\vec{\mathbf{r}} \in \Re_2 | s_2(t)]P[s_2(t)] \\ &\quad + P[\vec{\mathbf{r}} \in \Re_3 | s_3(t)]P[s_3(t)] + P[\vec{\mathbf{r}} \in \Re_4 | s_4(t)]P[s_4(t)], \end{aligned} \tag{7.28}$$

where $\vec{\mathbf{r}} = (\mathbf{r}_1, \mathbf{r}_2, \mathbf{r}_3, \ldots, \mathbf{r}_m)$ and $P[\vec{\mathbf{r}} \in \Re_i | s_i(t)]$ is the probability that the observation vector $\vec{\mathbf{r}}$ falls into the ith region when signal $s_i(t)$ (or message m_i) is transmitted. $P[s_i(t)] \equiv P_i$ is the *a priori* probability of message m_i being transmitted. Therefore,

$$\begin{aligned} P[\text{correct}] &= \int_{\Re_1} P_1 f(\vec{r}|s_1(t))\text{d}\vec{r} + \int_{\Re_2} P_2 f(\vec{r}|s_2(t))\text{d}\vec{r} \\ &\quad + \int_{\Re_3} P_3 f(\vec{r}|s_3(t))\text{d}\vec{r} + \int_{\Re_4} P_4 f(\vec{r}|s_4(t))\text{d}\vec{r}. \end{aligned} \tag{7.29}$$

To maximize the probability of a correct decision it is readily seen from the above that the decision rule becomes

assign the observation vector $\vec{r} = (r_1, r_2, \ldots, r_m)$ in
the m-dimensional signal space to the region for which
the integrand $P_i f(\vec{r}|s_i(t))$ is the *largest*.

Or stated in another way:

$$\text{choose } s_i(t) \text{ if } P_i f(\vec{r}|s_i(t)) > P_j f(\vec{r}|s_j(t)),\ j = 1, 2, 3, 4;\ j \neq i. \tag{7.30}$$

Letting $m \to \infty$, the conditional pdf given the ith message is transmitted can be written as

$$f(\vec{r}|s_i(t)) = \frac{1}{\sqrt{\pi N_0}} \exp\left\{-\frac{(r_1 - s_{i1})^2}{N_0}\right\} \frac{1}{\sqrt{\pi N_0}} \exp\left\{-\frac{(r_2 - s_{i2})^2}{N_0}\right\} \times \prod_{k=3}^{\infty} \frac{1}{\sqrt{\pi N_0}} \exp\left\{-\frac{r_k^2}{N_0}\right\}. \tag{7.31}$$

The terms due to $r_3, r_4, \ldots$ are common to each conditional density and therefore can be ignored in the decision rule of (7.30), which can be rewritten as

$$\begin{gathered} \text{choose } s_i(t) \text{ if} \\ P_i f(r_1, r_2|s_i(t)) > P_j f(r_1, r_2|s_j(t)), \\ j = 1, 2, 3, 4;\ j \neq i. \end{gathered} \tag{7.32}$$

In essence, the projections of $r(t)$ onto $\phi_j(t)$, $j > 2$ are ignored since they do not provide us with any information concerning the transmitted signal (i.e., the input message). The expression $P_j f(r_1, r_2|s_j(t))$ is given by

$$P_j f(r_1, r_2|s_j(t)) = P_j \frac{1}{\sqrt{\pi N_0}} \exp\left\{-\frac{(r_1 - s_{j1})^2}{N_0}\right\} \frac{1}{\sqrt{\pi N_0}} \exp\left\{-\frac{(r_2 - s_{j2})^2}{N_0}\right\}. \tag{7.33}$$

Taking the natural logarithm of (7.33) still preserves the relative values, i.e., the signal $s_j(t)$ for which (7.33) is maximum does not change. The resulting expression is

$$\ln P_j - \ln(\pi N_0) - \frac{(r_1 - s_{j1})^2}{N_0} - \frac{(r_2 - s_{j2})^2}{N_0}. \tag{7.34}$$

Ignoring the term $\ln(\pi N_0)$ and multiplying through by N_0 (since again this does not affect the maximum), the decision rule can be written as:

$$\begin{gathered} \text{choose } s_i(t) \text{ if} \\ N_0 \ln P_i - (r_1 - s_{i1})^2 - (r_2 - s_{i2})^2 > \\ N_0 \ln P_j - (r_1 - s_{j1})^2 - (r_2 - s_{j2})^2, \\ j = 1, 2, 3, 4;\ j \neq i. \end{gathered} \tag{7.35}$$

Multiplying the terms out, ignoring the r_1^2, r_2^2 terms that are common to all expressions and dividing through by 2, the decision rule is

$$\begin{gathered} \text{choose } s_i(t) \text{ if} \\ \frac{N_0}{2} \ln P_i + r_1 s_{i1} + r_2 s_{i2} - \frac{\left(s_{i1}^2 + s_{i2}^2\right)}{2} > \\ \frac{N_0}{2} \ln P_j + r_1 s_{j1} + r_2 s_{j2} - \frac{\left(s_{j1}^2 + s_{j2}^2\right)}{2}, \\ j = 1, 2, 3, 4;\ j \neq i. \end{gathered} \tag{7.36}$$

Recognizing that $s_{j1}^2 + s_{j2}^2$ is the energy of signal $s_j(t)$ and that all the signals are of equal energy, the final decision rule becomes

$$\text{choose } s_i(t) \text{ if} \quad \frac{N_0}{2}\ln P_i + r_1 s_{i1} + r_2 s_{i2} > \frac{N_0}{2}\ln P_j + r_1 s_{j1} + r_2 s_{j2}, \quad j = 1,2,3,4;\ j \neq i. \tag{7.37}$$

The block diagram of the above receiver is shown in Figure 7.12.

For the special case where the messages are equally likely, i.e., $P_1 = P_2 = P_3 = P_4 = 0.25$, it is simple to see the decision rule becomes:

$$\text{choose } s_i(t) \text{ if } (r_1 - s_{i1})^2 + (r_2 - s_{i2})^2 \text{ is } \textit{the smallest}, \tag{7.38}$$

which is simply interpreted as the *minimum-distance receiver*. Its decision region therefore looks as shown in Figure 7.13.

The symbol (message) error probability of the minimum-distance receiver is determined by first changing variables or rotating coordinates and then finding the volume under the appropriate pdfs. Because of the symmetry and equal *a priori* probabilities, one has

$$P[\text{error}] = P[\text{error}|s_i(t)] = 1 - P[\text{correct}|s_i(t)]. \tag{7.39}$$

One computes $P[\text{correct}|s_1(t)]$ by finding the volume of $f(r_1, r_2|s_1(t))$ over the shaded quadrant in Figure 7.14. This volume is readily found by a change of variables to produce two new axes, $\hat{r}_1, \hat{r}_2$, as shown. This change of variables is a rotation of the r_1, r_2 axes by $\theta = \pi/4$, i.e.,

$$\begin{bmatrix} \hat{r}_1 \\ \hat{r}_2 \end{bmatrix} = \begin{bmatrix} \cos\theta & \sin\theta \\ -\sin\theta & \cos\theta \end{bmatrix} \begin{bmatrix} r_1 \\ r_2 \end{bmatrix}.$$

The new variables $\hat{\mathbf{r}}_1$, $\hat{\mathbf{r}}_2$ are still statistically independent Gaussian random variables of variance $N_0/2$ and means of $\left(\sqrt{E_s/2}, -\sqrt{E_s/2}\right)$. The volume under $f(\hat{r}_1, \hat{r}_2|s_1(t)) = f(\hat{r}_1|s_1(t)) \cdot f(\hat{r}_2|s_1(t))$ is now readily found to be

$$P[\text{correct}|s_1(t)] = \left[1 - Q\left(\sqrt{\frac{E_s}{N_0}}\right)\right]^2. \tag{7.40}$$

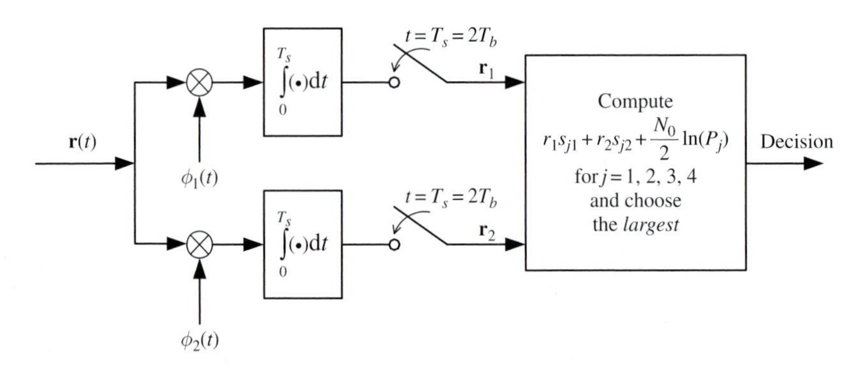

Fig. 7.12 Receiver implementation for QPSK signaling.

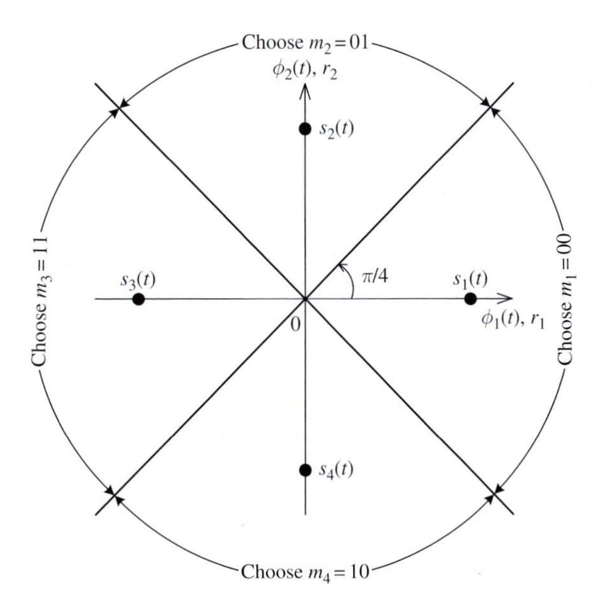

Fig. 7.13 Decision regions of the minimum-distance receiver of QPSK.

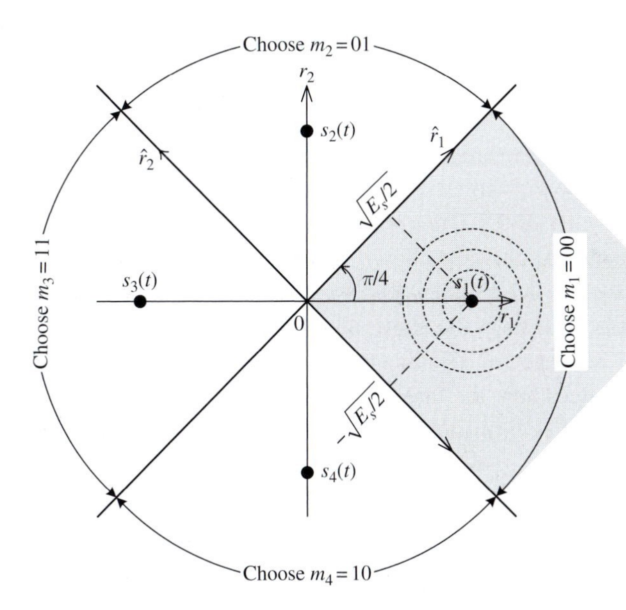

Fig. 7.14 Signal space diagram of QPSK: to compute $P[\text{correct}|s_1(t)]$ one finds the volume of $f(\hat{r}_1, \hat{r}_2|s_1(t))$ over the shaded quadrant.

This, however, would give the symbol (message) error probability and not the bit error probability. Even though a message error has been made it does not mean that a specific bit is in error. As an example, if the receiver decides on message m_2 when m_1 is the true message, then only the *second* bit is in error. To determine the bit error probability it is necessary to distinguish between the different message errors. Again, because of the symmetry, it is sufficient to consider only a specific message, say m_1. Then the different errors are

$$\begin{aligned} m_1(00) &\Rightarrow m_2(01): && \text{second bit is in error,} \\ m_1(00) &\Rightarrow m_3(11): && \text{both bits are in error,} \\ m_1(00) &\Rightarrow m_4(10): && \text{first bit is in error.} \end{aligned}$$

The different error probabilities are given by (see Problem 7.3):

$$P[m_2|m_1] = Q\left(\sqrt{\frac{E_s}{N_0}}\right)\left[1 - Q\left(\sqrt{\frac{E_s}{N_0}}\right)\right], \tag{7.41}$$

$$P[m_3|m_1] = Q^2\left(\sqrt{\frac{E_s}{N_0}}\right), \tag{7.42}$$

$$P[m_4|m_1] = Q\left(\sqrt{\frac{E_s}{N_0}}\right)\left[1 - Q\left(\sqrt{\frac{E_s}{N_0}}\right)\right]. \tag{7.43}$$

The bit error probability is then calculated as

$$\begin{aligned} P[\text{bit error}] &= 0.5P[m_2|m_1] + 0.5P[m_4|m_1] + 1.0P[m_3|m_1] \\ &= Q\left(\sqrt{\frac{E_s}{N_0}}\right), \end{aligned} \tag{7.44}$$

where the viewpoint is taken that one of the two bits is chosen at random, i.e., with a probability of 0.5. Thus when the message errors of m_2 or m_4 occur (given that m_1 was transmitted), then the chosen bit is in error with a probability of 0.5. When the message error of m_3 occurs, then the chosen bit is certain to be in error, i.e., has a probability of 1.

It is clear from the above derivation that the way each information bit pair is mapped to a message (or signal) influences the bit error probability. Specifically, the mapping determines how the three probabilities $\{0.5, 0.5, 1.0\}$ are associated with the three error probabilities $\{P(m_2|m_1), P(m_3|m_1), P(m_4|m_1)$. Recognizing that

$$P(m_3|m_1) < P(m_2|m_1) = P(m_4|m_1), \tag{7.45}$$

it is desirable to associate the probability 1.0 with $P(m_3|m_1)$ in order to minimize the overall bit error probability. This means that the signals $s_1(t)$ and $s_3(t)$, which are separated by the largest Euclidean distance, should be mapped to the bit pairs that differ in both bits. Equivalently, the signals that are closest to each other, known as the *nearest neighbors*, should be mapped to the bit pairs that differ in only one bit. Such a mapping is called *Gray* mapping. For example, the mapping scheme in Table 7.1 is a Gray mapping.

In general, there is a total of $4! = 24$ possible ways to map the bit pairs to QPSK signals. However, due to the symmetry of the QPSK constellation, it is not hard to verify that all 24 mappings are equally divided into two types of mapping as far as the bit error probability is concerned. These are Gray mappings and anti-Gray mappings (i.e., any mapping that is not a Gray mapping). The bit error probability of QPSK with an anti-Gray mapping is examined in Problem 7.4.

For a fair comparison with the error performance of binary modulation schemes considered previously, again it is necessary to express (7.44) in terms of E_b, the *average energy per bit*. Since each signal of QPSK carries two bits and the energy of each signal is $E_s = V^2 T_b$, the average energy per bit is $E_b = E_s/2 = V^2 T_b/2$. Therefore the bit error probability of QPSK with a Gray mapping is

$$P[\text{bit error}] = Q\left(\sqrt{\frac{2E_b}{N_0}}\right), \tag{7.46}$$

which is exactly the same as that of BPSK. This clearly demonstrates the advantage of QPSK over BPSK. With QPSK modulation, the bit rate can be *doubled* without requiring any additional transmission bandwidth or sacrificing the error performance. Interpreted differently, to deliver the same transmission rate at the same bit error performance, using QPSK reduces the transmission bandwidth to half of that required by BPSK.

7.6.2 An alternative representation of QPSK

In this representation, the information bit stream is first converted to an NRZ-L waveform $a(t)$ with ± 1 levels. The waveform $a(t)$ is then demultiplexed into even, $a_I(t)$, and odd, $a_Q(t)$, bit streams (waveforms) where I and Q are mnemonics for inphase and quadrature, respectively. The individual bits in each stream occupy $T_s = 2T_b$ seconds and modulate the inphase carrier, $V\cos(2\pi f_c t)$, and quadrature carrier, $V\sin(2\pi f_c t)$, respectively. A block diagram of such a QPSK modulator is illustrated in Figure 7.15 and examples of various waveforms are shown in Figure 7.16.

The transmitted signal is

$$s(t) = a_I(t)V\cos(2\pi f_c t) + a_Q(t)V\sin(2\pi f_c t), \tag{7.47}$$

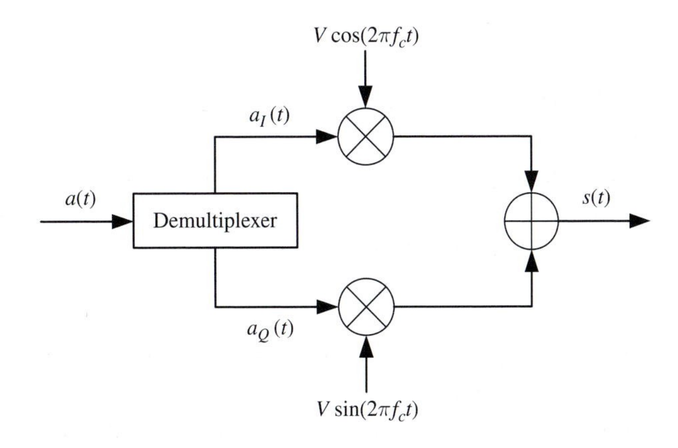

Fig. 7.15 A different block diagram of a QPSK modulator.

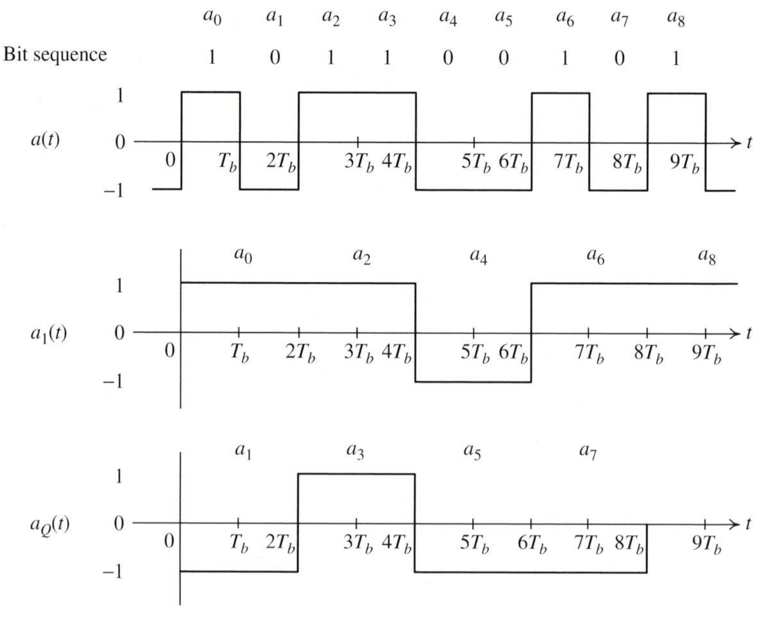

Fig. 7.16 Examples of $a_I(t)$ and $a_Q(t)$ in QPSK modulation.

which can be rewritten as

$$\begin{aligned} s(t) &= \sqrt{a_I^2(t) + a_Q^2(t)} V \cos\left(2\pi f_c t - \tan^{-1}\left(\frac{a_Q(t)}{a_I(t)}\right)\right) \\ &= \sqrt{2} V \cos[2\pi f_c t - \theta(t)], \end{aligned} \tag{7.48}$$

where the phase $\theta(t)$ is determined as follows:

$$\theta(t) = \begin{cases} \pi/4, & \text{if } a_I = +1, a_Q = +1 \text{ (bits are 11)} \\ -\pi/4, & \text{if } a_I = +1, a_Q = -1 \text{ (bits are 10)} \\ 3\pi/4, & \text{if } a_I = -1, a_Q = +1 \text{ (bits are 01)} \\ -3\pi/4, & \text{if } a_I = -1, a_Q = -1 \text{ (bits are 00)} \end{cases}. \tag{7.49}$$

Figure 7.17 shows how the QPSK transmitted waveform is generated from its inphase and quadrature components for the bit sequence given in Figure 7.16.

As can be seen from (7.49) the transmitted signal is a QPSK signal and depends upon the specific even (inphase) and odd (quadrature) bits which select the phase $\theta(t)$ of the sinusoidal carrier $\sqrt{2}V\cos[2\pi f_c t - \theta(t)]$. These four signals can be represented in terms of the following orthonormal basis functions:

$$\begin{cases} \phi_1(t) = \dfrac{V\cos(2\pi f_c t)}{\sqrt{V^2 T_b}} \\ \phi_2(t) = \dfrac{V\sin(2\pi f_c t)}{\sqrt{V^2 T_b}} \end{cases}, \quad 0 < t < T_s = 2T_b, \tag{7.50}$$

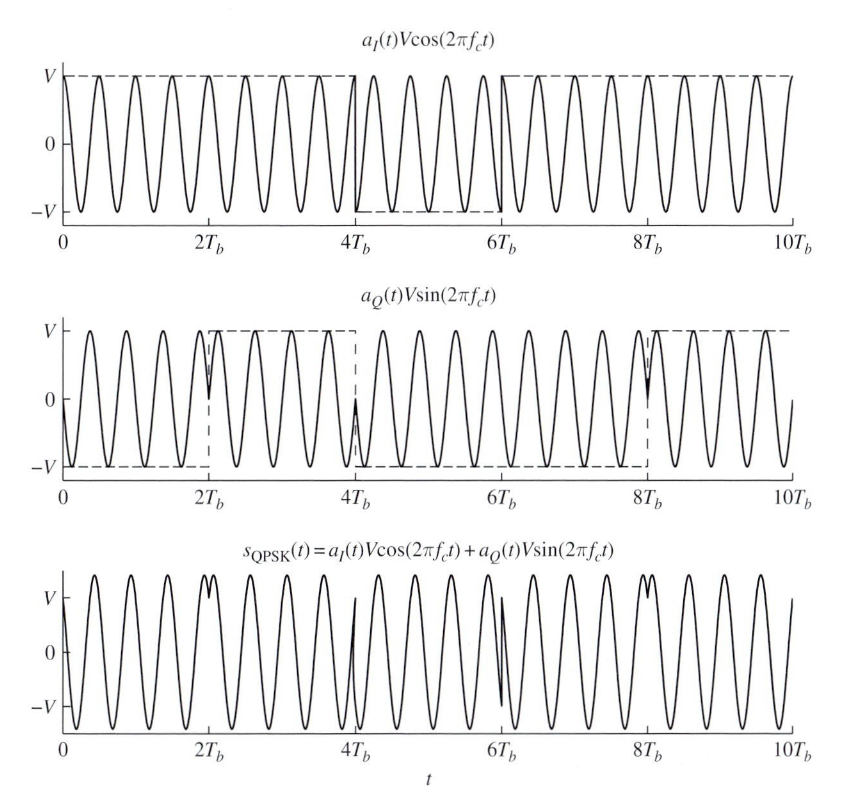

Fig. 7.17 An example of a QPSK signal, viewed as a sum of inphase and quadrature components.

as shown in Figure 7.18. Note that, the phase of each signal point is relative to the $\phi_1(t)$ axis, which is proportional to $\sqrt{2}V\cos(2\pi f_c t)$. Further, the energy of each QPSK signal in this representation is $E_s = V^2 T_s$.

To derive the minimum-error-probability receiver, observe that the even and odd bit streams can be treated separately since $a_I(t)$ does not have a component along $\phi_2(t)$, and $a_Q(t)$ does not have a component along $\phi_1(t)$. Thus the QPSK signal can be considered to consist of two separate (noninterfering) BPSK signals.

The signal space plots for the inphase and quadrature bit streams are shown in Figure 7.19, where $s_1^{(I)}(t) = V\cos(2\pi f_c t)$, $s_2^{(I)}(t) = -V\cos(2\pi f_c t)$, $s_1^{(Q)}(t) = V\sin(2\pi f_c t)$ and $s_2^{(Q)}(t) = -V\sin(2\pi f_c t)$. The receiver looks as shown in Figure 7.20. For equally likely signals, the bit error probability is that of BPSK modulation and is given by

$$P[\text{bit error}] = Q\left(\sqrt{\frac{V^2 T_s}{N_0}}\right). \tag{7.51}$$

Of course, one expects that the above expression for bit error probability is identical to that of (7.46) when expressed in terms of E_b, the energy per information bit. To see this, as noted earlier, the energy of each QPSK signal is $V^2 T_s$ (see Figure 7.18),

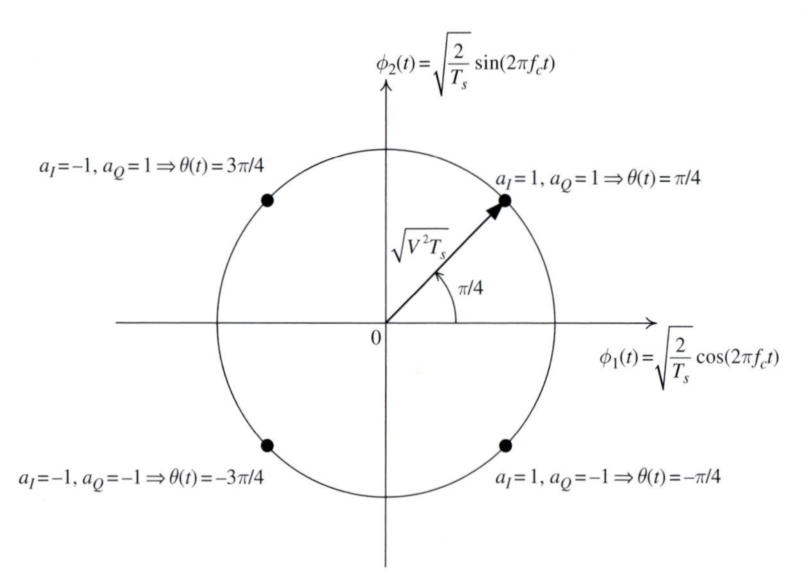

Fig. 7.18 Representing QPSK signals as phases of the sinusoidal carrier.

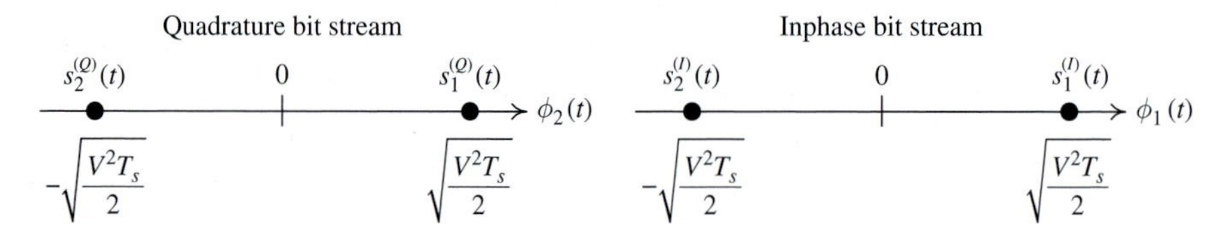

Fig. 7.19 Signal space plots for inphase and quadrature bit streams.

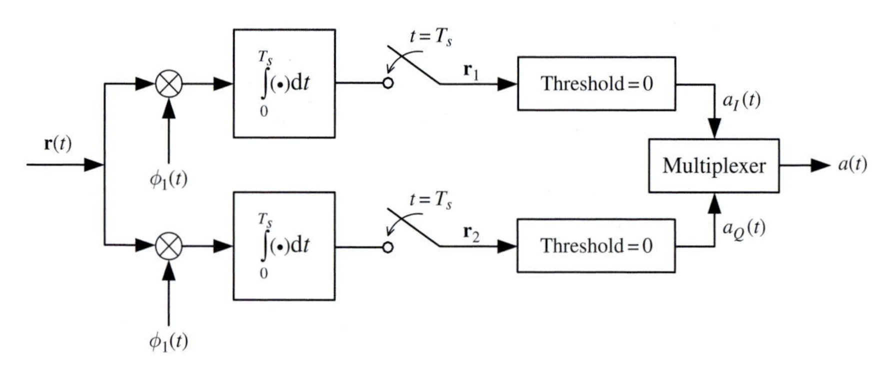

Fig. 7.20 Receiver implementation for QPSK.

hence $E_b = E_s/2 = V^2T_s/2$, or $V^2T_s = 2E_b$. Substituting this expression into (7.51) gives $P[\text{bit error}] = Q\left(\sqrt{2E_b/N_0}\right)$, which shows that (7.51) is indeed the same as (7.46). This is not a surprising fact if one recognizes from Figure 7.18 that the mapping from two bits to a signal point is exactly a Gray mapping.

Finally, viewing the QPSK signal as a sum of two BPSK signals facilitates the determination of its PSD. Specifically, due to the statistical independence of the even and odd bits, the two BPSK signals that make up the QPSK signal are *uncorrelated*. It follows that the PSD of the QPSK signal is twice the PSD of each BPSK signal. It is given as in (7.8), but scaled by a factor of 2 and with T_s substituted for T_b.

7.6.3 Offset quadrature phase-shift keying (OQPSK)

One of the attractive properties of the QPSK signal is that its envelope is ideally constant. However, in many applications (such as satellite communications), the QPSK signals must be bandlimited by a bandpass filter in order to conform to out-of-band emission standards. The filtering degrades the constant-envelope property of QPSK, and the occasional phase shifts of π in QPSK signals cause the envelope to pass through zero momentarily. When this signal is amplified by the final stage, usually a highly nonlinear power amplifier, in the transmitter, the filtered sidelobes in the signal spectrum are recreated (the nonlinearity causes signal energy to reappear in the spectrum where the sidelobes are located). To prevent spectral widening, it is necessary for the QPSK signal to be amplified with *linear* amplifiers, which usually have very low efficiency. To prevent the phase change of π in QPSK, *offset* (or staggered) quadrature phase-shift keying (OQPSK) is used so that signal amplification can be done more efficiently.

OQPSK differs from QPSK only in that in OQPSK the $a_I(t)$ and $a_Q(t)$ bit streams are offset by one bit interval T_b as shown in Figure 7.21. The resultant signal $s(t) = a_I(t)V\cos(2\pi f_c t) + a_Q(t)V\sin(2\pi f_c t)$ cannot undergo a change of π radians since the $a_I(t)$ bit stream has a transition in the middle of the $a_Q(t)$ bit stream. The possible changes are 0 or $\pm\pi/2$ and phase changes occur more frequently, namely every T_b, as compared to every $T_s = 2T_b$ in QPSK. This is illustrated in Figure 7.22.

The optimum receiver for OQPSK is identical to that of QPSK except that the time shift of T_b seconds for $a_I(t)$ must be taken into account by the correlator and sampler. The error probability is the same. And because the PSD of a signal does not depend on the phase, the PSDs of OQPSK and QPSK are identical.

7.6.4 Minimum shift keying (MSK)

The typical QPSK and OQPSK waveforms in Figures 7.17 and 7.22 show that the transmitted signals $s(t)$ in QPSK and OQPSK have sudden jumps at multiples of symbol duration (for QPSK), or multiples of bit duration (for OQPSK). The maximum value of the jump is $2V$ for both schemes. However, it is not the magnitude of the jump but the rapid transition that influences the PSD of the signal. Intuitively, if these jumps at the transitions are

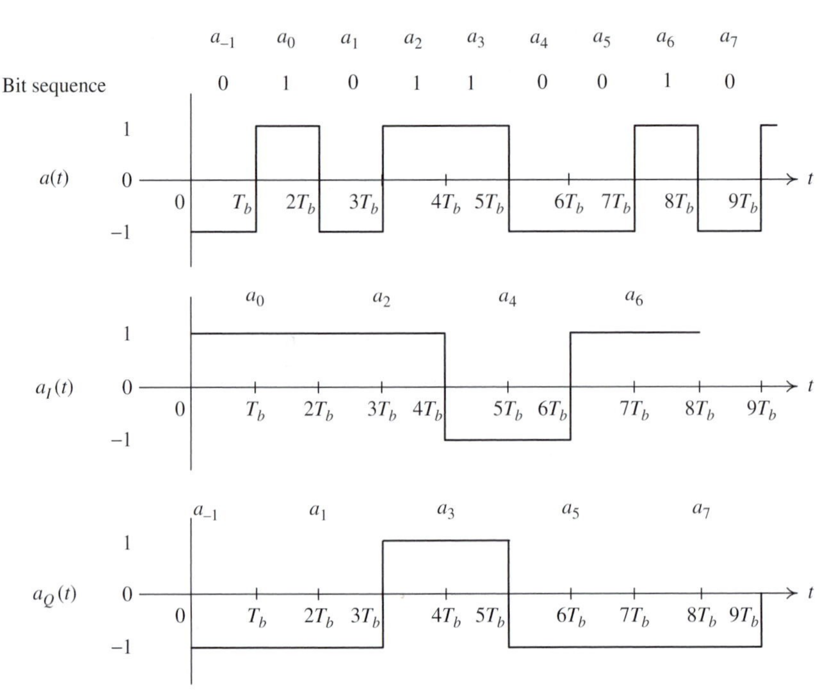

Fig. 7.21 Examples of $a_I(t)$ and $a_Q(t)$ in OQPSK.

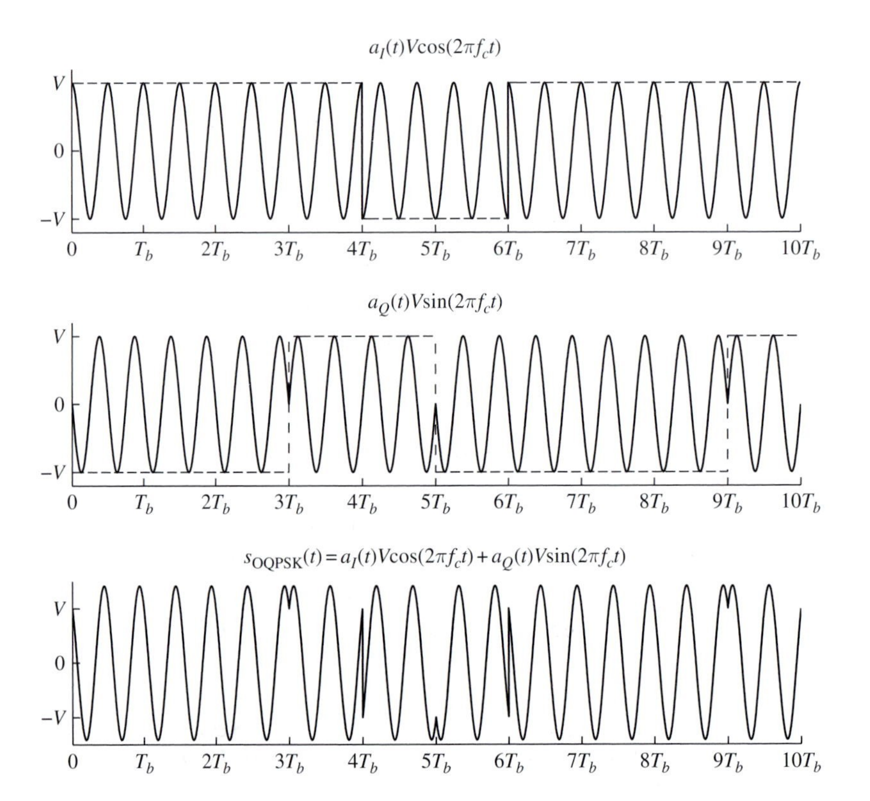

Fig. 7.22 Example of an OQPSK signal.

eliminated, then the PSD will not occupy as large a frequency band. This is the basic idea behind *minimum shift keying* (MSK). Here, the two carriers $V\cos(2\pi f_c t)$ and $V\sin(2\pi f_c t)$ are weighted by sinusoids of frequency $1/(4T_b)$ as follows:[2]

$$\sqrt{2}\cos\left(\frac{\pi t}{2T_b}\right)V\cos(2\pi f_c t), \tag{7.52}$$

$$\sqrt{2}\sin\left(\frac{\pi t}{2T_b}\right)V\sin(2\pi f_c t). \tag{7.53}$$

The signals represented by (7.52) and (7.53) are orthogonal over the interval of T_b seconds, or any integer multiple of T_b. This is shown as follows:

$$\begin{aligned}
&\int_0^{T_b} \cos\left(\frac{\pi t}{2T_b}\right)\sin\left(\frac{\pi t}{2T_b}\right)\cos(2\pi f_c t)\sin(2\pi f_c t)\mathrm{d}t \\
&\quad = \frac{1}{4}\int_0^{T_b} \sin\left(\frac{\pi t}{T_b}\right)\sin(4\pi f_c t)\mathrm{d}t \\
&\quad = \frac{1}{8}\int_0^{T_b} \left\{\cos\left[\left(4\pi f_c - \frac{\pi}{T_b}\right)t\right] - \cos\left[\left(4\pi f_c + \frac{\pi}{T_b}\right)t\right]\right\}\mathrm{d}t \\
&\quad = \frac{1}{8}\left\{\frac{\sin\left[\left(4\pi f_c - \frac{\pi}{T_b}\right)t\right]}{\left(4\pi f_c - \frac{\pi}{T_b}\right)}\Bigg|_0^{T_b} - \frac{\sin\left[\left(4\pi f_c + \frac{\pi}{T_b}\right)t\right]}{\left(4\pi f_c + \frac{\pi}{T_b}\right)}\Bigg|_0^{T_b}\right\} \\
&\quad = 0,
\end{aligned} \tag{7.54}$$

where f_c, as usual, is an integer multiple of $1/T_b$. The energy in each signal over the interval $(0, 2T_b)$ is

$$\begin{aligned}
&2V^2\int_0^{2T_b} \cos^2\left(\frac{\pi t}{2T_b}\right)\cos^2(2\pi f_c t)\mathrm{d}t \\
&\quad = 2V^2\int_0^{2T_b} \left[\frac{1}{2} + \frac{1}{2}\cos\left(\frac{\pi t}{T_b}\right)\right]\left[\frac{1}{2} + \frac{1}{2}\cos(4\pi f_c t)\right]\mathrm{d}t \\
&\quad = \frac{V^2}{2}\int_0^{2T_b} \left[1 + \cos\left(\frac{\pi t}{T_b}\right) + \cos(4\pi f_c t) + \cos\left(\frac{\pi t}{T_b}\right)\cos(4\pi f_c t)\right]\mathrm{d}t \\
&\quad = V^2 T_b \text{ (joules)}.
\end{aligned} \tag{7.55}$$

The two weighted carriers therefore are orthogonal over the interval $[0, 2T_b]$ and have the same energy. When they are modulated by the odd and even bit streams, these bit streams can be separately demodulated at the receiver. It follows that the bit error probability of MSK is the same as that of BPSK, QPSK, and OQPSK, namely

[2] Other weighting functions or pulse shaping functions, as they are commonly called, are possible. The challenge is to still preserve the orthogonality of the two *weighted carriers*. The "half-cosine" pulses here are most commonly used.

$$P[\text{bit error}] = Q\left(\sqrt{\frac{2E_b}{N_0}}\right), \tag{7.56}$$

where $E_b = V^2 T_b$ is the energy per bit.

The MSK modulator is shown in Figure 7.23, while Figure 7.24 is a sketch of the receiver. As in OQPSK, the odd bit stream in MSK is shifted by one bit period, T_b, without affecting the bit error probability or PSD. This shifting is necessary to produce the constant envelope and continuity of the phase in MSK. The two *orthonormal* basis functions used in the receiver of Figure 7.24 are simply

$$\phi_1(t) = \left[\sqrt{2}\sin\left(\frac{\pi t}{2T_b}\right) V\sin(2\pi f_c t)\right] \Big/ \sqrt{V^2 T_b}, \tag{7.57}$$

$$\phi_2(t) = \left[\sqrt{2}\cos\left(\frac{\pi t}{2T_b}\right) V\cos(2\pi f_c t)\right] \Big/ \sqrt{V^2 T_b}. \tag{7.58}$$

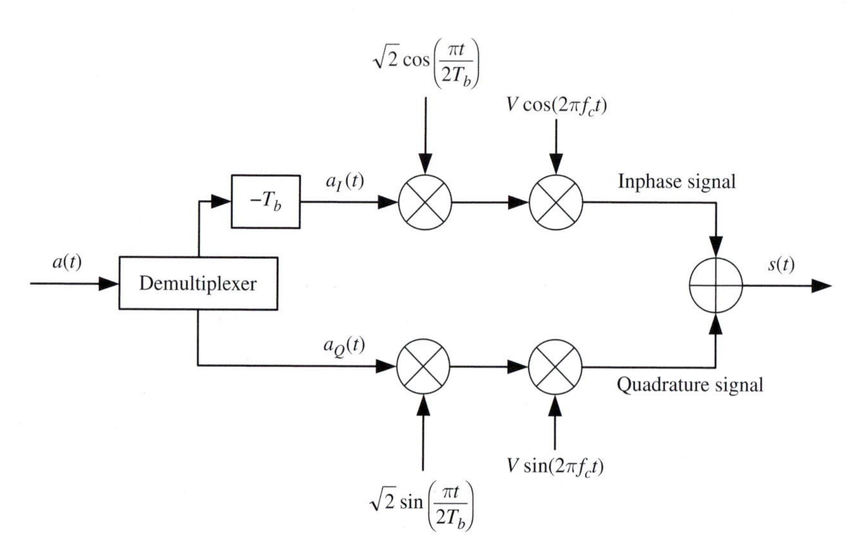

Fig. 7.23 Block diagram of the MSK modulator.

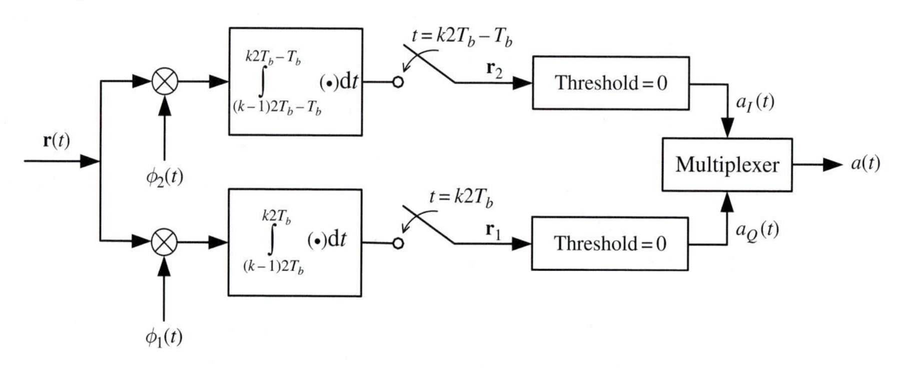

Fig. 7.24 Block diagram of the MSK receiver.

To gain a further insight into the transmitted signal, $s(t)$, of MSK, write

$$s(t) = a_I(t)\sqrt{2}\cos\left(\frac{\pi t}{2T_b}\right)V\cos(2\pi f_c t) + a_Q(t)\sqrt{2}\sin\left(\frac{\pi t}{2T_b}\right)V\sin(2\pi f_c t) \quad (7.59)$$

$$= A\cos(2\pi f_c t - \theta). \quad (7.60)$$

In the representation of (7.60), the amplitude is

$$A = \left[a_I^2(t)2V^2\cos^2\left(\frac{\pi t}{2T_b}\right) + a_Q^2(t)2V^2\sin^2\left(\frac{\pi t}{2T_b}\right)\right]^{\frac{1}{2}} = \sqrt{2}V \quad (7.61)$$

since $a_I(t) = \pm 1$ and $a_Q(t) = \pm 1$. The phase is given by

$$\theta = \tan^{-1}\left\{\frac{a_Q(t)\sin(\pi t/2T_b)}{a_I(t)\cos(\pi t/2T_b)}\right\}. \quad (7.62)$$

Since $a_I(t) = \pm 1$ and $a_Q(t) = \pm 1$, then

$$\theta = \tan^{-1}\left\{\pm\tan\left(\frac{\pi t}{2T_b}\right)\right\} = \tan^{-1}\left\{\tan\left(\pm\frac{\pi t}{2T_b}\right)\right\} = \pm\frac{\pi t}{2T_b}. \quad (7.63)$$

Therefore

$$s(t) = \sqrt{2}V\cos\left[2\pi\left(f_c \pm \frac{1}{4T_b}\right)t\right]. \quad (7.64)$$

The expression in (7.64) shows that $s(t)$ not only has a constant envelope, but also a continuous phase. Furthermore, the transmitted signal is of either frequency $f_2 = f_c + 1/4T_b$ or frequency $f_1 = f_c - 1/4T_b$ depending on the ratio $a_Q(t)/a_I(t)$. It is f_1 if $a_Q(t)$ and $a_I(t)$ are of the same sign and f_2 if they are of the opposite signs. Due to the offset alignment of the inphase and quadrature bit streams, the switching between frequencies f_1 and f_2 can occur every T_b seconds. Thus the transmitted signal may be considered to be a *frequency-shift keying* signal with continuous phase (CPFSK). Note also that the frequency separation is $f_2 - f_1 = 1/2T_b$, which is the minimum separation possible for the two sinusoidal carriers to be "coherently" orthogonal. This explains the name "minimum shift keying" of the modulation scheme.

The waveforms involved in generating the MSK signal are illustrated in Figure 7.25. Figures 7.25(a) and 7.25(b) show the inphase and the sinusoidally shaped inphase bit stream waveforms, respectively. The inphase carrier (the first term in (7.60)), obtained by multiplying the waveform in Figure 7.25(b) by $V\cos(2\pi f_c t)$, is shown in Figure 7.25(e). Similarly, the sinusoidally shaped odd-bit stream and the quadrature carrier are shown in Figures 7.25(d) and 7.25(f), respectively. The MSK signal, the addition of the waveforms of Figures 7.25(e) and 7.25(f), is shown in Figure 7.25(g). As expected, the MSK signal appears and is continuous in phase and takes on one of the two frequencies over each bit duration.

The demodulation of MSK can be accomplished differently, namely by considering it to be an FSK modulation technique. During the interval of T_b seconds the receiver decides whether frequency f_2 or frequency f_1 is sent as shown in Figure 7.26. Given the decision

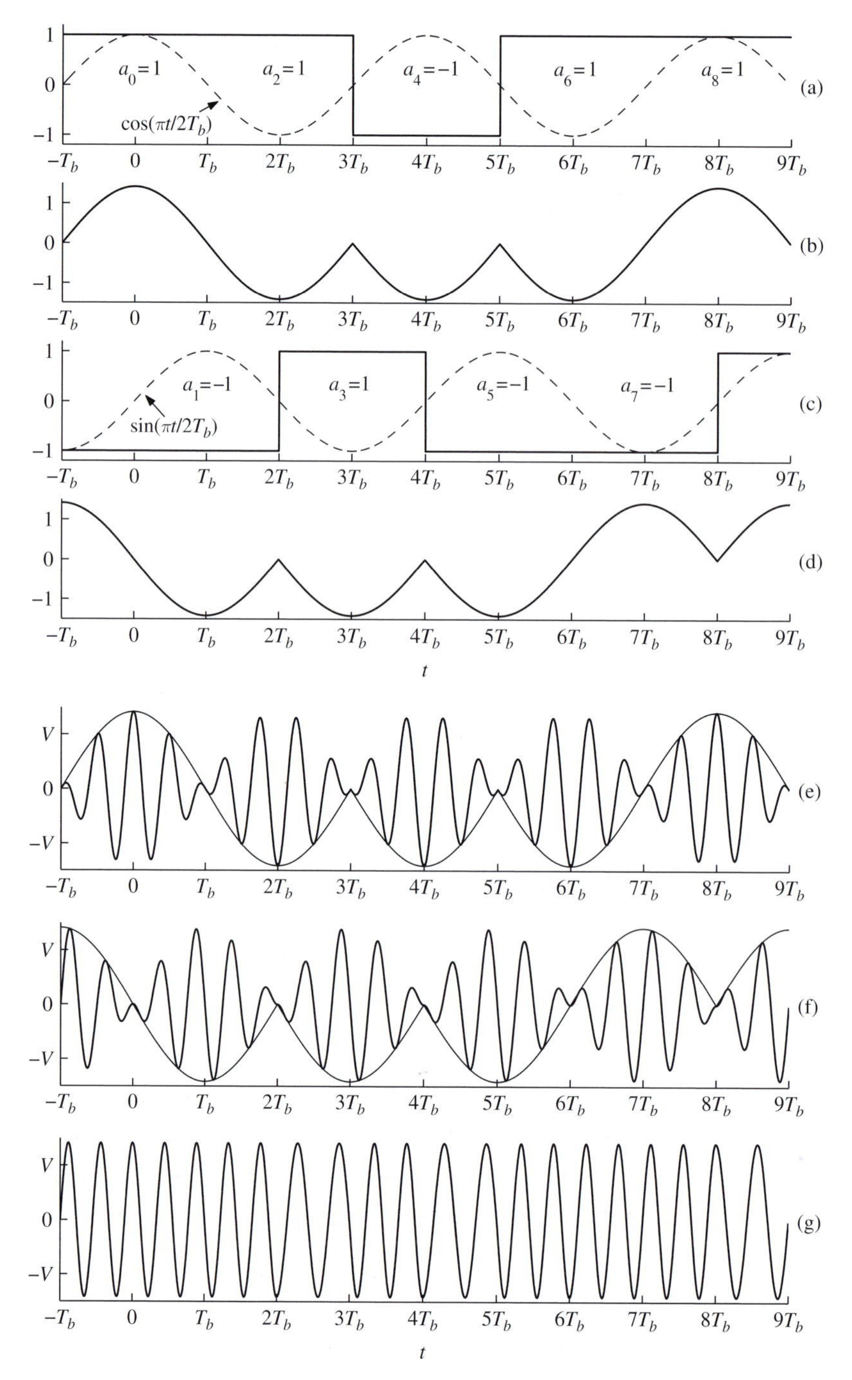

Fig. 7.25 Various components of the MSK signal defined in (7.60): (a) $a_I(t)$; (b) $a_I(t)\sqrt{2}\cos(\pi t/2T_b)$; (c) $a_Q(t)$; (d) $a_Q(t)\sqrt{2}\sin(\pi t/2T_b)$; (e) $a_I(t)\sqrt{2}\cos(\pi t/2T_b)V\cos(2\pi f_c t)$; (f) $a_Q(t)\sqrt{2}\sin(\pi t/2T_b)V\sin(2\pi f_c t)$; and (g) $s_{\text{MSK}}(t) = a_I(t)\sqrt{2}\cos(\pi t/2T_b)V\cos(2\pi f_c t) + a_Q(t)\sqrt{2}\sin(\pi t/2T_b)V\sin(2\pi f_c t)$.

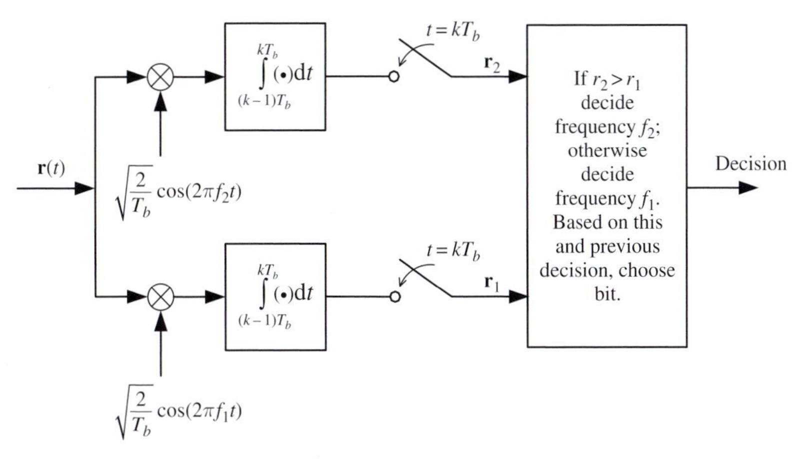

Fig. 7.26 Demodulation of MSK by viewing it as continuous-phase BFSK.

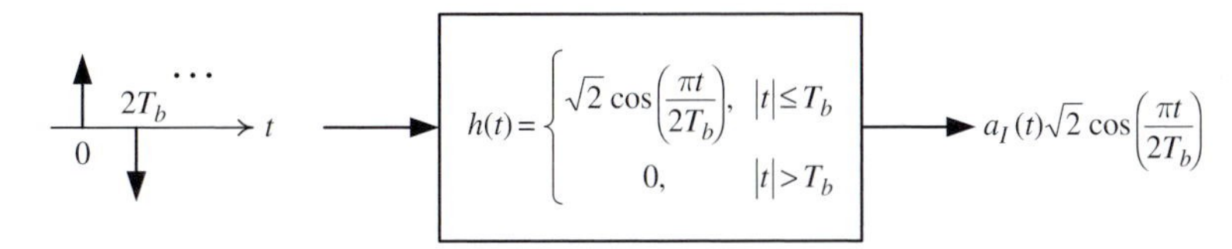

Fig. 7.27 Generation of the even bit stream in MSK.

during the previous bit duration then one can decide on the transmitted bit during the present duration. Note that the decision alternates between the even and odd bits.

The above demodulation method suffers, however, from error propagation since the present bit decision depends on the previous one. This could be overcome by precoding at the transmitter (this technique is discussed in Chapter 10). Another disadvantage is that, being considered as FSK, it is 3 decibels poorer in error performance than BPSK or QPSK/OQPSK.

The PSD of the MSK signal is most easily obtained by applying the impulse modulator technique presented in Section 5.7. Consider the even bit stream and the function $a_I(t)\sqrt{2}\cos(\pi t/2T_b)$. The signal can be modeled as being generated by a linear system whose impulse response is $\sqrt{2}\cos(\pi t/2T_b)$, $|t| \le T_b$. The input to the system is a random impulse train with the impulses occurring every $2T_b$ seconds. This is illustrated in Figure 7.27. Note that the PSD of the input process is $1/2T_b$, $\forall f$, which follows from the result in Section 5.7 after substituting $R_c(m) = 1$ for $m = 0$ and $R_c(m) = 0$ for $m \neq 0$.

The transfer function of the linear system is

$$H(f) = \int_{-T_b}^{T_b} \sqrt{2}\cos\left(\frac{\pi t}{2T_b}\right) e^{-j2\pi ft} dt = -\sqrt{2}\left(\frac{\pi}{T_b}\right)\left[\frac{\cos(2\pi f T_b)}{4\pi^2 f^2 - \frac{\pi^2}{4T_b^2}}\right]. \tag{7.65}$$

Therefore the PSD of $a_I(t)$ is given by

$$S_I(f) = \frac{1}{2T_b}|H(f)|^2 = \frac{1}{T_b}\left(\frac{\pi}{T_b}\right)^2 \left[\frac{\cos(2\pi f T_b)}{4\pi^2 f^2 - \frac{\pi^2}{4T_b^2}}\right]^2. \quad (7.66)$$

The PSD for the odd bit stream, $a_Q(t)\sqrt{2}\sin(\pi t/2T_b)$ is also given by (7.66). This is because, as a random signal, the odd bit stream is generated in an identical manner to the even bit stream except that there is a time shift of T_b seconds. This leaves the PSD unchanged. Finally, each bit stream is modulated by a carrier, $V\cos(2\pi f_c t)$ and $V\sin(2\pi f_c t)$. This effectively translates the PSD to be centered at the carrier frequency f_c. About the only question that might arise is "what is the effect of the crosscorrelation between the even and odd bit streams?" This is zero. The reason is that the statistical independence of the even and odd bits makes the inphase and quadrature bit streams, $a_I(t)$ and $a_Q(t)$, uncorrelated.

Finally, the PSD of the transmitted MSK signal is given by

$$S_{\text{MSK}}(f) = K\left\{\left[\frac{\cos[2\pi(f-f_c)T_b]}{4\pi^2(f-f_c)^2 - \pi^2/(4T_b^2)}\right]^2 + \left[\frac{\cos[2\pi(f+f_c)T_b]}{4\pi^2(f+f_c)^2 - \pi^2/(4T_b^2)}\right]^2\right\}, \quad (7.67)$$

where K is a scaling factor.

It should be noted that the PSD of QPSK/OQPSK can be found in the same manner but with the impulse response in Figure 7.27 set to a rectangular pulse, i.e., $h(t) = 1$, $|t| \leq T_b$. As mentioned before, except for a scaling factor of 2, the PSD of QPSK/OQPSK has the same expression as the PSD of BPSK in (7.8). It can be seen from (7.67) that the MSK PSD decays as $1/f^4$, which is considerably faster than the $1/f^2$ decay behavior of the PSDs of BPSK and QPSK/OQPSK (see (7.8)). This decay reflects the fact that there are no discontinuities in the transmitted signal.

The *normalized* PSDs for QPSK/OQPSK and MSK are sketched in Figure 7.28. The PSD of BPSK is also included for comparison. The fact that BPSK requires more bandwidth than the others for any reasonable definitions of bandwidth, such as the null-to-null or fractional out-of-band power definitions[3] is obvious. The theoretical null-to-null bandwidth efficiency of BPSK is half that of QPSK. As predicted by the decay rate discussed above, it is observed from Figure 7.28 that MSK has lower sidelobes than QPSK/OQPSK. This is a consequence of multiplying the binary bit stream with a sinusoid, yielding more gradual phase transitions. The more gradual the transition, the faster the spectral tails drop to zero. Consider a bandwidth that contains 99% of the total power. Then for the same bit rate of $r_b = 1/T_b$ MSK has a bandwidth of $\approx 1.18/T_b$, QPSK/OQPSK has a bandwidth of

[3] The null-to-null bandwidth is determined based on the first frequency value where the PSD equals zero, while the fractional out-of-band power bandwidth is defined based on a required percentage of the total power in the bandwidth.

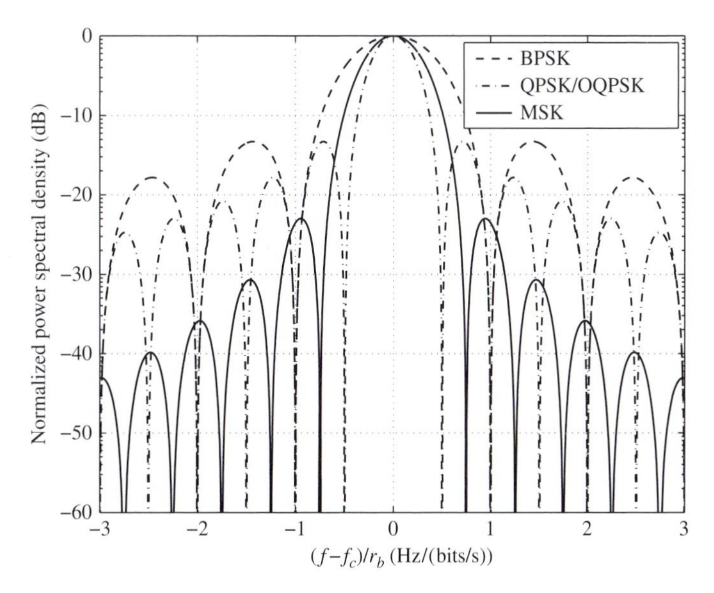

Fig. 7.28 PSDs of different modulation schemes.

$\approx 10.28/T_b$, while BPSK has a bandwidth of $\approx 20.56/T_b$. However, the MSK spectrum has a wider mainlobe than QPSK/OQPSK, which means that when compared in terms of null-to-null bandwidth, MSK is less spectrally efficient than QPSK/OQPSK.

Since there are no discontinuities at the bit transitions in the MSK signal, bandlimiting the MSK signal to meet the required out-of-band power specifications does not cause its envelope to go to zero. The envelope of an MSK signal is basically kept constant after bandlimiting. Even if there are small variations of the envelope after bandlimiting at the transmitter, these can be removed by hardlimiting at the receiver without significantly raising the sidelobe levels. Due to its constant envelope, continuous phase, low sidelobe levels, and the fact that it can be demodulated as easily as FSK, MSK is a very popular modulation technique in mobile communications.

7.7 Summary

Basic passband modulations for digital communications have been discussed in this chapter. As with baseband, the focus was on the power needed to achieve a certain performance, measured as the bit error probability, and on the bandwidth requirements of the specific modulations. The channel model was one of AWGN and the mantra was the same as in baseband modulation. Find a signal space representation of the modulator's signal set, project the received signal onto the signal space basis to generate a set of sufficient statistics, and process them to obtain the decision. Indeed this mantra will be invoked in subsequent chapters where other channel models and other modulations are studied.

Besides the basic binary passband modulations, the chapter also discussed higher-level modulations, namely, quadrature phase-shift keying (QPSK) and the related minimum shift keying (MSK). They were introduced as methods to conserve bandwidth without sacrificing error performance. The next chapter investigates higher-level modulations, known as M-ary ($M > 2$) more fully. Minimum shift keying belongs to the class of modulations known as continuous-phase modulation (CPM). The interested reader is referred to the classic text by Anderson *et al.* [1] for further reading on this subject.

Lastly, the reader might have observed that the basic analysis is very similar for both baseband and passband modulation. Indeed, though not done in the chapter, one can develop an *equivalent baseband model*, and then proceed to develop the optimum demodulator and its performance at baseband. The equivalent baseband model is pursued in the problems at the end of the chapter. However, though a passband digital communication system may be represented mathematically by an equivalent baseband system, one should be aware that there are excellent engineering reasons for passband modulation. The two major reasons are: (i) the decreased size of antennas for efficient electromagnetic transmission at higher frequencies, (ii) the ability for different users to share the same transmission medium, be it cable, fiber optics, or the air waves.

7.8 Problems

7.1 (*Why modulate?*) In passband modulation, the message signal modulates a sinusoidal *carrier*. One may ask why it is necessary to use a carrier for radio transmission. The answer is examined in this question.

Radio transmission is achieved with the *electromagnetic* field. The transmission of electromagnetic fields through space is accomplished with the use of an antenna. The size of the antenna depends on the wavelength λ and the application. For a cellular phone, the antenna is typically $\lambda/4$ in size (diameter), where wavelength is equal to c/f, with $c = 3 \times 10^8$ (m/s) the speed of light.

(a) Consider sending a message signal, say with $f = 4$ kilohertz, by coupling it to an antenna directly without a carrier. How large would the antenna have to be?

(b) Suppose that the message signal is first modulated onto a carrier of 1.2 gigahertz before coupling to the antenna. What is the size of the antenna in this case? Comment.

7.2 (*Autocorrelation of the NRZ waveform*) Show that the autocorrelation function of the baseband NRZ waveform is given as

$$R_{\mathbf{m}}(\tau) = \begin{cases} \dfrac{1}{4} + \dfrac{1}{4}\left(1 - \dfrac{|\tau|}{T_b}\right), & |\tau| < T_b \\ \dfrac{1}{4}, & |\tau| > T_b \end{cases}. \tag{P7.1}$$

7.3 Verify the expressions in (7.41), (7.42), and (7.43).

7.4 (*Anti-Gray mapping*) Recall that a Gray mapping is any mapping with the property that the nearest symbols differ in only one bit. We have shown that with a Gray mapping QPSK modulation has the same bit error probability as that of BPSK. Here consider the mapping from two bits to one QPSK symbol as in Table 7.2, called an anti-Gray mapping.

Table 7.2 An anti-Gray mapping of QPSK

Bit pattern	Message	Signal transmitted	
00	m_1	$s_1(t) = V\cos(2\pi f_c t)$,	$0 \leq t \leq T_s = 2T_b$
11	m_2	$s_2(t) = V\sin(2\pi f_c t)$,	$0 \leq t \leq T_s = 2T_b$
10	m_3	$s_3(t) = -V\cos(2\pi f_c t)$,	$0 \leq t \leq T_s = 2T_b$
01	m_4	$s_4(t) = -V\sin(2\pi f_c t)$,	$0 \leq t \leq T_s = 2T_b$

(a) Find, in terms of E_b/N_0, the expression for the *bit error probability* of QPSK modulation using the above anti-Gray mapping.

(b) Compare the result with that of QPSK employing Gray mapping. Plot the two bit error probabilities on the same graph versus E_b/N_0 and comment.

7.5 (*Asymmetric QPSK*) Asymmetric constellations provide a simple solution for unequal error protection, where bits that are deemed to be important can be protected more than bits of lesser importance (e.g., image and voice signals in multimedia applications). Consider an asymmetric QPSK constellation as shown in Figure 7.29, where the mapping of two bits b_1b_2 to each signal point is also shown. The information bits are equally likely. As usual, the signal set is used for communications over an ideal AWGN channel with two-sided PSD $N_0/2$ (watts/hertz).

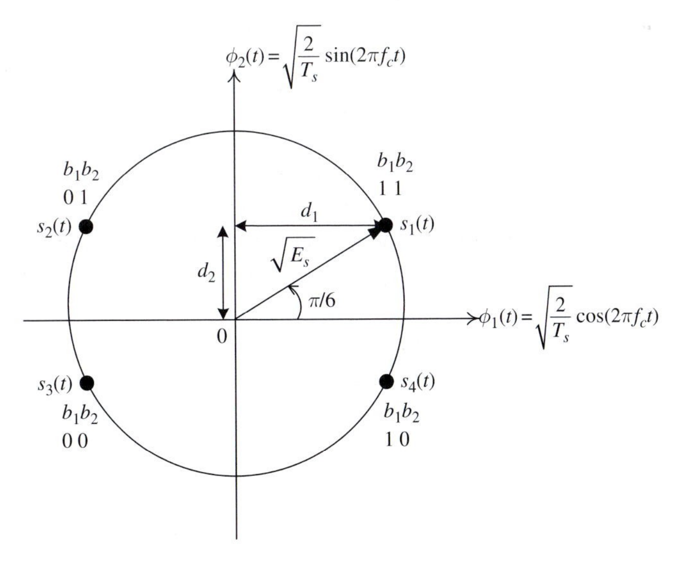

Fig. 7.29 Asymmetric QPSK constellation.

(a) Draw the decision boundary and the decision regions of the receiver that minimizes the *symbol* error probability.
(b) Determine the *symbol* error probability of the receiver in (a) as a function of E_s/N_0.
(c) Determine the *bit* error probabilities for b_1 and b_2 *separately*. Which bit is more protected and why? *Hint* Argue that $P[b_k \text{ is in error}] = P[b_k \text{ is in error}|s_1(t)]$, $k = 1, 2$.
(d) Assume that $N_0 = 10^{-6}$ watts/hertz. How large does E_s need to be set to achieve $P[b_1 \text{ is in error}] \leq 10^{-3}$?

7.6 (*Propagation delay and phase shift*) Consider the mobile radio link illustrated in Figure 7.30. In the figure a user stands at point A, at a distance d from the base station. The propagation delay from point A to the base station is T_d. Consider that a single tone $s(t) = \cos(2\pi f_c t)$ is transmitted from the user and let $f_c = 1.2$ gigahertz. Neglecting the noise, the waveform received at the base station is $r(t) = \cos[2\pi f_c(t + T_d)]$.
(a) If the user moves away from the base station to point B, or toward the base station to point C, what is the minimum distance of the movement that will cause a 2π rotation of the received waveform?
(b) Do we really care about a 2π phase rotation? What are the minimum distances of the user's movement that cause $\pi/2$ and π phase rotations? Comment.

7.7 (*BASK with phase uncertainty*) Consider BASK with the two elementary signals:

$$\begin{cases} s_1(t) = 0, & \text{if "}0_T\text{"} \\ s_2(t) = \sqrt{E}\sqrt{\dfrac{2}{T_b}}\cos(2\pi f_c t), & \text{if "}1_T\text{"} \end{cases}, \quad 0 < t \leq T_b, \tag{P7.2}$$

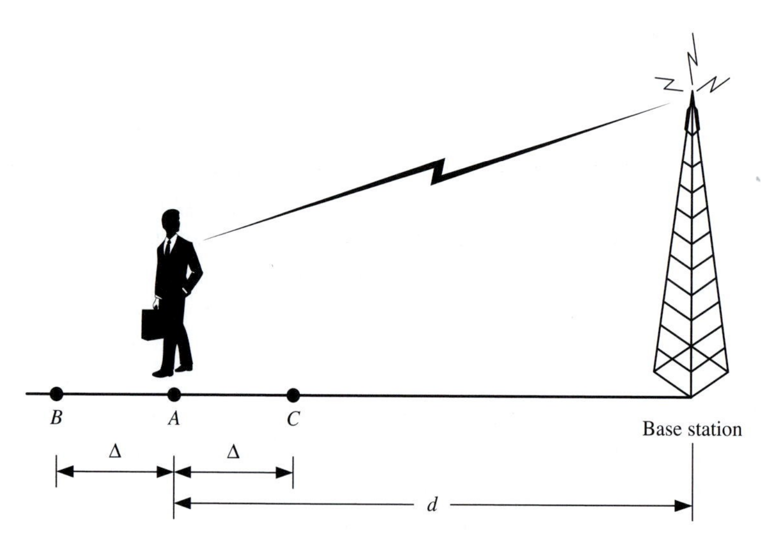

Fig. 7.30 Example of a radio link.

where $f_c = n/T_b$, $n \gg 1$. Upon transmission through the channel, the signal undergoes a *phase shift*. The received signal is therefore given by

$$\mathbf{r}(t) = \begin{cases} s_1^R(t) + \mathbf{w}(t) \\ s_2^R(t) + \mathbf{w}(t) \end{cases}$$
$$= \begin{cases} 0 + \mathbf{w}(t), & \text{if "}0_T\text{"} \\ \sqrt{E}\sqrt{\dfrac{2}{T_b}}\cos(2\pi f_c t + \theta) + \mathbf{w}(t), & \text{if "}1_T\text{"} \end{cases},$$

where, as usual, $\mathbf{w}(t)$ is AWGN with a two-sided PSD of $N_0/2$, $s_1^R(t)$ and $s_2^R(t)$ are the two possible received signals in the absence of noise. The phase θ is assumed to be *unknown* to the receiver since either the company cannot afford a phase-locked loop (PLL), or their engineers do not understand the circuit and cannot design one. Thus they are attempting to overcome this problem.

(a) One proposed idea is as follows. Forget about the phase shift (i.e., assume $\theta = 0$) and just use the conventional "optimum" receiver. Calculate the error probabilities for this method when $\theta = 30°, 60°, 90°$. Plot the error probability curves as functions of E_b/N_0 on the same graph and comment.

(b) After testing the performance of the receiver in (a) and not being satisfied with it, the company comes to you for a suggestion. What should be a reasonable receiver in the face of this phase uncertainty? Explain. *Hint* As a first step, determine the signal space of the received signals $s_1^R(t)$ and $s_2^R(t)$. Assume the phase can be any value between 0 and 2π and plot the locus of $s_2^R(t)$ in the signal space.

7.8 (*BPSK with phase and attenuation uncertainty*) Consider BPSK modulation, where two antipodal signals

$$s_1^T(t) = -s_2^T(t) = \sqrt{E_b}\sqrt{\frac{2}{T_b}}\cos(2\pi f_c t)$$

are used for the transmission of equally likely bits 0 and 1 every T_b seconds, respectively. Besides the usual AWGN of spectral strength $N_0/2$ (watts/hertz), the channel introduces a phase shift of θ radians, $-\pi/4 < \theta < \pi/4$, and an attenuation of k (volts/volts), $0 < k \leq 1$, to the transmitted signal.

(a) On the same graph, show the signal space diagrams at the transmitter and the receiver for some arbitrary values of θ and k within the ranges specified above.

(b) When both θ and k are *unknown* to the receiver, the receiver just employs the minimum distance rule as if there were no phase shift nor attenuation. Determine the resultant bit error probability of this receiver in terms of E_b/N_0, θ, and k.

(c) Assume now that the receiver can perfectly estimate the phase shift θ and the attenuation k. Draw the decision regions and the block diagram of the optimal receiver in this situation. Determine the error probability of this optimal receiver. Compare it with the performance of the receiver in (b) and comment.

7.9 (*QPSK with phase and attenuation uncertainty*) Consider QPSK modulation where the transmitted signal is

$$s_i^T(t) = \sqrt{E_s}\sqrt{\frac{2}{T_s}}\cos\left[2\pi f_c t + (2i+1)\frac{\pi}{4}\right],$$
$$i = 0, 1, 2, 3;\ 0 \le t \le T_s.$$

Besides the usual AWGN of spectral strength $N_0/2$ (watts/hertz), the channel introduces a phase shift of θ radians, $-\pi/4 < \theta < \pi/4$, and an attenuation of k (volts/volts), $0 < k \le 1$, to the transmitted signal. Both θ and k are *fixed* but *unknown* to the receiver.

(a) On the same graph, show the signal space diagrams at the transmitter and the receiver.

(b) Assuming that the four transmitted signals are equally likely, determine the resultant *symbol* error probability of the "conventional optimum receiver" (i.e., the receiver that assumes $\theta = 0$ and $k = 1$) under the phase and gain uncertainty.

(c) Consider the following Gray mapping: $00 \to s_0^T(t)$, $01 \to s_1^T(t)$, $11 \to s_2^T(t)$ and $10 \to s_3^T(t)$. Obtain the *bit* error probability and compare it to the case when there is no phase and attenuation uncertainty.

7.10 (*BPSK with two different signal sets*) Consider the following two signal sets proposed for BPSK:

$$\text{signal set \#1:}\quad \begin{cases} 0_T: & -\sqrt{E_b}\sqrt{2/T_b}\cos(2\pi f_c t) \\ 1_T: & \sqrt{E_b}\sqrt{2/T_b}\cos(2\pi f_c t) \end{cases},\quad 0 \le t \le T_b,$$

$$\text{signal set \#2:}\quad \begin{cases} 0_T: & -\sqrt{E_b}\sqrt{2/T_b}\sin(2\pi f_c t) \\ 1_T: & \sqrt{E_b}\sqrt{2/T_b}\sin(2\pi f_c t) \end{cases},\quad 0 \le t \le T_b.$$

(a) Sketch the signals of the two signal sets. Without any computation, what would you say about the PSD of each signal set.

(b) Determine the PSD of the two signal sets.

(c) Plot the respective PSDs. The plots should be versus fT_b with f_cT_b a parameter. Plot for $f_cT_b = 1, 2, 5, 20$. Compare and comment.

(d) A general remark that is often made (and indeed has been made in the text) is that the PSD ignores the phase information of a signal or that phase information is unimportant. Try to reconcile this statement with what you have observed above.

7.11 (*Sensitivity of error performance to incorrect* a priori *probability*) Consider BPSK, where $P[1_T] = p$. The "optimum" demodulator is designed on the assumption that $p = 1/2$.

(a) Fix the demodulator with the $p = 1/2$ assumption. What happens to the error performance of this demodulator if $p \neq 1/2$.

(b) Design the optimum demodulator for general p and derive the expression for its error performance.

(c) Plot the error performance based on the demodulator of (a) and that of (b) for $p = 0.6, 0.7, 0.8, 0.9$. Comment on the sensitivity of the error performance to an incorrect assumption of the *a priori* probability.

7.12 Import your favorite music into Matlab as a `wav` file. Determine the "probability" of a one occurring (or a zero occurring). Is the equally probable assumption reasonable? *Remark* The format of your music file may or may not be important. You may wish to import different formats of the music file.

7.13 (*QPSK simulation*) This problem uses Matlab to simulate QPSK modulation and demodulation. To this end, let the duration of each QPSK signal be normalized to $T_s = 1$ second, and the energy of each signal be $E_s = 1$ joule. With these normalizations, the coordinates of QPSK signals in the signal space diagram are as shown in Table 7.3.

Table 7.3 QPSK signal coordinates

Bit pattern	QPSK signal	Signal coordinates
00	$s_1(t)$	(1, 0)
01	$s_2(t)$	(0, 1)
11	$s_3(t)$	(−1, 0)
10	$s_4(t)$	(0, −1)

Perform the following steps in Matlab:

(a) Generate a random binary information sequence of length $L = 1000$ bits, whose elements are drawn from $\{0, 1\}$. Take two bits at a time and *Gray* map them to one of the four QPSK signals according to Table 7.3. This yields the sequence of transmitted QPSK signals.

(b) Add to each transmitted QPSK signal zero-mean AWGN of variance $\sigma^2 = N_0/2$. This can be done by independently adding to each coordinate of a QPSK signal a Gaussian random variable of variance $N_0/2$. In Matlab this means adding σ`*randn(1,1)` to each coordinate.

(c) Plot the received signals in the signal space diagram. Also, plot the optimum decision boundary.

Remark It is convenient to use four different "markers" for the received signals, each corresponding to one of the four transmitted signals. In this way you can associate a particular received signal with each transmitted QPSK signal. To carry out (b) and (c), first determine the values of σ^2 so that the bit error probabilities of QPSK are $10^{-1}, 10^{-2}, 10^{-3}, 10^{-4}$. Then run your program for each value of σ. (d) Comment on the four plots obtained in (c).

Though not used in the text, it is common to develop an equivalent baseband model for the various passband modulation schemes and then develop the optimum demodulator and analysis. The approach has the advantage of being applicable to any carrier

frequency, f_c, provided of course that the model (in our case here that of an AWGN channel) holds. It has the disadvantage that it somewhat abstracts the physical reality. Although the Hilbert transform might be used to develop the equivalent baseband model, here we take a more direct, intuitive approach.

7.14 First we develop the equivalent baseband signals for the passband signals used in the chapter. In baseband the signal should be such that when shifted up (and down) by f_s hertz we should get the passband signal. The choice of the frequency f_s can be made arbitrarily but in most modulation there is a natural or logical choice. In general then

$$s_{\text{PB}}(t) = \mathcal{R}\left\{s_{\text{BB}}(t)\text{e}^{\text{j}2\pi f_s t}\right\}, \tag{P7.3}$$

where $s_{\text{PB}}(t)$ is the passband signal, $s_{\text{BB}}(t)$ is the baseband signal, $\text{e}^{\text{j}2\pi f_s t}$ represents a *shift* of f_s hertz in the frequency domain, and $\mathcal{R}\{\cdot\}$ takes the real part of the signal because we wish to transmit real signals. With all this, identify and sketch the baseband signals for:

(a) the BASK signal set;
(b) the BPSK signal set;
(c) the BFSK signal set.

Remark In (a) and (b) there is a natural choice for the shift frequency, namely $f_s = f_c$. For BFSK one has some choices. Three obvious choices are $f_s = f_1, f_s = f_2$ and $f_s = (f_1 + f_2)/2$.

7.15 The equivalent baseband signals in Problem 7.14(c) are complex, at least some of them, while in (a) and (b) they are real. Can you deduce when you would get a real equivalent baseband signal and when you would have a complex equivalent baseband signal. *Hint* A time signal $s(t)$ is real if and only if $|S(f)|$ is even and $\angle S(f)$ is odd.

7.16 What are the equivalent baseband signals for QPSK?

We now develop an equivalent baseband representation of a passband random process or, in our context, noise. By passband we mean that most of power of the noise is concentrated around some frequency f_s and the PSD becomes negligible for f sufficiently far from f_s. In essence given a passband process $n(t)$ with PSD $S_n(f)$, $S_n(f)$ is related to the equivalent baseband process $S_B(f)$ by

$$S_n(f) = S_B(f - f_s) + S_B(-f - f_s), \tag{P7.4}$$

where f_s is the chosen shift or "representation" frequency. This relationship is illustrated in Figure 7.31.

7.17 As in the deterministic signal case, the passband process, in terms of the equivalent baseband process, $\mathbf{n}_{\text{B}}(t)$, is to be expressed in the form $\mathbf{n}(t) = \mathcal{R}\left\{\mathbf{n}_{\text{B}}(t)\text{e}^{\text{j}2\pi f_s t}\right\}$ and the aim is to find $\mathbf{n}_{\text{B}}(t)$. Since we are interested only in the first- and second-order moments, i.e., mean (which for us is 0) and autocorrelation/crosscorrelation, we start by observing that $\mathbf{n}(t)$ can be generated by passing a white process through an LTI filter (known as a shaping filter) as shown in Figure 7.32.

(a) Given that $S_{\mathbf{n}}(f)$ lies in the passband, show that $H(f)$ can be written as $H(f) = H_{\text{B}}(f - f_s) + H_{\text{B}}^*(-f - f_s)$, where $H_{\text{B}}(f)$ is lowpass and negligible for $|f| > f_s$. Some sketches might be helpful.

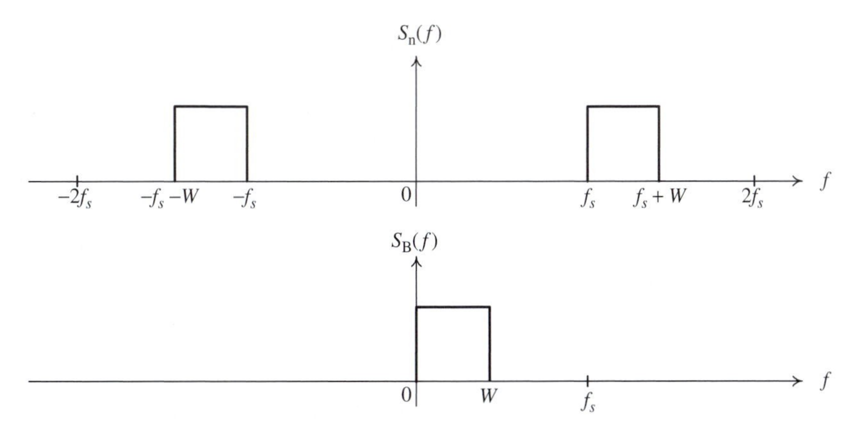

Fig. 7.31 Relationship between PSDs of passband and equivalent baseband processes.

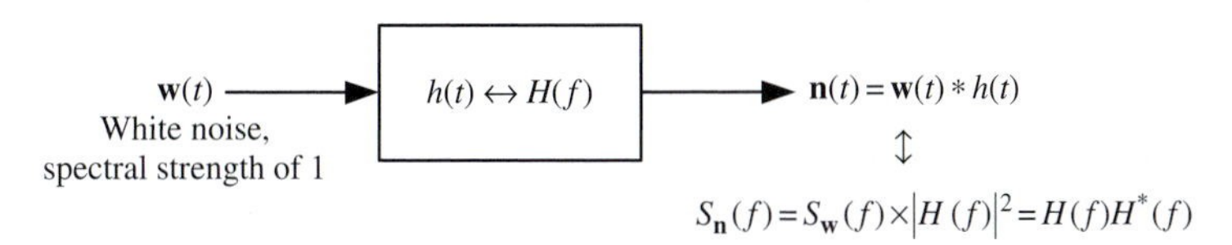

Fig. 7.32 Producing a passband random process from white noise and a shaping filter.

(b) Using $S_{\mathbf{n}}(f) = S_{\mathrm{B}}(f - f_s) + S_{\mathrm{B}}(-f - f_s)$, $S_{\mathbf{n}}(f) = |H(f)|^2$, $H(f) = H_{\mathrm{B}}(f - f_s) + H_{\mathrm{B}}^*(-f - f_s)$, show that $S_{\mathrm{B}}(f) = |H_{\mathrm{B}}(f)|^2$.

(c) Now show that the relationship between the impulse response of the shaping filter, $h(t)$, and the equivalent baseband filter impulse response, $h_{\mathrm{B}}(t)$ is

$$h(t) = 2\mathcal{R}\left\{h_{\mathrm{B}}(t)\mathrm{e}^{\mathrm{j}2\pi f_s t}\right\}, \qquad \text{(P7.5)}$$

where $h_{\mathrm{B}}(t) = \int_{-\infty}^{\infty} H_{\mathrm{B}}(f)\mathrm{e}^{\mathrm{j}2\pi f t}\mathrm{d}f$.

(d) Now show that

$$\mathbf{n}(t) = \mathcal{R}\left\{\int_{-\infty}^{\infty} \mathbf{w}(t-\lambda)h_{\mathrm{B}}(\lambda)\mathrm{e}^{\mathrm{j}2\pi f_s \lambda}\mathrm{d}\lambda\right\}. \qquad \text{(P7.6)}$$

(e) Observe that we want to express $\mathbf{n}(t)$ as $\mathbf{n}(t) = \mathcal{R}\left\{\mathbf{n}_{\mathrm{B}}(t)\mathrm{e}^{j2\pi f_s t}\right\}$. To put the expression in (d) in this form, multiply the quantity in the curly brackets by 1 (changes nothing), but express 1 as $1 = \mathrm{e}^{-\mathrm{j}2\pi f_s t}\mathrm{e}^{\mathrm{j}2\pi f_s t}$. Therefore

$$\mathbf{n}(t) = \mathcal{R}\left\{\left[\mathrm{e}^{-\mathrm{j}2\pi f_s t}\int_{-\infty}^{\infty} \mathbf{w}(t-\lambda)h_{\mathrm{B}}(\lambda)\mathrm{e}^{\mathrm{j}2\pi f_s \lambda}\mathrm{d}\lambda\right]\mathrm{e}^{\mathrm{j}2\pi f_s t}\right\}. \qquad \text{(P7.7)}$$

What is $\mathbf{n}_{\mathrm{B}}(t)$? Would you say it is a baseband or lowpass process? Why?

(f) Express $\mathbf{n}_{\mathrm{B}}(t) = \mathbf{n}_I(t) - \mathrm{j}\mathbf{n}_Q(t)$. What is $\mathbf{n}_I(t)$? $\mathbf{n}_Q(t)$? Show that $\mathbf{n}(t) = \mathbf{n}_I(t)\cos(2\pi f_s t) + \mathbf{n}_Q(t)\sin(2\pi f_s t)$.

Remark The choice of f_s is somewhat arbitrary but as in the case of deterministic signals, there are bad, good, and better choices. More about this in the next problem.

We now investigate the statistical properties of $\mathbf{n}_I(t)$ and $\mathbf{n}_Q(t)$, the baseband processes needed to represent the passband process.

7.18 Up till now we have made no assumptions about $\mathbf{w}(t)$, except that it should be white. Now assume that it is also Gaussian. Then $\mathbf{n}(t)$, $\mathbf{n}_I(t)$, and $\mathbf{n}_Q(t)$ are Gaussian. Why? If Gaussian, then what is needed to completely describe the two baseband processes is the autocorrelation and crosscorrelation functions, i.e., $R_I(\tau)$, $R_Q(\tau)$, $R_{I,Q}(\tau)$ and $R_{Q,I}(-\tau)$.

(a) Find $E\{\mathbf{n}_B(t)\mathbf{n}_B(t+\tau)\} = R_{\mathbf{n}_B}(t,\tau)$. Fourier transform $R_{\mathbf{n}_B}(t,\tau)$ with respect to the variable τ. The result you get should be of the form $e^{-j4\pi f_s t}H_B(-f-2f_s)H_B(f)$. Argue that this product is zero and therefore that $R_I(\tau) = R_Q(\tau)$ and $R_{I,Q}(\tau) = -R_{Q,I}(\tau)$.

(b) As in (a) find $E\left\{\mathbf{n}_B^*(t)\mathbf{n}_B(t+\tau)\right\}$. Show that it equals $\int_0^\infty h_B^*(\lambda)h_B(\tau+\lambda)d\lambda$ and that the Fourier transform is $|H_B(f)|^2 = S_B(f)$.

(c) Use the results of (a) and (b) to conclude that

$$R_I(\tau) = R_Q(\tau) = \frac{1}{2}\int_{-\infty}^{\infty} S_B(f)\cos(2\pi f\tau)df, \tag{P7.8}$$

$$R_{I,Q}(\tau) = -R_{Q,I}(\tau) = -\frac{1}{2}\int_{-\infty}^{\infty} S_B(f)\sin(2\pi f\tau)df. \tag{P7.9}$$

(d) If the PSD of $\mathbf{n}(t)$ is locally symmetric about f_s then what sort of symmetry does $S_B(f)$ possess? What can you say about the crosscorrelation functions? What can you say about the two processes $\mathbf{n}_I(t)$ and $\mathbf{n}_Q(t)$? Remember that if two Gaussian random variables are uncorrelated, then they are ________________.

7.19 To finish with the representation of a communication system in equivalent baseband consider the representation of the convolution operation in equivalent baseband. The system is shown in Figure 7.33. Here both $X(f)$ and $H(f)$ are passband signals whose

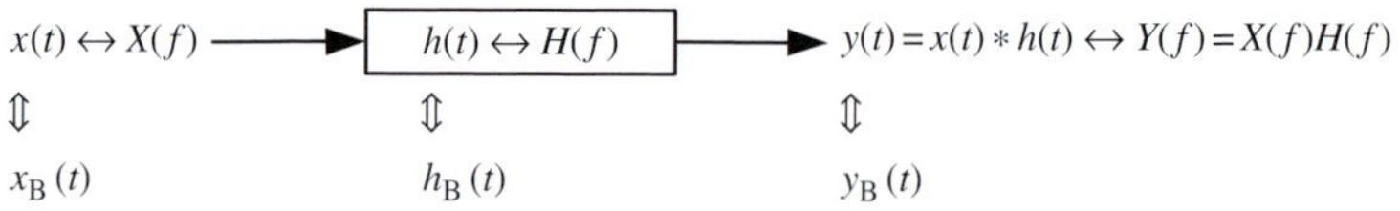

$x(t) \leftrightarrow X(f)$ → $h(t) \leftrightarrow H(f)$ → $y(t) = x(t) * h(t) \leftrightarrow Y(f) = X(f)H(f)$

$\Updownarrow$ $\Updownarrow$ $\Updownarrow$

$x_B(t)$ $h_B(t)$ $y_B(t)$

Fig. 7.33 Equivalent baseband representation of the convolution operation.

spectral content lies around or near $f_s \gg 0$. Express $X(f)$ and $H(f)$ as $X(f) = X_B(f-f_s) + X_B^*(-f-f_s)$ and $H(f) = H_B(f-f_s) + H_B^*(-f-f_s)$, where $X_B(f)$ and $H_B(f)$ are the respective baseband representations. Show that

$$y_B(t) = x_B(t) * h_B(t). \tag{P7.10}$$

7.20 As we have seen in Chapter 5, the matched filter, $h(t)$, is of the form $s(T_b - t)$, perhaps normalized to have unit energy and $s(t)$ is a signal chosen from the signal constellation. If $s_B(t)$ is the equivalent baseband signal for $s(t)$, what is $h_B(t)$, the equivalent baseband impulse response of $h(t)$?

References

[1] J. B. Anderson, T. Aulin, and C. Sundberg, *Digital Phase Modulation*. Kluwer Academic/Plenum Publishers, 1986.

8 M-ary signaling techniques

8.1 Introduction

The previous chapter shows that there are benefits to be gained when M-ary ($M = 4$) signaling methods are used rather than straightforward binary signaling. In general, M-ary communication is used when one needs to design a communication system that is bandwidth efficient. It is based on the observation that as the time duration of a signal, T_s, increases, the bandwidth requirement decreases. See Examples 2.11, 2.16, and Problem 2.38, which illustrate this. Typically, unlike QPSK and its variations in the previous chapter, the gain in bandwidth is accomplished at the expense of error performance. M-ary modulation is also a natural choice when the source is inherently M-ary, for example, the transmission of the English alphabet or when error control coding is used.

However, even when the source is inherently M-ary, the usual scenario is that the M messages are mapped to a sequence of bits, e.g., the ASCII code used for text. Therefore, even in these situations the final source output is binary and from the perspective of the modulator looks like a binary source. The typical application of M-ary modulation is one where a binary source has its bit stream blocked into groups of λ bits. The number of different bit patterns is 2^λ, which means $M = 2^\lambda$, where each bit pattern is mapped (modulated) into a distinct signal. Each block of λ bits is a symbol and given that the source's bit rate is $r_b = 1/T_b$ bits/second, the symbol transmission rate is $r_s = 1/T_s = 1/(\lambda T_b) = r_b/\lambda$ symbols/second. A signal from the modulator occupies $T_s = \lambda T_b$ seconds and the implication is that the bandwidth requirement is on the order of $1/T_s$, or that there is a bandwidth saving of $1/\lambda$ compared to binary modulation.

In this chapter M-ary ASK, PSK, QAM (quadrature amplitude modulation) and FSK signaling methods are discussed. In particular the demodulator which minimizes the *message error* probability (or symbol error probability) is derived and applied to the different signaling methods. The error performance of these modulation methods and bandwidth saving are analyzed and contrasted. It should be noted that though a message will usually represent a block of binary digits or bits, minimizing message error probability is not the same as minimizing *bit error* probability. In general, there is no simple relationship between message error probability and bit error probability. Relationships, however, are derived for the discussed modulations, albeit only in terms of bounds for some of them. Only M-ary signaling techniques using passband modulation are discussed. M-ary communication using baseband modulation is a fairly straightforward application of the ideas presented for the passband modulation signals. Before discussing the different modulation methods the optimum decision rule of Chapter 5 is generalized to the M-ary case.

8.2 Optimum receiver for M-ary signaling

Consider the problem of trying to transmit one of M messages, say m_1, m_2, ..., m_M, through a communication channel every T_s seconds as shown in Figure 8.1. The messages are represented by M signals $s_1(t)$, $s_2(t)$, ..., $s_M(t)$. As usual, the channel is assumed to be wideband enough to transmit the signals without any distortion. The only effect of the channel is to corrupt the signal with zero-mean, white Gaussian noise of strength $N_0/2$ (watts/hertz).

The problem now is: having received the signal $\mathbf{r}(t) = s_i(t) + \mathbf{w}(t)$ over the time interval of $[0, T_s]$ seconds how does one decide on the transmitted signal with minimum error? The determination of the optimum receiver proceeds in a manner analogous to that for the binary case. The M signals are first represented by an orthonormal basis set, $\phi_1(t), \phi_2(t), \ldots, \phi_N(t)$, where $N \leq M$. The set can be determined as follows (recall the *Gram–Schmidt* orthogonalization procedure):

(i) Choose one of the signals, say $s_1(t)$, normalize it by its energy and let this normalized signal be the first orthonormal function, $\phi_1(t)$.
(ii) Choose a second signal, $s_2(t)$, find its component along $\phi_1(t)$ and subtract this component from $s_2(t)$. Normalize the resultant signal and let it be $\phi_2(t)$.
(iii) Choose a third signal, $s_3(t)$. Find the projections of $s_3(t)$ along the $\phi_1(t)$ and $\phi_2(t)$ axes and subtract the two projections from $s_3(t)$. Normalize the resultant signal and call it $\phi_3(t)$. Note, if $s_3(t)$ is a linear combination of $s_1(t)$ and $s_2(t)$, then the subtraction will yield zero resulting in a zero $\phi_3(t)$ at this step. In this case go on to the fourth signal $s_4(t)$, project onto $\phi_1(t), \phi_2(t)$, subtract, normalize, and call this signal $\phi_3(t)$.
(iv) Repeat the procedure for all signals. In each case, project the considered signal onto the previously determined orthonormal functions. Subtract the components along these axes from the considered signal and if the resultant signal is nonzero, normalize so that it has unit energy and include it as part of the orthonormal basis.

The procedure thus yields $N \leq M$ orthonormal basis functions. The number of basis functions, N, equals the number of signals, M, if no signal is a linear combination

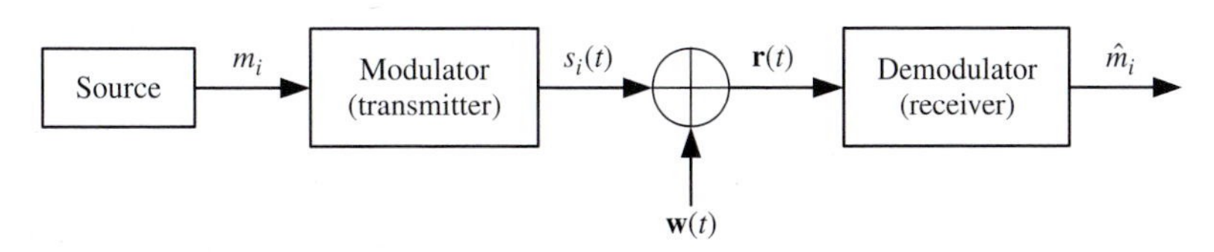

Fig. 8.1 Block diagram of an M-ary signaling system.

of some set of the other signals. Otherwise $N < M$. Signal $s_i(t)$ is then represented as follows:

$$s_i(t) = s_{i1}\phi_1(t) + s_{i2}\phi_2(t) + \cdots + s_{iN}\phi_N(t), \tag{8.1}$$

where

$$s_{ik} = \int_0^{T_s} s_i(t)\phi_k(t)\mathrm{d}t. \tag{8.2}$$

Typically a lot of the coefficients will be zero.

To determine the minimum message error probability receiver the received signal $\mathbf{r}(t)$ is expanded into the series

$$\begin{aligned}\mathbf{r}(t) &= s_i(t) + \mathbf{w}(t)\\ &= \mathbf{r}_1\phi_1(t) + \mathbf{r}_2\phi_2(t) + \cdots + \mathbf{r}_N\phi_N(t) + \mathbf{r}_{N+1}\phi_{N+1}(t) + \cdots\\ &= (s_{i1} + \mathbf{w}_1)\phi_1(t) + \cdots + (s_{iN} + \mathbf{w}_N)\phi_N(t) + \mathbf{w}_{N+1}\phi_{N+1}(t) + \cdots,\end{aligned} \tag{8.3}$$

where

$$\mathbf{r}_k = \int_0^{T_s} \mathbf{r}(t)\phi_k(t)\mathrm{d}t. \tag{8.4}$$

As before, for $k > N$, the coefficients $\mathbf{r}_k$ are the same regardless of which message, m_i (or signal $s_i(t)$), is being transmitted and they are statistically independent of the coefficients $\mathbf{r}_j, j \leq N$. Therefore they can be discarded.

The decision problem is to partition the N-dimensional space formed by $\vec{\mathbf{r}} = (\mathbf{r}_1, \mathbf{r}_2, \ldots, \mathbf{r}_N)$ into M regions so that the message error probability is minimized or equivalently the probability of a correct decision is maximized. This is illustrated in Figure 8.2.

The probability of making a correct decision is:

$$P[\text{correct}] = \int_{\Re_1} P_1 f(\vec{r}|s_1(t))\mathrm{d}\vec{r} + \cdots + \int_{\Re_M} P_M f(\vec{r}|s_M(t))\mathrm{d}\vec{r}. \tag{8.5}$$

To maximize $P[\text{correct}]$ the observed vector $\vec{r} = (r_1, r_2, \ldots, r_N)$ is assigned to region $\Re_i$ if

$$P_i f(\vec{r}|s_i(t)) \text{ is larger than } P_j f(\vec{r}|s_j(t)), \quad j = 1, 2, \ldots, M;\ j \neq i \tag{8.6}$$

where P_i is the *a priori* probability of the ith message and $f(\vec{r}|s_i(t))$ is the conditional pdf of the observations $(r_1, r_2, \ldots, r_N)$. The conditional pdf is given by

$$f(\vec{r}|s_i(t)) = \prod_{k=1}^{N} \frac{1}{\sqrt{\pi N_0}} \exp\left\{-\frac{1}{N_0}(r_k - s_{ik})^2\right\}. \tag{8.7}$$

The decision rule expressed by (8.6) can be rewritten, by using (8.7) and taking the natural logarithm of $P_i f(\vec{r}|s_i(t))$, as

$$\begin{gathered}\text{choose } m_i \text{ if}\\ \ln P_i - \frac{1}{N_0}\textstyle\sum_{k=1}^{N}(r_k - s_{ik})^2 > \ln P_j - \frac{1}{N_0}\sum_{k=1}^{N}(r_k - s_{jk})^2;\\ j = 1, 2, \ldots, M;\ j \neq i,\end{gathered} \tag{8.8}$$

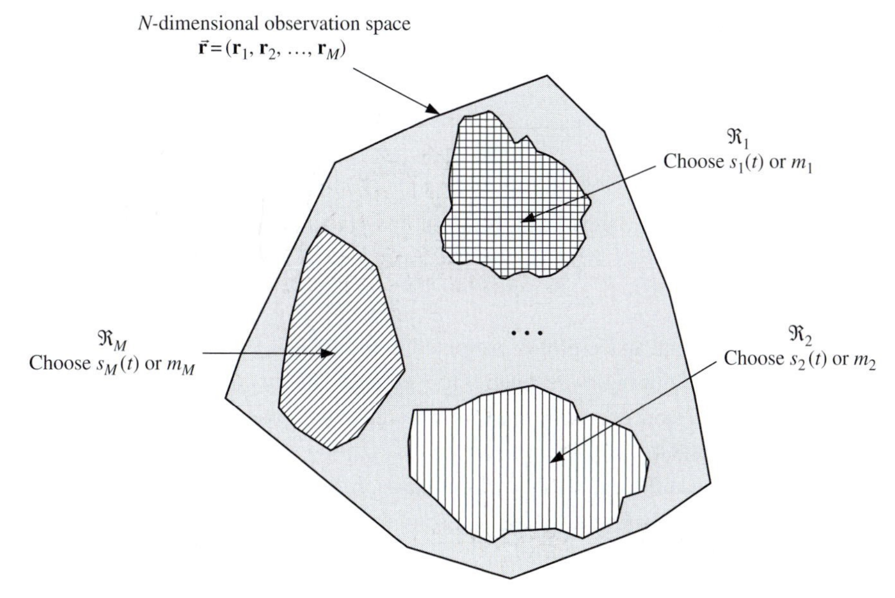

Fig. 8.2 Decision regions of M-ary signaling.

where the common term $-0.5N\ln(\pi N_0)$ is discarded. If the messages are equally likely, i.e., $P_i = P_j$, then the decision rule simplifies to

$$\begin{gathered}\text{choose } m_i \text{ if} \\ \sum_{k=1}^{N}(r_k - s_{ik})^2 < \sum_{k=1}^{N}(r_k - s_{jk})^2; \\ j = 1, 2, \ldots, M;\ j \neq i.\end{gathered} \tag{8.9}$$

Again the quantity $\sum_{k=1}^{N}(r_k - s_{ik})^2$ can be interpreted as the distance (squared) from the observations $(r_1, r_2, \ldots, r_N)$ to the ith signal $s_i(t)$, which is represented by the components $(s_{i1}, s_{i2}, \ldots, s_{iN})$. Therefore the intuitive interpretation of a *minimum-distance receiver* also applies in the M-ary situation. A general block diagram of the minimum-distance receiver that implements the decision rule in (8.9) is obvious. The decision rule and its implementation can be greatly simplified for the specific signaling techniques discussed in the following sections.

8.3 M-ary coherent amplitude-shift keying (M-ASK)

The transmitted signal is one of

$$s_i(t) = V_i\sqrt{\frac{2}{T_s}}\cos(2\pi f_c t),\ 0 \le t \le T_s$$
$$i = 1, 2, \ldots, M;\ f_c = k/T_s,\ k \text{ integer}, \tag{8.10}$$

where the amplitude takes on one of M values and usually is increased in equal increments. Obviously the above signal set is represented by only one orthonormal function, namely,

$$\phi_1(t) = \sqrt{\frac{2}{T_s}}\cos(2\pi f_c t), \quad 0 \le t \le T_s. \tag{8.11}$$

Let the signal amplitude be $V_i = (i-1)\Delta$. Then

$$s_i(t) = [(i-1)\Delta]\phi_1(t), \quad i = 1, 2, \ldots, M. \tag{8.12}$$

The signal space plot is presented in Figure 8.3.

The optimum receiver, for equally likely messages, computes $\mathbf{r}_1 = \int_{(k-1)T_s}^{kT_s} \mathbf{r}(t)\phi_1(t)\mathrm{d}t$ and determines to which signal the observed r_1 is closest. This is illustrated in Figure 8.4.

The sufficient statistic $\mathbf{r}_1$ is Gaussian, with variance $N_0/2$ and a mean value determined by the transmitted signal. The conditional pdf of $\mathbf{r}_1$ given that signal $s_i(t)$ was transmitted is

$$f(r_1|s_i(t)) = \frac{1}{\sqrt{\pi N_0}}\exp\left\{-\frac{1}{N_0}[r_1 - (i-1)\Delta]^2\right\}. \tag{8.13}$$

The decision regions are shown graphically in Figure 8.5. Mathematically they are:

$$\text{choose}\begin{cases} s_k(t), & \text{if } \left(k-\frac{3}{2}\right)\Delta < r_1 < \left(k-\frac{1}{2}\right)\Delta,\ k = 2, 3, \ldots, M-1 \\ s_1(t), & \text{if } r_1 < \Delta/2 \\ s_M(t), & \text{if } r_1 > \left(M-\frac{3}{2}\right)\Delta \end{cases}. \tag{8.14}$$

The message error probability is given by

$$P[\text{error}] = \sum_{i=1}^{M} P[s_i(t)]P[\text{error}|s_i(t)]. \tag{8.15}$$

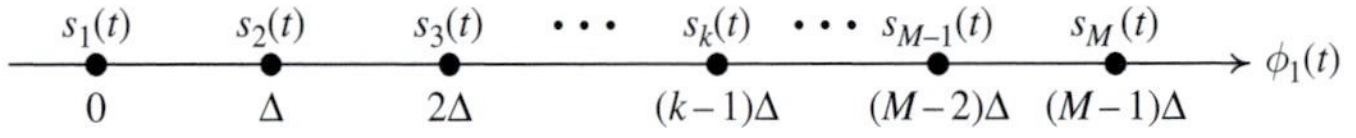

$s_1(t)$ $s_2(t)$ $s_3(t)$ … $s_k(t)$ … $s_{M-1}(t)$ $s_M(t)$ $\phi_1(t)$

0 Δ 2Δ $(k-1)\Delta$ $(M-2)\Delta$ $(M-1)\Delta$

Fig. 8.3 Signal space diagram for M-ASK signaling.

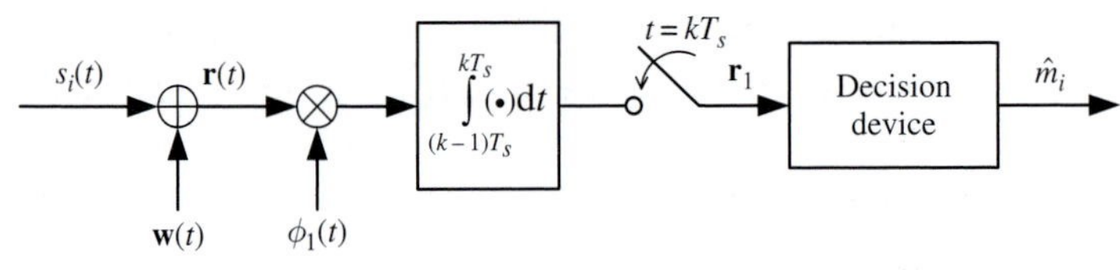

Fig. 8.4 Receiver implementation for M-ASK signaling.

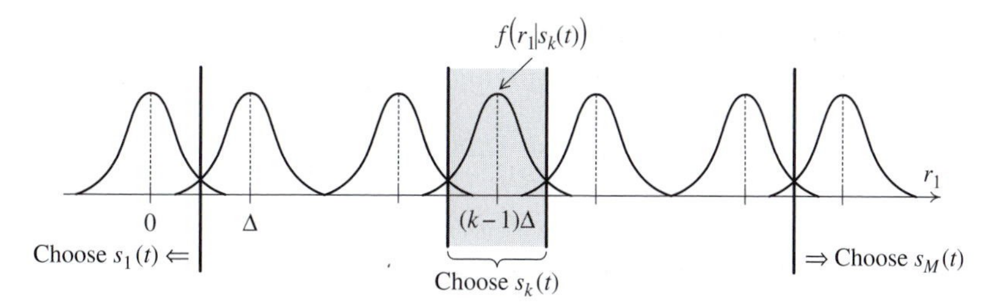

Fig. 8.5 Decision regions for M-ASK.

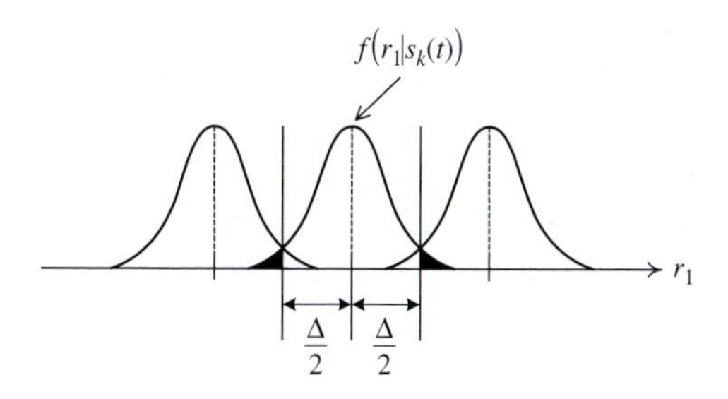

Fig. 8.6 Calculating the error probability for M-ASK.

For the $M-2$ inner signals, $s_i(t)$, $i=2,3,\ldots,M-1$, $P[\text{error}|s_i(t)]$ is equal to the two shaded areas as shown in Figure 8.6. Each area is equal to $Q\left(\Delta/\sqrt{2N_0}\right)$. Therefore

$$P[\text{error}|s_i(t)] = 2Q\left(\Delta/\sqrt{2N_0}\right), \quad i=2,3,\ldots,M-1. \tag{8.16}$$

The two end signals have an error probability of

$$P[\text{error}|s_i(t)] = Q\left(\Delta/\sqrt{2N_0}\right), \quad i=1,M. \tag{8.17}$$

Combining the above and using the fact (or assumption) that $P[s_i(t)] = 1/M$ results in

$$P[\text{error}] = \frac{2(M-1)}{M} Q\left(\Delta/\sqrt{2N_0}\right). \tag{8.18}$$

The maximum amplitude of a transmitted signal is $(M-1)\Delta$ and therefore the maximum transmitted energy is $[(M-1)\Delta]^2$. This can be reduced, without any sacrifice in error probability, by changing the signal set to one which includes the negative version of each signal. In essence the modulation is now a combination of amplitude and phase where the phase is either 0 or π. It is, however, the practice to call this M-ASK; M-ary phase modulation is discussed in the next section. The resultant signals are

$$s_i(t) = \underbrace{(2i-1-M)\frac{\Delta}{2}}_{V_i}\sqrt{\frac{2}{T_s}}\cos(2\pi f_c t), \quad 0 \le t \le T_s, \quad i=1,2,\ldots,M. \tag{8.19}$$

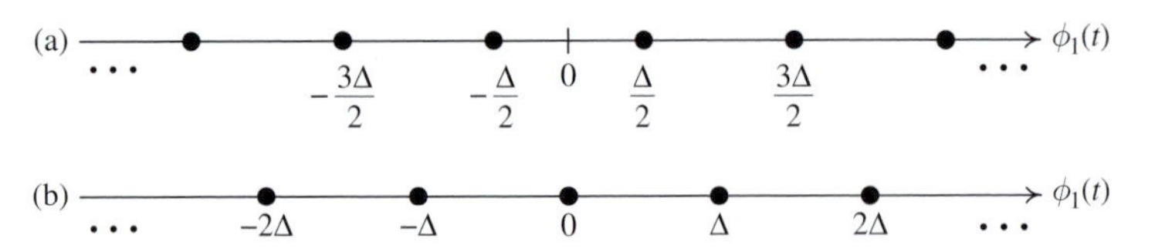

Fig. 8.7 Signal space plot when (a) M is even, and (b) M is odd.

With the above signal set, the maximum transmitted energy is

$$E_{\text{max}} = \left[\frac{(M-1)\Delta}{2}\right]^2 \text{ (joules)}, \tag{8.20}$$

which is reduced by four times compared to the previous signal set. The signal space plots are shown in Figure 8.7 for even and odd M.

Note that with the above modified signal constellation, not only is the maximum transmitted energy reduced, but also the average transmitted energy. As before, for a fair comparison among different signal constellations (with either the same or different M), it is desired to express the error probability in terms of E_b/N_0, where E_b is the average transmitted energy per bit. To do this, compute the average transmitted energy per message (or symbol) as follows:

$$\begin{aligned} E_s = \frac{\sum_{i=1}^{M} E_i}{M} &= \frac{\Delta^2}{4M}\sum_{i=1}^{M}(2i-1-M)^2 \\ &= \frac{\Delta^2}{4M}\left[\frac{1}{3}M(M^2-1)\right] = \frac{(M^2-1)\Delta^2}{12}. \end{aligned} \tag{8.21}$$

Thus the average transmitted energy per bit[1] is

$$E_b = \frac{E_s}{\log_2 M} = \frac{(M^2-1)\Delta^2}{12\log_2 M}, \tag{8.22}$$

which also implies that

$$\Delta = \sqrt{\frac{12E_s}{M^2-1}} \tag{8.23}$$

$$= \sqrt{\frac{(12\log_2 M)E_b}{M^2-1}}. \tag{8.24}$$

[1] The number of bits per symbol here is λ, i.e., $\log_2 M = \lambda$. However, even when M is not a power of 2, $\log_2 M$ is still taken as the number of "equivalent" bits that the message represents.

Substituting (8.23) and (8.24) into (8.18) yields

$$P[\text{error}] = \frac{2(M-1)}{M} Q\left(\sqrt{\frac{6E_s}{(M^2-1)N_0}}\right) \tag{8.25}$$

$$= \frac{2(M-1)}{M} Q\left(\sqrt{\frac{6\log_2 M}{M^2-1}\frac{E_b}{N_0}}\right). \tag{8.26}$$

The above probability of error is the symbol error probability. For M-ary modulations, the relation between the symbol error probability and the bit error probability is often tedious due to its dependence on the mapping from λ-bit patterns into the signal points. If Gray mapping is used, two adjacent symbols differ in only a single bit. Since the most probable errors due to noise result in the erroneous selection of a signal adjacent to the true signal, most symbol errors contain only a single-bit error. Hence the equivalent bit error probability for M-ASK modulation is well approximated by

$$P[\text{bit error}] \approx \frac{1}{\lambda} P[\text{symbol error}]. \tag{8.27}$$

Therefore for M-ASK with Gray mapping one has

$$P[\text{bit error}] = \frac{P[\text{symbol error}]}{\log_2 M} \approx \frac{2(M-1)}{M\log_2 M} Q\left(\sqrt{\frac{6\log_2 M}{M^2-1}\frac{E_b}{N_0}}\right). \tag{8.28}$$

The symbol error probability of M-ASK is plotted in Figure 8.8 for $M = 2$, $M = 4$, $M = 8$, and $M = 16$. This figure clearly illustrates how one can trade power for bandwidth

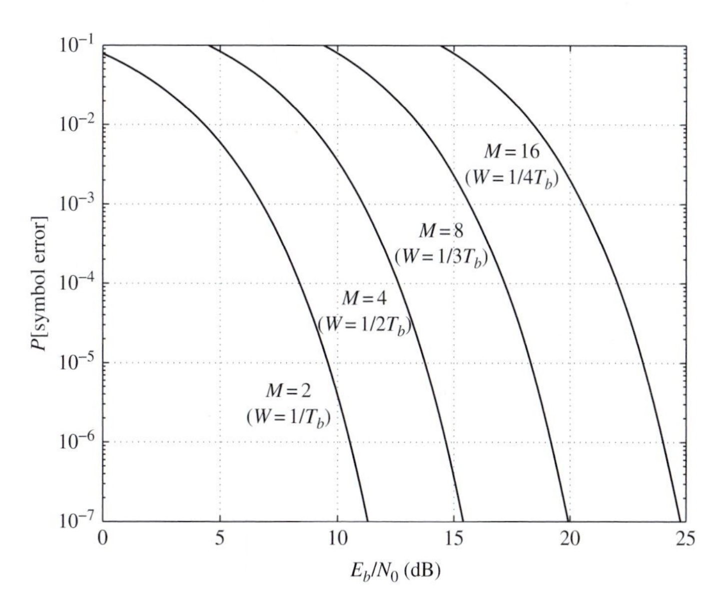

Fig. 8.8 Probability of symbol error for M-ASK signaling. The bandwidth, W, is obtained by using the $WT_s = 1$ rule-of-thumb. Here $1/T_b$ is the bit rate (bits/second).

by using a higher-order modulation. In particular, observe that the SNR per bit needs to increase by over 4 decibels when M is doubled. For large M, the additional power per bit needed when M increases by a factor of 2 approaches 6 decibels.

Though M-ASK gives a bandwidth saving, it comes at the expense of either error performance or an increased transmission power if the same error performance is required. The problem arises from the fact that the energy grows linearly with λ, while the number of signals one needs to place on the single axis, $\phi_1(t)$, grows exponentially (as 2^{λ}). Perhaps if one went to two dimensions, this effect, if not overcome completely, could at least be ameliorated. M-ary phase-shift keying (M-PSK), which uses two orthonormal basis functions, may be a possible solution. It was certainly successful for the $M = 4$ (QPSK) situation (see Section 7.6.1). The next section discusses general M-PSK.

8.4 M-ary phase-shift keying (M-PSK)

The signal set is given by

$$s_i(t) = V\cos\left[2\pi f_c t - \frac{(i-1)2\pi}{M}\right], \quad 0 \le t \le T_s,$$
$$i = 1, 2, \ldots, M;\ f_c = k/T_s,\ k \text{ integer};\ E_s = V^2 T_s/2 \text{ joules} \tag{8.29}$$

which can be written as

$$s_i(t) = V\cos\left[\frac{(i-1)2\pi}{M}\right]\cos(2\pi f_c t) + V\sin\left[\frac{(i-1)2\pi}{M}\right]\sin(2\pi f_c t). \tag{8.30}$$

From the above it is seen that each signal is expressed in terms of the following two orthonormal functions:

$$\phi_1(t) = \frac{V\cos(2\pi f_c t)}{\sqrt{E_s}},\ \phi_2(t) = \frac{V\sin(2\pi f_c t)}{\sqrt{E_s}}. \tag{8.31}$$

The coefficients of the ith signal are then

$$s_{i1} = \sqrt{E_s}\cos\left[\frac{(i-1)2\pi}{M}\right],\ s_{i2} = \sqrt{E_s}\sin\left[\frac{(i-1)2\pi}{M}\right]. \tag{8.32}$$

The signals therefore lie on a circle of radius $\sqrt{E_s}$, and are spaced every $2\pi/M$ radians around the circle. This is illustrated in Figure 8.9 for $M = 8$ and Figure 8.10 for an arbitrary M (only the first two and the last signals are shown). Also shown in Figure 8.9 is a Gray mapping from three-bit patterns to the signal points.

The receiver for minimum message error probability computes

$$P_i f(r_1, r_2|s_i(t)), \quad i = 1, 2, \ldots, M \text{ and chooses the largest,}$$

where

$$r_1 = \int_0^{T_s} r(t)\phi_1(t)\mathrm{d}t,\ r_2 = \int_0^{T_s} r(t)\phi_2(t)\mathrm{d}t. \tag{8.33}$$

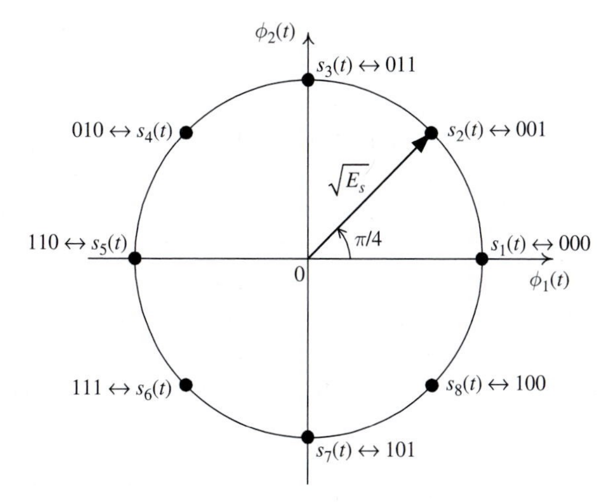

Fig. 8.9 Signal space plot for 8-PSK.

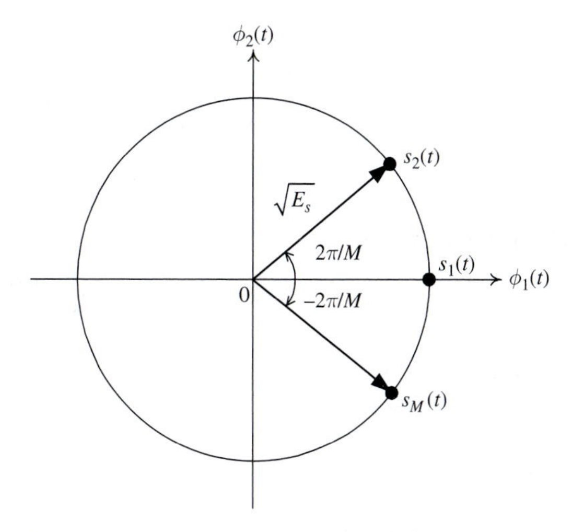

Fig. 8.10 Signal space plot for M-PSK.

When the messages are *equally likely*, the receiver is a *minimum-distance receiver*, which in block diagram form looks as in Figure 8.11. Graphically the decision regions are depicted in Figure 8.12.

To determine the error probability, note that due to the symmetry

$$\begin{aligned} P[\text{error}] &= P[\text{error}|s_1(t)] \\ &= P[r_1, r_2 \text{ fall outside Region } 1|s_1(t) \text{ transmitted}] \\ &= 1 - P[r_1, r_2 \text{ fall in Region } 1|s_1(t) \text{ transmitted}] \\ &= 1 - \iint_{r_1, r_2 \in \text{Region } 1} f(r_1, r_2|s_1(t)) \mathrm{d}r_1 \mathrm{d}r_2, \end{aligned} \tag{8.34}$$

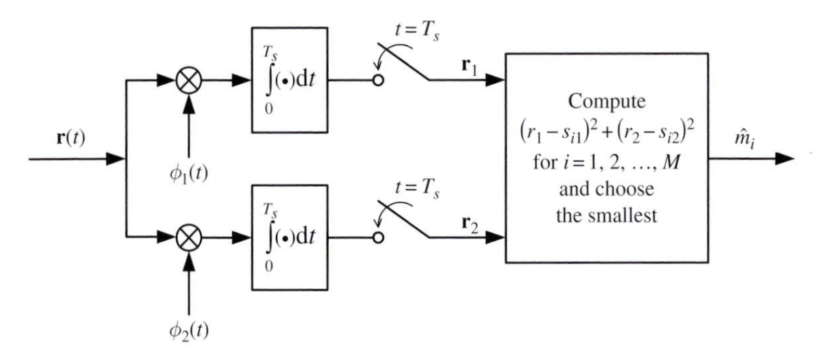

Fig. 8.11 Receiver implementation for M-PSK signaling.

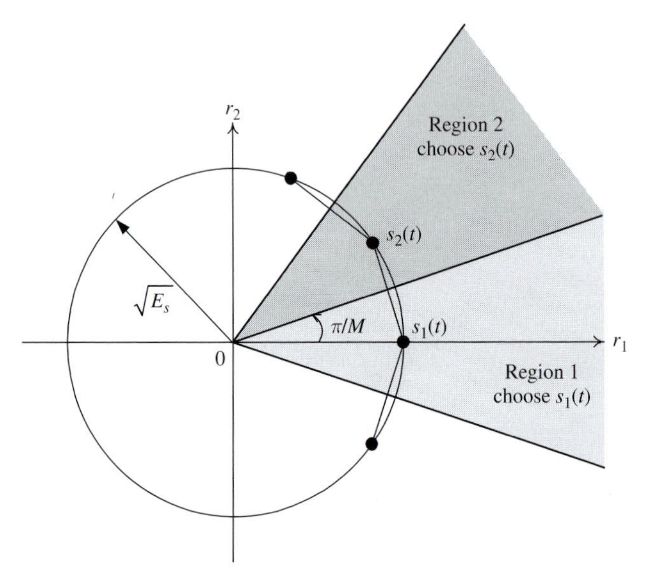

Fig. 8.12 Decision regions for M-PSK signaling.

where

$$f(r_1, r_2|s_1(t)) = \frac{1}{\sqrt{\pi N_0}} \exp\left\{-\frac{1}{N_0}\left(r_1 - \sqrt{E_s}\right)^2\right\} \frac{1}{\sqrt{\pi N_0}} \exp\left\{-\frac{r_2^2}{N_0}\right\}. \tag{8.35}$$

Unfortunately the above integral cannot be evaluated in a closed form. Thus to determine how the error probability behaves with M and the SNR, the error probability is bounded as follows.

For a lower bound, consider region $\Re_1$ as shown in Figure 8.13. Obviously,

$$P[\text{error}|s_1(t)] > P[r_1, r_2 \text{ fall in } \Re_1|s_1(t)], \text{ or}$$
$$P[\text{error}|s_1(t)] > Q\left\{\sin\left(\frac{\pi}{M}\right)\sqrt{2E_s/N_0}\right\}. \tag{8.36}$$

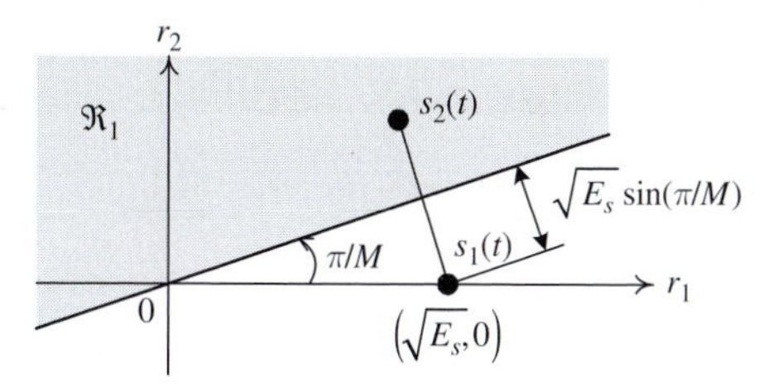

Fig. 8.13 Region $\Re_1$.

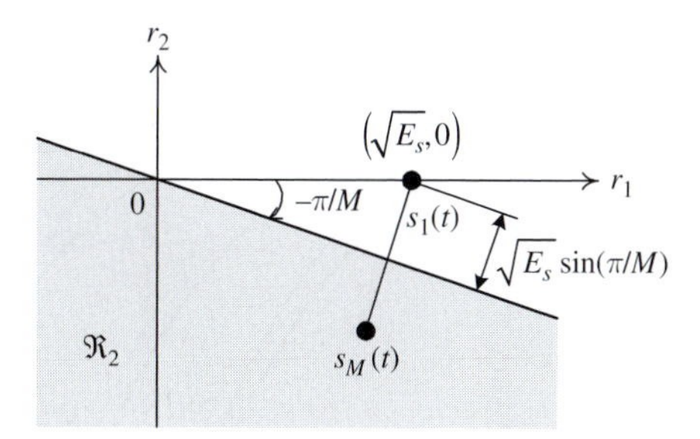

Fig. 8.14 Region $\Re_2$.

The upper bound is obtained by simply considering region $\Re_1$ as above and region $\Re_2$ shown in Figure 8.14. Then

$$P[\text{error}] < P[r_1, r_2 \text{ fall in } \Re_1 | s_1(t)] + P[r_1, r_2 \text{ fall in } \Re_2 | s_1(t)], \text{ or}$$
$$P[\text{error}] < 2Q\left(\sin\left(\frac{\pi}{M}\right)\sqrt{2E_s/N_0}\right), \tag{8.37}$$

where the common volume under $f(r_1, r_2|s_1(t))$ where regions $\Re_1, \Re_2$ intersect has been counted twice.

To compare M-PSK with BPSK let $M = 2^\lambda$. Thus each message represents a bit pattern of λ bits. The message occupies $T_s = \lambda T_b$ seconds and therefore the bandwidth requirement for the M-ary scheme is $1/\lambda$ that of the binary case. With regard to error probability note that the transmitted energy in the M-ary case is

$$E_s = \frac{V^2 T_s}{2} = \frac{V^2 \lambda T_b}{2} = \lambda E_b \quad \text{(joules/message)}, \tag{8.38}$$

where E_b is the energy transmitted in the binary case. Assuming that the SNR is high, the message error probability can be approximated as

$$P[\text{error}]_{M\text{-PSK}} \simeq Q\left(\sqrt{\lambda \sin^2\left(\frac{\pi}{M}\right)\frac{2E_b}{N_0}}\right). \tag{8.39}$$

The bit error probability for BPSK is

$$P[\text{error}]_{\text{BPSK}} = Q(\sqrt{2E_b/N_0}). \tag{8.40}$$

Table 8.1 Performance comparison of M-PSK and BPSK

λ	M	$\frac{M\text{-ary bandwidth}}{\text{binary bandwidth}}$	$\lambda \sin^2(\pi/M)$	$\frac{M\text{-ary energy}}{\text{binary energy}}$
3	8	1/3	0.44	3.6 dB
4	16	1/4	0.15	8.2 dB
5	32	1/5	0.05	13.0 dB
6	64	1/6	0.0144	17.0 dB

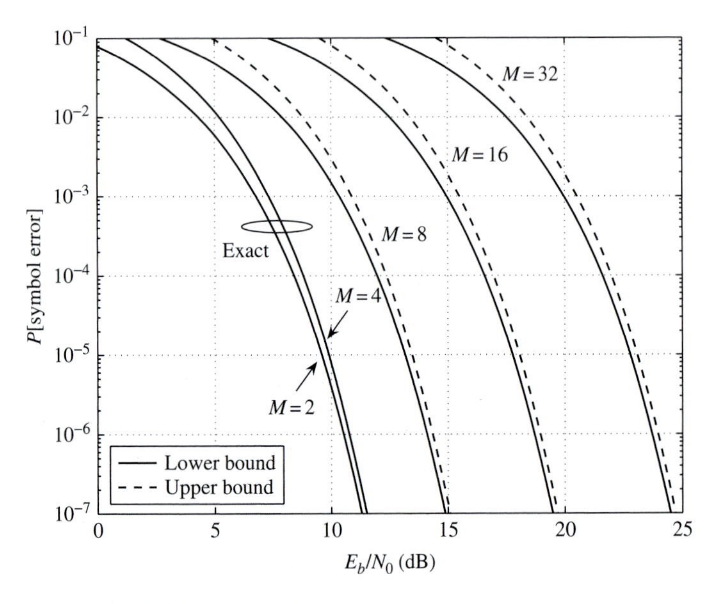

Fig. 8.15 Probability of symbol error for M-PSK signaling.

For the two error probabilities to be equal the M-PSK signal energy needs to be increased by the reciprocal of the factor $\lambda \sin^2(\pi/M)$. Table 8.1 summarizes the performances of M-PSK and BPSK. It should be noted that the above compares the symbol error of M-PSK to the bit error of BPSK.

The Gray approximation of (8.27) applies also to M-PSK. Therefore the bit error probability of M-PSK can be obtained from (8.39) as follows:

$$P[\text{bit error}]_{M\text{-PSK}} \simeq \frac{1}{\log_2 M} Q\left(\sqrt{\lambda \sin^2\left(\frac{\pi}{M}\right)\frac{2E_b}{N_0}}\right). \tag{8.41}$$

Finally, the symbol error probability of M-PSK is illustrated in Figure 8.15 for various values of M. In particular, upper and lower bounds on the symbol error probability are plotted for $M = 8,\ 16,\ 32$, while the exact results are shown for the cases of $M = 2,\ 4$.

Observe the tightness of the lower and upper bounds in the high SNR region. The performance curves clearly illustrate the penalty in SNR per bit as M increases beyond $M = 4$. For example, at $P[\text{error}] = 10^{-5}$, the difference between $M = 4$ and $M = 8$ is approximately 4 decibels, and the difference between $M = 8$ and $M = 16$ is approximately 5 decibels as in the case of M-ASK. For large values of M, doubling the number of messages requires an additional 6 decibels of power to maintain the same error performance.

M-PSK therefore behaves much the same as M-ASK in terms of bandwidth and error performance. In retrospect this might have been expected. Since the energy, E_s, grows linearly with λ, the radius of the circle that the signals are mapped onto grows as $\sqrt{E_s}$ and hence so does the circumference. But the number of signals mapped onto the circumference grows exponentially with λ. The distance between the signals therefore becomes smaller and smaller. It appears that a straight line and a circle are "topologically" the same. However, as the circle grows larger there is a lot of real estate inside it that should be able to house signal points. This is what quadrature amplitude modulation (QAM), discussed next, does.

8.5 M-ary quadrature amplitude modulation (M-QAM)

In M-ASK and M-PSK, the messages (patterns of $\lambda = \log_2 M$ binary bits) are encoded either into amplitudes or phases of a sinusoidal carrier. M-QAM is a more general modulation that includes M-ASK and M-PSK as special cases. In QAM, the messages are encoded into both the amplitude and phase of the carrier. M-QAM constellations are two-dimensional and they involve two orthonormal basis functions, given by

$$\phi_I(t) = \sqrt{\frac{2}{T_s}}\cos(2\pi f_c t), \quad 0 \le t \le T_s, \tag{8.42}$$

$$\phi_Q(t) = \sqrt{\frac{2}{T_s}}\sin(2\pi f_c t), \quad 0 \le t \le T_s, \tag{8.43}$$

where the subscripts I and Q refer to the inphase and quadrature carriers.[2] Examples of QAM constellations are shown in Figures 8.16 and 8.17 for various values of M.

The ith transmitted M-QAM signal is defined as follows:

$$s_i(t) = V_{I,i}\sqrt{\frac{2}{T_s}}\cos(2\pi f_c t) + V_{Q,i}\sqrt{\frac{2}{T_s}}\sin(2\pi f_c t), \quad \begin{array}{l} 0 \le t \le T_s, \\ i = 1, 2, \ldots, M, \end{array} \tag{8.44}$$

where $V_{I,i}$ and $V_{Q,i}$ are the information-bearing discrete amplitudes of the two quadrature carriers. The above shows that the signal $s_i(t)$ consists of two phase-quadrature carriers,

[2] Note that $\phi_I(t)$ and $\phi_Q(t)$ are exactly the same as $\phi_1(t)$ and $\phi_2(t)$ used in the representation of M-PSK. As will be seen shortly, it is more informative to rename them in the representation of M-QAM.

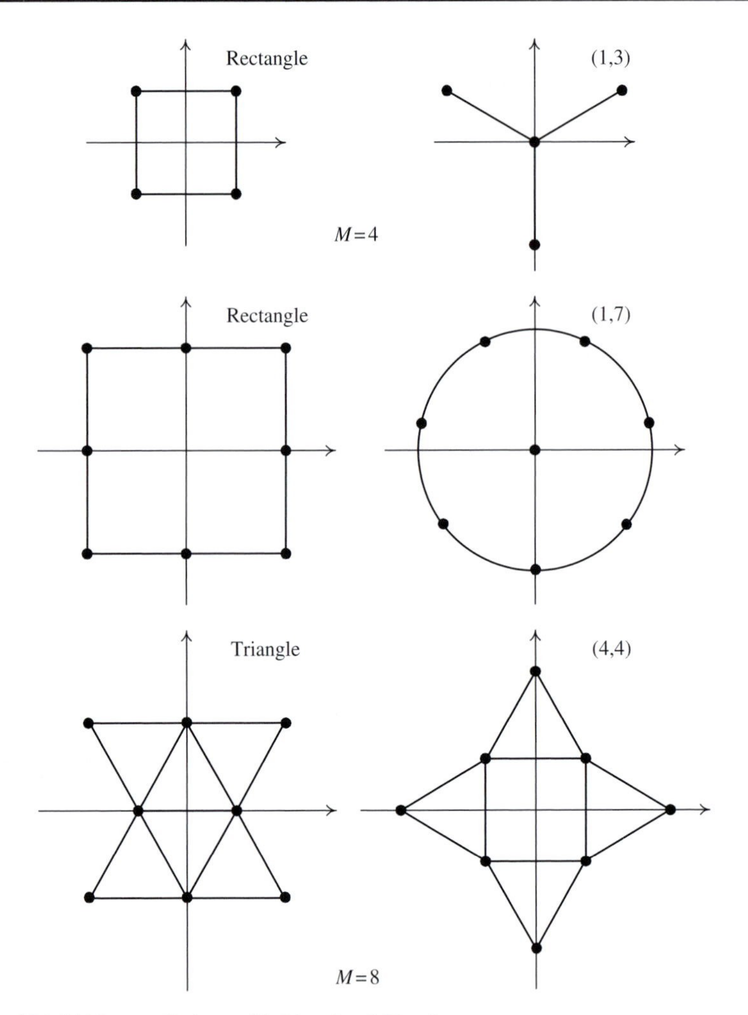

Fig. 8.16 Examples of M-QAM constellations with $M = 4$ and $M = 8$.

each being modulated by a set of discrete amplitudes, hence the name *quadrature amplitude modulation*. Also note that $V_{I,i}$, $V_{Q,i}$ in (8.44) are precisely the coefficients s_{i1}, s_{i2} in the usual representation $s_i(t) = s_{i1}\phi_1(t) + s_{i2}\phi_2(t)$.

Alternatively, the QAM signal waveforms may be expressed as

$$s_i(t) = \sqrt{E_i}\sqrt{\frac{2}{T_s}}\cos(2\pi f_c t - \theta_i), \tag{8.45}$$

where $E_i = V_{I,i}^2 + V_{Q,i}^2$ and $\theta_i = \tan^{-1}(V_{Q,i}/V_{I,i})$. From this expression, it is apparent that the QAM signal waveforms may be viewed as *combined* amplitude and phase modulation.

Since any M-QAM constellation is two-dimensional, just like M-PSK, the optimum receiver for the case of equally likely signal points is the minimum-distance receiver and can be implemented exactly as shown in Figure 8.11.

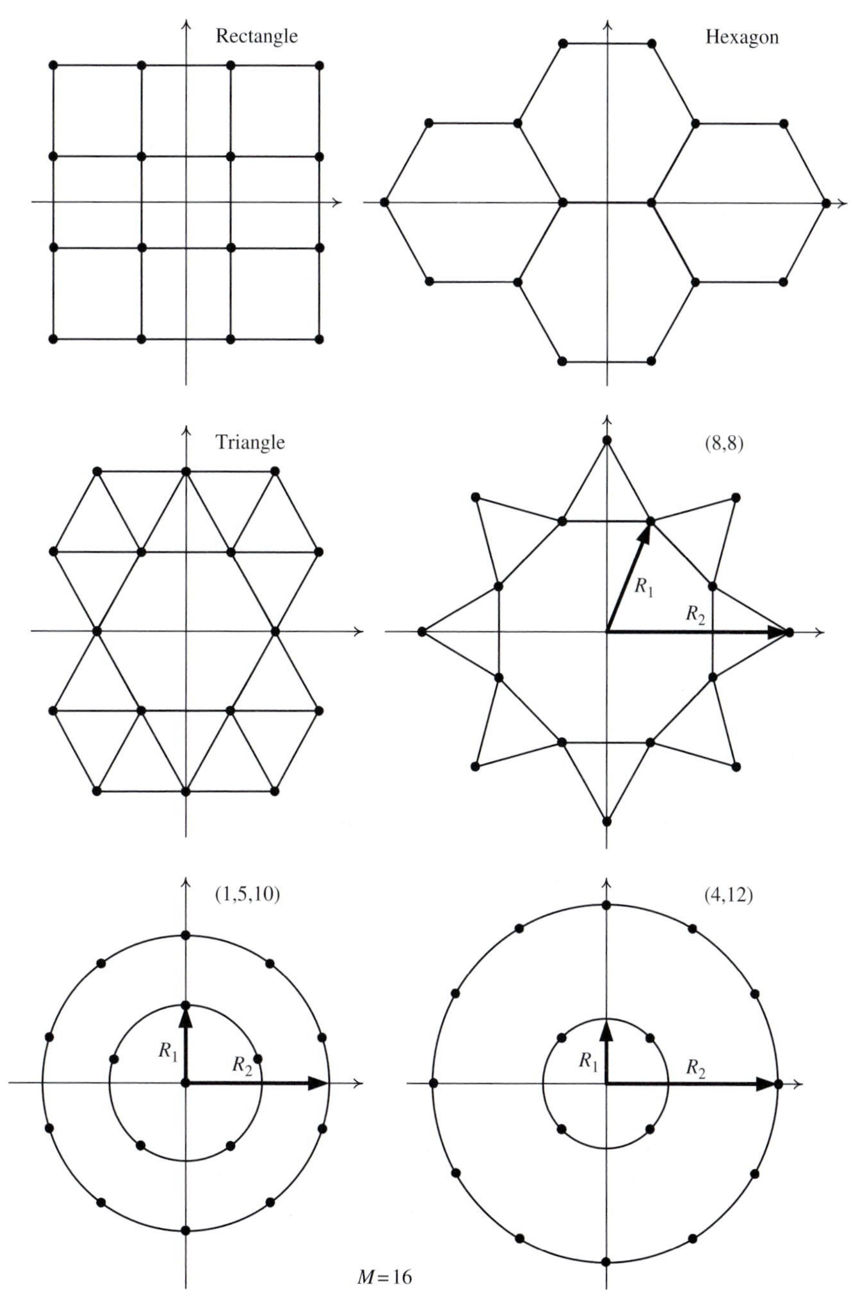

Fig. 8.17 Examples of 16-QAM constellations.

Depending on the number of possible symbols M and the set of amplitudes $\{V_{I,i}, V_{Q,i}\}$, a large variety of QAM constellations can be realized (see Figures 8.16 and 8.17 for example). Obviously, for a given M, an important question is how to compare different M-QAM constellations in terms of error performance. To answer this question, observe that for signaling over an AWGN channel the most likely error event is the one where the transmitted signal is confused with its *nearest neighbors*. Therefore, in order to (roughly) maintain the same symbol error probability, the distance between the *nearest* neighbors in all the signal constellations should be kept the same. With this constraint, the more efficient signal constellation is the one that has smaller average transmitted energy. For the special but also common case of equally likely signal points, the average transmitted energy is simply $E_s = (1/M)\sum_{i=1}^{M}(V_{I,i}^2 + V_{Q,i}^2)$.

As an example, consider the 8-QAM signal constellations of Figure 8.16. Let Δ be the minimum distance between signal points in any constellation. Assuming that the signal points are equally probable, it is simple to find that the average transmitted energies for the rectangular, triangular, (1,7), and (4,4) constellations are $1.50\Delta^2$, $1.125\Delta^2$, $1.162\Delta^2$, and $1.183\Delta^2$, respectively. Therefore, among the four 8-QAM constellations, the triangular constellation is most efficient, while the rectangular one is least efficient. In particular, the triangular 8-QAM requires approximately 1.25 decibels less energy than the rectangular 8-QAM to achieve approximately the same error probability.

The most important configuration of signal points is rectangular. Known naturally as *rectangular* QAM, the signal points are placed on a rectangular grid spaced equally in

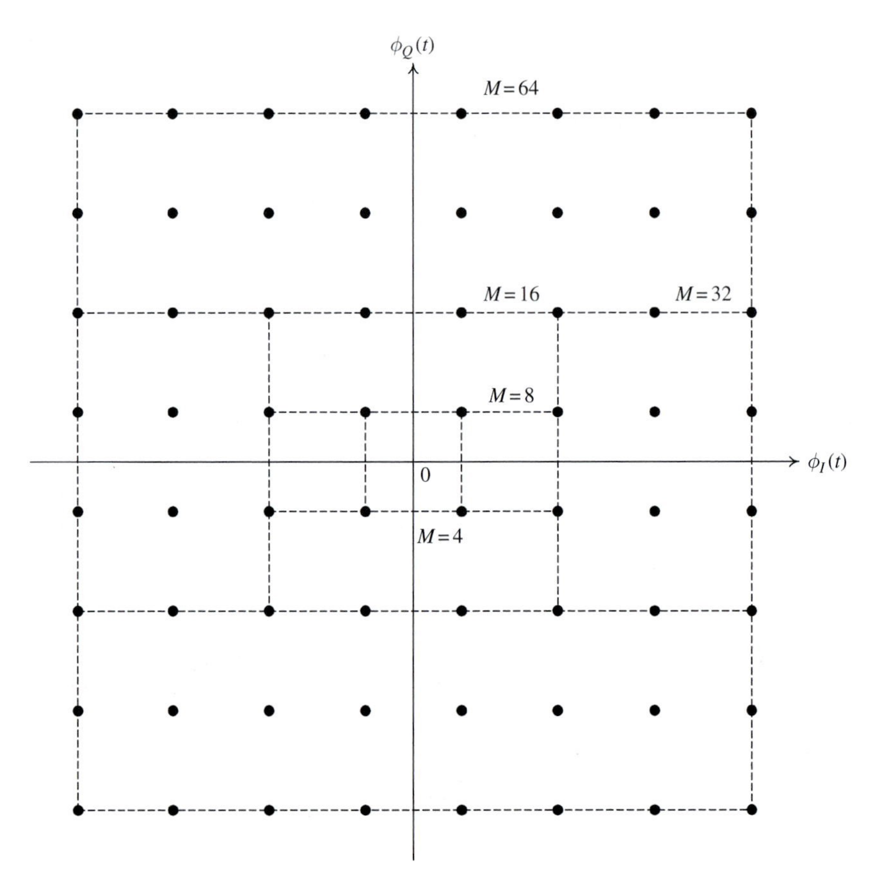

Fig. 8.18 Rectangular M-QAM constellations.

amplitude by Δ in each direction (dimension). Figure 8.18 shows rectangular QAM signal constellations for $M = 4, 8, 16, 32, 64$. The signal *components*, $V_{I,i}$ and $V_{Q,i}$, take their values from the set of discrete values $\{(2i - 1 - M)\Delta/2\}$, $i = 1, 2, \ldots, M/2$. The *minimum* Euclidean distance between signal points is therefore Δ, as in M-ASK.

The major advantage of rectangular QAM is its simple modulation and demodulation. This is because each group of $\lambda = \log_2 M$ bits can be divided into λ_I inphase bits and λ_Q quadrature bits, where $\lambda_I + \lambda_Q = \lambda$. The inphase bits and quadrature bits then modulate the inphase and quadrature carriers *independently*. These inphase and quadrature ASK signals are then added to form the QAM signal before being transmitted. At the receiver, due to the orthogonality of the inphase and quadrature signals, the two ASK signals can be *independently* detected to give the decisions on the inphase and quadrature bits. The above discussion on the modulator and demodulator of a rectangular M-QAM is illustrated in Figure 8.19.

Although in terms of energy, rectangular QAM constellations are not the best M-QAM signal constellations for $M \geq 16$, the average transmitted energy required to achieve a

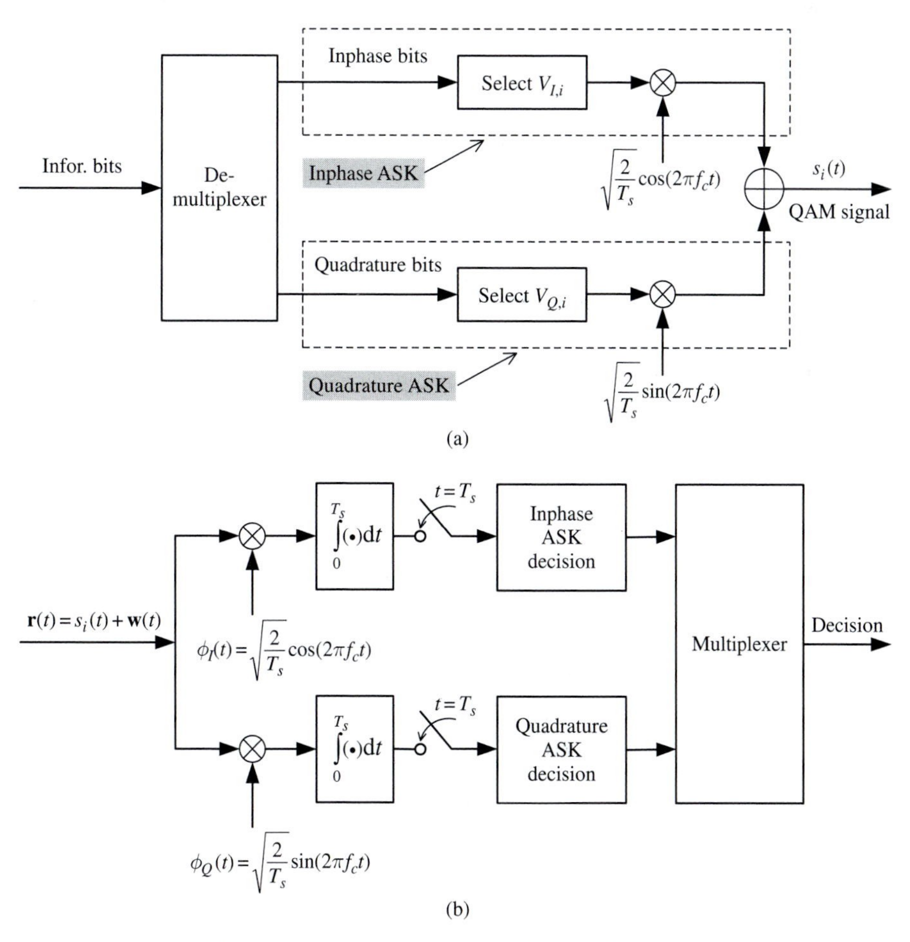

Fig. 8.19 Implementations of modulator and demodulator for rectangular M-QAM: (a) transmitter; (b) receiver.

given minimum distance is negligibly greater than the average energy required for the best M-QAM signal constellation. For these reasons, rectangular M-QAM signals are those most frequently used in practice. The most practical rectangular QAM constellation is one which "carries" an equal number of bits on each axis, i.e., λ is even, 2, 4, 6, 8, etc., and therefore M is a perfect square, 4, 16, 64, 256, etc., i.e., the rectangle is a square.

Any rectangular QAM signal constellation is equivalent to two ASK signals modulating quadrature carriers. The inphase carrier, $\phi_I(t)$, carries λ_I bits and has 2^{λ_I} signal points, while the quadrature carrier $\phi_Q(t)$ has λ_Q bits assigned to it and 2^{λ_Q} signal points. For square constellations, each carrier has $\sqrt{M} = 2^{\lambda/2}$ signal points. Since the signals in the phase-quadrature components can be perfectly separated at the demodulator, the probability of error for QAM can be easily determined from the probability of error for ASK. Specifically, the probability of symbol error for a square M-QAM system is

$$P[\text{error}] = 1 - P[\text{correct}] = 1 - \left(1 - P_{\sqrt{M}}[\text{error}]\right)^2, \tag{8.46}$$

where $P_{\sqrt{M}}[\text{error}]$ is the probability of error of a $\sqrt{M}$-ary ASK with one-half the average energy in each quadrature signal of the equivalent QAM system. By appropriately modifying the probability of error for M-ASK, we obtain

$$P_{\sqrt{M}}[\text{error}] = 2\left(1 - \frac{1}{\sqrt{M}}\right) Q\left(\sqrt{\frac{3E_s}{(M-1)N_0}}\right), \tag{8.47}$$

where E_s/N_0 is the average SNR per M-QAM symbol.

The above result is for $M = 2^{\lambda}$, λ even. However, even when λ is odd, an exact expression for the symbol error probability can be derived for rectangular QAM (see Problem 8.15). Furthermore, by upper bounding the conditional error probability (conditioned on the transmission of *any* signal point in the constellation) by the *worst case* conditional error probability (which corresponds to a signal point whose decision region is a square of size Δ centered at the signal point), the symbol error probability of a rectangular QAM is tightly upper-bounded by (see Problem 8.16)

$$\begin{aligned} P[\text{error}] &\leq 1 - \left[1 - 2Q\left(\sqrt{\frac{3E_s}{(M-1)N_0}}\right)\right]^2 \\ &\leq 4Q\left(\sqrt{\frac{3\lambda E_b}{(M-1)N_0}}\right) \end{aligned} \tag{8.48}$$

for any $\lambda \geq 1$, where E_b/N_0 is the average SNR per bit. The probability of a symbol error for M-QAM is plotted in Figure 8.20 as a function of E_b/N_0 for various values of M. The upper bounds plotted in this figure are based on (8.48) and can be seen to be very tight at high SNR.

For nonrectangular QAM signal constellations, the error probability can be upper-bounded by using the following union bound:

$$P[\text{error}] < (M-1)Q\left(\sqrt{\Delta^2/2N_0}\right), \tag{8.49}$$

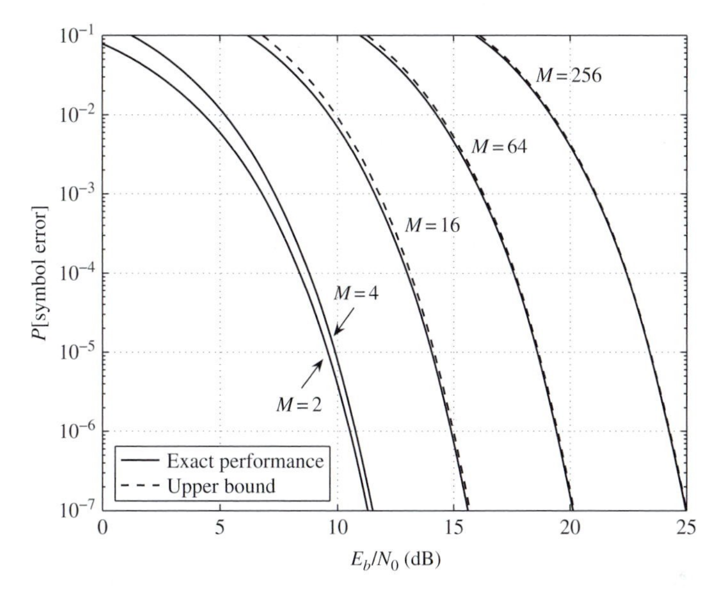

Fig. 8.20 Probability of symbol error for M-QAM signaling.

where Δ is the minimum Euclidean distance between signal points. This bound may be quite loose when M is large. In such a case, one may approximate P[error] by replacing $(M-1)$ by M_n, where M_n is the largest number of neighboring points that are at distance Δ from any constellation point, i.e., $P[\text{error}] < M_n Q(\sqrt{\Delta^2/2N_0})$.

Since both QAM and PSK are two-dimensional signal sets, it is of interest to compare their error performance for any given signal size M. Recall that for M-PSK, the probability of a symbol error is approximated as

$$P[\text{error}] \approx Q\left(\sqrt{\frac{2\lambda E_b}{N_0}} \sin\frac{\pi}{M}\right). \tag{8.50}$$

For M-QAM, one may use the expression (8.48). Since the error probability is dominated by the argument of the Q function, one may simply compare the squared arguments of $Q(\cdot)$ for the two modulation formats. The ratio of these two arguments is

$$\kappa_M = \frac{3/(M-1)}{2\sin^2(\pi/M)}. \tag{8.51}$$

When $M = 4$, $\kappa_M = 1$, which is not surprising since 4-PSK and 4-QAM are the same modulation format. However, when $M > 4$ it is seen that $\kappa_M > 1$, which means that M-QAM yields better performance than M-PSK. Table 8.2 illustrates the SNR advantage of QAM over PSK for several values of M. For example, one observes that 64-QAM has about 10 decibels SNR advantage over 64-PSK. The advantage, of course, is gained at the expense of increased sensitivity to amplitude and phase degradation in the transmission.

Table 8.2 SNR advantage of M-QAM over M-PSK

M	$10\log_{10}\kappa_M$
8	1.65 dB
16	4.20 dB
32	7.02 dB
64	9.95 dB
256	15.92 dB
1024	21.93 dB

8.6 M-ary coherent frequency-shift keying (M-FSK)

M-ary frequency-shift keying takes an essentially different approach to the previously discussed modulations. It gives each individual signal its own orthogonal axis and gains real estate in this manner. This, is done however, at the expense of bandwidth.

The signal set is given by

$$s_i(t) = \begin{cases} V\cos(2\pi f_i t), & 0 \le t \le T_s \\ 0, & \text{elsewhere} \end{cases}, \quad i = 1, 2, \ldots, M, \tag{8.52}$$

where the frequencies are chosen so that the signals are orthogonal over the interval $[0, T_s]$. A possible choice with minimum separation between the frequencies is

$$f_i = (k \pm i)\left(\frac{1}{T_s}\right), \qquad i = 0, 1, 2, \ldots \tag{8.53}$$

or

$$f_i = (k \pm i)\left(\frac{1}{2T_s}\right), \qquad i = 0, 1, 2, \ldots. \tag{8.54}$$

In the first case the signals are "noncoherently" orthogonal and in the second case they are "coherently" orthogonal. The energy in each signal is $E_s = V^2 T_s/2$. Since the signals are orthogonal they represent a natural choice for the orthonormal basis; all that is required is for the energy to be normalized to unity. Thus

$$\phi_i(t) = \frac{s_i(t)}{\sqrt{E_s}}, \qquad i = 1, 2, \ldots, M. \tag{8.55}$$

A typical signal space diagram for $M = 3$ is shown in Figure 8.21.

For M-FSK, the optimum (minimum message error probability) receiver computes the M projections, $r_i = \int_0^{T_s} r(t)\phi_i(t)\mathrm{d}t$, and then uses the decision rule of (8.8). As usual, $\mathbf{r}(t) = s_i(t) + \mathbf{w}(t)$, $\mathbf{w}(t)$ being AWGN and therefore the $\mathbf{r}_i$ are Gaussian random variables, with a variance of $N_0/2$ and the mean determined by $s_i(t)$. And, of course, they

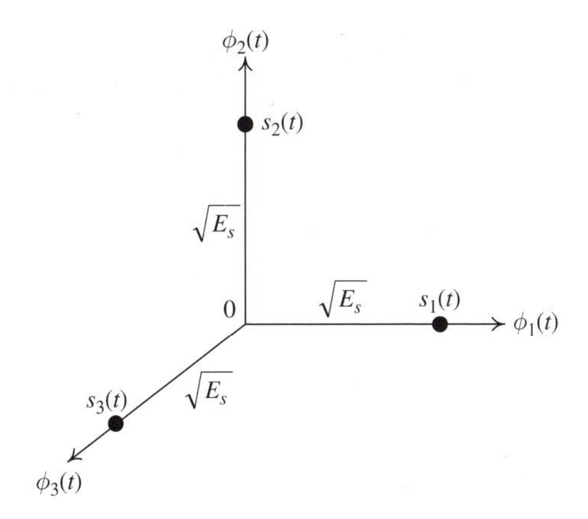

Fig. 8.21 Signal space plot for 3-FSK.

are also uncorrelated and statistically independent. If the messages are equally likely, i.e., $P[s_i(t)] = 1/M$, $i = 1, 2, \ldots, M$, then the receiver becomes that of the *minimum-distance receiver*. This is expressed as follows:

$$\text{choose } m_i \text{ if} \quad \sum_{k=1}^{M}(r_k - s_{ik})^2 < \sum_{k=1}^{M}(r_k - s_{jk})^2 \quad j = 1, 2, \ldots, M;\ j \neq i, \tag{8.56}$$

where $N = M$, i.e., the number of basis functions equals the number of signals.

Note that

$$s_{ik} = \begin{cases} 0, & k \neq i \\ \sqrt{E_s}, & k = i \end{cases} \tag{8.57}$$

and similarly

$$s_{jk} = \begin{cases} 0, & k \neq j \\ \sqrt{E_s}, & k = j \end{cases}. \tag{8.58}$$

Therefore the decision rule can be rewritten as

$$\text{choose } m_i \text{ if} \quad (r_i - \sqrt{E_s})^2 + r_j^2 < r_i^2 + (r_j - \sqrt{E_s})^2 \quad j = 1, 2, \ldots, M;\ j \neq i, \tag{8.59}$$

which reduces to

$$\text{choose } m_i \text{ if} \quad r_i > r_j, \quad j = 1, 2, \ldots, M;\ j \neq i. \tag{8.60}$$

In block diagram form the receiver looks as shown in Figure 8.22. Note that instead of $\phi_i(t)$ one may use $s_i(t)$. This only scales the correlator output by the factor $\sqrt{E_s}$ and does not in any way change the ordering of the sampled outputs. To determine the message error

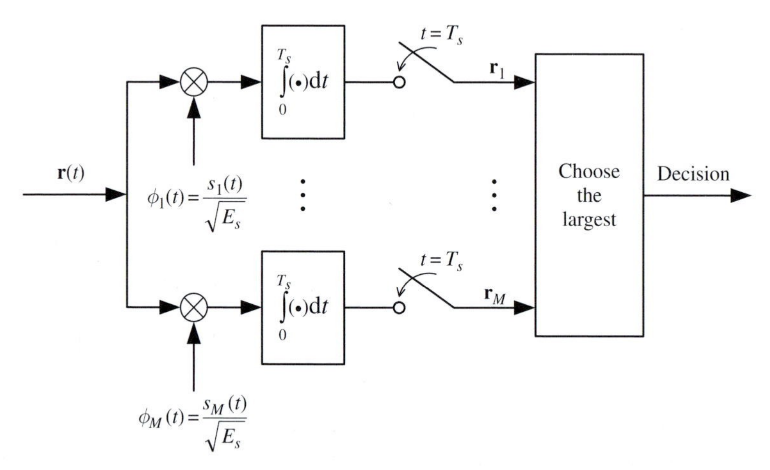

Fig. 8.22 Receiver implementation for M-FSK.

probability consider that message $s_1(t)$ is transmitted. Due to the symmetry of the signal space and because the messages are equally likely

$$P[\text{error}] = P[\text{error}|s_1(t)] = 1 - P[\text{correct}|s_1(t)]. \tag{8.61}$$

The conditional probability of being correct can be expressed as follows:

$$P[\text{correct}|s_1(t)] = \\ P[(\mathbf{r}_2 < \mathbf{r}_1) \text{ and } (\mathbf{r}_3 < \mathbf{r}_1) \text{ and } \ldots \text{ and } (\mathbf{r}_M < \mathbf{r}_1)|s_1(t) \text{ sent}]. \tag{8.62}$$

The sampled outputs $\mathbf{r}_1, \mathbf{r}_2, \ldots, \mathbf{r}_M$ are random variables and thus the above is concerned basically with the probability that one random variable is smaller than another. To proceed further, fix the random variable $\mathbf{r}_1$ at some specific value, say r_1. Then (8.62) can be rewritten as

$$P[\text{correct}|s_1(t)] = \\ \int_{r_1=-\infty}^{\infty} P[(\mathbf{r}_2 < r_1) \text{ and} \ldots \text{and } (\mathbf{r}_M < r_1)|\{\mathbf{r}_1 = r_1, s_1(t)\}]f(r_1|s_1(t))\mathrm{d}r_1, \tag{8.63}$$

where $f(r_1|s_1(t))\mathrm{d}r$ is interpreted as the probability that the random variable $\mathbf{r}_1$ is in the infinitesimal range r_1 to $r_1 + \mathrm{d}r_1$. Given $s_1(t)$, $f(r_1|s_1(t))$ is Gaussian, with mean $\sqrt{E_s}$ and variance $N_0/2$. Furthermore, given $\mathbf{r}_1 = r_1$, the random variables $\mathbf{r}_2$, $\mathbf{r}_3$, ..., $\mathbf{r}_M$ are statistically independent Gaussian random variables with zero mean and variance $N_0/2$. Therefore

$$P[(\mathbf{r}_2 < r_1) \text{ and} \ldots \text{and } (\mathbf{r}_M < r_1)|\{\mathbf{r}_1 = r_1, s_1(t)\}] = \\ \prod_{j=2}^{M} P[(\mathbf{r}_j < r_1)|\{\mathbf{r}_1 = r_1, s_1(t)\}]. \tag{8.64}$$

The term $P[\mathbf{r}_j < r_1|\{\mathbf{r}_1 = r_1, s_1(t)\}]$ is

$$P[\mathbf{r}_j < r_1|\{\mathbf{r}_1 = r_1, s_1(t)\}] = \int_{-\infty}^{r_1} \frac{1}{\sqrt{\pi N_0}} \exp\left\{-\frac{\lambda^2}{N_0}\right\} d\lambda. \tag{8.65}$$

It follows that

$$P[\text{correct}] = \int_{r_1=-\infty}^{\infty} \left[\int_{\lambda=-\infty}^{r_1} \frac{1}{\sqrt{\pi N_0}} \exp\left\{-\frac{\lambda^2}{N_0}\right\} d\lambda\right]^{M-1} \times \frac{1}{\sqrt{\pi N_0}} \exp\left\{-\frac{(r_1 - \sqrt{E_s})^2}{N_0}\right\} dr_1. \tag{8.66}$$

The above integral can only be evaluated numerically. It can be normalized so that only two parameters, namely M (the number of messages) and E_b/N_0 (the SNR per bit), enter into the numerical integration (see Problem 8.17) as

$$P[\text{error}] = 1 - \frac{1}{\sqrt{2\pi}} \int_{-\infty}^{\infty} \left[\frac{1}{\sqrt{2\pi}} \int_{-\infty}^{y} \exp\left(\frac{-x^2}{2}\right) dx\right]^{M-1} \times \exp\left[-\frac{1}{2}\left(y - \sqrt{\frac{2\log_2 ME_b}{N_0}}\right)^2\right] dy. \tag{8.67}$$

Due to the symmetry of the M-FSK constellation, all mappings from sequences of λ bits to signal points yield the same bit error probability. The exact relationship between the probability of bit error and the probability of symbol error for M-FSK can be found as follows. For equally likely signals, all the conditional error events, each conditioned on the transmission of a specific signal, are equiprobable and occur with probability $P[\text{symbol error}]/(M-1) = P[\text{symbol error}]/(2^{\lambda}-1)$. There are $\binom{\lambda}{k}$ ways in which k bits out of λ may be in error. Hence the average number of bit errors per λ-bit symbol is

$$\sum_{k=1}^{\lambda} k \binom{\lambda}{k} \frac{P[\text{symbol error}]}{2^{\lambda}-1} = \lambda \frac{2^{\lambda-1}}{2^{\lambda}-1} P[\text{symbol error}]. \tag{8.68}$$

The probability of bit error is simply the above quantity divided by λ, i.e.,

$$P[\text{bit error}] = \frac{2^{\lambda-1}}{2^{\lambda}-1} P[\text{symbol error}]. \tag{8.69}$$

Note that the ratio $P[\text{bit error}]/P[\text{symbol error}] = 2^{\lambda-1}/(2^{\lambda}-1)$ is precisely the ratio between the number of ways that a bit error can be made and the number of ways that a symbol error can be made. Furthermore, this ratio approaches 1/2 as $\lambda \to \infty$.

Figure 8.23 plots the exact symbol error probability of M-FSK as a function of E_b/N_0 for different values of M. In a completely opposite behavior as compared to M-ASK, M-PSK, and M-QAM, the required E_b/N_0 to achieve a given error probability decreases as M increases in M-FSK signaling. It should be pointed out that this happens at the expense

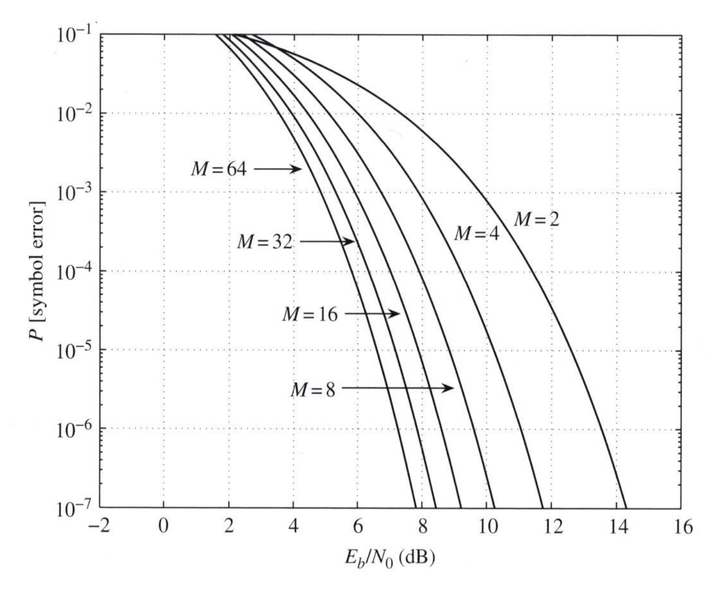

Fig. 8.23 Probability of symbol error for M-FSK signaling.

of a larger transmission bandwidth in order to accommodate a higher number of orthogonal carriers.

Although (8.67) and (8.69) provide *exact* symbol and bit error probabilities, respectively, for M-FSK, they do not clearly show how the error probabilities behave with SNR, E_b/N_0, nor with the constellation size, M. To overcome this difficulty, consider upper-bounding the error probability by using the union bound. The basic expression for the error is:

$$P[\text{error}] = P[(\mathbf{r}_1 < \mathbf{r}_2) \text{ or } (\mathbf{r}_1 < \mathbf{r}_3) \text{ or} \ldots \text{or } (\mathbf{r}_1 < \mathbf{r}_M)|s_1(t)]. \tag{8.70}$$

Since the events are not mutually exclusive, the error probability is obviously bounded by

$$\begin{aligned} P[\text{error}] &< P[(\mathbf{r}_1 < \mathbf{r}_2)|s_1(t)] \\ &\quad + P[(\mathbf{r}_1 < \mathbf{r}_3)|s_1(t)] + \cdots + P[(\mathbf{r}_1 < \mathbf{r}_M)|s_1(t)]. \end{aligned} \tag{8.71}$$

The quantity $P[(\mathbf{r}_1 < \mathbf{r}_2)|s_1(t)]$ is the same as the probability of making an error when only $s_1(t)$ and $s_2(t)$ are considered (i.e., binary FSK). This is given by (see Figure 8.24)

$$P[(\mathbf{r}_1 < \mathbf{r}_2)|s_1(t)] = Q\left(\sqrt{E_s/N_0}\right). \tag{8.72}$$

The other probabilities, $P[(\mathbf{r}_1 < \mathbf{r}_j)|s_1(t)]$, $j = 3, 4, \ldots, M$, are equal to $P[(\mathbf{r}_1 < \mathbf{r}_2)|s_1(t)]$. Therefore

$$P[\text{error}] < (M-1)Q\left(\sqrt{E_s/N_0}\right) < MQ\left(\sqrt{E_s/N_0}\right). \tag{8.73}$$

An even simpler expression is obtained by further applying the following upper bound on $Q(x)$:

$$Q(x) = \int_x^{\infty} \frac{1}{\sqrt{2\pi}} \mathrm{e}^{-\lambda^2/2}\, \mathrm{d}\lambda < \mathrm{e}^{-x^2/2}. \tag{8.74}$$

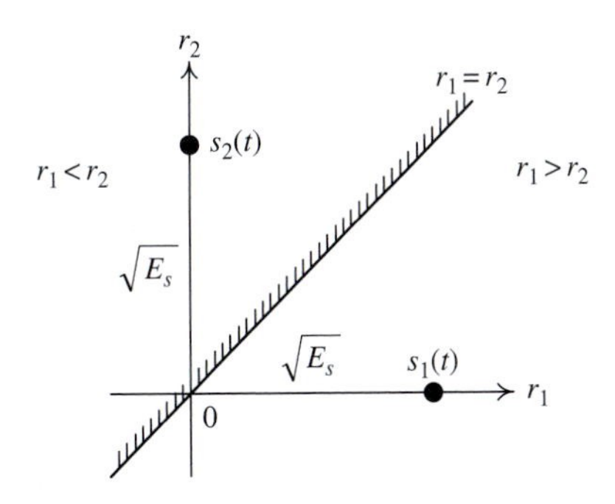

Fig. 8.24 Determining $P[(\mathbf{r}_1 < \mathbf{r}_2)|s_1(t)]$.

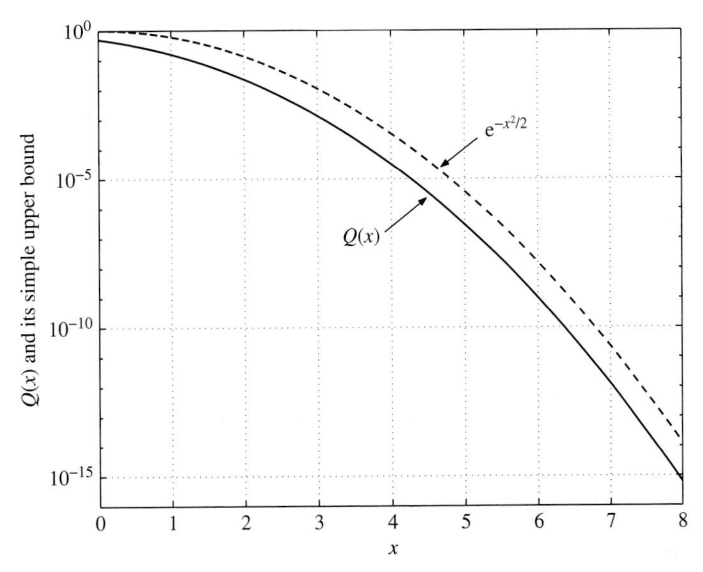

Fig. 8.25 Plot of $Q(x)$ and $e^{-x^2/2}$.

The above bound is graphically illustrated in Figure 8.25.

Therefore, the symbol error probability is upper-bounded by

$$P[\text{error}] < M e^{-E_s/(2N_0)}. \tag{8.75}$$

If the number of messages is a power of 2, i.e., $M = 2^\lambda$, where each message represents a pattern of λ bits occurring at a rate of $r_b = 1/T_b$ (bits/second), then $M = e^{\lambda \ln 2}$ and $E_s = \lambda E_b$. Thus the error probability can be expressed as

$$P[\text{error}] < e^{\lambda \ln 2} e^{-\lambda E_b/(2N_0)} = e^{-\lambda(E_b/N_0 - 2\ln 2)/2}. \tag{8.76}$$

Observe that, as $\lambda \to \infty$, or equivalently, as $M \to \infty$, the probability of error *approaches zero* exponentially, *provided that*

$$\frac{E_b}{N_0} > 2\ln 2 = 1.39 = 1.42 \text{ decibels}. \tag{8.77}$$

The upper bound on the probability of error given by (8.76) implies that, as long as SNR per bit > 1.42 decibels, we can achieve *an arbitrarily low probability of error*!

A different interpretation of the upper bound in (8.76) can be obtained as follows. Since $E_s = \lambda E_b = V^2 T_s/2$, (8.76) can be rewritten as

$$P[\text{error}] < e^{\lambda \ln 2} e^{-V^2 T_s/(4N_0)} = e^{-T_s[-r_b \ln 2 + V^2/(4N_0)]}. \tag{8.78}$$

The above implies that if $-r_b \ln 2 + V^2/(4N_0) > 0$, or $r_b < V^2/(4N_0 \ln 2)$ the probability or error tends to zero as T_s or M becomes larger and larger. This behavior of the error probability is surprising since what it shows is that, provided the bit rate, r_b, is small enough, the error probability can be made arbitrarily small even though the SNR is finite. Equivalently the transmitter power can be finite and still one can achieve as small an error as desired when the bit stream is transmitted by means of M-FSK.

8.7 Comparison of M-ary signaling techniques

Important parameters of a signaling (modulation/demodulation) technique include: (i) transmission bit rate, r_b (bits/second), (ii) bandwidth requirement, W (hertz), (iii) error performance (bit or symbol error probability), and (iv) transmitted power (usually quantified by the SNR per bit, E_b/N_0) to achieve a certain error performance. To have a meaningful comparison of different passband signaling techniques described in this chapter, these important parameters need to be taken into account. In fact, this has been partially done in the comparisons between M-PSK and BPSK in Section 8.4, and between M-QAM and M-PSK in Section 8.5.

A more compact and meaningful comparison of different modulation techniques is the one based on the bit-rate-to-bandwidth ratio, r_b/W (bits per second per hertz of bandwidth) versus the SNR per bit (E_b/N_0) required to achieve a given error probability. The ratio r_b/W, commonly called the *normalized* bit rate, measures the *bandwidth efficiency* of a signaling scheme. Let us determine this ratio for different signaling techniques.

Since M-ASK is amplitude modulation, the bandwidth-efficient transmission method is *single-sideband* (SSB), which is identical to SSB amplitude modulation in analog communications. With SSB transmission, the required channel bandwidth is approximately equal to half of the reciprocal of the symbol duration T_s, i.e., $W = 1/2T_s$. Since $T_s = \lambda T_b = \lambda/r_b = \log_2 M/r_b$, it follows that

$$W = \frac{r_b}{2\log_2 M}. \tag{8.79}$$

The above expression shows that, when the bit rate r_b is fixed, the channel bandwidth required decreases as M is increased. Consequently, the bandwidth efficiency of SSB-ASK increases with M as follows:

$$\left(\frac{r_b}{W}\right)_{\text{SSB-ASK}} = 2\log_2 M \quad \text{((bits/second)/hertz)}. \tag{8.80}$$

For M-PSK ($M > 2$), the signals must be transmitted via *double sidebands*. Therefore the required channel bandwidth is $W = 1/T_s$. Similarly to the case of M-ASK, one has $T_s = \log_2 M/r_b$. It then follows that

$$\left(\frac{r_b}{W}\right)_{\text{PSK}} = \log_2 M \quad \text{((bits/second)/hertz)}, \tag{8.81}$$

which is a factor of 2 less than that of M-ASK with SSB transmission.

In the case of (rectangular) QAM, since the transmitted signal consists of two *independent* ASK signals on orthogonal quadrature carriers, the transmission rate is twice that of ASK. However, like M-PSK the QAM signals must be transmitted via double sidebands. Consequently, QAM and SSB-ASK have the same bandwidth efficiency.

Orthogonal M-FSK has a totally different bandwidth requirement, and hence a totally different bandwidth efficiency. If the orthogonal carriers in M-FSK have the minimum frequency separation of $1/2T_s$, the bandwidth required for transmission of $\lambda = \log_2 M$ information bits is approximately

$$W = \frac{M}{2T_s} = \frac{M}{2(\lambda/r_b)} = \frac{M}{2\log_2 M} r_b. \tag{8.82}$$

The above expression shows that, for a fixed transmission rate r_b, the bandwidth increases as M increases. As a consequence, the bandwidth efficiency of M-FSK reduces in the following manner as M increases:

$$\left(\frac{r_b}{W}\right)_{\text{FSK}} = \frac{2\log_2 M}{M}. \tag{8.83}$$

Figure 8.26 shows the plots of r_b/W versus E_b/N_0 for SSB-ASK, PSK, QAM, and FSK when $P[\text{symbol error}] = 10^{-5}$. Observe that, in the cases of ASK, PSK, and QAM, increasing M results in a higher bandwidth efficiency r_b/W. However, the cost of achieving the higher data rate per unit of bandwidth is an increase in SNR per bit. Consequently, these modulation techniques are appropriate for communication channels that are *bandwidth-limited*, where it is desired to have a bit-rate-to-bandwidth ratio $r_b/W > 1$ and where there is sufficiently high SNR to support increases in M. Telephone channels and digital microwave channels are examples of such bandwidth-limited channels.

In contrast, M-FSK modulation provides a bit-rate-to-bandwidth ratio of $r_b/W \leq 1$. As M increases, r_b/W decreases due to the larger increase in required channel bandwidth. However, the SNR per bit required to achieve a given error probability (in the case of Figure 8.26, $P[\text{symbol error}] = 10^{-5}$) decreases as M increases. Therefore, the M-FSK signaling technique is appropriate for *power-limited* channels that have sufficiently large bandwidth to accommodate a large number of signals but cannot afford a large SNR per bit. For the case of M-FSK, as $M \to \infty$, the error probability can be made as small as desired, provided that $E_b/N_0 > 1.39 = 1.42$ decibels (see (8.77)).

Note that also shown in Figure 8.26 is the graph for the normalized *channel capacity* of the bandlimited AWGN channel, which is due to Shannon [1]. The ratio C/W, where C is the capacity in bits/second, represents the *highest achievable* bit-rate-to-bandwidth ratio on this channel, an AWGN channel. Hence, it serves as the *upper bound* on the bandwidth efficiency of any modulation technique. This bound is discussed in more detail in the next section.

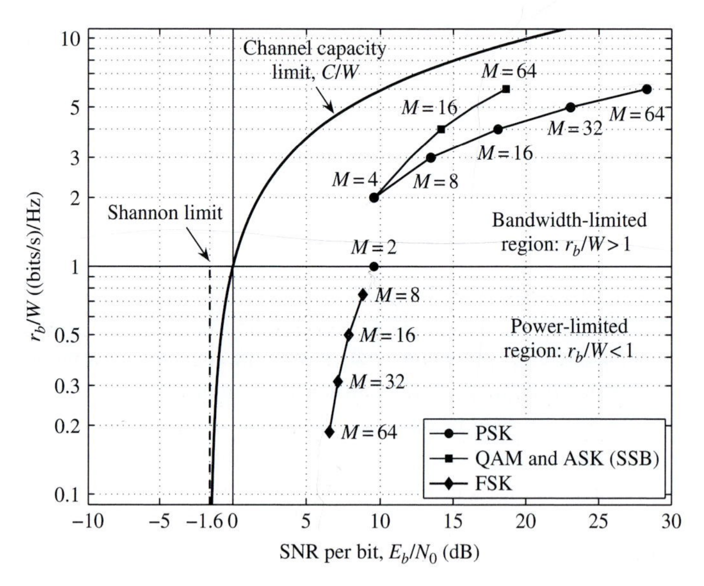

Fig. 8.26 Comparison of different M-ary signaling techniques at a *symbol error probability* of 10^{-5}. Note that the values of M shown next to square markers are for QAM. For SSB-ASK, take the square roots of these values.

8.8 Shannon channel capacity theorem

Shannon [1] showed that the system capacity (in bits/second) of an AWGN channel is a function of the average signal power P_{av}, the average noise power, and the bandwidth W. The Shannon capacity theorem can be stated as

$$C = W \log_2 \left(1 + \frac{P_{\text{av}}}{W N_0} \right), \tag{8.84}$$

where W is in hertz and, as usual, $N_0/2$ is the two-sided PSD of the noise.

Shannon proved that it is theoretically possible to transmit information over such a channel at any rate r_b, as long as $r_b \leq C$, with an *arbitrarily small* error probability by using a sufficiently complicated modulation scheme. For an information rate $r_b > C$, it is not possible to find a modulation that can achieve an arbitrarily small error probability. Thus Shannon's work showed that the values of P_{av}, N_0, and W *set a limit on transmission rate, not on error probability*!

It follows from (8.84) that the normalized channel capacity C/W ((bits/second)/hertz) is given by

$$\frac{C}{W} = \log_2 \left(1 + \frac{P_{\text{av}}}{W N_0} \right). \tag{8.85}$$

It is instructive to express the normalized channel capacity as a function of the SNR per bit. Since P_{av} represents the average transmitted power and C is the rate in bits/second, one has $P_{av} = CE_b$, where E_b is the energy per bit. Hence (8.85) can be expressed as

$$\frac{C}{W} = \log_2\left(1 + \frac{C}{W}\frac{E_b}{N_0}\right). \tag{8.86}$$

Consequently,

$$\frac{E_b}{N_0} = \frac{2^{C/W} - 1}{C/W}. \tag{8.87}$$

This relation is plotted in Figure 8.26.

An important observation from (8.87) is that there exists a limiting value of E_b/N_0 below which there can be no error-free communication at any information rate (and no matter how much bandwidth one may have). This is shown below.

Let $x = (C/W)(E_b/N_0)$. Then from (8.87) one has

$$\frac{C}{W} = x \log_2(1 + x)^{1/x} \tag{8.88}$$

and

$$\left(\frac{E_b}{N_0}\right)^{-1} = \log_2(1 + x)^{1/x}. \tag{8.89}$$

Now apply the identity $\lim_{x \to 0}(1 + x)^{1/x} = \mathrm{e}$ to (8.89). It follows that when $C/W \to 0$, one has

$$\frac{E_b}{N_0} = \frac{1}{\log_2 \mathrm{e}} = \ln 2 = 0.693 = -1.6 \text{ decibels}. \tag{8.90}$$

This value of E_b/N_0 is called the *Shannon limit*.

Finally, it should be mentioned that Shannon's work provides a theoretical proof for the *existence* of a coding/modulation technique that could achieve the channel capacity. It does not tell us how to construct such coding/modulation. For example, for a bit error probability of 10^{-5}, BPSK[3] modulation requires an E_b/N_0 of 9.6 decibels in order to transmit 1 ((bits/second)/hertz). For this spectral efficiency, Shannon's proof promises the existence of a theoretical improvement of 9.6 decibels over the performance of BPSK, through the use of some coding/modulation techniques. Today, most of that promised improvement has been realized with *turbo codes* [2], *low-density parity check (LDPC) codes* [3], and the *iterative processing principle*.

8.9 Summary

Basic M-ary modulation methods have been described and analyzed in this chapter. Bandwidth reduction is the main motivation for M-ary modulation and accounts for the popularity of QAM modulation. The reduction eliminates or at least mitigates intersymbol interference

[3] BPSK is the optimum (uncoded) binary modulation technique.

(ISI) effects which occur with bandlimited channels. Bandlimited channels are the subject of the next chapter.

However, even the M-ary modulation of FSK, which involves an exponential expansion of bandwidth, has important applications. It is the basis of the modulation technique known as OFDM (orthogonal frequency-division multiplexing) used over channels where fading, amplitude and phase distortion, are experienced. Typically it is combined with CDMA (code-division multiple access). CDMA is a topic in a later chapter on advanced modulation techniques where TCM (trellis-coded modulation), a very bandwidth efficient modulation, is also discussed.

8.10 Problems

As mentioned in the text, the most important neighbors are the nearest ones. The first set of problems explores how to ensure that nearest neighbors differ as little as possible. This might make for a dull neighborhood but it does reduce bit error probability. It is also shown just how influential the near neighbors are. Finally, the problems explore how for certain important M-ary constellations one can demodulate directly to the bits.

8.1 (*Gray coding*) Two different methods of generating a Gray code (mapping) are presented. The first can be called an "inductive" technique. The second a direct method.

(a) *Inductive technique* To Gray code a single bit is, of course, trivial. It is

$$\frac{0}{1}.$$

However, use this as a starting point to *Gray* code two-bit sequences as

$$\frac{\frac{00}{01}}{\frac{11}{10}}.$$

With this pattern in mind, what is the Gray code for three-bit sequences? For the ambitious reader, what is the Gray code for four-bit sequences?

(b) *Direct method* Arrange the 2^n n-bit sequences in *natural order* from $00\cdots0$ to $11\cdots1$. Then map each n-bit sequence $a_1, a_2, \ldots, a_n$ to a corresponding Gray sequence, $b_1, b_2, \ldots, b_n$ as follows. Let $b_1 = a_1$ and $b_k = a_k \oplus a_{k-1}$, where $\oplus$ is an *exclusive or* operation. Apply the method to obtain a Gray code for five-bit sequences. Note you should only need to do the algebra for the first half, i.e., sequences 00000 to 01111, the other half of the table should be simple to fill in. But describe how you would do it.

(c) Which approach do you prefer and why? Can one adapt the inductive technique to produce a Gray code "as easily" as the direct method, whatever easy means.

8.2 To explore the influence of nearest neighbors consider the simple 4-ASK modulation with Gray mapping shown in Figure 8.27. The probability of symbol error is

$$P[\text{symbol error}] = \frac{2(M-1)}{M} Q\left(\frac{\Delta}{\sqrt{2N_0}}\right).$$

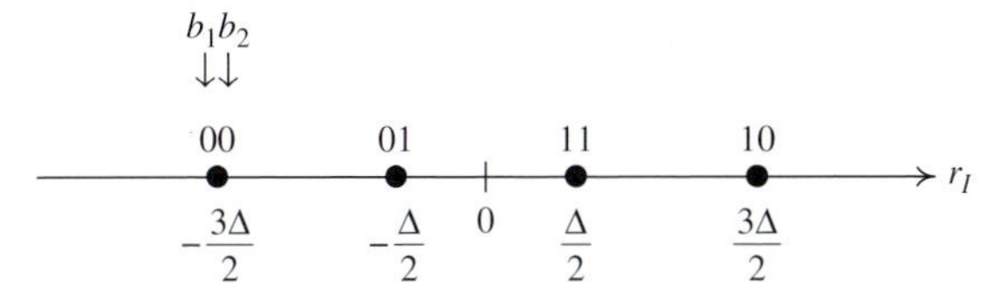

Fig. 8.27 Signal constellation of 4-ASK with a Gray mapping.

(a) Determine the ratio $\Delta/\sqrt{2N_0}$ (which is related to the SNR $= E_b/N_0$) so that $P[\text{symbol error}] = 10^{-1},\ 10^{-2},\ 10^{-3},\ 10^{-4}$.

(b) For each $P[\text{symbol error}]$ determine the following:

$$Q\left(\frac{\Delta}{\sqrt{2N_0}}\right),\ Q\left(\frac{3\Delta}{\sqrt{2N_0}}\right),\ Q\left(\frac{5\Delta}{\sqrt{2N_0}}\right).$$

(c) Consider that symbol 00 was transmitted. Determine the following error probabilities:

$$P\left[\{01\}_D | \{00\}_T\right],\ P\left[\{11\}_D | \{00\}_T\right],\ P\left[\{10\}_D | \{00\}_T\right].$$

(d) How would the answer in (c) change if one of the other symbols was chosen to be the transmitted one?

Therefore, in terms of neighbors the most important ones are the ________________ ones.

8.3 Though in general it is difficult to determine bit error probabilities for M-ary modulation, it is very feasible for the constellation shown in Figure 8.27.

(a) Determine the bit error probability for the constellation shown in Figure 8.27.

(b) Determine the *individual* bit error probabilities, i.e., $P[b_1 \text{ error}]$, $P[b_2 \text{ error}]$. Compare the three bit error probabilities, as well as the approximate expression for bit error probability in Equation (8.27). Comment.

8.4 Given r_I, the sufficient statistic, and the constellation in Figure 8.27 one can readily demodulate directly to the bits b_1, b_2. As an example, the decision rule for bit b_1 is

$$r_I \underset{b_1=0}{\overset{b_1=1}{\gtrless}} 0.$$

(a) What is the corresponding decision rule for bit b_2?

(b) Based on these decision rules, determine $P[\text{bit error}]$, $P[b_1 \text{ error}]$, and $P[b_2 \text{ error}]$. Compare with the answers in Problem 8.3.

8.5 Instead of Gray coding let the mapping be a *natural* one, i.e., $-3\Delta/2 \leftrightarrow 00$, $-\Delta/2 \leftrightarrow 01$, $\Delta/2 \leftrightarrow 10$, $3\Delta/2 \leftrightarrow 11$. Repeat Problems 8.3 and 8.4 for this mapping and compare the results.

8.6 Consider 8-ASK with Gray coding. As done for 4-ASK in Problem 8.4:
(a) Develop a set of rules that would demodulate directly to the bits b_1, b_2, b_3.
(b) Determine P [bit error], P [b_i error], $i = 1, 2, 3$ and compare the results.

8.7 (*Two-dimensional Gray coding*) Consider *rectangular* QAM and develop a Gray mapping technique for the bit sequences. *Hint* Recall rectangular QAM is in essence two separate ASK modulations. Illustrate the technique for 16-QAM.

8.8 Consider 16-QAM with Gray coding.
(a) Develop a set of decision rules that will directly demodulate to the bits b_1, b_2, b_3, b_4.
(b) Determine P [bit error], P [b_i error], $i = 1, 2, 3, 4$.
(c) If you have an application where one bit is deemed to be more valuable than the others, how would you exploit the result of (b)?

8.9 (*8-ary constellations*) Consider the four 8-ary signal constellations in Figure 8.28, where all the signal points in each constellation are equally probable.
(a) Compute the average energies for the four constellations and rank the signal constellations in terms of energy efficiency.

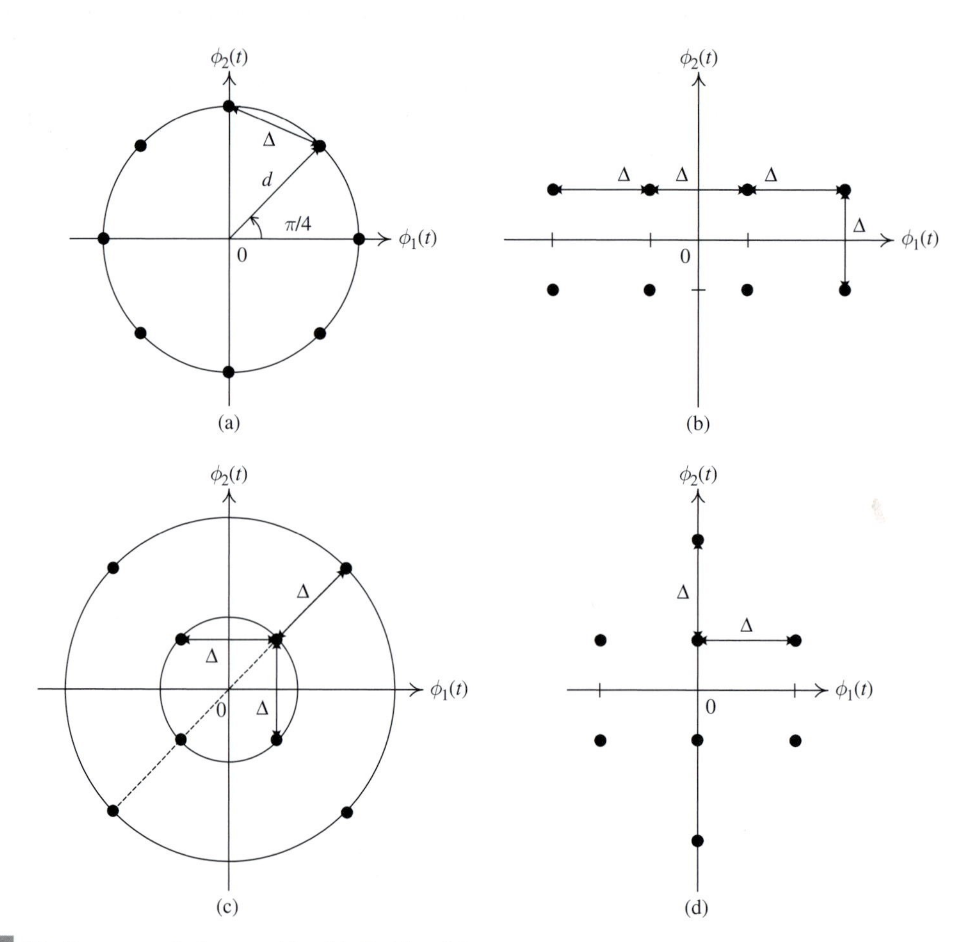

Fig. 8.28 The 8-ary constellations considered in Problem 8.9.

(b) Specify Gray mapping for the constellation in Figure 8.28(b).

(c) Draw the minimum-distance decision boundaries for the signal constellation in Figure 8.28(d). Which signals in this constellation are *most* susceptible to error and why?

(d) How does the best constellation found in (a) compare with the best 8-ary constellation in Figure 8.16.

8.10 (*V.29 constellation*) The 16-QAM signal constellation shown in Figure 8.29 is an international standard for telephone-line modems, called V.29.

(a) Ignoring the four corner points at $(\pm 1.5\Delta, \pm 1.5\Delta)$, specify a Gray mapping of the constellation.

(b) Assume that all the 16 signal points are equally likely. Sketch the optimum decision boundaries of the minimum-distance receiver.

8.11 (*16-QAM constellations*) Figure 8.30 shows two 16-QAM constellations.

(a) What can you say about the error performance of the two constellations? Which constellation is more energy-efficient? Explain.

(b) Specify a Gray mapping for constellation (b).

(c) Draw the minimum-distance decision boundaries for constellation (b). Which signals in this constellation are *least* susceptible to error and why?

8.12 (*8-QAM*) You are asked to design an 8-QAM modulator with a *peak* energy constraint of E_s joules. Due to its regular structure you decide on the signal constellation shown in Figure 8.31. The signal set is such that the signal points are symmetrical about the $\phi_1(t)$ and $\phi_2(t)$ axes.

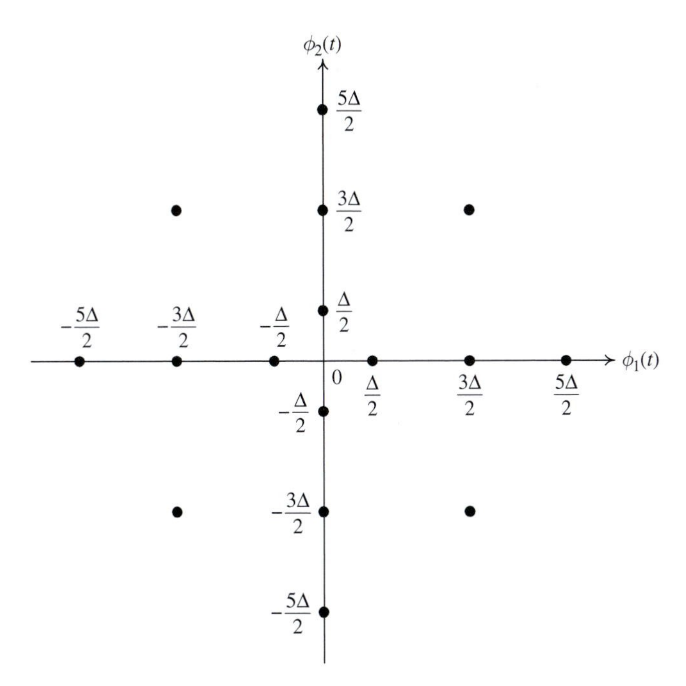

Fig. 8.29 V.29 constellation.

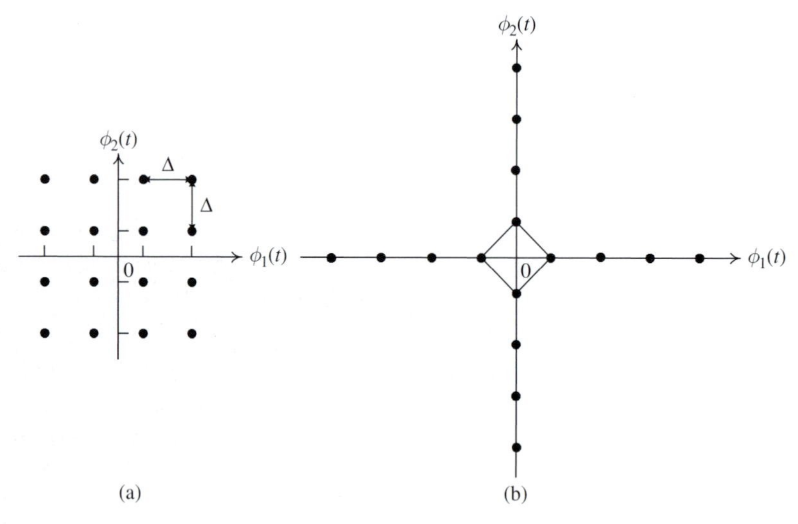

Fig. 8.30 The two 16-QAM constellations considered in Problem 8.11.

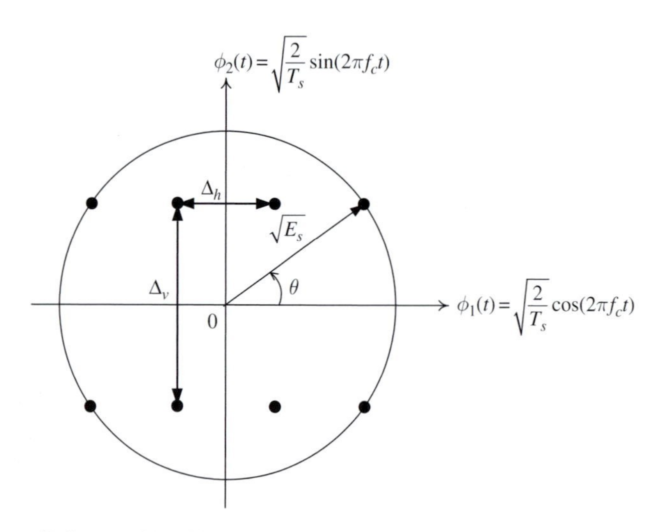

Fig. 8.31 The 8-QAM constellation considered in Problem 8.12.

(a) Determine the angle θ so that the minimum Euclidean distance, $d_{\min}$, is *maximized*.

(b) Specify a Gray mapping for the signal constellation obtained in (a).

(c) How does the constellation obtained in (a) compare with the 8-PSK constellation with the same peak energy E_s in terms of error performance? Explain.

8.13 (*Another design of 8-QAM*) You have been asked to design a modulation scheme for a communication system, and to conserve bandwidth it has been decided to use an 8-QAM constellation. Unhappy with the 8-PSK and 8-QAM you have learned because you feel that they do not use the available energy very efficiently, you decide

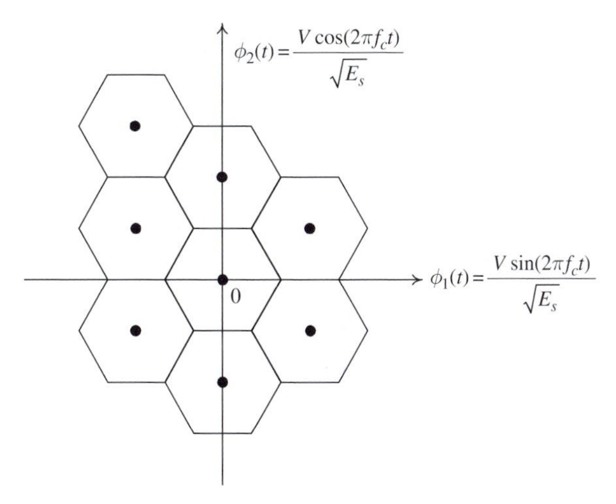

Fig. 8.32 The 8-QAM constellation considered in Problem 8.13.

to attempt a different signal constellation. Inspired by a tile design you notice in the local shopping mall, you propose the signal constellation in Figure 8.32.

Assume that the side of each hexagon is of length Δ. Determine:

(a) The *minimum* distance between the signals in terms of Δ.
(b) The average transmitted energy per bit in terms of Δ.
(c) Assuming that all the signal points are equally likely, draw the decision regions of the minimum-error-probability receiver.
(d) Is it possible to do a Gray mapping for this constellation? Explain.

8.14 (*16-QAM in DVB-S2*) The 16-QAM constellation labeled with (4,12) in Figure 8.17 is used in the DVB-S2 (Digital Video Broadcasting via Satellite–Second Generation) standard. Since the constellation combines ASK and PSK, it is widely known as 16-APSK. There are two concentric circles, where four signals are equally spaced on the inner circle and twelve signals are equally spaced on the outer circle. Assume that the radii R_1 and R_2 are adjusted so that the minimum distances on both inner and outer circles are Δ.

(a) Determine the *average* symbol energy of this 16-APSK constellation in terms of Δ.
(b) Your friend is not totally convinced about the performance of the DVB-S2 16-APSK constellation over an AWGN channel. She thus proposes to use the constellation labeled with (8,8) in Figure 8.17, where there are eight signal points equally spaced on both inner and outer circles. Note that the radii R_1 and R_2 of this constellation are also adjusted so that the distance Δ between the adjacent signal points on the inner circle is equal to the distance between the nearest points on the two circles (note the equilateral triangles). Determine the *average* symbol energy of this (8,8) constellation and compare it with that of the DVB-S2 16-APSK constellation. Comment.
(c) Is it possible to perform a Gray mapping for the (8,8) or (4,12) constellation? Explain.

8.15 Consider a rectangular M-QAM constellation where $M = 2^\lambda$. Assume that λ_I bits independently modulate the inphase carrier, $\cos(2\pi f_c t)$, while $\lambda_Q = \lambda - \lambda_I$ bits independently modulate the quadrature carrier, $\sin(2\pi f_c t)$, by means of ASK.
   (a) Obtain the exact symbol error probability in terms of the average energy per bit E_b and the two-sided PSD $N_0/2$ of AWGN.
   (b) Obtain a union upper bound on the symbol error probability.

8.16 Consider a rectangular M-QAM constellation where $M = 2^\lambda$. Verify the upper bound on the symbol error probability in (8.48).

8.17 Verify that (8.66) can be written as (8.67).

8.18 (*Hypercube constellation*) Consider an M-ary signal constellation where $M = 2^\lambda$ and λ is the dimension of the signal space. Suppose that the M signals lie on the vertices of a hypercube that is centered at the origin. Determine the symbol error probability as a function of E_b/N_0, where E_b is the energy per bit and $N_0/2$ is the two-sided PSD of the AWGN. Also assume that all the signals are equally probable.

8.19 (*Biorthogonal modulation*) An M-ary *biorthogonal* signal constellation is obtained from a set of $M/2$ orthogonal signals (such as $M/2$-FSK), $\{s_i(t)\}_{i=1}^{M/2}$, and their negatives, $\{-s_i(t)\}_{i=1}^{M/2}$. As an example, a 4-QAM (or a QPSK) is a 4-ary biorthogonal modulation.
   (a) What is the transmission bandwidth requirement of an M-ary biorthogonal set compared to that of an M-ary orthogonal set? Explain.
   (b) Also assume that all the signals are equally probable. Obtain the expression of the decision rule for the minimum-error-probability receiver of the M-ary biorthogonal modulation.
   (c) Based on (b), obtain the expression to compute the exact symbol error probability of M-ary biorthogonal modulation in terms of E_b/N_0, where E_b is the energy per bit and $N_0/2$ is the two-sided power spectral density of the AWGN.

8.20 (*SSB transmission of BPSK*) In determining the bandwidth efficiency of ASK signaling, SSB transmission was assumed. This problem examines the transmitted waveform of BPSK with SSB (note that BPSK can be viewed as an amplitude modulation).
   (a) Assume that the information bit sequence is $\{0, 1, 1, 0, 1, 0\}$. Use Matlab to obtain and plot the BPSK waveform with upper single sideband (USSB). *Remark* Refer to Figure 2.42 for the block diagram of a USSB modulator. In Matlab the Hilbert transform can be performed with the command `hilbert`. To make the plot visible, let $f_c = T_b/4$, where T_b is the bit interval.
   (b) How does the USSB waveform in (a) compare to the conventional BPSK waveform? Does it still have a constant envelope?

8.21 (*Bandwidth efficiency*)
   (a) Consider the bandwidth–power plane of Figure 8.26. Determine the locations of 2-FSK and 4-FSK schemes. For 4-FSK, estimate the E_b/N_0 value from Figure 8.23.
   (b) Obtain the plots similar to Figure 8.26 but for $P[\text{error}] = 10^{-2}$ and $P[\text{error}] = 10^{-7}$. Comment.

In the 1200s a major philosophical debate was concerned with the question, "How many angels can fit on the head of a pin?" Variants on this question are: "How many angels can dance on the head of a pin?" and "How many angels can dance on the point of a needle?" No less a person than St. Thomas Aquinas was involved in the debate. However, in the twentieth century a more relevant question for communication engineers became: "How many orthogonal signals of time duration T seconds can one fit in a bandwidth of W hertz?" The answer, $2WT + 1 \approx 2WT$, was provided by Landau, Pollack, and Slepian [4, 5, 6]. They used functional analysis to obtain the result. The next problem provides a very heuristic derivation of the answer.

8.22 Consider a signal bandlimited to W hertz and of duration T seconds.

(a) What is the minimum number of samples per unit time needed to represent the signal?

(b) In the time interval T, based on (a), how many independent time samples are there?

(c) But the time signal can also be represented by a set of orthogonal time functions (which are linearly independent). Reason that the number of orthogonal functions needed is $2WT$.

The difficulty in the above argument is that a time-limited signal cannot be bandlimited and vice versa. One needs to have a reasonable definition of bandwidth. Landau, Pollack, and Slepian used a fractional out-of-band *energy definition. Here we use a* null-to-null *bandwidth.*

(d) Consider a rectangular pulse centered at the origin and of duration T seconds. What is the *null-to-null* bandwidth after it is modulated up to f_c hertz, i.e., of the signal

$$s(t) = V\left[u\left(t + T/2\right) - u\left(t - T/2\right)\right]\cos\left(2\pi f_c t\right).$$

Based on this show that the number of orthogonal functions is four.

8.23 (*Q^2PSK*) QPSK has two orthogonal functions, time duration T and *null-to-null* bandwidth $2/T$ hertz. The previous question promises four possible orthogonal functions. This is what *quadrature–quadrature phase-shift keying* (Q^2PSK) achieves.

Consider the following basis set:

$$\left\{\begin{array}{l} \phi_1'(t) = \cos\left(\pi t/T\right)\cos\left(2\pi f_c t\right) \\ \phi_2'(t) = \sin\left(\pi t/T\right)\cos\left(2\pi f_c t\right) \\ \phi_3'(t) = \cos\left(\pi t/T\right)\sin\left(2\pi f_c t\right) \\ \phi_4'(t) = \sin\left(\pi t/T\right)\sin\left(2\pi f_c t\right) \end{array}\right., \quad |t| \le T/2 \tag{P8.1}$$

and $\phi_i'(t) = 0$, $i = 1, 2, 3, 4$, for $|t| > T/2$; $f_c = k/2T$, k integer and typically $\gg 1$. Note that the basis set is a combination of two orthogonal carriers, $c_1(t) = \cos(2\pi f_c t)$, $c_2(t) = \sin(2\pi f_c t)$ and two orthogonal *pulse shaping functions*, $p_1(t) = \cos(\pi t/T)$,

$p_2(t) = \sin(\pi t/T)$. This is somewhat reminiscent of MSK but there is no time offset.

(a) Show that $\{\phi_i^{'}(t)\}_{i=1}^{4}$ form a set of orthogonal functions.

(b) Normalize the set so that each basis function has unit energy over the T-second interval.

(c) The four orthogonal basis functions suggest one can use an individual bit to modulate each basis function. This is what is done in Q^{2}PSK. The binary input is split into four bit streams. Draw a block diagram of the Q^{2}PSK modulator. If T_b is the bit interval of the primary source, what is the symbol interval T_s? And how are they related to T?

(d) Assuming that the energy per bit is E_b joules, draw the signal space (or state what it is). How many signals are there? What is the minimum distance between two signals in the constellation? How many nearest neighbors does a signal have?

8.24 (*Q^{2}PSK demodulation*) One has two choices for demodulation. Either demodulate to the signal (symbol) representing the four-bit sequence or directly to the individual bits.

(a) Draw a block diagram of each demodulator.

(b) For the symbol demodulator determine P[symbol error], and for the bit demodulator determine P[bit error]. Is the bit error probability of the symbol demodulator different from that of the bit demodulator? Compare the bit error probability with that of BPSK and QPSK/OQPSK/MSK.

8.25 (*Q^{2}PSK envelope*) The transmitted signal in an interval of T_s is $s(t) = b_1\sqrt{E_b}\phi_1(t) + b_2\sqrt{E_b}\phi_2(t) + b_3\sqrt{E_b}\phi_3(t) + b_4\sqrt{E_b}\phi_4(t)$, where $b_i = \pm 1,\ i = 1, 2, 3, 4,$ and the $\{\phi_i(t)\}_{i=1}^{4}$ are the properly normalized basis functions.

(a) Write $s(t)$ as $s(t) = e(t)\cos(2\pi f_c t + \theta(t))$. The envelope is of particular interest. Show that

$$e(t) = \left[2 + (b_1 b_2 + b_3 b_4)\sin\left(\frac{\pi t}{4T_b}\right)\right]^{1/2}. \tag{P8.2}$$

Plot $e(t)$ for the various combinations of $(b_1 b_2 + b_3 b_4)$.

(b) Since a constant envelope is beneficial when saturating nonlinearities are encountered, devise a coding scheme that would make it constant, i.e., make one of the bits dependent on the other bits. What is the price paid to have a constant envelope?

(c) When the coding of (b) is used, how many signals are there in the signal space? What is the $d_{\min}$ between signals?

(d) Draw a block diagram of a demodulator that demodulates directly to the bits. What is the bit error probability?

(e) Can the coding of (b) be used for *error detection*? If so, how?

8.26 (Q^2PSK *PSD*) To determine the PSD proceed as follows. Write the transmitted signal as

$$s(t) = \underbrace{\sqrt{E_b}\left\{\sum_{k=-\infty}^{\infty} b_{1k}\cos\left(\frac{\pi t}{4T_b}\right)\right\}}_{s_1(t)}\cos(2\pi f_c t)$$
$$+ \underbrace{\sqrt{E_b}\left\{\sum_{k=-\infty}^{\infty} b_{2k}\sin\left(\frac{\pi t}{4T_b}\right)\right\}}_{s_2(t)}\cos(2\pi f_c t)$$
$$+ \underbrace{\sqrt{E_b}\left\{\sum_{k=-\infty}^{\infty} b_{3k}\cos\left(\frac{\pi t}{4T_b}\right)\right\}}_{s_3(t)}\sin(2\pi f_c t)$$
$$+ \underbrace{\sqrt{E_b}\left\{\sum_{k=-\infty}^{\infty} b_{4k}\sin\left(\frac{\pi t}{4T_b}\right)\right\}}_{s_4(t)}\sin(2\pi f_c t). \quad \text{(P8.3)}$$

Since $\sqrt{E_b}$ does not affect the signal shape, we set it equal to 1. Further, the carriers $\cos(2\pi f_c t)$ and $\sin(2\pi f_c t)$ simply translate the PSD to lie around $\pm f_c$. Therefore we concentrate on the PSD of $s_1(t)$, $s_2(t)$, $s_3(t)$, $s_4(t)$.

(a) Under the assumption that the bits are *statistically independent* what is the crosscorrelation between any pair of the four signals $s_1(t)$, $s_2(t)$, $s_3(t)$, $s_4(t)$?

(b) Argue that $R_{s_1}(\tau) = R_{s_3}(\tau)$ and $R_{s_2}(\tau) = R_{s_4}(\tau)$.

(c) Based on (c), the PSD at baseband is $2\mathcal{F}\{R_{s_1}(\tau)\} + 2\mathcal{F}\{R_{s_2}(\tau)\}$. Determine the two PSDs. *Hint* Treat $s_1(t)$ generation as a train of impulses, weight ± 1, which is input into a filter with impulse response

$$h(t) = \cos\left(\frac{\pi t}{4T_b}\right)[u(t+2T_b) - u(t-2T_b)].$$

Similarly with $s_2(t)$.

(d) Plot the resultant PSD and comment.

References

[1] C. E. Shannon, "A mathematical theory of communications," *Bell System Technical Journal*, vol. 27, pp. 379–423, 623–657, 1948.

[2] C. Berrou, A. Glavieux, and P. Thitimajshima, "Near Shannon limit error-correcting coding and decoding: Turbo-codes," in *Proceedings of the IEEE International Conference on Communications*, pp. 1064–1070, May 1993.

[3] S. Chung, G. D. Forney Jr., T. J. Richardson, and R. Urbanke, "On the design of low-density parity-check codes within 0.0045 dB of the Shannon limit," *IEEE Commununications Letters*, vol. 5, pp. 58–60, Feb. 2001.

[4] D. Slepian and H. O. Pollak, "Prolate spheroidal wave functions, Fourier analysis, and uncertainty – I," *Bell System Technical Journal*, vol. 40, pp. 43–64, Jan. 1961.

[5] H. J. Landau and H. O. Pollak, "Prolate spheroidal wave functions, Fourier analysis and uncertainty – II," *Bell System Technical Journal*, vol. 40, pp. 65–84, Jan. 1961.

[6] D. Slepian and H. O. Pollak, "Prolate spheroidal wave functions, Fourier analysis, and uncertainty – III. The dimension of essentially time and band-limited signals," *Bell System Technical Journal*, vol. 41, pp. 1295–1336, July 1962.

9 Signaling over bandlimited channels

9.1 Introduction

Up to now we have considered only the detection (or demodulation) of signals transmitted over channels of infinite bandwidth, or at least a large enough bandwidth that any signal distortion is negligible and can be ignored. Though in some situations this assumption is reasonable, satellite communications is a common example, bandlimitation is also common. The classical example is the telephone channel where the twisted-pair wires used as the transmission medium have a bandwidth on the order of kilohertz. But even a medium such as optical fiber exhibits a phenomenon called dispersion which results in an effect very analogous to bandlimitation.

It is important to realize that bandlimitation depends not only on the channel medium but also on the source, specifically the source rate, R_s (symbols/second). One common measure of the bandwidth needed or occupied by a source is $W = 1/T_s = R_s$ (hertz). As source rates keep increasing to accommodate more data eventually any channel starts to look bandlimited. Bandlimitation can also be imposed on a communication system by regulatory requirements. A user is usually allotted only so much bandwidth in which to transmit her/his information.

The general effect of bandlimitation on a transmitted signal of finite time duration, T_s seconds, is to *disperse* it or to *spread* it out. Therefore the signal transmitted in a particular time slot (or symbol interval) will interfere with signals in other time slots resulting in what is called *intersymbol interference* (each signal represents a data symbol) or ISI. Thus in this chapter we consider the demodulation of signals which are not only corrupted by AWGN but also by ISI.

Three major approaches are developed. The first approach deals with the problem[1] by answering the question: "Under what conditions is it possible to achieve zero ISI even in the presence of bandlimitation?" This leads to what is called Nyquist's first criterion and to zero-forcing equalization. The second approach is to allow some ISI but in a controlled manner, resulting in what is known as partial response signaling. The final approach is to live with what you are given, i.e., with the ISI present, and design the best demodulation for the situation. The demodulation then determines the entire transmitted sequence using a maximum likelihood sequence criterion. It is usually realized by a Viterbi algorithm.

[1] To paraphrase the great philosopher, Charlie Brown: "No problem is too big or too complex that cannot be run away from."

9.2 The communication system model

Figure 9.1 shows the communication system model that will be used. Regarding the model it is logical enough to assume that the bandlimitation imposed on the system is due to the channel. However, as mentioned, it may also come about at the transmitter due to regulatory constraint. Though for practical purposes the modulator and transmit filter may be separate blocks, for our purposes they are lumped together into one block. Finally, antipodal modulation is represented by a train of impulses resulting in an impulse modulator. The resultant system to be considered is shown in Figure 9.2.

Before proceeding to investigate ISI let us consider a simple example to illustrate it. Let the modulator be NRZ-L, signaling rate of $r_b = 1/T_b$ (bits/second), and let the channel be a simple lowpass filter as shown in Figure 9.3, where the time constant RC is on order of T_b. The signals at points (A) and (B) are shown in Figure 9.4 for a specific transmitted bit sequence. The received signal $\mathbf{r}(t)$ during the interval, say $[3T_b, 4T_b]$, is

$$\mathbf{r}(t) = \mathbf{b}_3 s_B(t - 3T_b) + \mathbf{b}_2 s_B(t - 2T_b) + \mathbf{b}_1 s_B(t - T_b) + \mathbf{b}_0 s_B(t) + \mathbf{w}(t). \tag{9.1}$$

It is composed of three components: (i) $\mathbf{b}_3 s_B(t - 3T_b)$, which represents the bit transmitted in the interval $[3T_b, 4T_b]$, (ii) $\mathbf{w}(t)$, the additive random noise, (iii) $\mathbf{b}_2 s_B(t - 2T_b) + \mathbf{b}_1 s_B(t - T_b) + \mathbf{b}_0 s_B(t)$ due to the previous transmitted bits. Since to the receiver the $\mathbf{b}_i$ are unknown, $\mathbf{b}_0, \mathbf{b}_1, \mathbf{b}_2$ are binary random variables to it and therefore the receiver sees

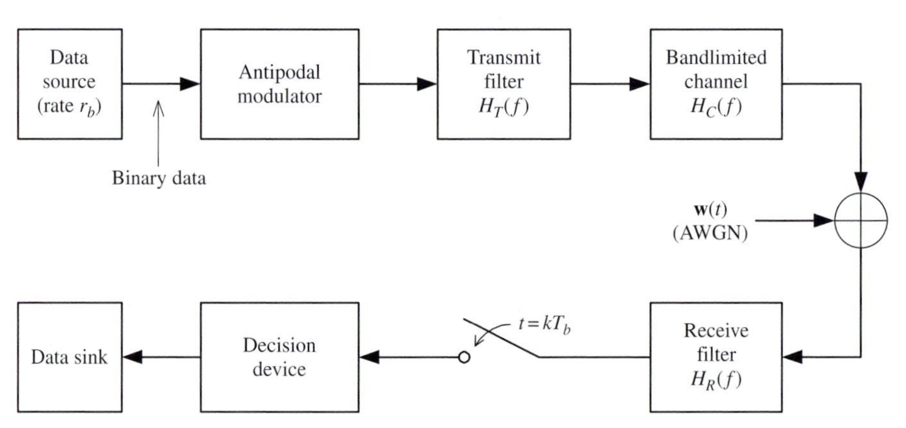

Fig. 9.1 Communication system model.

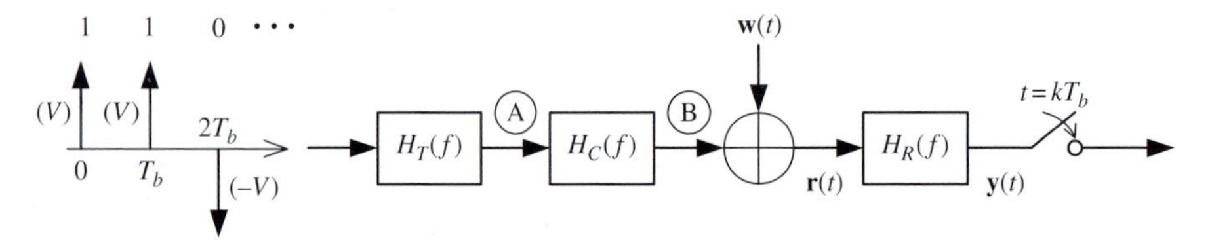

Fig. 9.2 Impulse modulator.

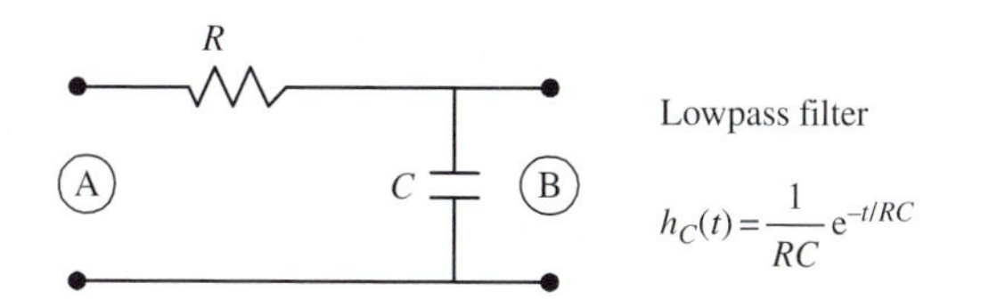

Fig. 9.3 A lowpass filter.

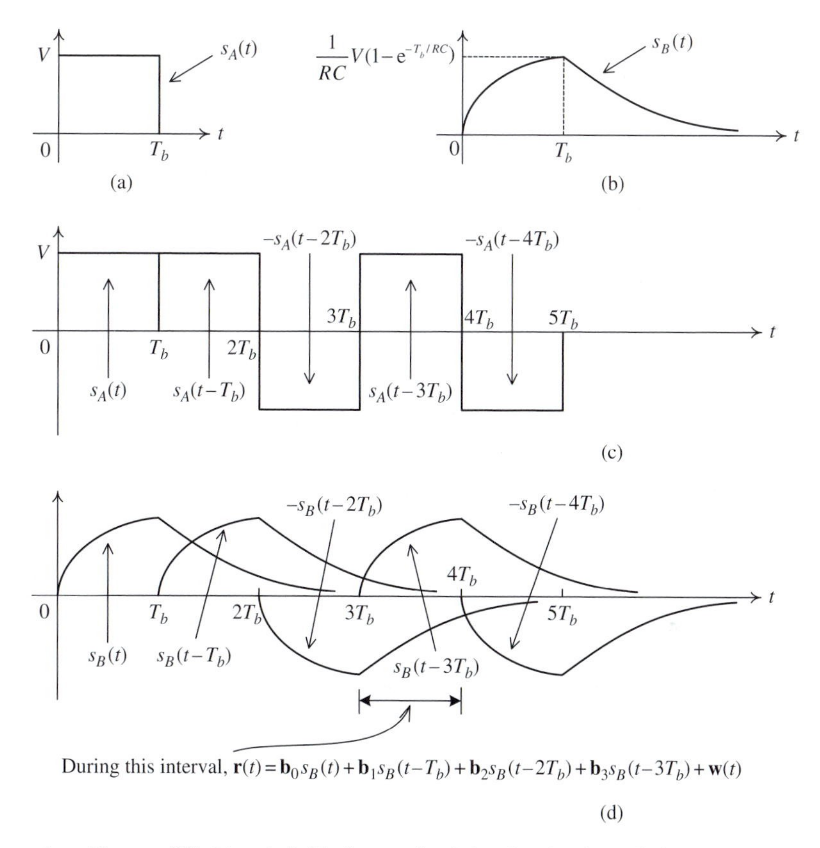

Fig. 9.4 An example to illustrate ISI: (a) an individual transmitted signal at the channel's input; (b) an individual received signal at the channel's output; (c) a sequence of transmitted signals at the channel's input; (d) received signal components at the channel's output.

component (iii) as noise. This noise is not Gaussian and as such is difficult to handle in an optimum, i.e., minimum error probability, sense. One possible approach is to ignore it completely and proceed as in previous chapters. This, however, usually results in severe performance degradation. Another approach is to determine under what conditions on $H_T(f)$, $H_C(f)$, $H_R(f)$ will the sample at $t = 4T_b$ have no contribution from the previous transmitted bits, i.e., in the term $\mathbf{b}_2 s_B(t-2T_b) + \mathbf{b}_1 s_B(t-T_b) + \mathbf{b}_0 s_B(t)$ each of the signals goes through zero at $t = 4T_b$. This leads to Nyquist's first criterion which is discussed next.

9.3 Nyquist criterion for zero ISI

In general, the received signal at the output of the receiver filter is given by

$$\mathbf{y}(t)=\sum_{k=-\infty}^{\infty}\mathbf{b}_k s_R(t-kT_b)+\mathbf{w}_{\text{out}}(t), \tag{9.2}$$

where $s_R(t)=h_T(t)*h_C(t)*h_R(t)$ is the overall response of the system due to a unit impulse at the input,

$$\mathbf{b}_k=\begin{cases} V & \text{if the } k\text{th bit is "1"} \\ -V & \text{if the } k\text{th bit is "0"} \end{cases}. \tag{9.3}$$

Without loss of generality, it is assumed that the overall response $s_R(t)$ is normalized so that $s_R(0)=1$. At sampling time $t=mT_b$ the sampler's output is

$$\mathbf{y}(mT_b)=\mathbf{b}_m+\sum_{\substack{k=-\infty\\k\neq m}}^{\infty}\mathbf{b}_k s_R(mT_b-kT_b)+\mathbf{w}_{\text{out}}(mT_b). \tag{9.4}$$

The second term represents ISI and now we look into the conditions on the overall transfer function $S_R(f)=H_T(f)H_C(f)H_R(f)$ which would make it zero. To this end consider a unit impulse applied to the system at $t=0$ and look at the sampled output (see Figure 9.5(a)). Ideally the samples of $s_R(t)$ due to this input should be 1 at $t=0$ and zero at all other sampling times kT_b ($k\neq 0$) as shown in Figure 9.5. If $s_R(t)$ is such that this is the case, then the system will experience no ISI since an impulse say at $T=mT_b$ will produce a nonzero value at that time instance and zero at all other sampling points.

To determine what implication this has for $S_R(f)$ recall the sampling theorem, which states that the continuous-time signal is uniquely specified by its samples provided the

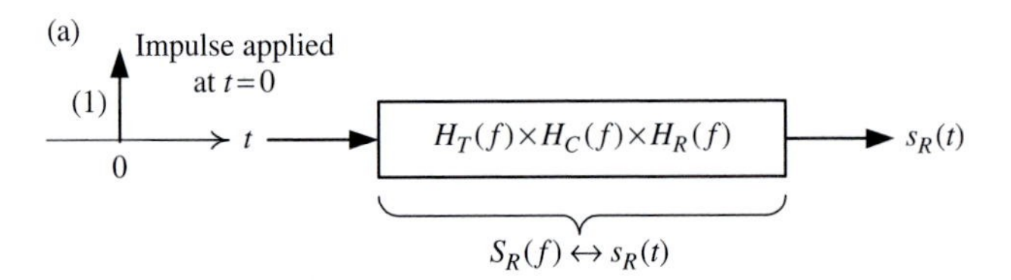

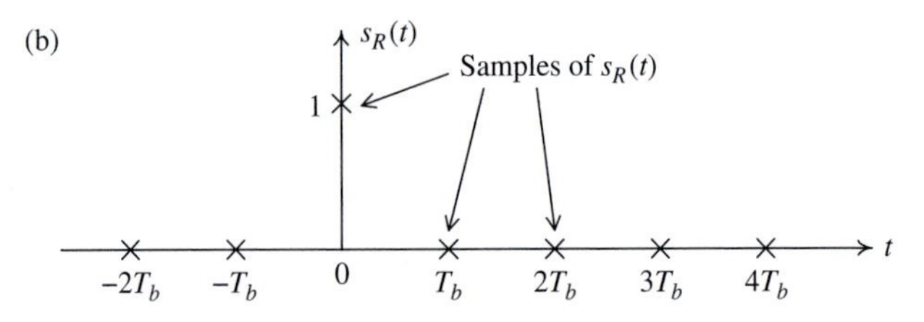

Fig. 9.5 Condition on $s_R(t)$ for zero ISI: (a) obtain the overall impulse response $s_R(t)$; (b) conditions on the samples of $s_R(t)$ at the sampling times.

sampling frequency is at least twice the maximum bandwidth of the signal. Here $s_R(t)$ is sampled at the signaling rate, every T_b seconds, and therefore for the samples in Figure 9.5(b) to uniquely specify $s_R(t)$ it follows that $S_R(f) = 0$ for frequencies $f > 1/2T_b$ hertz. To determine what $S_R(f)$ is in the interval $-1/2T_b \leq f \leq 1/2T_b$ we are faced with the problem of finding a time function of bandwidth $1/2T_b$ which in the time domain goes through zero at multiples of T_b. Some reflection indicates that $s_R(t) = \sin(\pi t/T_b)/(\pi t/T_b) = \mathrm{sinc}(t/T_b)$ satisfies this condition. The function $S_R(f)$ is shown in Figure 9.6.

The above discussion indicates that if the overall bandwidth of the system is less than $1/2T_b$ hertz then the ISI terms *cannot be made zero* at the sampling instances. Typically, however, the available bandwidth is greater than $1/2T_b$. In this case, because the signal is undersampled, aliasing occurs and many different $S_R(f)$, indeed an infinite number, result in the same set of desired samples shown in Figure 9.5(b). We claim that if the sampled spectrum is such that $S_R(f)$ and all its aliases add up to a constant value in the frequency band $|f| \leq 1/2T_b$, then no ISI occurs. Stated formally:

Claim If $\sum_{k=-\infty}^{\infty} S_R(f + k/T_b) = T_b$ for $|f| \leq 1/2T_b$ then

$$s_R(kT_b) = \begin{cases} 1, & k = 0 \\ 0, & \text{otherwise} \end{cases}. \tag{9.5}$$

Graphically the condition looks as shown in Figure 9.7.

Proof $s_R(t) = \int_{-\infty}^{\infty} S_R(f) e^{j2\pi ft} df$. Split up the integral into intervals of $1/T_b$ and look at the value of $s_R(t)$ at $t = kT_b$:

$$s_R(kT_b) = \sum_{m=-\infty}^{\infty} \int_{(2m-1)/2T_b}^{(2m+1)/2T_b} S_R(f) e^{j2\pi fkT_b} df. \tag{9.6}$$

Change the integration variable to $\lambda = f - m/T_b$. This effectively shifts the integrand into one range, namely, $[-1/2T_b, 1/2T_b]$. Therefore,

$$s_R(kT_b) = \sum_{m=-\infty}^{\infty} \int_{-1/2T_b}^{1/2T_b} S_R\left(\lambda + \frac{m}{T_b}\right) e^{j2\pi(\lambda + m/T_b)kT_b} d\lambda. \tag{9.7}$$

Using the fact that $e^{j2\pi(\lambda+m/T_b)kT_b} = e^{j2\pi\lambda kT_b}$ and interchanging the summation and integration operations gives

$$s_R(kT_b) = \int_{-1/2T_b}^{1/2T_b} \left[\sum_{m=-\infty}^{\infty} S_R\left(f + \frac{m}{T_b}\right) \right] e^{j2\pi fkT_b} df. \tag{9.8}$$

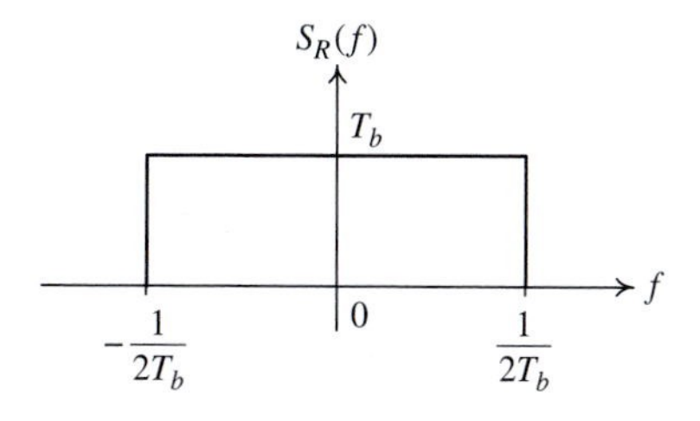

Fig. 9.6 An example of $S_R(f)$ which satisfies the zero-ISI condition.

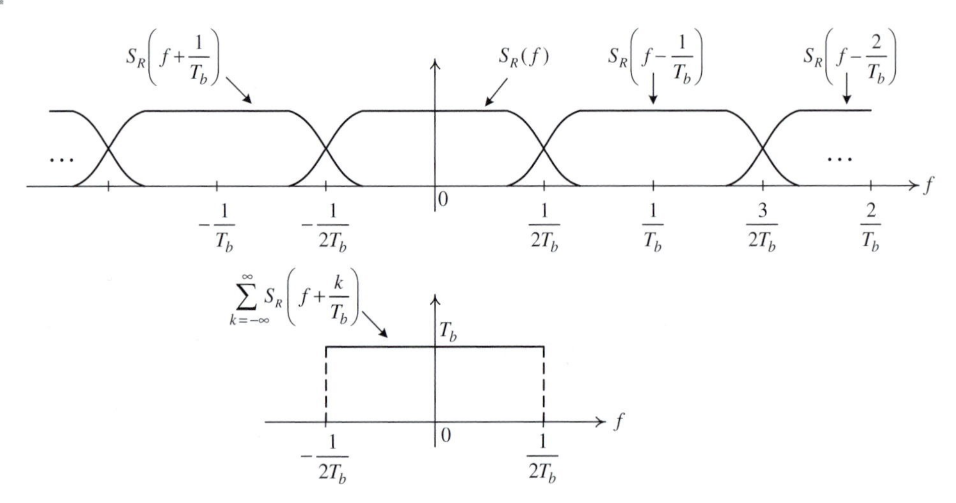

Fig. 9.7 Condition on $S_R(f)$ to achieve zero ISI.

But again the frequency function $\sum_{m=-\infty}^{\infty} S_R(f + m/T_b)$ must be such that the sample values of the continuous time function that it represents satisfy (9.5). This means that

$$\sum_{m=-\infty}^{\infty} S_R\left(f + \frac{m}{T_b}\right) = T_b \text{ for } -\frac{1}{2T_b} \le f \le \frac{1}{2T_b}, \tag{9.9}$$

which is Nyquist's first criterion.

If the available bandwidth is exactly equal to $1/2T_b$, then the overall response must be flat over $-1/2T_b \le f \le 1/2T_b$ hertz. The corresponding time function $s_R(t)$ is $\sin(\pi t/T_b)/(\pi t/T_b)$. It goes through zero at every sampling point, except $t = 0$ as required. The function is plotted in Figure 9.8. Besides going through zero at the sampling point as required, an important characteristic is that it decays rather slowly as $1/t$ which means that if the sampler is not perfectly synchronized in time, considerable ISI can be encountered.

Usually, however, the available bandwidth is greater than $1/2T_b$ hertz. Since the sampling rate of $1/T_b$ now results in aliasing, then as mentioned an infinite number of $S_R(f)$ is available to satisfy Nyquist's criterion. One can exploit this by choosing $S_R(f)$ to meet other criteria. One criterion is to attempt to increase the rate of decay. Practically, though the available bandwidth is greater than $1/2T_b$, it is less than $1/T_b$ which means only one alias is present in the interval $[-1/2T_b, 1/2T_b]$ hertz as shown in Figure 9.7. To satisfy Nyquist's criterion, $S_R(f)$ must have a certain symmetry about the point $1/2T_b$. Namely, it should have the following form:

$$S_R(f) = \begin{cases} T_b, & 0 \le |f| \le \frac{1-\beta}{2T_b} \\ T_b - X\left(-|f| + \frac{1}{2T_b}\right), & \frac{1-\beta}{2T_b} \le |f| \le \frac{1}{2T_b} \\ X\left(|f| - \frac{1}{2T_b}\right), & \frac{1}{2T_b} \le |f| \le \frac{1+\beta}{2T_b} \\ 0, & \frac{1+\beta}{2T_b} \le |f| \end{cases}, \tag{9.10}$$

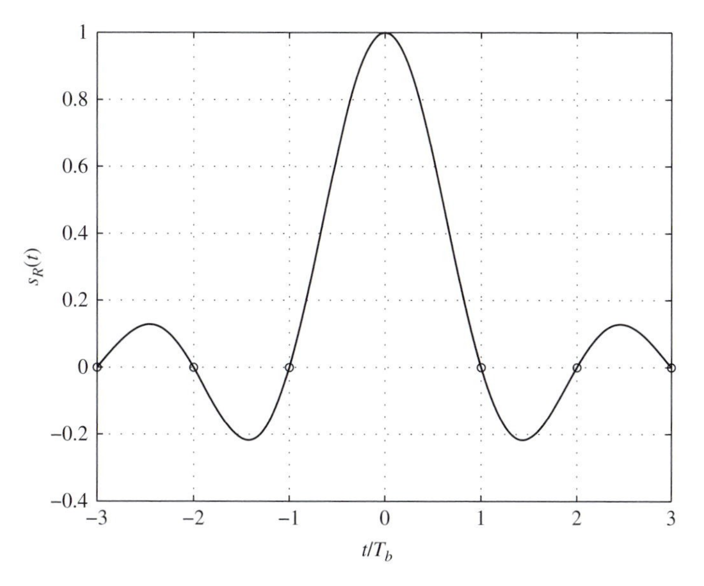

Fig. 9.8 Time domain function of the rectangular spectrum.

where $0 \le \beta \le 1$ and $X(f)$ is any function that satisfies the following conditions:

$$\begin{cases} 0 \le X(f) \le T_b \text{ for } 0 \le f \le \beta/(2T_b), \\ X(0) = T_b/2 \text{ and } X(\beta/(2T_b)) = 0, \\ X(f) = 0 \text{ for } f < 0 \text{ and } f > \beta/(2T_b). \end{cases} \tag{9.11}$$

In the above, β is a parameter that controls the excess bandwidth, over $1/2T_b$, that the system response occupies. This parameter is also commonly referred to as the *roll-off* factor.

A popular overall system response, $S_R(f)$, is the *raised-cosine* function shown in Figure 9.9 with the corresponding time function shown in Figure 9.10. It is defined as follows:

$$S_R(f) = S_{RC}(f) = \begin{cases} T_b, & |f| \le \frac{1-\beta}{2T_b} \\ T_b \cos^2\left[\frac{\pi T_b}{2\beta}\left(|f| - \frac{1-\beta}{2T_b}\right)\right], & \frac{1-\beta}{2T_b} \le |f| \le \frac{1+\beta}{2T_b} \\ 0, & |f| \ge \frac{1+\beta}{2T_b} \end{cases}.$$

The time domain response is given by

$$s_R(t) = s_{RC}(t) = \frac{\sin(\pi t/T_b)}{(\pi t/T_b)} \frac{\cos(\pi \beta t/T_b)}{1 - 4\beta^2 t^2/T_b^2} = \text{sinc}(t/T_b) \frac{\cos(\pi \beta t/T_b)}{1 - 4\beta^2 t^2/T_b^2}, \tag{9.12}$$

which decays as $1/t^3$, as expected, since $S_R(f)$ must be differentiated twice before a discontinuity is created. This makes it less sensitive to mistiming errors.

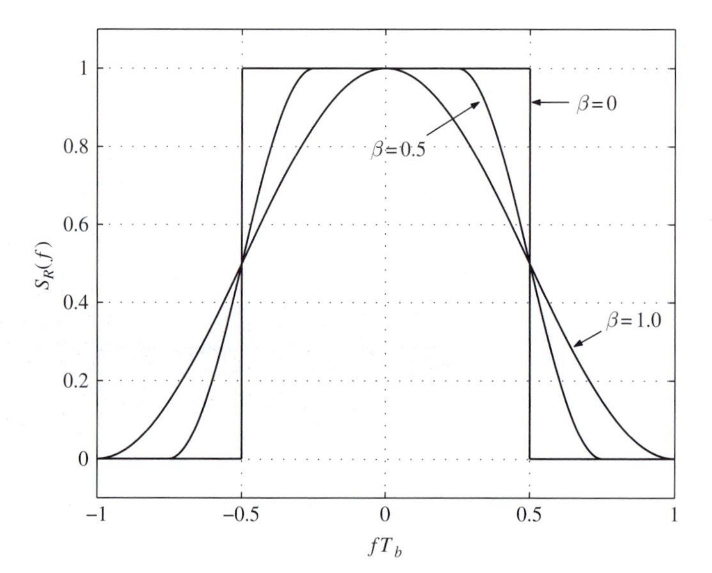

Fig. 9.9 The raised-cosine spectrum.

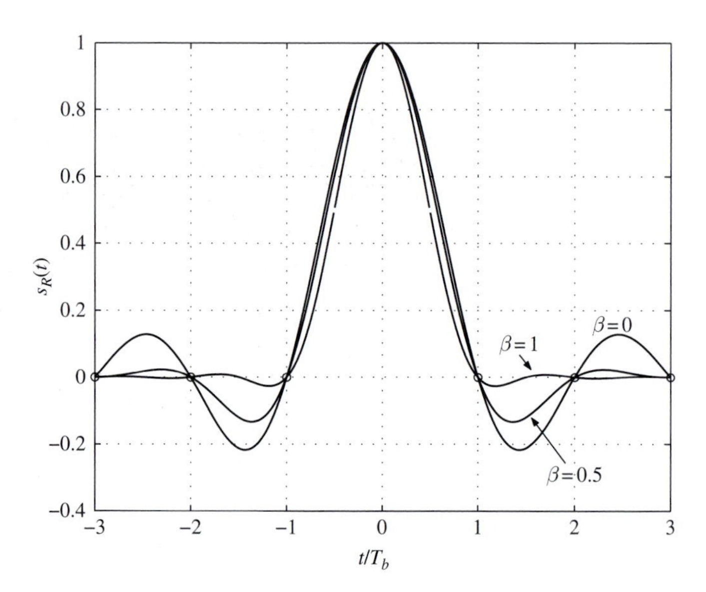

Fig. 9.10 Time domain function of the raised-cosine spectrum.

One common way to observe and measure (qualitatively) the effect of ISI is to look at the *eye diagram* of the received signal. The effect of ISI and other noise can be observed on an oscilloscope by displaying the output of the receiver filter on the vertical input with the horizontal sweep rate set at multiples of $1/T_b$. Such a display is called an eye diagram. For illustration, Figures 9.11 and 9.12 show the eye diagrams (without the additive random noise component) for two different overall impulse responses, namely the ideal lowpass filter and a raised-cosine filter with roll-off factor $\beta = 0.35$. The effect of ISI is to cause

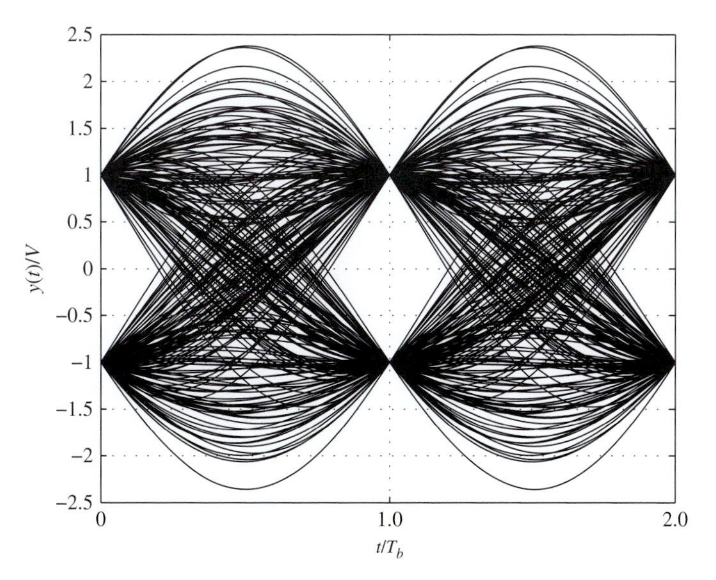

Fig. 9.11 Eye diagram: ideal lowpass filter of the overall frequency response and no AWGN.

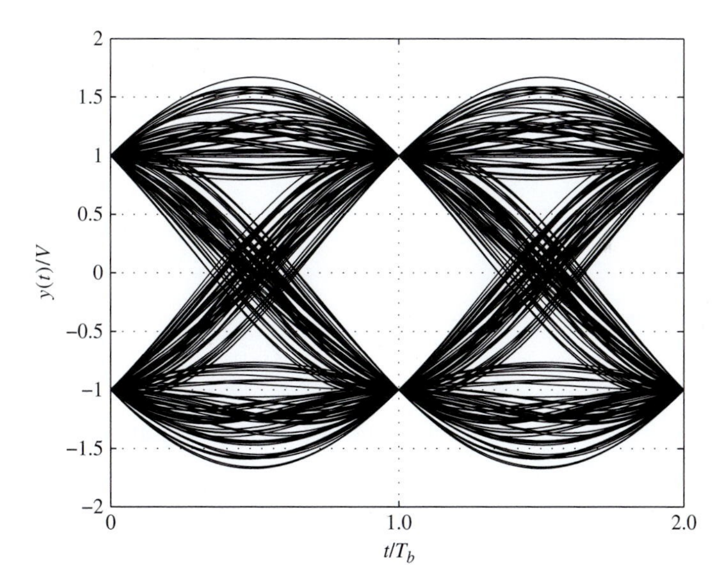

Fig. 9.12 Eye diagram: raised-cosine filter of the overall frequency response with $\beta = 0.35$ and no AWGN.

a closing of the eye, which makes the system more sensitive to a synchronization error. Observe from Figures 9.11 and 9.12 that, compared to the ideal lowpass filter, the raised-cosine filter has a larger eye opening. This is to be expected based on the decay behaviors of the impulse responses of the two filters just discussed. Finally, the eye diagram in the presence of AWGN is also shown in Figure 9.13 for the case of a raised-cosine filter with $\beta = 0.35$ and an SNR of 20 decibels.

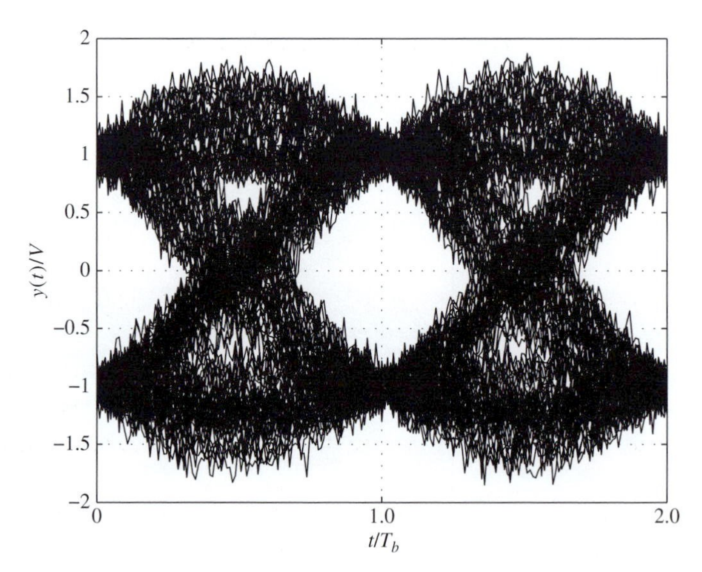

Fig. 9.13 Eye diagram: raised-cosine filter of the overall frequency response with $\beta = 0.35$ and with the presence of AWGN ($V^2/\sigma_{\mathbf{w}}^2$=20 decibels).

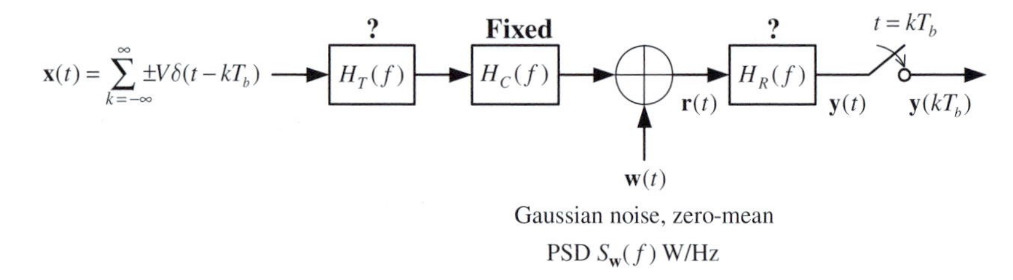

Fig. 9.14 Communication system model.

9.3.1 Design of transmitting and receiving filters

It is customary and logical to assume that the channel transfer function, $H_C(f)$, is *fixed*. However, all that has been shown thus far is that the overall response

$$S_R(f) = H_T(f)H_C(f)H_R(f) \tag{9.13}$$

must be such that it satisfies the zero-ISI criterion. Though $H_C(f)$ is fixed, one still has flexibility in the design of $H_T(f)$ and $H_R(f)$ to satisfy the zero-ISI criterion. A natural approach to the design is to attempt to minimize the probability of error. To develop the design equations for $H_T(f)$, $H_R(f)$ let us revisit the overall block diagram of the system as shown in Figure 9.14. Note that the noise here is assumed to be Gaussian (as usual) but *does not necessarily have to be white*.

It was shown earlier that the sampler's output is given by

$$\mathbf{y}(mT_b) = \pm V + \mathbf{w}_{\text{out}}(mT_b). \tag{9.14}$$

The output noise sample $\mathbf{w}_{\text{out}}(mT_b)$ is Gaussian, zero-mean with variance

$$\sigma_{\mathbf{w}}^2 = \int_{-\infty}^{\infty} S_{\mathbf{w}}(f)|H_R(f)|^2 \mathrm{d}f, \tag{9.15}$$

where $S_{\mathbf{w}}(f)$ is the PSD of the noise. The conditional pdfs of sample $\mathbf{y}(mT_b)$ are shown in Figure 9.15.

Since the bits are assumed to be equally likely the minimum error probability receiver sets the threshold at zero. Its error probability is given by $P[\text{error}] = Q(V/\sigma_{\mathbf{w}})$. To make this as small as possible the argument $V/\sigma_{\mathbf{w}}$, or equivalently $V^2/\sigma_{\mathbf{w}}^2$, needs to be maximized.

The design problem therefore becomes: *given the transmitted power P_T, the channel's frequency response $H_C(f)$, and the additive noise's PSD $S_{\mathbf{w}}(f)$, choose $H_T(f)$ and $H_R(f)$ so that the zero-ISI criterion is satisfied and the* SNR $= V^2/\sigma_{\mathbf{w}}^2$ *is maximized.* To proceed we need to obtain an expression for the SNR in terms of $H_T(f), H_R(f)$. This is accomplished as follows.

(i) The average transmitted power (of the transmitter) is given by

$$P_T = \frac{V^2}{T_b}\int_{-\infty}^{\infty} |H_T(f)|^2 \mathrm{d}f \quad \text{(watts)}, \tag{9.16}$$

where the average power of a train of impulses, occurring every T_b seconds and weight $\pm V$ is V^2/T_b (watts).

(ii) Using (9.15) and (9.16) we obtain

$$\frac{V^2}{\sigma_{\mathbf{w}}^2} = P_T T_b \left[\int_{-\infty}^{\infty} |H_T(f)|^2 \mathrm{d}f\right]^{-1} \left[\int_{-\infty}^{\infty} S_{\mathbf{w}}(f)|H_R(f)|^2 \mathrm{d}f\right]^{-1}, \tag{9.17}$$

or equivalently, the inverse of the SNR is

$$\frac{\sigma_{\mathbf{w}}^2}{V^2} = \frac{1}{P_T T_b}\left[\int_{-\infty}^{\infty} |H_T(f)|^2 \mathrm{d}f\right]\left[\int_{-\infty}^{\infty} S_{\mathbf{w}}(f)|H_R(f)|^2 \mathrm{d}f\right]. \tag{9.18}$$

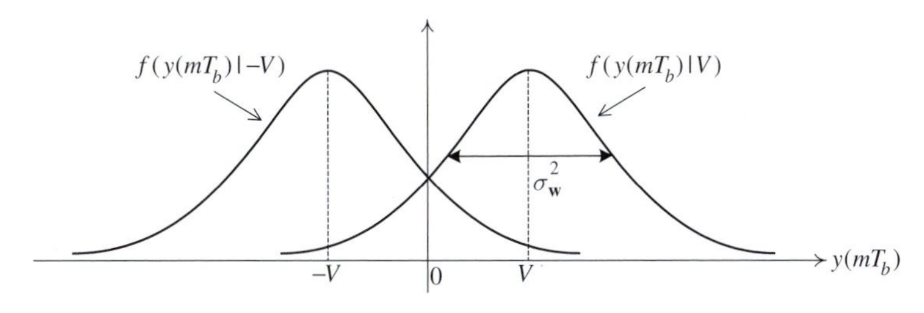

Fig. 9.15 The conditional pdfs of $\mathbf{y}(mT_b)$.

(iii) Eliminate one of the variables, say $H_T(f)$ by using (9.13). One has

$$\frac{\sigma_{\mathrm{w}}^2}{V^2} = \frac{1}{P_T T_b}\left[\int_{-\infty}^{\infty} \frac{|S_R(f)|^2}{|H_C(f)|^2 |H_R(f)|^2}\mathrm{d}f\right]\left[\int_{-\infty}^{\infty} S_{\mathrm{w}}(f)|H_R(f)|^2 \mathrm{d}f\right]. \tag{9.19}$$

(iv) Apply the Cauchy–Schwartz inequality to *minimize* the above. It states that $|\int_{-\infty}^{\infty} A(f)B^*(f)\mathrm{d}f|^2 \le [\int_{-\infty}^{\infty} |A(f)|^2 \mathrm{d}f][\int_{-\infty}^{\infty} |B(f)|^2 \mathrm{d}f]$ and holds with equality if and only if $A(f) = KB(f)$, where K is an arbitrary constant. Now identify $|A(f)| = \sqrt{S_{\mathrm{w}}(f)}|H_R(f)|$, $|B(f)| = |S_R(f)|/|H_C(f)||H_R(f)|$. To minimize (9.19) we want equality to hold and therefore it follows that

$$|H_R(f)|^2 = \frac{K|S_R(f)|}{\sqrt{S_{\mathrm{w}}(f)}|H_C(f)|}, \tag{9.20}$$

and from (9.13)

$$|H_T(f)|^2 = \frac{|S_R(f)|\sqrt{S_{\mathrm{w}}(f)}}{K|H_C(f)|}. \tag{9.21}$$

If the noise is white, or at least has a flat PSD over the channel bandwidth, then the design equations simplify to

$$|H_R(f)|^2 = K_1 \frac{|S_R(f)|}{|H_C(f)|}, \tag{9.22}$$

$$|H_T(f)|^2 = K_2 \frac{|S_R(f)|}{|H_C(f)|} = \frac{K_2}{K_1}|H_R(f)|^2, \tag{9.23}$$

where K_1, K_2 are arbitrary constants which set the power levels at the transmitter and the receiver. To complete the design the phases of the filters need to be specified. They are arbitrary but the phase functions must cancel each other. Therefore

$$\begin{aligned} H_R(f) &= |H_R(f)|\mathrm{e}^{\mathrm{j}\angle H_R(f)}, \\ H_T(f) &= K|H_R(f)|\mathrm{e}^{\mathrm{j}\angle -H_R(f)}, \end{aligned} \tag{9.24}$$

which tells us that the transmit and receive filters are a *matched-filter pair*.

With the optimal filters in (9.21) and (9.20) the maximum output SNR is given by

$$\left(\frac{V^2}{\sigma_{\mathrm{w}}^2}\right)_{\max} = P_T T_b \left[\int_{-\infty}^{\infty} \frac{|S_R(f)|\sqrt{S_{\mathrm{w}}(f)}}{|H_C(f)|}\mathrm{d}f\right]^{-2}. \tag{9.25}$$

In the special case where the channel is ideal, i.e., $H_C(f) = 1$ for $|f| \le W$ and $K_1 = K_2$, one has $|H_T(f)| = |H_R(f)| = \sqrt{|S_R(f)|}/\sqrt{K_1}$. If $S_R(f)$ is a raised-cosine spectrum (and $K_1 = 1$), then both $H_T(f)$ and $H_R(f)$ assume the following *square-root raised-cosine* (SRRC) spectrum:

$$H_T(f) = H_R(f) = \begin{cases} \sqrt{T_b}, & |f| \le \dfrac{1-\beta}{2T_b} \\ \sqrt{T_b}\cos\left[\dfrac{\pi T_b}{2\beta}\left(|f| - \dfrac{1-\beta}{2T_b}\right)\right], & \dfrac{1-\beta}{2T_b} \le |f| \le \dfrac{1+\beta}{2T_b} \\ 0, & |f| \ge \dfrac{1+\beta}{2T_b} \end{cases}. \tag{9.26}$$

It can be shown that the impulse response of a filter having SRRC spectral characteristic is given by

$$h_T(t) = h_R(t) = s_{\text{SRRC}}(t) = \frac{(4\beta t/T_b)\cos[\pi(1+\beta)t/T_b] + \sin[\pi(1-\beta)t/T_b]}{(\pi t/T_b)[1-(4\beta t/T_b)^2]}. \tag{9.27}$$

The time waveform of the above impulse response is plotted in Figure 9.16 along with the time waveform having the raised-cosine spectrum for $\beta = 0.5$. Note that the waveform having SRRC spectrum does not go to zero at nonzero multiples of $1/T_b$.

The following example illustrates the design procedure.

Example 9.1 Design a binary NRZ-L system with the following specifications:

(i) Transmission rate $r_b = 3600$ bits/second.
(ii) $P[\text{bit error}] \leq 10^{-4}$.
(iii) Channel model: $H_C(f) = 10^{-2}$ for $|f| \leq 2400$ hertz and $H_C(f) = 0$ for $|f| > 2400$ hertz.
(iv) Noise model: $S_{\mathbf{w}}(f) = 10^{-14}$ watts/Hz, $\forall f$ (i.e., white noise).

Solution

(a) The transmission rate of 3600 bits/second tells us that we need a bandwidth of at least 1800 hertz to transmit without ISI. Since the available bandwidth is 2400 hertz, choose a raised-cosine spectrum with $\beta r_b/2 = 600$ or $\beta = \frac{1}{3}$. That is,

$$S_R(f) = \begin{cases} \frac{1}{3600}, & |f| < (1-\beta)\frac{r_b}{2} = 1200 \text{ hertz} \\ \frac{1}{3600}\cos^2\left[\frac{\pi}{2400}(|f| - 1200)\right], & 1200 \text{ hertz} \leq |f| \leq 2400 \text{ hertz} \\ 0, & \text{elsewhere} \end{cases}. \tag{9.28}$$

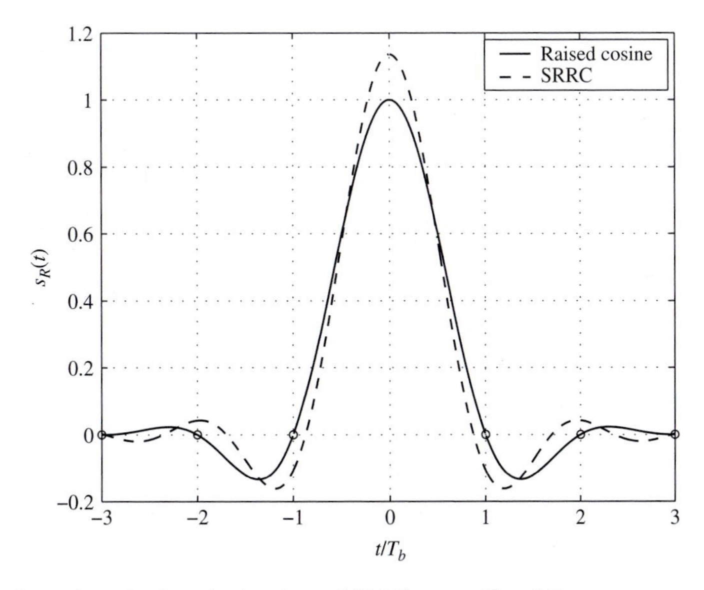

Fig. 9.16 An example of waveforms having raised-cosine and SRRC spectra ($\beta = 0.5$).

(b) The transmit and receive filters are $|H_T(f)| = K_1|S_R(f)|^{1/2}$ and $|H_R(f)| = |S_R(f)|^{1/2}$. The gain constant K_1 is found from $|H_T(f)||H_C(f)||H_R(f)| = |S_R(f)|$ which at $f = 0$ gives $\frac{1}{\sqrt{3600}}K_1(10^{-2})\frac{1}{\sqrt{3600}} = \frac{1}{3600}$, or $K_1 = 100$.

(c) To determine the transmitted power needed the specification on the required bit error probability is used as follows:

$$Q\left(\sqrt{\left(\frac{V^2}{\sigma_{\mathrm{w}}^2}\right)_{\max}}\right) \leq 10^{-4} \Rightarrow \left(\frac{V^2}{\sigma_{\mathrm{w}}^2}\right)_{\max} \geq 14.04 \approx 14. \tag{9.29}$$

The transmitted power P_T is given by

$$\begin{aligned} P_T &= \frac{1}{T_b}\left(\frac{V^2}{\sigma_{\mathrm{w}}^2}\right)_{\max}\left[\int_{-\infty}^{\infty}\frac{|S_R(f)|\sqrt{S_{\mathrm{w}}(f)}}{|H_C(f)|}\mathrm{d}f\right]^2 \\ &= 3600 \times 14 \times \frac{10^{-14}}{10^{-4}}\left[\int_{-\infty}^{\infty}|S_R(f)|\mathrm{d}f\right]^2 \quad \text{(watts)}, \end{aligned} \tag{9.30}$$

but $\int_{-\infty}^{\infty}|S_R(f)|\mathrm{d}f = s_R(t)|_{t=0} = 1$. Therefore $P_T = 5$ microwatts.

■

All of the above discussion assumes that one has adequate bandwidth to achieve zero ISI. However, if bandwidth is very limited, i.e., one cannot afford to use more than $1/2T_b$ hertz, then it would appear that to achieve zero ISI, the overall response $S_R(f)$ must be flat in this frequency band. Besides being sensitive to mistiming, a more severe problem is the design of the transmit filter to achieve this brickwall response. Among other issues, Gibb's phenomenon starts to come into effect. One approach to circumvent these difficulties is to allow a certain amount of ISI but in a controlled manner. This is what *duobinary* modulation, which is discussed next, achieves.

9.3.2 Duobinary modulation

Duobinary modulation falls in the general class that is known as *partial response signaling* (PRS). The terminology "partial" reflects that the effect of a symbol is not confined to only one symbol interval but is allowed to appear in other intervals as well, albeit as ISI. To develop duobinary modulation consider restricting the ISI to only one term, namely that due to the previous symbol. With impulse sampling the sampled overall response would look like in Figure 9.17.

To determine $s_R(t)$, find $S_R(f)$ by first determining $S_R^{[\text{sampled}]}(f)$. Throw away all the aliases of the sampled spectrum by restricting $S_R^{[\text{sampled}]}(f)$ to the band $\left[-1/2T_b, 1/2T_b\right]$ hertz, and then perform the inverse transform. Therefore,

$$\begin{aligned} S_R^{[\text{sampled}]}(f) &= \int_{-\infty}^{\infty}[\delta(t-T_b)+\delta(t)]\,\mathrm{e}^{-\mathrm{j}2\pi ft}\mathrm{d}t \\ &= 1+\mathrm{e}^{-\mathrm{j}2\pi fT_b} = 2\mathrm{e}^{-\mathrm{j}\pi fT_b}\cos(\pi fT_b). \end{aligned} \tag{9.31}$$

The factor $e^{-j\pi f T_b} = e^{-j2\pi f(T_b/2)}$ represents a shift of the sampled signal by $T_b/2$ seconds to the right. The sampling time is under our control, therefore we shift the signal back by $T_b/2$ seconds and now sample it at times $t = kT_b + T_b/2$, $k = 0, \pm 1, \pm 2, \ldots$; still every T_b seconds but offset by $T_b/2$. Thus the sampled signal of Figure 9.17 is shifted by $T_b/2$ seconds and the corresponding sampled spectrum $S_R^{[\text{sampled}]}(f)$ becomes $2\cos(\pi f T_b)$. The overall spectrum $S_R(f)$, known as duobinary modulation, is obtained by scaling $S_R^{[\text{sampled}]}(f)$ by T_b (see (4.7)). It is given as

$$S_R(f) = \begin{cases} 2T_b \cos(\pi f T_b), & -\frac{1}{2T_b} \le f \le \frac{1}{2T_b} \\ 0, & \text{elsewhere} \end{cases} \tag{9.32}$$

and is shown in Figure 9.18.

The corresponding impulse response can be shown to be

$$s_R(t) = \frac{\cos(\pi t/T_b)}{\pi\left[\frac{1}{4} - (t/T_b)^2\right]}. \tag{9.33}$$

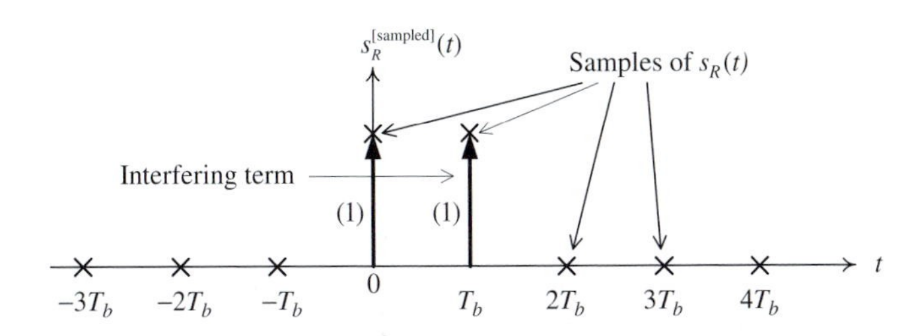

Fig. 9.17 Samples of the overall response in duobinary modulation.

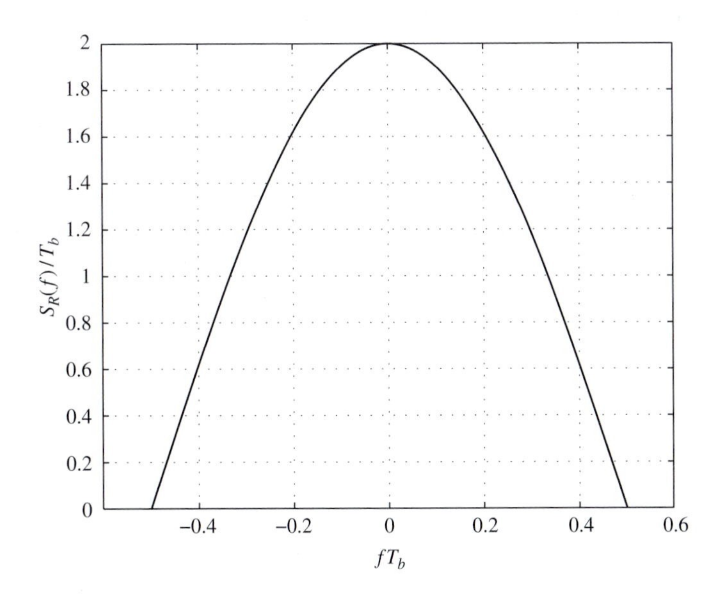

Fig. 9.18 Plot of $S_R(f)$.

A plot of the above waveform is shown in Figure 9.19. Note that it decays as $1/t^2$, as expected since $S_R(f)$ must be differentiated twice before an impulse appears. From (9.33) or Figure 9.19 it is readily seen or determined that at $t = -T_b/2$ the sample value contains information about the present bit and also has a component due to the previous bit. All other bits, however, contribute zero to it.

Thus

$$\mathbf{y}(t_k) = \mathbf{V}_k + \mathbf{V}_{k-1} + \mathbf{w}_{\text{out}}(t_k), \tag{9.34}$$

where $t_k = kT_b - \frac{T_b}{2}$, $k = 0, \pm 1, \pm 2, \ldots$. The signal sample, $\mathbf{V}_k$, at the output is $\pm V$ and therefore

$$\mathbf{y}(t_k) = \begin{cases} 2V + \mathbf{w}_{\text{out}}(t_k), & \text{if bits } k \text{ and } (k-1) \text{ are both "1"} \\ 0 + \mathbf{w}_{\text{out}}(t_k), & \text{if bits } k \text{ and } (k-1) \text{ are different} \\ -2V + \mathbf{w}_{\text{out}}(t_k), & \text{if bits } k \text{ and } (k-1) \text{ are both "0"} \end{cases}. \tag{9.35}$$

Observe that if the previous bit is known, the ISI is known and can be subtracted from the present observed signal $\mathbf{y}(t_k)$. One is then left with standard antipodal signaling. One, of course, needs to know the previous bits, something that the receiver is supposed to determine from the received signal. One possible escape from this dilemma is to make a decision on the previous bit, assume (more like hope) it is correct, and subtract out the interference that it causes in the present bit. Make a decision on the present bit and proceed. This procedure, known as *decision-feedback equalization*, works very well as long as correct decisions are made. However, when an error is made, the wrong value ISI is subtracted. This increases the probability of the next bit being in error with the implication that the errors will propagate. To circumvent the need for knowledge of the previous bit a precoder can be used at the transmitter as shown in Figure 9.20.

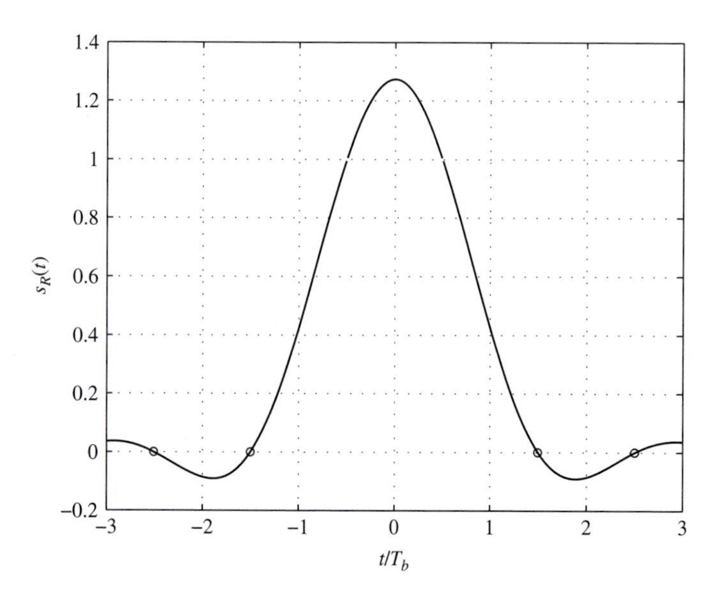

Fig. 9.19 Plot of $s_R(t)$ given by (9.33).

Observe that sample output, $\mathbf{y}(t_k)$ is now unambiguously related to the information bit. In essence what the precoder does is to eliminate the memory of the channel which is what the ISI represents. Figure 9.21 shows the signal space at the sampler output along with the conditional densities and decision thresholds.

Note that $P[\mathbf{y}(t_k) = -2V] = P[\mathbf{y}(t_k) = 2V] = \frac{1}{4}$; $P[\mathbf{y}(t_k) = 0] = \frac{1}{2}$. The error probability is well approximated by

$$P[\text{error}] \approx \frac{1}{4}\text{ area }④ + \frac{1}{2}\left[\text{area }① + \text{area }②\right] + \frac{1}{4}\text{ area }③ = \frac{3}{2}Q\left(\frac{V}{\sigma_{\mathbf{w}}}\right), \tag{9.36}$$

where $\sigma_{\mathbf{w}}^2$ is the noise variance at the output of the receive filter $H_R(f)$. The transmit and receive filters, $H_T(f), H_R(f)$, again are chosen to maximize $(V/\sigma_{\mathbf{w}})$ subject to the constraint that $S_R(f)$ is the duobinary shape and also that the transmitted power level is fixed.

It is of interest to compare the duobinary system with the zero-forcing approach. To do this let $H_C(f) = 1$ over $-1/2T_b \leq f \leq 1/2T_b$ hertz. Further, let the noise be white Gaussian with spectral density of $N_0/2$ watts/hertz. Then for binary PAM with zero ISI we have

$$\begin{aligned}\left(\frac{V^2}{\sigma_{\mathbf{w}}^2}\right)_{\max} &= P_T T_b\left[\int_{-\infty}^{\infty}\frac{|S_R(f)|\sqrt{S_{\mathbf{w}}(f)}}{|H_C(f)|}\mathrm{d}f\right]^{-2} \\ &= P_T T_b\left[\sqrt{N_0/2}\int_{-\frac{1}{2T_b}}^{\frac{1}{2T_b}}|S_R(f)|\mathrm{d}f\right]^{-2}.\end{aligned} \tag{9.37}$$

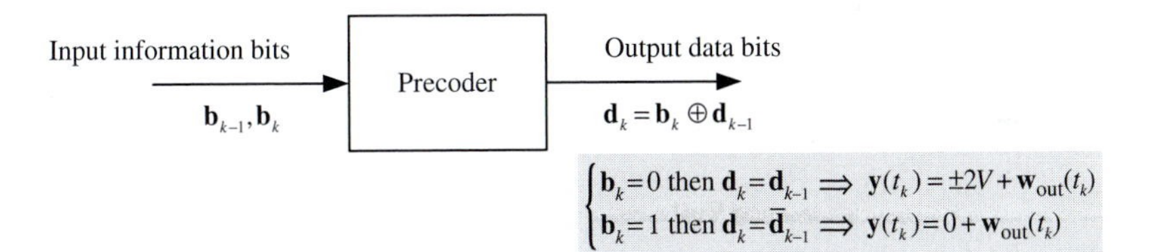

Fig. 9.20 Block diagram of a precoder.

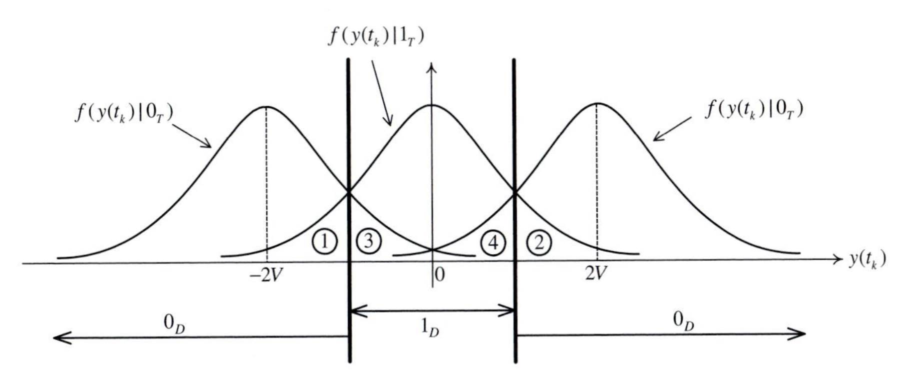

Fig. 9.21 Signal space and conditional probability density functions of $\mathbf{y}(t_k)$.

But $\int_{-1/2T_b}^{1/2T_b} |S_R(f)|\mathrm{d}f = 1$. Therefore $\left(V^2/\sigma_{\mathbf{w}}^2\right)_{\max} = P_T T_b \,(2/N_0)$ and

$$P[\text{error}]_{\text{binary}} = Q\left[\sqrt{\frac{2P_T T_b}{N_0}}\right]. \tag{9.38}$$

For the duobinary case the SNR is

$$\begin{aligned}\left(\frac{V^2}{\sigma_{\mathbf{w}}^2}\right)_{\max} &= P_T T_b\left[\sqrt{N_0/2}\int_{-1/2T_b}^{1/2T_b} 2T_b\cos(\pi f T_b)\mathrm{d}f\right]^{-2}\\ &= (P_T T_b)\left(\frac{2}{N_0}\right)\left(\frac{\pi}{4}\right)^2\end{aligned} \tag{9.39}$$

and

$$P[\text{error}]_{\text{duobinary}} = \frac{3}{2}Q\left(\frac{\pi}{4}\sqrt{\frac{2P_T T_b}{N_0}}\right). \tag{9.40}$$

Ignoring the 3/2 factor, the above shows that for duobinary modulation the transmitted power must be increased by $(4/\pi)^2$ or $10\log_{10}(4/\pi)^2 = 2.1$ decibels to achieve the same error probability as the zero-forcing approach.

9.4 Maximum likelihood sequence estimation

Up till now the ISI caused by bandlimitation has been combated either by designing the transmitter/receiver filter so that the overall response satisfies Nyquist's criterion or by allowing a controlled amount of interference. A disadvantage of these approaches is that the receiver filter invariably is such that it enhances the noise power at its output which degrades the potential performance. Here an approach is presented that does not attempt to eliminate the ISI but rather takes it into account in the demodulator. The criterion changes from minimizing the bit error probability to minimizing the sequence error probability. The demodulator becomes a *maximum likelihood sequence estimation* (MLSE).

To begin consider a sequence of N equally likely bits transmitted over a bandlimited channel where the transmission begins at $t = 0$ and ends at $t = NT_b$, with T_b the bit interval. The system block diagram is shown in Figure 9.22, where $h(t)$ is the impulse response of the overall chain: modulator/transmitter filter/channel. It extends over more than one bit interval but we shall assume that $h(t)$ is nonzero over a finite time interval of LT_b or can be reasonably approximated as such. Therefore the number of ISI terms is L.

Since there are N transmitted bits the receiver sees one of the $M = 2^N$ possible signals, $s_i(t) = \sum_{k=0}^{N-1} b_{i,k}h(t - kT_b)$, $i = 1, 2, \ldots, M = 2^N$, corrupted by $\mathbf{w}(t)$. Consider this as a humongous M-ary problem of determining which of the M signals (or symbols or

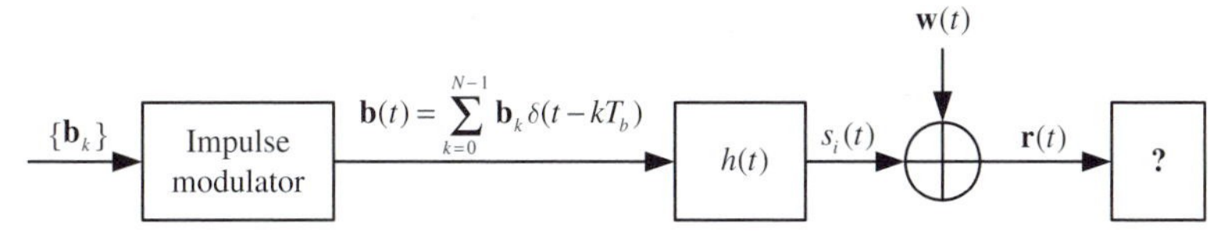

Fig. 9.22 System block diagram.

sequences) was transmitted with the criterion of minimizing the *probability of sequence error*. The procedure is one that we have used before for M-ary modulation and is outlined below.

(i) Take the M signals, $s_i(t)$ and determine a set of orthonormal functions using the Gram–Schmidt procedure that will represent the signals exactly i.e., find the orthonormal set $\phi_1(t), \ldots, \phi_K(t)$, where $K \leq M$. Now

$$s_i(t) = \sum_{j=1}^{K} s_{ij}\phi_j(t), \; i = 1, 2, \ldots, M, \; \text{with} \; s_{ij} = \int_{-\infty}^{\infty} s_i(t)\phi_j(t)\mathrm{d}t. \tag{9.41}$$

(ii) Project the received signal $\mathbf{r}(t) = s_i(t) + \mathbf{w}(t)$ onto the orthonormal basis to generate the set of sufficient statistics (or observables).

$$\mathbf{r}_1 = \int_{-\infty}^{\infty} \mathbf{r}(t)\phi_1(t)\mathrm{d}t = \int_{-\infty}^{\infty} [s_i(t) + \mathbf{w}(t)]\phi_1(t)\mathrm{d}t = s_{i1} + \mathbf{w}_1,$$
$$\mathbf{r}_2 = \int_{-\infty}^{\infty} [s_i(t) + \mathbf{w}(t)]\phi_2(t)\mathrm{d}t = s_{i2} + \mathbf{w}_2,$$
$$\vdots$$
$$\mathbf{r}_j = s_{ij} + \mathbf{w}_j,$$
$$\vdots$$
$$\mathbf{r}_K = s_{iK} + \mathbf{w}_K.$$

Note that the jth sufficient statistic is Gaussian, mean value s_{ij}, and variance $N_0/2$ (watts).

(iii) Determine the decision rule to minimize the sequence error probability. Assuming the bit sequences are equally probable, which is the case if the individual bits are equally probable and statistically independent, the decision rule is very simply

$$\begin{gathered}\text{compute:}\\ f(r_1, r_2, \ldots, r_K|s_i(t)) \text{ for } i = 1, 2, \ldots, M\\ \text{and choose the } s_i(t) \text{ or sequence that gives the } \textit{largest} \text{ value.}\end{gathered} \tag{9.42}$$

The conditional density function is given by

$$f(r_1, r_2, \ldots, r_K | s_i(t)) = \prod_{j=1}^{K} \frac{1}{\sqrt{\pi N_0}} \mathrm{e}^{-(r_j - s_{ij})^2/N_0}. \tag{9.43}$$

Take the natural logarithm, ignoring the constant term $K \ln\left(1/\sqrt{\pi N_0}\right)$ to obtain the decision rule:

$$\begin{gathered}\text{compute:}\\ -\frac{1}{N_0}\sum_{j=1}^{K} r_j^2 + \frac{2}{N_0}\sum_{j=1}^{K} r_j s_{ij} - \frac{1}{N_0}\sum_{j=1}^{K} s_{ij}^2, \quad \text{for } i = 1, 2, \ldots, M\\ \text{and choose the } \textit{largest} \text{ value.}\end{gathered} \tag{9.44}$$

The first term is independent of i and can be ignored. The second and third sums can be written as $\int_{-\infty}^{\infty} r(t)s_i(t)\mathrm{d}t$, the correlation between the received signal and test signal $s_i(t)$, and $\int_{-\infty}^{\infty} s_i^2(t)\mathrm{d}t$, the energy in the ith signal, respectively. The decision rule becomes

$$\begin{gathered}\text{compute:}\\ \gamma_i = \frac{2}{N_0}\int_{-\infty}^{\infty} r(t)s_i(t)\mathrm{d}t - \frac{1}{N_0}\int_{-\infty}^{\infty} s_i^2(t)\mathrm{d}t, \quad i = 1, 2, \ldots, M\\ \text{and choose the } \textit{largest} \text{ value.}\end{gathered} \tag{9.45}$$

To bring the impulse response, $h(t)$, into the decision rule, express $s_i(t)$ in terms of it. Very simply, it is

$$s_i(t) = \sum_{k=0}^{N-1} b_{i,k} h(t - kT_b). \tag{9.46}$$

The decision rule now becomes

$$\begin{gathered}\text{compute:}\\ \gamma_i = \frac{2}{N_0}\sum_{k=0}^{N-1} b_{i,k}\int_{-\infty}^{\infty} r(t)h(t - kT_b)\mathrm{d}t\\ -\frac{1}{N_0}\sum_{k=0}^{N-1}\sum_{j=0}^{N-1} b_{i,k}b_{i,j}\int_{-\infty}^{\infty} h(t - kT_b)h(t - jT_b)\mathrm{d}t\\ \text{and choose the } \textit{largest} \text{ value.}\end{gathered} \tag{9.47}$$

The integral $\int_{-\infty}^{\infty} h(t - kT_b)h(t - jT_b)\mathrm{d}t$ is the autocorrelation function of $h(t)$, with the usual properties of the autocorrelation function, namely that the output is a function of the time difference $kT_b - jT_b$ and even. Call the output h_{k-j}. Consider now the integral $\int_{-\infty}^{\infty} r(t)h(t - kT_b)\mathrm{d}t = \int_{-\infty}^{\infty} r(t)h(t - \tau)\mathrm{d}t|_{\tau = kT_b}$. It can be looked upon as the output of a filter with impulse response $h(-t)$ and input $r(t)$ sampled at $t = kT_b$ as shown in Figure 9.23.

The decision rule can be written now in terms of these sampled outputs, r_k, and the autocorrelation coefficients, h_{k-j}. Note that the r_k here are not the same as the ones obtained

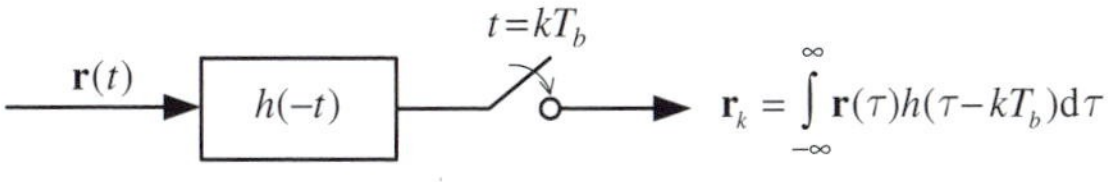

Fig. 9.23 Computation of $\int_{-\infty}^{\infty} \mathbf{r}(t)h(t-kT_b)\mathrm{d}t$.

by projecting $r(t)$ onto $\phi_k(t)$. However, they serve as a set of sufficient statistics for the decision rule which now is

$$\text{compute:}\quad \gamma_i = \frac{2}{N_0}\sum_{k=0}^{N-1} b_{i,k}r_k - \frac{1}{N_0}\sum_{k=0}^{N-1}\sum_{j=0}^{N-1} b_{i,k}b_{i,j}h_{k-j},\ i=1,2,\ldots,M \quad \text{and choose the } \textit{largest} \text{ value.} \tag{9.48}$$

Though, in principle, the above gives the decision rule which minimizes the sequence error probability, in practice the number M makes it infeasible. Consider a sequence length of 50 bits. Then the number of computations is $M = 2^{50} \approx 2^{(3.3)(16)} \approx 10^{16}$ (a very large number). The computation of γ_i is simplified considerably and made practical by using the Viterbi algorithm.

Consider the RHS of (9.48), which is known as a *path metric* for reasons that will become apparent later. For simplicity let the index i which indicates a specific sequence be understood. The term $\sum_{k=0}^{N-1}\sum_{j=0}^{N-1} b_k b_j h_{k-j}$ is a quadratic in the sequence bits b_k and therefore can be written as a matrix multiplication.

$$\underbrace{[b_0, b_1, \ldots, b_{N-1}]}_{\vec{b}} \underbrace{\begin{bmatrix} h_0 & h_{-1} & \ldots & h_{-N} \\ h_1 & h_0 & \ldots & h_{-N+1} \\ \vdots & \vdots & & \vdots \\ h_{N-1} & h_{N-2} & \ldots & h_{-1} \\ h_N & h_{N-1} & \ldots & h_0 \end{bmatrix}}_{H} \underbrace{\begin{bmatrix} b_0 \\ b_1 \\ \vdots \\ b_{N-2} \\ b_{N-1} \end{bmatrix}}_{\vec{b}^\top} = \vec{b}H\vec{b}^\top. \tag{9.49}$$

The correlation matrix H is symmetric, i.e., element $h_{kj} = h_{k-j} = h_{j-k}$, and we write it as a sum of three matrices

$$H = \begin{bmatrix} h_0 & 0 & \ldots & 0 \\ 0 & h_0 & \ldots & 0 \\ \vdots & \vdots & & \vdots \\ 0 & 0 & \ldots & 0 \\ 0 & 0 & \ldots & h_0 \end{bmatrix} + \begin{bmatrix} 0 & 0 & \ldots & 0 \\ h_1 & 0 & \ldots & 0 \\ \vdots & \vdots & & \vdots \\ h_{N-1} & h_{N-2} & \ldots & 0 \\ h_N & h_{N-1} & \ldots & 0 \end{bmatrix} + \begin{bmatrix} 0 & h_1 & \ldots & h_N \\ 0 & 0 & \ldots & h_{N-1} \\ \vdots & \vdots & & \vdots \\ 0 & 0 & \ldots & h_1 \\ 0 & 0 & \ldots & 0 \end{bmatrix}. \tag{9.50}$$

Then (9.49) becomes

$$\vec{b}H\vec{b}^{\top} = \vec{b}\begin{bmatrix} h_0 & 0 & \dots & 0 \\ 0 & h_0 & \dots & 0 \\ \vdots & \vdots & & \vdots \\ 0 & 0 & \dots & 0 \\ 0 & 0 & \dots & h_0 \end{bmatrix}\vec{b}^{\top} + \vec{b}\begin{bmatrix} 0 & 0 & \dots & 0 \\ h_1 & 0 & \dots & 0 \\ \vdots & \vdots & & \vdots \\ h_{N-1} & h_{N-2} & \dots & 0 \\ h_N & h_{N-1} & \dots & 0 \end{bmatrix}\vec{b}^{\top} + \vec{b}\begin{bmatrix} 0 & h_1 & \dots & h_N \\ 0 & 0 & \dots & h_{N-1} \\ \vdots & \vdots & & \vdots \\ 0 & 0 & \dots & h_1 \\ 0 & 0 & \dots & 0 \end{bmatrix}\vec{b}^{\top}. \tag{9.51}$$

The first term above is $\sum_{k=0}^{N-1} b_k^2 h_0$. Observe that the third term is equal to the second term. The $(k+1)$th element of the vector produced by the product

$$\begin{bmatrix} 0 & 0 & \dots & 0 \\ h_1 & 0 & \dots & 0 \\ \vdots & \vdots & & \vdots \\ h_{N-1} & h_{N-2} & \dots & 0 \\ h_N & h_{N-1} & \dots & 0 \end{bmatrix}\begin{pmatrix} b_0 \\ b_1 \\ \vdots \\ b_{N-2} \\ b_{N-1} \end{pmatrix} \tag{9.52}$$

is given by $h_k b_0 + h_{k-1} b_1 + \dots + h_1 b_{k-1} = \sum_{j=1}^{k} b_{k-j} h_j$. This element is then weighted by b_k when the vector is premultiplied by $\vec{b}$. The last two terms can therefore be written as

$$2\sum_{k=0}^{N-1} b_k \sum_{j=1}^{k} b_{k-j} h_j. \tag{9.53}$$

Combining all of the above, the path metric can be expressed as

$$\gamma_i = \frac{2}{N_0}\sum_{k=0}^{N-1} b_{i,k} r_k - \frac{1}{N_0}\sum_{k=0}^{N-1} b_{i,k}^2 h_0 - \frac{2}{N_0}\sum_{k=0}^{N-1} b_{i,k}\sum_{j=1}^{k} b_{i,k-j} h_j. \tag{9.54}$$

Since $b_{i,k} = \pm 1$ the quantity $b_{i,k}^2 h_0/2 = h_0/2$ is a constant independent of i which means that the second term can be ignored. Then under the assumption that $h(t) = 0$ for $t \geq LT_b$, which means $h_j = 0$ for $j \geq L$, the path metric becomes

$$\gamma_i = \frac{2}{N_0}\sum_{k=0}^{N-1}\left\{ b_{i,k} r_k - b_{i,k}\sum_{j=1}^{L-1} b_{i,k-j} h_j \right\}. \tag{9.55}$$

The term in the brackets is called a *branch metric*, for reasons which, as for the path metric, will soon be clear. It depends on three quantities: (i) the present output of the matched filter, r_k; (ii) the present value of the considered bit pattern, $b_{i,k}$; (iii) the previous $L-1$ values of the considered bit pattern $b_{i,k-1}, b_{i,k-2}, \dots, b_{i,k-(L-1)}$.

The system thus has memory, namely the ISI terms. The finite memory aspect of the branch metric computation suggests that it and hence the path metric generation can be done by using a finite state diagram and the trellis associated with the finite state diagram. The situation is directly analogous to that of Miller modulation encountered in Chapter 6. Though the memory in the two cases is due to different reasons, the approach to determining the maximum likelihood (most probable) sequence is the same. Namely, represent the memory by a finite state diagram and/or by the associated trellis graph. The states are the previous $L-1$ bits, where as usual a state is what is needed from the past which along with the present input allows one to compute the present branch metric, $b_{i,k}r_k - b_{i,k}\sum_{j=1}^{L-1} b_{i,k-j}h_j$, for the branch in the trellis along the ith path. The determination of the best path through the trellis, in this case the path that yields the maximum γ_i, can therefore be accomplished as a graph search procedure. This is done most efficiently using the *Viterbi algorithm*.

Because of the importance of the trellis graph and the Viterbi algorithm in communications,[2] an example of their application to the ISI channel is discussed in some detail next. Let the impulse response, $h(t)$, be as shown in Figure 9.24(a). The autocorrelation of $h(t)$ is $R_h(\tau) = \int_{-\infty}^{\infty} h(t)h(t-\tau)\mathrm{d}t$ and is shown in Figure 9.24(b). Note that there are two ISI terms, which are due to $h_1 = 0.6$ and $h_2 = 0.2$, which means that $L = 3$ or $L - 1 = 2$, i.e., the memory is two bits in length (as can be seen from the impulse response). The branch metric term of (9.55) becomes $b_{i,k}r_k - 0.6b_{i,k}b_{i,k-1} - 0.2b_{i,k}b_{i,k-2}$. The inputs $b_{i,k-1}$, $b_{i,k-2}$ represent the system memory and hence are states of the state diagram. The state diagram and corresponding trellis are shown in Figure 9.25.

Before proceeding to illustrate the Viterbi algorithm some remarks are pertinent to the state diagram and/or the trellis:

(1) The subscript i in $b_{i,k}$, which refers to a specific bit sequence, has been suppressed; however, any bit sequence pattern of length N can be traced on either figure.
(2) b_k is represented by logic 0 or 1, which at the transmitter is modulated to an impulse of strength -1 or $+1$, respectively. In the computation of the branch metric(s) one should use -1 and $+1$ as appropriate.

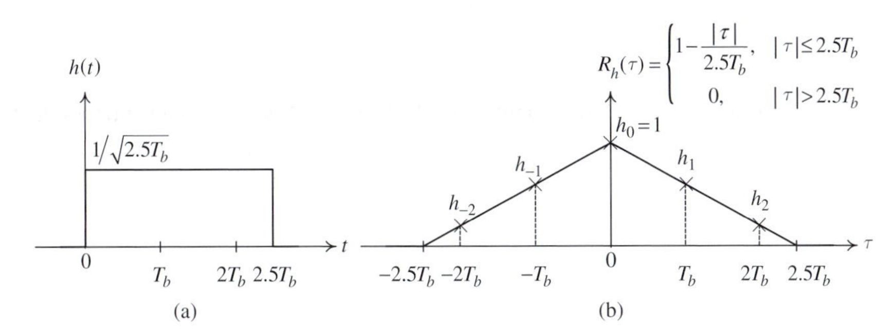

Fig. 9.24 (a) Overall impulse response, $h(t)$, of a bandlimited channel, and (b) the autocorrelation of $h(t)$. Here T_b is the bit interval.

[2] The trellis graph and the Viterbi algorithm arose originally in the demodulation of the convolutional codes. They have since been used in TCM, discussed in Chapter 11, in turbo coding/decoding, and even in the decoding of block codes.

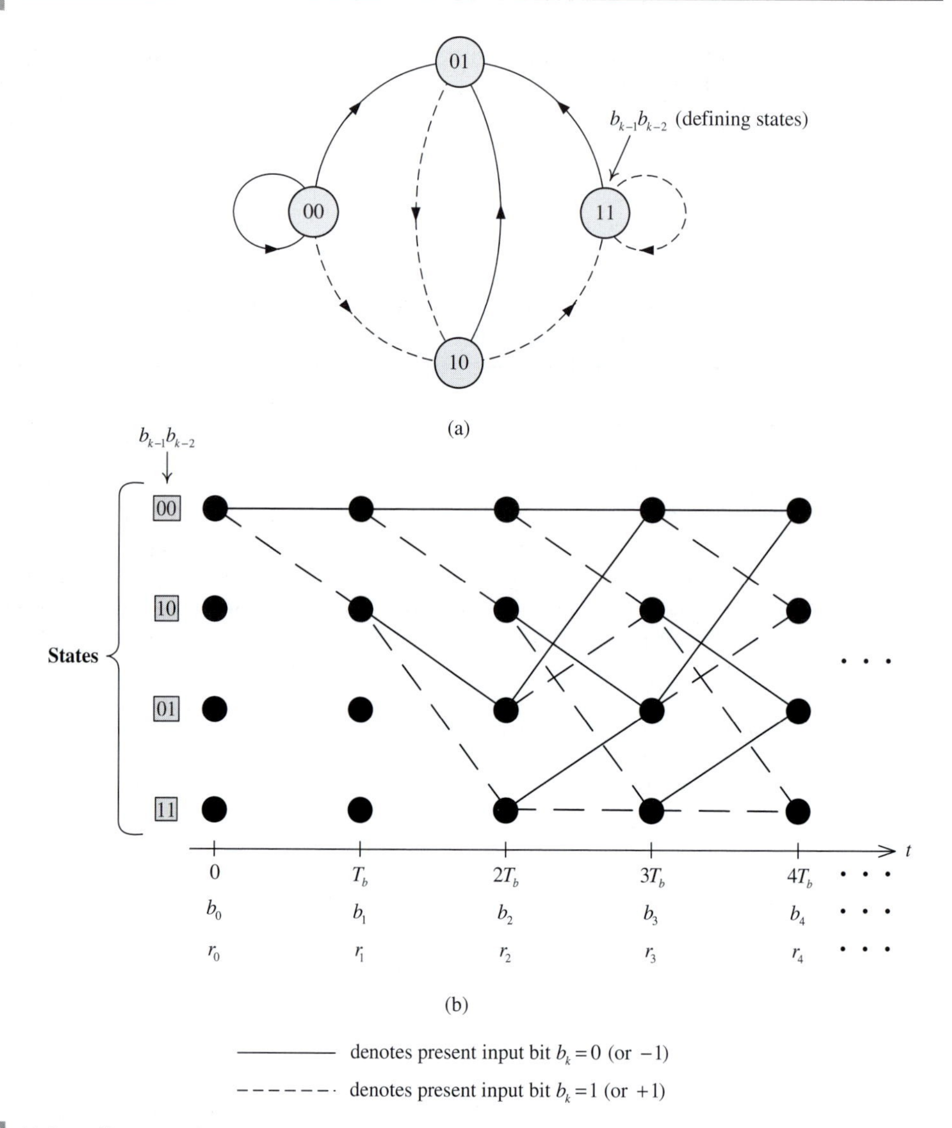

Fig. 9.25 (a) State diagram and (b) trellis for the ISI example.

(3) The starting state of the trellis was chosen to be 00. The choice is arbitrary, but the transmitter and the receiver should agree on what it is. Before $t = 0$ it is assumed that everything is zero, which means that there is no interference at $t = 0$.
(4) After two bit intervals, $t = 2T_b$, the trellis is fully developed and it is the same from bit interval to bit interval. In general the trellis becomes fully developed at $(L - 1)T_b$.
(5) If two paths diverge at a state, it takes at least three bit intervals ($3T_b$) before they can meet at a state. In general this would be L bit intervals.

To proceed with the example, consider the input bit sequence given in Table 9.1, a sequence of length 20. The table also contains a sequence of 20 zero-mean Gaussian

Table 9.1 Examples of transmitted sequence, noise sequence, and matched filter's sample outputs

k	0	1	2	3	4	5	6
b_k	0	1	0	1	0	0	1
b_k	−1	+1	−1	+1	−1	−1	+1
y_k	−1.0	0.4	−0.6	0.6	−0.6	−1.4	0.2
w_k	−0.074	0.059	0.116	0.336	−0.216	−0.163	0.165
r_k	−1.074	0.459	−0.484	0.936	−0.816	−1.563	0.365

k	7	8	9	10	11	12	13
b_k	1	0	1	0	0	0	0
b_k	+1	−1	+1	−1	−1	−1	−1
y_k	1.4	−0.2	0.6	−0.6	−1.4	−1.8	−1.8
w_k	−0.062	0.220	0.050	0.247	0.113	0.311	0.080
r_k	1.338	0.020	0.650	−0.353	−1.287	−1.489	−1.720

k	14	15	16	17	18	19
b_k	1	1	0	0	1	0
b_k	+1	+1	−1	−1	+1	−1
y_k	0.2	1.4	−0.2	−1.4	0.2	−0.6
w_k	−0.296	−0.054	−0.181	−0.034	0.189	−0.177
r_k	−0.096	1.346	−0.381	−1.434	0.389	−0.777

samples, generated by Matlab. The noise variance, σ^2, was set so that the SNR $= E_b/\sigma^2$ was 16 dB. This means that, if the transmitted energy per bit is set at $E_b = 1$ joule, then $\sigma = 10^{-\text{SNR}/20} = 0.158$. The sample output is $r_k = y_k + w_k$, where $y_k = b_k + 0.6b_{k-1} + 0.2b_{k-2}$. Both sequences $\{r_k\}$ and $\{y_k\}$ are also given in Table 9.1.

The branch metric, $b_k r_k - b_k \sum_{j=1}^{2} b_{k-j}h_j$, becomes $b_k r_k - 0.6b_k b_{k-1} - 0.2b_k b_{k-2}$, where as mentioned earlier subscript i, which refers to a specific path through the trellis, is suppressed and b_k is ± 1. Table 9.2 sets up the systematic computation of the metric for branches emanating from the trellis states.

Branch metrics are now computed for the first three bit transmissions. Figure 9.26 shows the results along with the (*partial*) path metrics for all possible paths for the first three bit transmissions.

It is at this point that the Viterbi algorithm comes into play. Consider the two paths 000 and 100, which merge at state 00 at $k = 3$. Their path metrics are -0.301 and -0.849, respectively. Regardless of what happens in succeeding bit transmissions, path 100 can never be a part of an overall path where γ_i is maximum. Therefore, it can be discarded. At this stage 000 is a survivor path and -0.301 is the corresponding survivor metric, which is also known as a *state metric*. Similar statements hold for the other three states. This procedure is repeated for each succeeding bit transmission where a branch metric is computed for that bit interval, *added* to the state (survivor) metric from which the branch emanates,

Table 9.2 Branch metric computation for the example under consideration

b_{k-2}	b_{k-1}	b_k	Branch metric
0 (−1)	0 (−1)	0 (−1)	$-r_k - 0.6 - 0.2 = -r_k - 0.8$
0 (−1)	0 (−1)	1 (+1)	$+r_k + 0.6 + 0.2 = +r_k + 0.8$
0 (−1)	1 (+1)	0 (−1)	$-r_k + 0.6 - 0.2 = -r_k + 0.4$
0 (−1)	1 (+1)	1 (+1)	$+r_k - 0.6 + 0.2 = +r_k - 0.4$
1 (+1)	0 (−1)	0 (−1)	$-r_k - 0.6 + 0.2 = -r_k - 0.4$
1 (+1)	0 (−1)	1 (+1)	$+r_k + 0.6 - 0.2 = +r_k + 0.4$
1 (+1)	1 (+1)	0 (−1)	$-r_k + 0.6 + 0.2 = -r_k + 0.8$
1 (+1)	1 (+1)	1 (+1)	$+r_k - 0.6 - 0.2 = +r_k - 0.8$

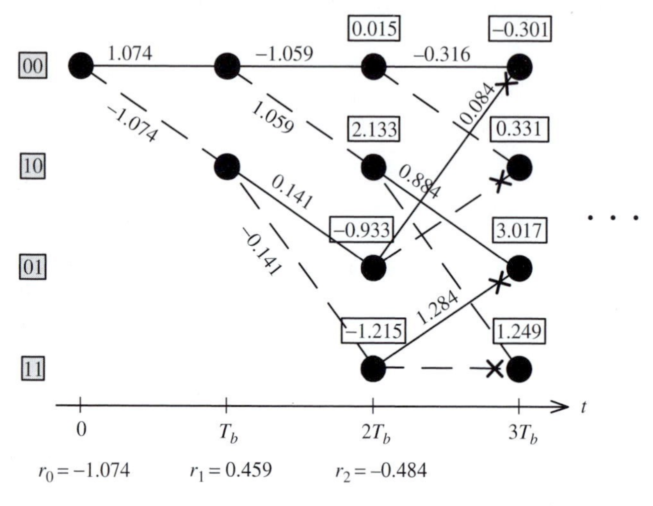

Fig. 9.26 Branch metrics and partial path metrics for the first three bit transmissions. Note that the two branch metrics coming out of any state are the negative of each other.

and then *compared* with the competing path metric at the state where the branch terminates. The trellis is "*pruned*" by retaining only the survivor path at each stage. Figure 9.27 shows the pruned trellis up to $k = 12$.

The use of the Viterbi algorithm to search the trellis graph reduces in this case the number of paths that need to be considered from 2^N, where N is the length of the transmitted bit sequence to just 4, the number of states. One could therefore search the trellis to the end and at that point choose, out of the four paths, the one that has the maximum survivor metric to be the *maximum likelihood sequence*. Though N is taken to be 20 here, typically it would be on the order of hundreds or thousands of bits. This leads to two problems: (i) a large storage requirement in the electronic circuitry, (ii) more fundamentally, an unacceptable delay in outputting the information bits.

These problems can be alleviated by realizing that as one progresses "deep" into the trellis the survivor paths will exhibit a backward merge, i.e., they will all share a common tail. This can be seen in the present example at $k = 4$ (Figure 9.27) where the four survivor

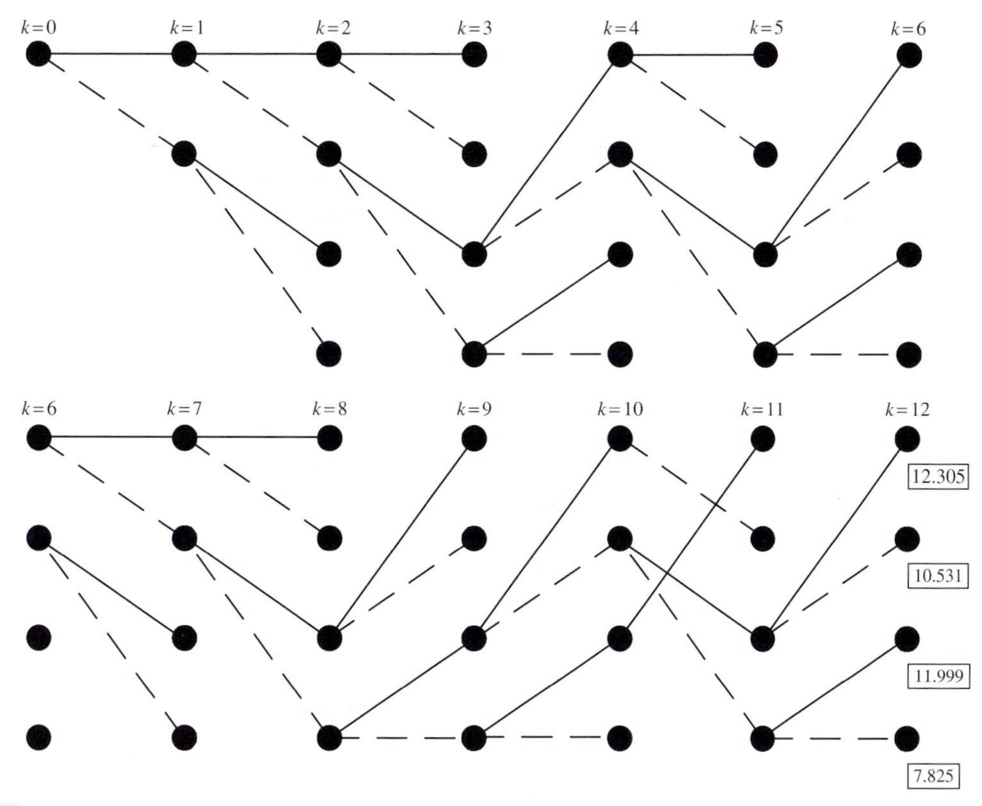

Fig. 9.27 Survivor paths and survivor metrics after 12 bit transmissions.

paths share a common tail of 01. These two bits can then be outputted. The difficulty with this is that backward merges are dependent on the specific noise sequence and hence are random. The number of steps into the trellis needed before a backward merge occurs depends on the severity of the noise, but the further along the trellis one goes, the more likely it is that there will be a backwards merge.

Practically, a rule of thumb is to assume that a length of five times the memory length ($5 \times 2 = 10$ in this example) is sufficient to ensure a backward merge, at least on the bit interval furthest back from the present. One can then output this bit irrespective of whether a merge actually occurred. There are several approaches to deciding on how to choose the value of the outputted bit:

(i) One can take the survivor path with the largest metric and output its "oldest" bit.
(ii) One can take any survivor path and output its "tail" bit.
(iii) One can take a majority vote among the tail end bits of all the survivor paths (four in the current example) to determine the output bit value. Ties would be broken by flipping a coin.

Note that if a merge has occurred, all the approaches result in the same output bit value.

After the decision is made on the tail bit, one goes to the next bit interval and repeats the procedure.

Observe in the present example that if one ignores the ISI and uses simple *symbol-by-symbol* demodulation based on the decision rule:

$$r_k \underset{0_D}{\overset{1_D}{\gtrless}} 0, \tag{9.56}$$

then two bit errors are made. One at $k = 8$ and one at $k = 14$. Applying the Viterbi algorithm for the remainder of the trellis to determine the maximum path metric, γ_i, it can be shown (left as a problem) that the correct bit sequence is demodulated.

The poor performance of the symbol-by-symbol demodulation can be explained by looking at the eye diagram at the output of $h(-t)$ in the absence of AWGN. This is shown in Figure 9.28. As can be seen from Figure 9.28, the eye is almost closed. There are signal points that are only 0.2 units away from the threshold. Based on this worst case scenario (which can happen quite often) and given that $\sigma = 0.158$, the bit error probability is $P[\text{bit error}] = Q\left(\frac{0.2}{0.158}\right) \approx 0.1028$, which agrees well with having two bit errors out of 20 information bits. Alternatively the SNR in the worst case is $20\log_{10}\left(\frac{0.2}{0.158}\right) \approx$ 2 decibels, a far cry from the 16 decibels based on the transmitted energy. Therefore, while the ISI severely degrades the performance of the symbol-by-symbol demodulation, it does not impair the performance of the maximum likelihood sequence demodulation (MLSD). Qualitatively, MLSD takes advantage of the signal energy, i.e., it works with an SNR of 16 decibels.

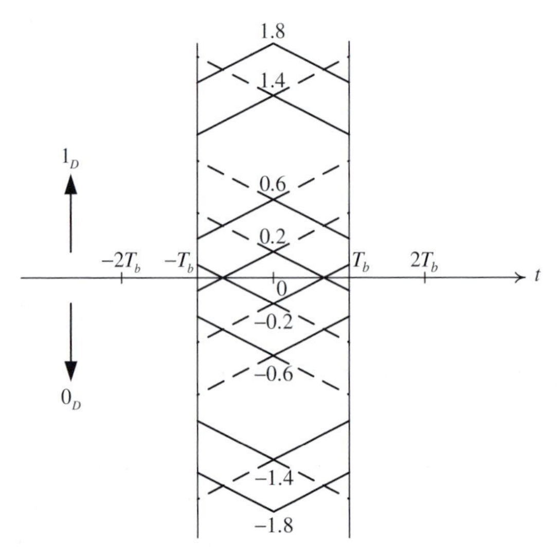

Fig. 9.28 Eye diagram in the absence of AWGN for the overall impulse response considered in the ISI example.

9.5 Summary

Bandlimitation and the subsequent effect of intersymbol interference (ISI) have been discussed in this chapter. Various methods to combat ISI, from eliminating it (Nyquist's criterion) to allowing a controlled amount (partial response signaling) to using a Viterbi algorithm (equalizer) to accomplish a maximum likelihood sequence detection in the presence of any inherent ISI have been described. The phenomenon of ISI is not only present in what maybe considered to be traditional communication systems, ISI is also a problem in digital storage devices, in particular magnetic disk recording systems. The ISI in this application is limited to a few terms with a Viterbi algorithm used to demodulate the recorded bit sequence.

9.6 Problems

9.1 Consider a *bandpass* channel with a bandwidth of 4 kilohertz. Assume that the PSD of AWGN is $N_0/2 = 10^{-8}$ watts/hertz. Design a QAM modulator to deliver a transmission rate of 9600 bits/second. Use a raised-cosine spectrum having a roll-off factor of at least 50% for the combined transmit and receive filters. Also determine the minimum transmitted power to achieve a bit error probability of 10^{-6}. *Hint* The result on excess bandwidth to achieve zero ISI for the binary baseband system can be extended to an M-ary passband system by replacing the bit rate with the symbol rate.

9.2 Figure 9.29 illustrates the frequency response of a typical voice-band telephone channel. You are asked to design a modem that transmits at a symbol rate of 2400 symbols/second, with the objective of achieving 9600 bits/second. It is also decided that the raised-cosine spectrum is used for the overall system.

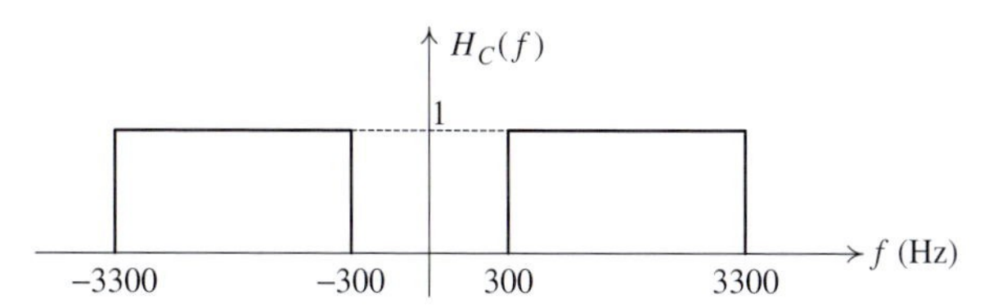

Fig. 9.29 Frequency response considered in Problem 9.2.

(a) Select an appropriate QAM signal constellation and the roll-off factor β of the raised-cosine spectrum that utilizes the entire frequency band.

(b) Sketch the spectrum of the transmitted signal and label all the relevant frequencies.

9.3 Consider a channel with the following frequency response $H_C(f) = 1 + \alpha\cos(2\pi f T_s)$. Determine the frequency response characteristic of the optimum transmit and receive filters that yields zero ISI at a rate of $1/T_s$ symbols/second and has 75% access bandwidth. Assume that the additive Gaussian noise is *white*.

9.4 Consider a lowpass channel with bandwidth W and the following frequency response:

$$H_C(f) = \begin{cases} 1 + \alpha\cos(2\pi f t_0), & -W \le f \le W,\ |\alpha| < 1 \\ 0, & \text{otherwise.} \end{cases} \tag{P9.1}$$

An input signal $s(t)$ whose spectrum is bandlimited to W (hertz) is passed through the channel.

(a) Show that, in the absence of AWGN, the output of the channel is

$$y(t) = s(t) + \tfrac{1}{2}\alpha[s(t - t_0) + s(t + t_0)], \tag{P9.2}$$

i.e., the channel produces a pair of echoes.

(b) The received signal $y(t)$ is passed through a filter matched to $s(t)$. Determine the output of the matched filter at $t = kT$, $k = 0, \pm 1, \pm 2, \ldots$. *Hint* The output of the matched filter is $y(t) * s(T - t) = \int_{-\infty}^{\infty} y(\lambda)s(t - (T - \lambda))\mathrm{d}\lambda$.

(c) What is the ISI pattern at the output of the matched filter if $t_0 = T$? Comment.

9.5 Consider transmitting binary antipodal signals, $\pm s(t)$, over a nonideal bandlimited channel. Let E_b and T_b be the signal energy (i.e., energy per bit) and signaling interval, respectively. The channel introduces ISI over two adjacent symbols. In the absence of AWGN, the output of the matched filter is $\sqrt{E_b}$ at $t = T_b$, $\sqrt{E_b}/4$ at $t = 2T_b$, and zero for $t = kT_b$, $k > 2$.

(a) Assume that the two signals are equally probable and the two-sided PSD of AWGN is $N_0/2$. Determine the average probability of bit error.

(b) Plot the bit error probability obtained in (a) and that for the case of no ISI. Determine the relative difference in SNR $= E_b/N_0$ (decibels) for the bit error probability of 10^{-5}.

9.6 Figure 9.30 shows the system model for binary baseband signaling over a bandlimited channel. The source rate is $r_b = 1/T_b$ and the overall response $S_R(f) = H_T(f) \cdot H_C(f) \cdot H_R(f)$ of the system is shown in Figure 9.31 in which K is an arbitrary constant.

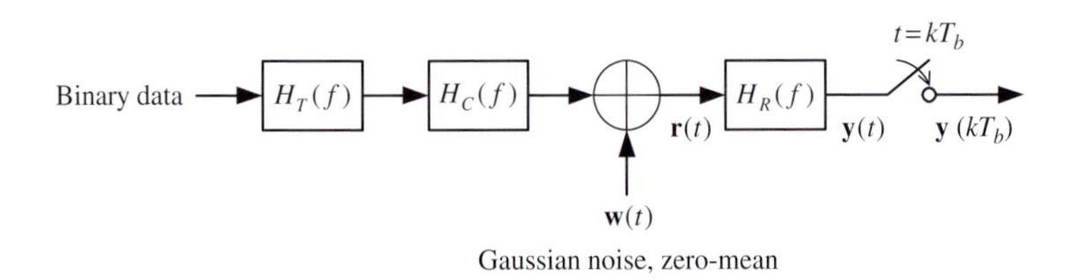

Fig. 9.30 System model considered in Problem 9.6.

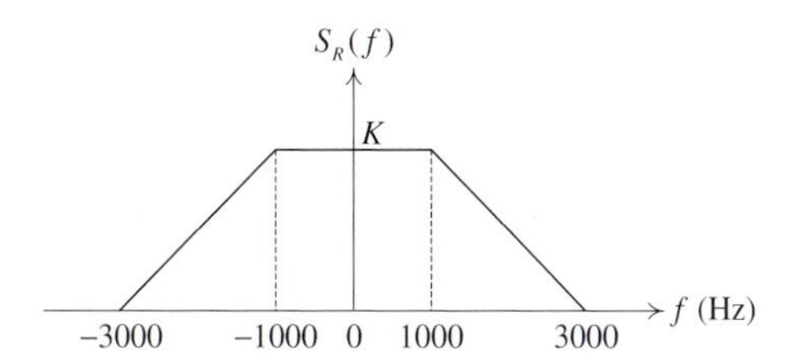

Fig. 9.31 The overall response considered in Problem 9.6.

(a) At what rate would you transmit if you desire the ISI terms to be zero at the sampling instants? Explain.

(b) If you signal at rates slower than 4×10^3 hertz does it follow that the ISI is zero? Explain.

(c) What is the excess bandwidth of the above system.

9.7 Your friend has been asked to design a binary digital communication system for a baseband channel whose bandwidth is $W = 3000$ hertz. The block diagram of the system is shown in Figure 9.32. Having learnt that ISI generally occurs for a

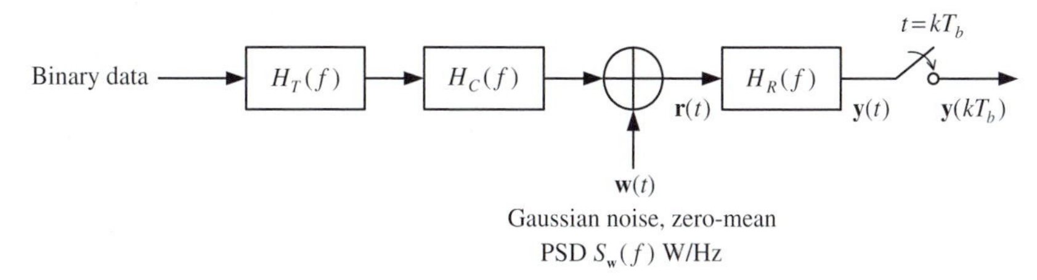

Fig. 9.32 System model considered in Problem 9.7.

bandlimited channel, which can severely degrade the system performance, she proposes two possible choices for the overall system response $S_R(f) = H_T(f) \cdot H_C(f) \cdot H_R(f)$ in Figures 9.33(a) and 9.33(b).

(a) For each choice of the overall response, at what rate should your friend transmit if she desires the ISI terms to be zero at the sampling instants? Explain.

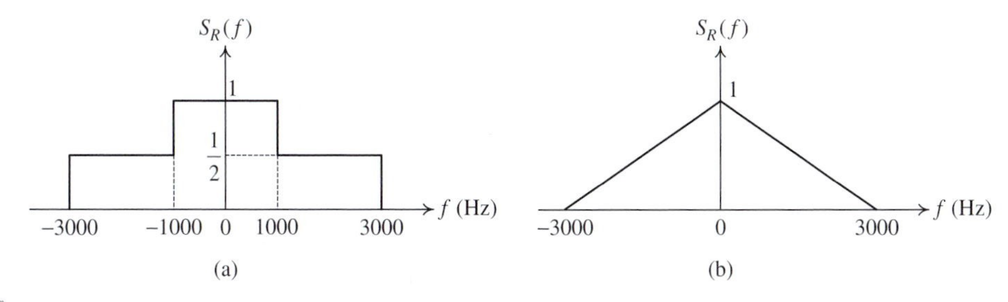

Fig. 9.33 The two overall responses, (a) and (b), considered in Problem 9.7.

(b) From your answers in (a), what is the excess bandwidth of the system for each choice of the overall response?

(c) Being aware of the raised-cosine spectrum, you suggest it to your friend as an alternative design for the spectrum in Figure 9.33(a). Neatly sketch the raised-cosine spectrum on top of the spectrum in Figure 9.33(a).

(d) Your friend considers your suggestion by examining the eye diagrams in Figure 9.34. What should be your friend's choice for the overall spectrum? Explain.

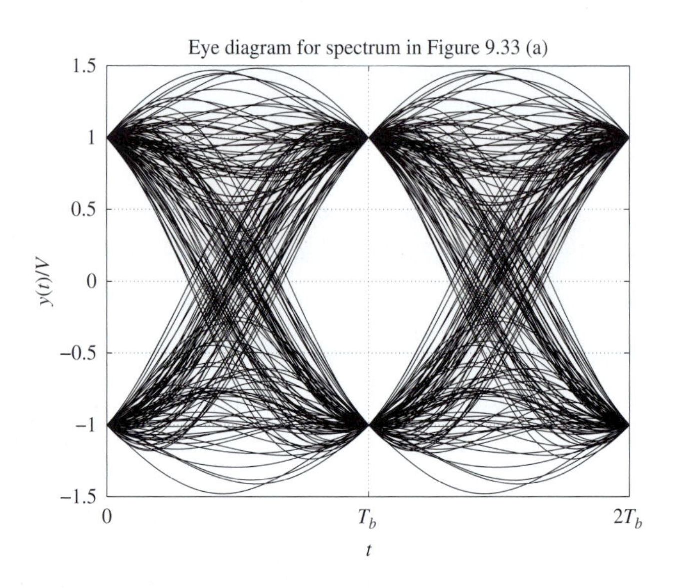

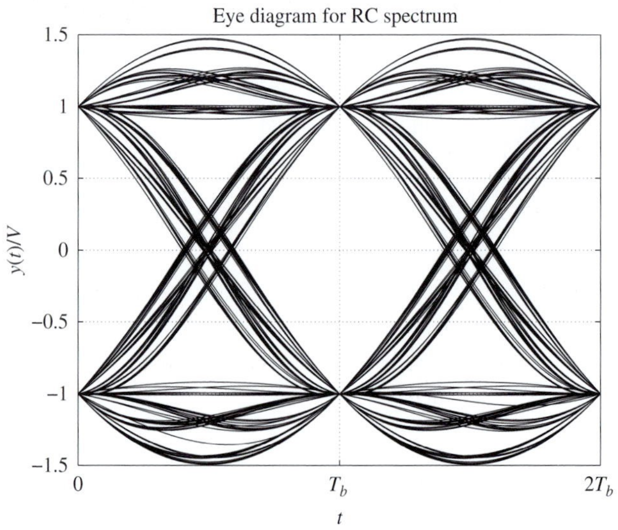

Fig. 9.34 Two eye diagrams considered in Problem 9.7.

9.8 For each of the overall system responses $S_R(f) = H_T(f) \cdot H_C(f) \cdot H_R(f)$ in Figure 9.35 determine if one can signal at a rate which would result in zero ISI, and if so what the signaling rate r_b (bits/second) is.

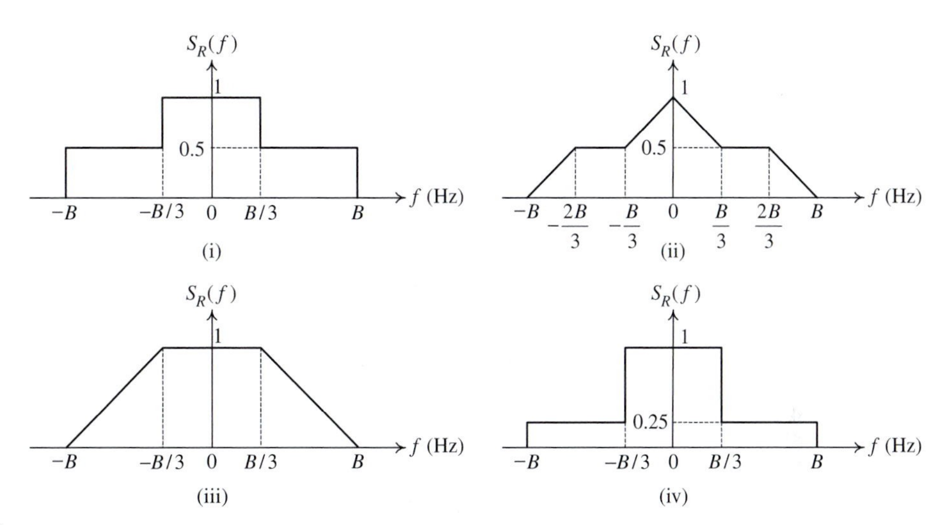

Fig. 9.35 The four overall system responses considered in Problem 9.8.

9.9 The binary sequence 110010101101 is the input to a precoder whose output is used to modulate a duobinary transmit filter. Complete Table 9.3, which shows the precoded sequence, the transmitted amplitude levels, the received signal levels, and the decoded sequence (all in the absence of the channel noise).

Table 9.3 Binary signaling with duobinary pulses

Time index k	0	1	2	3	4	5	6	7	8	9	10	11	12
Information bit $\{b_k\}$		1	1	0	0	1	0	1	0	1	1	0	1
Precoded bit $\{d_k\}$	0												
Transmitted level													
Received level													
Decoded bit													

9.10 Consider a generalization of the discussion of duobinary to PRS. Let $s_R(t)$ be nonzero only at a finite number of sampled values, say N, i.e., $s_R(kT_s) = s_k$, for $k = 0, 1, \ldots, N-1$ and $s_R(kT_s) = 0$ otherwise.

(a) Show that

$$S_R(f) = \begin{cases} \sum_{k=0}^{N-1} T_b s_k \mathrm{e}^{-\mathrm{j}2\pi k T_b}, & -1/2T_b \leq f \leq 1/2T_b \\ 0, & \text{otherwise} \end{cases}. \tag{P9.3}$$

(b) Show that $s_R(t) = \sum_{k=0}^{N-1} s_k \sin(\pi(t - kT_b))/\pi(t - kT_b)$.

Remark It is common to assume that the samples are nonzero at values kT_b, where k ranges in a symmetrical range around zero, i.e., $-(N-1)/2 \le k \le (N-1)/2$.

9.11 The spectrum $S_R(f)$ of duobinary modulation has most of its energy concentrated around $f = 0$, a feature that is undesirable in communication systems that have AC coupling to eliminate DC and low frequency drift. To overcome this you propose, instead of a "cosine" spectrum, a "sine" spectrum on the basis that it will be zero at $f = 0$. The spectrum $S_R(f)$ is then of the form $\sin(2\pi f T_b)$ which ensures that it is also zero at $f = \pm 1/2T_b$. Since $s_R(t)$ is a real-time function, $S_R(f)$ must have a magnitude spectrum that is even and an odd phase spectrum. Bearing this in mind you arrive at the following:

$$S_R(f) = \begin{cases} \mathrm{j}2T_b \sin(2\pi f T_b), & -1/2T_b \le f \le 1/2T_b \\ 0, & \text{otherwise} \end{cases}. \qquad \text{(P9.4)}$$

(a) Determine the overall response $s_R(t)$ and plot the sampled values $s_R(kT_b)$. How does $s_R(t)$ decay with t?

(b) Determine the equation for the sampled output, $\mathbf{y}_k$. How many interfering terms are there? What is the number of received levels?

(c) Design a precoder to eliminate the memory and show the decision space when this precoder is used.

(d) What is the SNR degradation, in decibels, (with precoding) over the ideal binary case?

9.12 A partial response system has the sampled overall impulse response shown in Figure 9.36.

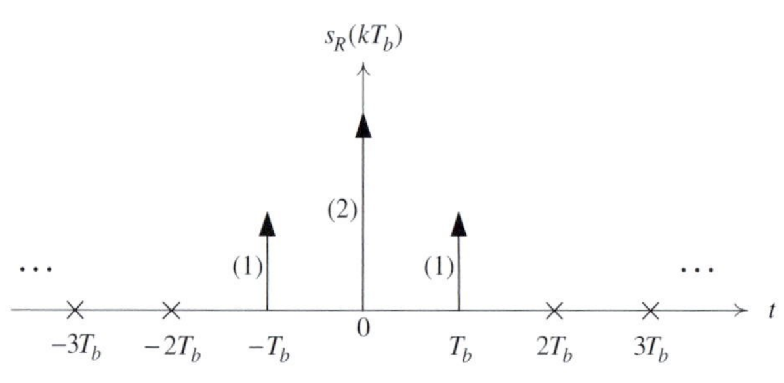

Fig. 9.36 The sampled overall impulse response considered in Problem 9.12.

(a) Show that the frequency spectrum $S_R(f)$ is

$$S_R(f) = \begin{cases} 4T_b \cos^2(\pi f T_b), & -1/2T_b \le f \le 1/2T_b \\ 0, & \text{otherwise} \end{cases}. \qquad \text{(P9.5)}$$

(b) How does the time domain response decay with t?

(c) Derive the equation for the sampled output.

(d) How many received levels are there?

(e) Design a precoder to eliminate the ISI and show the decision regions when this precoder is used.

(f) What is the SNR degradation, in decibels, over the ideal binary case?

9.13 Repeat Problem 9.12 for the sampled overall impulse response shown in Figure 9.37.

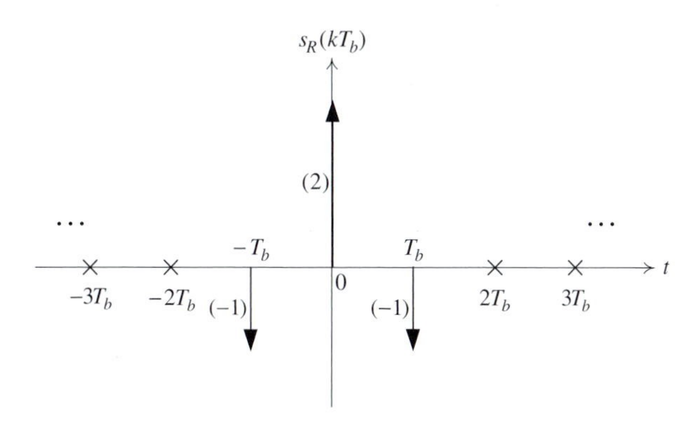

Fig. 9.37 The sampled overall impulse response considered in Problem 9.13.

9.14 Use Matlab to obtain the eye diagrams for the duobinary modulation and for the PRS of Problem 9.11. Discuss the differences, if significant.

9.15 Use Matlab to obtain the eye diagrams for the lowpass raised-cosine filter considered as an example in the chapter (see Figure 9.3). Let $RC = 0.2T_b,\ T_b,\ 2T_b$.

10 Signaling over fading channels

10.1 Introduction

Up to now we have assumed that the transmitted signal is only degraded by AWGN. Even when it is subjected to filtering, as in the previous chapter, the filtering characteristics are known precisely by the receiver. This knowledge is exploited in the design of the modulator/demodulator by employing Nyquist's criterion to avoid intersymbol interference (ISI), or by allowing a certain amount of ISI as in the case of partial response systems, or by using a maximum likelihood sequence detection based on the unavoidable ISI.

In practice, however, there arise communication channels where the received signal is not subjected to a known transformation or filtering. In particular the gain and/or phase of a digitally modulated transmitted signal is not known precisely at the receiver. These parameters can be modeled as either unknown but fixed over the period of transmission or as random. In the former case, one could transmit a known signal briefly at the beginning of transmission to estimate the parameter(s) and then use the estimate(s) for the remainder of the transmission, which would be the message of interest. However, in the more typical application, the parameters do change in time, so though they may remain reasonably constant over a bit interval, or several bit intervals, they do change over the course of the entire message transmission, typically unpredictably. It is therefore common to model these parameters, which in this context are also known as *nuisance parameters*, as *random*.

This chapter considers channel models where the amplitude and/or phase of the received signal is random. It starts by considering amplitude effects on binary digital modulations. This is followed by a channel model where the phase is random which introduces an additional complication into the analysis and design of the optimum receiver. The last section of this chapter considers the most important channel model: one where both amplitude and phase are random. The model applies to channels where "scattering" occurs; these are found in ionospheric point-to-point transmission and in wireless communications. Fading due to scattering causes a severe degradation in error performance due to "deep" fades where the signal is essentially obliterated. Diversity transmission as a means of overcoming deep fades concludes the chapter.

10.2 Demodulation with random amplitude

This is a pure fading model in which only the amplitude of the received signal is changed. In general the received signal is

$$\mathbf{r}(t) = \mathbf{a}s(t) + \mathbf{w}(t), \tag{10.1}$$

where $s(t)$ is the transmitted signal, $\mathbf{w}(t)$ is the usual ubiquitous AWGN, and $\mathbf{a}$ is a random variable with known pdf, $f_{\mathbf{a}}(a)$. It is assumed that $\mathbf{a}$ is statistically independent of both $s(t)$ and $\mathbf{w}(t)$. Further, $\mathbf{a} \geq 0$, since a negative value implies not only gain distortion but also phase distortion, namely one of $180°$. Logically one would feel that $\mathbf{a} \leq 1$ since values larger than 1 would mean that the received signal energy is greater than the transmitted signal energy. However $\mathbf{a} > 1$ does not mean that an increase in the received signal energy is beneficial. Depending on the modulation it can lead to a degradation of performance.

Consider now binary modulation and the three basic schemes, on–off keying, antipodal and orthogonal, which in passband are known as BASK, BPSK, and BFSK. The signal space plots and decision regions are shown in Figure 10.1. For simplicity it is assumed here and for the rest of the chapter that the bits (or symbols) are equally probable. Also the transmitted energy is E for a *nonzero* signal. Since the two signals used in BPSK and BFSK are of equal energy, the transmitted energy per bit, commonly denoted by E_b, is simply E in these two modulation schemes. On the other hand $E_b = E/2$ in BASK since one signal in this scheme is zero.

If the transmitted energy is E joules, then the received energy is $\mathbf{a}^2E$ or the signal point is at a distance of $\mathbf{a}\sqrt{E}$ from the origin. Figure 10.1 shows the signal points for the three signal sets for a specific value of $\mathbf{a}$. The following observations are readily made from Figure 10.1. The optimum receiver remains the same for antipodal and orthogonal signaling for any $\mathbf{a} \geq 0$. This is a not unexpected result, since the information is not carried by the amplitude. The error performance, however, is affected. It is given by

$$\mathbf{P}[\text{error}] = Q\left(\mathbf{a}\sqrt{\frac{2E_b}{N_0}}\right) \quad \text{(antipodal)}, \tag{10.2}$$

$$\mathbf{P}[\text{error}] = Q\left(\mathbf{a}\sqrt{\frac{E_b}{N_0}}\right) \quad \text{(orthogonal)}. \tag{10.3}$$

In essence, the $\mathbf{P}$[error] is a *random variable*. To determine the average error probability one needs to determine

$$E\{\mathbf{P}[\text{error}]\} = \int_0^\infty Q\left(a\sqrt{\frac{2E_b}{N_0}}\right) f_{\mathbf{a}}(a)\mathrm{d}a \quad \text{(antipodal)}, \tag{10.4}$$

$$E\{\mathbf{P}[\text{error}]\} = \int_0^\infty Q\left(a\sqrt{\frac{E_b}{N_0}}\right) f_{\mathbf{a}}(a)\mathrm{d}a \quad \text{(orthogonal)}. \tag{10.5}$$

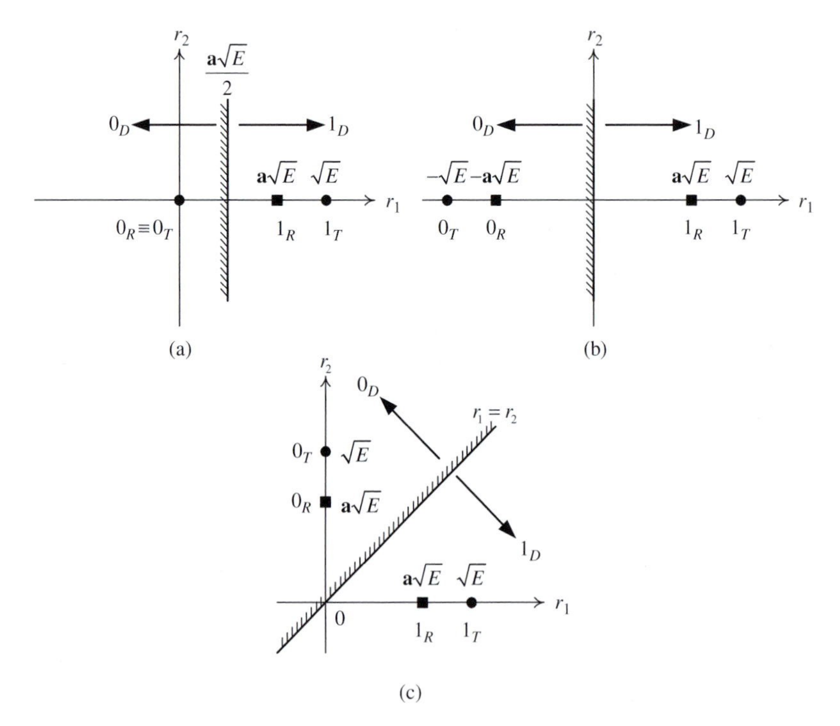

Fig. 10.1 Signal space plots and decision regions of three binary modulations with random amplitude: (a) BASK; (b) BPSK; (c) BFSK.

Though the evaluation of the integrals can be complicated or only done numerically, it can be shown that

$$\int_0^{\infty} Q(ax) f_{\mathbf{a}}(a) \mathrm{d}a \geq Q(m_{\mathbf{a}} x), \tag{10.6}$$

where $m_{\mathbf{a}} = E\{\mathbf{a}\}$. The equality holds if and only if $\mathbf{a}$ is not random, which means $f_{\mathbf{a}}(a) = \delta(a - m_{\mathbf{a}})$.

Consider now BASK. The optimum receiver is changed since if $\mathbf{a} = a$ then the receiver threshold becomes $a\sqrt{E}/2$. The minimum bit error probability and the optimum receiver are still given by determining the following likelihood ratio:

$$\frac{f_{\mathbf{r}_1}(r_1|1_T)}{f_{\mathbf{r}_1}(r_1|0_T)} \underset{0_D}{\overset{1_D}{\gtrless}} 1,$$

which means that the two pdfs need to be determined. The conditional pdf, $f_{\mathbf{r}_1}(r_1|0_T)$, is obtained by observation. It is Gaussian, zero-mean, variance $N_0/2$, the usual mantra. The conditional pdf $f_{\mathbf{r}_1}(r_1|1_T)$ is not as straightforward since when a “1” is transmitted the sufficient statistic becomes $\mathbf{r}_1 = \mathbf{a}\sqrt{E} + \mathbf{w}_1$, where $\mathbf{w}_1$ is Gaussian, zero-mean, variance $N_0/2$. The conditional pdf $f_{\mathbf{r}_1}(r_1|1_T)$ can then be determined in two ways. One method is to determine the pdf of $\mathbf{a}\sqrt{E}$ which is $(1/\sqrt{E}) f_{\mathbf{a}}\left(a/\sqrt{E}\right)$, and convolve this density with that of $f_{\mathbf{w}_1}(w_1)$ to obtain $f_{\mathbf{r}_1}(r_1|1_T)$. Another approach is used here. Assume that $\mathbf{a} = a$, where

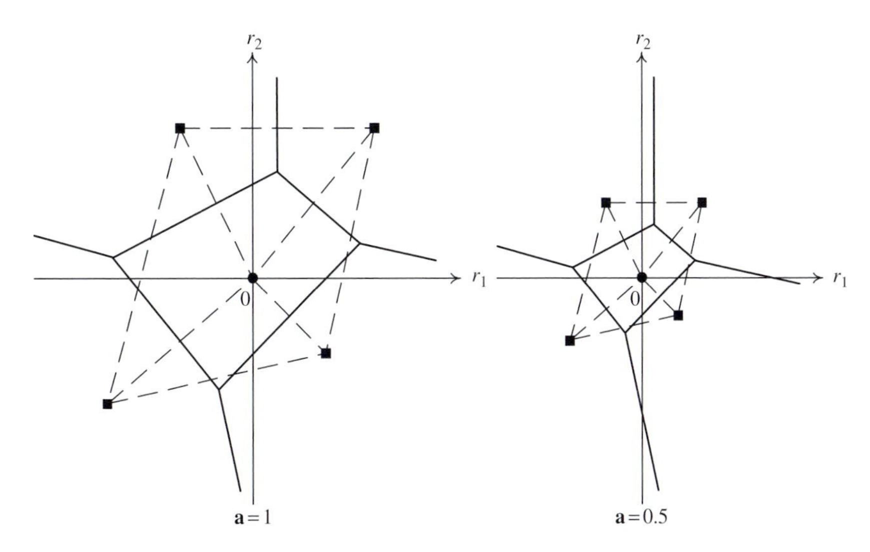

Fig. 10.2 Illustration of random amplitude effect on the decision regions.

a is a specific value. Then $f_{\mathbf{r}_1}(r_1|1_T, \mathbf{a}=a)$ is Gaussian, mean value $a\sqrt{E}$, variance $N_0/2$. Average this pdf over all possible values of $\mathbf{a}$ where the weights in the average are given by $f_{\mathbf{a}}(a)\mathrm{d}a$. Therefore:

$$f_{\mathbf{r}_1}(r_1|1_T) = \int_0^{\infty} f_{\mathbf{r}_1}(r_1|1_T, \mathbf{a}=a) f_{\mathbf{a}}(a)\mathrm{d}a \tag{10.7}$$

$$= E\left\{f_{\mathbf{r}_1}(r_1|1_T, \mathbf{a}=a)\right\}. \tag{10.8}$$

Whether the above expectation can be evaluated and meaningfully interpreted depends on $f_{\mathbf{a}}(a)$. The threshold (and hence the decision regions) is therefore a balance between the different regions given by the values that $\mathbf{a}$ takes, weighted by the probabilities that $\mathbf{a}$ takes these values. This is true, even in the general M-ary case, as illustrated in Figure 10.2, for a two-dimensional signal plot.

Note, however, that if all the signal points lie at distance of $\sqrt{E_s}$ from the origin, i.e., they are of equal energy E_s, then the optimum decision regions are *invariant* to any scaling by $\mathbf{a}$, provided of course that $\mathbf{a} \geq 0$. This is exactly what was observed for the binary antipodal and orthogonal signaling. The matched-filter or correlation receiver structure is still optimum, one does not even need to know $f_{\mathbf{a}}(a)$. The error performance, as noted earlier, does depend crucially on $\mathbf{a}$ and $f_{\mathbf{a}}(a)$.

10.3 Demodulation with random phase

Phase uncertainty can arise due to different effects. One possibility is slow drift in the receiver's local oscillator that is used to demodulate the incoming signal to baseband.

Another common effect is changes in the propagation time of the signal between the transmitter and receiver. To illustrate this consider BPSK modulation used at a carrier frequency[1] of 1 gigahertz. The transmitted signal is $\pm V\cos(2\pi f_c t)$. The received signal is $\pm V\cos(2\pi f_c(t - t_d))$, where t_d is the propagation delay from the transmitter to the receiver. If constant, then the receiver oscillator can adjust and therefore the reference signal used at the demodulator would be $\sqrt{2/T_b}\cos(2\pi f_c t - \theta_d)$, where $\theta_d = 2\pi f_c t_d$. However, typically t_d varies due to changes in the propagation path and looks like $t_d = t_{\text{nominal}} \pm \Delta t$. The signal at the receiver is now $\pm V\cos(2\pi f_c(t - t_{\text{nominal}} \pm \Delta t)) = \pm V\cos(2\pi f_c t - \theta_{\text{nominal}} \pm \Delta\theta)$, where $\theta_{\text{nominal}} = 2\pi f_c t_{\text{nominal}}$ and $\Delta\theta = 2\pi f_c \Delta t$. Though the receiver oscillator may adjust to the phase θ_{nominal}, it is typically very difficult to track the $\Delta\theta$ perturbations in phase which can be quite large. Indeed, if Δt is on the order of $1/f_c$, which at 1 gigahertz is 10^{-9} seconds, the phase perturbation can vary from 0 to 2π as the propagation time Δt changes from 0 to 10^{-3} microseconds. Significant phase uncertainly can therefore be introduced quite readily into the received signal.

Based on the above discussion, we shall model the phase uncertainty as a *random variable*, with a uniform pdf in the range 0 to 2π or ($-\pi$ to π). It would be expected that using BPSK as a modulation technique in the face of this phase uncertainty is a dead-end street. However, BASK and BFSK may have potential for communications and we explore this next. In what follows we ignore the known delay t_{nominal} and set it to zero. As usual the bits are assumed to be equally probable and we have AWGN. Observe also that the phase uncertainty does not change the energy of the received signal, i.e., $\sqrt{E}\sqrt{2/T_b}\cos(2\pi f_c t + \alpha)$ has the same energy for any α. Figure 10.3(a) shows the signal space plots at the receiver. For completeness BPSK is also shown.

10.3.1 Optimum receiver for noncoherent BASK

Consider the optimum receiver for BASK. The signal space of Figure 10.3(a) shows that when a "1" is transmitted the signal lies on the circumference of a circle. One would therefore conjecture that the optimum decision boundary consists of a circle enclosing the signal point at the origin, radius to be determined. A radius of $\sqrt{E}/2$ seems plausible. To check these conjectures and intuitions the receiver is derived next.

The received signal is:

$$\mathbf{r}(t) = \begin{cases} \mathbf{w}(t), & \text{if “}0_T\text{”} \\ \sqrt{E}\sqrt{2/T_b}\cos(2\pi f_c t - \boldsymbol{\theta}) + \mathbf{w}(t), & \text{if “}1_T\text{”} \end{cases}. \tag{10.9}$$

It is relatively straightforward to show that a set of *sufficient statistics* is generated by projecting the received signal onto the two basis functions $\phi_I(t) = \sqrt{2/T_b}\cos(2\pi f_c t)$ and $\phi_Q(t) = \sqrt{2/T_b}\sin(2\pi f_c t)$. The sufficient statistics become

[1] Mobile communication systems are in the region of 800 MHz; personal communication systems lie in the region of 1.6 gigahertz so the chosen carrier frequency is quite representative.

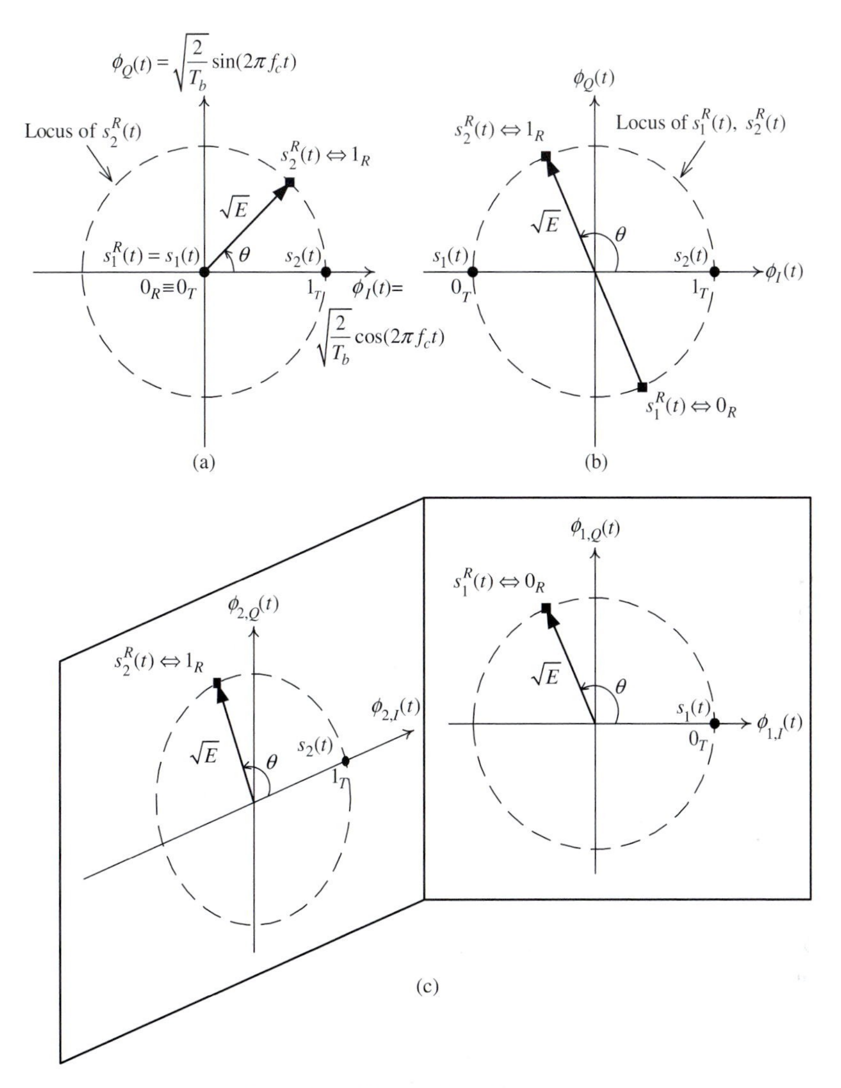

Fig. 10.3 Signal space plots of (a) BASK, (b) BPSK, and (c) BFSK with random phase.

$$\mathbf{r}_I = \begin{cases} \mathbf{w}_I, & \text{if "}0_T\text{"} \\ \sqrt{E}\cos\boldsymbol{\theta} + \mathbf{w}_I, & \text{if "}1_T\text{"} \end{cases}, \quad \mathbf{r}_Q = \begin{cases} \mathbf{w}_Q, & \text{if "}0_T\text{"} \\ \sqrt{E}\sin\boldsymbol{\theta} + \mathbf{w}_Q, & \text{if "}1_T\text{"} \end{cases}, \tag{10.10}$$

where $\mathbf{w}_I$ and $\mathbf{w}_Q$ are statistically independent Gaussian random variables, with zero-mean, variance $N_0/2$. As before one needs to determine the likelihood ratio, $f(r_I, r_Q|1_T)/f(r_I, r_Q|0_T)$. The conditional pdf $f(r_I, r_Q|0_T)$ is easily determined. It equals $f(r_I|0_T)f(r_Q|0_T) = (1/\pi N_0)\exp\left(-(r_I^2 + r_Q^2)/N_0\right)$.

To determine $f(r_I, r_Q|1_T)$ the same approach as in Section 10.2 is used. Namely we assume a specific value for $\boldsymbol{\theta}$, determine $f(r_I, r_Q|1_T, \boldsymbol{\theta} = \theta)$, and then average over all values of $\boldsymbol{\theta}$. The result is

$$f(r_I, r_Q|1_T) = \int_0^{2\pi} \frac{1}{\pi N_0} \exp\left[-\frac{\left(r_I - \sqrt{E}\cos\theta\right)^2 + \left(r_Q - \sqrt{E}\sin\theta\right)^2}{N_0}\right] \frac{1}{2\pi} \mathrm{d}\theta, \tag{10.11}$$

which simplifies to

$$f(r_I, r_Q|1_T) = \left[\frac{1}{\pi N_0} \mathrm{e}^{-(r_I^2+r_Q^2)/N_0} \mathrm{e}^{-E/N_0}\right] \left[\frac{1}{2\pi} \int_0^{2\pi} \mathrm{e}^{(2\sqrt{E}/N_0)(r_I\cos\theta + r_Q\sin\theta)} \mathrm{d}\theta\right]. \tag{10.12}$$

The integral can be written as

$$\frac{1}{2\pi} \int_0^{2\pi} \mathrm{e}^{(2\sqrt{E}/N_0)\sqrt{r_I^2+r_Q^2}\cos(\theta - \tan^{-1}(r_Q/r_I))} \mathrm{d}\theta$$

and recognized to be a *modified Bessel function of the first kind*,[2] namely $I_0\left((2\sqrt{E}/N_0)\sqrt{r_I^2 + r_Q^2}\right)$. Therefore

$$f(r_I, r_Q|1_T) = \frac{1}{\pi N_0} \mathrm{e}^{-(r_I^2+r_Q^2)/N_0} \mathrm{e}^{-E/N_0} I_0\left(\frac{2\sqrt{E}}{N_0}\sqrt{r_I^2 + r_Q^2}\right). \tag{10.13}$$

The likelihood ratio becomes:

$$\mathrm{e}^{-E/N_0} I_0\left(\frac{2\sqrt{E}}{N_0}\sqrt{r_I^2 + r_Q^2}\right) \underset{0_D}{\overset{1_D}{\gtrless}} 1, \tag{10.14}$$

which, since $I_0(\cdot)$ is a monotonic (increasing) function, can be written as

$$\sqrt{r_I^2 + r_Q^2} \underset{0_D}{\overset{1_D}{\gtrless}} \frac{N_0}{2\sqrt{E}} I_0^{-1}\left(\mathrm{e}^{E/N_0}\right). \tag{10.15}$$

From (10.15) it is readily seen that the decision region is a circle in the sufficient statistic space, (r_I, r_Q), of radius $T_h = (N_0/2\sqrt{E}) I_0^{-1}\left(\mathrm{e}^{E/N_0}\right)$. This is shown in Figure 10.4.

It is interesting to observe that the optimum threshold depends not only on the energy E of the nonzero signal, but also on the noise level, N_0. This means that the optimum threshold is different from the radius $\sqrt{E}/2$ that we conjectured before. It can be shown, however, that if the signal-to-noise ratio E/N_0 is sufficiently high (which is typically the case in practice), the optimum threshold approaches $\sqrt{E}/2$ quite closely. Nevertheless, there is a noticeable performance degradation due to the use of the simpler suboptimum threshold of $\sqrt{E}/2$, as illustrated later in this section.

A block diagram of the optimum receiver is shown in Figure 10.5. The generation of the test statistic $\sqrt{r_I^2 + r_Q^2}$ can be done differently. Consider the received signal applied to the filter shown in Figure 10.6. The output is given by

[2] This function is defined as $I_0(x) \triangleq \frac{1}{2\pi} \int_0^{2\pi} \mathrm{e}^{x\cos\theta} \mathrm{d}\theta$.

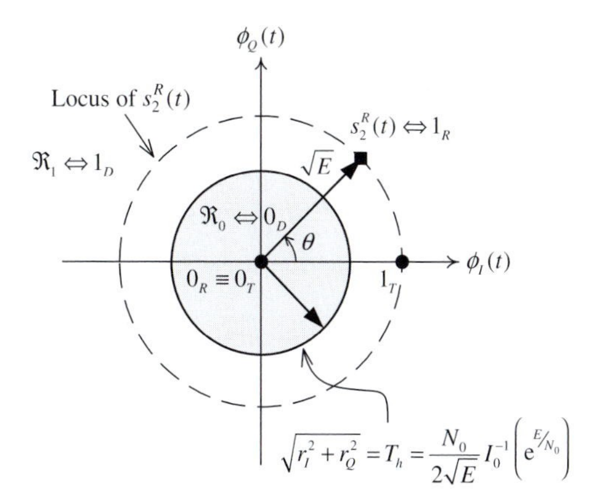

Fig. 10.4 Decision regions of BASK with random phase.

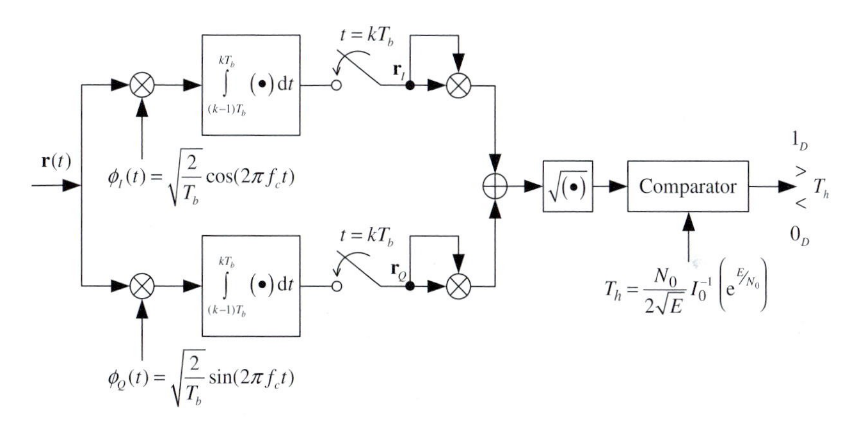

Fig. 10.5 Block diagram of the optimum receiver for BASK under random phase.

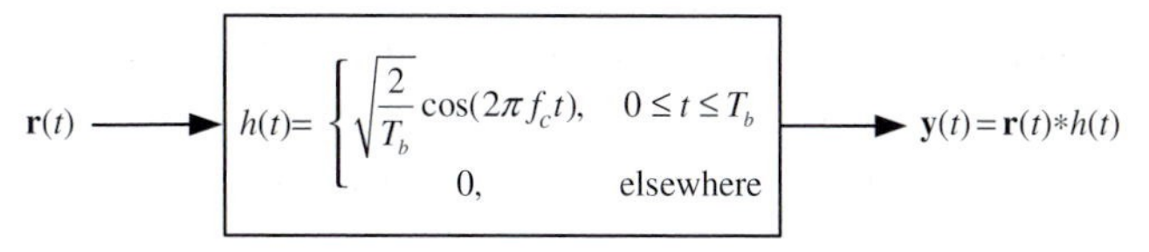

Fig. 10.6 Computation of $y(t)$.

$$\begin{aligned}
y(t) &= \int_{-\infty}^{\infty} r(\lambda)\sqrt{\frac{2}{T_b}}\cos(2\pi f_c(t-\lambda))[u(t-\lambda)-u(t-\lambda-T_b)]\mathrm{d}\lambda \\
&= \underbrace{\left[\int_{t-T_b}^{t} r(\lambda)\sqrt{\frac{2}{T_b}}\cos(2\pi f_c\lambda)\mathrm{d}\lambda\right]}_{y_I(t)}\cos(2\pi f_c t) \\
&\quad + \underbrace{\left[\int_{t-T_b}^{t} r(\lambda)\sqrt{\frac{2}{T_b}}\sin(2\pi f_c\lambda)\mathrm{d}\lambda\right]}_{y_Q(t)}\sin(2\pi f_c t) \\
&= y_I(t)\cos(2\pi f_c t) + y_Q(t)\sin(2\pi f_c t) \\
&= \sqrt{y_I^2(t)+y_Q^2(t)}\cos\left[2\pi f_c\left(t-\tan^{-1}\frac{y_Q(t)}{y_I(t)}\right)\right],
\end{aligned} \tag{10.16}$$

where $\sqrt{y_I^2(t)+y_Q^2(t)}$ is the envelope of $y(t)$. But at the sampling instant, $t = kT_b$, one has $y_I(kT_b) = r_I$ and $y_Q(kT_b) = r_Q$, the sufficient statistics for the kth bit. Therefore the envelope is precisely $\sqrt{r_I^2+r_Q^2}$, the test statistic. Therefore the optimum receiver can be realized by the block diagram of Figure 10.7. This form of optimum demodulator realization lends itself to the following interpretation. The matched filter, as usual, extracts the maximum energy from the transmitted signal. However, because of the phase uncertainty, at the filter output, only the envelope is looked at. For this reason the receiver is frequently known as a *noncoherent demodulator*. Note also that the filter is a bandpass filter centered at the carrier frequency f_c.

Turning now to the error performance, $P[\text{error}|0_T]$ is computed quite easily. It is given by

$$P[\text{error}|0_T] = \iint_{\Re_1} f(r_I, r_Q|0_T)\mathrm{d}r_I\mathrm{d}r_Q = \iint_{\Re_1} \frac{1}{\pi N_0}\mathrm{e}^{-(r_I^2+r_Q^2)/N_0}\mathrm{d}r_I\mathrm{d}r_Q, \tag{10.17}$$

which, using polar coordinates, becomes:

$$P[\text{error}|0_T] = \frac{1}{\pi N_0}\int_{\alpha=0}^{2\pi}\int_{\rho=T_h}^{\infty}\rho\mathrm{e}^{-\rho^2/N_0}\mathrm{d}\rho\mathrm{d}\alpha = \mathrm{e}^{-T_h^2/N_0}. \tag{10.18}$$

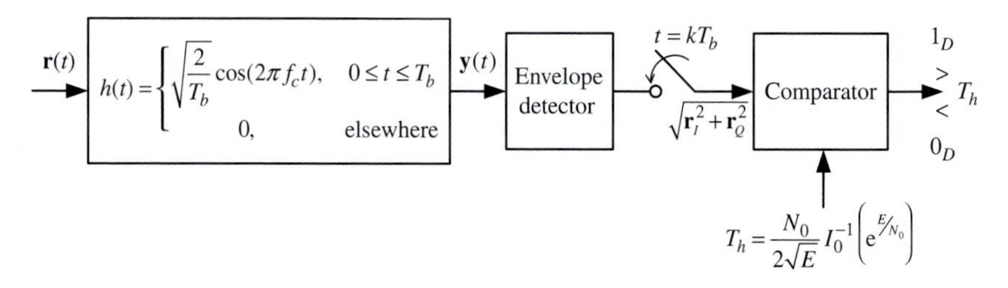

Fig. 10.7 A different implementation of the optimum noncoherent demodulation of BASK.

The error probability when a "1" is transmitted is not as readily evaluated. We have

$$P[\text{error}|1_T] = 1 - P[\text{correct}|1_T] = 1 - \iint_{\Re_1} f(r_I, r_Q|1_T)\mathrm{d}r_I\mathrm{d}r_Q. \tag{10.19}$$

Using (10.13) for $f(r_I, r_Q|1_T)$ and expressing the variables in polar coordinates the integral becomes

$$\begin{aligned}\frac{1}{\pi N_0}\int_{\alpha=0}^{2\pi}\int_{\rho=T_h}^{\infty} \rho \mathrm{e}^{-\rho^2/N_0}\mathrm{e}^{-E/N_0} I_0\left(\frac{2\sqrt{E}}{N_0}\rho\right)\mathrm{d}\rho\mathrm{d}\alpha \\ = \frac{2}{N_0}\int_{\rho=T_h}^{\infty} \rho \mathrm{e}^{-(\rho^2+E)/N_0} I_0\left(\frac{2\sqrt{E}}{N_0}\rho\right)\mathrm{d}\rho.\end{aligned} \tag{10.20}$$

The integral cannot be evaluated analytically, but to recognize it as a tabulated function, let $\lambda = \sqrt{2/N_0}\rho$. Then the integral becomes $\int_{\sqrt{2/N_0}T_h}^{\infty} \lambda \mathrm{e}^{-(\lambda^2+2E/N_0)/2} I_0\left(\sqrt{2E/N_0}\lambda\right)\mathrm{d}\lambda$. This is now in the form of a function commonly called *Marcum's Q*-function, defined as

$$Q(\alpha, \beta) = \int_{\beta}^{\infty} x\mathrm{e}^{-(x^2+\alpha^2)/2} I_0(\alpha x)\mathrm{d}x. \tag{10.21}$$

Identify $\alpha = \sqrt{2E/N_0}$ and $\beta = \sqrt{2/N_0}T_h$. The error probability is then

$$P[\text{error}|1_T] = 1 - Q\left(\sqrt{\frac{2E}{N_0}}, \sqrt{\frac{2}{N_0}}T_h\right). \tag{10.22}$$

Finally the overall error probability is given by

$$P[\text{error}] = \frac{1}{2}\mathrm{e}^{-T_h^2/N_0} + \frac{1}{2}\left[1 - Q\left(\sqrt{\frac{2E}{N_0}}, \sqrt{\frac{2}{N_0}}T_h\right)\right]. \tag{10.23}$$

Figure 10.8 plots the above error performance as a function of E_b/N_0, where $E_b = E/2$ is the average energy per bit, along with the coherent case, i.e., when the phase θ is known at the receiver. It can be seen that, over the error probability range of 10^{-2}–10^{-6}, noncoherent BASK is only about 0.5–1.0 decibels less efficient in power than its coherent version. Also plotted in Figure 10.8 is the error performance of noncoherent BASK when the "intuitive," suboptimum threshold of $\sqrt{E}/2 = \sqrt{E_b/2}$ is applied. Observe that there is about a 0.3 decibel penalty in power when using such a simpler suboptimum threshold instead of the optimum one.

10.3.2 Optimum receiver for noncoherent BFSK

Consider now BFSK in which the frequency separation between two signals is such that they are "noncoherently" orthogonal as discussed in Chapter 7. The two possible transmitted signals are

$$s(t) = \begin{cases}\sqrt{E}\sqrt{2/T_b}\cos(2\pi f_1 t), & \text{if "}0_T\text{"} \\ \sqrt{E}\sqrt{2/T_b}\cos(2\pi f_2 t), & \text{if "}1_T\text{"}\end{cases}. \tag{10.24}$$

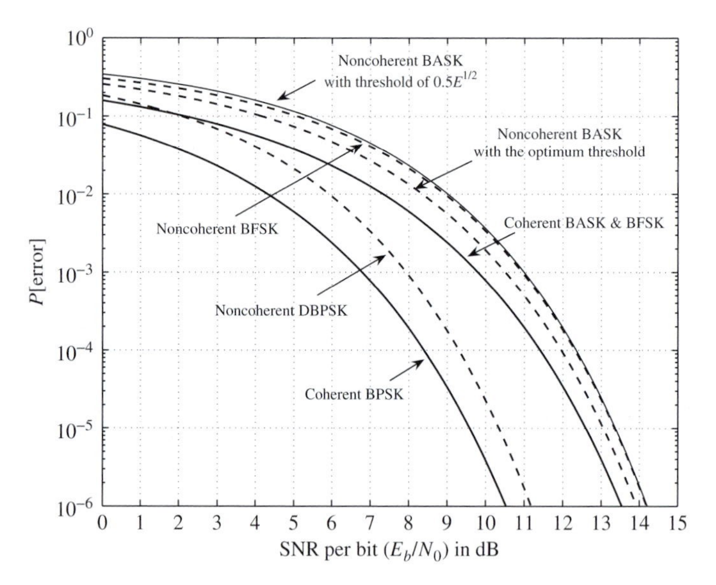

Fig. 10.8 Error performance of coherent and noncoherent demodulations of binary signaling schemes.

The received signal is

$$\mathbf{r}(t) = \begin{cases} \sqrt{E}\sqrt{2/T_b}\cos(2\pi f_1 t - \boldsymbol{\theta}) + \mathbf{w}(t), & \text{if "}0_T\text{"} \\ \sqrt{E}\sqrt{2/T_b}\cos(2\pi f_2 t - \boldsymbol{\theta}) + \mathbf{w}(t), & \text{if "}1_T\text{"} \end{cases}. \tag{10.25}$$

The relevant signal space at the receiver is 4-dimensional with basis functions $\sqrt{2/T_b}\cos(2\pi f_1 t)$, $\sqrt{2/T_b}\sin(2\pi f_1 t)$, $\sqrt{2/T_b}\cos(2\pi f_2 t)$, $\sqrt{2/T_b}\sin(2\pi f_2 t)$. Projecting $\mathbf{r}(t)$ onto these basis functions one gets the following sufficient statistics:

$$\underline{0_T} \qquad\qquad\qquad\qquad \underline{1_T}$$

$$\begin{aligned} \mathbf{r}_{1,I} &= \sqrt{E}\cos\boldsymbol{\theta} + \mathbf{w}_{1,I} \\ \mathbf{r}_{1,Q} &= \sqrt{E}\sin\boldsymbol{\theta} + \mathbf{w}_{1,Q} \\ \mathbf{r}_{2,I} &= \mathbf{w}_{2,I} \\ \mathbf{r}_{2,Q} &= \mathbf{w}_{2,Q} \end{aligned} \tag{10.26}$$

$$\begin{aligned} \mathbf{r}_{1,I} &= \mathbf{w}_{1,I} \\ \mathbf{r}_{1,Q} &= \mathbf{w}_{1,Q} \\ \mathbf{r}_{2,I} &= \sqrt{E}\cos\boldsymbol{\theta} + \mathbf{w}_{2,I} \\ \mathbf{r}_{2,Q} &= \sqrt{E}\sin\boldsymbol{\theta} + \mathbf{w}_{2,Q} \end{aligned} \tag{10.27}$$

The random variables $\mathbf{w}_{1,I}$, $\mathbf{w}_{1,Q}$, $\mathbf{w}_{2,I}$, $\mathbf{w}_{2,Q}$ are Gaussian with zero mean and variance $N_0/2$, uncorrelated, and therefore statistically independent. The conditional pdf of, say, $f(r_{1,I}, r_{1,Q}, r_{2,I}, r_{2,Q}|0_T)$ can therefore be written as $f(r_{1,I}, r_{1,Q}|0_T)f(r_{2,I}|0_T)f(r_{2,Q}|0_T)$. The pdfs $f(r_{2,I}|0_T)$ and $f(r_{2,Q}|0_T)$ are each $\mathcal{N}(0, N_0/2)$ while the pdf $f(r_{1,I}, r_{1,Q}|0_T)$ is determined using the same approach as was done in the BASK case when $f(r_I, r_Q|1_T)$ was evaluated. The final pdfs are

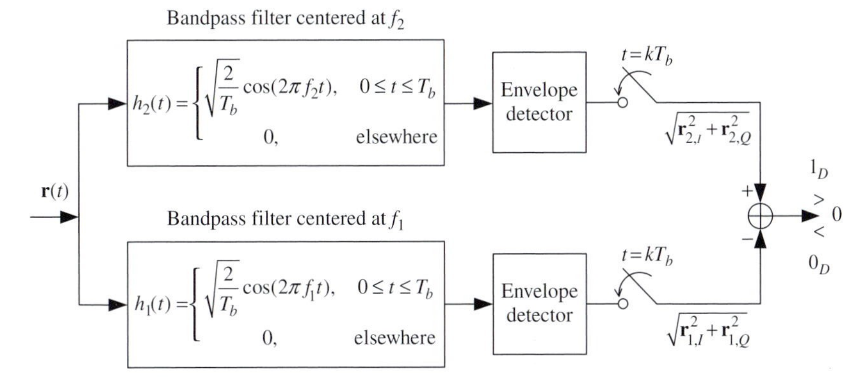

Fig. 10.9 Block diagram of the optimum demodulator for BFSK with random phase.

$$f(r_{1,I}, r_{1,Q}, r_{2,I}, r_{2,Q}|0_T) = \frac{1}{\pi N_0} \mathrm{e}^{-\left(r_{1,I}^2 + r_{1,Q}^2\right)/N_0} \times \mathrm{e}^{-\sqrt{E}/N_0} I_0\left(\frac{2\sqrt{E}}{N_0}\sqrt{r_{1,I}^2 + r_{1,Q}^2}\right) \frac{1}{\pi N_0} \mathrm{e}^{-\left(r_{2,I}^2 + r_{2,Q}^2\right)/N_0}, \tag{10.28}$$

$$f(r_{1,I}, r_{1,Q}, r_{2,I}, r_{2,Q}|1_T) = \frac{1}{\pi N_0} \mathrm{e}^{-\left(r_{1,I}^2 + r_{1,Q}^2\right)/N_0} \times \frac{1}{\pi N_0} \mathrm{e}^{-\left(r_{2,I}^2 + r_{2,Q}^2\right)/N_0} \mathrm{e}^{-\sqrt{E}/N_0} I_0\left(\frac{2\sqrt{E}}{N_0}\sqrt{r_{2,I}^2 + r_{2,Q}^2}\right). \tag{10.29}$$

The likelihood ratio becomes

$$\frac{I_0\left((2\sqrt{E}/N_0)\sqrt{r_{2,I}^2 + r_{2,Q}^2}\right)}{I_0\left((2\sqrt{E}/N_0)\sqrt{r_{1,I}^2 + r_{1,Q}^2}\right)} \underset{0_D}{\overset{1_D}{\gtrless}} 1, \tag{10.30}$$

or since $I_0(\cdot)$ is a monotonically increasing function the decision rule can be simplified to

$$\sqrt{r_{2,I}^2 + r_{2,Q}^2} \underset{0_D}{\overset{1_D}{\gtrless}} \sqrt{r_{1,I}^2 + r_{1,Q}^2}. \tag{10.31}$$

A block diagram of the receiver using a matched filter–envelope detector realization is shown in Figure 10.9. Put simply, the demodulator finds the envelope at the two frequencies and chooses the larger one at the sampling instant.

Consider now the error performance of "noncoherent" BFSK. By symmetry $P[\text{error}] = P[\text{error}|0_T]$. To determine $P[\text{error}]$ observe that the decision rule can be expressed as $\mathbf{r}_{2,I}^2 + \mathbf{r}_{2,Q}^2 \underset{0_D}{\overset{1_D}{\gtrless}} \mathbf{r}_{1,I}^2 + \mathbf{r}_{1,Q}^2$. Fix the random variable $\mathbf{r}_{1,I}^2 + \mathbf{r}_{1,Q}^2$ at a specific value, R^2, and determine

$$P\left(\mathbf{r}_{2,I}^2+\mathbf{r}_{2,Q}^2 \geq R^2 \middle| 0_T, \mathbf{r}_{1,I}^2+\mathbf{r}_{1,Q}^2=R^2\right)$$
$$= P\left[(\mathbf{r}_{2,I}, \mathbf{r}_{2,Q}) \text{ falls outside the circle of radius } R \middle| 0_T\right]. \qquad (10.32)$$

Expressing the resulting integral in polar coordinates the conditional probability becomes

$$P\left[\text{error}\middle| 0_T, \mathbf{r}_{1,I}^2+\mathbf{r}_{1,Q}^2=R^2\right]$$
$$= \frac{1}{\pi N_0}\int_{\alpha=0}^{2\pi}\int_{\rho=R}^{\infty} \rho e^{-\rho^2/N_0} d\rho d\alpha = e^{-R^2/N_0} = e^{-\left(r_{1,I}^2+r_{1,Q}^2\right)/N_0}. \qquad (10.33)$$

Now average $e^{-\left(r_{1,I}^2+r_{1,Q}^2\right)/N_0}$ over all possible values, i.e., find

$$E\left\{e^{-\left(r_{1,I}^2+r_{1,Q}^2\right)/N_0}\middle| 0_T\right\} = \int_{r_{1,I}=-\infty}^{\infty}\int_{r_{1,Q}=-\infty}^{\infty} e^{-\left(r_{1,I}^2+r_{1,Q}^2\right)/N_0} f(r_{1,I}, r_{1,Q}|0_T) dr_{1,I} dr_{1,Q}. \qquad (10.34)$$

To find $f(r_{1,I}, r_{1,Q}|0_T)$ fix the random variable $\boldsymbol{\theta}$ at $\boldsymbol{\theta}=\alpha$, find $f(r_{1,I}, r_{1,Q}|0_T, \boldsymbol{\theta}=\alpha)$ and average over all values of $\boldsymbol{\theta}$. Note that given $\boldsymbol{\theta}$ (and 0_T), $\mathbf{r}_{1,I}$ and $\mathbf{r}_{1,Q}$ are statistically independent random variables, with means $\sqrt{E}\cos\alpha$ and $\sqrt{E}\sin\alpha$, respectively, and variance $N_0/2$. The pdf $f(r_{1,I}, r_{1,Q}|0_T, \boldsymbol{\theta}=\alpha)$ becomes a product of two pdfs, namely $f(r_{1,I}, r_{1,Q}|0_T, \boldsymbol{\theta}=\alpha) = \mathcal{N}\left(\sqrt{E}\cos\alpha, N_0/2\right)\mathcal{N}\left(\sqrt{E}\sin\alpha, N_0/2\right)$. The expectation becomes

$$E\left\{e^{-\left(r_{1,I}^2+r_{1,Q}^2\right)/N_0}\middle| 0_T\right\} = \int_{\alpha=0}^{2\pi}\left[\int_{r_{1,I}=-\infty}^{\infty} e^{-r_{1,I}^2/N_0}\mathcal{N}\left(\sqrt{E}\cos\alpha, \frac{N_0}{2}\right) dr_{1,I}\right]$$
$$\times\left[\int_{r_{1,Q}=-\infty}^{\infty} e^{-r_{1,Q}^2/N_0}\mathcal{N}\left(\sqrt{E}\sin\alpha, \frac{N_0}{2}\right) dr_{1,Q}\right] f_{\boldsymbol{\theta}}(\alpha) d\alpha. \qquad (10.35)$$

The integrations in the square brackets are accomplished by completing the square in the exponent, and manipulating the result so that one has an integral which is the area under a Gaussian function. The two integrals become $(1/\sqrt{2})e^{-E\cos^2\alpha/2N_0}$ and $(1/\sqrt{2})e^{-E\sin^2\alpha/2N_0}$, respectively. Therefore:

$$P[\text{error}] = E\left\{e^{-\left(r_{1,I}^2+r_{1,Q}^2\right)/N_0}\middle| 0_T\right\}$$
$$= \int_{\alpha=0}^{2\pi}\frac{1}{\sqrt{2}}e^{-E\cos^2\alpha/2N_0}\frac{1}{\sqrt{2}}e^{-E\sin^2\alpha/2N_0}\frac{1}{2\pi}d\alpha = \frac{1}{2}e^{-E/2N_0} = \frac{1}{2}e^{-E_b/2N_0}. \qquad (10.36)$$

The error performance of coherent and noncoherent demodulation of BFSK is shown in Figure 10.8, where $E_b = E$ is the energy per bit in BFSK. The performance advantage of coherent BFSK over noncoherent BFSK can be seen clearly. It is more interesting to observe that, although coherent BASK and BFSK perform identically in terms of

E_b/N_0, noncoherent BASK is about 0.3 decibels more power efficient than noncoherent BFSK. This observation can be explained by the fact that phase and frequency are related while amplitude and phase are not. Thus phase uncertainty degrades the performance of BFSK more than BASK.

10.3.3 Differential BPSK and its optimum demodulation

Knowing that coherent BPSK is 3 decibels better than either coherent BASK or BFSK, one wonders if there is any hope of using BPSK on a channel with a uniform phase uncertainty. At first glance, this seems like an oxymoron but a little thought establishes that what is needed is a *phase reference* at the receiver that is matched to the received signal, which for BPSK is $\pm\sqrt{E_b}\sqrt{2/T_b}\cos(2\pi f_c t - \boldsymbol{\theta}) + \mathbf{w}(t)$. Assuming that the phase uncertainty changes relatively slowly with time, indeed let us assume that the change from one bit interval to another is negligible, then the received signal in one bit interval can act as a phase reference for the succeeding bit interval. This is shown by the block diagram of Figure 10.10.

The sampled output is $\mathbf{r}_k = \pm E_b + \mathbf{w}_k + \mathbf{w}_{k-1} + \mathbf{w}_{k,k-1}$, where $\mathbf{w}_{k,k-1} = \left(1/\sqrt{E_b}\right) \int_{(k-1)T_b}^{kT_b} \mathbf{w}(t)\mathbf{w}(t - T_b)\mathrm{d}t$ is a random variable with a zero-mean symmetrical pdf. Based on this the decision rule is as given in Figure 10.10. Note that the decision rule depends on knowledge of the previous bit. This can be circumvented by initially transmitting a known bit, using it to make a decision on the first message (information) bit, and using this decision to make a decision on the next message bit, etc. The difficulty with this approach, however, is that if noise causes an error, then all succeeding bits will be in error, at least until the noise causes another error, which is not a very satisfactory situation. Some (or considerable) thought leads to the observation that the difficulty arises because at the transmitter the bits are mapped into an *absolute phase*, irrespective of what the preceding phase is. Instead of this, let the present phase depend not only on the present bit but also on the

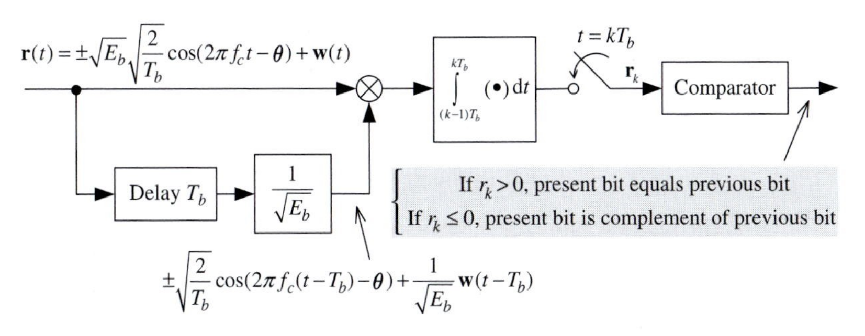

Fig. 10.10 Block diagram of the noncoherent demodulator for BPSK with random phase.

previous phase. This leads to the following modulation:

$$\begin{array}{l}\text{if present bit} = 0\text{, let present phase} = \text{previous phase,}\\ \text{if present bit} = 1\text{, let present phase} = \text{previous phase plus } \pi,\end{array} \tag{10.37}$$

or more succinctly,

$$\begin{array}{l}0_T\text{: no phase change,}\\ 1_T\text{: } \pi \text{ phase change.}\end{array} \tag{10.38}$$

The described modulation is known as *differential* binary phase-shift keying (DBPSK). With DBPSK the sufficient statistic in Figure 10.10 is equal to $\mathbf{r}_k = E_b + \mathbf{w}_k + \mathbf{w}_{k-1} + \mathbf{w}_{k,k-1}$ when a "0" is transmitted and $\mathbf{r}_k = -E_b + \mathbf{w}_k + \mathbf{w}_{k-1} + \mathbf{w}_{k,k-1}$ when a "1" is transmitted. The decision rule is

$$r_k \underset{0_D}{\overset{1_D}{\gtrless}} 0 \tag{10.39}$$

and, most importantly, is independent of the previous decision.

A different demodulator for DBPSK can be obtained by considering the trellis which arises from the memory in the modulator. This is shown in Figure 10.11, where the states are given by the previous phase.

Over the two bit intervals, observe that there are only four distinct transmitted signals, namely:

$$\begin{array}{lcc} & (k-1)\text{th bit interval} & k\text{th bit interval}\\ & \Downarrow & \Downarrow \\ 0_T: & \left\{\begin{array}{r}\sqrt{E_b}\sqrt{2/T_b}\cos(2\pi f_c t)\\ -\sqrt{E_b}\sqrt{2/T_b}\cos(2\pi f_c t)\end{array}\right., & \begin{array}{r}\sqrt{E_b}\sqrt{2/T_b}\cos(2\pi f_c t)\\ -\sqrt{E_b}\sqrt{2/T_b}\cos(2\pi f_c t)\end{array}, \\ 1_T: & \left\{\begin{array}{r}-\sqrt{E_b}\sqrt{2/T_b}\cos(2\pi f_c t)\\ \sqrt{E_b}\sqrt{2/T_b}\cos(2\pi f_c t)\end{array}\right. & \begin{array}{r}\sqrt{E_b}\sqrt{2/T_b}\cos(2\pi f_c t)\\ -\sqrt{E_b}\sqrt{2/T_b}\cos(2\pi f_c t)\end{array}. \end{array} \tag{10.40}$$

The two signals that represent 0_T are antipodal as are the two signals representing 1_T. Further, and most importantly, the two signals that represent 0_T are *orthogonal* to the two signals that represent 1_T. Putting this all together the four signals, at the transmitter, can be represented by the following orthonormal basis set:

$$\left\{\begin{array}{l}\phi_1(t) = \sqrt{1/T_b}\cos(2\pi f_c t)\,[u(t) - u(t-T_b)]\\ \qquad\qquad + \sqrt{1/T_b}\cos(2\pi f_c t)\,[u(t-T_b) - u(t-2T_b)]\\ \phi_2(t) = \sqrt{1/T_b}\cos(2\pi f_c t)\,[u(t) - u(t-T_b)]\\ \qquad\qquad - \sqrt{1/T_b}\cos(2\pi f_c t)\,[u(t-T_b) - u(t-2T_b)]\end{array}\right. . \tag{10.41}$$

At the receiver, the received signals over the two bit intervals are

$$\mathbf{r}(t) = \left\{\begin{array}{ll}\pm\sqrt{2E_b/T_b}\cos(2\pi f_c t - \boldsymbol{\theta})\,[u(t) - u(t-2T_b)] + \mathbf{w}(t), & \text{if "}0_T\text{"}\\ \pm\sqrt{2E_b/T_b}\cos(2\pi f_c t - \boldsymbol{\theta})\{[u(t) - u(t-T_b)] & \\ \qquad\qquad - [u(t-T_b) - u(t-2T_b)]\} + \mathbf{w}(t), & \text{if "}1_T\text{"}\end{array}\right. . \tag{10.42}$$

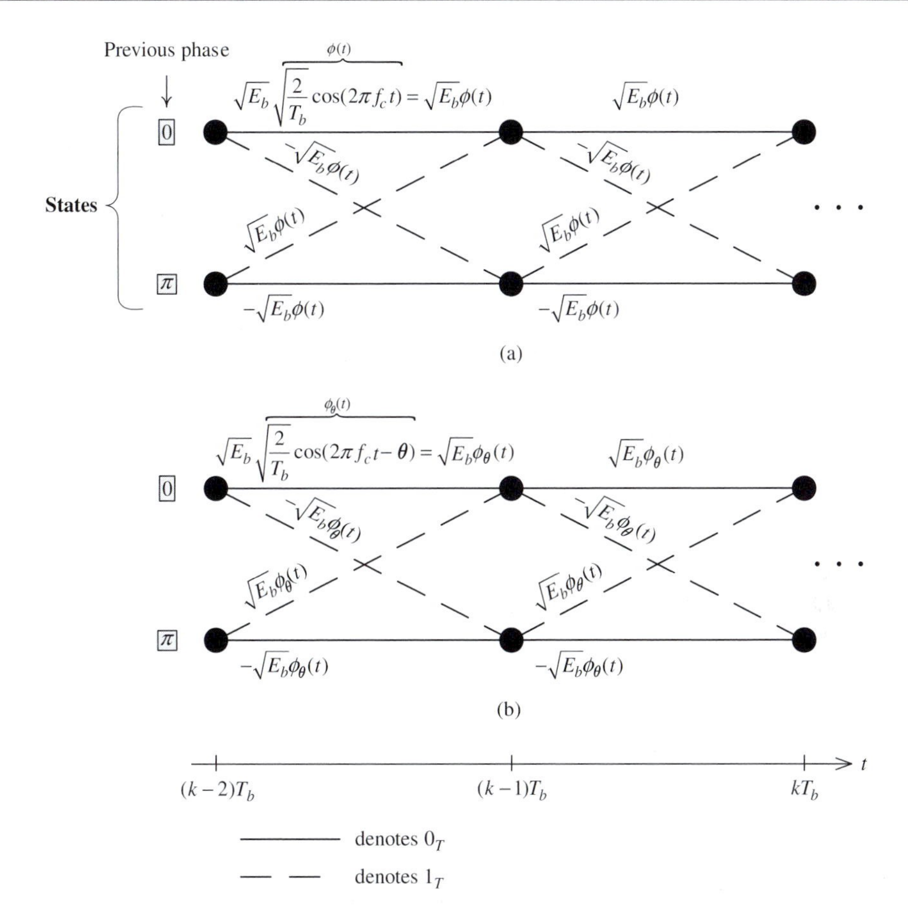

Fig. 10.11 Trellis representation of DBPSK: (a) at the transmitter; (b) at the receiver.

Note that, in both (10.41) and (10.42), for simplicity and without loss of generality, only the first two bit intervals are considered, i.e., $k = 2$. Because of the phase uncertainty at the receiver we shall need four basis functions to represent the signal part of $\mathbf{r}(t)$. These are

$$\begin{cases} \phi_{1,I}(t) = \sqrt{1/T_b}\cos(2\pi f_c t)\,[u(t) - u(t-2T_b)]\,, \\ \phi_{1,Q}(t) = \sqrt{1/T_b}\sin(2\pi f_c t)\,[u(t) - u(t-2T_b)]\,, \\ \phi_{2,I}(t) = \sqrt{1/T_b}\cos(2\pi f_c t)\{[u(t) - u(t-T_b)] \\ \qquad\qquad - [u(t-T_b) - u(t-2T_b)]\}, \\ \phi_{2,Q}(t) = \sqrt{1/T_b}\sin(2\pi f_c t)\{[u(t) - u(t-T_b)] \\ \qquad\qquad - [u(t-T_b) - u(t-2T_b)]\}. \end{cases} \tag{10.43}$$

Since at the receiver the four basis functions completely represent the signal that is due to the transmitted signal, projecting $\mathbf{r}(t)$ onto them generates the sufficient statistics on which to make a decision. These projections are

$$0_T: \begin{cases} \mathbf{r}_{1,I} = \sqrt{2E_b}\cos\boldsymbol{\theta} + \mathbf{w}_{1,I}, \\ \mathbf{r}_{1,Q} = \sqrt{2E_b}\sin\boldsymbol{\theta} + \mathbf{w}_{1,Q}, \\ \mathbf{r}_{2,I} = \mathbf{w}_{2,I}, \\ \mathbf{r}_{2,Q} = \mathbf{w}_{2,Q}, \end{cases} \underline{\text{OR}} \quad \begin{matrix} \mathbf{r}_{1,I} = -\sqrt{2E_b}\cos\boldsymbol{\theta} + \mathbf{w}_{1,I}, \\ \mathbf{r}_{1,Q} = -\sqrt{2E_b}\sin\boldsymbol{\theta} + \mathbf{w}_{1,Q}, \\ \mathbf{r}_{2,I} = \mathbf{w}_{2,I}, \\ \mathbf{r}_{2,Q} = \mathbf{w}_{2,Q}, \end{matrix} \tag{10.44}$$

$$1_T: \begin{cases} \mathbf{r}_{1,I} = \mathbf{w}_{1,I}, \\ \mathbf{r}_{1,Q} = \mathbf{w}_{1,Q}, \\ \mathbf{r}_{2,I} = \sqrt{2E_b}\cos\boldsymbol{\theta} + \mathbf{w}_{2,I}, \\ \mathbf{r}_{2,Q} = \sqrt{2E_b}\sin\boldsymbol{\theta} + \mathbf{w}_{2,Q}, \end{cases} \underline{\text{OR}} \quad \begin{matrix} \mathbf{r}_{1,I} = \mathbf{w}_{1,I}, \\ \mathbf{r}_{1,Q} = \mathbf{w}_{1,Q}, \\ \mathbf{r}_{2,I} = -\sqrt{2E_b}\cos\boldsymbol{\theta} + \mathbf{w}_{2,I}, \\ \mathbf{r}_{2,Q} = -\sqrt{2E_b}\sin\boldsymbol{\theta} + \mathbf{w}_{2,Q}, \end{matrix} \tag{10.45}$$

where $\mathbf{w}_{1,I}$, $\mathbf{w}_{1,Q}$, $\mathbf{w}_{2,I}$, $\mathbf{w}_{2,Q}$ are statistically independent Gaussian random variables, zero-mean, variance $N_0/2$. To form the likelihood ratio $f(r_{1,I}, r_{1,Q}, r_{2,I}, r_{2,Q}|1_T)/f(r_{1,I}, r_{1,Q}, r_{2,I}, r_{2,Q}|0_T)$ one proceeds exactly as before, namely letting $\boldsymbol{\theta} = \theta$ and then averaging the two pdfs over $f_{\boldsymbol{\theta}}(\theta)$. The only concern might be that for either 0_T or 1_T there are two sets of sufficient statistics. However, it turns out that regardless of which set is considered, the conditional pdfs are identical. The optimum decision rule becomes

$$\sqrt{r_{2,I}^2 + r_{2,Q}^2} \underset{0_D}{\overset{1_D}{\gtrless}} \sqrt{r_{1,I}^2 + r_{1,Q}^2}, \tag{10.46}$$

or, equivalently,

$$r_{2,I}^2 + r_{2,Q}^2 \underset{0_D}{\overset{1_D}{\gtrless}} r_{1,I}^2 + r_{1,Q}^2. \tag{10.47}$$

The implementation of the demodulator is shown in Figure 10.12. The demodulator also has a bandpass filter–envelope detector implementation. This is shown in Figure 10.13.

The determination of the error performance of DBPSK follows readily by observing that DBPSK is orthogonal signaling and therefore, in principle, identical to noncoherent BFSK. The error analysis for noncoherent BFSK therefore applies to DBPSK. The only difference is that rather than E_b joules/bit the energy in DBPSK becomes $2E_b$. The factor 2 reflects the fact that the signal over two bit intervals is used to make a decision. The error performance is given by (10.36) and is equal to

$$P[\text{error}]_{\text{DBPSK}} = \tfrac{1}{2}\mathrm{e}^{-E_b/N_0}. \tag{10.48}$$

This is plotted in Figure 10.8 and shows minimal degradation (about 1 decibel) over coherent BPSK.

Communication systems where the channel model exhibits a random phase, though they have a performance loss, still have an error performance that behaves exponentially with the SNR, i.e., $P[\text{error}] \sim \mathrm{e}^{-E_b/N_0}$. The next section discusses a channel model where *both* the amplitude and phase are random. The considered channel is commonly called a Rayleigh fading channel. As shall be seen, its error performance behaves dramatically different.

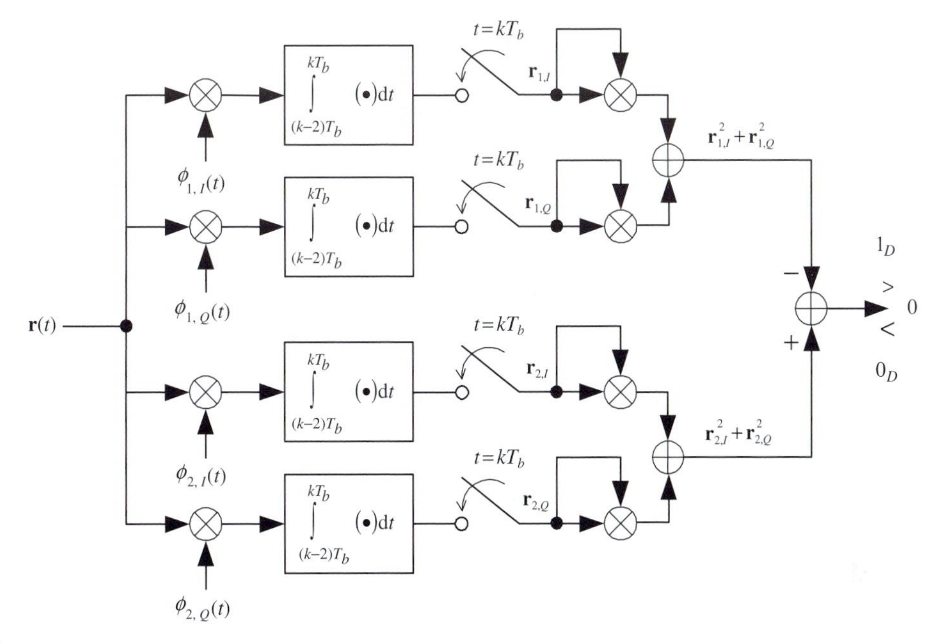

Fig. 10.12 Implementation of the demodulator for DBPSK.

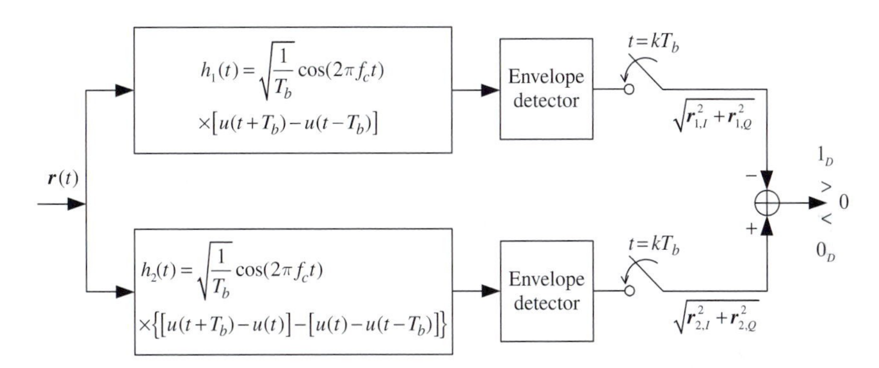

Fig. 10.13 Implementation of the demodulator for DBPSK with a bandpass filter–envelope detector.

10.4 Detection with random amplitude and random phase: Rayleigh fading channel

Nature is seldom kind. The fading channel model presents one of the most severe environments for communications. It arises when there are multiple transmission paths, i.e., multipaths, from the transmitter to the receiver. Two typical situations where this happens are in ionospheric or tropospheric communications and in mobile wireless communications. Figure 10.14 shows these scenarios diagrammatically. In both scenarios it is

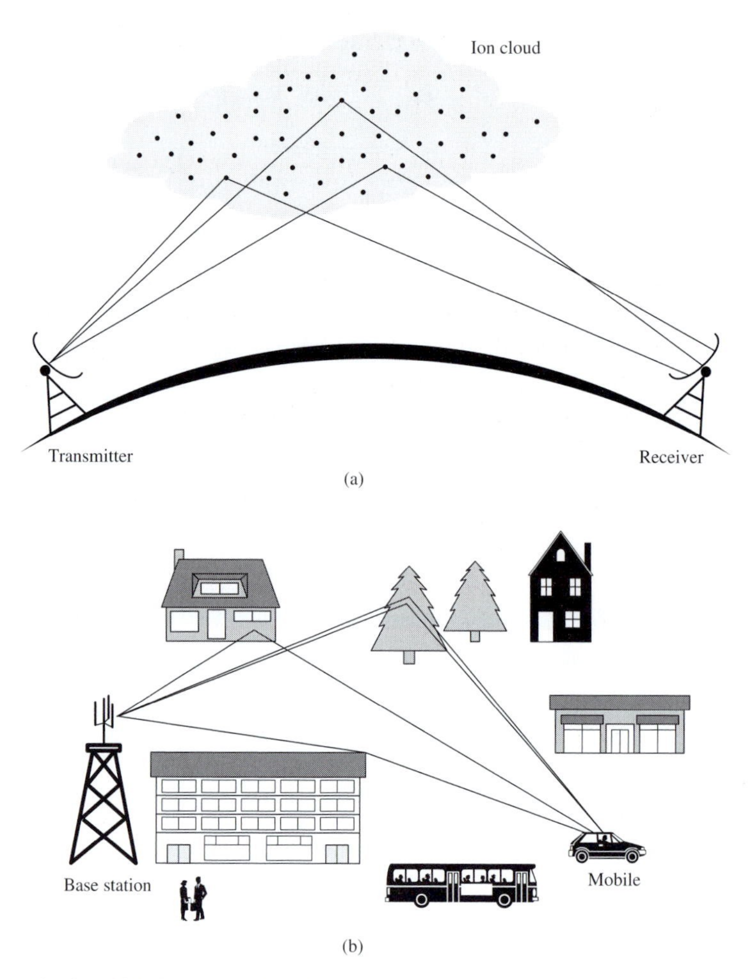

Fig. 10.14 Typical scenarios in which there are multipaths between the transmitter and receiver: (a) ionospheric/tropospheric scattering channel; (b) mobile wireless channel.

assumed that there is no line-of-sight (LOS) path between the transmitter and receiver. This assumption can be relaxed and the effect of having a direct path is pursued in the chapter problems. Having no direct path is the more common situation and unfortunately the more severe situation. A model for the channel is developed next.

10.4.1 Fading channel model

To develop the model consider the following transmitted signal:

$$s_T(t) = s(t)\cos(2\pi f_c t), \tag{10.49}$$

where $s(t)$ is a *lowpass* signal. For us $s(t)$ will be $\pm\sqrt{E_b}\sqrt{2/T_b}$ over the bit interval with bit rate $r_b \ll f_c$. As seen from Figure 10.14 the transmitted signal arrives at the receiver over many different paths and therefore

$$r(t) = \sum_j r_j(t) = \sum_j s(t - t_j)\alpha_j \cos(2\pi f_c(t - t_j)), \tag{10.50}$$

where α_j represents the attenuation and t_j the delay along the jth path. In general, α_j and t_j are time varying due to movement of ions in the cloud, swaying of buildings, wind through the trees, vehicles moving, etc. However, we shall assume that at least over a bit interval they are reasonably constant. Further, the t_j represents variations around an average delay, t_d, which is accounted for by the synchronization at the receiver. Finally it is reasonable to assume that the attenuations and delays are unpredictable, i.e., *random*. The received signal is then

$$\mathbf{r}(t) = s(t)\sum_j \boldsymbol{\alpha}_j \cos\left(2\pi f_c t - 2\pi f_c \mathbf{t}_j\right) = s(t)\sum_j \boldsymbol{\alpha}_j \cos\left(2\pi f_c t - \boldsymbol{\theta}_j\right), \tag{10.51}$$

where, because $s(t)$ is lowpass, the approximation $s(t) \approx s(t - \mathbf{t}_j)$ is used. Since $\mathbf{t}_j$ is on the order of $1/f_c$ we assume that the random phase $\boldsymbol{\theta}_j$ lies in the range $[0, 2\pi)$. The received signal becomes

$$\mathbf{r}(t) = s(t)\left[\left(\sum_j \boldsymbol{\alpha}_j \cos\boldsymbol{\theta}_j\right)\cos(2\pi f_c t) + \left(\sum_j \boldsymbol{\alpha}_j \sin\boldsymbol{\theta}_j\right)\sin(2\pi f_c t)\right]. \tag{10.52}$$

Consider the random variables $(\Sigma_j \boldsymbol{\alpha}_j \cos\boldsymbol{\theta}_j)$ and $(\Sigma_j \boldsymbol{\alpha}_j \sin\boldsymbol{\theta}_j)$. Call them $\mathbf{n}_{F,I}$, $\mathbf{n}_{F,Q}$, respectively, where F denotes fading, and I and Q as usual denote inphase and quadrature. It is reasonable to assume that the attenuation and delay on any path are not related. Further, they are not related between paths. By not related we mean *statistically independent*. Lastly the phase $\boldsymbol{\theta}_j$ is assumed to be uniform over the interval $[0, 2\pi)$. The first and second moments of $\mathbf{n}_{F,I}$, $\mathbf{n}_{F,Q}$ are

$$E\left\{\mathbf{n}_{F,I}\right\} = \sum_j E\{\boldsymbol{\alpha}_j\}E\{\cos\boldsymbol{\theta}_j\} = 0, \tag{10.53}$$

$$E\left\{\mathbf{n}_{F,Q}\right\} = \sum_j E\{\boldsymbol{\alpha}_j\}E\{\sin\boldsymbol{\theta}_j\} = 0, \tag{10.54}$$

$$E\left\{\mathbf{n}_{F,I}^2\right\} = \sum_j E\left\{\boldsymbol{\alpha}_j^2\right\}E\left\{\cos^2\boldsymbol{\theta}_j\right\} = \sigma_F^2/2, \tag{10.55}$$

$$E\left\{\mathbf{n}_{F,Q}^2\right\} = \sum_j E\left\{\boldsymbol{\alpha}_j^2\right\}E\left\{\sin^2\boldsymbol{\theta}_j\right\} = \sigma_F^2/2, \tag{10.56}$$

$$\begin{aligned} E\left\{\mathbf{n}_{F,I}\mathbf{n}_{F,Q}\right\} &= E\left\{\sum_j \boldsymbol{\alpha}_j \cos\boldsymbol{\theta}_j \sum_k \boldsymbol{\alpha}_k \sin\boldsymbol{\theta}_k\right\} \\ &= \sum_j\sum_k E\left\{\boldsymbol{\alpha}_j\boldsymbol{\alpha}_k\right\}\underbrace{E\left\{\cos\boldsymbol{\theta}_j \sin\boldsymbol{\theta}_k\right\}}_{=0} = 0, \end{aligned} \tag{10.57}$$

where, with a slight abuse of notation, $\sigma_F^2 = \Sigma_j E\{\boldsymbol{\alpha}_j^2\}$ is the mean-squared value of the attenuation effect.

Finally, since the number of multipaths is large, we invoke the *central limit theorem* (see Section 10.6), which means that $\mathbf{n}_{F,I}$, $\mathbf{n}_{F,Q}$ are Gaussian random variables, zero-mean, variance $\sigma_F^2/2$, and uncorrelated and hence statistically independent. The joint pdf is

$$f_{\mathbf{n}_I,\mathbf{n}_Q}(n_I, n_Q) = f_{\mathbf{n}_I}(n_I) f_{\mathbf{n}_Q}(n_Q) = \mathcal{N}\left(0, \frac{\sigma_F^2}{2}\right) \mathcal{N}\left(0, \frac{\sigma_F^2}{2}\right). \tag{10.58}$$

The received signal is therefore

$$\mathbf{r}(t) = s(t)\left[\mathbf{n}_{F,I}\cos(2\pi f_c t) + \mathbf{n}_{F,Q}\sin(2\pi f_c t)\right] \tag{10.59}$$

$$= s(t)\left[\boldsymbol{\alpha}\cos(2\pi f_c t - \boldsymbol{\theta})\right], \tag{10.60}$$

where the random variables $\boldsymbol{\alpha} = \sqrt{\mathbf{n}_{F,I}^2 + \mathbf{n}_{F,Q}^2}$ and $\boldsymbol{\theta} = \tan^{-1}\left(\mathbf{n}_{F,Q}/\mathbf{n}_{F,I}\right)$ have the following pdfs (left as an exercise):

$$f_{\boldsymbol{\theta}}(\theta) = \frac{1}{2\pi} \quad \text{(uniform)}, \tag{10.61}$$

$$f_{\boldsymbol{\alpha}}(\alpha) = \frac{2\alpha}{\sigma_F^2} \mathrm{e}^{-\alpha^2/\sigma_F^2} u(\alpha) \quad \text{(Rayleigh)}. \tag{10.62}$$

Note that the term "Rayleigh fading" comes from the envelope distribution (statistics) given in (10.62) of the received signal.

Equation (10.62) shows that not only is the phase of the received signal severely degraded but that the amplitude is affected as well. The incoming signals add not only constructively but also destructively. In the case where two paths have approximately the same attenuation but are 180° out of phase they tend to cancel each other out. Since both amplitude and phase are affected by the fading channel it seems prudent to look at FSK as the modulation to be used over this channel. This is what we consider next.

10.4.2 Binary FSK with noncoherent demodulation in Rayleigh fading

The transmitted signal is

$$s(t) = \begin{cases} \sqrt{E_b}\sqrt{2/T_b}\cos(2\pi f_1 t), & \text{if "}0_T\text{"} \\ \sqrt{E_b}\sqrt{2/T_b}\cos(2\pi f_2 t), & \text{if "}1_T\text{"} \end{cases}, \tag{10.63}$$

while the received signal is

$$\mathbf{r}(t) = \begin{cases} \sqrt{E_b}\sqrt{2/T_b}\boldsymbol{\alpha}\cos(2\pi f_1 t - \boldsymbol{\theta}) + \mathbf{w}(t), & \text{if "}0_T\text{"} \\ \sqrt{E_b}\sqrt{2/T_b}\boldsymbol{\alpha}\cos(2\pi f_2 t - \boldsymbol{\theta}) + \mathbf{w}(t), & \text{if "}1_T\text{"} \end{cases}. \tag{10.64}$$

In the following development of the optimum receiver, both the random amplitude, $\boldsymbol{\alpha}$, and phase, $\boldsymbol{\theta}$ are completely unknown at the receiver. This type of *noncoherent* demodulation is commonly used due to its simple implementation. If the channel fading is sufficiently slow, then it is possible to estimate the random phase $\boldsymbol{\theta}$ from the received signal with a small error (typically via a pilot or training signal). In that case the ideal *coherent* demodulation can be implemented and one expects a better performance to be achieved. Coherent demodulation of BFSK as well as BPSK is considered in Section 10.4.3.

The received signal can be rewritten as

$$\mathbf{r}(t) = \begin{cases} \sqrt{E_b}\mathbf{n}_{F,I}\underbrace{\sqrt{2/T_b}\cos(2\pi f_1 t)}_{\phi_{1,I}(t)} \\ \quad +\sqrt{E_b}\mathbf{n}_{F,Q}\underbrace{\sqrt{2/T_b}\sin(2\pi f_1 t)}_{\phi_{1,Q}(t)} + \mathbf{w}(t), & \text{if “}0_T\text{”}, \\ \sqrt{E_b}\mathbf{n}_{F,I}\underbrace{\sqrt{2/T_b}\cos(2\pi f_2 t)}_{\phi_{2,I}(t)} \\ \quad +\sqrt{E_b}\mathbf{n}_{F,Q}\underbrace{\sqrt{2/T_b}\sin(2\pi f_2 t)}_{\phi_{2,Q}(t)} + \mathbf{w}(t), & \text{if “}1_T\text{”}, \end{cases} \tag{10.65}$$

where $\mathbf{n}_{F,I}$, $\mathbf{n}_{F,Q}$ are due to the fading and $\mathbf{w}(t)$ is the ever present thermal noise. To derive the optimum demodulator it is simpler to work with (10.65) which shows that at the receiver the transmitted signal lies entirely within the signal space spanned by $\phi_{1,I}(t)$, $\phi_{1,Q}(t)$, $\phi_{2,I}(t)$, and $\phi_{2,Q}(t)$. Projecting $\mathbf{r}(t)$ onto these basis functions yields the following set of sufficient statistics:

$$\underline{0_T} \qquad\qquad \underline{1_T}$$

$$\begin{aligned} \mathbf{r}_{1,I} &= \sqrt{E_b}\mathbf{n}_{F,I} + \mathbf{w}_{1,I} \\ \mathbf{r}_{1,Q} &= \sqrt{E_b}\mathbf{n}_{F,Q} + \mathbf{w}_{1,Q} \\ \mathbf{r}_{2,I} &= \mathbf{w}_{2,I} \\ \mathbf{r}_{2,Q} &= \mathbf{w}_{2,Q} \end{aligned} \tag{10.66}$$

$$\begin{aligned} \mathbf{r}_{1,I} &= \mathbf{w}_{1,I} \\ \mathbf{r}_{1,Q} &= \mathbf{w}_{1,Q} \\ \mathbf{r}_{2,I} &= \sqrt{E_b}\mathbf{n}_{F,I} + \mathbf{w}_{2,I} \\ \mathbf{r}_{2,Q} &= \sqrt{E_b}\mathbf{n}_{F,Q} + \mathbf{w}_{2,Q} \end{aligned} \tag{10.67}$$

where the $\mathbf{w}_{1,I}$, $\mathbf{w}_{1,Q}$, $\mathbf{w}_{2,I}$, $\mathbf{w}_{2,Q}$ are due to thermal noise and are Gaussian, statistically independent, zero-mean, with variance $N_0/2$. As developed in the previous section, the terms due to fading, $\mathbf{n}_{F,I}$ and $\mathbf{n}_{F,Q}$, are Gaussian, statistically independent, zero-mean with variance $\sigma_F^2/2$. The sufficient statistics are therefore Gaussian, statistically independent, zero-mean, with a variance of either $N_0/2$ or $E_b\sigma_F^2/2 + N_0/2$, depending on whether a “0” or “1” was transmitted and which sufficient statistic we are considering. Note that, given a “0” or “1” is transmitted, the difference between the sufficient statistics lies in the *received power* and it will be on the basis of received power that a decision is made. This is reasonable, since the amplitude and phase information is “obliterated” by the channel.

It is straightforward to form the likelihood ratio of

$$\frac{f(r_{1,I}, r_{1,Q}, r_{2,I}, r_{2,Q}|1_T)}{f(r_{1,I}, r_{1,Q}, r_{2,I}, r_{2,Q}|0_T)},$$

take the natural logarithm, and simplify to get the following decision rule:

$$r_{2,I}^2 + r_{2,Q}^2 \underset{0_D}{\overset{1_D}{\gtrless}} r_{1,I}^2 + r_{1,Q}^2, \tag{10.68}$$

which says that if the received power is greater on $(\phi_{2,I},\phi_{2,Q})$ than on $(\phi_{1,I},\phi_{1,Q})$, then choose "1" transmitted and the converse. Equivalently the decision rule can be expressed as

$$\sqrt{r_{2,I}^2 + r_{2,Q}^2} \underset{0_D}{\overset{1_D}{\gtrless}} \sqrt{r_{1,I}^2 + r_{1,Q}^2}. \tag{10.69}$$

The decision rule is identical to that of (10.31) for noncoherent BFSK and therefore the demodulator implementation is that of Figure 10.9. In retrospect one might have anticipated this. Compare the received signal when fading occurs, (10.64), with that when only the phase is random, (10.25). Since α is positive, one can incorporate it into the energy term, i.e., the received energy term is $\sqrt{\alpha^2 E_b}$. For any value of positive α, the decision rule is that of (10.31). Therefore, α does not affect the form of the optimum demodulator. It does, however, have a dramatic effect on the error performance. We turn our attention to the error performance next.

From symmetry it follows that $P[\text{error}] = P[\text{error}|0_T]$. To determine $P[\text{error}|0_T]$ consider decision rule (10.69):

$$P[\text{error}|0_T] = P\left[\sqrt{\mathbf{r}_{2,I}^2 + \mathbf{r}_{2,Q}^2} \geq \sqrt{\mathbf{r}_{1,I}^2 + \mathbf{r}_{1,Q}^2}\,\middle|\, 0_T\right]. \tag{10.70}$$

To proceed, fix the value of $\mathbf{r}_{1,I}^2 + \mathbf{r}_{1,Q}^2$ at a specific value, say R^2, determine $P\left[\sqrt{\mathbf{r}_{2,I}^2 + \mathbf{r}_{2,Q}^2} \geq R|0_T, \sqrt{\mathbf{r}_{1,I}^2 + \mathbf{r}_{1,Q}^2} = R\right]$ and then average this probability over all possible values of R, or equivalently over all possible values of $\mathbf{r}_{1,I}$, $\mathbf{r}_{1,Q}$. Since $P\left[\sqrt{\mathbf{r}_{2,I}^2 + \mathbf{r}_{2,Q}^2} \geq R|0_T, \sqrt{\mathbf{r}_{1,I}^2 + \mathbf{r}_{1,Q}^2} = R\right]$ is the volume under $f(r_{2,I}, r_{2,Q}|0_T)$ in the region shown in Figure 10.15, it can be computed as

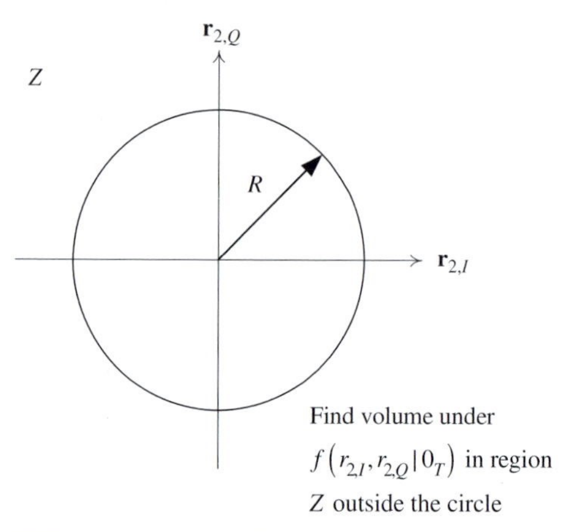

Fig. 10.15 Computation of $P\left[\sqrt{\mathbf{r}_{2,I}^2 + \mathbf{r}_{2,Q}^2} \geq R|0_T, \sqrt{\mathbf{r}_{1,I}^2 + \mathbf{r}_{1,Q}^2} = R\right]$.

$$P\left[\sqrt{\mathbf{r}_{2,I}^2+\mathbf{r}_{2,Q}^2}\geq R\,\middle|\,0_T,\sqrt{\mathbf{r}_{1,I}^2+\mathbf{r}_{1,Q}^2}=R\right]$$
$$=\iint_Z \frac{1}{\pi N_0}\mathrm{e}^{-(r_{2,I}^2+r_{2,Q}^2)/N_0}\mathrm{d}r_{2,I}\mathrm{d}r_{2,Q}$$
$$=\int_{\lambda=0}^{2\pi}\int_{\rho=R}^{\infty}\frac{1}{\pi N_0}\rho\mathrm{e}^{-\rho^2/N_0}\mathrm{d}\rho\mathrm{d}\lambda=\mathrm{e}^{-\left(r_{1,I}^2+r_{1,Q}^2\right)/N_0}. \tag{10.71}$$

Now average the error probability over all possible values of $\mathbf{r}_{1,I}$, $\mathbf{r}_{1,Q}$, i.e., find:

$$E\left\{\mathrm{e}^{-\left(r_{1,I}^2+r_{1,Q}^2\right)/N_0}\,\middle|\,0_T\right\}=\int_{r_{1,I}=-\infty}^{\infty}\int_{r_{1,Q}=-\infty}^{\infty}\mathrm{e}^{-\left(r_{1,I}^2+r_{1,Q}^2\right)/N_0}f(r_{1,I},r_{1,Q}|0_T)\mathrm{d}r_{1,I}\mathrm{d}r_{1,Q}. \tag{10.72}$$

Given 0_T, the random variables $\mathbf{r}_{1,I}$, $\mathbf{r}_{1,Q}$ are statistically independent Gaussian random variables, zero-mean, with equal variances of $E_b\sigma_F^2/2+N_0/2$. The integration in (10.72) proceeds employing an approach that has been used several times before. Namely, (i) observe that the integral splits into a product of two identical integrals, (ii) in either integral complete the square in the exponent, (iii) manipulate the integral so that one recognizes it as being the area under a Gaussian pdf (which by the way is still equal to 1) along with whatever multiplicative factors arise. The algebra is left as an exercise but the result is

$$P[\text{error}]=\frac{1}{2+\sigma_F^2E_b/N_0}. \tag{10.73}$$

The quantity $E_b\sigma_F^2$ in the above expression can be interpreted as the received energy per bit.

The most important feature of the error performance is its behavior with respect to the SNR. Up to now, for all modulations and channel models considered, the behavior has been that $P[\text{error}]\propto \mathrm{e}^{-\text{SNR}}$. Here, however, the error probability is $P[\text{error}]\propto 1/\text{SNR}$, a much much slower rate of decay. In particular, this relationship tells us that, in the log–log plot of the $P[\text{error}]$ versus SNR in decibels, the error performance curve appears to be a straight line of slope -1 in the high SNR region. Figure 10.16 shows the performance in terms of received SNR per bit, i.e., $E_b\sigma_F^2/N_0$, together with the error performance of other schemes. For example, compared to noncoherent demodulation of BFSK in random phase only, at an error probability of 10^{-3} about 19 decibels more power is needed for noncoherent demodulation of BFSK in Rayleigh fading to achieve the same performance.

10.4.3 BFSK and BPSK with coherent demodulation

As mentioned before, if the random phase introduced by fading can be perfectly estimated at the receiver, then coherent demodulation can be achieved. Assuming a perfect phase estimation, only the random amplitude needs to be dealt with and the situation is exactly the same as that considered in Section 10.2. Since the Rayleigh fading channel is considered, we know that the random amplitude $\boldsymbol{\alpha}$ is a Rayleigh random variable.

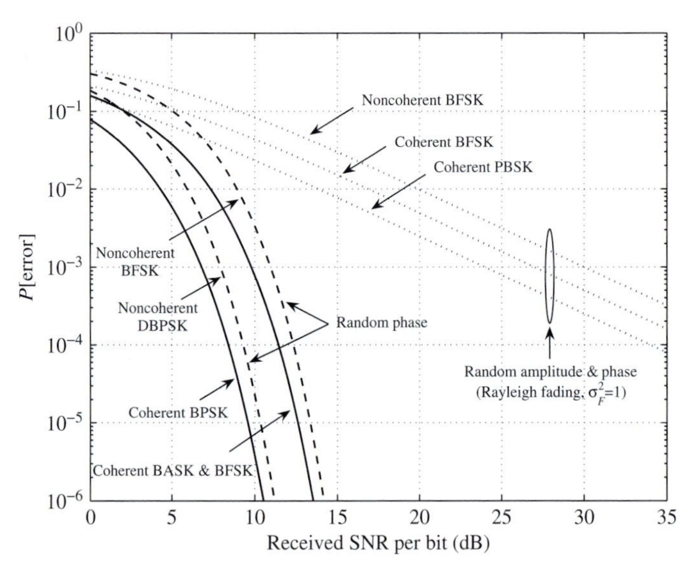

Fig. 10.16 Error performance of noncoherent and coherent BFSK and coherent BPSK over a Rayleigh fading channel.

For BFSK, projecting the received signal $\mathbf{r}(t)$ over the two basis functions, $\phi_1(t) = \sqrt{2/T_b}\cos(2\pi f_1 t - \boldsymbol{\theta})$ and $\phi_2(t) = \sqrt{2/T_b}\cos(2\pi f_2 t - \boldsymbol{\theta})$, yields the following sufficient statistics:

$\underline{0_T}$	$\underline{1_T}$
$\mathbf{r}_1 = \sqrt{E_b}\boldsymbol{\alpha} + \mathbf{w}_1$	$\mathbf{r}_1 = \mathbf{w}_1$
$\mathbf{r}_2 = \mathbf{w}_2$	$\mathbf{r}_2 = \sqrt{E_b}\boldsymbol{\alpha} + \mathbf{w}_2$

where $\mathbf{w}_1$ and $\mathbf{w}_2$ are Gaussian, statistically independent, zero-mean, with variance $N_0/2$. By equating $E = E_b$ and $\mathbf{a} = \boldsymbol{\alpha}$, the signal space representation of Figure 10.1(c) applies here. The optimum decision rule is simply

$$r_1 \underset{1_D}{\overset{0_D}{\gtrless}} r_2. \tag{10.74}$$

To determine the error performance, start with (10.5), which is $P[\text{error}] = E\left\{Q\left(\boldsymbol{\alpha}\sqrt{E_b/N_0}\right)\right\}$, where the expectation is over the Rayleigh random variable $\boldsymbol{\alpha}$. By changing to the variable $\boldsymbol{\beta} = \boldsymbol{\alpha}^2 E_b/N_0$, one has $P[\text{error}] = E\left\{Q\left(\sqrt{\boldsymbol{\beta}}\right)\right\}$, where the expectation is over the random variable $\boldsymbol{\beta}$. It is simple to show that if $\boldsymbol{\alpha}$ is a Rayleigh random variable with mean-squared value σ_F^2, then $\boldsymbol{\beta}$ has a decaying exponential distribution with mean value $\sigma_F^2 E_b/N_0$. That is,

$$f_{\boldsymbol{\beta}}(\beta) = \frac{1}{\sigma_F^2 E_b/N_0}\exp\left(-\frac{\beta}{\sigma_F^2 E_b/N_0}\right)u(\beta). \tag{10.75}$$

Furthermore, the expectation $E\left\{Q\left(\sqrt{\beta}\right)\right\}$ can be evaluated conveniently by making use of the following alternative form of the Q function, known as Craig's formula [1]:

$$Q(x) = \frac{1}{\pi}\int_0^{\pi/2} \exp\left(-\frac{x^2}{2\sin^2\theta}\right) \mathrm{d}\theta. \tag{10.76}$$

Using (10.76) and (10.75) one has

$$P[\text{error}] \\ = \frac{1}{\pi}\frac{1}{\sigma_F^2 E_b/N_0}\int_0^{\pi/2}\int_0^{\infty} \exp\left(-\frac{\beta}{2\sin^2\theta}\right)\exp\left\{-\frac{\beta}{\sigma_F^2 E_b/N_0}\right\}\mathrm{d}\beta\mathrm{d}\theta. \tag{10.77}$$

Performing the inner integration first gives

$$P[\text{error}] = \frac{1}{2} - \frac{1}{\pi}\int_0^{\pi/2} \frac{\sigma_F^2 E_b/N_0}{\left(\sigma_F^2 E_b/N_0 + 1\right) - \cos(2\theta)}\mathrm{d}\theta. \tag{10.78}$$

The integral in (10.78) can be evaluated by making use of the following identity:

$$\int_0^{\pi} \frac{\mathrm{d}x}{1 + a\cos x} = \frac{\pi}{\sqrt{1-a^2}}, \quad a^2 < 1. \tag{10.79}$$

The final expression for the error probability is

$$P[\text{error}] = \frac{1}{2}\left[1 - \sqrt{\frac{\sigma_F^2 E_b/N_0}{2 + \sigma_F^2 E_b/N_0}}\right]. \tag{10.80}$$

For the case of BPSK, the transmitted signal is $\pm\sqrt{E_b}\sqrt{2/T_b}\cos(2\pi f_c t)$ and the received signal is $\pm\sqrt{E_b}\sqrt{2/T_b}\boldsymbol{\alpha}\cos(2\pi f_c t - \boldsymbol{\theta}) + \mathbf{w}(t)$. Since we assume that the phase $\boldsymbol{\theta}$ is perfectly estimated, only one basis function, namely $\phi_1(t) = \sqrt{2/T_b}\cos(2\pi f_c t - \boldsymbol{\theta})$, is needed. The sufficient statistic is $\mathbf{r}_1 = \pm\sqrt{E_b}\boldsymbol{\alpha} + \mathbf{w}_1$ and the optimum decision rule is

$$r_1 \underset{0_D}{\overset{1_D}{\gtrless}} 0$$

(see Figure 10.1(b)). The error probability of the coherent demodulation of BPSK can be obtained in the same manner as for BFSK. It is given by

$$P[\text{error}] = \frac{1}{2}\left[1 - \sqrt{\frac{\sigma_F^2 E_b/N_0}{1 + \sigma_F^2 E_b/N_0}}\right]. \tag{10.81}$$

The above error probabilities of coherent BFSK and BPSK are plotted in Figure 10.16. As can be seen, coherent BPSK is 3 decibels more efficient that coherent BFSK, which in turn is 3 decibels more efficient than the noncoherent BFSK. The more important observation, however, is that all three schemes have the same discouraging performance behavior of $P[\text{error}] \propto 1/\text{SNR}$. Indeed, nature is not benign. The reason for the large performance degradation in Rayleigh fading is that it is very probable for the channel to exhibit what is called a *deep fade*, i.e., the received signal amplitude becomes very small. To circumvent this what is resorted to is a technique called *diversity* where multiple copies of the same message are transmitted in the hope that at least one of them will not experience a deep fade. This is the topic of the next section.

10.5 Diversity

Diversity is a technique in which multiple copies of a message are transmitted to avoid a deep fade. If the fading experienced by each transmission is statistically independent of the others, then the probability of all of them experiencing a deep fade simultaneously is low. Diversity can be accomplished in several ways. The basic forms are *time* diversity, *frequency* diversity, and *space* diversity but one can use some combination of these.

Time diversity is achieved by transmitting the same message in different time slots. For the transmissions to experience statistically independent fades the time slot spacing should be greater than what is called the channel *coherence time*, i.e., the responses in the two time slots are uncorrelated. Though not necessarily requiring an increase in the transmitted power or bandwidth, time diversity does result in a lowering of the data rate.

Frequency diversity is accomplished by sending the message copies in different frequency slots. For statistical independence, the carriers in the different slots should be separated by the *coherence bandwidth* of the channel. Since the transmissions occur simultaneously in time the data rate is not affected. However, the transmitted power and required bandwidth are increased.

Antenna arrays are used to realize space diversity. The multiple transmit and/or receive antennas should be spaced far enough apart so that independent fading is achieved. For an omnidirectional antenna in a uniformly scattering environment, this spacing is on the order of one half-wavelength. More directional antennas require a larger separation. As with frequency diversity, space diversity does not affect the data rate and the required bandwidth remains unchanged. A somewhat specialized form of space diversity can be achieved with a single antenna by using the vertical and horizontal polarizations as two separate channels.

The three forms of diversity can be viewed as a form of coding, namely, the simplest type of code, a repetition code. Standard codes, such as block and convolutional codes, can be used to achieve diversity. The next chapter discusses Alamouti space-time coding as a form of (transmit) space diversity. Spread spectrum techniques such as code-division multiple access (CDMA) (also discussed in the next chapter) are a form of diversity as well. In this section we develop the optimum demodulator for BFSK when diversity is used. An analysis of error performance is also presented. The discussion is quite general in that it applies to any of the basic diversity methods. Extensions to other modulation schemes are explored in the problems.

10.5.1 Optimum demodulation of binary FSK with diversity

Consider N transmissions of BFSK over a fading channel. At the transmitter the basic signal set over a duration of T_b seconds is

$$s(t) = \begin{cases} \sqrt{E_b'}\sqrt{2/T_b}\cos(2\pi f_1 t), & \text{if "}0_T\text{"} \\ \sqrt{E_b'}\sqrt{2/T_b}\cos(2\pi f_2 t), & \text{if "}1_T\text{"} \end{cases}, \tag{10.82}$$

where, as usual, the bits are equally likely, the frequencies are chosen so that $f_2 - f_1 = k/T_b$ and E_b' is either E_b or E_b/N. The first case is typical of time diversity and NE_b joules/bit are expended. In the second situation the energy expended per bit is E_b; it arises when space or frequency diversity is used.

Assume that time diversity is used and the signal $s(t)$ is transmitted N times. At the receiver the following signals are received:

$$\mathbf{r}_j(t) = \begin{cases} \sqrt{E_b'}\sqrt{2/T_b}\boldsymbol{\alpha}_j\cos(2\pi f_1 t - \boldsymbol{\theta}_j) + \mathbf{w}(t), & \text{if "}0_T\text{"} \\ \sqrt{E_b'}\sqrt{2/T_b}\boldsymbol{\alpha}_j\cos(2\pi f_2 t - \boldsymbol{\theta}_j) + \mathbf{w}(t), & \text{if "}1_T\text{"} \end{cases} \tag{10.83}$$

$$= \begin{cases} \sqrt{E_b'}\mathbf{n}_{j,I}\underbrace{\sqrt{2/T_b}\cos(2\pi f_1 t)}_{\phi_{j,I}^{(1)}(t)} \\ +\sqrt{E_b'}\mathbf{n}_{j,Q}\underbrace{\sqrt{2/T_b}\sin(2\pi f_1 t)}_{\phi_{j,Q}^{(1)}(t)} + \mathbf{w}(t), & \text{if "}0_T\text{"} \\ \sqrt{E_b'}\mathbf{n}_{j,I}\underbrace{\sqrt{2/T_b}\cos(2\pi f_2 t)}_{\phi_{j,I}^{(2)}(t)} \\ +\sqrt{E_b'}\mathbf{n}_{j,Q}\underbrace{\sqrt{2/T_b}\sin(2\pi f_2 t)}_{\phi_{j,Q}^{(2)}(t)} + \mathbf{w}(t), & \text{if "}1_T\text{"} \end{cases} \tag{10.84}$$

for $(j-1)T_b \le t \le jT_b$ and $j = 1, \ldots, N$. To generate the sufficient statistics, project the received signals, $\mathbf{r}_j(t), j = 1, \ldots, N$ onto the following set of $2N$ basis functions:

$$\begin{cases} \phi_{j,I}^{(1)}(t) = \sqrt{2/T_b}\cos(2\pi f_1 t), \\ \phi_{j,Q}^{(1)}(t) = \sqrt{2/T_b}\sin(2\pi f_1 t), \\ \phi_{j,I}^{(2)}(t) = \sqrt{2/T_b}\cos(2\pi f_2 t), \\ \phi_{j,Q}^{(2)}(t) = \sqrt{2/T_b}\sin(2\pi f_2 t), \end{cases} \quad \text{for } (j-1)T_b \le t \le jT_b,\ j = 1, \ldots, N. \tag{10.85}$$

Outside the signal space spanned by the above basis set there is nothing but white Gaussian noise, $\mathbf{w}(t)$. The $4N$ sufficient statistics are summarized in Table 10.1.

The sufficient statistics are Gaussian random variables, zero-mean and have a variance of either $\sigma_t^2 = E_b'\sigma_F^2/2 + N_0/2$ or $\sigma_w^2 = N_0/2$. Given that each transmission experiences a statistically independent fade (indeed uncorrelated is sufficient) the sufficient statistics are also statistically independent. Number the sufficient statistics corresponding to f_1 from

Table 10.1 Sufficient statistics for optimum demodulation of BFSK with diversity.

0_T	
$\mathbf{r}_{1,I}^{(1)} = \sqrt{E_b'}\mathbf{n}_{1,I}^{(1)} + \mathbf{w}_{1,I}^{(1)}$	$\mathbf{r}_{1,Q}^{(1)} = \sqrt{E_b'}\mathbf{n}_{1,Q}^{(1)} + \mathbf{w}_{1,Q}^{(1)}$
$\vdots$	$\vdots$
$\mathbf{r}_{N,I}^{(1)} = \sqrt{E_b'}\mathbf{n}_{N,I}^{(1)} + \mathbf{w}_{N,I}^{(1)}$	$\mathbf{r}_{N,Q}^{(1)} = \sqrt{E_b'}\mathbf{n}_{N,Q}^{(1)} + \mathbf{w}_{N,Q}^{(1)}$
$\mathbf{r}_{1,I}^{(2)} = \mathbf{w}_{1,I}^{(2)}$	$\mathbf{r}_{1,Q}^{(2)} = \mathbf{w}_{1,Q}^{(2)}$
$\vdots$	$\vdots$
$\mathbf{r}_{N,I}^{(2)} = \mathbf{w}_{N,I}^{(2)}$	$\mathbf{r}_{N,Q}^{(2)} = \mathbf{w}_{N,Q}^{(2)}$
1_T	
$\mathbf{r}_{1,I}^{(1)} = \mathbf{w}_{1,I}^{(1)}$	$\mathbf{r}_{1,Q}^{(1)} = \mathbf{w}_{1,Q}^{(1)}$
$\vdots$	$\vdots$
$\mathbf{r}_{N,I}^{(1)} = \mathbf{w}_{N,I}^{(1)}$	$\mathbf{r}_{N,Q}^{(1)} = \mathbf{w}_{N,Q}^{(1)}$
$\mathbf{r}_{1,I}^{(2)} = \sqrt{E_b'}\mathbf{n}_{1,I}^{(2)} + \mathbf{w}_{1,I}^{(2)}$	$\mathbf{r}_{1,Q}^{(2)} = \sqrt{E_b'}\mathbf{n}_{1,Q}^{(2)} + \mathbf{w}_{1,Q}^{(2)}$
$\vdots$	$\vdots$
$\mathbf{r}_{N,I}^{(2)} = \sqrt{E_b'}\mathbf{n}_{N,I}^{(2)} + \mathbf{w}_{N,I}^{(2)}$	$\mathbf{r}_{N,Q}^{(2)} = \sqrt{E_b'}\mathbf{n}_{N,Q}^{(2)} + \mathbf{w}_{N,Q}^{(2)}$

1 to $2N$ and sufficient statistics associated with f_2 from $2N+1$ to $4N$. The likelihood ratio is then

$$\frac{f(r_1,\ldots,r_N;r_{2N+1},\ldots,r_{4N}|1_T)}{f(r_1,\ldots,r_N;r_{2N+1},\ldots,r_{4N}|0_T)} = \frac{\prod_{j=1}^{2N}\left(1/\sqrt{2\pi}\sigma_w\right)\mathrm{e}^{-r_j^2/(2\sigma_w^2)}\prod_{j=2N+1}^{4N}\left(1/\sqrt{2\pi}\sigma_t\right)\mathrm{e}^{-r_j^2/(2\sigma_t^2)}}{\prod_{j=1}^{2N}\left(1/\sqrt{2\pi}\sigma_t\right)\mathrm{e}^{-r_j^2/(2\sigma_t^2)}\prod_{j=2N+1}^{4N}\left(1/\sqrt{2\pi}\sigma_w\right)\mathrm{e}^{-r_j^2/(2\sigma_w^2)}} \underset{0_D}{\overset{1_D}{\gtrless}} 1. \tag{10.86}$$

Canceling, taking natural logarithm, and rearranging gives the following decision rule:

$$\sum_{j=2N+1}^{4N} r_j^2 \underset{0_D}{\overset{1_D}{\gtrless}} \sum_{j=1}^{2N} r_j^2. \tag{10.87}$$

Again the decision rule is intuitively pleasing. Under the assumption, at the receiver, that a “1” was transmitted if the received power is greater than that under the assumption that a “0” was transmitted, then make the decision that a “1” was transmitted and vice versa.

Consider now the error performance. By symmetry one has $P[\text{error}] = P[\text{error}|1_T] = P[\text{error}|0_T]$. Define $\boldsymbol{\ell}_1 = \sum_{j=2N+1}^{4N} \mathbf{r}_j^2$ and $\boldsymbol{\ell}_0 = \sum_{j=1}^{2N} \mathbf{r}_j^2$. The decision rule is then

$$\boldsymbol{\ell}_1 \underset{0_D}{\overset{1_D}{\gtrless}} \boldsymbol{\ell}_0$$

and the determination of $P[\text{error}|0_T]$ follows the procedure already used in this chapter. Fix $\boldsymbol{\ell}_0$ at a specific value $\boldsymbol{\ell}_0 = \ell_0$, determine the probability that $\boldsymbol{\ell}_1 \geq \ell_0$, given by $\int_{\ell_0}^{\infty} f(\ell_1|0_T)\mathrm{d}\ell_1$ and then average over all possible values of $\boldsymbol{\ell}_0$. Therefore

$$P[\text{error}|0_T] = \int_0^{\infty} f(\ell_0|0_T)\left[\int_{\ell_0}^{\infty} f(\ell_1|0_T)\mathrm{d}\ell_1\right]\mathrm{d}\ell_0. \tag{10.88}$$

The sufficient statistics have a chi-square pdf with $2N$ degrees of freedom. The chi-square pdf is discussed next.

Chi-square probability density function Consider $\mathbf{y} = \mathbf{x}_1^2 + \mathbf{x}_2^2 + \cdots + \mathbf{x}_N^2$ where the $\mathbf{x}_i$ are zero-mean, statistically independent Gaussian random variables with identical variances, σ^2. To find $f_{\mathbf{y}}(y)$ determine the characteristic function $\Phi_{\mathbf{y}}(f)$ and then inverse transform it. The characteristic function of $\mathbf{y}$ is

$$\begin{aligned}\Phi_{\mathbf{y}}(f) &= E\left\{e^{j2\pi f\mathbf{y}}\right\} = E\left\{e^{j2\pi \sum_{k=1}^{N}\mathbf{x}_k^2}\right\} \\ &= E\left\{\prod_{k=1}^{N} e^{j2\pi f\mathbf{x}_k^2}\right\} = \prod_{k=1}^{N} E\left\{e^{j2\pi f\mathbf{x}_k^2}\right\}.\end{aligned} \tag{10.89}$$

Now $E\left\{e^{j2\pi f\mathbf{x}_k^2}\right\} = (1/\sqrt{2\pi}\sigma)\int_{-\infty}^{\infty} e^{j2\pi f x_k^2}e^{-x_k^2/(2\sigma^2)}\mathrm{d}x_k = 1/\sqrt{1-j4\pi\sigma^2 f}$. Therefore

$$\Phi_{\mathbf{y}}(f) = \frac{1}{(1-j4\pi\sigma^2 f)^{N/2}}$$

and

$$f_{\mathbf{y}}(y) = \int_{-\infty}^{\infty} \frac{1}{(1-j4\pi\sigma^2 f)^{N/2}} e^{-j2\pi y f}\mathrm{d}f,$$

where $y \geq 0$. From the identity[3] provided by Gradshteyn & Ryzhik [2, p. 343, Eqn 3.382-7], the pdf is

$$f_{\mathbf{y}}(y) = \frac{y^{N/2-1}e^{-y/(2\sigma^2)}}{2^{N/2}\sigma^N\Gamma(N/2)}u(y), \tag{10.90}$$

where the Gamma function, $\Gamma(x)$, is defined as

$$\Gamma(x) = \int_0^{\infty} t^{x-1}e^{-t}\mathrm{d}t \tag{10.91}$$

$$= (x-1)! \text{ for } x \text{ integer.} \tag{10.92}$$

[3] $\int_{-\infty}^{\infty}(\beta - \mathrm{i}x)^{-\nu}e^{-\mathrm{i}px}\mathrm{d}x = \frac{2\pi p^{\nu-1}e^{-\beta p}}{\Gamma(\nu)}u(p)$, where $\mathcal{R}(\nu) > 0$ and $\mathcal{R}(\beta) > 0$.

It follows from the chi-square distribution in (10.90) that

$$f(\ell_1|0_T) = \frac{\ell_1^{N-1}\mathrm{e}^{-\ell_1/(2\sigma_w^2)}}{2^N\sigma_w^{2N}\Gamma(N)}u(\ell_1),\ f(\ell_0|0_T) = \frac{\ell_0^{N-1}\mathrm{e}^{-\ell_0/(2\sigma_t^2)}}{2^N\sigma_t^{2N}\Gamma(N)}u(\ell_0).$$

The inner integral in (10.88) evaluates to

$$\frac{1}{2^N\sigma_w^{2N}\Gamma(N)}\int_{\ell_0}^{\infty}\ell_1^{N-1}\mathrm{e}^{-\ell_1/(2\sigma_w^2)}\mathrm{d}\ell_1 \overset{x=\ell_1/2\sigma_w^2}{=} \frac{1}{\Gamma(N)}\int_{\ell_0/(2\sigma_w^2)}^{\infty} x^{N-1}\mathrm{e}^{-x}\mathrm{d}x. \tag{10.93}$$

Integrating by parts $N-1$ times gives

$$\int_{\ell_0}^{\infty} f(\ell_1|0_T)\mathrm{d}\ell_1 = \mathrm{e}^{-\ell_0/(2\sigma_w^2)}\sum_{j=1}^{N}\frac{\ell_0^{N-j}}{\Gamma(N-j+1)\left(2\sigma_w^2\right)^{N-j}}. \tag{10.94}$$

The integration with respect to ℓ_0, after some algebra, becomes

$$\sum_{j=1}^{N}\frac{1}{2^{2N-j}\sigma_t^{2N}\sigma_w^{2N-2j}\Gamma(N)(N-j)!}\int_0^{\infty}\ell_0^{2N-j-1}\mathrm{e}^{-\ell_0\left[(\sigma_t^2+\sigma_w^2)/2\sigma_t^2\sigma_w^2\right]}\mathrm{d}\ell_0. \tag{10.95}$$

Again Gradshteyn & Ryzhik [2] is consulted: Equation 3.381-4 on page 342 states $\int_0^{\infty} x^{\nu-1}\mathrm{e}^{-\mu x}\mathrm{d}x = (1/\mu^{\nu})\Gamma(\nu)$, where $\mathcal{R}(\mu) > 0$ and $\mathcal{R}(\nu) > 0$. Identifying the appropriate terms the integral becomes

$$\frac{\left(2\sigma_t^2\sigma_w^2\right)^{2N-j}}{\left(\sigma_t^2+\sigma_w^2\right)^{2N-j}}\Gamma(2N-j).$$

More algebra yields:

$$P[\text{error}] = \sum_{j=1}^{N}\left(\frac{\sigma_t^2}{\sigma_w^2}\right)^{N-j}\frac{1}{\left(1+\sigma_t^2/\sigma_w^2\right)^{2N-j}}\frac{\Gamma(2N-j)}{\Gamma(N)\Gamma(N-j+1)}. \tag{10.96}$$

Define $\gamma_T = E_b'\sigma_F^2/N_0$ as the averaged SNR *per transmission*. Recognize that $\sigma_t^2/\sigma_w^2 = 1+\gamma_T$ and $\Gamma(x) = (x-1)!$ for integer x. By changing the index variable to $k = N-j$, the final expression for the error probability is[4]:

$$P[\text{error}] = \frac{1}{(2+\gamma_T)^N}\sum_{k=0}^{N-1}\binom{N-1+k}{k}\left(\frac{1+\gamma_T}{2+\gamma_T}\right)^k. \tag{10.97}$$

The error performance is plotted in Figure 10.17 versus the averaged SNR *per transmission*, i.e., γ_T. Observe that, compared with no diversity, there is a significant improvement

[4] $\binom{n}{k} = \frac{n!}{k!(n-k)!}$.

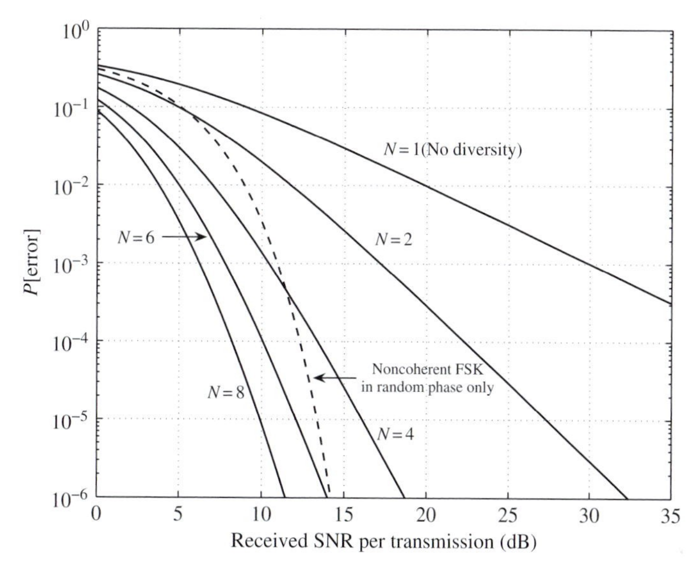

Fig. 10.17 Error performance of noncoherent BFSK with diversity in Rayleigh fading versus the average received SNR *per transmission*, $10\log_{10}\left(E_b^{'}\sigma_F^2/N_0\right)$.

in performance. Indeed, for large values of SNR, one can approximate $1+\gamma_T \approx 2+\gamma_T \approx \gamma_T$. Then (10.97) simplifies to

$$P[\text{error}] \approx \frac{1}{(\gamma_T)^N}\sum_{k=0}^{N-1}\binom{N-1+k}{k} = \frac{1}{(\gamma_T)^N}\binom{2N-1}{N}, \tag{10.98}$$

which shows that the error performance now decays inversely with the Nth power of the received SNR. The exponent N of the SNR is generally referred to as the *diversity order* of the modulation scheme. Note that the diversity order N can also be verified from Figure 10.17 by examining the slopes of the performance curves.

10.5.2 Optimum diversity

The error performance curves with diversity show that as N, the diversity order, increases the error performance improves. However, as mentioned, this improvement comes at the expense of a reduced data rate in the case of time diversity, or an increase in the transmitted power for the case of frequency or space diversity. If the transmitter's power or equivalently the energy expended per information bit is constrained to E_b joules, then increasing N does not necessarily lead to a better error performance.

Qualitatively this is explained by realizing that, though with increased N we increase the probability of avoiding a deep fade, at the same time the energy, $E_b^{'}$, of each transmission is reduced. Therefore the SNR of each transmission is reduced, which in turn

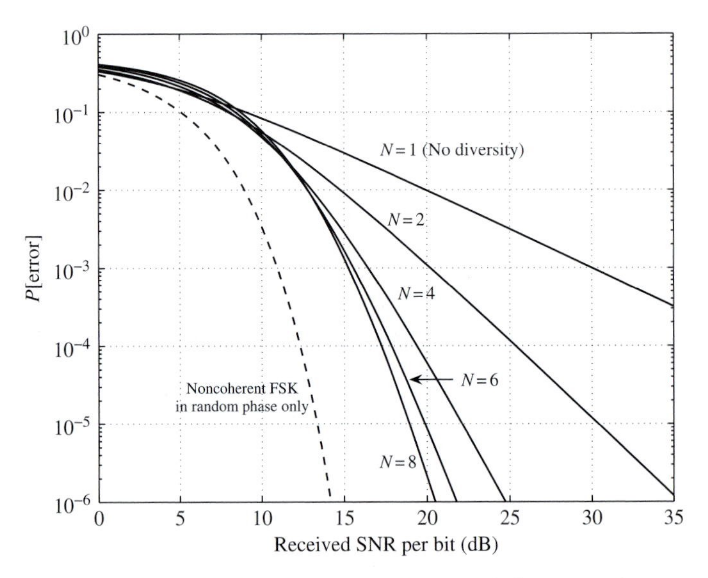

Fig. 10.18 Error performance of noncoherent BFSK with diversity in Rayleigh fading versus the average received SNR *per bit*, $10\log_{10}\left(E_b\sigma_F^2/N_0\right)$.

increases the error probability. All of this implies that there is an optimum value for the diversity order N at each level of error probability. This can also be seen from Figure 10.18, which plots the error performance in (10.97) versus the average received SNR *per bit*, i.e., $E_b\sigma_F^2/N_0$. Observe that the higher slope of the error performance curve corresponding to a larger value of N in the high-SNR region comes with an increase in the error probability in the low-SNR region. This implies that, at each level of P[error], increasing the diversity order is not necessarily helpful. Obtaining the optimum value analytically is an intractable task due to the complicated dependence of error performance on N. Though approximations to the expression of error performance can be used to obtain a feel for the optimum diversity [3] a numerical approach is used here.

Specifically divide the energy equally between each transmission, i.e., $E_b^{'} = E_b/N$. Then at each SNR $= E_b\sigma_F^2/N_0$ the error probability is determined using (10.97) as a function of N and plotted. These plots are shown in Figure 10.19(a). A minimum can be observed but the plots are quite "shallow," i.e., the minimum is not that critically dependent on N. Another observation is that the minimum N increases with SNR. Figure 10.19(b) plots optimum N versus SNR (in decibels). From this plot the following empirical relationship can be verified:

$$N_{\text{opt}} = Ke^{10\log_{10}\gamma_T}. \tag{10.99}$$

where K is some constant.

However, to reiterate, the most important aspect is that the error performance is reasonably insensitive over a broad range to the diversity order used. Therefore other engineering considerations should also be considered when N is specified.

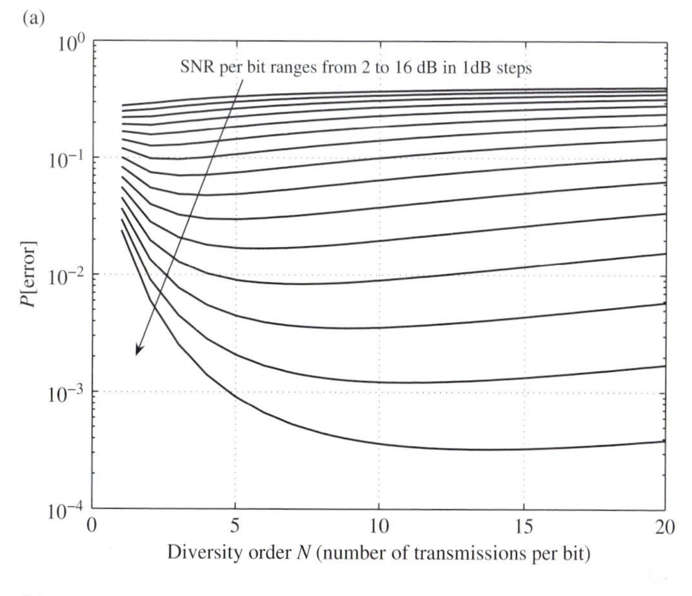

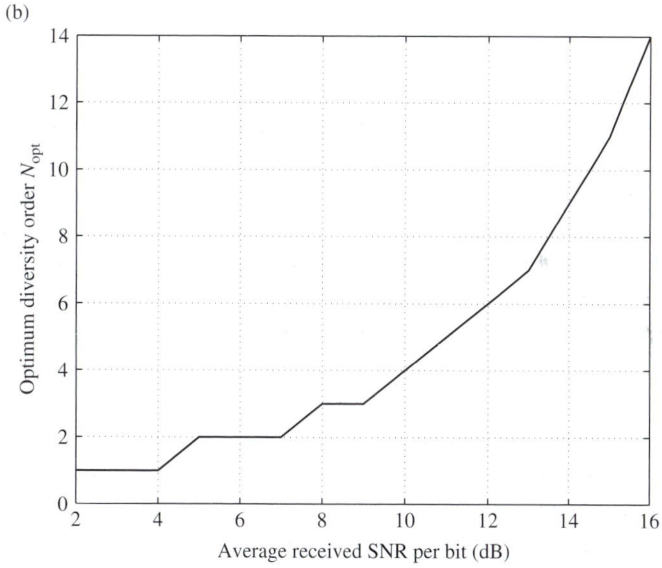

Fig. 10.19 Determining the optimum diversity order: (a) P [error] versus N with varying SNR, (b) plot of the optimum diversity order versus the average received SNR per bit.

10.6 Central limit theorem

The central limit theorem states that under certain general conditions the sum of n statistically independent continuous random variables has a pdf that approaches a Gaussian pdf as n increases. Let $\mathbf{x} = \sum_{i=1}^{n} \mathbf{x}_i$. where the $\mathbf{x}_i$ are statistically independent random variables with mean, $E\{\mathbf{x}_i\} = m_i$, and variance, $E\left\{(\mathbf{x}_i - m_i)^2\right\} = \sigma_i^2$. Then $\mathbf{x}$ is a random variable with mean $m_{\mathbf{x}} = \sum_{i=1}^{n} m_i$, variance $\sigma_{\mathbf{x}}^2 = \sum_{i=1}^{n} \sigma_i^2$ and a pdf of

$$f_{\mathbf{x}}(x) = f_{\mathbf{x}_1}(x) * f_{\mathbf{x}_2}(x) * \cdots * f_{\mathbf{x}_n}(x). \tag{10.100}$$

By the central limit theorem $f_{\mathbf{x}}(x)$ approaches a Gaussian pdf as n increases, i.e.,

$$f_{\mathbf{x}}(x) \sim \frac{1}{\sqrt{2\pi}\sigma_{\mathbf{x}}} \mathrm{e}^{-(x-m_{\mathbf{x}})^2/2\sigma_{\mathbf{x}}^2}. \tag{10.101}$$

If $\mathbf{x}$ is scaled so that its variance in the limit is finite, then (10.101) becomes an equality as $n \to \infty$. Two conditions, not the most general ones but applicable in many situations, for this to occur are:

(1) $\sum_{i=1}^{n} \sigma_i^2 \to \infty$ as $n \to \infty$.
(2) For some $k > 2$, $\int_{-\infty}^{\infty} x_i^k f_{\mathbf{x}_i}(x_i)\mathrm{d}x_i < C$, a constant.

Though not the most general conditions, the above are satisfied in our applications. The first condition is satisfied if the given random variables have equal variances, while the second is true if all pdfs, $f_{\mathbf{x}_i}(x_i)$, are nonzero over a finite interval only or if they decay fast enough.

Rather than prove the theorem[5], which is lengthy, the central limit theorem is illustrated by two examples. In both examples the $f_{\mathbf{x}_i}(x_i)$ are identical. In the first the $f_{\mathbf{x}_i}(x_i)$ are zero-mean and uniform, while in the second the underlying pdfs are one-sided exponentials. The pdfs of the resultant sum for $n = 2, 4, 6, 10$ are shown in Figures 10.20(a), (b) along with Gaussian fits of the same mean and variance. As can be seen, even for "small" values of n the pdf of $\mathbf{x}$ approaches a Gaussian pdf. However, one should realize that the close approximation is in the main body of the pdf, not in the tails, which are of most interest to us. Nonetheless, as n becomes large, on the order of hundreds or thousands[6] the Gaussian approximation is an appropriate one, hence the preponderance of Gaussian models in nature. Given that Gaussian problems are the most tractable analytically perhaps nature is not all that malevolent.

To conclude consider the filter system of Figure 10.21, which is a system of n identical first-order lowpass filters in cascade. The impulse response of a section is $h(t) = (1/RC)\mathrm{e}^{-t/(RC)}u(t)$. The overall impulse response is n convolutions,

$$h_{\mathrm{o}}(t) = \underbrace{h(t) * h(t) * \cdots * h(t)}_{n \text{ times}}.$$

By the central limit theorem as n increases, $h_{\mathrm{o}}(t)$ tends to a Gaussian curve. In other words the central limit theorem is simply a property of convolving a large number of positive functions. As such, it does not depend on probabilistic considerations.

[5] A sketch of the proof is given in [3, pp. 109–111]. For a proof the reader is referred to [4].
[6] Which brings us back to the question: how many free electrons are there in a 1 ohm resistor?

(a)

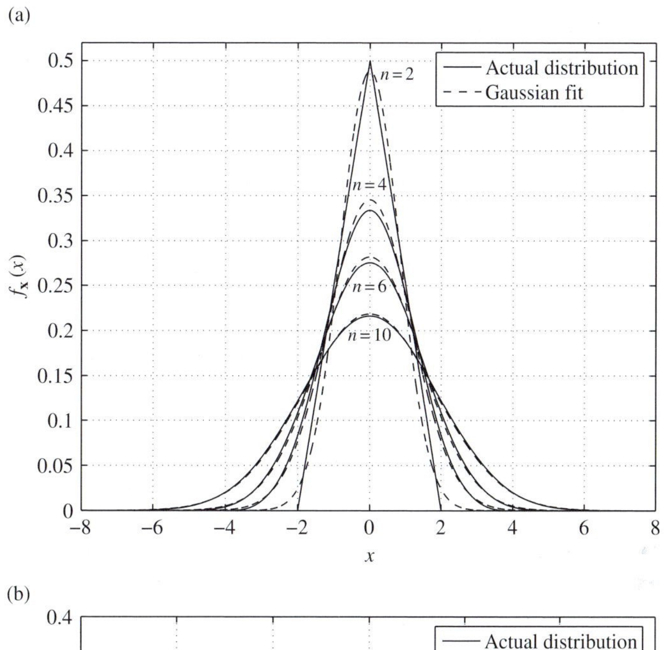

(b)

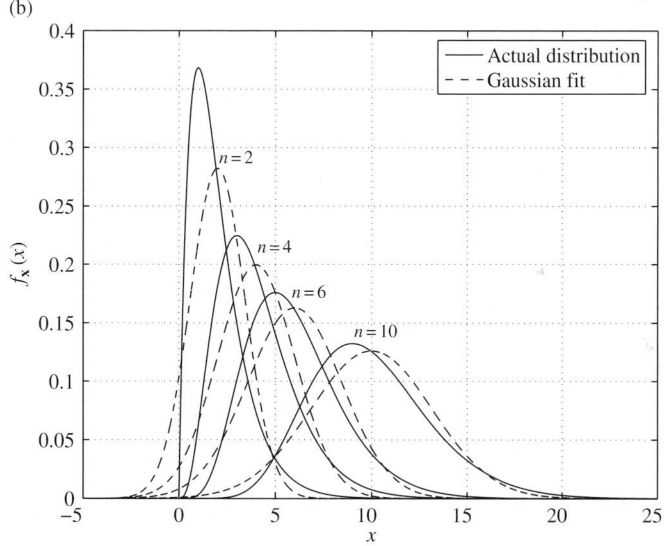

Fig. 10.20 Illustration of the central limit theorem: (a) i.i.d. uniform random variables; (b) i.i.d. Laplacian random variables.

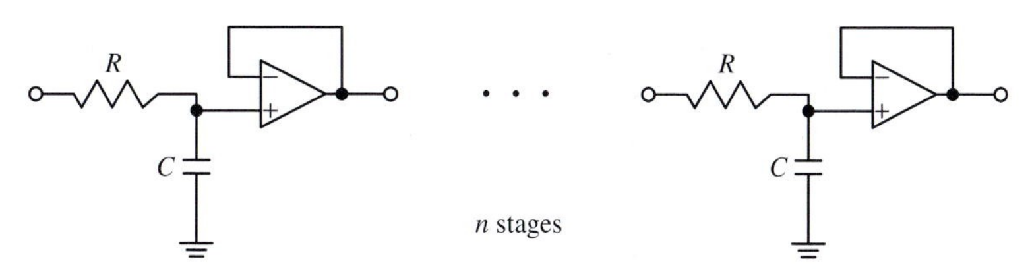

Fig. 10.21 A cascade of n identical lowpass filters.

10.7 Summary

This chapter has emphasized Rayleigh fading, a model that was originally developed in the 1950s and 1960s for non-line-of-sight communications. The typical applications were for ionospheric propagation in the 3–30 megahertz (HF) band and for the tropospheric scatter experienced in the 300 megahertz–3 gigahertz (UHF) and 3–30 gigahertz (SHF) bands. Though present day fading effects experienced by mobile digital communication systems are somewhat different, the Rayleigh model is still relevant for design purposes. Particularly for mobile systems such as cellular and personal communication systems (PCS) that operate in the UHF band. Other models, besides Rayleigh, are appropriate for the envelope statistics of received signal. When, besides the multipath components, a line-of-sight (LOS) component is present, also referred to as a specular component, the envelope statistics become *Rician* [5], basically noncentral chi-square. The optimum demodulator is a straightforward extension of the approach used in this chapter. Indeed, it turns out to be the same. The error performance analysis, due to the noncentral chi-square pdfs, is considerably more involved. Experimental data have shown situations where neither the Rayleigh nor the Rician model is entirely appropriate. An empirical model called the *Nakagami-m* pdf (see Problem 10.7) has been developed for the envelope where the parameter m, $0.5 \leq m < \infty$, can be chosen to fit a variety of experimental situations. For $m = 1$ the Nakagami-m pdf becomes Rayleigh and as $m \to \infty$ it approaches the Rician pdf.

The fading described by all of these models is termed *small-scale* fading where small changes in path length result in dramatic changes in the received signal's amplitude and phase. Large-scale fading, not discussed in this chapter, occurs when the receiver is "shadowed" by prominent geographic features such as hills, forests, a cluster of buildings, etc. It results in an attenuation of the received power due to movement over "large" distances. The two classes of fading lead to two types of diversity techniques. The diversity technique discussed in this chapter is used to combat multipath. It is called microdiversity. Macrodiversity is used for large-scale fading. It typically involves the combining of signals received by several base stations or access points. This implies coordination between the different stations, which is implemented as part of the network protocols governing an infrastructure based wireless network. There are different approaches to combining the received signals in both micro- and macrodiversity. These are *selection* combining, *threshold* combining, *maximal-ratio* combining and *equal-gain* combining.

The fading model described in this chapter, besides Rayleigh, is a *narrowband* model. This refers to the assumption that the information signal $s(t) \approx s(t - \tau_n)$, i.e., the transmitted signal is essentially confined to the symbol interval. If the delay spread (range of τ_n) is such that there is significant signal energy overlapping into other symbol intervals, then the system starts to experience ISI. When this happens the channel is called frequency selective as opposed to frequency nonselective (or flat fading). The demodulation then typically involves maximum likelihood sequence estimation and a Viterbi equalization type of approach. Another popular approach to deal with a frequency-selective fading channel is OFDM (orthogonal frequency-division multiplexing). In OFDM the entire bandwidth is divided into overlapping, but orthogonal, narrowband channels. Since the information symbols are transmitted and demodulated independently over these narrowband channels the ISI effects are considerably reduced.

Lastly, the modulation investigated in this chapter was binary. M-ary modulation is also possible and desirable. This and some of the other topics mentioned in the summary are pursued in the chapter problems. In an introductory text, such as this, it is not possible to do full justice to the rich area of wireless communications. The reader is directed to references [6, 5, 7, 8] for more advanced reading.

10.8 Problems

10.1 Figure 10.8 shows that the error performance of noncoherent BASK with threshold of $\sqrt{E}/2$ is slightly worse than that of the noncoherent BFSK. Based on (10.23) and (10.36), verify this observation analytically.

10.2 Plot the optimum threshold, T_h, in noncoherent BASK versus E_b/N_0 over the range 0–20 decibels. Compare it with the "intuitive" threshold of $\sqrt{E}/2 = \sqrt{E_b/2}$. Comment.

10.3 Develop *coherent* demodulation of BASK in Rayleigh fading, assuming that both the random phase and amplitude can be perfectly estimated at the receiver. Then obtain its error performance in terms of the average received SNR per bit. Compare the result with that of coherent BFSK and BPSK in Section 10.4.3.

10.4 Develop *noncoherent* demodulation of DBPSK in Rayleigh fading. Then obtain its error performance in terms of the average received SNR per bit. Compare the result with that of coherent BFSK and BPSK in Section 10.4.3.

10.5 Section 10.5.1 considers noncoherent demodulation of BFSK with diversity in Rayleigh fading. If the phase can be perfectly estimated at the receiver, one can use BPSK with diversity and its coherent demodulation.

(a) Develop the optimum coherent demodulation of BPSK with diversity order N in Rayleigh fading.

(b) Show that the error performance is given by

$$P[\text{error}] = \left(\frac{1-\mu}{2}\right)^N \sum_{k=0}^{N-1} \binom{N-1+k}{k} \left(\frac{1+\mu}{2}\right)^k, \tag{P10.1}$$

where $\mu = \sqrt{\gamma_T/(1+\gamma_T)}$, $\gamma_T = E_b^{'}\sigma_F^2/N_0$ and $E_b^{'}$ is the average energy *per transmission*.

(c) To see the influence of the diversity order N at high SNR, use the following approximations $(1+\mu)/2 \approx 1$, $(1-\mu)/2 \approx 1/4\gamma_T$ and

$$\sum_{k=0}^{N-1} \binom{N-1+k}{k} = \binom{2N-1}{N}$$

to show that

$$P[\text{error}] \approx \binom{2N-1}{N} \frac{1}{(4\gamma_T)^N}. \tag{P10.2}$$

Comment.

10.6 Repeat the above problem for *noncoherent* demodulation of DBPSK in Rayleigh fading with N diversity receptions.

10.7 (*Amount of fading*) For coherent demodulation in fading, it is assumed that the random phase can be perfectly estimated and accounted for at the receiver. This means that only the random amplitude affects the quality of the demodulation. Let $f_{\boldsymbol{\alpha}}(\alpha)$ be the pdf of the random amplitude $\boldsymbol{\alpha}$. Then, the severity of fading due to random amplitude can be quantified through a single parameter, called the "amount of fading" or fading figure, defined as

$$\text{AF} = \frac{\text{var}\{\boldsymbol{\alpha}^2\}}{(E\{\boldsymbol{\alpha}^2\})^2}. \tag{P10.3}$$

In general, the smaller the AF, the less severe the fading is.

(a) What is the value of AF when the amplitude is a constant.

(b) Show that, for a Rayleigh amplitude, AF $= 1$. Recall that the Rayleigh distribution is

$$f_{\boldsymbol{\alpha}}(\alpha) = \frac{2\alpha}{\sigma_F^2} \mathrm{e}^{-\alpha^2 / \sigma_F^2} u(\alpha). \tag{P10.4}$$

(c) Consider the following Nakagami-m distribution:

$$f_{\boldsymbol{\alpha}}(\alpha) = \frac{2m^m \alpha^{2m-1}}{\Gamma(m)\sigma^{2m}} \mathrm{e}^{-m\alpha^2/\sigma^2} u(\alpha), \quad m \geq 0.5. \tag{P10.5}$$

Show that AF $= 1/m$.

In the chapter, differential encoding and differential demodulation were introduced to overcome phase uncertainty in the received signal at the receiver. The phase uncertainty model assumed was quite severe in that it was assumed to be equally probable over $[0, 2\pi]$. However, there are practical applications where the phase uncertainty is quite restricted, namely $0°$ or $180°$. A phase of $180°$ can occur simply due to inadvertent insertion of an inverting amplifier in the chain. More commonly it occurs in synchronization where phase-locked loop (PLL) circuits (discussed in Chapter 12) can lock on, falsely, to a phase that is out by π radians. The next set of problems explores differential encoding and differential modulation when the phase model is thus restricted.

10.8 (*Coherently demodulated DBPSK*) The terminology now begins to become a bit confusing. What "coherently demodulated DBPSK" means is that differential mapping (or encoding) is used at the modulator but coherent demodulation is done at the receiver. This is possible because the phase uncertainty, as noted above, is very restricted. How the demodulation is done and the resultant error performance are investigated here.

(a) Consider the differential mapping rule of (10.38). How do the signals of (10.40) change if a phase uncertainty of π radians is present.

Table 10.2 Differential phase mapping in DQPSK

Present symbol	Previous phase	Present phase = previous phase $+\Delta\phi$
00	ϕ_{k-1}	$\phi_k = \phi_{k-1} +$ ______
01	ϕ_{k-1}	$\phi_k = \phi_{k-1} +$ ______
11	ϕ_{k-1}	$\phi_k = \phi_{k-1} +$ ______
10	ϕ_{k-1}	$\phi_k = \phi_{k-1} +$ ______

(b) Consider the block diagram of the modulator for DBPSK in Figure 10.22. The purpose of the logic circuit is to implement mapping rule (10.38). Thus, if $b_k = 0$ then $d_k = d_{k-1}$, and if $b_k = 1$ then $d_k = \overline{d_{k-1}}$ (complement of d_{k-1}). Therefore d_k is a Boolean function of b_k and d_{k-1}, i.e., $d_k = g(d_{k-1}, b_k)$. What is $g(\cdot)$? How would you implement $g(\cdot)$?

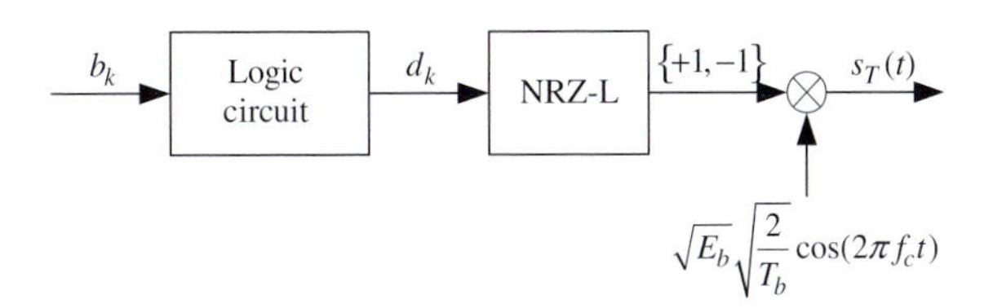

Fig. 10.22 Block diagram of the modulator for DBPSK.

(c) Draw the signal space plot of (10.40) which shows the possible transmitted and received signals (in the absence of AWGN) with the phase uncertainty model with which we are dealing. Compare with the signal space plot of the Miller modulation (see Chapter 6).

(d) Based on (b) draw a block diagram of the coherent demodulator.

(e) Determine the error performance of this demodulator and compare it with that of DBPSK in (10.48).

(f) Come up with a mnemonic for this modulation and demodulation.

10.9 (*DQPSK – differential quadrature phase-shift keying*) Since QPSK offered, at seemingly no cost, benefits over BPSK for the AWGN, we attempt here to develop a differential version of it for the phase model of Problem 10.8. The phase model is extended slightly in that not only is the phase uncertainty 0 or π, it can also be multiples of $\pi/2$ radians. The first step is to develop a differential mapping.

(a) As in differential BPSK we do not map the two bits in any signaling interval of $T_s = 2T_b$ seconds to an absolute phase but to a *relative* one, one that depends on the previous phase. With this in mind complete Table 10.2.

(b) Consider the following transmitted bit sequence:

$$01111100011010.$$

What is the corresponding phase sequence of the transmitted signal?

(c) Since the modulator has (finite) memory, there is a corresponding trellis. Define the states for the trellis and draw it.

(d) Discuss the signal space that represents the output of the modulator. What is its dimensionality? How many signal points are there in the constellation? Where are they located? *Hint* The dimensionality is greater than 3.

(e) Consider the block diagram of the DQPSK modulator. It should be similar to the DBPSK modulator since one view of QPSK is that of two BPSK modulators (one on the inphase, I axis, the other on the quadrature, Q axis). Thus a possible block diagram is shown in Figure 10.23. The differential mapper maps the

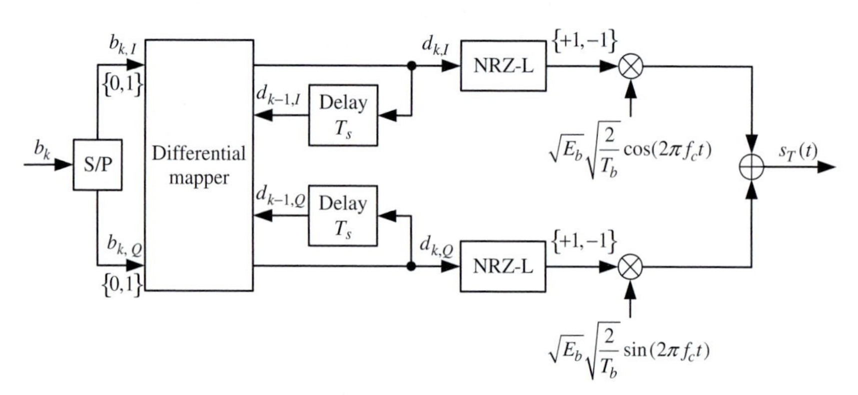

Fig. 10.23 Block diagram of the modulator for DQPSK.

information bits, $b_{k,I}$ and $b_{k,Q}$ to $d_{k,I}$ and $d_{k,Q}$, respectively, so that the transmitted signal satisfies the map given by Table 10.2. What are the two Boolean functions $d_{k,I} = g\left(b_{k,I}, b_{k,Q}, d_{k-1,I}, d_{k-1,Q}\right)$ and $d_{k,Q} = h\left(b_{k,I}, b_{k,Q}, d_{k-1,I}, d_{k-1,Q}\right)$ that would accomplish this?

10.10 (*Coherent demodulation of DQPSK*)

(a) As usual we need to generate a set of sufficient statistics. The received signal (ignoring the AWGN for now) depends on both the previous two bits and the present two bits. Each set is represented by an inphase component and a quadrature component, i.e., $\sqrt{2/T_b}\cos(2\pi f_c t)$ and $\sqrt{2/T_b}\sin(2\pi f_c t)$. Therefore one concludes that there are ______ basis functions that can completely represent the transmitted signal set.

(b) Since a transmitted signal depends on four bits, the number of signals in the transmitted signal set is ________.

(c) In one dimension we have $2^1 = 2$ quadrants, in two, $2^2 = 4$ quadrants, in three, $2^3 = 8$ quadrants. Therefore in four dimensions we have ________ quadrants.

(d) To get some idea of what the signal space looks like, consider that the bits to be transmitted in the present interval are $b_{k-1}b_k = 00$. Let the previous two bits be 00 as well, i.e., $b_{k-3}b_{k-2} = 00$. Complete Table 10.3.

Note that, there are four possible signals corresponding to the bits $b_{k-1}b_k$ depending on the previous phase. To see where these signals lie in the signal

Table 10.3 Table to complete in Problem 10.10

b_{k-3}	b_{k-2}	b_{k-1}	b_k		Previous phase	Present phase
0	0	0	0	(i)	$45°(\pi/4)$	______
				(ii)	$135°(3\pi/4)$	______
				(iii)	$225°(5\pi/4)$	______
				(iv)	$315°(7\pi/4)$	______

space, let $\{\phi_{k-1,I}, \phi_{k-1,Q}\}$ be the basis functions for the previous interval and $\{\phi_{k,I}, \phi_{k,Q}\}$ that for the present interval. Thus signal (i) (see Table 10.3) lies in the signal space illustrated in Figure 10.24.

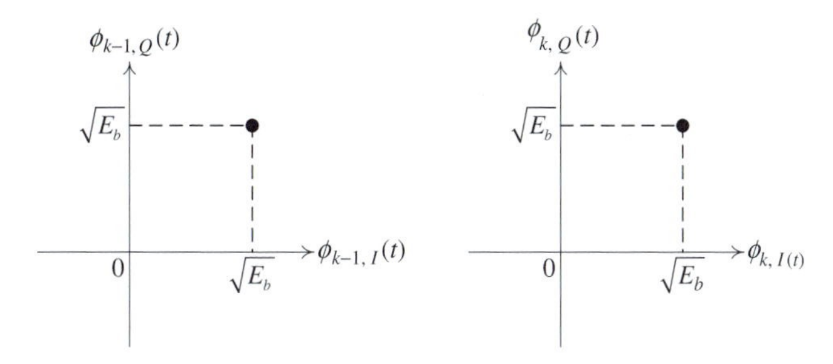

Fig. 10.24 Signal space representation for case (i) in Table 10.3.

Here we are using two cross-sections to visualize the four-dimensional signal space. To denote the quadrant in which signal (i) falls, let us use the notation $(+1, +1, +1, +1)$. Plot the other three signals, (ii), (iii), (iv), and determine in which quadrants they fall.

(e) But, of course, the previous bits, $b_{k-3}b_{k-2}$, were just as likely to have been 01, 10, 11. Show that if the present two bits are 00, then we still have the same signal points.

(f) Repeat (d) and (c) for the situations where $b_{k-1}b_k = 01, 11, 10$.

(g) If you have come this far, congratulate yourself. Now develop a block diagram of the demodulator.

Remark The observant reader might note that in this problem we have in essence answered some of the previous questions. Consider that a benefit of being a widely read person.

10.11 (*Error performance of coherent demodulation of DQPSK*) Show that the probability of symbol error is given by

$$P[\text{error}] = 4Q\left(\sqrt{\frac{2E_b}{N_0}}\right) - 8Q^2\left(\sqrt{\frac{2E_b}{N_0}}\right) + 8Q^3\left(\sqrt{\frac{2E_b}{N_0}}\right) - 4Q^4\left(\sqrt{\frac{2E_b}{N_0}}\right). \quad \text{(P10.6)}$$

Plot the error probability and compare the result with that of coherent QPSK. What penalty is paid in SNR (decibels) at, say, 10^{-4} for this phase uncertainty? *Hint* Use symmetry to argue that $P\,[\text{error}] = P\,[\text{error}\,|00\,]$. Note also that only one of the signals corresponding to 00 transmitted needs to be considered.

10.12 (*Differential encoding of square QAM*) First, consider as a specific example 16-QAM. Note that any phase rotation of $k\pi/2$ results in the same signal constellation, i.e., the signal constellation has a four-fold symmetry.

(a) Based on the above observation come up with a differential encoding rule for the signal constellation. *Hint* You should need to differentially encode only two of the bits.

(b) Map the remaining two bits to the signals of a quadrant in a manner that you feel (intuitively) minimizes the bit error probability.

(c) Repeat the above for 64-QAM.

(d) Generalize to any square QAM.

(e) For differentially encoded 16-QAM, draw block diagrams of the modulator and demodulator.

10.13 Consider a channel model where there is a direct LOS path as well as the scattering (i.e., Rician fading). The modulation is BFSK. The received signal is

$$\mathbf{r}(t) = \begin{cases} \sqrt{E_1}\sqrt{2/T_b}\cos(2\pi f_1 t) \\ \qquad +\sqrt{E_2}\boldsymbol{\alpha}\sqrt{2/T_b}\cos(2\pi f_1 t - \boldsymbol{\theta}) + \mathbf{w}(t), & \text{if “}0_T\text{”} \\ \sqrt{E_1}\sqrt{2/T_b}\cos(2\pi f_2 t) \\ \qquad +\sqrt{E_2}\boldsymbol{\alpha}\sqrt{2/T_b}\cos(2\pi f_2 t - \boldsymbol{\theta}) + \mathbf{w}(t), & \text{if “}1_T\text{”} \end{cases} \tag{P10.7}$$

where f_1 and f_2 are orthogonal carriers, $\boldsymbol{\alpha}$ is a Rayleigh random variable of mean-squared value σ_F^2, $\boldsymbol{\theta}$ is uniform over $[0, 2\pi]$ and $\mathbf{w}(t)$ is white Gaussian noise of two-sided PSD $N_0/2$ (watts/hertz).

(a) Come up with a set of orthonormal basis functions and obtain the sufficient statistics.

(b) Determine the likelihood ratio test, take natural logarithm (ln) and simplify. Express the decision rule in terms of $\sigma^2 \equiv N_0/2$ and $\sigma_t^2 \equiv E_2\sigma_F^2/2 + N_0/2$.

10.14 (*Rician pdf*) In Problem 10.13 the LOS component results in a test statistic that has a Rician pdf, which comes from what is termed a noncentral chi-square density. It is called noncentral because the Gaussian random variables have a nonzero mean. Here we derive the Rician pdf, plot it, and compare the plots with the Rayleigh pdf which arises from a central chi-square density.

Consider the following random variable:

$$\mathbf{z} = \sqrt{\mathbf{x}^2 + \mathbf{y}^2}, \tag{P10.8}$$

where $\mathbf{x}$ and $\mathbf{y}$ are statistically independent, Gaussian, random variables of equal variance σ^2. The random variable $\mathbf{x}$ has a *nonzero* mean, m, whereas $\mathbf{y}$ has zero mean.

(a) To find the pdf of $\mathbf{z}$, determine the *characteristic functions* of $\mathbf{x}$ and $\mathbf{y}$, i.e., determine $\Phi_{\mathbf{x}}(f) = E\left\{e^{j2\pi f\mathbf{x}}\right\}$ and $\Phi_{\mathbf{y}}(f) = E\left\{e^{j2\pi f\mathbf{y}}\right\}$. *Hint* The usual trick of completing the square, multiplying (and dividing) by an appropriate constant, recognizing the integral as the area under a Gaussian pdf should avoid any integration as such.

(b) Multiply the two characteristic functions together to obtain the characteristic function of the random variable $\mathbf{w} = \mathbf{x}^2 + \mathbf{y}^2$, i.e., $\Phi_{\mathbf{w}}(f) = \Phi_{\mathbf{x}}(f) \cdot \Phi_{\mathbf{y}}(f)$.

(c) Find the inverse transform of $\Phi_{\mathbf{w}}(f)$, i.e., $\int_{-\infty}^{\infty} \Phi_{\mathbf{w}}(f) e^{-j2\pi f w} df$ to obtain

$$f_{\mathbf{w}}(w) = \frac{1}{2\sigma^2} e^{-(w+m^2)/2\sigma^2} I_0\left(\sqrt{w}\frac{m}{\sigma^2}\right). \tag{P10.9}$$

(d) Now $\mathbf{z} = \sqrt{\mathbf{w}}$. Determine $f_{\mathbf{z}}(z)$ from $f_{\mathbf{w}}(w)$ and show that it is given by

$$f_{\mathbf{z}}(z) = \frac{z}{\sigma^2} e^{-(z^2+m^2)/2\sigma^2} I_0\left(\frac{zm}{\sigma^2}\right). \tag{P10.10}$$

(e) In the Rician fading channel model, the random variable $\mathbf{z}$ plays the role of the channel "gain" (perhaps more accurately, the channel attenuation). The parameters m^2 and $2\sigma^2$ reflect the received powers in the LOS component and the non-LOS multipath components, respectively. The Rician distribution is often described in terms of a fading parameter κ, defined as $\kappa = m^2/2\sigma^2$. In essence, κ is the ratio of the power in the LOS component to the power in the non-LOS multipath components. Note that $\kappa = 0$ corresponds to Rayleigh fading, while $\kappa = \infty$ means no fading. Let $\sigma_F^2 = m^2 + 2\sigma^2$ be the total received power. Express $f_{\mathbf{z}}(z)$ in terms of κ and σ_F^2. Then plot $\sigma_F f_{\mathbf{z}}(z/\sigma_F)$ for various values of κ. On the same figure, also plot the scaled Rayleigh pdf, i.e., $\sigma_F f_{\mathbf{z}}(z/\sigma_F)$ where $\kappa = 0$. Comment.

References

[1] J. W. Craig, "A new, simple and exact result for caculating the probability of error for two-dimension signal constellations," in *Proc. IEEE Vehicular Technology Conference*, pp. 571–575, Oct. 1991.

[2] L. S. Gradshteyn and L. M. Ryzhik, *Table of Integrals, Series, and Products*. Academic Press, 6th edn, 2000.

[3] J. M. Wozencraft and I. M. Jacobs, *Principles of Communications Engineering*. John Wiley & Sons, 1965.

[4] J. V. Uspensky, *Introduction to Mathematical Probability*. McGraw-Hill, 1937.

[5] M. K. Simon and M. S. Alouini, *Digital Communication over Fading Channels*. John Wiley & Sons, 2nd edn, 2005.

[6] J. G. Proakis, *Digital Communications*. McGraw-Hill, 4th edn, 2001.

[7] A. Goldsmith, *Wireless Communications*. Cambridge University Press, 2005.

[8] D. Tse and P. Viswanath, *Fundamentals of Wireless Communication*. Cambridge University Press, 2005.

11 Advanced modulation techniques

This chapter looks at three important modulation paradigms. *Trellis-coded modulation* (TCM) is considered first. Developed in the late 1970s as a method to conserve bandwidth without sacrificing error performance [1, 2, 3], it has become an extremely important modulation.

The second technique is called *code-division multiple access* (CDMA). It falls within the broad class of multiple access methods and is an administrator's favorite. It makes the addition (or deletion) of new users essentially transparent, easing the administrator's work. However, the technique does have technical merits that warrant its study by communication engineers. It forms the basis for the so-called 3G (third generation) and beyond wireless communication systems. In contrast to TCM, CDMA is a very wideband modulation technique.

The last modulation method studied uses space-time codes which provide diversity gain for the fading channel through the use of multiple *transmit* antennas. The important Alamouti's space-time block code (STBC) is discussed in some detail. It was first described in 1998 [4], almost at the same time Tarokh *et al.* published their paper [5] on space-time trellis codes.

11.1 Trellis-coded modulation (TCM)

By now one can appreciate that over an AWGN channel, the basic idea in digital communications is to find signal constellations with as large a distance between signals as possible without increasing the energy E_b (joules) expended per bit inordinately and with as small a bandwidth as possible. To decrease bandwidth one went to M-ary modulation such as M-PSK and M-QAM. However, as more and more signal points were packed into a circle of radius $\sqrt{E}$ (volts) the minimum distance between signals suffered.

On reflection, however, one realizes that the signal transmitted in symbol period $[kT_s, (k+1)T_s]$ is chosen *independently* of that transmitted in the time interval $[(k-1)T_s, kT_s]$. Perhaps if one imposed some sort of constraint on the transmitted signal patterns one could increase the minimum distance between the signals. But note that one is talking about the minimum distance between *signal sequences*, i.e., between signals for time periods greater than one symbol period.[1] This is the basic idea behind TCM, a

[1] This is very similar to the discussion of sequence demodulation for Miller signaling in Section 6.4.

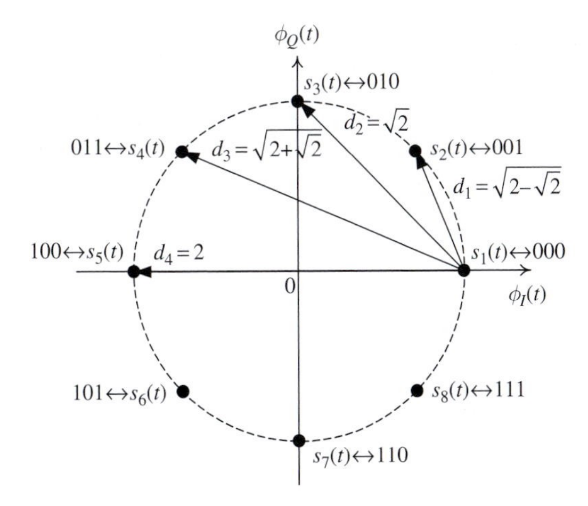

Fig. 11.1 8-PSK signal set and a natural mapping.

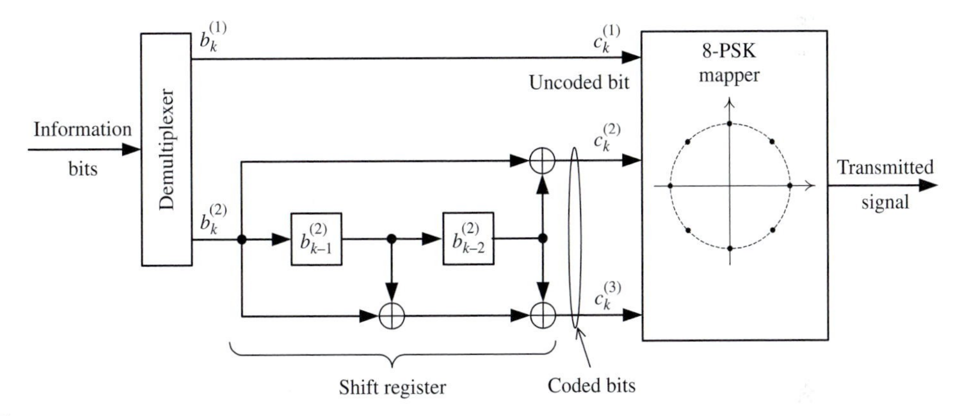

Fig. 11.2 Mapping from the two input information bits to the 8-PSK signal points in TCM.

modulation/coding technique first proposed by Ungerboeck[2] in the late 1970s. To introduce the method let us consider a specific example.

The signal set used by the modulator is 8-PSK as shown in Figure 11.1 along with the Euclidean distances between the signal points. For convenience the signals have been normalized to lie on a *unit* circle. Eventually it will be informative to look at the signal set as being composed of two QPSK sets, $\{s_1(t), s_3(t), s_5(t), s_7(t)\}$ and $\{s_2(t), s_4(t), s_6(t), s_8(t)\}$.

To impose a constraint on the *allowable* transmitted signal patterns implies that the signal transmitted in the present signaling interval depends on what happened previously, in other words, the modulator needs to have *memory*. A shift register circuit provides memory and this is what is used in the modulator as shown in Figure 11.2.

[2] Although not yet published in the literature, Ungerboeck's invention was already well known in the late 1970s.

For historical reasons the shift register circuit is called a *convolutional encoder*[3] but the important aspect of it is that it provides memory. The output bits, $c_k^{(2)}$ and $c_k^{(3)}$, not only depend on the present input bit, $b_k^{(2)}$, but also on the two previous bits, $b_{k-1}^{(2)}$ and $b_{k-2}^{(2)}$. Observe that there are three bits applied to the signal selector (mapper) block and these three bits are used to choose one of the 8-PSK signals. The shift register has the state diagram and trellis shown in Figure 11.3. Observe that the trellis structure is determined completely by the shift register's memory cells, i.e., $b_{k-1}^{(2)}$, $b_{k-2}^{(2)}$ along with $b_k^{(1)}$.

However, what we are interested in is the *transmitted signal* and for this one needs to specify the *mapping* from the three output bits of the encoder to the signal constellation.

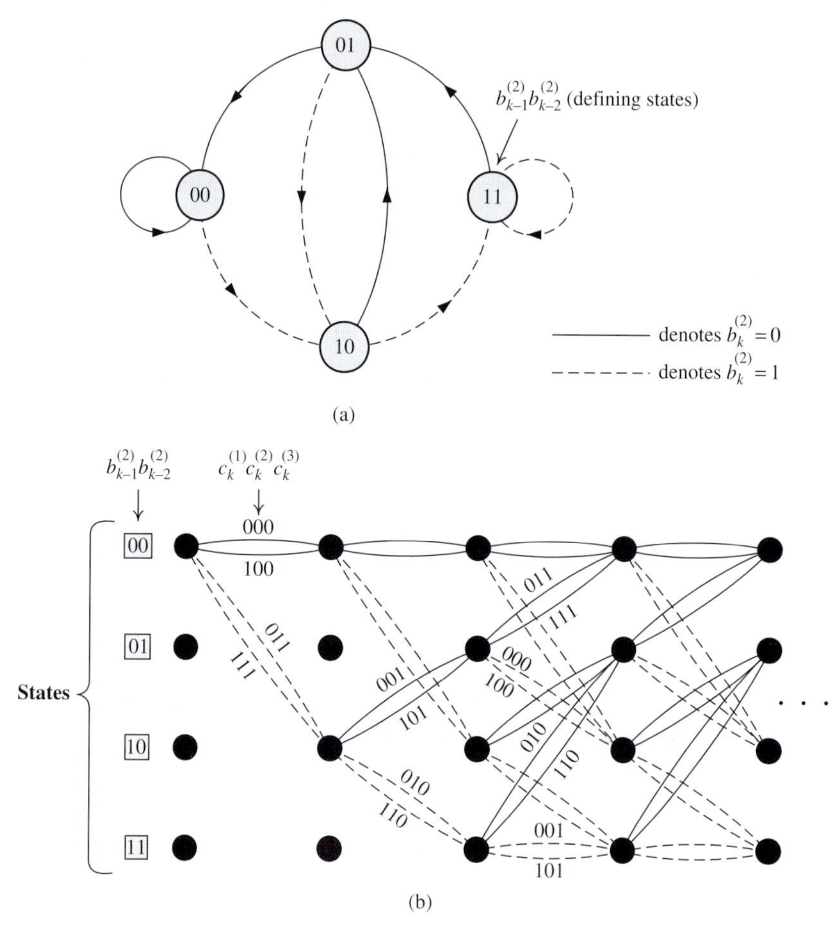

Fig. 11.3 (a) State diagram and (b) trellis of the shift register of Figure 11.2.

[3] The adjective "convolutional" comes from the fact that, when represented in binary sequences, the output of the register is a convolution of the input and a sequence representing the configuration of the shift register circuit.

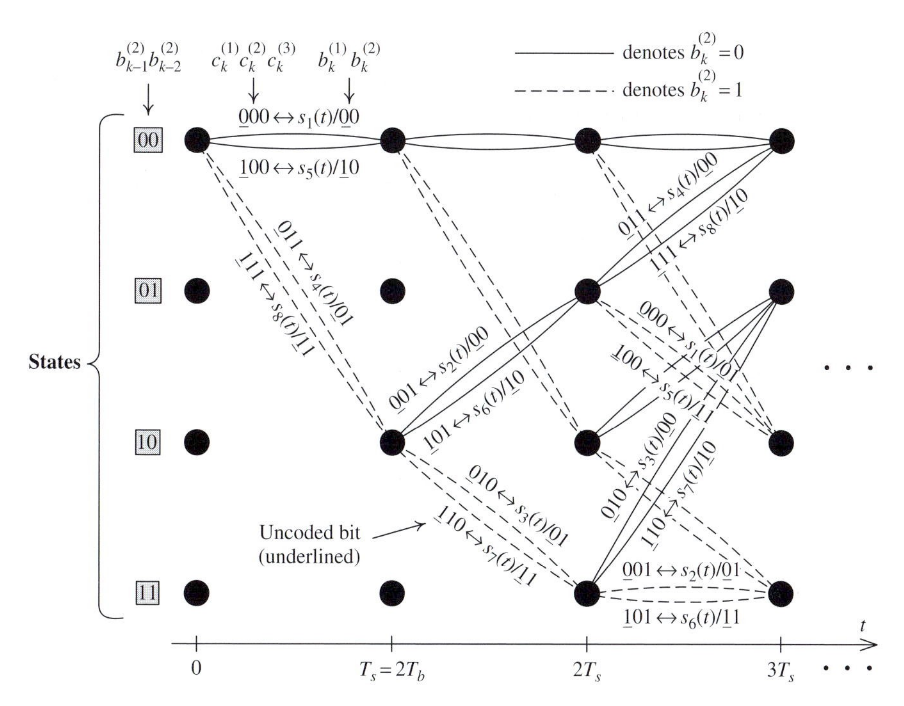

Fig. 11.4 Trellis diagram showing the transmitted signal sequences.

For this the *natural mapping*,[4] also shown in Figure 11.1, is chosen and Figure 11.4 shows the trellis diagram of this TCM scheme. Due to the uncoded bit, $b_k^{(1)}$, there are two *parallel branches* from every state. Consider now the minimum Euclidean distance between two signal sequences (subsequences is a more appropriate terminology) that start in a common state, diverge, and then merge and end in a common state. For simplicity, only sequences of one and three signaling intervals in length are considered. Using Figure 11.1, the minimum distances are:

$$\text{sequences of length 1: } d^2_{\min} = d_4^2 = 4,$$
$$\text{sequences of length 3: } d^2_{\min} = d_1^2 + d_1^2 + d_1^2 = 3d_1^2 = 6 - 3\sqrt{2} = 1.757.$$

Therefore sequences of length 3 determine the *minimum* distance as it is easily observed that longer sequences will have a larger distance.

Having designed the modulation, it is of interest and necessary to see how it compares with other modulations. To this end, take as a reference the straightforward QPSK of unit symbol energy. Both modulations have the same energy of $E_b = 1/2$ (joules) expended per bit and also the same bandwidth requirement. We know that QPSK has a $d^2_{\min} = 2$. Thus it appears that all that has been achieved is a more complex modulator and a reduced $d^2_{\min}$!

[4] The natural mapping assigns bits based on the binary representation of the symbols' indices. In particular, since the constellation symbols are indexed from 1 to M (the same convention used in previous chapters), the label of $s_k(t)$ is the binary representation of the integer $(k-1)$. For example, in an 8-ary constellation, $s_1(t)$ is labeled with 000, while the label of $s_7(t)$ is 110.

However, upon reflection, one realizes that the problem is *competing sequences* of length 3 and that the signal assignment to the branches of the trellis may not be the best possible, i.e., the natural mapping may be inappropriate. Let us attempt to map the signal set *directly* to the branches. Since signals along the *parallel* branches are in immediate direct competition to determine $d^2_{\min}$ it seems prudent to assign signals that are 180 degrees apart, i.e., *antipodal*. In fact this was done previously and is shown in Figure 11.5(a), where signals $\{s_1(t), s_5(t)\}$ are assigned to the parallel branches emanating from state 00. The other pair of parallel branches from state 00 will eventually compete with set $\{s_1(t), s_5(t)\}$ to determine $d^2_{\min}$. Therefore let us make them not only antipodal but *as far away as possible* from set $\{s_1(t), s_5(t)\}$. From the signal space diagram of Figure 11.1, it follows that this pair should be set $\{s_3(t), s_7(t)\}$ as shown in Figure 11.5(a). The same reasoning is used for the two sets of parallel branches that merge in a common state. They should also be as far

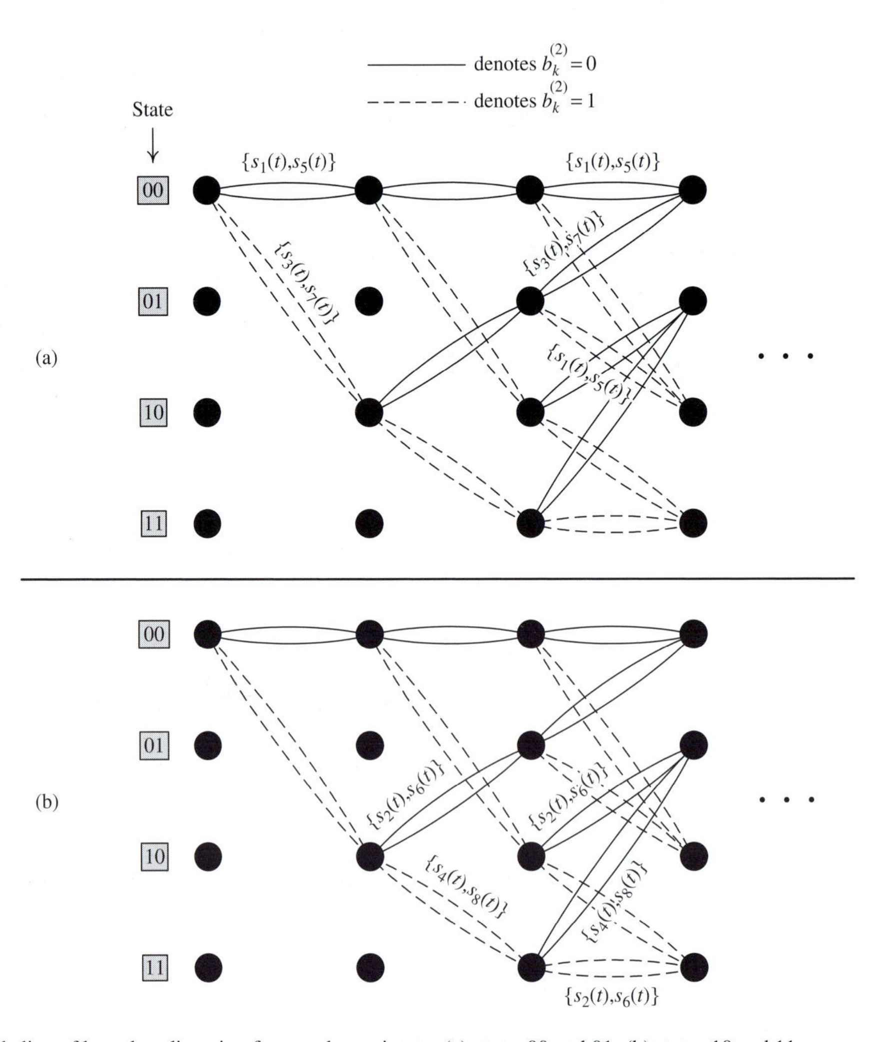

Fig. 11.5 Labeling of branches diverging from and merging to: (a) states 00 and 01; (b) states 10 and 11.

apart as possible. As Figure 11.5(a) shows this dictates that the branches from state 01 to state 00 should be assigned signals $\{s_3(t), s_7(t)\}$.

The branches out of states 00 and 01 have been fully labeled with the transmitted signals. It remains to label the branches out of states 10 and 11. For these branches we use the other subset of signals $\{s_2(t), s_4(t), s_6(t), s_8(t)\}$ and follow the same reasoning as above. Figure 11.5(b) shows the signal labeling out of state 10. Once this labeling is chosen the rest of the labeling falls into place. Figure 11.6 shows the final mapping with two sequences of length 3 highlighted. The squared Euclidean distance between them is

$$d_2^2 + d_1^2 + d_2^2 = 2 + (2 - \sqrt{2}) + 2 = 6 - \sqrt{2}, \tag{11.1}$$

which is *larger* than $d_4^2 = 4$, the squared distance between two parallel paths. It is left as an exercise to show that any two sequences of length 3 that start in any common state and end in a common state have at least this squared distance.

Therefore with this mapping, it is the parallel paths that determine the minimum distance, which is $d_{\min} = 2$. Compared with QPSK, a power saving of $10\log_{10}\left(d_{\min}^2(\text{TCM})/d_{\min}^2(\text{QPSK})\right) = 10\log_{10}\left(2^2/(\sqrt{2})^2\right) = 3.01$ decibels has been achieved *without any bandwidth expansion*. The modest increase in the complexity of the modulator may well be worth the price now.

It is very important to emphasize that the above improved modulator design is obtained by mapping the 8-PSK signals directly to the trellis branches, i.e., the labeling of the trellis branches in Figure 11.2 provided by the bits $c_k^{(1)}, c_k^{(2)}, c_k^{(3)}$ is not used. What 8-PSK signal is to be transmitted over the interval of $[kT_s, (k+1)T_s]$ depends on the trellis state at time kT_s, namely $b_{k-1}^{(2)}, b_{k-2}^{(2)}$, and the two information bits to be transmitted over that symbol

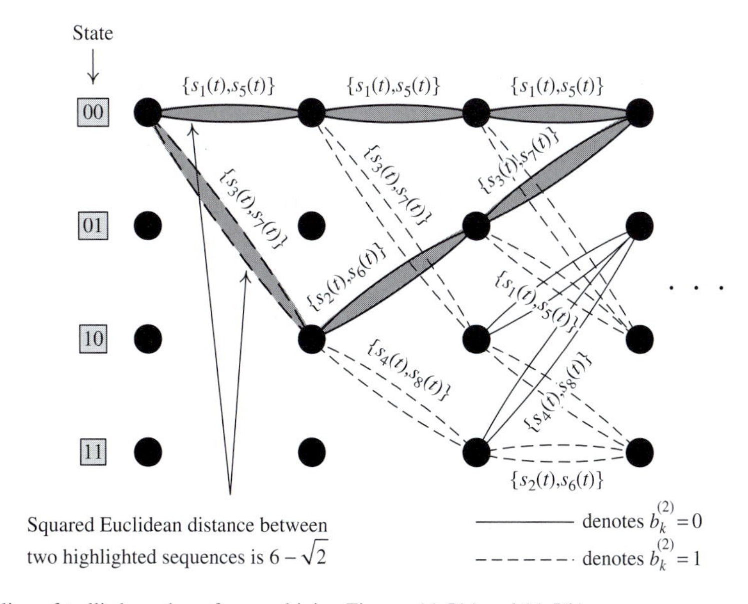

Fig. 11.6 Final labeling of trellis branches after combining Figures 11.5(a) and 11.5(b).

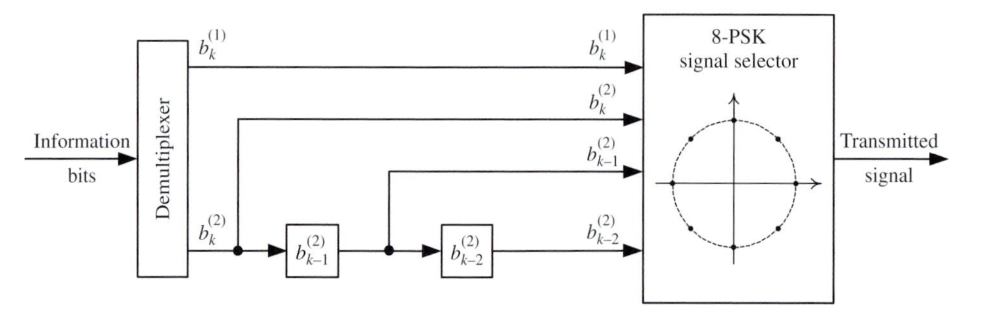

Fig. 11.7 A general modulator of four-state 8-PSK/TCM: The "8-PSK signal selector" block follows Figure 11.6, where two bits, $b_{k-1}^{(2)}, b_{k-2}^{(2)}$, choose a state and two bits $b_k^{(1)}, b_k^{(2)}$, select a signal on the trellis branch emanating from that state.

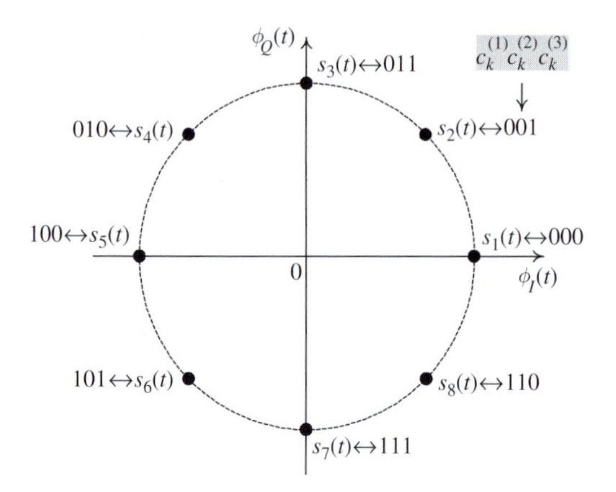

Fig. 11.8 8-PSK mapping for the circuitry of Figure 11.2 and TCM trellis of Figure 11.6.

interval, i.e., $b_k^{(1)}, b_k^{(2)}$. In essence the four bits $b_k^{(1)}, b_k^{(2)}, b_{k-1}^{(2)}, b_{k-2}^{(2)}$ are used to select the transmitted 8-PSK signal over $[kT_s, (k+1)T_s]$. This is illustrated in Figure 11.7.

From the practical implementation point of view, it is desired to process the three bits $b_k^{(2)}, b_{k-1}^{(2)}, b_{k-2}^{(2)}$ with simple XOR gates to produce two "coded" bits $c_k^{(2)}, c_k^{(3)}$, so that these coded bits together with $c_k^{(1)} = b_k^{(1)}$ uniquely select an 8-PSK signal according to some constellation mapping scheme. For the 8-PSK/TCM scheme of Figures 11.6 and 11.7, it is simple to verify that the same circuitry as in Figure 11.2 works with the constellation mapping of Figure 11.8.

The natural question then arises: what if other signal constellations such as QAM or other trellises, perhaps of eight or more states are considered? How does one go about designing the modulator to achieve an increase in $d_{\min}^2$? To do this the above procedure needs to be made more systematic and this is what is done next.

The starting point is to expand the signal set from the minimum needed to represent the source bits uniquely to the number needed after some of the source bits have been passed through the shift register circuit which provides the trellis. An explanation for this expansion is given later. Usually, though other expansions are possible, the signal set is expanded by a factor of 2 as was done in the previous example. After the signal constellation is decided on and a memory arrangement has been chosen, one then needs to assign the

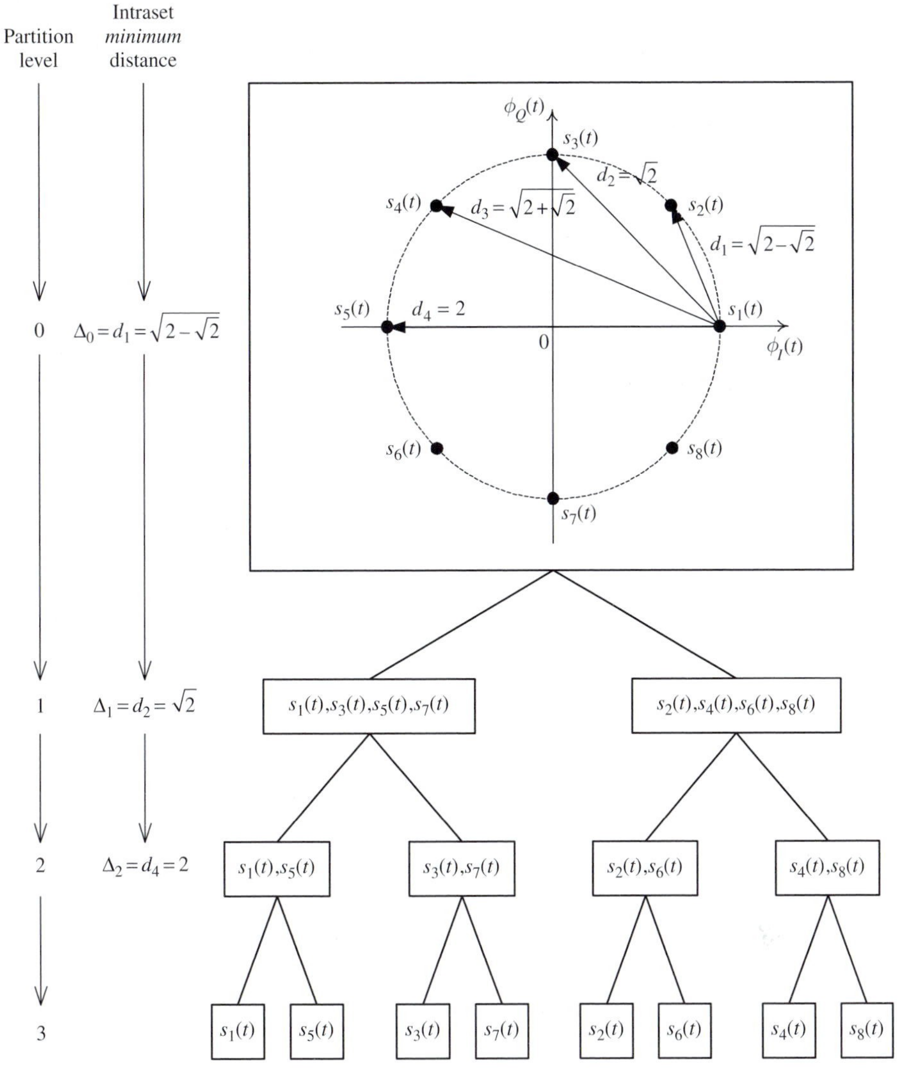

Fig. 11.9 Set partitioning applied to the 8-PSK signal constellation.

signals to the trellis branches. To do this a concept known as *set partitioning* is used. In this technique, the signal constellation is respectedly partitioned into subsets. At each step a subset is partitioned into two subsets where the Euclidean distance between signals *within* each subset is made as large as possible. Figure 11.9 illustrates this for the signal constellation of 8-PSK.

The mapping of the signals to the trellis branches then proceeds by observing three, essentially heuristic, guidelines:

(1) Parallel transitions, due to the uncoded bit(s), are assigned signals from within a subset at the lowest level. Since these signal sequences are immediate candidates to determine the minimum distance, this ensures that it is as large as possible.

(2) Transitions that diverge or merge from a common state are assigned signal subsets with as large an *intersubset* distance as possible. This also helps to ensure that the distance between two signal sequences that diverge from a common state and later merge in a common state is as large as possible.
(3) All signals and subsets are used with equal frequency in the trellis diagram.

Consider two further examples to illustrate the above guidelines. In the example just considered the minimum distance was determined by the parallel branches, which in turn arose due to the uncoded bit. The next example eliminates the parallel branches by applying both information bits to memory cells. It also increases the number of memory cells to three, which means that the trellis will have eight states. Figure 11.10 shows the general modulator, whereas Figure 11.11 shows the trellis along with the signal labeling.

The mapping (trellis labeling) proceeds as follows. Call the subsets $\{s_1(t), s_3(t), s_5(t), s_7(t)\}$ and $\{s_2(t), s_4(t), s_6(t), s_8(t)\}$ the *odd* and *even* subsets, respectively.

(a) First, the branches out of state 000 are labeled with signals from the lowest level of the partition tree of Figure 11.9. Choosing the four needed signals to have the largest possible distance between any two forces one to consider the signals in either the odd or even subset. Choose (choice is arbitrary) the odd subset.
(b) After this, attention is turned to labeling the branches from the states that the four branches terminate on, namely, 010, 100, 110 (state 000 is ignored since we already know its branch labeling). The second guideline dictates the labeling of the branches out of these states. Start with state 100 since the branches from this state merge at states that already have a labeled branch. Choose the branch signal to be as far as possible from the signal on the already labeled branch. In essence the signals are chosen to be antipodal, which means that they are chosen from the odd subset. For example, the branch from state 100 to state 000 is chosen to be $s_5(t)$ which is antipodal to $s_1(t)$, etc.
(c) It follows from the third guideline that four of the states have signals from the odd subset and the other four have signals from the even subset. We have established that states 000 and 100 are associated with the odd subset. Nature loves symmetry and therefore we assign the two subsets to the states as follows: (i) even subset

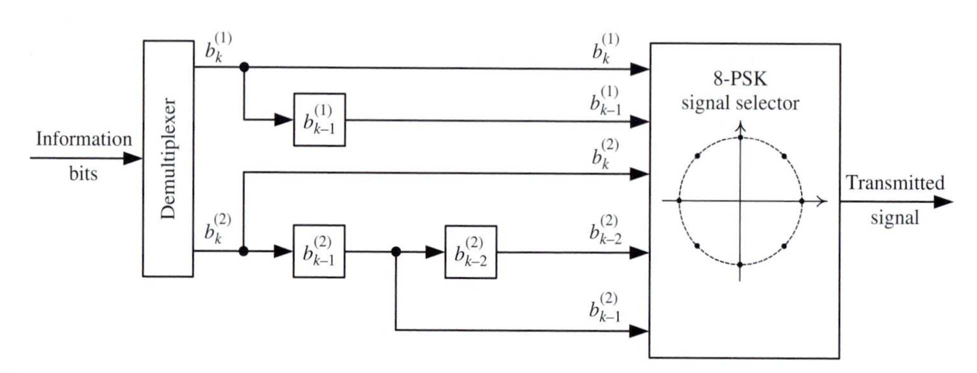

Fig. 11.10 A general modulator for eight-state 8-PSK/TCM.

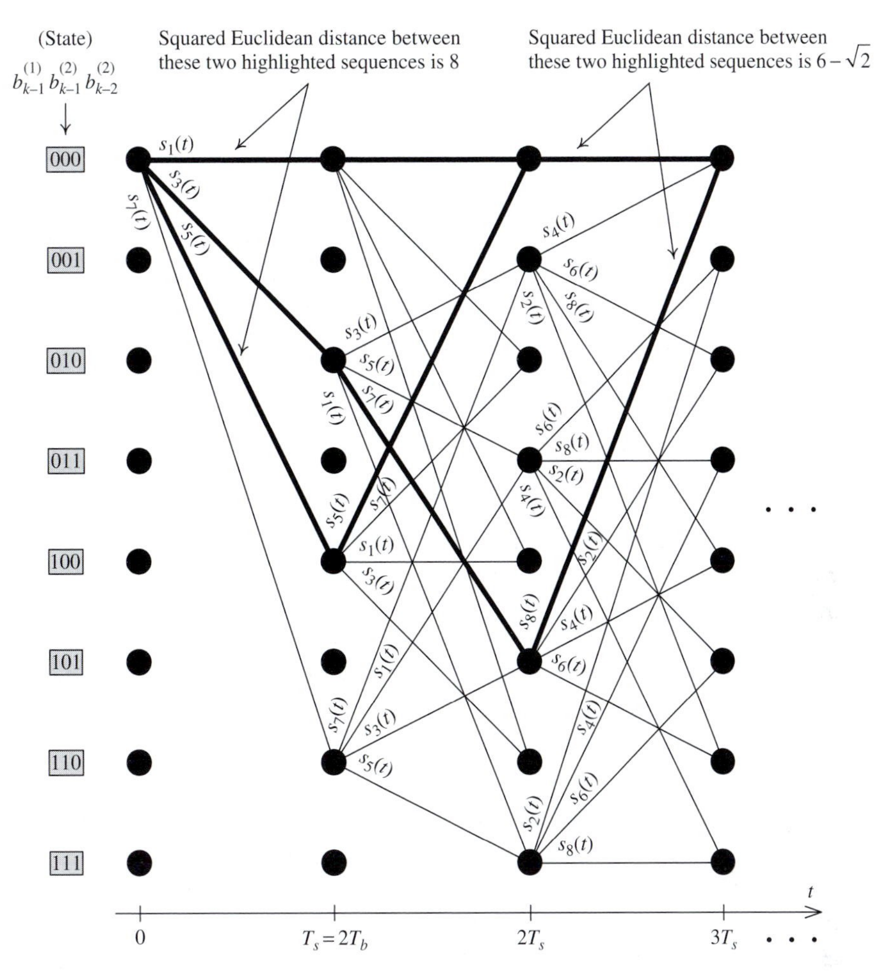

Fig. 11.11 Labeling of trellis branches for an eight-state trellis and 8-PSK. Note that for clarity not all connections are shown.

to states 001, 011, 101, 111; and (ii) odd subset to 000, 010, 100, 110. It should be pointed out that other assignments are possible.

(d) One needs to label the branches coming out of these states. To do this turn your attention to state 001 at time $2T_s$. The signal along the branch from state 001 to state 000 competes with $s_1(t)$ and comes from the even subset. Choose it for maximum distance. Either $s_4(t)$ or $s_6(t)$ could be chosen. Choose $s_4(t)$ and then label the other three branches as shown in Figure 11.11. Though this latter labeling, at this point, can be made arbitrarily, for aesthetic purposes, the labels are chosen to be in cyclic order.

(e) Once the branches out of state 001 have been determined, look at state 101 and label it as shown in Figure 11.11. The branch from state 101 to state 000 is labeled with signal $s_8(t)$ to make it antipodal to $s_4(t)$ and similarly for the other three branches. Note how the cyclic order of branch labeling nicely falls into place.

(f) Finish the labeling for all other states.

Table 11.1 Power savings of 8-PSK/TCM over QPSK

Number of trellis states	4	8	16	32	64	128	256
Power saving (dB)	3.0	3.6	4.1	4.6	5.0	5.2	5.8

Figure 11.11 shows the minimum squared Euclidean distances between competing paths of length 2 and length 3. The length 3 paths determine d^2_{min} which is $6-\sqrt{2}$. Compared with QPSK this means a power saving of $10\log_{10}\frac{6-\sqrt{2}}{2}=3.6$ decibels. It is left as an exercise (see Problem 11.1) to come up with a shift register circuit, principally the XOR gates and tap connections, and 8-PSK mapping such that the three coded bits (produced by the circuit) can uniquely select the 8-PSK signal point in accordance with the trellis labelings of Figure 11.11.

One could attempt to increase d^2_{min} by increasing the number of states of the trellis. What this does is increase (with a proper arrangement of the memory cells) the length of paths that start in a common state, diverge, and merge later in a common state. This has been done and the gain in power savings over QPSK is shown in Table 11.1 [6].

Of course, as the number of states becomes larger the modulator and especially the demodulator (discussed in Section 11.1.3) become more and more complex. Even for the examples just discussed one needs to design the circuit to do the mapping from the source bits to the modulator's output signal. In general, the maximum saving that can be achieved with this signal constellation (8-PSK) is on the order of 6.0 decibels.

The next example uses rectangular 16-QAM as the expanded signal constellation. Using the factor of 2 rule-of-thumb for signal expansion this means that three information bits are eventually mapped to the signal constellation. Since in the example one of the information bits is chosen to be uncoded (which means that there will be parallel transitions in the trellis) the other two information bits are input to memory cells. The memory cell configuration is the same as that of the previous example, i.e., there are three cells and an 8-state trellis. Figure 11.12 shows the set partitioning of the 16-QAM signal constellation, Figure 11.13 the general modulator, and Figure 11.14 the trellis and mapping. Again, the design of a specific shift register circuit and 16-QAM constellation mapping to work with the memory arrangement in Figure 11.13 and the trellis labelings of Figure 11.14 is left as an exercise (see Problem 11.2).

To determine what power saving, or *coding gain* as it is commonly called, has been achieved by this example one, of course, needs to compare it with a reference modulation. The comparison is not quite as straightforward as it was for the M-PSK signals because in going from 8-QAM signals to the expanded signal set of 16-QAM the average energy expenditure per bit (or per symbol) has changed. To account for this define the power saving or coding gain as

$$\text{coding gain} = 10\log_{10}\frac{d^2_{\text{min}}(\text{coded})/E_s(\text{coded})}{d^2_{\text{min}}(\text{uncoded})/E_s(\text{uncoded})}, \tag{11.2}$$

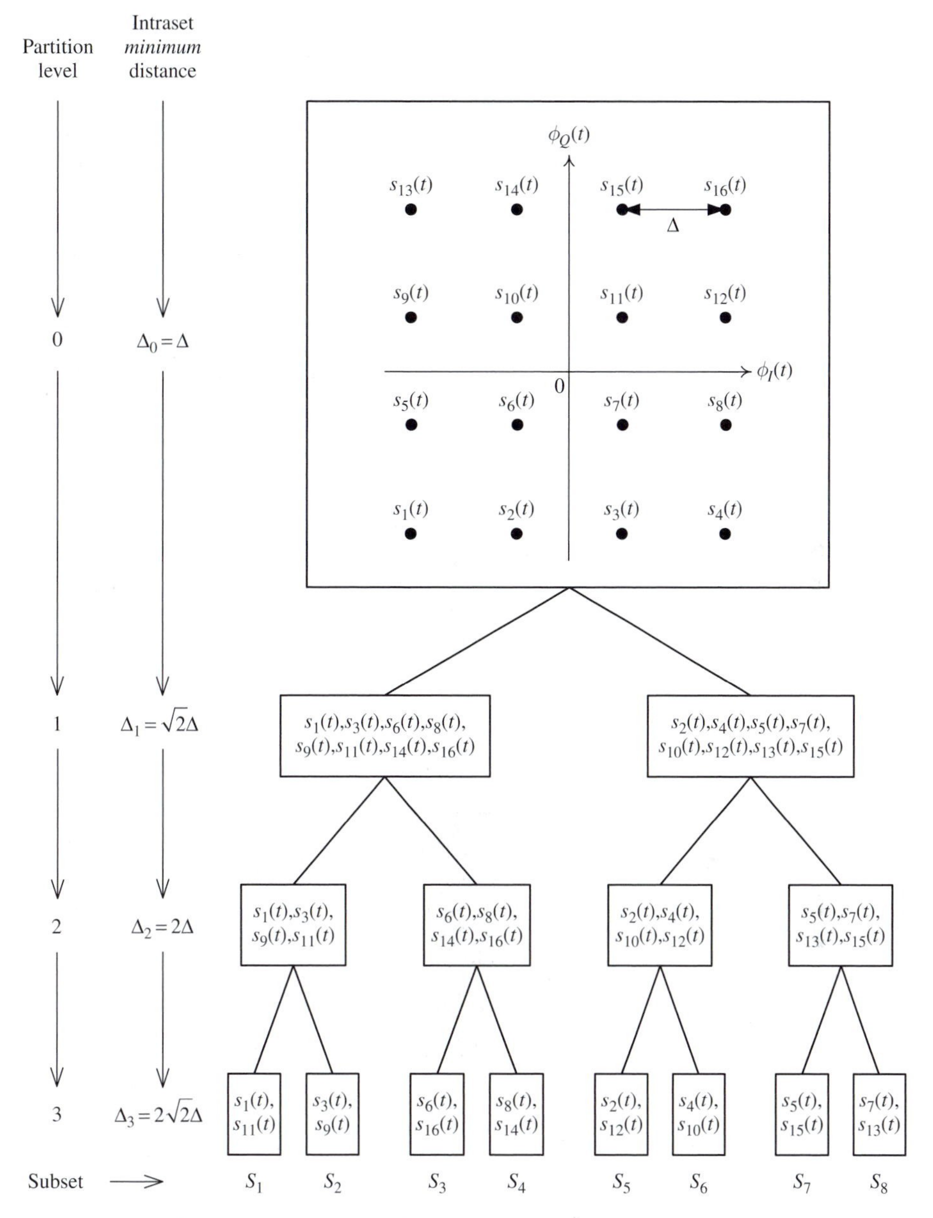

Fig. 11.12 Set partitioning applied to the 16-QAM signal constellation.

where E_s(coded) and E_s(uncoded) are the average symbol energies of the signal constellations used in the coded and uncoded systems, respectively.

One still needs to specify the *uncoded* signal constellation. A simple choice is that of 8-QAM as shown in Figure 11.15(a). From Figure 11.14 it is sequences of length 3 that determine the minimum distance, d^2_{min}(coded). The squared distance between the subset sequence paths $\{S_1, S_1, S_1\}$ and $\{S_3, S_5, S_3\}$, which diverge at

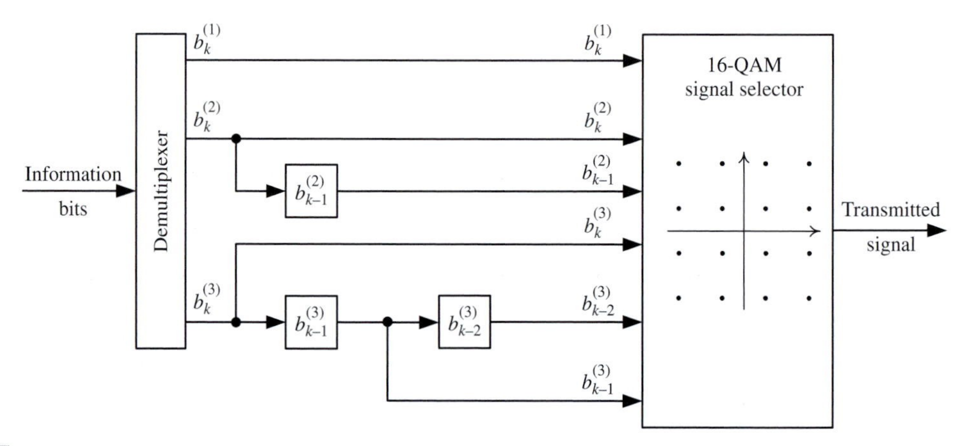

Fig. 11.13 A general modulator for an eight-state 16-QAM/TCM.

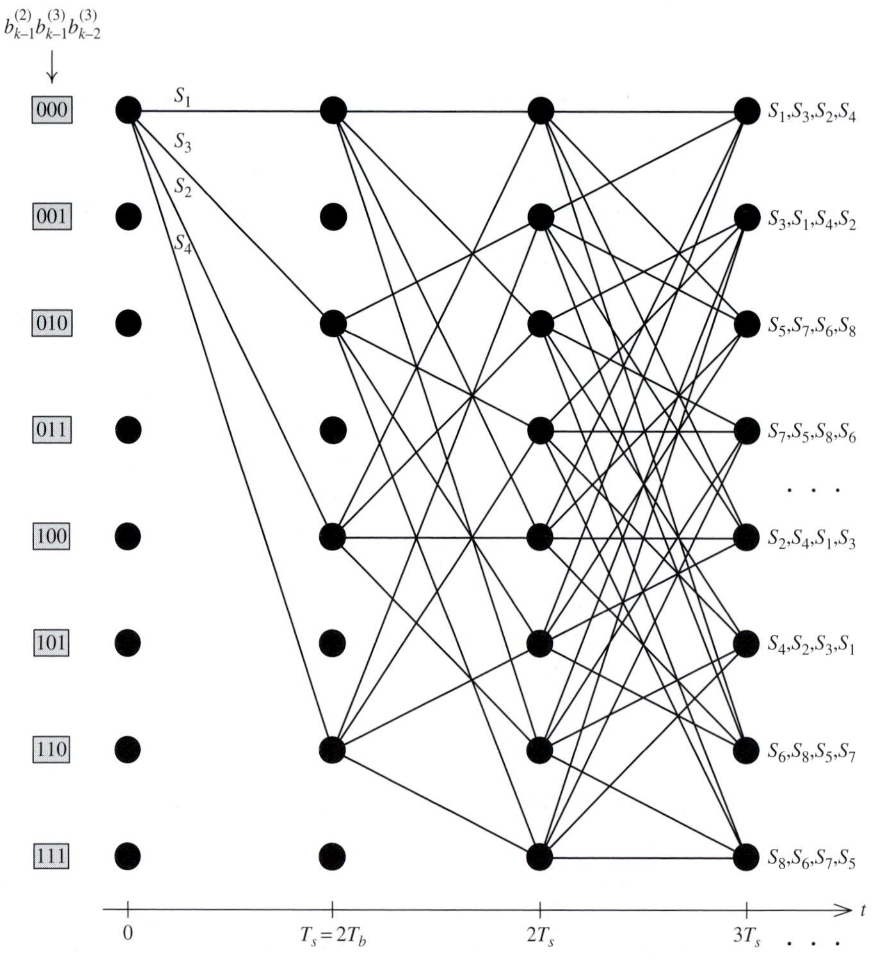

Fig. 11.14 Labeling of trellis branches for an eight-state trellis and 16-QAM. The subset notation is that of Figure 11.12. The first subset is mapped to the top (parallel) branch(es), the second one to the next branch and so on.

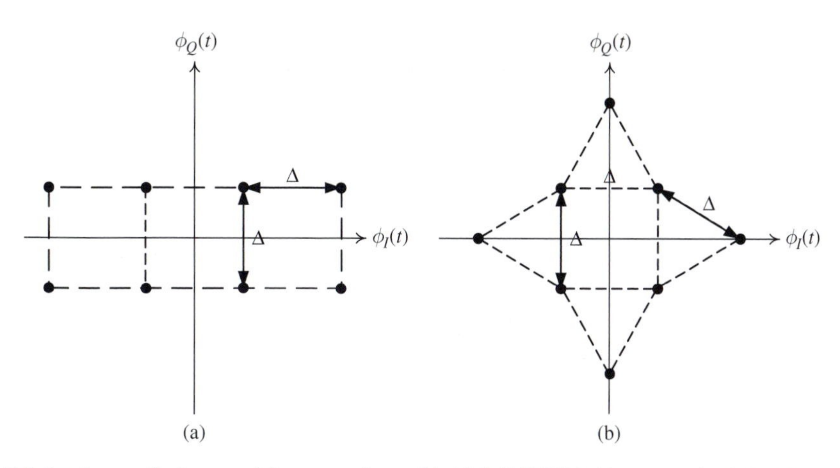

Fig. 11.15 8-QAM signal constellations used for comparison with 16-QAM/TCM: (a) rectangular 8-QAM, and (b) a more efficient 8-QAM.

state 000 and also merge in 000 is $2\Delta^2 + \Delta^2 + 2\Delta^2 = 5\Delta^2$. The average energy for the coded system is given by $E_s(\text{coded}) = 4\left(\frac{1}{2}\Delta^2 + 5\Delta^2 + \frac{9}{2}\Delta^2\right)/16 = \frac{5}{2}\Delta^2$ joules/symbol. For the 8-QAM constellation of Figure 11.15(a) the minimum squared distance $d^2_{\text{min}}(\text{uncoded}) = \Delta^2$, while the average energy is $E_s(\text{uncoded}) = 4\left(\frac{1}{2}\Delta^2 + \frac{5}{2}\Delta^2\right)/8 = \frac{3}{2}\Delta^2$. The power saving is therefore equal to

$$10\log_{10}\left[\left(\frac{5\Delta^2}{5\Delta^2/2}\right)\bigg/\left(\frac{\Delta^2}{3\Delta^2/2}\right)\right] = 4.77 \text{ decibels.} \tag{11.3}$$

The saving, of course, depends on the 8-QAM constellation chosen for comparison. Consider the constellation of Figure 11.15(b) which still has the same minimum squared distance of Δ^2 but is more power efficient. Its average energy is $\frac{3+\sqrt{3}}{4}\Delta^2$ joules/symbol. The power saving of the TCM scheme is now 3.74 decibels, still substantial. Note that the 8-QAM constellation of Figure 11.15(b) would also require a more complicated demodulator than that of Figure 11.15(a).

Having illustrated the concepts of TCM through the above examples we give next an explanation of the performance improvement offered by TCM followed by a brief presentation of a generalization which allows one to deal with more complex signal constellations, particularly those of higher dimensions than two.

11.1.1 Explanation of performance improvement achieved with TCM

Consider each transmitted signal in an interval to be chosen from a signal point in a signal constellation. In the examples above two-dimensional signal constellations were used but this is not necessary. Further, assume that some of the signal points do not need to be used. When the signal set is expanded by a factor of 2 this means that strictly speaking only half the signal points need to be used, at least coming out of any state.

More generally, let the number of signals in the constellation be M and let $M_0 < M$ be actually used for the transmitted signal in a signaling interval, T_s. Now consider a transmitted sequence of signals corresponding to N signaling intervals. The total number of available sequences from the overall constellation is M^N, of which only M_0^N are used. The ratio of used signal sequences to the number available is $(M_0/M)^N$. As N becomes larger the ratio $(M_0/M)^N$ tends to zero since $(M_0/M) < 1$. Or, proportionally, fewer and fewer of the available sequences are used as N increases. This, in turn, suggests that the distance between the transmitted sequences can be made larger with increasing N. Not all choices of the M_0^N sequences out of the M^N possible sequences are good but the use of a trellis as provided by the shift register circuit along with set partitioning and the associated mapping guidelines allow one to select a good set of signal sequences.

The most popular TCM schemes have $(M_0/M) = \frac{1}{2} < 1$, i.e., the signal set is expanded by a factor of 2 as was done in the examples. Other expansions are possible and this generalization is described briefly in the next section. However, the actual performance of TCM depends not only on the signal expansion, but just as crucially on the introduction of memory by the shift register. What this suggests is that the design of the shift register circuit is also very important. For this design the theory of convolutional coding (or block coding) needs to be used, a topic beyond the scope of the present discussion.

11.1.2 A (reasonably) general approach to TCM

A common structure for a TCM modulator using a convolutional encoder is as shown in Figure 11.16. Basically a group of λ bits is mapped onto one of $2^{\lambda+1}$ signal points every signaling interval. The convolutional encoder takes $\tilde{\lambda} \leq \lambda$ bits in and puts out $\tilde{\lambda} + 1$ bits. Since there is one extra bit this implies that the signal constellation has been expanded by a factor of 2. The $\tilde{\lambda} + 1$ coded bits select a sublattice of the signal constellation. A sublattice (the lattice concept is explained below) in essence is one of the subsets we have seen in the set partitioning method. Using lattice concepts is a useful and convenient way to describe the process of set partitioning mathematically. Once the sublattice has been selected the uncoded bits select the signal from within the sublattice. This signal point is then transmitted.

The uncoded bits result in $2^{(\lambda-\tilde{\lambda})}$ parallel transitions or branches from any state to any other state of the trellis. Therefore the sublattices should be chosen such that, all other parameters being equal, the minimum distance between signal points in the sublattice is *maximized*. The memory cells of the convolutional encoder determine the trellis structure. The number of shift register cells determines the number of states and in turn the minimum length of a path pair which diverge from a common state and later merge in a common state. As a general rule the longer it takes for a merge to occur the larger the distance between these paths is. Thus increasing the memory usually improves the minimum distance but, of course, at the expense of a more complex modulator and demodulator. Finally, the connections from the shift register cells to the *exclusive-or* gates to form the coded bits provide a convenient method to select a sublattice.

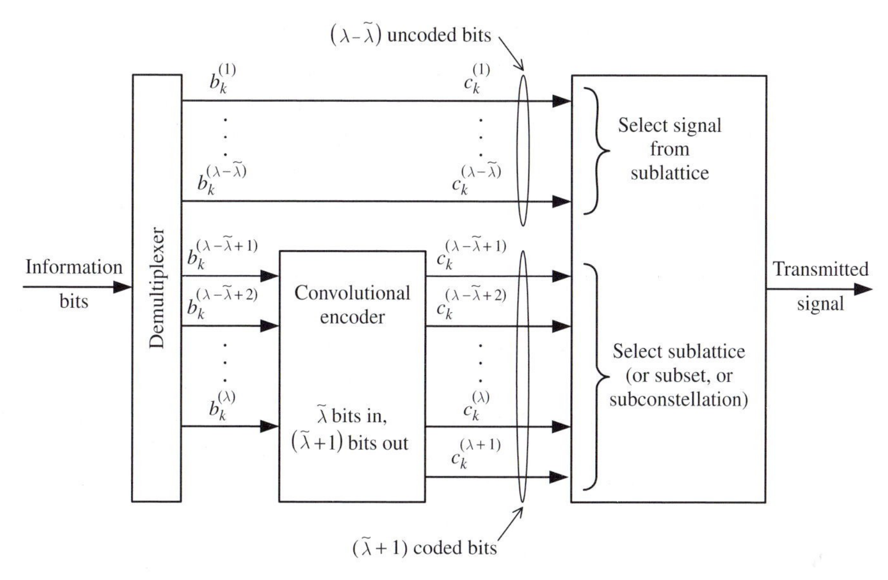

Fig. 11.16 General structure of modulation for TCM.

Since lattices play an important part in TCM (and other aspects of communications) this section concludes with a discussion of them. Generally, a lattice is a *regular* spacing of points over the *entire extent* of a n-dimensional space. The most familiar lattice is that of regularly spaced grid points in one or two dimensions, which were used for M-ASK or rectangular M-QAM. We discuss two-dimensional lattices next since they are the most important ones for communications.

A two-dimensional lattice is generated by the following equation:

$$\begin{bmatrix} s_1 \\ s_2 \end{bmatrix} = \begin{bmatrix} b_{11} & b_{21} \\ b_{12} & b_{22} \end{bmatrix} \begin{bmatrix} k_1 \\ k_2 \end{bmatrix}, \tag{11.4}$$

where vector $\begin{bmatrix} s_1 \\ s_2 \end{bmatrix}$ is a lattice point, vectors $\begin{bmatrix} b_{11} \\ b_{12} \end{bmatrix}$ and $\begin{bmatrix} b_{21} \\ b_{22} \end{bmatrix}$ are basis vectors, and vector $\begin{bmatrix} k_1 \\ k_2 \end{bmatrix}$ is an integer vector, i.e., k_1 and k_2 are any values drawn from the set of integers. Note that a lattice point can be written as

$$\begin{bmatrix} s_1 \\ s_2 \end{bmatrix} = k_1 \begin{bmatrix} b_{11} \\ b_{12} \end{bmatrix} + k_2 \begin{bmatrix} b_{21} \\ b_{22} \end{bmatrix}, \tag{11.5}$$

i.e., as a *linear integer combination* of the two vectors $\vec{b_1} = \begin{bmatrix} b_{11} \\ b_{12} \end{bmatrix}$ and $\vec{b_2} = \begin{bmatrix} b_{21} \\ b_{22} \end{bmatrix}$, hence the name basis. The choice of the basis vectors determines the appearance of the lattice. The most familiar is the simple $\begin{bmatrix} b_{11} \\ b_{12} \end{bmatrix} = \begin{bmatrix} 1 \\ 0 \end{bmatrix}, \begin{bmatrix} b_{21} \\ b_{22} \end{bmatrix} = \begin{bmatrix} 0 \\ 1 \end{bmatrix}$, which results in a rectangular grid but other choices are possible (actually an infinite number of choices

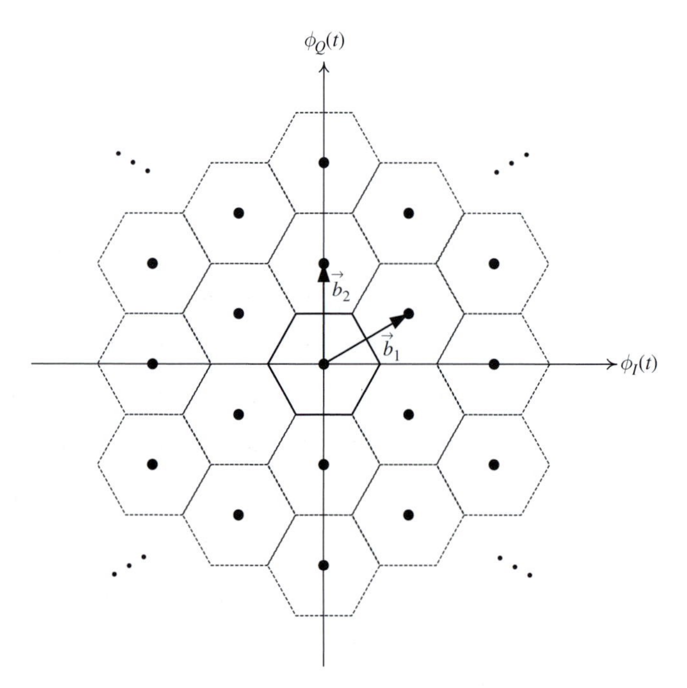

Fig. 11.17 Hexagonal lattice.

is available). As an example the choice of $\begin{bmatrix} b_{11} \\ b_{12} \end{bmatrix} = \begin{bmatrix} \frac{\sqrt{3}}{2} \\ \frac{1}{2} \end{bmatrix}$ and $\begin{bmatrix} b_{21} \\ b_{22} \end{bmatrix} = \begin{bmatrix} 0 \\ 1 \end{bmatrix}$ results in a hexagonal lattice as shown in Figure 11.17.

The signal points used in, say M-ary modulation, are then drawn from a finite set of the lattice points, located inside a circle centered at the origin of maximum radius $\sqrt{E_s}$. The subsets are determined by utilizing a sublattice.

A sublattice is a subset of the lattice points with the property that any linear, integer combination of two points in this subset results in a lattice point that is also in this subset. Consider the simple and most frequently encountered lattice, the rectangular lattice. It is commonly denoted as $\mathbb{Z}_2$ and we have used it for rectangular QAM. Figure 11.18 shows the lattice along with a sublattice, which is defined as the set of all lattice points with the property that the coordinates of the point are even. Obviously any linear, integer combination of two points in this sublattice has coordinates that are even and therefore is some point in the sublattice.

The important thing about a sublattice is that it partitions the lattice into mutually exclusive subsets.[5] The sublattice just described partitions the lattice into four subsets (counting itself). The three other subsets are obtained by adding $(1, 0)$, $(0, 1)$, $(1, 1)$ respectively to every point in the sublattice. The coded bits of Figure 11.16 select the subset. In this case since there are four subsets this means that two coded bits are needed.

[5] For readers familiar with finite or abstract algebra and the algebraic structures of *groups*, *rings*, *fields* the lattice is a group and the sublattice is a subgroup. One of the first properties that is established for a subgroup is that it partitions the group into mutually exclusive subsets called *cosets*.

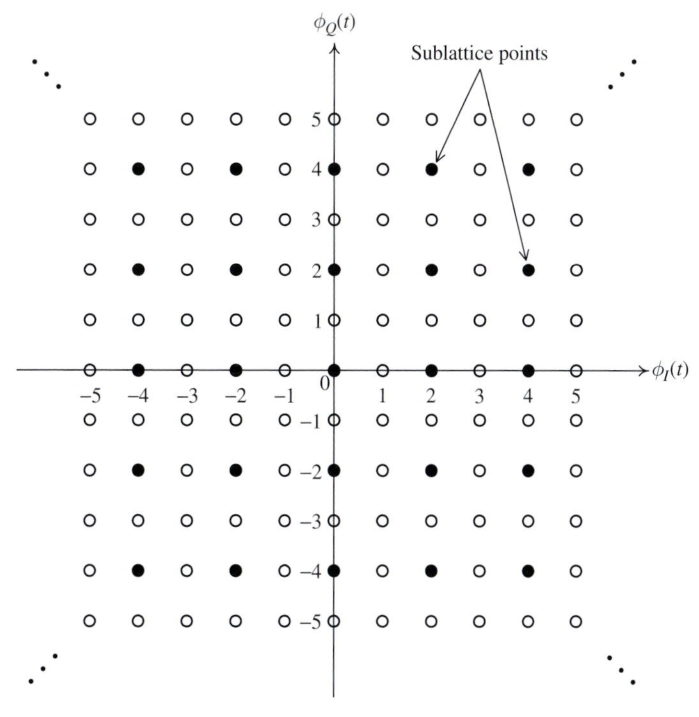

Fig. 11.18 Rectangular or $\mathbb{Z}_2$ lattice and a sublattice.

As an exercise the reader may attempt to determine other sublattices for the given lattice. Possibilities are: (i) all lattice points whose coordinates sum to an even number, (ii) all lattice points whose coordinates are a multiple of 4, etc. In each case you should check that a proper sublattice has been created and, if so, how many subsets does the sublattice partition the lattice into.

The sublattice in essence does the set partitioning mentioned earlier. However, a constellation such as M-PSK does not form a lattice and therefore for it one must still rely on set partitioning as described earlier. The sublattice concept though is very useful when constellations of higher dimensions are considered. Using pencil and paper does not work when one goes to a signal space dimension of 4, 8 or higher.

To illustrate the design of a TCM modulator using lattice concepts the following example is given.

Example 11.1 TCM design with hexagonal lattice Consider the hexagonal lattice, which is repeated in Figure 11.19. As a sublattice we choose all points whose coordinates are even multipliers of the lattice basis, i.e., the basis vectors for the sublattice are $\begin{bmatrix}\sqrt{3}\\1\end{bmatrix},\begin{bmatrix}0\\2\end{bmatrix}$. The sublattice partitions the lattice space into four cosets. These are the

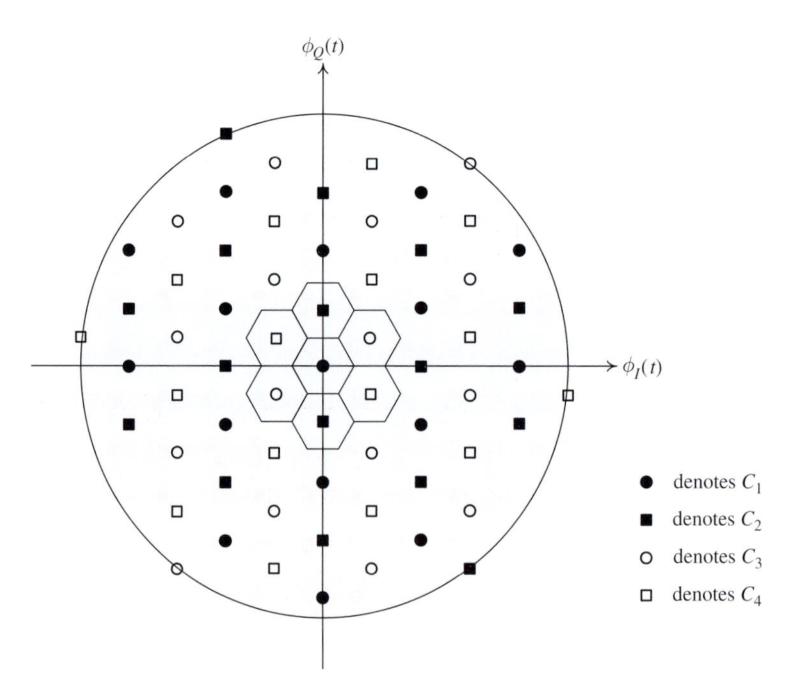

Fig. 11.19 Lattice and sublattice considered in Example 11.1.

sublattice itself, called C_1, and three other cosets with coset leaders $\begin{bmatrix} 0 \\ 1 \end{bmatrix}$, $\begin{bmatrix} \frac{\sqrt{3}}{2} \\ \frac{1}{2} \end{bmatrix}$, $\begin{bmatrix} -\frac{\sqrt{3}}{2} \\ \frac{1}{2} \end{bmatrix}$, called C_2, C_3, C_4.

Having chosen the lattice and sublattice, the next choice is that of the shift register to provide a coset selector. Since there are four cosets, we need two coded bits. The memory is chosen to make the trellis simple, i.e., there are two memory cells. The circuit is shown in Figure 11.20 along with trellis and the mapping of the cosets to the trellis branches. The figure shows that there are four uncoded bits, which means that choosing a factor of 2 signal expansion, we have to map 2^5 patterns of information bits onto $2^6 = 64$ signal points. There are four cosets and therefore we have $64/4 = 16$ points per coset. The uncoded bits select which of these 16 signal points is to be transmitted (after a coset is selected). To determine the actual signal points to be included from the lattice, draw a circle centered at the origin of radius large enough to include 16 signals from each coset. The final constellation is shown in Figure 11.19.

The last step in this preliminary design is to determine the coding gain. To do this we need to determine two distances, one between points in a coset, $d^{(1)}_{\text{min}}$, the other between cosets, $d^{(2)}_{\text{min}}$. The distance $d^{(1)}_{\text{min}}$ is necessary because there are parallel branches in the trellis, each branch being a signal point in a coset. The distance $d^{(2)}_{\text{min}}$ is due to the distance between two sequences that diverge at a common state and then merge at some common state. For the purpose of distance calculation, the lattice may be scaled arbitrarily since the squared distance is divided by the average energy and the scaling factor cancels out.

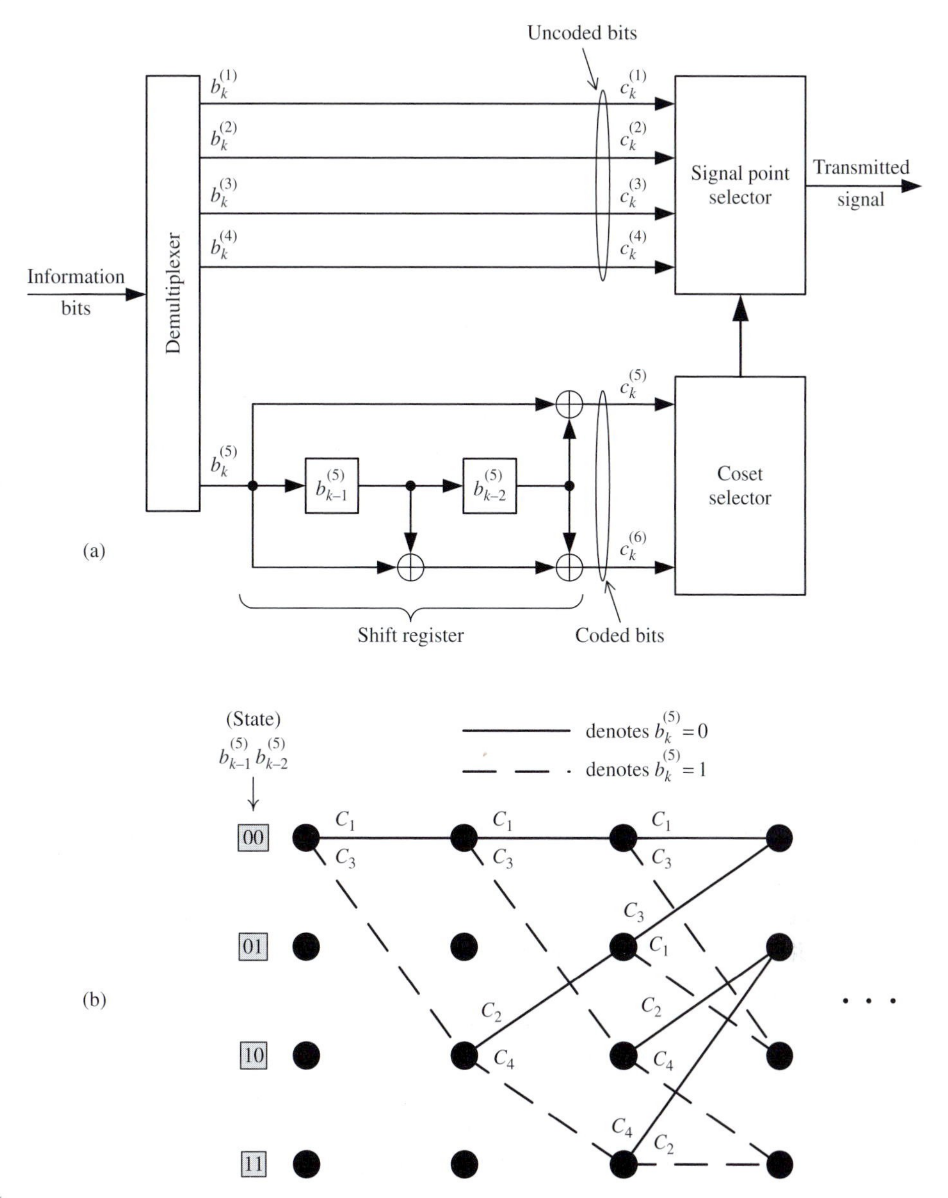

Fig. 11.20 (a) Encoder (modulator), and (b) trellis diagram for TCM design in Example 11.1.

Based on this, it is readily seen from Figure 11.19 that $d_{\min}^{(1)} = 2$ and $d(C_i, C_j) = 1$, $i, j = 1, 2, 3, 4$, $i \neq j$, where $d(C_i, C_j)$ is the minimum distance between cosets i and j. The distance $d_{\min}^{(2)}$ is determined by the two (sub)sequences $\{C_1, C_1, C_1\}$ and $\{C_3, C_2, C_3\}$ and is

$$\left[d_{\min}^{(2)}\right]^2 = d^2(C_1, C_3) + d^2(C_1, C_2) + d^2(C_1, C_3) = 3. \qquad (11.6)$$

Therefore $d_{\min}(\text{coded}) = \min\left\{d_{\min}^{(1)}, d_{\min}^{(2)}\right\} = \sqrt{3} = 1.73$. The average transmitted energy is found to be 9.0 joules/symbol.

To determine the coding gain we need to specify a reference uncoded constellation. For simplicity, a rectangular 32-QAM is chosen with $d_{\min}(\text{uncoded}) = 1$. Its average energy is 6.5 joules/symbol. Therefore the coding gain achieved by the designed TCM is

$$\begin{aligned}\text{coding gain} &= 10\log_{10}\frac{d_{\min}^2(\text{coded})/E_s(\text{coded})}{d_{\min}^2(\text{uncoded})/E_s(\text{uncoded})}\\ &= 10\log_{10}\left(\frac{3}{9.0}\times\frac{6.5}{1}\right) = 3.36\text{ decibels.}\end{aligned}\tag{11.7}$$

■

Finally, other reference constellations can be chosen (see Problem 11.3), some of which are more energy efficient, which would reduce the coding gain. Note that the demodulator (or modulator for that matter) for the hexagonal signal set is not as simple as that for rectangular QAM. It is of interest, therefore, to see what happens with the same modulator, one coded bit, four uncoded bits, and a QAM lattice. This is pursued in Problem 11.4.

Having discussed the design of TCM we now turn our attention to the demodulator and analysis of the error performance of TCM.

11.1.3 Demodulation of TCM

The fact that the transmitted signal in TCM is a sequence of symbols (signal points of the expanded constellation) along a path in the trellis used to define the modulation implies that the demodulator can proceed by determining the best path (transmitted signal sequence) through the trellis. The situation is directly analogous to that encountered in Miller baseband modulation (Chapter 6) and in Chapter 9 when ISI was encountered. There the best path through the trellis was determined using a *Viterbi algorithm*. This is also done by the TCM demodulator. The difference is in the details: namely in how the branch metric is computed, the facts that the energy of the transmitted symbol in any signaling interval may be important and that there may be *parallel branches* in the trellis. But once the branch metric is established the principle of adding it to the survivor metric that resides in the state from which the branch emanates, then comparing the resultant (partial) path metric with other (partial) path metrics at the state on which the branch terminates, choosing the one that is most favorable, calling it the survivor (state) metric at this state, etc., etc., is the same.

The derivation of the branch metric proceeds similarly to that of Miller demodulation and ISI demodulation. It is included here for completeness and on the principle that practice makes perfect (the P^3 approach to learning). Consider the situation where λ information bits are mapped every signaling interval to a signal point, drawn typically from a $2^{\lambda+1}$ signal constellation. The constellation size is not important for the discussion here, but it should be larger than 2^{λ} in order to achieve a coding gain. Further, let the information bits be equally probable and statistically independent. This means that any transmitted signal is equally likely. Finally, the degradation is AWGN and we are synchronized.

To determine the branch metric, start by considering N signaling intervals. There are $2^{\lambda N}$ possible transmitted signals, each signal representing a path through the TCM trellis. The received signal is

$$\mathbf{r}(t) = S_i(t) + \mathbf{w}(t), \quad 0 \le t \le (N-1)T_s, \; T_s = \lambda T_b, \quad i = 1, 2, \ldots, 2^{\lambda N}, \tag{11.8}$$

where $S_i(t)$ is the *signal sequence* corresponding to the ith path through the trellis, which will be simply referred to as the ith signal. Viewed as an M-ary modulation with equally likely signals and over an AWGN channel, the demodulator would choose the signal that the received signal $r(t)$ is closest to in the Euclidean distance sense, i.e., the decision rule is

compute

$$\int_0^{(N-1)T_s} [(r(t) - S_i(t)]^2 \, \mathrm{d}t, \quad i = 1, 2, \ldots, 2^{\lambda N}, \tag{11.9}$$

and choose the signal that has the *smallest* distance.

The above is fine in principle but useless in practice since $2^{\lambda N}$ is a large, large number, particularly as $N \to \infty$. The problem can be overcome by rewriting the integral as:

$$\int_0^{(N-1)T_s} [(r(t) - S_i(t)]^2 \, \mathrm{d}t = \int_0^{(N-1)T_s} r^2(t)\mathrm{d}t - 2 \int_0^{(N-1)T_s} r(t)S_i(t)\mathrm{d}t + \int_0^{(N-1)T_s} S_i^2(t)\mathrm{d}t. \tag{11.10}$$

The first term is the same for all i and can be discarded. The last term is the energy of the ith signal, call it E_i. By multiplying through by -1 and dividing by 2 the decision rule becomes

compute

$$\int_0^{(N-1)T_s} r(t)S_i(t)\mathrm{d}t - \frac{E_i}{2}, \quad i = 1, 2, \ldots, 2^{\lambda N}, \tag{11.11}$$

and choose the signal that has the *maximum* value.

However, this is still no better from a computational complexity point of view than (11.9).

To simplify the computational effort write $S_i(t)$ as $S_i(t) = \sum_{k=1}^{N} S_{ik}(t)$, where $S_{ik}(t)$ is the constellation symbol transmitted in the time interval $[(k-1)T_s, kT_s]$ along the ith path through the trellis. Then (11.11) becomes

$$\sum_{k=1}^{N} \underbrace{\int_0^{(N-1)T_s} r(t)S_{ik}(t)\mathrm{d}t}_{= \int_{(k-1)T_s}^{kT_s} r(t)S_{ik}(t)\mathrm{d}t} - \frac{1}{2}\sum_{k=1}^{N}\sum_{j=1}^{N} \underbrace{\int_0^{(N-1)T_s} S_{ik}(t)S_{ij}(t)\mathrm{d}t}_{\substack{=0, \; k \neq j \\ = E_{ik}, \; k = j}}, \tag{11.12}$$

where E_{ik} is the energy of the ith signal sequence in the kth signaling interval. The decision rule is therefore

$$\text{compute} \quad \sum_{k=1}^{N}\left[\int_{(k-1)T_s}^{kT_s} r(t)S_{ik}(t)\mathrm{d}t - \frac{E_{ik}}{2}\right], \quad i = 1, 2, \ldots, 2^{\lambda N}, \tag{11.13}$$

and choose the signal that has the *maximum* value.

Now we address how to reduce the computational complexity that is still present in (11.13). The term in the brackets is a branch metric for the kth signaling interval. If we knew the best (partial) path to each state and the corresponding path (or survivor or state) metric at time $(k-1)T_s$ then this knowledge and the branch metrics for all branches in the kth signaling interval could be used to compute the best (partial) path for each state at $t = kT_s$. The algorithm, known generically as the Viterbi algorithm, is as follows:

(1) Start in a known state at $t = 0$, typically the all-zero state. Compute the branch metrics for all branches emanating from this state.
(2) Proceed to the states at $t = T_s$. If at this stage there are two (or more) branches merging into a state, retain only the one with the *largest* branch metric and call the metric the state metric (also called the survivor metric). If only one branch terminates at a state, then the branch metric becomes the state metric.
(3) Compute the branch metrics for all the trellis branches for the interval $[T_s, 2T_s]$ based on the received signal $r(t)$ in this interval. Now ADD the branch metric to the state metric present at the state the branch emanates from, COMPARE the resultant path metric with all other path metrics at the state at which the branch terminates, and SELECT the largest path metric to be the state (survivor) metric at this state.
(4) Do Step (3) for all the states at $t = 2T_s$ and then repeat Step (3) for the next signaling interval $[T_s, 2T_s]$ and so on.

Implicit in the above is that one also retains the survivor paths at each trellis stage. The number of paths to be saved is equal to the number of trellis states, a much smaller number than the total number of possible transmitted signals. Besides the branch metric computation one needs to do an ADD–COMPARE–SELECT operation at each state. Though in theory one searches the trellis to the end of time or at least to the end of transmission, in practice one uses a sliding finite block length to make a decision as per the discussion in Chapter 9.

Having discussed the demodulator for TCM we now consider how one would evaluate its error performance.

11.1.4 Error performance of TCM

The demodulator just described minimizes the probability of *sequence error*. As in the M-ary case in Chapter 8 where we distinguished between symbol error and bit error here we need to distinguish between sequence error and symbol error. Nonetheless we would expect (or hope) that a demodulator that minimizes sequence error will also perform well in regard to symbol error, and eventually in regard to the bit error. There is another, more subtle, aspect with sequence detection. Consider a demodulator where $N \to \infty$, i.e., a very, very, long sequence. One would expect that occasionally the demodulator might stray from the correct trellis path. However, the demodulated sequence would agree with the transmitted sequence in a large number of places, indeed an overwhelming majority of places. This implies that though the probability of sequence error is 1 the *meaningful* error probability is quite small.

So what is a meaningful error probability? One that allows us to make an analysis and is a useful measure of performance. To obtain an error analysis that is useful we need to look at the concept of what is called an *error event*. An error event occurs when the demodulated path deviates from the correct (transmitted) path and then rejoins it later. Figure 11.21 illustrates it.

In an error event the symbols between the correct path and the error event disagree at the first step. Within the error event there is some chance that the symbols might agree with those of the correct path but in general they disagree. Even correct symbols within the error event are typically not that useful. The result of all this is that the probability of an error event is, in general, a more useful way to describe the error performance of TCM than the symbol or sequence error. However, even to compute the probability of an error event is difficult, somewhat analogously to the M-ary case. What is presented are bounds on the event error probability. They indicate what parameter(s) are important in TCM design. In the final analysis one resorts to simulation to determine the error performance of a designed TCM scheme.

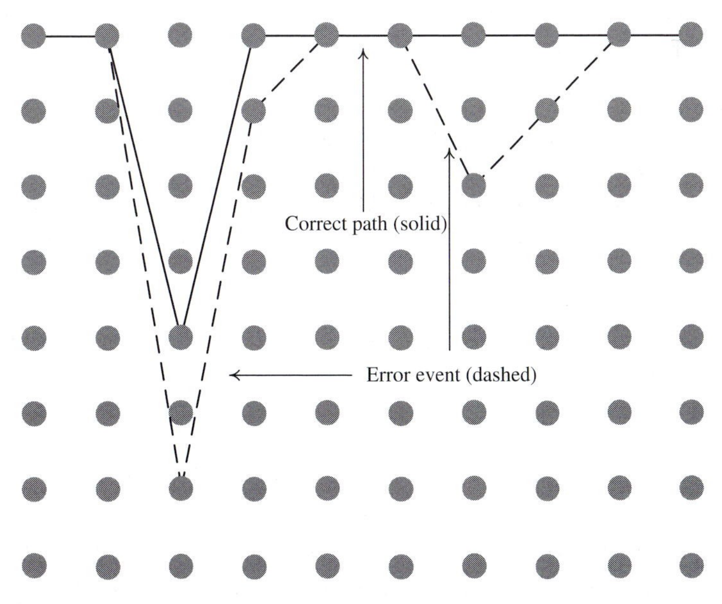

Fig. 11.21 Illustration of an error event.

The first bound we establish is a *lower* bound. Observe that an error event is actually part of the correct signal path through the trellis. It lies at Euclidean distance d from the correct signal path and since an AWGN channel is considered, the probability of this error event is $Q\left(d/2\sigma\right)$, where $\sigma^2 = N_0/2$ is the two-sided PSD of white Gaussian noise. The lower bound follows immediately:

$$P[\text{error event}] \geq Q\left(\frac{d_{\min}}{2\sigma}\right), \tag{11.14}$$

where $d_{\min}$ is the minimum distance between two signal paths. Though correct as a lower bound, and better than 0, (11.14) is not that tight a lower bound. However, based on it one can get an approximate lower bound that is reasonable. Observe that if the correct path has $N_{d_{\min}}$ neighbors at distance $d_{\min}$ and if the decision regions for these neighbors have minimal intersection with respect to the probability density centered at the correct signal, then

$$P[\text{error event}] \gtrsim N_{d_{\min}} Q\left(\frac{d_{\min}}{2\sigma}\right). \tag{11.15}$$

To get an *upper* bound we use the pairwise union bound approach where the correct path is compared with every other path starting at the state and containing a (*single*) error event. Comparing all these error event paths in turn with the correct path we have:

$$P[\text{error event}] \leq \sum_{\substack{\text{all single error}\\ \text{event paths}}} Q\left(\frac{d(\text{correct path, error event path})}{2\sigma}\right). \tag{11.16}$$

Grouping the error event paths into sets where the paths in a set are each at a common distance d_i from the correct path with N_{d_i} paths in the set we have:

$$P[\text{error event}] \leq \sum_{i=0}^{\infty} N_{d_i} Q\left(\frac{d_i}{2\sigma}\right), \tag{11.17}$$

where the indexing is such that $d_0 = d_{\min} < d_1 < d_2 < d_3 < \cdots$. Rewrite (11.17) as

$$P[\text{error event}] \leq N_{d_{\min}} Q\left(\frac{d_{\min}}{2\sigma}\right) + \sum_{i=1}^{\infty} N_{d_i} Q\left(\frac{d_i}{2\sigma}\right). \tag{11.18}$$

Note that $Q(\cdot)$ is a function that decreases rapidly with its argument, particularly in the "waterfall" region – the region the system typically operates in. Therefore the first term in (11.18) is the dominant one. Ignoring the second term with respect to the first one[6] an approximate upper bound is:

$$P[\text{error event}] \lesssim N_{d_{\min}} Q\left(\frac{d_{\min}}{2\sigma}\right). \tag{11.19}$$

[6] Care should be exercised here. Though the terms are small, there are many of them and discarding them may not be appropriate, i.e., the argument is not that rigorous mathematically. However, simulations show it is reasonable.

Accepting the approximations to the lower and upper bounds we conclude that

$$P[\text{error event}] \approx N_{d_{\min}} Q\left(\frac{d_{\min}}{2\sigma}\right). \qquad (11.20)$$

The above expression clearly shows that maximizing $d_{\min}$, the minimum Euclidean distance between any two allowable signal sequences (paths), is the main objective in the design of a good TCM scheme. This is precisely what Ungerboeck's design rules try to achieve.

The above analysis of error events was carried out by taking an arbitrary path through the trellis as a correct (reference) path. If the reference path changes, then the set of parameters $\{d_i, N_{d_i}\}_{i=0}^{\infty}$ in (11.17) also changes in general. This fact can be taken into account by viewing N_{d_i} as the *average* number of paths at distance d_i from some reference path. This means that the parameter $N_{d_{\min}}$ in (11.20) need not be an integer. One method to determine $\{d_i, N_{d_i}\}_{i=0}^{\infty}$ in general and $\{d_{\min}, N_{d_{\min}}\}$ in particular is to work with the "transfer function" of the TCM encoder, but this is quite complicated and beyond the scope of this book.

For the four-state 8-PSK/TCM presented earlier, it can be shown that $N_{d_{\min}} = 1$. Since $d_{\min} = 2$ when E_s is normalized to 1, it follows that $d_{\min} = 2\sqrt{E_s} = \sqrt{8E_b}$, where $E_b = E_s/2$ is the average transmitted energy per information bit. Substituting this and $\sigma = \sqrt{N_0/2}$ into (11.20) gives $P[\text{error event}] \approx Q(2\sqrt{E_b/N_0})$. Similarly, the 8-state 8-PSK/TCM has $N_{d_{\min}} = 2$ and $d_{\min} = \sqrt{(6-\sqrt{2})E_b}$, which yields $P[\text{error event}] \approx 2Q(\sqrt{6-\sqrt{2}}\sqrt{E_b/N_0})$. These error event probabilities are plotted in Figure 11.22 together with the probability of bit error, obtained by computer simulation, for each 8-PSK/TCM scheme. Also shown in this figure is the probability of bit error of Gray-mapped QPSK, which is $Q\left(\sqrt{2E_b/N_0}\right)$. Observe that the probability of an error event is lower than the probability of bit error in low to medium SNR regions, but the two probabilities are very close at high SNR. This clearly shows the usefulness of adopting $P[\text{error event}]$ and $d_{\min}$ in the design of TCM. The simulation results in Figure 11.22 also confirm the power savings (or coding gains) of 3.0 and 3.6 decibels at the high SNR region ($E_b/N_0 \geq 7.0$ decibels) achieved by the four-state and eight-state 8-PSK/TCM over the uncoded QPSK, respectively. It should be pointed out, however, that the two TCM schemes perform worse than QPSK in the low SNR region. This is due to the fact that there are typically many more sequences at the distance $d_{\min}$ and also at other distances which are close to $d_{\min}$ in TCM than in the reference uncoded scheme. In the low SNR region, errors often occur not only because of the sequences at $d_{\min}$ away from the correct sequence, but also because of the sequences at larger distances.

11.1.5 Differential TCM for rotational invariance

The signals constellations of TCM are invariably I/Q signals, i.e., $\sqrt{2/T_s}\cos(2\pi f_c t)$ and $\sqrt{2/T_s}\sin(2\pi f_c t)$ are the basis set. As discussed in the next chapter on synchronization, the synchronizer for QAM signals typically may have a phase ambiguity that is a multiple of $\pi/2$ radians. Therefore, as in Chapter 10, where differential modulation was used to make

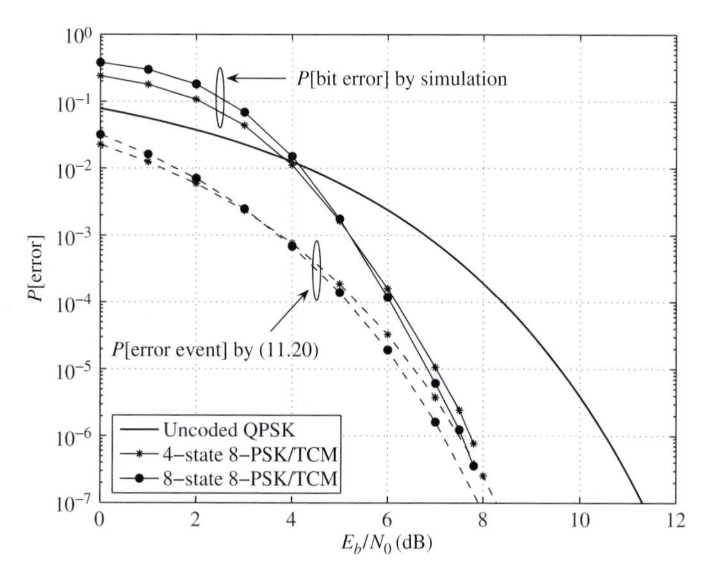

Fig. 11.22 Performance comparison between 8-PSK/TCM and uncoded QPSK.

phase ambiguities of this type transparent, one can pose the question of whether this is possible for TCM. The answer is yes, but not for all TCM signal constellations or formats. Since an encompassing general approach is not available we illustrate the ideas behind differential TCM by specific examples.

Consider the TCM scheme of Figure 11.2 repeated here in Figure 11.23(b) along with the signal constellation and signal subsets or cosets.[7] Note that the uncoded bit selects which signal is transmitted from the chosen coset and that each coset looks the same if rotated by π radians. This suggests that if we *differentially encode* the uncoded bit as shown in Figure 11.23(b) the resultant TCM signal and demodulation will be transparent to any phase ambiguity of π radians. Further, the demodulator is as shown in Figure 11.23(c). First the trellis is demodulated, typically using the Viterbi algorithm, to produce $\{\hat{b}_k^{(2)}, \hat{d}_k\}$, and then $\{\hat{d}_k\}$ is passed through the inverse of the differential encoding operation to obtain $\{\hat{b}_k^{(1)}\}$.

To obtain a feel for how the modulator/demodulator works let us consider a specific information input sequence and the different signals both with and without an ambiguity of π radians. The (partial) trellis is shown in Figure 11.24, while Table 11.2 shows the different sequences. Only the demodulated $\hat{b}_k^{(1)}$ bit is shown since a π shift does not affect which sequence of cosets is chosen by the demodulator. The table shows the effect of a phase shift occurring both at the beginning and also in the middle of transmission. Also shown for illustrative purposes is the effect of a $\pi/2$ phase shift (occurring at the beginning of the transmission). Note that we assume no AWGN since the objective is to show that a phase shift of π still results in a valid demodulated sequence (or transmitted sequence for

[7] We shall use the terminology "subsets" and "cosets" interchangeably in this section.

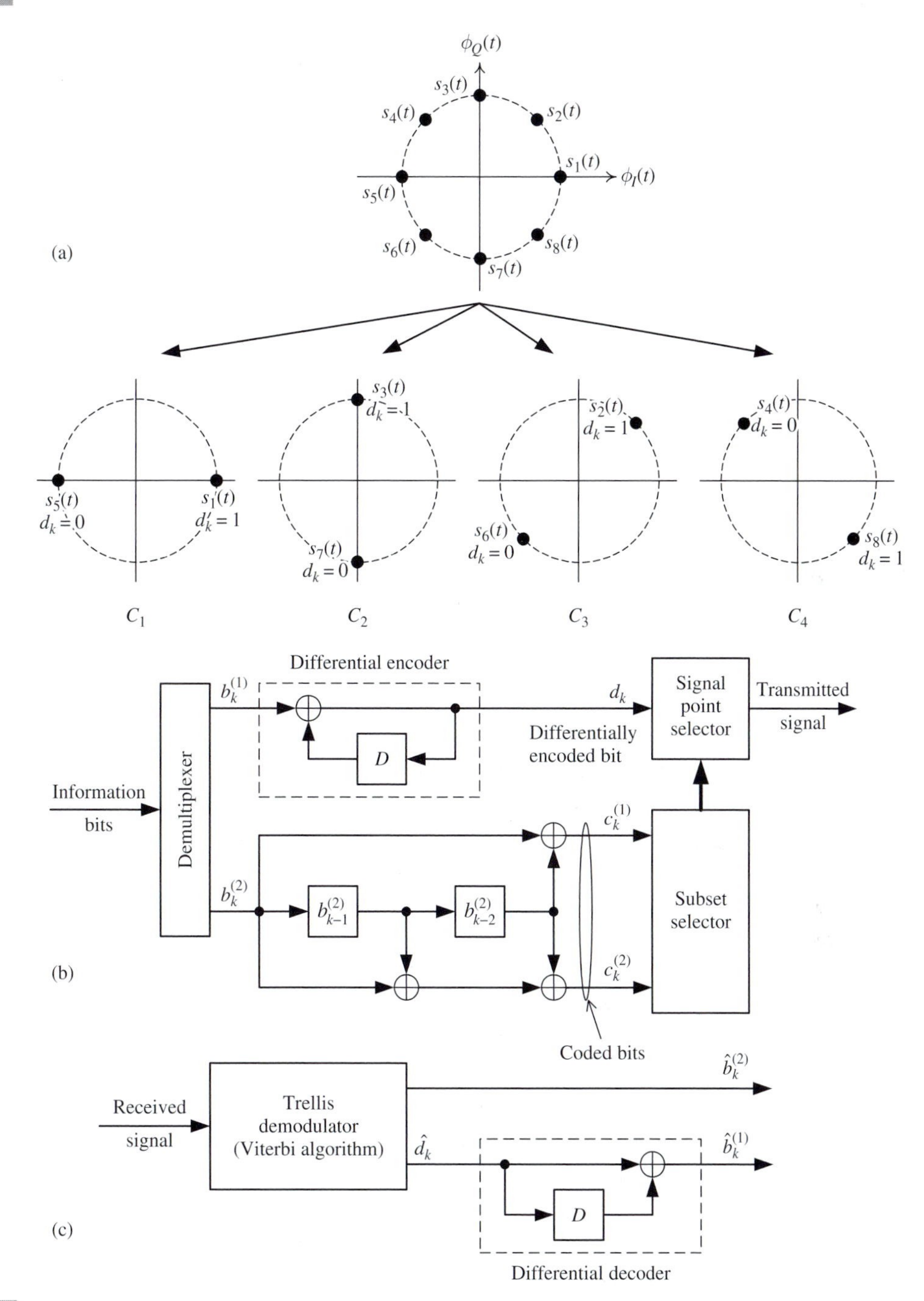

Fig. 11.23 A differential TCM scheme with 8-PSK for $k\pi$ rotational invariance: (a) signal constellation and subsets, (b) modulator, and (c) demodulator.

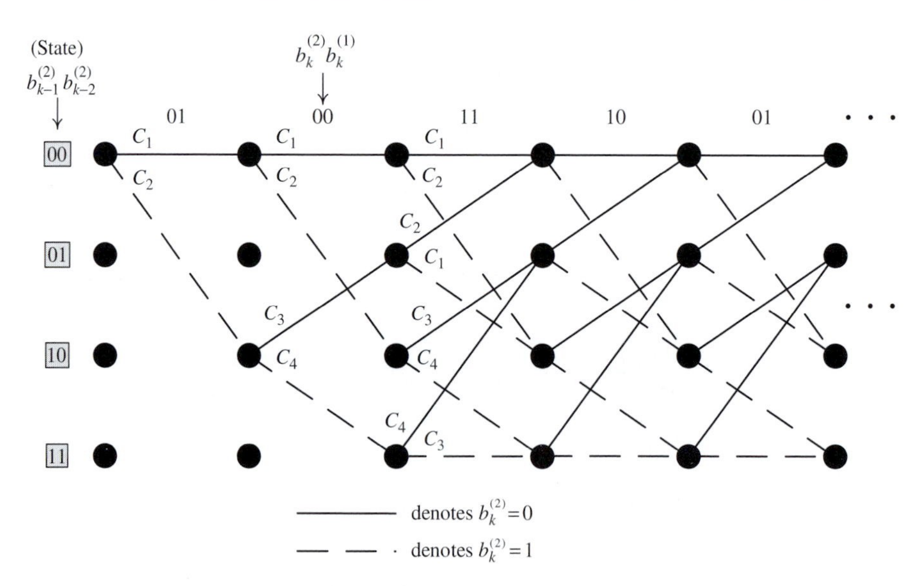

Fig. 11.24 The trellis diagram of the differential TCM in Figure 11.23(b).

that matter). Not only is the demodulated sequence valid but, except for errors where the phase shift occurs, the sequence is *correct*. The errors just after a phase shift are due to the memory of the differential coder. As seen from the last row of the table a $\pi/2$ phase shift results in an invalid sequence and many errors.

However, as stated in the opening paragraph of this subsection, it is of greater importance to have rotational invariance to $k\pi/2$ phase rotations, k integer. To achieve this one must go to signal constellations where the *cosets* are invariant under a rotation of $k\pi/2$ radians. Figure 11.25(a) shows a 16-PSK constellation partitioned into four QPSK subsets. Each is invariant to a $k\pi/2$ phase shift and therefore there is a potential to design a $k\pi/2$ rotationally invariant TCM using it. Observing the factor of 2 signal expansion guideline means that we shall map three information bits to the signal constellation. Two of these information bits will select the signals within a subset which means they are uncoded. However, to achieve rotational invariance they are differentially encoded. This differential encoder is the standard one found for any QAM constellation that lends itself to differential encoding (see Problem 10.12). The remaining information bit selects the subset and since there are four of them it is mapped through a shift register (or a convolutional encoder) to produce two coded bits. The shift register, as always, determines the trellis. Block diagrams of the modulator and demodulator are shown in Figures 11.25(b) and 11.25(c), respectively.

An excellent example of a TCM that is $k\pi/2$ rotationally invariant is the one found in the V.32 modem.[8] It is based on a 32-QAM signal constellation which is partitioned into eight cosets of four signals each. This means that four information bits are mapped to one

[8] V.32 modem standard is for digital transmission over dial-up telephone lines. It offers a maximum transmission rate of 9600 bits/second.

Table 11.2 Effect of π and $\pi/2$ phase shifts on the demodulation of $b_k^{(1)}$.

$b_k^{(2)}$	0	0	1	1	1	0	1	0	0	1	1	1
Coset	C_1	C_1	C_2	C_4	C_3	C_4	C_1	C_3	C_2	C_2	C_4	C_3
$b_k^{(1)}$	1	0	1	0	0	1	0	0	0	0	0	1
$d_k \to 0$	1	1	0	0	0	1	1	1	1	1	1	0
Signal	$s_1(t)$	$s_1(t)$	$s_7(t)$	$s_4(t)$	$s_6(t)$	$s_8(t)$	$s_1(t)$	$s_2(t)$	$s_3(t)$	$s_3(t)$	$s_8(t)$	$s_6(t)$
Signal with π shift	$\boxed{s_5(t)}$	$s_5(t)$	$s_3(t)$	$s_8(t)$	$s_2(t)$	$s_4(t)$	$s_5(t)$	$s_6(t)$	$s_7(t)$	$s_7(t)$	$s_4(t)$	$s_2(t)$
$\hat{d}_k \to 0$	0	0	1	1	1	0	0	0	0	0	0	1
$\hat{b}_k^{(1)}$	$\boxed{0}$	0	1	0	0	1	0	0	0	0	0	1
Signal with π shift	$s_1(t)$	$s_1(t)$	$s_7(t)$	$s_4(t)$	$\boxed{s_2(t)}$	$s_4(t)$	$s_5(t)$	$s_6(t)$	$\boxed{s_3(t)}$	$s_3(t)$	$s_8(t)$	$s_6(t)$
$\hat{d}_k \to 0$	1	1	0	0	1	0	0	0	1	1	1	0
$\hat{b}_k^{(1)}$	1	0	1	0	$\boxed{1}$	1	0	0	$\boxed{1}$	0	0	1
Signal with $+\pi/2$ shift	$\boxed{s_3(t)}$	$s_3(t)$	$s_1(t)$	$s_6(t)$	$s_8(t)$	$s_2(t)$	$s_3(t)$	$s_4(t)$	$\boxed{s_7(t)}$	$s_7(t)$	$s_4(t)$	$s_2(t)$
$\hat{d}_k \to 0$	1	1	1	0	1	1	1	0	0	0	0	1
$\hat{b}_k^{(1)}$	1	0	$\boxed{0}$	$\boxed{1}$	$\boxed{1}$	$\boxed{0}$	0	$\boxed{1}$	0	0	0	1

$\boxed{\cdot}$ indicates where the phase shift occurs and/or where $b_k^{(1)}$ bit error occurs.

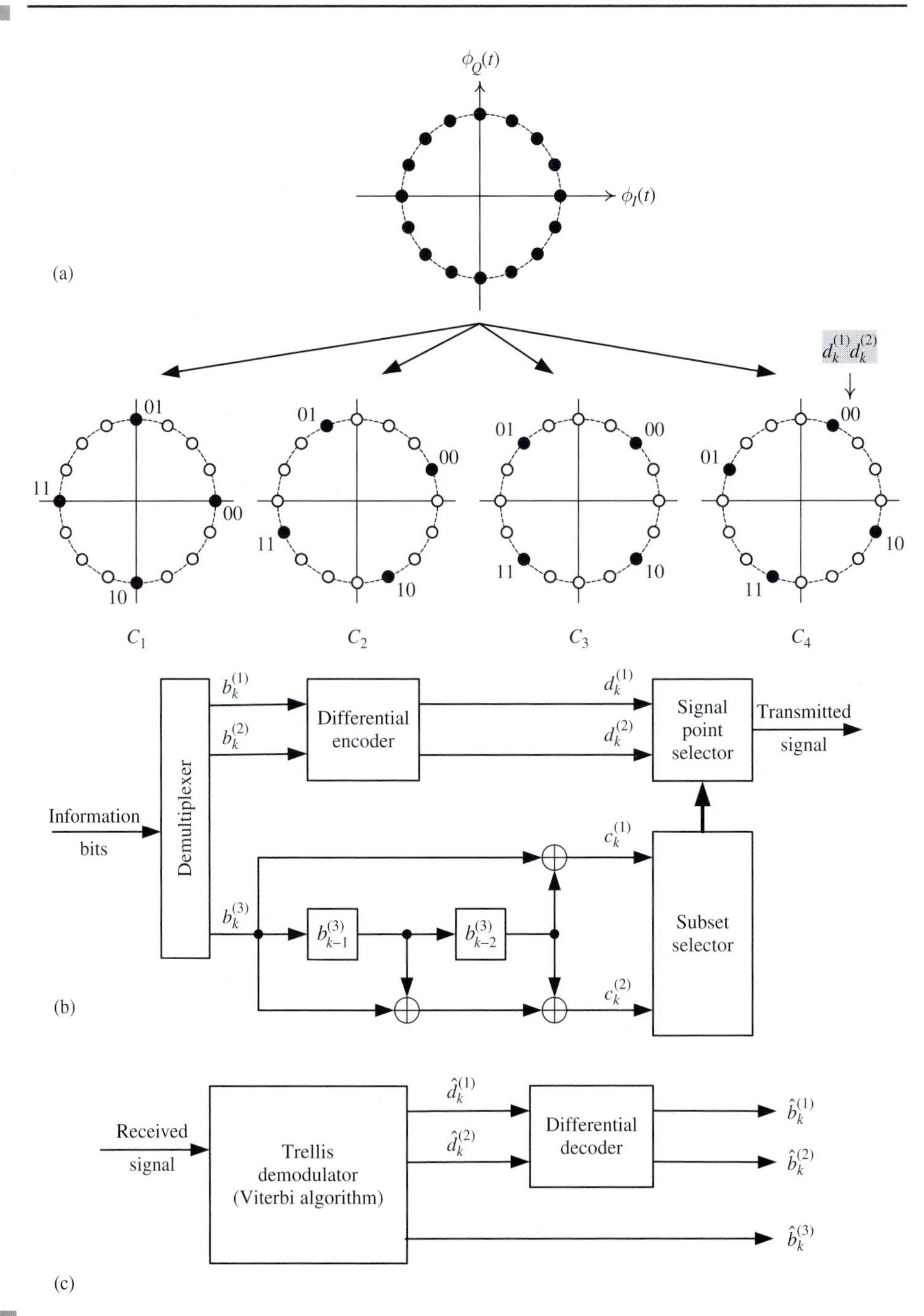

Fig. 11.25 A differential TCM scheme with 16-PSK for $k\pi/2$ rotational invariance: (a) signal constellation and subsets, (b) modulator, and (c) demodulator.

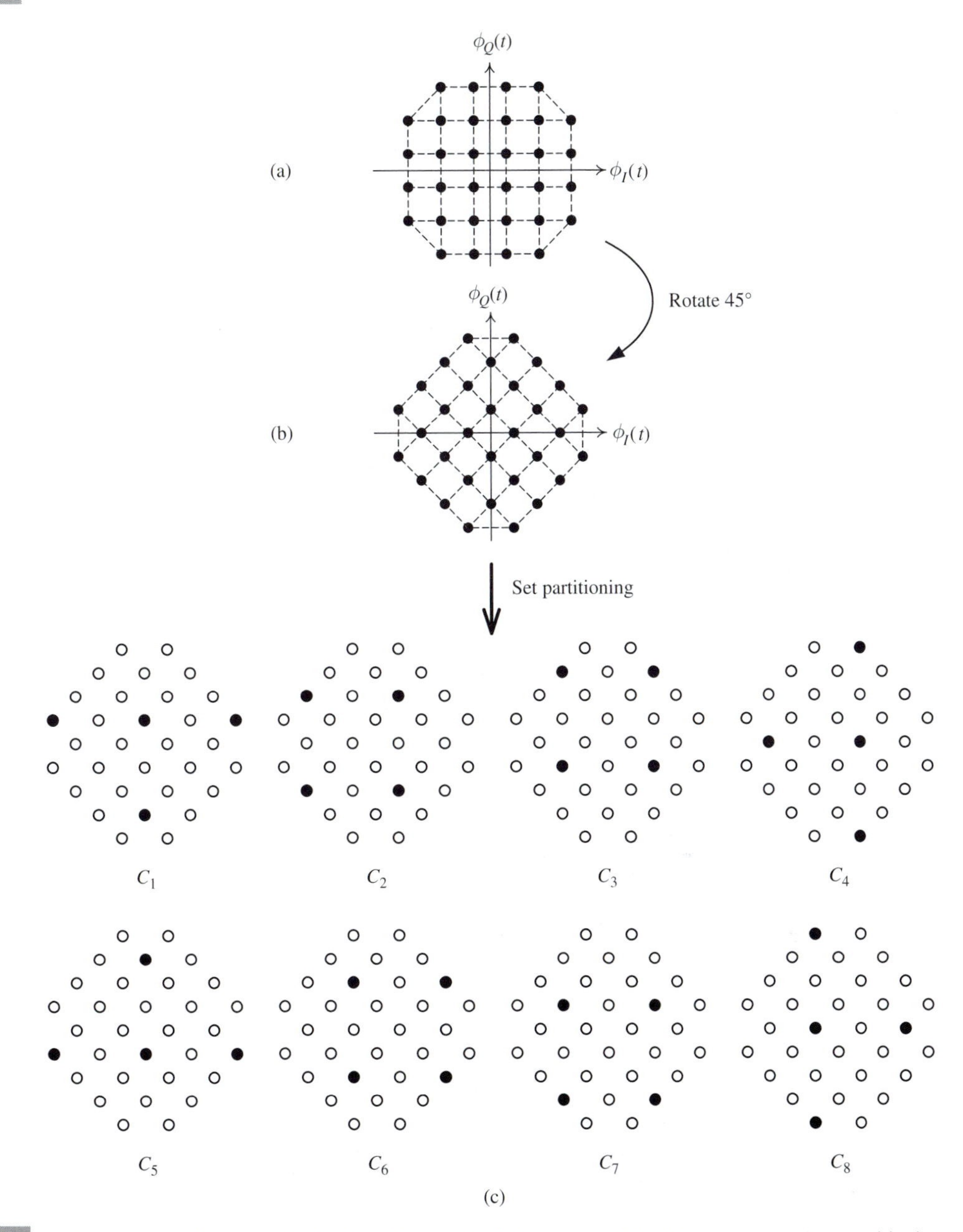

Fig. 11.26 (a) Master constellation for the V.32 modem standard, (b) its rotated (by 45°) version, and (c) partitioning the rotated constellation into eight subsets.

of 32 signal points. Two of these information bits choose a signal point within a coset and are uncoded. The other two bits select one of the eight cosets which means they are mapped by a subset selector to three coded bits. Note that the term "subset selector" is used. This is because it is not the standard linear shift register circuit we have seen till now but a *nonlinear* logic circuit (because it contains AND gates). It has three memory cells and therefore a state diagram of eight states and a corresponding trellis. Another difference is that it is the coded information bits that are differentially encoded. This is because it is the cosets that display a rotational invariance, not the signals within the coset. We now proceed to give a more detailed, but still very brief, description of the modulator/demodulator. Detailed explanations can be found in [7, 8, 9].

We start with the 32-QAM master constellation, which is the cross 32-QAM (Figure 11.26(a)), also known as the 32-CROSS constellation. It is then rotated by 45 degrees to yield the constellation in Figure 11.26(b). This constellation is partitioned into eight subsets, Figure 11.26(c). Note that subsets $\{C_1, C_4, C_5, C_8\}$ are $k\pi/2$ rotationally invariant within themselves as are subsets $\{C_2, C_3, C_6, C_7\}$. A block diagram of the modulator is shown in Figure 11.27(a). The trellis of the subset selector is given in Figure 11.28 along with the corresponding subset mapping. Note that the sequence of

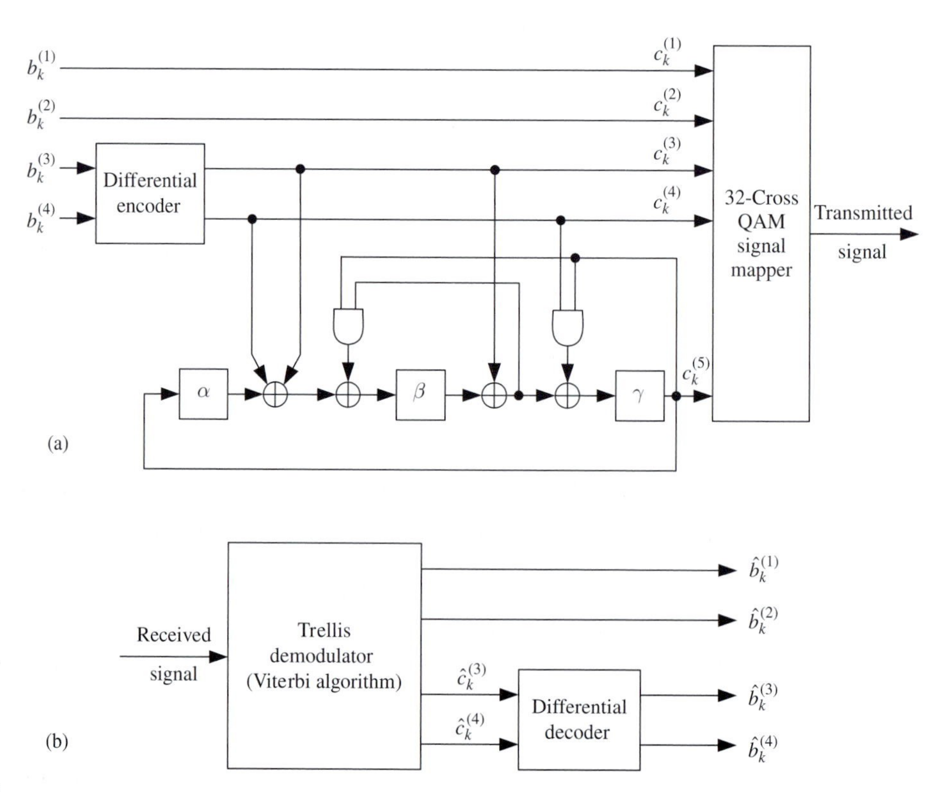

Fig. 11.27 Block diagrams of: (a) the modulator (or encoder), and (b) the demodulator (or decoder) for V.32 differential TCM.

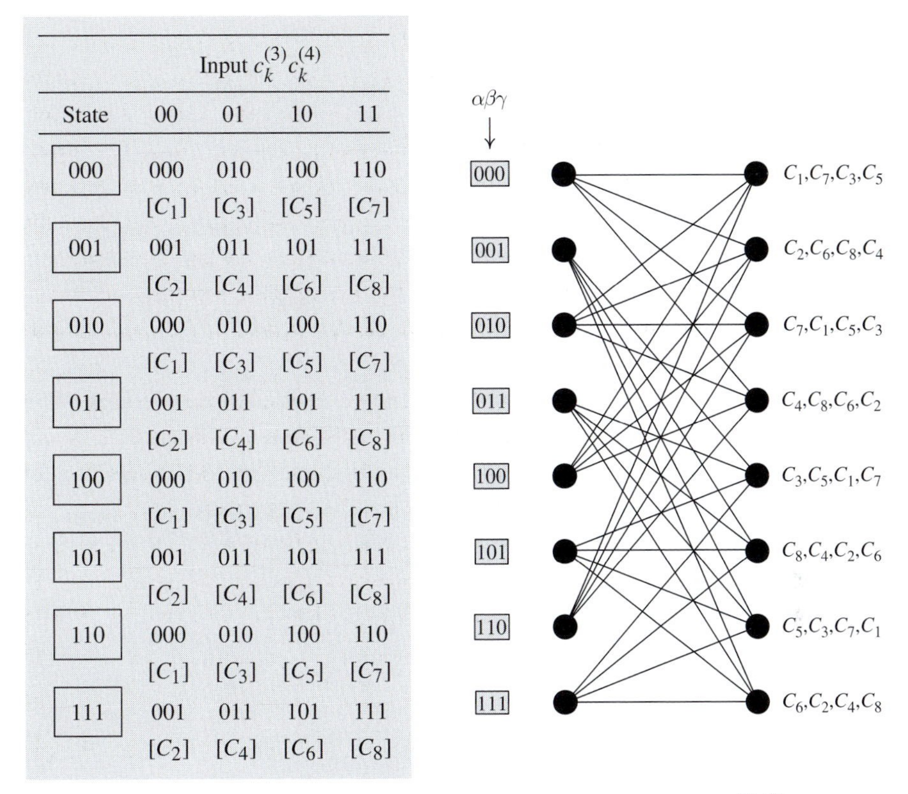

State	Input $c_k^{(3)}c_k^{(4)}$ 00	01	10	11
000	000 [C_1]	010 [C_3]	100 [C_5]	110 [C_7]
001	001 [C_2]	011 [C_4]	101 [C_6]	111 [C_8]
010	000 [C_1]	010 [C_3]	100 [C_5]	110 [C_7]
011	001 [C_2]	011 [C_4]	101 [C_6]	111 [C_8]
100	000 [C_1]	010 [C_3]	100 [C_5]	110 [C_7]
101	001 [C_2]	011 [C_4]	101 [C_6]	111 [C_8]
110	000 [C_1]	010 [C_3]	100 [C_5]	110 [C_7]
111	001 [C_2]	011 [C_4]	101 [C_6]	111 [C_8]

Fig. 11.28 Trellis diagram of V.32 differential TCM. The table on the left shows the "coded" bits $c_k^{(3)}c_k^{(4)}$ used to select the coset, which is written in the squared brackets underneath.

subsets (or cosets) that appears next to each trellis state applies to the branches, either coming from or going into that state, *from top to bottom*. Figure 11.29 shows the mapping of the bits to the signal constellation points. Finally, Figure 11.27(b) shows in block diagram form the demodulator.[9]

11.2 Code-division multiple access (CDMA)

"You can fool some of the people all of the time, all of the people some of the time, but you cannot fool all of the people all of the time." This quotation is attributed to Abraham Lincoln, sixteenth president of the USA and arguably its greatest. In the context of communication engineering the appropriate statement is: You can use some of the bandwidth

[9] It should be pointed out that the V.33 standard uses the same code as the V.32 standard, along with four uncoded bits and a 128-CROSS constellation. Its maximum data rate with full-duplex operation over four-wire leased lines is 14.4 kilobits/second.

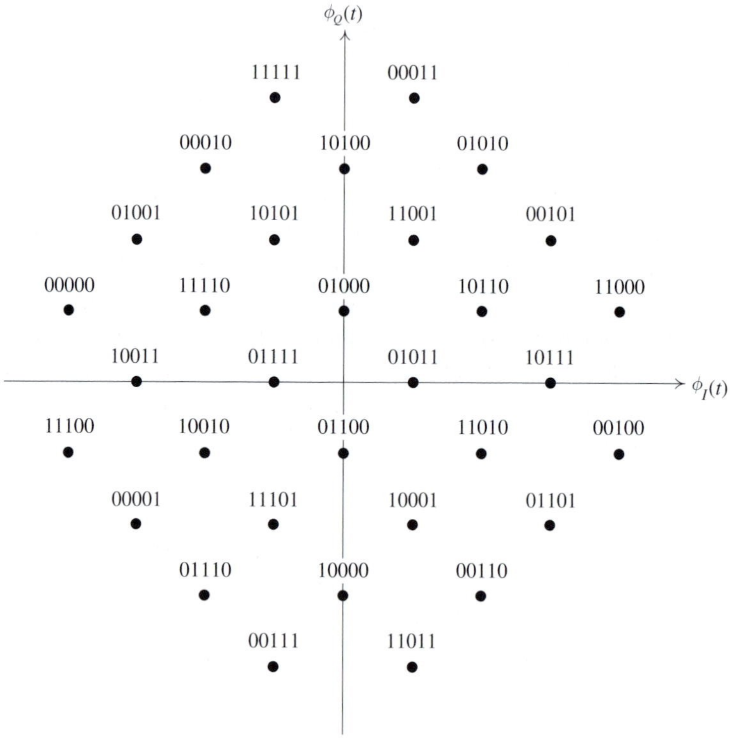

Fig. 11.29 Signal mapping, $c_k^{(1)}c_k^{(2)}c_k^{(3)}c_k^{(4)}c_k^{(5)}$, of the 32-CROSS constellation used in V.32 differential TCM. Note that the first two bits of a label are the uncoded bits, $c_k^{(1)}c_k^{(2)}$.

all of the time (FDMA), all of the bandwidth some of the time (TDMA), and you can also use all of the bandwidth all of the time (CDMA)!

Multiple access schemes can be classified into three broad categories. Perhaps the earliest multiple access scheme was frequency-division multiple access (FDMA) in which each user is assigned a frequency band. The assigned bands typically do not overlap and the users transmit their messages simultaneously in time but over disjoint frequency bands. Commercial AM and FM radio, and television are classical examples of this multiple access scheme as well as the first generation of mobile communications known as *Advanced Mobile Phone System* (AMPS).[10] In the second generation of mobile communication as exemplified by the GSM standard,[11] time-division multiple access (TDMA) is used. Users now occupy the same frequency band simultaneously but send their messages in different time slots, i.e., they are disjoint in time.

[10] AMPS is the analog mobile phone system standard developed by Bell Labs., and officially introduced in the Americas in 1983. It was the primary analog mobile phone system in North America through the 1980s and into the 2000s, and is still widely available today (see http://en.wikipedia.org/wiki/Advanced_Mobile_Phone_System).

[11] The abbreviation "GSM" originally came from the French phrase *Groupe Spécial Mobile*. It is also interpreted as the abbreviation for "Global System for Mobile communications."

CDMA is different from the above in that users now occupy the same frequency band and transmit/receive simultaneously in time. Different users are separated or distinguished by distinct *codes* assigned to them. CDMA relies on a technique called spread spectrum. As the terminology suggests the spectrum of a user's message is spread to occupy a much wider frequency band than necessary for the given user's signaling rate. As such the modulation philosophy is different from that encountered thus far where the emphasis was on developing spectrally efficient modulation methods. Historically spread spectrum arose from attempts by the military to mitigate attempts by hostile forces to jam (interfere with), disrupt, or intercept communications. Now, however, it has been incorporated into civilian applications and the so-called third (and beyond) generations of mobile communications have adopted CDMA as the multiple access scheme. Since the terminology from this initial application of spread spectrum and many of the concepts has carried over to CDMA, the spread spectrum concept is discussed next.

11.2.1 Spread spectrum

Consider a user sending a signal using binary NRZ-L modulation at baseband and signaling rate $r_b = 1/T_b$ bits/second, as illustrated in Figure 11.30. The modulator output, $m(t)$, is multiplied by another NRZ-L signal, $c(t)$, but one of a *much higher rate* as shown in the figure. This signal is the coding part of the process and its clock period is T_c seconds, where $T_c \ll T_b$ and $T_b/T_c = N$ (an integer). Note that the two signals are aligned (synchronized) in time. The bandwidth of $m(t)$ is on the order of $1/T_b$ (hertz) while that of $c(t)$ is on the order of $1/T_c$ (hertz). Since T_c is much smaller than T_b, then compared to $m(t)$, $c(t)$ is a wideband signal. The resultant spectrum of the product $s_T(t) = c(t)m(t)$ is a convolution of the two spectra $C(f)$ and $M(f)$ and essentially it will occupy a bandwidth that is practically the same as that of $c(t)$. Further, $c(t)$ is usually considered to be a random sequence of level changes every T_c seconds; practically it is chosen to be pseudorandom which means that the sequence is known to the user and at the desired receiver but looks random to everyone else, in particular to any interfering signal, $i(t)$. The signal $c(t)$ can be written as

$$c(t) = \sum_{n=0}^{N-1} c_n p(t - nT_c), \quad 0 \le t \le T_b, \tag{11.21}$$

where

$$p(t) = \begin{cases} 1, & 0 \le t \le T_c \\ 0, & \text{elsewhere} \end{cases}. \tag{11.22}$$

The transmitted signal, $s_T(t)$ is sent across the channel as shown in Figure 11.31. The channel model shows an interfering signal, $i(t)$, added to the transmitted signal. Consider now the received signal, $r(t)$. AWGN, though it may be present, is ignored in this discussion. At the receiver the user can recover the desired signal by *despreading* the received signal as shown in Figure 11.31. It is a modulation identical to that at the transmitter. The output is $r_{\text{out}}(t) = c(t)r(t) = c^2(t)m(t) + c(t)i(t)$. But $c^2(t) = 1$ and therefore

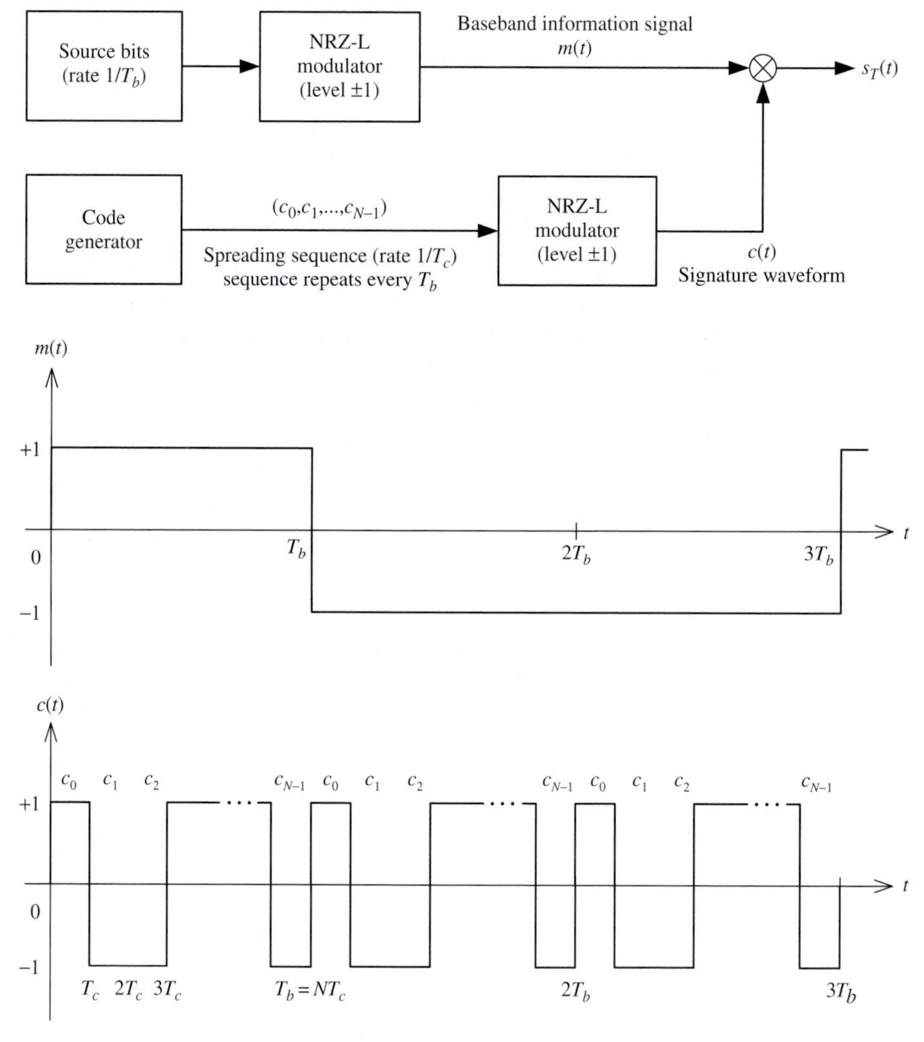

Fig. 11.30 Spread spectrum transmitter.

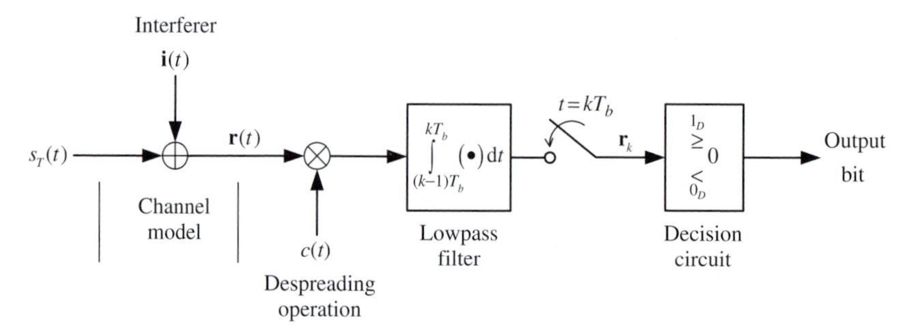

Fig. 11.31 Spread spectrum receiver.

$r_{\text{out}}(t) = m(t) + c(t)i(t)$. The important thing about the above is that $m(t)$ is narrowband. If the interfering signal, $i(t)$, is chosen to occupy the same band as $m(t)$, multiplication of it by $c(t)$ spreads its spectrum, i.e., energy, over a wide band in the frequency domain. Following the product modulator by a lowpass filter whose bandwidth is that of $m(t)$ filters out most of the interference energy. Normally the lowpass filter is an integrator followed by sampler and threshold device as shown in Figure 11.31.

The technique described is known as *direct-sequence spread spectrum* (DSSS). To summarize a pseudorandom spreading code, $c(t)$, is used to directly (in time) spread the message spectrum so that the transmitted message appears noise-like to any receiver that does not know $c(t)$. Therefore, if an interfering signal of the same bandwidth as the message is injected, then at the receiver one simply spreads the interference energy over the wide bandwidth while simultaneously despreading the desired signal energy into the original bandwidth. Conceptually this is shown in Figure 11.32.

However, what if the potential interferer knows the spread bandwidth and decides to place the interfering signal energy over the entire bandwidth occupied by the transmitted signal? What then are the benefits of spreading? This analysis is given next.

Analysis of benefits of spread spectrum Since for transmission purposes it is necessary to shift the baseband signal spectrum the modulator of Figure 11.30 is changed to that shown in Figure 11.33. In effect there are two stages of modulation: the signal spectrum is first spread and then the spread signal undergoes BPSK modulation. The transmitted output is

$$s_T(t) = m(t)c(t)\sqrt{E_b}\sqrt{\frac{2}{T_b}}\cos(2\pi f_c t), \quad 0 \le t \le T_b = NT_c, \tag{11.23}$$

where $f_c = k/T_c$, for some integer $k \gg 1$. The only impairment that is considered is that due to the interferer, denoted as $i(t)$ in Figure 11.34.

The demodulator then has two stages of demodulation as shown in Figure 11.34. Though the two stages can be interchanged usually the received signal is first (de)modulated back to baseband where a lowpass filter removes the spectral components around $2f_c$. The output of the lowpass filter is then despread in the next stage and passed through a matched filter followed by the threshold device.

The signal space representation of the transmitted signal is straightforward:

$$s_T(t) = \sum_{n=0}^{N-1} s_n \phi_{n,I}(t), \quad s_n = \pm\sqrt{E_b}\frac{T_c}{T_b} \tag{11.24}$$

and

$$\phi_{n,I}(t) = \begin{cases} \sqrt{\frac{2}{T_b}}\cos(2\pi f_c t), & nT_c \le t \le (n+1)T_c \\ 0, & \text{elsewhere} \end{cases}, \quad n = 0, 1, \ldots, N-1 \tag{11.25}$$

are a set of orthonormal functions. The transmitted signal therefore lies in an N-dimensional signal space.

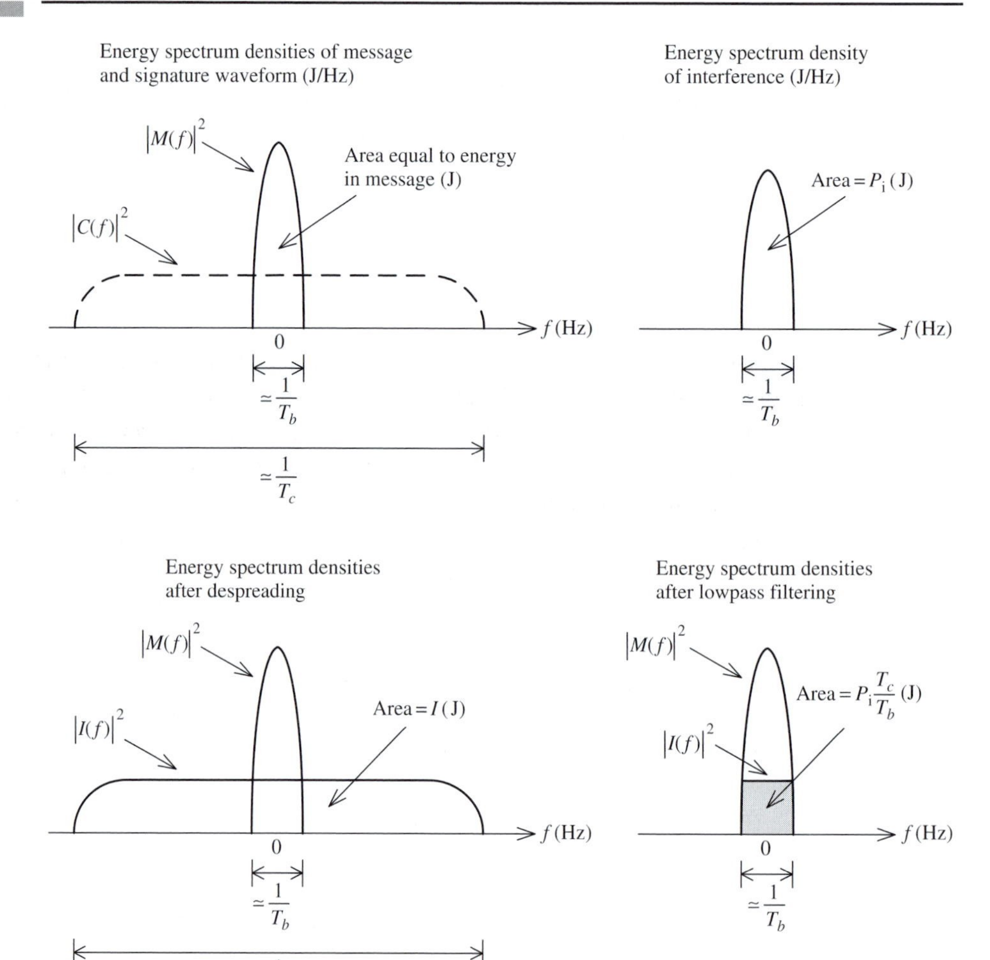

Fig. 11.32 Illustration of spreading and despreading in combating narrowband interference.

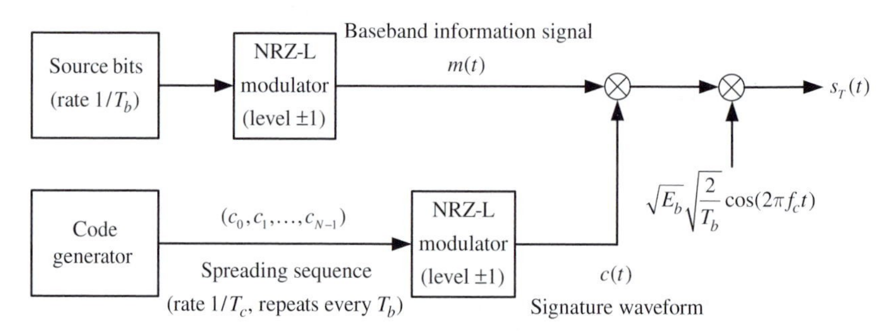

Fig. 11.33 Spread spectrum with BPSK modulation.

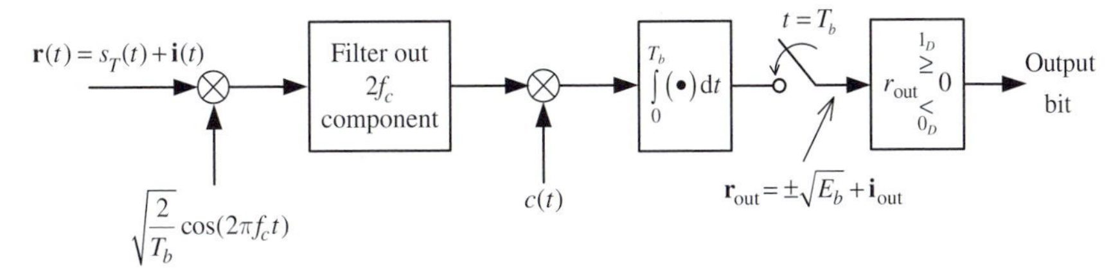

Fig. 11.34 Demodulation of the spread spectrum signal.

Consider now the interferer. The only knowledge available to him/her is that the transmitted signal lies in the N-dimensional space. It would seem prudent to place all of the available power at hand in exactly the same N-dimensional space. To do this the interferer would have to know the phase of the transmitted signal and even this is not available. To estimate the phase would require knowledge of the spreading code, $c(t)$ and it is safe to assume this is not readily available. The interferer therefore has to place the power available in a $2N$-dimensional space, namely that spanned by the orthonormal basis:

$$\begin{cases} \phi_{n,I}(t) = \sqrt{2/T_b}\cos(2\pi f_c t), & nT_c \le t \le (n+1)T_c \\ \phi_{n,Q}(t) = \sqrt{2/T_b}\sin(2\pi f_c t), & nT_c \le t \le (n+1)T_c \\ \phi_{n,I}(t) = \phi_{n,Q}(t) = 0, & \text{elsewhere} \end{cases}, \quad n = 0, 1, \ldots, N-1. \tag{11.26}$$

The interfering signal is then represented as

$$i(t) = \sum_{n=0}^{N-1} i_{n,I}\phi_{n,I}(t) + \sum_{n=0}^{N-1} i_{n,Q}\phi_{n,Q}(t), \quad 0 \le t \le T_b, \tag{11.27}$$

where

$$\begin{cases} i_{n,I} = \int_0^{T_b} i(t)\phi_{n,I}(t)\mathrm{d}t \\ i_{n,Q} = \int_0^{T_b} i(t)\phi_{n,Q}(t)\mathrm{d}t \end{cases}, \quad n = 0, 1, \ldots, N-1. \tag{11.28}$$

The average power of the interference is given by[12]:

$$P_\mathrm{i} = \frac{1}{T_b}\int_0^{T_b} i^2(t)\mathrm{d}t = \frac{1}{T_b}\left[\sum_{n=0}^{N-1} i_{n,I}^2 + \sum_{n=0}^{N-1} i_{n,Q}^2\right] \quad \text{(watts)}. \tag{11.29}$$

Further, in the absence of phase information, the best strategy for the interferer would be to place equal power into the inphase and quadrature components, i.e., $(1/T_b)\sum_{n=0}^{N-1} i_{n,I}^2 =$

[12] Here the subscript "i" of P_i refers to "interferer", not a dummy index as typically used.

$(1/T_b)\sum_{n=0}^{N-1} i_{n,Q}^2$ is a reasonable assumption. Therefore the average interfering power is:

$$P_{\mathrm{i}} = \frac{2}{T_b}\sum_{n=0}^{N-1} i_{n,I}^2 \quad \text{(watts)}. \tag{11.30}$$

Since the average signal power at the input to the matched filter is E_b/T_b watts, the signal to interference ratio (SIR), at the input is $\mathrm{SIR_{in}} = E_b/T_bP_{\mathrm{i}}$.

Though the relevant criterion at the output of the matched filter is *bit error probability* we first obtain an expression for the signal to interference ratio at the output, $\mathrm{SIR_{out}}$. The output due to the despread BPSK signal $\sqrt{E_b}\sqrt{2/T_b}\cos(2\pi f_c t)$ is given by $s_{\mathrm{out}} = \pm\sqrt{E_b}$ (volts). That due to the interference is given by $i_{\mathrm{out}} = \sqrt{2/T_b}\int_0^{T_b} c(t)i(t)\cos(2\pi f_c t)\mathrm{d}t$. Using the expression for $c(t)$ given in (11.21) one has

$$i_{\mathrm{out}} = \sqrt{\frac{2}{T_b}}\sum_{n=0}^{N-1} c_n \int_{nT_c}^{(n+1)T_c} i(t)\cos(2\pi f_c t)\mathrm{d}t. \tag{11.31}$$

Substituting (11.27) into (11.31) gives

$$i_{\mathrm{out}} = \sqrt{\frac{2}{T_b}}\sum_{n=0}^{N-1} c_n \int_{nT_c}^{(n+1)T_c} \sum_{l=0}^{N-1} i_{l,I}\sqrt{\frac{2}{T_c}}\cos^2(2\pi f_c t)\mathrm{d}t = \sqrt{\frac{T_c}{T_b}}\sum_{n=0}^{N-1} c_n i_{n,I}. \tag{11.32}$$

To proceed we assume that the coefficients of the spreading sequence, c_n, are a set of statistically independent and i.i.d. equally probable binary random variables. A reasonable assumption from the point of view of the matched filter. Further, an assumption is made that the interferer's coefficients, $i_{n,I}$, though unknown, are fixed. Therefore the output interference is a random variable

$$\mathbf{i}_{\mathrm{out}} = \sqrt{\frac{T_c}{T_b}}\sum_{n=0}^{N-1} \mathbf{c}_n i_{n,I}. \tag{11.33}$$

To determine the output interference power, one needs to compute the mean and variance of $\mathbf{i}_{\mathrm{out}}$. The mean value of $\mathbf{i}_{\mathrm{out}}$ is

$$E\{\mathbf{i}_{\mathrm{out}}\} = \sqrt{\frac{T_c}{T_b}}\sum_{n=0}^{N-1} E\{\mathbf{c}_n\} i_{n,I} = 0, \tag{11.34}$$

since $E\{\mathbf{c}_n\} = 0$. The variance of $\mathbf{i}_{\mathrm{out}}$ can therefore be computed as

$$\begin{aligned}\mathrm{var}\{\mathbf{i}_{\mathrm{out}}\} = E\{\mathbf{i}_{\mathrm{out}}^2\} &= E\left\{\left(\sqrt{\frac{T_c}{T_b}}\sum_{n=0}^{N-1}\mathbf{c}_n i_{n,I}\right)\left(\sqrt{\frac{T_c}{T_b}}\sum_{l=0}^{N-1}\mathbf{c}_l i_{l,I}\right)\right\} \\ &= \frac{T_c}{T_b}\sum_{n=0}^{N-1}\sum_{l=0}^{N-1} E\{\mathbf{c}_n\mathbf{c}_l\} i_{n,I} i_{l,I} = \frac{T_c}{T_b}\sum_{n=0}^{N-1} i_{n,I}^2,\end{aligned} \tag{11.35}$$

where we have used the fact that

$$E\{\mathbf{c}_n\mathbf{c}_l\} = \begin{cases} 1, & n = l \\ 0, & n \neq l \end{cases}. \tag{11.36}$$

In terms of the average interference power, P_{i} in (11.30), the average interference power at the output is

$$\text{var}\{\mathbf{i}_{\text{out}}\} = \frac{P_{\text{i}} T_c}{2} \quad \text{(watts)}. \tag{11.37}$$

Define the output SIR as the ratio of the peak instantaneous output signal power (which is E_b watts across a 1 ohm resistor) to the interference (average) power, i.e.,

$$\text{SIR}_{\text{out}} = \frac{2E_b}{P_{\text{i}} T_c}. \tag{11.38}$$

From this it follows that

$$\frac{\text{SIR}_{\text{out}}}{\text{SIR}_{\text{in}}} = \frac{2E_b/(P_{\text{i}} T_c)}{E_b/(P_{\text{i}} T_b)} = \frac{2T_b}{T_c}, \tag{11.39}$$

which states that the matched filter has improved the signal to noise ratio by a factor of $2T_b/T_c$ or in decibels by $10\log_{10}(2T_b/T_c) = 3 + 10\log_{10}(T_b/T_c)$ decibels. The 3 decibels improvement comes from the fact that the interferer had to put power into a $2N$-dimensional space even though the signal lay in only an N-dimensional space, i.e., equal power in the inphase and quadrature components of the signal space. The second term, however, is due to the *spreading of the spectrum*. The ratio T_b/T_c is referred to as the *processing gain*, G. Obviously $G = T_b/T_c = N$, i.e., the processing gain is directly related to the bandwidth expansion factor N. The parameter N is also commonly referred to as the *spreading factor*.

To complete the discussion on spread spectrum consider the probability of bit error of the system. For this the pdf of the interfering term at the output, $\mathbf{i}_{\text{out}}$, is needed. The usual approximation that is made is to invoke the *central limit theorem*, since $\mathbf{i}_{\text{out}}$ is the sum of many (here N) i.i.d. random variables and to approximate its probability density as a Gaussian density whose mean and variance have been determined already.

The probability of bit error is then given by

$$P[\text{bit error}] = Q\left(\sqrt{\frac{E_b}{\text{var}\{\mathbf{i}_{\text{out}}\}}}\right). \tag{11.40}$$

The argument of the Q-function can be manipulated to bring the processing gain explicitly into the bit error probability expression as follows:

$$P[\text{bit error}] = Q\left(\sqrt{\frac{2E_b}{P_{\text{i}} T_c}}\right) = Q\left(\sqrt{\frac{2E_b}{T_b P_{\text{i}}}\frac{T_b}{T_c}}\right) = Q\left(\sqrt{2G\frac{P_b}{P_{\text{i}}}}\right), \tag{11.41}$$

where $P_b = E_b/T_b$ is the average power (watts) per bit, G the processing gain, and P_{i} the interferer's average power. Note that if the interference is well modeled as Gaussian, then the demodulator is the optimum one (assuming as usual that the bits are equally probable).

With this as background, attention is turned to CDMA. As mentioned CDMA relies on the concept of spread spectrum with some major differences. The two major differences are that in CDMA there is no deliberate interference as such and that there is more than one user.

11.2.2 CDMA

Figure 11.35 shows a block diagram of a typical CDMA transmitter at the base station. The station relays (transmits) K message signals simultaneously in time and frequency to the users. Each user's bit is spread by a distinct *signature waveform*, $c^{(k)}(t)$, with the spreading

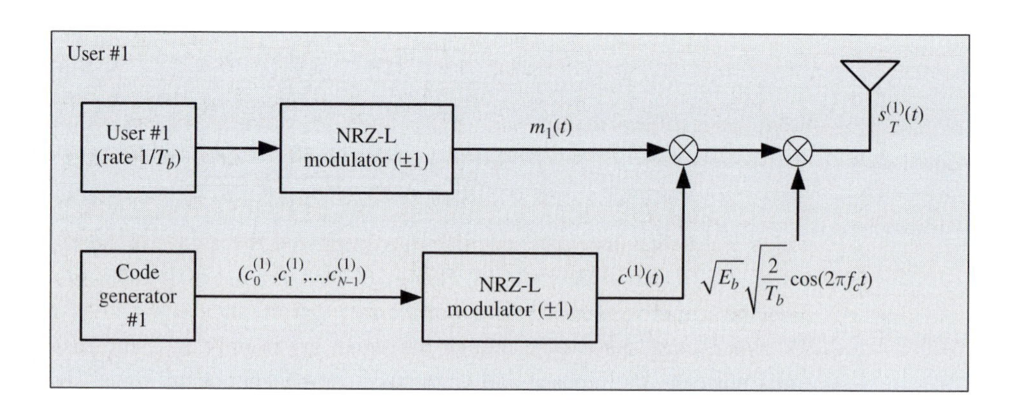

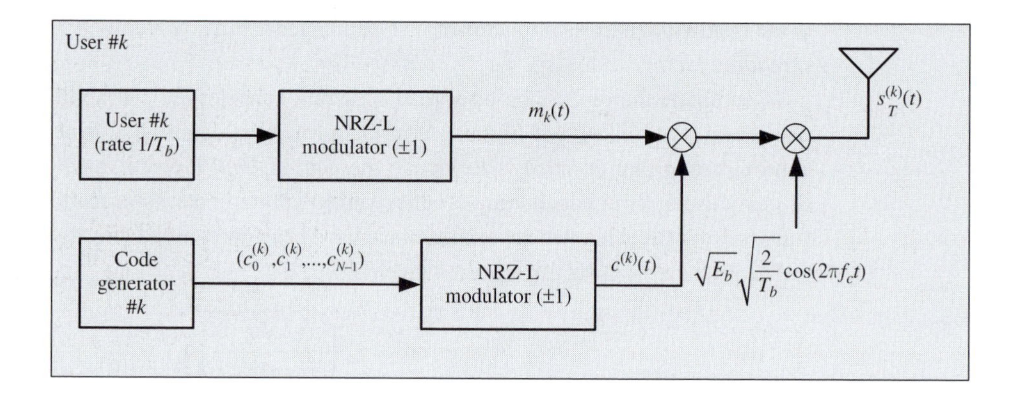

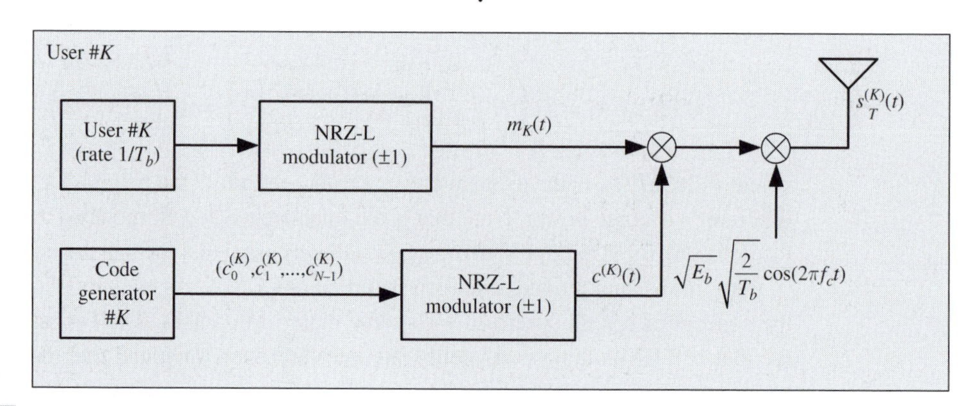

Fig. 11.35 CDMA transmitter with K users.

sequences,[13] $c_0^{(k)}, c_1^{(k)}, \ldots, c_{N-1}^{(k)}$, $k = 1, \ldots, K$, chosen so that the signature waveforms are orthogonal, at least ideally, i.e.,

$$\int_0^{T_b} c^{(k)}(t)c^{(j)}(t)\mathrm{d}t = 0, \quad k \neq j. \tag{11.42}$$

In essence each user is given an orthogonal axis to play around in.

The kth user receives the entire transmitted signal and proceeds to demodulate the desired message as shown in Figure 11.36. The signals in the time interval $[0, T_b]$ at the various points of the demodulator are as follows. The received signal $\mathbf{r}(t)$ at carrier frequency f_c is

$$\mathbf{r}(t) = \sum_{j=1}^{K} c^{(j)}(t)m_j(t)\sqrt{E_b}\sqrt{\frac{2}{T_b}}\cos(2\pi f_c t) + \mathbf{w}(t), \quad 0 \le t \le T_b, \tag{11.43}$$

where $\mathbf{w}(t)$ is the usual thermal noise modeled as white Gaussian noise, zero-mean, spectral strength, $N_0/2$ watts/hertz.

After the double frequency filter the baseband signal is

$$\mathbf{r}_{BB}(t) = \frac{\sqrt{E_b}}{T_b}\sum_{j=1}^{K} c^{(j)}(t)m_j(t) + \mathbf{w}_{BB}(t), \quad 0 \le t \le T_b, \tag{11.44}$$

where $\mathbf{w}_{BB}(t)$ is baseband noise. The signal $\mathbf{r}_{BB}(t)$ is now despread by $c^{(k)}(t)$. The output of the despreading operation is

$$\mathbf{r}_{DS}(t) = \frac{\sqrt{E_b}}{T_b}\sum_{j=1}^{K} c^{(k)}(t)c^{(j)}(t)m_j(t) + c^{(k)}(t)\mathbf{w}_{BB}(t), \quad 0 \le t \le T_b. \tag{11.45}$$

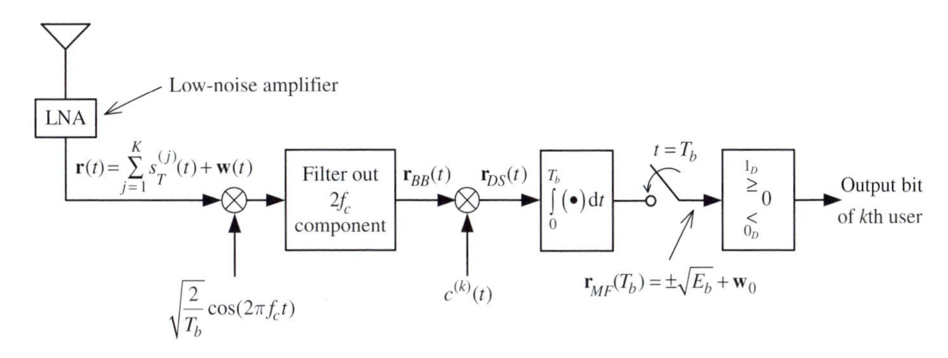

Fig. 11.36 CDMA demodulator for the kth user.

[13] Note the meanings of the superscript and subscript: For each element $c_j^{(k)}$ of the spreading sequence, the superscript indexes the user, while the subscript indexes the position of the element in the sequence.

The output of the integrator and sampler is

$$\mathbf{r}_{MF}(T_b) = \frac{\sqrt{E_b}}{T_b} \sum_{j=1}^{K} \int_0^{T_b} c^{(k)}(t)c^{(j)}(t)m_j(t)\mathrm{d}t + \int_0^{T_b} c^{(k)}(t)\mathbf{w}_{BB}(t)\mathrm{d}t. \quad (11.46)$$

However, $m_j(t) = \pm 1$ for any of the users. Using this and the fact that $c^{(j)}(t)$ and $c^{(k)}(t)$ are orthogonal gives

$$\mathbf{r}_{MF}(T_b) = \frac{\pm\sqrt{E_b}}{T_b} \int_0^{T_b} \left[c^{(k)}(t)\right]^2 \mathrm{d}t + \mathbf{w}_0 = \pm\sqrt{E_b} + \mathbf{w}_0, \quad (11.47)$$

where $\mathbf{w}_0$ is Gaussian, zero-mean, variance $N_0/2$ watts and the signal component is $+\sqrt{E_b}$ if user k transmitted bit 1 and $-\sqrt{E_b}$ if she transmitted bit 0 (over the interval $0 \le t \le T_b$).

The bit error probability is given by the usual expression for equally probable, binary, antipodal signaling. That is,

$$P[\text{bit error}]_{\text{user } k} = Q\left(\sqrt{\frac{2E_b}{N_0}}\right). \quad (11.48)$$

The error probability for all the other users is the same, where it is assumed that each user is expending and receiving the same energy per bit which is not necessarily always, if ever, the case.

From the above one sees that the users who occupy the same bandwidth and same time are *transparent* to each other. This transparency is a result of the signature waveforms being chosen to be orthogonal over T_b seconds. One possible set of orthogonal waveforms can be generated using what are called the *Walsh–Hadamard* sequences.

Walsh–Hadamard sequences Walsh–Hadamard sequences can be generated for most integer values of N but here only the case $N = 2^n$ is discussed. They are constructed iteratively starting with $n = 1$ or $N = 2$. Then the two orthogonal signature waveforms are the first two waveforms shown in Figure 11.37. The corresponding spreading sequences are $\vec{c}^{(1)} = \begin{bmatrix} 1 \\ 1 \end{bmatrix}$ and $\vec{c}^{(2)} = \begin{bmatrix} 1 \\ -1 \end{bmatrix}$, or in matrix notation $H_2 = \begin{bmatrix} 1 & 1 \\ 1 & -1 \end{bmatrix}$. The matrix H_2 is then used to get the spreading sequences for $N = 4$ as follows:

$$H_4 = \begin{bmatrix} H_2 & H_2 \\ -H_2 & H_2 \end{bmatrix} = \begin{array}{c} \begin{array}{cccc} \vec{c}^{(1)} & \vec{c}^{(2)} & \vec{c}^{(3)} & \vec{c}^{(4)} \\ \downarrow & \downarrow & \downarrow & \downarrow \end{array} \\ \begin{bmatrix} 1 & 1 & 1 & 1 \\ 1 & -1 & 1 & 1 \\ -1 & -1 & 1 & 1 \\ -1 & 1 & 1 & -1 \end{bmatrix} \end{array} \begin{array}{l} \\ \\ \leftarrow c_0^{(k)} \\ \leftarrow c_1^{(k)} \\ \leftarrow c_2^{(k)} \\ \leftarrow c_3^{(k)} \end{array} \quad (11.49)$$

It is relatively easy to determine that the four sequences generate orthogonal signature waveforms. The construction procedure is then iterated to give

$$H_8 = \begin{bmatrix} H_4 & H_4 \\ -H_4 & H_4 \end{bmatrix}, \; H_{16} = \begin{bmatrix} H_8 & H_8 \\ -H_8 & H_8 \end{bmatrix}, \; H_{32} = \begin{bmatrix} H_{16} & H_{16} \\ -H_{16} & H_{16} \end{bmatrix}, \ldots. \tag{11.50}$$

In general,

$$H_{2N} = \begin{bmatrix} H_N & H_N \\ -H_N & H_N \end{bmatrix}. \tag{11.51}$$

The above demodulator's performance is based on the signature waveforms being orthogonal. This is the case if: (i) the spreading sequences are chosen such that the $\{c^{(k)}(t)\}_{k=1}^{K}$ are orthogonal, and (ii) the received signals are *time aligned*, i.e., perfect synchronization is established. Both conditions are difficult to achieve in practice. For the first this is because one can have only so many orthogonal signals in a bandwidth $W \approx 1/T_c$ and time duration $T_b = NT_c$. Therefore if more users than this are to be accommodated then one is forced to choose $\{c^{(k)}(t)\}_{k=1}^{K}$ that are nonorthogonal. The second condition can be reasonably achieved in the downlink of a wireless system but not at all in the uplink. CDMA is typically used in a cellular environment where a base station serves many users in a geographical area (cell). Downlink is transmission from the base station to the users while uplink is the reverse, transmission from the users to the base station (see Figure 11.38).

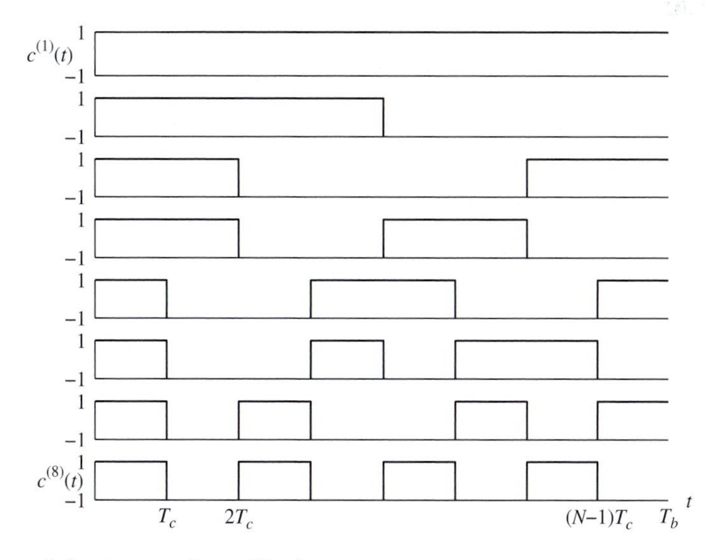

Fig. 11.37 Walsh–Hadamard signature waveforms, $N = 8$.

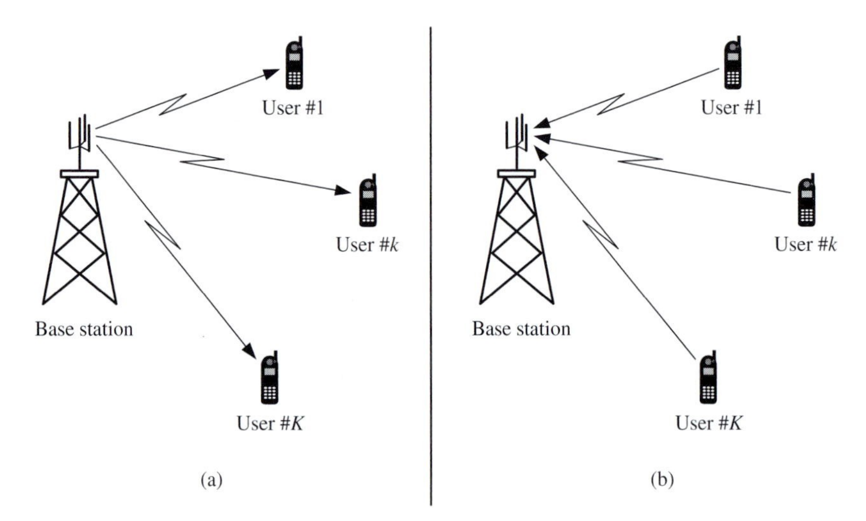

Fig. 11.38 (a) Downlink and (b) uplink transmissions.

Assume that the multipath delays are not significant so that the signal received by any user from the base station is time synchronized with regard to the signature waveforms, $\{c^{(k)}(t)\}_{k=1}^{K}$. Therefore Walsh–Hadamard sequences can be used at the base station to achieve orthogonality. However, in the uplink it is unreasonable, if not impossible, to expect the user signals to be time coordinated.

To see the effect of this, let the transmitted signals from the K users arrive at the base station at different times, i.e., they experience different delays $0 \leq \tau_j \leq T_b, j = 1, 2, \ldots, K$. Since the users are at different geographical locations, their local oscillators typically have different phases, θ_j, $j = 1, 2, \ldots, K$. Furthermore, let $b_i^{(j)} \in \{\pm 1\}$ denote the information bit of the jth user transmitted in the interval $iT_b \leq t \leq (i+1)T_b$. The received signal, over $-\infty \leq t \leq \infty$, is

$$\begin{aligned}\mathbf{r}(t) &= \sum_{j=1}^{K} \sum_{i=-\infty}^{\infty} b_i^{(j)} c^{(j)}(t - iT_b - \tau_j) \sqrt{E_b} \sqrt{\frac{2}{T_b}} \cos(2\pi f_c (t - \tau_j) + \theta_j) + \mathbf{w}(t) \\ &= \sum_{j=1}^{K} \sum_{i=-\infty}^{\infty} b_i^{(j)} c^{(j)}(t - iT_b - \tau_j) \sqrt{E_b} \sqrt{\frac{2}{T_b}} \cos(2\pi f_c t + \varphi_j) + \mathbf{w}(t),\end{aligned} \tag{11.52}$$

where $\varphi_j = \theta_j - 2\pi f_c \tau_j$. Consider the detection of the first information bit of the kth user, namely $b_0^{(k)}$. It is assumed that the base station can track the kth user's delay and phase shift. With this assumption, one can set $\tau_k = 0$ and $\varphi_k = 0$ and the other users' delays and phase shifts are interpreted relative to that of the kth user.

Now, the baseband signal is

$$\mathbf{r}_{BB}(t) = \frac{\sqrt{E_b}}{T_b} \sum_{j=1}^{K} \sum_{i=-\infty}^{\infty} b_i^{(j)} c^{(j)}(t - iT_b - \tau_j) \cos(\varphi_j) + \mathbf{w}_{BB}(t) \tag{11.53}$$

and the despread signal is $\mathbf{r}_{DS}(t) = \mathbf{r}_{BB}(t)c^{(k)}(t)$. The output of the integrator and sampler is

$$\begin{aligned}
\mathbf{r}_{MF}(T_b) &= \frac{\sqrt{E_b}}{T_b} \sum_{j=1}^{K} \int_0^{T_b} \sum_{i=-\infty}^{\infty} b_i^{(j)} c^{(j)}(t - iT_b - \tau_j) c^{(k)}(t) \cos(\varphi_j) \mathrm{d}t + \mathbf{w}_0 \\
&= b_0^{(k)} \sqrt{E_b} + \frac{\sqrt{E_b}}{T_b} \sum_{\substack{j=1 \\ j \neq k}}^{K} \left[\int_0^{\tau_j} b_{-1}^{(j)} c^{(j)}(t + T_b - \tau_j) c^{(k)}(t) \mathrm{d}t \right. \\
&\quad \left. + \int_{\tau_j}^{T_b} b_0^{(j)} c^{(j)}(t - \tau_j) c^{(k)}(t) \mathrm{d}t \right] \cos(\varphi_j) + \mathbf{w}_0 \\
&= b_0^{(k)} \sqrt{E_b} + \sqrt{E_b} \sum_{\substack{j=1 \\ j \neq k}}^{K} \left[b_{-1}^{(j)} \hat{R}_{kj}(\tau_j) + b_0^{(j)} \tilde{R}_{kj}(\tau_j) \right] \cos(\varphi_j) + \mathbf{w}_0.
\end{aligned} \tag{11.54}$$

The functions

$$\hat{R}_{kj}(\tau_j) = \frac{1}{T_b} \int_0^{\tau_j} c^{(k)}(t) c^{(j)}(t + T_b - \tau_j) \mathrm{d}t, \tag{11.55}$$

$$\tilde{R}_{kj}(\tau_j) = \frac{1}{T_b} \int_{\tau_j}^{T_b} c^{(k)}(t) c^{(j)}(t - \tau_j) \mathrm{d}t \tag{11.56}$$

are the *partial* crosscorrelations between the signature waveforms $c^{(j)}(t)$ and $c^{(k)}(t)$. Note that both correlations are time correlations.

The middle term of (11.54) is interference from the other users and is usually called *multiple access interference* or MAI. It depends on the crosscorrelation properties of the signature waveforms.

To simplify the study of correlation properties of the signature waveforms, let us extend each signature waveform toward $\pm\infty$ by simply repeating it every T_b. In essence, this gives the following *periodic* signature waveforms:

$$c_P^{(j)}(t) = \sum_{n=-\infty}^{\infty} c_n^{(j)} p(t - nT_c), \; c_n^{(j)} = c_{n+N}^{(j)}, \; j = 1, 2, \ldots, K. \tag{11.57}$$

Further, define the *full* correlation between $c_P^{(k)}(t)$ and $c_P^{(j)}(t)$ as

$$R_{kj}(\tau) = \frac{1}{T_b} \int_0^{T_b} c_P^{(k)}(t) c_P^{(j)}(t - \tau_j) \mathrm{d}t, \quad 0 \leq \tau \leq T_b. \tag{11.58}$$

Note that $R_{kj}(\tau)$ is also a periodic function of period T_b. Ideally the spreading sequences are ones where the time correlations satisfy

$$\int_0^{T_b} c_P^{(k)}(t)c_P^{(j)}(t-\tau)\mathrm{d}t = 0 \text{ (or } \approx 0), \ j \neq k, \tag{11.59}$$

$$\int_0^{T_b} c_P^{(k)}(t)c_P^{(k)}(t-\tau)\mathrm{d}t = 0 \text{ (or } \approx 0), \ \tau \neq 0. \tag{11.60}$$

The first condition eliminates or minimizes MAI while the second condition minimizes a number of channel effects such as a delayed multipath component interfering with an LOS signal. Unfortunately signature waveforms generated by Walsh–Hadamard sequences are poor in this regard since the crosscorrelation has large peaks in it.

One approach to achieve orthogonality, particularly for crosscorrelation, is to generate the spreading sequences as follows. Assume for the moment there are only two users, your friend and yourself. Each of you has a fair coin and each proceeds to flip the coin every T_c seconds to produce two sequences $\left\{\mathbf{c}_n^{(1)}\right\}_{n=0}^{N-1}$ and $\left\{\mathbf{c}_n^{(2)}\right\}_{n=0}^{N-1}$ with heads mapped to 1 and tails to 0. The corresponding periodic spreading waveforms are

$$\mathbf{c}_P^{(1)}(t) = \sum_{n=-\infty}^{\infty} \mathbf{c}_n^{(1)} g(t-kT_c), \tag{11.61a}$$

$$\mathbf{c}_P^{(2)}(t) = \sum_{n=-\infty}^{\infty} \mathbf{c}_n^{(2)} g(t-kT_c), \tag{11.61b}$$

where $g(t)$ is a pulse waveform (or the chip waveform). Though we have chosen it to be the output of an NRZ-L modulator, other pulse shapes have been investigated. The NRZ-L waveform (where $g(t) = p(t)$ as defined in (11.22)), however, is the simplest and the one most practical. Therefore in the ensuing discussion we shall stay with it.

Now it is reasonable to assume that: (i) the coin tosses are statistically independent of each other, (ii) individual coin flips are statistically independent. In other words, the two sequences are i.i.d. sequences. The correlation properties of the spreading signals are as follows.

(1) *Crosscorrelation*

$$\begin{aligned}
R_{12}(\tau) &= E\left\{\mathbf{c}_P^{(1)}(t)\mathbf{c}_P^{(2)}(t-\tau)\right\} \\
&= E\left\{\sum_{n=-\infty}^{\infty} \mathbf{c}_n^{(1)} g(t-nT_c) \sum_{l=-\infty}^{\infty} \mathbf{c}_l^{(2)} g(t-lT_c-\tau)\right\} \\
&= \sum_{n=-\infty}^{\infty}\sum_{l=-\infty}^{\infty} E\left\{\mathbf{c}_n^{(1)}\mathbf{c}_l^{(2)}\right\} g(t-nT_c)g(t-lT_c-\tau).
\end{aligned} \tag{11.62}$$

However, $E\left\{\mathbf{c}_n^{(1)}\mathbf{c}_l^{(2)}\right\} = 0$ for all n, l. Therefore $R_{12}(\tau) = 0$ for all τ.

(2) *Autocorrelation* The autocorrelation can be determined from (11.62) by making the superscript for both $\mathbf{c}_n^{(1)}$ and $\mathbf{c}_l^{(2)}$ equal to j, where $j = 1, 2$. Since $E\left\{\mathbf{c}_n^{(j)}\mathbf{c}_l^{(j)}\right\} = 0$ if $n \neq l$ and $E\left\{\mathbf{c}_n^{(j)}\mathbf{c}_l^{(j)}\right\} = 1$ if $n = l$, it follows that

$$R_{jj}(t,t-\tau) = E\left\{\mathbf{c}_P^{(j)}(t)\mathbf{c}_P^{(j)}(t-\tau)\right\}$$
$$= \sum_{n=-\infty}^{\infty} g(t-nT_c)g(t-lT_c-\tau). \tag{11.63}$$

The above autocorrelation depends on t, which tells us that the $\mathbf{c}_P^{(j)}(t)$ is a nonstationary process, a consequence of the fact that $\mathbf{c}_P^{(1)}(t)$ is a periodic signal. By averaging over time, it can be shown that (see the derivation in Section 11.2.4):

$$R_{jj}(\tau) = \frac{1}{T_b}\int_0^{T_b} R_{jj}(t,t-\tau)\mathrm{d}t$$
$$= \begin{cases} 1-\left(1+\frac{1}{N}\right)\frac{|\tau|}{T_c}, & |\tau| \le T_c, \\ -\frac{1}{N}, & |\tau| > T_c, \end{cases} \quad \text{for } |\tau| < (N-1)T_c. \tag{11.64}$$

Moreover, since the spreading waveform is periodic with period $T_b = NT_c$, $R_{jj}(\tau)$ is also periodic with the same period. Figure 11.39 plots the above autocorrelation.

Though only two users were considered above, the results hold for K users. One simply reaches into one's pocket and produces another $K-2$ pennies. The correlation properties of spreading signals produced by coin flipping, though ideal, suffer from two disadvantages. One is that the receiver needs to know the spreading sequence in order to despread the received signal. This could be overcome by having the coin flip occur ahead of transmission, telling the user what the spreading sequence is; the user would then store it for future use. This leads to the second disadvantage. In practice, the spreading waveform, $c^{(j)}(t)$, of the jth user need not be the same over every bit interval of T_b. Instead it can be a segment of duration T_b of the periodic waveform $c_P^{(j)}(t)$. The periodic waveform itself corresponds to a periodic sequence whose period N_P is on the order of 2^{20}–2^{40}. Storing such a long $c_P^{(j)}(t)$ is prohibitive even if the practical fact of executing 2^{20} coin flips is overlooked. A deterministic method of generating one period of $c_P^{(j)}(t)$ is therefore needed. This can be done by a simple shift register circuit with feedback. It generates what is called a *pseudorandom* sequence, the subject we discuss next.

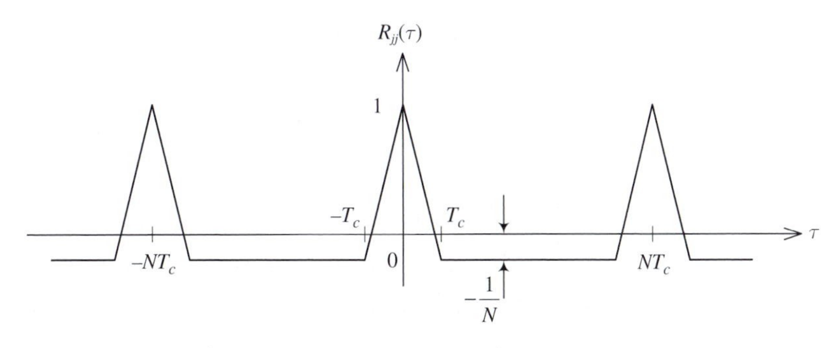

Fig. 11.39 Autocorrelation of the random spreading signal.

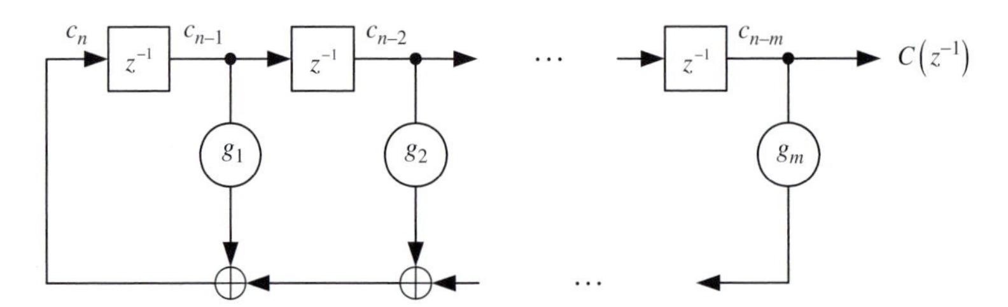

Fig. 11.40 Pseudorandom sequence generator.

11.2.3 Pseudorandom sequences: generation and properties

Pseudorandom sequences are in essence deterministic sequences but possess properties that make them appear to be random. The three key random properties that they should possess are [10]:

(1) The relative frequencies of 0 and 1 should be 1/2, i.e., $P[0] = P[1] = 1/2$.
(2) Each sequence should have the same run lengths of ones and zeros that one expects from a random sequence where half of all run lengths are unity, 1/4 are of length 2, 1/8 of length 3, $1/2^n$ are of length n, i.e., $P[n \text{ consecutive zeros}] = P[n \text{ consecutive ones}] = 1/2^n$.
(3) If an i.i.d. binary sequence is shifted by any nonzero number of indices, the shifted sequence has an equal number of agreements and disagreements with the original sequence. So should the pseudorandom sequence.

Sequences with these properties are generated by the linear binary shift register circuit of Figure 11.40. A word about terminology is in order. The delay element is depicted by z^{-1} to emphasize that z-transform concepts will be used subsequently. z-transform is a natural choice since the circuit is linear, time-invariant, and discrete. However, the algebra that the circuit performs is Boolean algebra or what is formally known as *Galois field* 2, GF(2). A common notation for the delay element is D (for delay); X is also used with the connotation that polynomial addition and multiplication are being performed. Regardless of the symbol the algebra is still GF(2).

Galois fields Galois fields are algebraic structures in which the number of elements is finite and the operations of addition/subtraction, multiplication/division are well defined. Well defined in the sense that they satisfy basic properties such as: (i) *closure* – operation on any two elements results in another element of the set; (ii) *associativity* – one can group elements and perform the operations in any order; (iii) *inverse* – each element has an inverse; (iv) *commutativity* – the operations commute. Galois fields exist only for sets that have a *prime* number of elements or when the number of elements is equal to the prime raised to some integer power. GF(2) has the following addition and multiplication tables

⊕	0	1
0	0	1
1	1	0

⊗	0	1
0	0	0
1	0	1

in which $\oplus$ can be recognized as an exclusive-or operation.

We now determine the response of the system to a set of nonzero initial conditions. From the circuit the following difference equation is easily written:

$$c_n = g_1 c_{n-1} + g_2 c_{n-2} + \cdots + g_m c_{n-m} = \sum_{i=1}^{m} g_i c_{n-i}. \tag{11.65}$$

Further the output sequence for $n \geq 0$ has the z-transform

$$C\left(z^{-1}\right) = \sum_{n=0}^{\infty} c_n z^{-n}. \tag{11.66}$$

Substituting (11.65) into (11.66) gives

$$\begin{aligned} C\left(z^{-1}\right) &= \sum_{n=0}^{\infty}\left[\sum_{i=1}^{m} g_i c_{n-i}\right] z^{-n} = \sum_{i=1}^{m} g_i z^{-i}\left[\sum_{n=0}^{\infty} c_{n-i} z^{-n+i}\right] \\ &\overset{l=n-i}{=} \sum_{i=1}^{m} g_i z^{-i}\left[\sum_{l=-i}^{\infty} c_l z^{-l}\right] \\ &= \sum_{i=1}^{m} g_i z^{-i}\left[\underbrace{c_{-i}z^{i} + \cdots + c_{-1}z^{1}}_{\text{initial condition}} + \underbrace{\sum_{l=0}^{\infty} c_l z^{-l}}_{C(z^{-1})}\right]. \end{aligned} \tag{11.67}$$

Therefore $C\left(z^{-1}\right)\left[1 - \sum_{i=1}^{m} g_i z^{-i}\right] = \sum_{i=1}^{m} g_i z^{-i}\left(c_{-i}z^{i} + \cdots + c_{-1}z^{1}\right)$. Equivalently

$$C\left(z^{-1}\right) = \frac{\sum_{i=1}^{m} g_i z^{-i}\left(c_{-i}z^{i} + \cdots + c_{-1}z^{1}\right)}{1 - \sum_{i=1}^{m} g_i z^{-i}} \triangleq \frac{C_{\text{ini}}\left(z^{-1}\right)}{f\left(z^{-1}\right)}, \tag{11.68}$$

where the feedback polynomial $f\left(z^{-1}\right) = 1 - \sum_{i=1}^{m} g_i z^{-i} = 1 + \sum_{i=1}^{m} g_i z^{-i}$ depends only on the tap gains $\{g_i\}_{i=1}^{m}$ and is known as the *characteristic polynomial* of the shift register sequence generator, $C_{\text{ini}}\left(z^{-1}\right)$ depends on the contents (initial condition) of the shift register when we start the clock (at $n = 0$).

We now establish the following property of the output sequence. It is periodic with a period that is less than $2^m - 1$. The all-zero initial condition vector is excluded since if this is the case the output is always zero. Consider then any nonzero initial condition vector. As the clock ticks the shift register moves from one state to another. When it reaches a particular state that it has been in previously, then the output will repeat itself. Since the maximum number of nonzero states is $2^m - 1$, the shift register must find itself in a previous state within this number of clock periods. The case of interest is when the period is exactly $2^m - 1$, i.e., it is of *maximum length*. Sequences with this period are also called m-sequences.

The period of the sequence is determined by the feedback polynomial. We now show that if the polynomial can be factored (keep in mind that the factorization must take place in GF(2)), then it cannot produce an m-sequence. The proof is by contradiction and the reasoning is as follows. Assume $f\left(z^{-1}\right)$ is a polynomial that produces an m-sequence. Take as an initial vector $C_{\text{ini}}\left(z^{-1}\right) = 1 = [1000 \ldots 00]$. There is no loss of generality since if $C\left(z^{-1}\right)$ is the maximum length every nonzero vector will be at some time in the shift register and can be considered to be an initial condition vector. Now let $f\left(z^{-1}\right)$ be factorable, i.e., $f\left(z^{-1}\right) = f_1\left(z^{-1}\right) \cdot f_2\left(z^{-1}\right)$, where $f_1\left(z^{-1}\right)$ is of degree m_1, $f_2\left(z^{-1}\right)$ is of degree m_2, $m_1 + m_2 = m$. Then

$$C\left(z^{-1}\right) = \frac{1}{f_1\left(z^{-1}\right) \cdot f_2\left(z^{-1}\right)} \overset{\text{partial}}{\underset{\text{fractions}}{=}} \frac{n_1\left(z^{-1}\right)}{f_1\left(z^{-1}\right)} + \frac{n_2\left(z^{-1}\right)}{f_2\left(z^{-1}\right)} \tag{11.69}$$

with period of $n_1\left(z^{-1}\right)/f_1\left(z^{-1}\right) \leq 2^{m_1} - 1$ and period of $n_2\left(z^{-1}\right)/f_2\left(z^{-1}\right) \leq 2^{m_2} - 1$. Since the period of $C\left(z^{-1}\right)$ is less than the product of the above two periods (think back to Fourier series), then

$$\begin{aligned} \text{period of } C\left(z^{-1}\right) &\leq \left(2^{m_1} - 1\right)\left(2^{m_2} - 1\right) \\ &= 2^m - 2^{m_1} - 2^{m_2} + 1 \leq 2^m - 3, \end{aligned} \tag{11.70}$$

which contradicts the fact that the sequence is maximum length. Hence the feedback polynomial, $f\left(z^{-1}\right)$ must be *irreducible*. However, this is just a *necessary* condition as the following counterexample shows.

Consider the irreducible polynomial $f\left(z^{-1}\right) = 1 + z^{-1} + z^{-2} + z^{-3} + z^{-4}$. Since $m = 4$, the maximum-length sequence is of period $2^4 - 1 = 15$. The polynomial $f\left(z^{-1}\right)$ divides $1 - z^{-5}$ and straightforward computation for the initial condition of 1,

$$\begin{aligned} C\left(z^{-1}\right) &= \frac{1}{f\left(z^{-1}\right)} = \frac{1}{1 + z^{-1} + z^{-2} + z^{-3} + z^{-4}} \\ &= 1 + z^{-1} + z^{-5} + z^{-6} + z^{-10} + z^{-11} + \cdots, \end{aligned} \tag{11.71}$$

shows that the sequence has period 5. The general statement is: if $f\left(z^{-1}\right)$ divides $\left(1 - z^{-p}\right)$, then sequence $C\left(z^{-1}\right)$ has period p. The feedback polynomial must not only be irreducible, it must also be what is called *primitive*, i.e., the smallest value of p is $2^m - 1$. Fortunately, primitive polynomials exist for every length m and just as fortunately, though they are not easy to recognize, mathematicians have found them for us. Table 11.3 gives a list of them. The list is complete for $m \leq 5$, but only the polynomial with the smallest number of terms is listed for each degree $m > 5$.

We now turn to proving that the sequence generated by a primitive polynomial satisfies the three properties for it to look like a random sequence.

Table 11.3 List of primitive polynomials

m	
1	$1+z^{-1}$
2	$1+z^{-1}+z^{-2}$
3	$1+z^{-1}+z^{-3}, 1+z^{-2}+z^{-3}$
4	$1+z^{-1}+z^{-4}, 1+z^{-3}+z^{-4}$
5	$1+z^{-2}+z^{-5}, 1+z^{-1}+z^{-2}+z^{-3}+z^{-5},$ $1+z^{-3}+z^{-5}, 1+z^{-1}+z^{-3}+z^{-4}+z^{-5},$ $1+z^{-2}+z^{-3}+z^{-4}+z^{-5}, 1+z^{-1}+z^{-2}+z^{-4}+z^{-5}$
6	$1+z^{-1}+z^{-6}$
7	$1+z^{-3}+z^{-7}$
8	$1+z^{-2}+z^{-3}+z^{-4}+z^{-8}$
9	$1+z^{-4}+z^{-9}$
10	$1+z^{-3}+z^{-10}$
11	$1+z^{-2}+z^{-11}$
12	$1+z^{-1}+z^{-4}+z^{-6}+z^{-12}$
13	$1+z^{-1}+z^{-3}+z^{-4}+z^{-13}$
14	$1+z^{-1}+z^{-6}+z^{-10}+z^{-14}$
15	$1+z^{-1}+z^{-15}$
16	$1+z^{-1}+z^{-3}+z^{-12}+z^{-16}$
17	$1+z^{-3}+z^{-17}$
18	$1+z^{-7}+z^{-18}$
19	$1+z^{-1}+z^{-2}+z^{-5}+z^{-19}$
20	$1+z^{-3}+z^{-20}$
21	$1+z^{-2}+z^{-21}$
22	$1+z^{-1}+z^{-22}$
23	$1+z^{-5}+z^{-23}$
24	$1+z^{-1}+z^{-2}+z^{-7}+z^{-24}$

(1) *Relative frequencies of ones and zeros* The maximal length shift register cycles through all the states except the all-zero one. Thus if we look at the output bit it is equivalent to writing all possible $2^m - 1$ binary m-tuples, one to a line, and looking at the last column (or any column). If the all-zero m-tuple was included we would see that exactly half the values are zero and half are one (see Problem 11.11). Because the all-zero vector is excluded we have 2^{m-1} ones and $2^{m-1} - 1$ zeros. If the initial condition vector is chosen randomly, which makes the output random, we can say

$$P[0] = \frac{2^{m-1} - 1}{2^m - 1} = \frac{1}{2}\left(1 - \frac{1}{2^m - 1}\right), \tag{11.72}$$

$$P[1] = \frac{2^{m-1}}{2^m - 1} = \frac{1}{2}\left(1 + \frac{1}{2^m - 1}\right). \tag{11.73}$$

The difference from the ideal situation is $1/(2^m - 1)$ which for m in the range of $20 - 40$ is from 10^{-6} to 10^{-12}.

(2) *Runlength property* To show that the runlengths are those of a random sequence examine $n + 2$ consecutive shift register cells where $n \le m - 2$. Consider contents that start and end with either a zero or a one:

$$0c_1c_2\cdots c_n0 \text{ or } 1c_1c_2\cdots c_n1.$$

Of the 2^n different bit patterns possible in each case only one will have a runlength of exactly n; n ones in the first case, n zeros in the second. The total number of sequences are 2^{n+1} but only two runlengths of length n. Therefore the fraction of subsequences that have a runlength of n for $1 \le n \le m - 2$ is $2/2^{n+1} = 2^{-n}$. Now consider runs of length $m - 1$. When they occur the shift register contents are either

$$\underbrace{000\cdots0}_{m-1}1 \text{ or } \underbrace{111\cdots1}_{m-1}0.$$

Therefore of the 2^{m-1} possible sequences only one has a runlength of $m - 1$ zeros and only one has a runlength of $m - 1$ ones. Therefore for $n \le m - 1$ the relative frequency of a runlength of length n is 2^{-n}. Lastly there is only one runlength of length m which occurs with relative frequency $1/2^{m-1}$.

(3) *Number of agreements and disagreements with respect to a shifted sequence* We now show that the third condition is also satisfied. Consider a maximum-length sequence generated by some initial condition vector, say $C_{\text{ini}}^{(1)}\left(z^{-1}\right)$, i.e., $C_1\left(z^{-1}\right) = C_{\text{ini}}^{(1)}\left(z^{-1}\right)/f\left(z^{-1}\right)$. Now shift the sequence $C_1\left(z^{-1}\right)$ by an arbitrary number of indices, $n \le 2^m - 1$. This results in a maximum-length sequence, $C_2(z^{-1})$ that was produced by a different initial condition vector, i.e., $C_2\left(z^{-1}\right) = C_{\text{ini}}^{(2)}\left(z^{-1}\right)/f\left(z^{-1}\right)$. Since the polynomial operations are linear the sum (modulo 2) of the two sequences is

$$C_1\left(z^{-1}\right) \oplus C_2\left(z^{-1}\right) = \frac{C_{\text{ini}}^{(1)}\left(z^{-1}\right) \oplus C_{\text{ini}}^{(2)}\left(z^{-1}\right)}{f\left(z^{-1}\right)}.$$

But $C_{\text{ini}}^{(1)}\left(z^{-1}\right) \oplus C_{\text{ini}}^{(2)}\left(z^{-1}\right)$ is a valid initial condition vector; it produces a sequence that is maximal length $(2^m - 1)$ and by the first property has 2^{m-1} ones and $2^{m-1} - 1$ zeros. But a one means that $C_1\left(z^{-1}\right)$ and $C_2\left(z^{-1}\right)$ disagree in that position and a zero means that they agree. Condition (3) is therefore satisfied.

11.2.4 Autocorrelation and crosscorrelation of the signature waveforms

Though one can generalize to the situation where the spreading sequences are any periodic sequences of the same period, here we are specifically interested in the auto- and crosscorrelation properties of signature waveforms produced by pseudorandom sequences. Consider two m-sequences of period $N_P = 2^m - 1$, or $T_P = N_P T_c$. The two periodic spreading signals are given by (11.61). Since the signals are periodic, so is the crosscorrelation with the same period. In reality the signals are also *deterministic*, therefore we define the crosscorrelation function of these two deterministic signals as

$$R_{c_P^{(1)},c_P^{(2)}}(\tau) = \frac{1}{T_P}\int_0^{T_P} c_P^{(1)}(t)c_P^{(2)}(t+\tau)\,dt, \quad -\infty \leq \tau \leq \infty. \tag{11.74}$$

Using (11.61), one has

$$R_{c_P^{(1)},c_P^{(2)}}(\tau) = \frac{1}{N_P}\sum_{n=-\infty}^{\infty}\sum_{l=-\infty}^{\infty} c_n^{(1)}c_l^{(2)}\frac{1}{T_c}\int_0^{T_P} g(t-nT_c)g(t+\tau-lT_c)\,dt. \tag{11.75}$$

Note that the integral is nonzero only if the two pulse waveforms do not overlap.

We now let $\tau = kT_c + \epsilon$, where $k = \lfloor \tau/T_c \rfloor$ and $0 \leq \epsilon \leq T_c$. Then the crosscorrelation becomes a function of k and ϵ and is

$$R_{c_P^{(1)},c_P^{(2)}}(k,\epsilon) = \frac{1}{N_P}\sum_{n=-\infty}^{\infty}\sum_{l=-\infty}^{\infty} c_n^{(1)}c_l^{(2)}\frac{1}{T_c}\int_0^{T_P} g(t-nT_c)g(t+\epsilon+(k-l)T_c)\,dt. \tag{11.76}$$

Change variable in the integral, $\lambda = t - nT_c$, to get

$$R_{c_P^{(1)},c_P^{(2)}}(k,\epsilon) = \frac{1}{N_P}\sum_{n=-\infty}^{\infty}\sum_{l=-\infty}^{\infty} c_n^{(1)}c_l^{(2)}\frac{1}{T_c}\int_{t=-nT_c}^{T_P-nT_c} g(\lambda)g(\lambda+\epsilon-(l-k-n)T_c)\,d\lambda. \tag{11.77}$$

Now $g(\lambda)$ sets the limits on λ to be between 0 and T_c, which means that n ranges from 0 to $N_P - 1$ since $T_P - nT_c = N_P T_c - nT_c = T_c$. Further, since ϵ ranges only over $(0, T_c)$, the integrand is nonzero only when (i) $l - k - n = 0$, which means $l = n + k$, or (ii) $l - k - n = 1$, which means $l = n + k + 1$. Based on these two observations, one obtains

$$\begin{aligned} R_{c_P^{(1)},c_P^{(2)}}(k,\epsilon) = {} & \frac{1}{N_P}\sum_{n=0}^{N_P-1} c_n^{(1)}c_{n+k}^{(2)}\frac{1}{T_c}\int_0^{T_c} g(\lambda)g(\lambda+\epsilon)d\lambda \\ & + \frac{1}{N_P}\sum_{n=0}^{N_P-1} c_n^{(1)}c_{n+k+1}^{(2)}\frac{1}{T_c}\int_0^{T_c} g(\lambda)g(\lambda+\epsilon-T_c)d\lambda. \end{aligned} \tag{11.78}$$

Let $g(t)$ be the standard rectangular chip waveform whose amplitude is unity over $[0, T_c]$ (i.e., $g(t)$ is the same as $p(t)$ given in (11.22)). Then it is easy to show that (see Problem 11.12):

$$R_{c_P^{(1)},c_P^{(2)}}(k,\epsilon) = \left(1 - \frac{\epsilon}{T_c}\right)\frac{1}{N_P}\sum_{n=0}^{N_P-1} c_n^{(1)}c_{n+k}^{(2)} + \frac{\epsilon}{T_c}\frac{1}{N_P}\sum_{n=0}^{N_P-1} c_n^{(1)}c_{n+k+1}^{(2)}. \tag{11.79}$$

Consider first the autocorrelation of an m-sequence, i.e., $c_n^{(1)} = c_n^{(2)}$, $n = 0, 1, \ldots, N_P - 1$. For $k = 0$, one has $(1/N_P)\Sigma_{n=0}^{N_P-1}[c_n^{(1)}]^2 = 1$ and $(1/N_P)\Sigma_{n=0}^{N_P-1} c_n^{(1)} c_{n+k}^{(1)} = -1/N_P$, which follows from the fact that, for m-sequences, the numbers of agreements and disagreements between $\left\{c_n^{(1)}\right\}$ and $\left\{c_{n+k}^{(1)}\right\}$ differ by 1. Then

$$R_{c_P^{(1)},c_P^{(2)}}(0, \epsilon) = 1 - \left(1 + \frac{1}{N_P}\right)\frac{\epsilon}{T_c}. \tag{11.80}$$

For $k = 1, 2, \ldots, N_P - 2$, both the discrete correlations are equal to $-1/N_P$ and the autocorrelation is

$$R_{c_P^{(1)},c_P^{(2)}}(k, \epsilon) = -\frac{1}{N_P}. \tag{11.81}$$

Using (11.80) and (11.81) and the property that the autocorrelation is an even function in τ, the plot of $R_{c_P^{(1)},c_P^{(1)}}(\tau)$ is shown in Figure 11.41.

The crosscorrelation depends on how the sequences of maximum length are produced. One way to produce two such sequences is to take the same pseudorandom sequence generator and choose two different initial condition vectors. The resultant outputs will be shifted versions of each other, say with a shift of T_c. This means that the crosscorrelation will simply be the autocorrelation just determined but shifted on the τ-axis by T_c.

There are $2^m - 1$ sequences that can be generated by one feedback polynomial, one for each shift register state and each sequence so generated could be assigned to a different user. The crosscorrelation between any two of such sequences would be $1/(2^m - 1)$, provided they are *synchronized*, which brings us back full circle. Otherwise, if the two sequences are shifted by only kT_c seconds, and the two transmitter clocks are out by kT_c seconds then the crosscorrelation between the two signature waveforms becomes an autocorrelation!

To overcome this problem, one can produce *distinct* m-sequences using different primitive polynomials of order m as the feedback polynomials and assign these m-sequences to different users. For a given m, the number of *primitive* polynomials of order m is $\Phi(2^m - 1)/m$. Here $\Phi(\cdot)$ is called the Euler totient function, defined as $\Phi(n) = n\prod_{p|n}^{n}\left(1 - \frac{1}{p}\right)$, where the product is over all primes p that divide n. For example, with

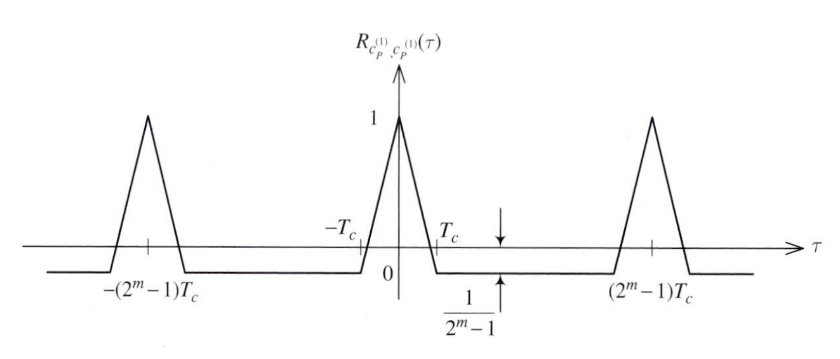

Fig. 11.41 Autocorrelation of periodic signature waveform constructed from an m-sequence.

$m = \{5, 6, 7, 8, 9, 10, 11, 12\}$ then $\Phi(2^m - 1)/m = \{6, 6, 18, 16, 48, 60, 176, 144\}$. Unfortunately, they also have poor crosscorrelation properties. Other sets of sequences have been proposed such as Gold and Kasami sequences that have better crosscorrelation properties. The interested reader is referred to the literature.

One should realize that regardless of the m-sequence chosen, what has been discussed so far is the behavior of the *full-period* correlation properties. In practice, the m-sequence period might last over many information symbol periods, i.e., $T_P = N_P T_c$, while $T_b = NT_c$, where N_P is on the order of 2^{20}–2^{50} while N (the processing gain) is on the order of 100–1000. In such a case, the correlation properties of different spreading sequences should be considered over T_b, leading to a *partial-period* correlation. Unfortunately, these are difficult to determine analytically.

Even if one has managed to design the spreading sequence set so that the MAI is less than $1/100$, i.e., 20 decibels, of the desired signal's power, there is still what is called the *near–far* problem in a CDMA environment. This typically occurs in the uplink and is due to two users transmitting to the base station from different geographical locations and hence different distances. The standard plane-earth propagation model shows that the receive power, P_R, is related to the transmit power, P_T, by an inverse fourth power law, i.e.,

$$P_R \propto \frac{P_T}{d^4} \quad \text{(watts)}, \tag{11.82}$$

where d is the distance between the transmitter and the receiver.

Assume that one user is 0.5 kilometers from the base station while the other is 10 kilometers away. The ratio of the two received powers is

$$\frac{P_R(0.5 \text{ kilometers})}{P_R(10 \text{ kilometers})} = 10 \log_{10} \frac{10^4}{0.5^4} = 52 \text{ decibels}. \tag{11.83}$$

Therefore if the processing gain is $N = 100$ or 20 decibels, the far user is completely swamped by the MAI from the near user. *Power control* is therefore very crucial to a CDMA system and it is briefly discussed next.

11.2.5 Power control

Power control can be and is done in two modes. The first mode is an *open-loop* control where the mobile user measures the received signal power and then adjusts its transmit power. This is fast but not that accurate because the correlation between the downlink received power and uplink received power is not that strong, especially when Rayleigh fading occurs. One link might be in a fade while the other is not.

With *closed-loop* power control the base station measures the received signal power from the mobile terminal and then sends a signal to the mobile instructing it to increase or decrease its power. Typically by a single bit command – for example, a 1 could signify an increase, a 0 a decrease by some predetermined amount, say 1 decibel. To minimize delay in response the power-control bits are sent at a rate in the range of 1 kilohertz.

11.2.6 Rake receiver

As mentioned, CDMA is typically used in wireless communications. Wireless channels invariably suffer from multipath and fading as discussed in Chapter 10. However, since CDMA is a wideband modulation, the approximation $s(t-\tau_j) \approx s(t)$, due to the narrow-band nature of the transmitted signal, used to develop the fading channel model, no longer holds. This assumption meant that the multipath signals were all "smeared" together. With a wideband signal one can resolve at least groups of paths at the receiver. This modifies the model and the resulting demodulator.

The resultant demodulator, called a *Rake receiver*, exploits the resolvability of the multipath. To develop the concepts behind the model and the Rake receiver, consider Figure 11.42.

There are L distinct scattering clusters. Each cluster has a mean delay of t_i with delay variations of $\boldsymbol{\tau}_{ij}$ around t_i. Assume that the delay spreads of individual clusters do not overlap in time. Finally, assume that the overall spread is less (much less) than the symbol period (which is T_b for binary modulation).

The development of the model proceeds identically to that in Chapter 10. Ignoring AWGN for now, the received signal is

$$\mathbf{r}(t) = \sum_{i=1}^{L} \sum_{j} \boldsymbol{\alpha}_{ij} s(t - t_i - \boldsymbol{\tau}_{ij}) \sqrt{\frac{2}{T_b}} \cos\left(2\pi f_c(t - t_i) - 2\pi f_c \boldsymbol{\tau}_{ij})\right). \tag{11.84}$$

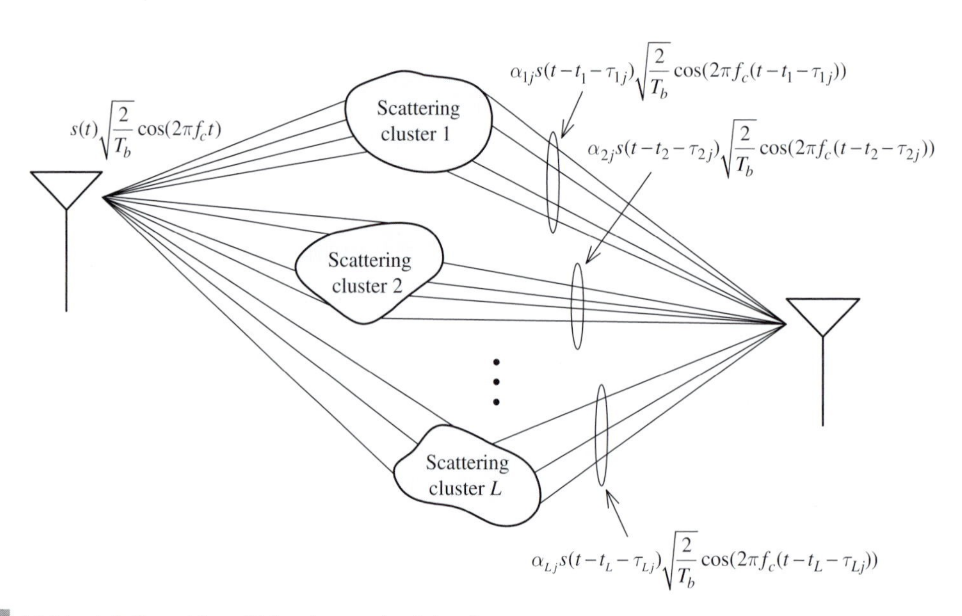

Fig. 11.42 Multipath fading with a wideband transmitted signal.

Though the overall $s(t)$ is not narrowband and $s(t - t_i - \boldsymbol{\tau}_{ij})$ is not approximately the same as $s(t)$, within the range of $\boldsymbol{\tau}_{ij}$ we can state that[14] $s(t - t_i - \boldsymbol{\tau}_{ij}) \approx s(t - t_i)$. Further, $2\pi f_c \boldsymbol{\tau}_{ij}$ ranges over $[0, 2\pi]$ and is uniform over the interval. Call it $\boldsymbol{\theta}_{ij}$. The received signal becomes

$$\mathbf{r}(t) = \sqrt{\frac{2}{T_b}} \sum_{i=1}^{L} s(t - t_i) \times \left\{ \left(\sum_j \boldsymbol{\alpha}_{ij} \cos \boldsymbol{\theta}_{ij} \right) \cos(2\pi f_c(t - t_i)) + \left(\sum_j \boldsymbol{\alpha}_{ij} \sin \boldsymbol{\theta}_{ij} \right) \sin(2\pi f_c(t - t_i)) \right\}. \tag{11.85}$$

Reasoning exactly as in Chapter 10, the quantities $\Sigma_j \boldsymbol{\alpha}_{ij} \cos \boldsymbol{\theta}_{ij}$ and $\Sigma_j \boldsymbol{\alpha}_{ij} \sin \boldsymbol{\theta}_{ij}$ are *statistically independent, zero-mean Gaussian random variables* with the same variance[15] $\sigma_F^2/2$. Call them $\mathbf{n}_{i,I}$ and $\mathbf{n}_{i,Q}$ respectively. Further, the random variables due to any one scattering cluster are statistically independent of any other. Restoring AWGN, the received signal is then

$$\begin{aligned}
\mathbf{r}(t) &= \sqrt{\frac{2}{T_b}} \sum_{i=1}^{L} s(t - t_i) \left[\mathbf{n}_{i,I} \cos(2\pi f_c(t - t_i)) + \mathbf{n}_{i,Q} \sin(2\pi f_c(t - t_i)) \right] + \mathbf{w}(t) \\
&= \sqrt{\frac{2}{T_b}} \sum_{i=1}^{L} s(t - t_i) \boldsymbol{\alpha}_i \cos(2\pi f_c(t - t_i) - \boldsymbol{\theta}_i) + \mathbf{w}(t),
\end{aligned} \tag{11.86}$$

where $\boldsymbol{\alpha}_i = \sqrt{\mathbf{n}_{i,I}^2 + \mathbf{n}_{i,Q}^2}$ is a Rayleigh random variable (see (10.62) for its pdf) and $\boldsymbol{\theta}_i = \tan^{-1}\left(\mathbf{n}_{i,Q}/\mathbf{n}_{i,I}\right)$ is a uniform random variable over $[0, 2\pi]$.

For simplicity, consider that there is only one user. Let $s(t)$ be a spread (i.e., wideband) signal with BPSK modulation, i.e.,

$$s(t) = bc(t), \quad 0 \le t \le T_b, \tag{11.87}$$

where $b = \pm\sqrt{E_b}$ (depending on the information bit to be transmitted) and $c(t) = \Sigma_{n=0}^{N-1} c_n p(t - nT_c)$ is the spreading waveform. Note that $\int_0^{T_b} s(t - t_i)s(t - t_j)\mathrm{d}t = E_b \int_0^{T_b} c(t - t_i)c(t - t_j)\mathrm{d}t \approx 0$ if $t_j - t_i > T_c$, which follows from the autocorrelation property of the spreading waveform. This means that the received signals are in essence orthogonal if the mean delays are separated enough. We assume this to be the case. The last assumption we make is that a genie tells us what $\boldsymbol{\alpha}_i$ and $\boldsymbol{\theta}_i$, $i = 1, 2, \ldots, L$ are during a transmission or if a genie is not to be found that we have a very good estimate of $\boldsymbol{\alpha}_i$, $\boldsymbol{\theta}_i$, so good, that we ignore the estimation error. With all this we now develop the minimum error probability demodulator.

[14] No man is an island and no band is so wide that it cannot be looked upon as being made up of many contiguous narrow bands.

[15] The assumption of equal variance for all scattering clusters, $i = 1, 2, \ldots, L$, is mainly for simplicity. The model and results of this section can be easily modified to accommodate the case of unequal variances over different scattering clusters.

The set of sufficient statistics is generated as shown in Figure 11.43. Observe that the integrator filters out the double frequency term and $\mathbf{w}_i$ is the usual Gaussian noise sample, statistically independent, zero-mean, variance $N_0/2$. The likelihood ratio is then

$$\frac{f(r_1, r_2, \ldots, r_L|1_T)}{f(r_1, r_2, \ldots, r_L|0_T)} = \frac{\prod_{i=1}^{L} \frac{1}{\sqrt{\pi N_0}} \mathrm{e}^{-\left(r_i - \alpha_i\sqrt{E_b}\right)^2/N_0}}{\prod_{i=1}^{L} \frac{1}{\sqrt{\pi N_0}} \mathrm{e}^{-\left(r_i + \alpha_i\sqrt{E_b}\right)^2/N_0}} \underset{0_D}{\overset{1_D}{\gtrless}} 1. \tag{11.88}$$

Taking the natural logarithm and doing the usual algebra we obtain the following decision rule:

$$\sum_{i=1}^{L} \alpha_i r_i \underset{0_D}{\overset{1_D}{\gtrless}} 0. \tag{11.89}$$

A block diagram of the receiver is shown in Figure 11.44.

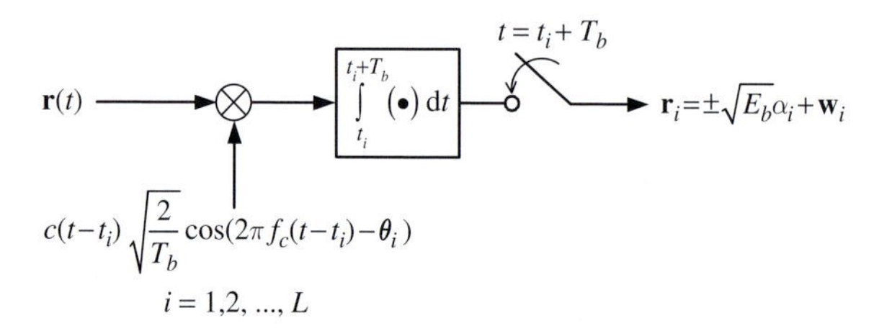

Fig. 11.43 Generation of the sufficient statistics for demodulation of the spread spectrum signal transmitted over a multipath fading channel.

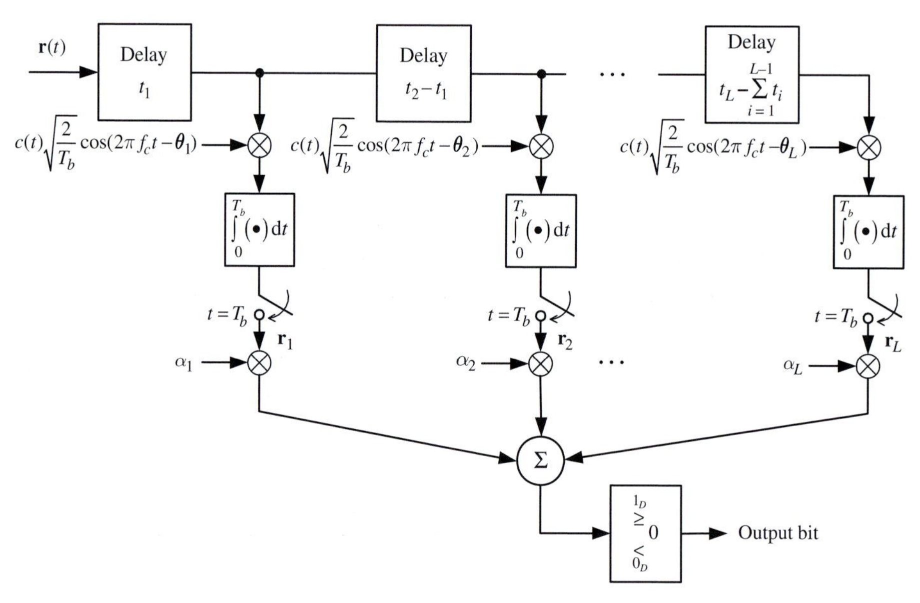

Fig. 11.44 Block diagram of the Rake receiver.

The error performance is that of Lth-order diversity with *coherent demodulation*. It is given by (see Problem 10.5):

$$P[\text{error}] = \left(\frac{1-\mu}{2}\right)^L \sum_{k=0}^{L-1} \binom{L-1+k}{k} \left(\frac{1+\mu}{2}\right)^k, \tag{11.90}$$

where $\mu = \sqrt{\gamma_T/1+\gamma_T}$ and $\gamma_T = E_b\sigma_F^2/N_0$.

The demodulator block diagram looks like a garden rake, and hence the name rake receiver. However, the fingers of the rake are very unevenly spaced and it would be hard to do a decent job of raking with it. Therefore let us revisit the model and revise it somewhat.

Let the delay spread in a CDMA environment be QT_c, where Q is considerably less than N, the processing gain. The delays are a continuum on the delay axis as shown in Figure 11.45. But because of the wideband nature of spread signal, we can discretize it into bins of width T_c and within each bin represent the multipath by a joint pdf in the random variables $\boldsymbol{\alpha}$ and $\boldsymbol{\theta}$. Typically the joint pdf is taken to be Rayleigh but other pdfs such as Rician (see Problem 10.14) or Nakagami-m (see Problem 10.7) are possible. Indeed, as the system becomes more wideband, it is able to resolve the signal on a finer scale implying that fewer and fewer multipaths fall in a bin, so that the central limit theorem works with fewer and fewer terms, implying that the Rayleigh model might not be that appropriate. The model may then need to be determined empirically, i.e., by experiment.

Be that as it may, we shall stay with the Rayleigh model. Crucially we assume that the random variables $\boldsymbol{\alpha}$ and $\boldsymbol{\theta}$ in one bin are uncorrelated (and hence statistically independent) from those of another bin. Assuming L bins, the model is exactly the same as discussed except the time delays are uniformly spaced at intervals of T_c. The block diagram of the demodulator then looks like that of Figure 11.46. The fingers of the rake are evenly spaced. The bit error probability of the receiver is also given as in (11.90). It is plotted in Figure 11.47 versus the average received SNR *per bit per branch*, $E_b\sigma_F^2/N_0$, for different values of L. Note how the error performance is dramatically improved with the number of rake fingers (branches). Thus, from the performance perspective, it is desired to have as many fingers as possible. However, it is important to realize that the number of available fingers depends on the wireless environment as well as the bandwidth of the spread transmitted signal.

Finally, the above analysis and result assume there is only one user in the system. In the case of multiple users, the Rake receiver still applies. However, the analysis of its error performance is more involved, because one needs to account for the MAI. A simple method is to treat MAI as Gaussian noise, in addition to and independent of the thermal Gaussian noise, and determine its power. Then use (11.90) but with $\gamma_T = E_b\sigma_F^2/(P_{\text{MAI}} + N_0)$, where

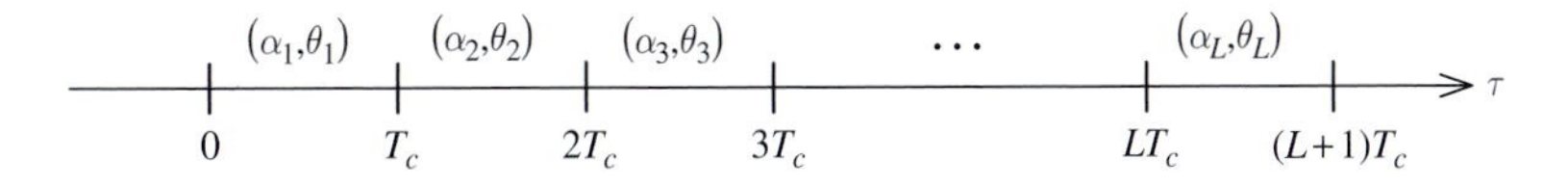

Fig. 11.45 Discrete-time approximation of the delay spread.

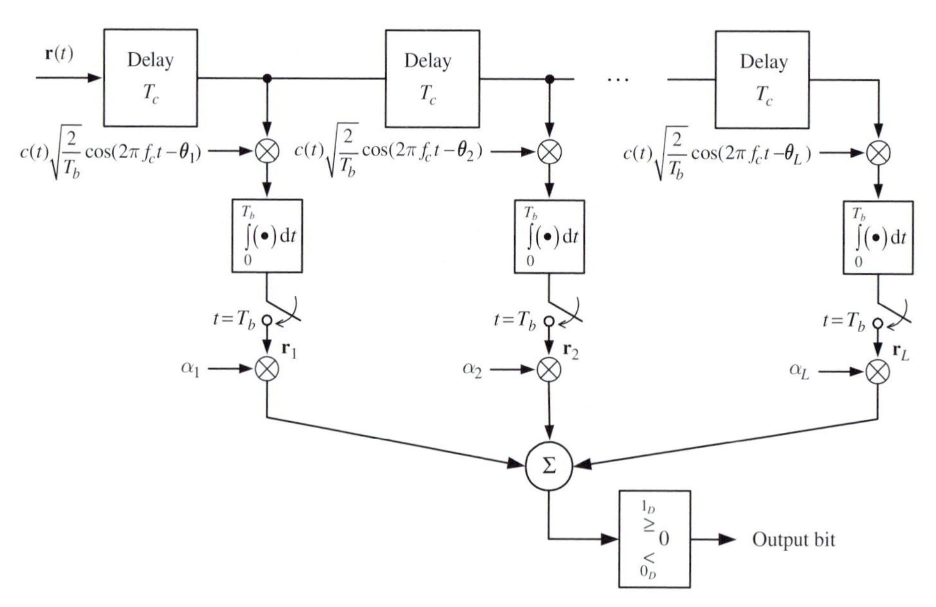

Fig. 11.46 Block diagram of the rake receiver with discrete-time approximation of the delay spread.

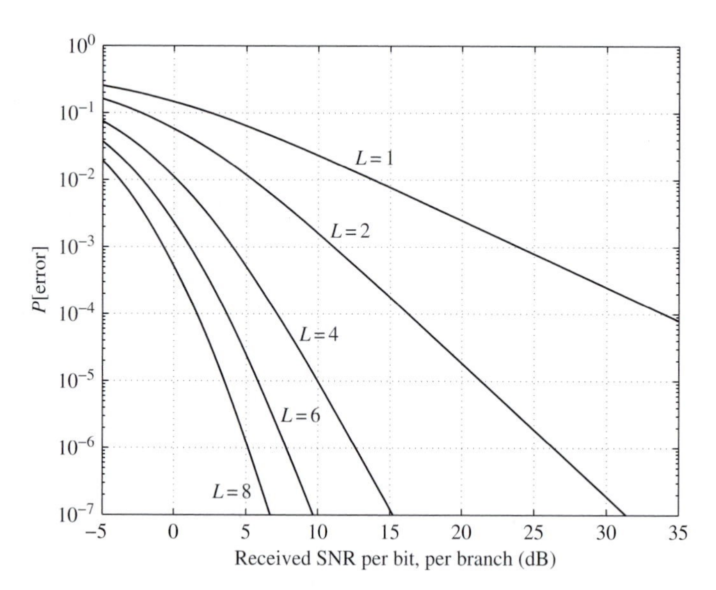

Fig. 11.47 Bit error probability of the rake receiver: single user and Rayleigh fading.

P_{MAI} is the power of MAI. Even with this simple approximation, the determination of P_{MAI} depends on the specific set of signature waveforms used and is still cumbersome. Another issue is that the MAI is not well approximated by Gaussian noise, especially when the number of users is small.

11.3 Space-time transmission

In Chapter 10 we saw that, in terms of the received SNR, the error performance of a communication system operating over a fading channel is severely degraded. To improve performance, the diversity concept and technique were introduced to combat the effect of a deep fade. Though effective, time and frequency diversity techniques have disadvantages and limitations.

Time diversity can be used if the fading channel changes reasonably fast compared with the delay requirement of the application. This allows one to rearrange (interleave) the multiple copies of the same symbol before transmitting them so that each copy experiences a statistically independent channel. If repetition coding is used in combination with interleaving to realize time diversity, then the information rate is decreased in proportion to the diversity order.

Frequency diversity relies on the fact that the channel response varies over frequency subbands and if the subbands are sufficiently far apart, the corresponding channels are more or less statistically independent. Typically frequency diversity is most suitable in wideband communication systems (such as CDMA and OFDM systems) where the channel bandwidth is much wider than the bandwidth of the information-bearing signal. Otherwise, frequency diversity also requires an increased bandwidth in proportion to the diversity order.

When neither time diversity nor frequency diversity is available, such as in a slowly varying, narrowband communication system, it is of interest to have another means of providing diversity. This can be achieved by *space* diversity which is the topic of this section.[16] In general, space diversity is achieved by having multiple antennas, at either the receiver or the transmitter or both, that are spaced sufficiently far apart (on the order of a wavelength). The case of one transmit antenna and multiple receive antennas is known as *receive diversity*. It is typically found in the uplink of a mobile communication system. Conversely, when there are multiple transmit antennas and a single receive antenna, we have *transmit diversity*. This is most often found in the downlink.

Here we discuss both transmit and receive diversity for the simplest cases of: (i) one transmit antenna and two receive antennas, and (ii) two transmit antennas and one receive antenna. Though simple, the discussion results in the introduction of the Alamouti space-time block code, originally developed in reference [4] for transmit diversity with two transmit antennas and one receive antenna. The coding aspect of this elegant and practical diversity scheme comes from not only using the space diversity offered by the two transmit antennas, but more importantly, by also introducing a clever constraint over time.

Before proceeding to the discussion, a word about the model with which we are working. We assume that the channel is slow fading so that at least over two symbol transmission

[16] Even when time diversity and/or frequency diversity are available, it might still be attractive and desirable to provide space diversity. In fact most modern wireless communication systems try to exploit all the available means of diversity in order to improve the system performance.

periods (also referred to as time slots) the fading coefficients (α_i, θ_i) of each space transmission path remain constant. The slow fading implies that one can obtain good estimates of the random variables (α_i, θ_i), so good that we assume they are perfect. This means that the BPSK, QPSK, or M-QAM signal set can be used at the modulator.

11.3.1 Receive diversity

Consider the receive diversity of Figure 11.48(a). Over the interval $[0, T_b]$ a single bit is transmitted with BPSK modulation. Assume that the receive antennas are spaced far enough apart so that the fading coefficients (α_1, θ_1), (α_2, θ_2) are statistically independent. Then it follows from Section 10.5 and Problem 10.5 that a second-order diversity is achieved without any decrease in throughput (information data transmission rate) or increase in bandwidth. One can easily visualize extending the above to the system with one transmit antenna and N receive antennas to achieve Nth-order diversity, as long as the (α_i, θ_i) pairs are statistically independent.

As mentioned earlier, to achieve statistically independent fading channels, the receive antennas should be spaced far apart, on the order of a wavelength. Consider a typical carrier frequency of 1 gigahertz. Then the wavelength is $c/f = \frac{3\times10^8}{10^9}$ m $= 30$ cm (≈ 12 inches). Though this requirement is quite reasonable at the base station, it is not possible to achieve with a pocket-size mobile receiver unless one is willing to strap the antennas onto body extremities. Thus, although simple in the concept, receive diversity might not be available in many situations, such as in the downlink of a mobile communication system. This practical limitation of receive diversity leads us to consider transmit diversity next.

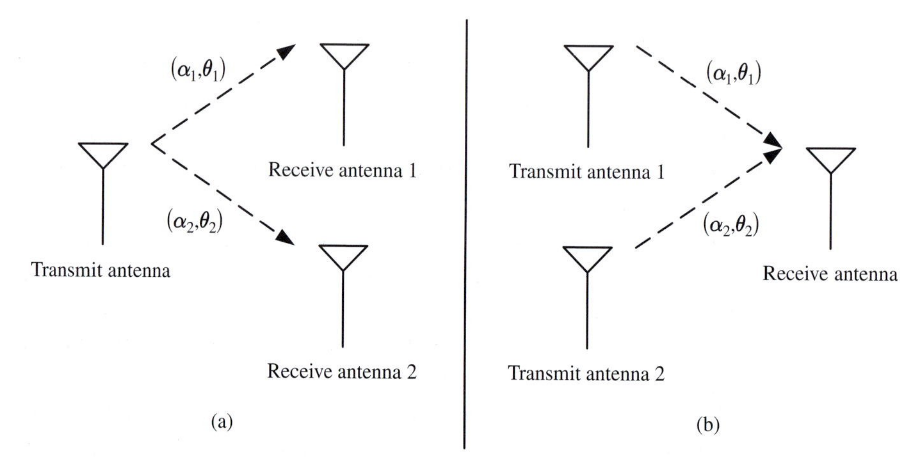

Fig. 11.48 Space diversity: (a) one transmit and two receive antennas and (b) two transmit and one receive antennas.

11.3.2 Transmit diversity

Consider the system setup in Figure 11.48(b), where the fading coefficients $(\boldsymbol{\alpha}_1, \boldsymbol{\theta}_1)$, $(\boldsymbol{\alpha}_2, \boldsymbol{\theta}_2)$ are statistically independent (visualize that the transmit antennas are at the base station and there should be no difficulty in spacing them far enough apart). Again, the objective is to transmit one information bit over the bit duration T_b by BPSK.

As the first attempt, in every bit interval, let us send the information bit *simultaneously* over both transmit antennas. Over the kth bit interval, $[(k-1)T_b, kT_b]$, the received signal is given by

$$\mathbf{r}(t) = \underbrace{b_k\sqrt{E_b'}\boldsymbol{\alpha}_1\sqrt{\frac{2}{T_b}}\cos(2\pi f_c t - \boldsymbol{\theta}_1)}_{\text{over transmit antenna 1}} + \underbrace{b_k\sqrt{E_b'}\boldsymbol{\alpha}_2\sqrt{\frac{2}{T_b}}\cos(2\pi f_c t - \boldsymbol{\theta}_2)}_{\text{over transmit antenna 2}} + \mathbf{w}(t), \quad (11.91)$$

where $b_k = \pm 1$ is the information bit. Note that E_b' is the transmitted energy per bit per antenna, hence the energy per bit is $E_b = 2E_b'$. Equation (11.91) can be rewritten as

$$\begin{aligned}\mathbf{r}(t) = &b_k\sqrt{E_b'}\sqrt{\frac{2}{T_b}}\big[(\boldsymbol{\alpha}_1\cos\boldsymbol{\theta}_1 + \boldsymbol{\alpha}_2\cos\boldsymbol{\theta}_2)\cos(2\pi f_c t) \\ &+ (\boldsymbol{\alpha}_1\sin\boldsymbol{\theta}_1 + \boldsymbol{\alpha}_2\sin\boldsymbol{\theta}_2)\sin(2\pi f_c t)\big] + \mathbf{w}(t).\end{aligned} \quad (11.92)$$

But $\boldsymbol{\alpha}_1\cos\boldsymbol{\theta}_1 = \mathbf{n}_{1,I}$, $\boldsymbol{\alpha}_2\cos\boldsymbol{\theta}_2 = \mathbf{n}_{2,I}$, $\boldsymbol{\alpha}_1\sin\boldsymbol{\theta}_1 = \mathbf{n}_{1,Q}$, $\boldsymbol{\alpha}_2\sin\boldsymbol{\theta}_2 = \mathbf{n}_{2,Q}$ are Gaussian random variables, zero-mean, variance $\sigma_F^2/2$ and statistically independent. For convenience, define $\mathbf{n}_I = \mathbf{n}_{1,I} + \mathbf{n}_{2,I}$ and $\mathbf{n}_Q = \mathbf{n}_{1,Q} + \mathbf{n}_{2,Q}$. They are two statistically independent Gaussian random variables with zero mean and variance σ_F^2. Therefore (11.92) can be written as

$$\mathbf{r}(t) = b_k\sqrt{E_b'}\boldsymbol{\alpha}\sqrt{\frac{2}{T_b}}\cos(2\pi f_c t - \boldsymbol{\theta}) + \mathbf{w}(t), \quad (11.93)$$

where $\boldsymbol{\alpha} = \sqrt{\mathbf{n}_I^2 + \mathbf{n}_Q^2}$ and $\boldsymbol{\theta} = \tan^{-1}(\mathbf{n}_Q/\mathbf{n}_I)$ are Rayleigh and uniform random variables, respectively. Unfortunately, the received signal in (11.93) is the same as that over a single fading channel (e.g., in a single transmit antenna and single receive antenna). This means that only a diversity order of 1 can be achieved with the above transmission scheme. We have not achieved anything with the two transmit antennas, except perhaps doubled the received signal power.

Upon reflection, the fact that the above transmission scheme can only achieve a first-order diversity can be explained as follows. Although we have two transmitted signals for the same information bit, these two signals are superimposed in the air before arriving at the single receive antenna. The superposition of the transmitted signals is beyond an engineer's control and therefore she cannot "constructively" combine the two signals to achieve a diversity order of 2. As another explanation, from the perspective of a transmission link from one transmit antenna to the single receive antenna, the second transmit antenna simply appears (or acts) as a scattering cluster! One can generalize the argument to the case of N transmit antennas and 1 receive antenna to see that, as long as the same signal is transmitted simultaneously over all N antennas, only a first-order diversity is achieved.

Recognizing that sending the same signal simultaneously over the two antennas is not helpful, another possibility is to turn one antenna on for T_b seconds, then switch it off and turn the other antenna on for the next T_b seconds. Here we have a choice. We can either transmit the next information bit, which would maintain the throughput (i.e., information data rate), but obviously the diversity order is still 1. Or we could retransmit the same information bit so that we obtain two independent received signals, over two consecutive bit intervals, for the same information bit. This is nothing more than a repetition code and though a diversity order 2 is achieved, the throughput has halved. The choice turned out to be somewhat of a Hobson's choice.[17] The transmission scheme just described can be represented in matrix form as

$$\begin{bmatrix} b_k\sqrt{E_b'}\sqrt{2/T_b}\cos(2\pi f_c t) & 0 \\ 0 & b_k\sqrt{E_b'}\sqrt{2/T_b}\cos(2\pi f_c t) \end{bmatrix} \begin{matrix} \text{antenna 1} \\ \text{antenna 2} \end{matrix}$$
$$\underbrace{\qquad\qquad}_{[(k-1)T_b, kT_b]} \quad \underbrace{\qquad\qquad}_{[kT_b,(k+1)T_b]} \quad \text{time} \longrightarrow \tag{11.94}$$

where $b_k = \pm 1$ is the information bit. Note also that the average transmitted energy per bit in this scheme is also $E_b = 2E_b'$.

The limitations of the previous two transmission schemes motivate one to seek a scheme that can transmit one information bit every T_b and achieve diversity order 2. Intuitively, such a scheme must use the two transmit antennas all the time and should be designed over multiple bit intervals. It is also desired that the transmitted signals from the two antennas be "decoupled" at the receiver. With this in mind, consider the following transmitted signals from the two antennas over two bit intervals:

$$\begin{bmatrix} b_k\sqrt{E_b'}\sqrt{2/T_b}\cos(2\pi f_c t) & ? \\ b_{k+1}\sqrt{E_b'}\sqrt{2/T_b}\cos(2\pi f_c t) & ? \end{bmatrix} \begin{matrix} \text{antenna 1} \\ \text{antenna 2} \end{matrix} \tag{11.95}$$
$$\underbrace{\qquad\qquad}_{[(k-1)T_b, kT_b]} \quad \underbrace{\qquad\qquad}_{[kT_b,(k+1)T_b]} \quad \text{time} \longrightarrow$$

We wish to fill in the question marks in such a manner that the signal transmitted over the first antenna (the first row) is always orthogonal to that transmitted over the second antenna (the second row) over $[(k-1)T_b, (k+1)T_b]$, regardless of the bit pattern (b_k, b_{k+1}). We also want b_k and b_{k+1} involved in both antennas (i.e., they should appear in both rows of the above matrix) so that each information bit is transmitted over both independent fading channels. This ensures that a diversity of 2 is achieved. Finally, the signals in

[17] See Wikipedia: "http://en.wikipedia.org/wiki/Hobson's_choice" and decide if this is a true Hobson's choice.

the second bit interval (time slot) should come from the BPSK signal set. Some thought (there are not that many possibilities) results in the following transmission scheme (see Problem 11.14):

$$\underbrace{\begin{bmatrix} b_k\sqrt{E_b'}\sqrt{2/T_b}\cos(2\pi f_c t) \\ b_{k+1}\sqrt{E_b'}\sqrt{2/T_b}\cos(2\pi f_c t) \end{bmatrix.}_{[(k-1)T_b,kT_b]} \underbrace{\begin{matrix} -b_{k+1}\sqrt{E_b'}\sqrt{2/T_b}\cos(2\pi f_c t) \\ b_k\sqrt{E_b'}\sqrt{2/T_b}\cos(2\pi f_c t) \end{bmatrix}}_{[kT_b,(k+1)T_b]} \begin{matrix}\text{antenna 1} \\ \text{antenna 2} \\ \text{time} \longrightarrow\end{matrix} \tag{11.96}$$

Note that E_b' is the transmitted energy on each antenna over one bit duration. The transmitted energy per bit is $E_b = 2E_b'$. The above scheme is an Alamouti space-time block code for the BPSK signal constellation.

Having designed the modulator, we turn our attention to the demodulator and its error performance. The received signal is

$$\mathbf{r}(t) = \mathbf{r}_1(t) + \mathbf{r}_2(t) + \mathbf{w}(t), \quad (k-1)T_b \leq t \leq (k+1)T_b, \tag{11.97}$$

where the bits b_k, $b_{k+1} \in \{-1, +1\}$ and $\mathbf{w}(t)$ is the usual zero-mean AWGN whose two-sided PSD is $N_0/2$ watts/hertz. The two components $\mathbf{r}_1(t)$ and $\mathbf{r}_2(t)$ are given by

$$\begin{bmatrix} \mathbf{r}_1(t) \\ \mathbf{r}_2(t) \end{bmatrix} = \begin{bmatrix} \underbrace{\begin{matrix} b_k\sqrt{E_b'}\boldsymbol{\alpha}_1\sqrt{2/T_b}\cos(2\pi f_c t - \boldsymbol{\theta}_1) \\ b_{k+1}\sqrt{E_b'}\boldsymbol{\alpha}_2\sqrt{2/T_b}\cos(2\pi f_c t - \boldsymbol{\theta}_2) \end{matrix}}_{[(k-1)T_b,kT_b]} & \underbrace{\begin{matrix} -b_{k+1}\sqrt{E_b'}\boldsymbol{\alpha}_1\sqrt{2/T_b}\cos(2\pi f_c t - \boldsymbol{\theta}_1) \\ b_k\sqrt{E_b'}\boldsymbol{\alpha}_2\sqrt{2/T_b}\cos(2\pi f_c t - \boldsymbol{\theta}_2) \end{matrix}}_{[kT_b,(k+1)T_b]} \end{bmatrix}. \tag{11.98}$$

To obtain the demodulator, we use the approach that has been a constant theme: find a signal space for the received signal, project $\mathbf{r}(t)$ onto it to get a set of sufficient statistics and the likelihood ratio to obtain a decision rule. As noted before, the receiver has perfect estimates of the fading coefficients. This means that the demodulator knows and works with specific coefficients (α_1, θ_1), (α_2, θ_2).

Because of the phase shifts in the channels, the signal space is four-dimensional with the following basis set:

$$\underbrace{\begin{matrix} \phi_{1,I}(t) = \sqrt{2/T_b}\cos(2\pi f_c t), \\ \phi_{1,Q}(t) = \sqrt{2/T_b}\sin(2\pi f_c t), \end{matrix}}_{t\in[(k-1)T_b,kT_b]} \; \underbrace{\begin{matrix} \phi_{2,I}(t) = \sqrt{2/T_b}\cos(2\pi f_c t), \\ \phi_{2,Q}(t) = \sqrt{2/T_b}\sin(2\pi f_c t). \end{matrix}}_{t\in[kT_b,(k+1)T_b]} \tag{11.99}$$

Projecting $\mathbf{r}(t)$ onto the basis functions results in the following set of sufficient statistics:

$$\begin{cases} \mathbf{r}_{1,I} = b_k\sqrt{E_b'}\alpha_1\cos\theta_1 + b_{k+1}\sqrt{E_b'}\alpha_2\cos\theta_2 + \mathbf{w}_{1,I} \\ \mathbf{r}_{1,Q} = b_k\sqrt{E_b'}\alpha_1\sin\theta_1 + b_{k+1}\sqrt{E_b'}\alpha_2\sin\theta_2 + \mathbf{w}_{1,Q} \\ \mathbf{r}_{2,I} = -b_{k+1}\sqrt{E_b'}\alpha_1\cos\theta_1 + b_k\sqrt{E_b'}\alpha_2\cos\theta_2 + \mathbf{w}_{2,I} \\ \mathbf{r}_{2,Q} = -b_{k+1}\sqrt{E_b'}\alpha_1\sin\theta_1 + b_k\sqrt{E_b'}\alpha_2\sin\theta_2 + \mathbf{w}_{2,Q} \end{cases}. \tag{11.100}$$

Write the above as two matrix equations:

$$\begin{bmatrix}\mathbf{r}_{1,I}\\ \mathbf{r}_{2,I}\end{bmatrix} = \sqrt{E_b'}\begin{bmatrix}\alpha_1\cos\theta_1 & \alpha_2\cos\theta_2\\ \alpha_2\cos\theta_2 & -\alpha_1\cos\theta_1\end{bmatrix}\begin{bmatrix}b_k\\ b_{k+1}\end{bmatrix} + \begin{bmatrix}\mathbf{w}_{1,I}\\ \mathbf{w}_{2,I}\end{bmatrix}, \tag{11.101a}$$

$$\begin{bmatrix}\mathbf{r}_{1,Q}\\ \mathbf{r}_{2,Q}\end{bmatrix} = \sqrt{E_b'}\begin{bmatrix}\alpha_1\sin\theta_1 & \alpha_2\sin\theta_2\\ \alpha_2\sin\theta_2 & -\alpha_1\sin\theta_1\end{bmatrix}\begin{bmatrix}b_k\\ b_{k+1}\end{bmatrix} + \begin{bmatrix}\mathbf{w}_{1,Q}\\ \mathbf{w}_{2,Q}\end{bmatrix}, \tag{11.101b}$$

where the noise terms $\mathbf{w}_{1,I}$, $\mathbf{w}_{1,Q}$, $\mathbf{w}_{2,I}$, $\mathbf{w}_{2,Q}$ are statistically independent Gaussian random variables, zero-mean, variance $N_0/2$ (watts). Premultiply (11.101a) by the transpose of the *inphase part* of the channel transmission matrix. Similarly, premultiply (11.101b) by the transpose of the *quadrature part* of the channel transmission matrix to obtain a new set of sufficient statistics:

$$\begin{bmatrix}\hat{\mathbf{r}}_{1,I}\\ \hat{\mathbf{r}}_{2,I}\end{bmatrix} = \sqrt{E_b'}\begin{bmatrix}\alpha_1^2\cos^2\theta_1 + \alpha_2^2\cos^2\theta_2 & 0\\ 0 & \alpha_1^2\cos^2\theta_1 + \alpha_2^2\cos^2\theta_2\end{bmatrix}\begin{bmatrix}b_k\\ b_{k+1}\end{bmatrix} + \begin{bmatrix}\hat{\mathbf{w}}_{1,I}\\ \hat{\mathbf{w}}_{2,I}\end{bmatrix}, \tag{11.102a}$$

$$\begin{bmatrix}\hat{\mathbf{r}}_{1,Q}\\ \hat{\mathbf{r}}_{2,Q}\end{bmatrix} = \sqrt{E_b'}\begin{bmatrix}\alpha_1^2\sin^2\theta_1 + \alpha_2^2\sin^2\theta_2 & 0\\ 0 & \alpha_1^2\sin^2\theta_1 + \alpha_2^2\sin^2\theta_2\end{bmatrix}\begin{bmatrix}b_k\\ b_{k+1}\end{bmatrix} + \begin{bmatrix}\hat{\mathbf{w}}_{1,Q}\\ \hat{\mathbf{w}}_{2,Q}\end{bmatrix}, \tag{11.102b}$$

where $\hat{\mathbf{w}}_{1,I}$, $\hat{\mathbf{w}}_{2,I}$, $\hat{\mathbf{w}}_{1,Q}$, $\hat{\mathbf{w}}_{2,Q}$ are still zero-mean Gaussian random variables which are uncorrelated and hence statistically independent. However, their variances are not equal. In particular one can show that (see Problem 11.15):

$$\mathrm{var}\{\hat{\mathbf{w}}_{1,I}\} = \mathrm{var}\{\hat{\mathbf{w}}_{2,I}\} = \left(\alpha_1^2\cos^2\theta_1 + \alpha_2^2\cos^2\theta_2\right)\frac{N_0}{2}, \tag{11.103a}$$

$$\mathrm{var}\{\hat{\mathbf{w}}_{1,Q}\} = \mathrm{var}\{\hat{\mathbf{w}}_{2,Q}\} = \left(\alpha_1^2\sin^2\theta_1 + \alpha_2^2\sin^2\theta_2\right)\frac{N_0}{2}. \tag{11.103b}$$

To find the decision rule(s), recognize that information about bit b_k is only provided by the sufficient statistics $\left(\hat{\mathbf{r}}_{1,I}, \hat{\mathbf{r}}_{1,Q}\right)$, while information for b_{k+1} is only provided by $\left(\hat{\mathbf{r}}_{2,I}, \hat{\mathbf{r}}_{2,Q}\right)$. Consider the derivation of the decision rule for b_k. Rewrite the sufficient statistics for it as

$$\hat{\mathbf{r}}_{1,I} = \sqrt{E_b'}\left(\alpha_1^2\cos^2\theta_1 + \alpha_2^2\cos^2\theta_2\right)b_k + \hat{\mathbf{w}}_{1,I}, \tag{11.104a}$$

$$\hat{\mathbf{r}}_{1,Q} = \sqrt{E_b'}\left(\alpha_1^2\sin^2\theta_1 + \alpha_2^2\sin^2\theta_2\right)b_k + \hat{\mathbf{w}}_{1,Q}. \tag{11.104b}$$

To make the noise variances equal, we make one final transformation of the sufficient statistics. Specifically, divide (11.104a) by $\sqrt{\alpha_1^2\cos^2\theta_1+\alpha_2^2\cos^2\theta_2}$ and (11.104b) by $\sqrt{\alpha_1^2\sin^2\theta_1+\alpha_2^2\sin^2\theta_2}$ to obtain

$$\mathbf{r}_I=\sqrt{E_b'}\sqrt{\alpha_1^2\cos^2\theta_1+\alpha_2^2\cos^2\theta_2}\cdot b_k+\mathbf{w}_I, \tag{11.105a}$$

$$\mathbf{r}_Q=\sqrt{E_b'}\sqrt{\alpha_1^2\sin^2\theta_1+\alpha_2^2\sin^2\theta_2}\cdot b_k+\mathbf{w}_Q, \tag{11.105b}$$

where $\mathbf{w}_I$ and $\mathbf{w}_Q$ are zero-mean statistically independent Gaussian random variables with the same variance $N_0/2$ (see Problem 11.16). Now form the likelihood ratio:

$$\frac{f(r_I,r_Q|b_k=1)}{f(r_I,r_Q|b_k=-1)}\underset{0_D}{\overset{1_D}{\gtrless}}1. \tag{11.106}$$

Taking the logarithm and simplifying results in the decision rule:

$$\sqrt{\alpha_1^2\cos^2\theta_1+\alpha_2^2\cos^2\theta_2}\cdot r_I+\sqrt{\alpha_1^2\sin^2\theta_1+\alpha_2^2\sin^2\theta_2}\cdot r_Q\underset{0_D}{\overset{1_D}{\gtrless}}0. \tag{11.107}$$

The decision rule can also be obtained from (11.105) in a different way, by using the geometry of Figure 11.49 and the fact that $\mathbf{w}_I$ and $\mathbf{w}_Q$ are the usual zero-mean statistically independent Gaussian random variables. The decision rule is a *minimum-distance* rule and since the information bits are equally likely, it partitions the (sufficient statistics) signal space as shown. From high-school geometry we recall that if $y=mx$ then the perpendicular is the line $y=-(1/m)x$. Here the equation of the line joining the two "signal" points is

$$r_Q=\frac{\sqrt{\alpha_1^2\sin^2\theta_1+\alpha_2^2\sin^2\theta_2}}{\sqrt{\alpha_1^2\cos^2\theta_1+\alpha_2^2\cos^2\theta_2}}r_I$$

and the equation of the perpendicular bisector is

$$r_Q=-\frac{\sqrt{\alpha_1^2\cos^2\theta_1+\alpha_2^2\cos^2\theta_2}}{\sqrt{\alpha_1^2\sin^2\theta_1+\alpha_2^2\sin^2\theta_2}}r_I.$$

From this one obtains the decision rule of (11.107).

The bit error performance of the receiver that implements the decision rule in (11.107) can be readily obtained from Figure 11.49. Given the fading coefficients (α_1,θ_1) and (α_2,θ_2), the Euclidean distance between the two "signal" points is $2\sqrt{E_b'}\sqrt{\alpha_1^2+\alpha_2^2}$. Therefore the *conditional* error probability is given by

$$P[\text{error}|\boldsymbol{\alpha}_1=\alpha_1,\boldsymbol{\theta}_1=\theta_1,\boldsymbol{\alpha}_2=\alpha_2,\boldsymbol{\theta}_2=\theta_2]=Q\left(\sqrt{\frac{2E_b'\left(\alpha_1^2+\alpha_2^2\right)}{N_0}}\right). \tag{11.108}$$

It should be pointed out that, although the above conditional error probability depends only on α_1 and α_2, it does not mean that the receiver does not need the perfect estimates

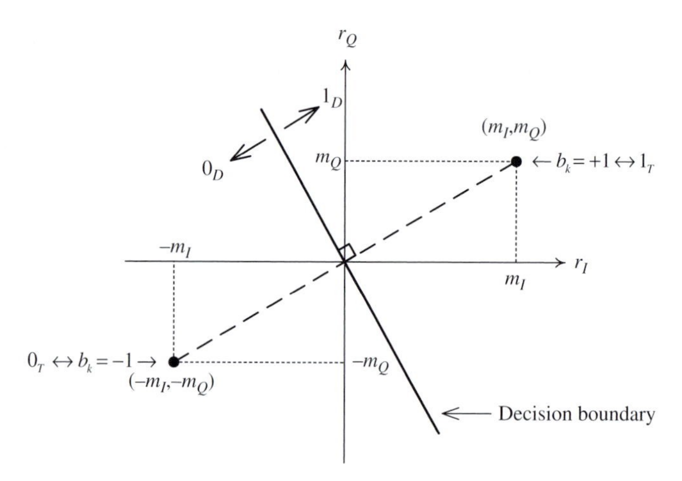

Fig. 11.49 The signal space representation of the sufficient statistics for the detection of b_k. Here $m_I = \sqrt{E_b'}\sqrt{\alpha_1^2 \cos^2\theta_1 + \alpha_2^2 \cos^2\theta_2}$ and $m_Q = \sqrt{E_b'}\sqrt{\alpha_1^2 \sin^2\theta_1 + \alpha_2^2 \sin^2\theta_2}$.

of the phases θ_1 and θ_2. Recall that knowledge of the phases is needed to perform various transformations on the sufficient statistics in order to arrive at the decision rule in (11.107).

The conditional error probability in (11.108) applies for a specific set of parameters α_1 and α_2. Though we assumed that α_1 and α_2 are known to the demodulator, we should not forget that they are *random variables* drawn from an underlying probabilistic sample space.[18] Therefore during the transmission of a long sequence of bits, (α_1, α_2) will vary according to two independent Rayleigh pdfs. Therefore the *average* error probability should be obtained by averaging (11.108) over the pdfs of $\boldsymbol{\alpha}_1$ and $\boldsymbol{\alpha}_2$, or equivalently over the pdf of $\mathbf{y} = \boldsymbol{\alpha}_1^2 + \boldsymbol{\alpha}_2^2$. This is basically what was considered in Problem 10.5 and it is summarized below.

Let σ_F^2 be the mean-squared value of both $\boldsymbol{\alpha}_1$ and $\boldsymbol{\alpha}_2$. Since $\boldsymbol{\alpha}_1$ and $\boldsymbol{\alpha}_2$ are independent Rayleigh random variables, the random variable $\mathbf{y}$ is a chi-square random variable with four degrees of freedom (see (10.90)). Its pdf is

$$f_{\mathbf{y}}(y) = \frac{y e^{-y/\sigma_F^2}}{\sigma_F^4} u(y). \tag{11.109}$$

Therefore,

$$\begin{aligned} P[\text{error}] &= E\left\{Q\left(\sqrt{\frac{E_b\left(\boldsymbol{\alpha}_1^2 + \boldsymbol{\alpha}_2^2\right)}{N_0}}\right)\right\} = E\left\{Q\left(\sqrt{\frac{E_b}{N_0}}\sqrt{\mathbf{y}}\right)\right\} \\ &= \frac{1}{\pi}\frac{1}{\sigma_F^4}\int_0^{\pi/2}\int_0^{\infty} \exp\left(-\frac{y}{2\sin^2\theta}\frac{E_b}{N_0}\right) y \exp\left\{-\frac{y}{\sigma_F^2}\right\} \mathrm{d}y\mathrm{d}\theta, \end{aligned} \tag{11.110}$$

[18] That is the nature of random events or variables. Once you see them or know them they are no longer random, but before that they are unpredictable.

where we have substituted $E_b = 2E_b^{'}$ and used Craig's formula for the Q-function (see (10.76)). Performing the inner integration first gives

$$P[\text{error}] = \frac{1}{\pi}\int_0^{\pi/2} \frac{1}{\left(1+\gamma_T/\sin^2\theta\right)^2}\text{d}\theta, \tag{11.111}$$

where $\gamma_T = E_b^{'}\sigma_F^2/N_0 = E_b\sigma_F^2/2N_0$. To evaluate the integral in (11.111), expand the square and use the identity in (10.79) and the following identity:

$$\int_0^{\pi} \frac{\text{d}x}{(b+a\cos x)^2} = \frac{\pi b}{(b^2-a^2)\sqrt{b^2-a^2}}, \quad |a| < b. \tag{11.112}$$

The final expression is

$$P[\text{error}] = \left(\frac{1}{2} - \frac{1}{2}\sqrt{\frac{\gamma_T}{1+\gamma_T}}\right)^2 \left(2 + \sqrt{\frac{\gamma_T}{1+\gamma_T}}\right). \tag{11.113}$$

At high SNR, i.e., $\gamma_T \gg 1$, we can use the approximations $\frac{1}{2} - \frac{1}{2}\sqrt{\gamma_T/(1+\gamma_T)} \approx 1/4\gamma_T$ and $2 + \sqrt{\gamma_T/(1+\gamma_T)} \approx 3$ to obtain

$$P[\text{error}] \approx \frac{3}{(4\gamma_T)^2} = \frac{3}{4}\frac{1}{\left(E_b\sigma_F^2/N_0\right)^2}, \tag{11.114}$$

which clearly shows that a diversity order of 2 is achieved.

Another, perhaps simpler, way to see that the diversity order is 2 is as follows. Start with (11.108) and use the approximation $Q(x) < \frac{1}{2}\text{e}^{-x^2/2}$ to obtain

$$\begin{aligned} P[\text{error}] &= E\left\{Q\left(\sqrt{\frac{E_b\left(\boldsymbol{\alpha}_1^2+\boldsymbol{\alpha}_2^2\right)}{N_0}}\right)\right\} < \frac{1}{2}E\left\{\exp\left[-\frac{E_b}{2N_0}\left(\boldsymbol{\alpha}_1^2+\boldsymbol{\alpha}_2^2\right)\right]\right\} \\ &= \frac{1}{2}\int_{\alpha_1=0}^{\infty}\int_{\alpha_2=0}^{\infty}\exp\left[-\frac{E_b}{2N_0}\left(\alpha_1^2+\alpha_2^2\right)\right]f_{\alpha_1}(\alpha_1)f_{\alpha_2}(\alpha_2)\text{d}\alpha_1\text{d}\alpha_2 \\ &= \frac{1}{2}\left[\int_{\alpha_1=0}^{\infty}\exp\left(-\frac{E_b}{2N_0}\alpha_1^2\right)f_{\alpha_1}(\alpha_1)\text{d}\alpha_1\right]^2, \end{aligned} \tag{11.115}$$

where the last equality follows from $f_{\alpha_1}(\alpha_1) = f_{\alpha_2}(\alpha_1) = \left(\alpha_1/\sigma_F^2\right)\text{e}^{-\alpha_1^2/2\sigma_F^2}$. Substituting the pdf of $\boldsymbol{\alpha}_1$ and doing the integration gives[19]

$$P[\text{error}] < \frac{1}{2}\frac{1}{\left(1+E_b\sigma_F^2/2N_0\right)^2} \approx \frac{2}{\left(E_b\sigma_F^2/N_0\right)^2}, \tag{11.116}$$

which means that a diversity order of 2 is achieved.

Figure 11.50 plots the bit error probability (11.113) and the upper bound (11.116) versus the average received SNR per bit, defined as $E_b\sigma_F^2/N_0$. Also shown for comparison is the bit error probability of BPSK with coherent detection over a Rayleigh fading

[19] Note that, instead of averaging over the pdf of $\boldsymbol{\alpha}_1$, one could also change the variable to $\boldsymbol{\beta} = \boldsymbol{\alpha}_1^2$ and average over the pdf of $\boldsymbol{\beta}$, which is a decaying exponential $f_{\boldsymbol{\beta}}(\beta) = (1/\sigma_F^2)\exp\left(-\beta/\sigma_F^2\right)$.

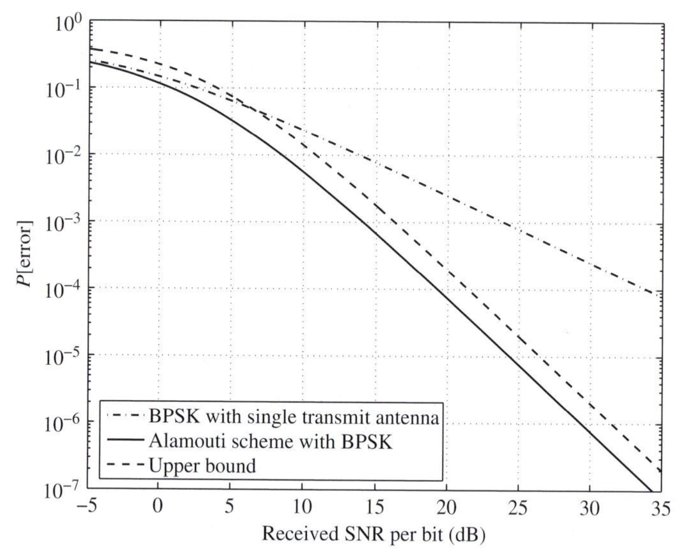

Fig. 11.50 Bit error probability of the Alamouti scheme with BPSK. Also shown for comparison are its upper bound and the bit error probability of BPSK with one transmit and one receive antennas.

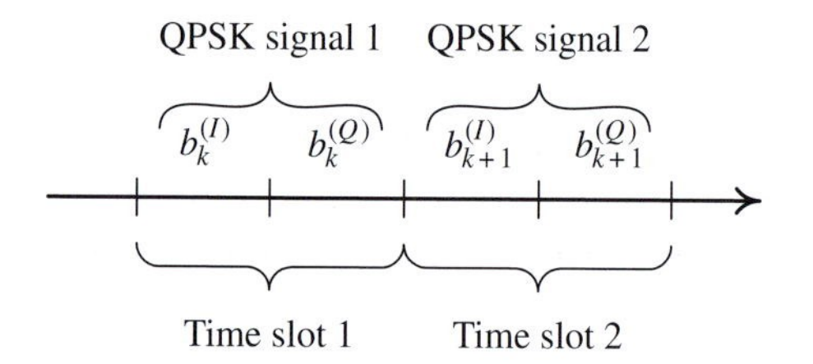

Fig. 11.51 Transmission of two consecutive QPSK symbols.

channel when there are single transmit and receive antennas, which can be shown to be $\frac{1}{2}\left[1 - \sqrt{\gamma_b/(1+\gamma_b)}\right]$, where $\gamma_b = E_b\sigma_F^2/N_0$. These performance curves clearly verify the diversity orders of the two schemes.

Upon examining the above derivations of the error performance of the Alamouti space-time code, one should realize that the achieved diversity order 2 is a direct consequence of the fact that transmission and reception of the Alamouti code yield an effective gain $\sqrt{\boldsymbol{\alpha}_1^2 + \boldsymbol{\alpha}_2^2}$. The pdf of $\sqrt{\boldsymbol{\alpha}_1^2 + \boldsymbol{\alpha}_2^2}$ is more favorable than the Rayleigh pdf of the individual channel gains $\boldsymbol{\alpha}_1$ and $\boldsymbol{\alpha}_2$ in the sense that the probability that the channel gain is small (i.e., that both channels are in a deep fade) becomes smaller.

The Alamouti space-time code just described can be extended to QAM signal constellations. Consider QPSK and two consecutive symbol transmissions as illustrated in Figure 11.51. In each symbol duration (time slot) of T_s a QPSK signal (symbol) is formed as usual by modulating the inphase and quadrature carriers separately using the respective bits. The signals to be transmitted are

$$b_k^{(I)}\sqrt{E_b'}\sqrt{\frac{2}{T_s}}\cos(2\pi f_c t) + b_k^{(Q)}\sqrt{E_b'}\sqrt{\frac{2}{T_s}}\sin(2\pi f_c t), \tag{11.117}$$

$$b_{k+1}^{(I)}\sqrt{E_b'}\sqrt{\frac{2}{T_s}}\cos(2\pi f_c t) + b_{k+1}^{(Q)}\sqrt{E_b'}\sqrt{\frac{2}{T_s}}\sin(2\pi f_c t). \tag{11.118}$$

These signals are transmitted over the two antennas over $[(k-1)T_s, kT_s]$, which are represented as the first column of the following transmission matrix:

$$\sqrt{E_b'}\sqrt{2/T_s}\Bigg[\underbrace{\begin{matrix} b_k^{(I)}\cos(2\pi f_c t) + b_k^{(Q)}\sin(2\pi f_c t) \\ b_{k+1}^{(I)}\cos(2\pi f_c t) + b_{k+1}^{(Q)}\sin(2\pi f_c t) \end{matrix}}_{[(k-1)T_s, kT_s]} \quad \underbrace{\begin{matrix} ? \\ ? \end{matrix}}_{[kT_s,(k+1)T_s]}\Bigg]. \tag{11.119}$$

Once again, the question marks in the second column are to be filled by signals from the QPSK constellation to "decouple" the two channels, i.e., make the transmitted signals over the two antennas orthogonal. Again, some thought leads to the following design:

$$\sqrt{E_b'}\sqrt{2/T_s} \times \Bigg[\underbrace{\begin{matrix} b_k^{(I)}\cos(2\pi f_c t) + b_k^{(Q)}\sin(2\pi f_c t) \\ b_{k+1}^{(I)}\cos(2\pi f_c t) + b_{k+1}^{(Q)}\sin(2\pi f_c t) \end{matrix}}_{[(k-1)T_s, kT_s]} \quad \underbrace{\begin{matrix} -b_{k+1}^{(I)}\cos(2\pi f_c t) + b_{k+1}^{(Q)}\sin(2\pi f_c t) \\ b_k^{(I)}\cos(2\pi f_c t) - b_k^{(Q)}\sin(2\pi f_c t) \end{matrix}}_{[kT_s,(k+1)T_s]}\Bigg]. \tag{11.120}$$

To develop the demodulator, write the received signal as

$$\mathbf{r}(t) = \mathbf{r}_1(t) + \mathbf{r}_2(t) + \mathbf{w}(t), \quad (k-1)T_s \le t \le (k+1)T_s, \tag{11.121}$$

where $\mathbf{w}(t)$ is the usual zero-mean AWGN whose two-sided PSD is $N_0/2$ watts/hertz. The two components $\mathbf{r}_1(t)$ and $\mathbf{r}_2(t)$ are

$$\begin{bmatrix}\mathbf{r}_1(t) \\ \mathbf{r}_2(t)\end{bmatrix} = \sqrt{E_b'}\sqrt{2/T_s}\Bigg[\underbrace{\begin{matrix} b_k^{(I)}\alpha_1\cos(2\pi f_c t - \theta_1) + b_k^{(Q)}\alpha_1\sin(2\pi f_c t - \theta_1) \\ b_{k+1}^{(I)}\alpha_1\cos(2\pi f_c t - \theta_1) + b_{k+1}^{(Q)}\alpha_1\sin(2\pi f_c t - \theta_1) \end{matrix}}_{\text{first column: } [(k-1)T_s, kT_s]}$$

$$\underbrace{\begin{matrix} -b_{k+1}^{(I)}\alpha_1\cos(2\pi f_c t - \theta_1) + b_{k+1}^{(Q)}\alpha_1\sin(2\pi f_c t - \theta_1) \\ b_k^{(I)}\alpha_1\cos(2\pi f_c t - \theta_1) - b_k^{(Q)}\alpha_1\sin(2\pi f_c t - \theta_1) \end{matrix}}_{\text{second column: } [kT_s,(k+1)T_s]}\Bigg]. \tag{11.122}$$

Then project $\mathbf{r}(t)$ onto the following basis set

$$\underbrace{\begin{array}{l}\phi_{1,I}(t) = \sqrt{2/T_s}\cos(2\pi f_c t), \\ \phi_{1,Q}(t) = \sqrt{2/T_s}\sin(2\pi f_c t),\end{array}}_{t\in[(k-1)T_s,kT_s]} \quad \underbrace{\begin{array}{l}\phi_{2,I}(t) = \sqrt{2/T_s}\cos(2\pi f_c t), \\ \phi_{2,Q}(t) = \sqrt{2/T_s}\sin(2\pi f_c t)\end{array}}_{t\in[kT_s,(k+1)T_s]} \tag{11.123}$$

and obtain the sufficient statistics:

$$\begin{bmatrix}\mathbf{r}_{1,I}\\ \mathbf{r}_{1,Q}\\ \mathbf{r}_{2,I}\\ \mathbf{r}_{2,Q}\end{bmatrix} = \sqrt{E_b'}\underbrace{\begin{bmatrix}\alpha_1\cos\theta_1 & -\alpha_1\sin\theta_1 & \alpha_2\cos\theta_2 & -\alpha_2\sin\theta_2\\ \alpha_1\sin\theta_1 & \alpha_1\cos\theta_1 & \alpha_2\sin\theta_2 & \alpha_2\cos\theta_2\\ \alpha_2\sin\theta_2 & \alpha_2\sin\theta_2 & -\alpha_1\cos\theta_1 & -\alpha_1\sin\theta_1\\ \alpha_2\sin\theta_2 & -\alpha_2\cos\theta_2 & -\alpha_1\sin\theta_1 & \alpha_1\cos\theta_1\end{bmatrix}}_{\text{“transmission matrix” } A}\begin{bmatrix}b_k^{(I)}\\ b_k^{(Q)}\\ b_{k+1}^{(I)}\\ b_{k+1}^{(Q)}\end{bmatrix} + \begin{bmatrix}\mathbf{w}_{1,I}\\ \mathbf{w}_{1,Q}\\ \mathbf{w}_{2,I}\\ \mathbf{w}_{2,Q}\end{bmatrix}. \tag{11.124}$$

Premultiply both sides of the above equation with $A^\top$ to get the following new set of the sufficient statistics:

$$\begin{aligned}\begin{bmatrix}\hat{\mathbf{r}}_{1,I}\\ \hat{\mathbf{r}}_{1,Q}\\ \hat{\mathbf{r}}_{2,I}\\ \hat{\mathbf{r}}_{2,Q}\end{bmatrix} &= \sqrt{E_b'}\left(\alpha_1^2+\alpha_2^2\right)\begin{bmatrix}1&0&0&0\\0&1&0&0\\0&0&1&0\\0&0&0&1\end{bmatrix}\begin{bmatrix}b_k^{(I)}\\ b_k^{(Q)}\\ b_{k+1}^{(I)}\\ b_{k+1}^{(Q)}\end{bmatrix} + \begin{bmatrix}\hat{\mathbf{w}}_{1,I}\\ \hat{\mathbf{w}}_{1,Q}\\ \hat{\mathbf{w}}_{2,I}\\ \hat{\mathbf{w}}_{2,Q}\end{bmatrix}\\ &= \sqrt{E_b'}\left(\alpha_1^2+\alpha_2^2\right)\begin{bmatrix}b_k^{(I)}\\ b_k^{(Q)}\\ b_{k+1}^{(I)}\\ b_{k+1}^{(Q)}\end{bmatrix} + \begin{bmatrix}\hat{\mathbf{w}}_{1,I}\\ \hat{\mathbf{w}}_{1,Q}\\ \hat{\mathbf{w}}_{2,I}\\ \hat{\mathbf{w}}_{2,Q}\end{bmatrix},\end{aligned} \tag{11.125}$$

where the noise variables are all i.i.d. zero-mean Gaussian random variables with variance $\left(\alpha_1^2+\alpha_2^2\right)N_0/2$. The above relationship shows that the four information bits are essentially decouped in a four-dimensional signal space with an individual bit lying (antipodally) on its own orthogonal axis.

The decision rule for each bit follows immediately. The conditional bit error probability is $Q\left(\sqrt{2E_b'\left(\alpha_1^2+\alpha_2^2\right)/N_0}\right) = Q\left(\sqrt{E_b\left(\alpha_1^2+\alpha_2^2\right)/N_0}\right)$, where $E_b = 2E_b'$ is also the average transmitted energy per information bit for the Alamouti scheme described in (11.120). This is exactly the same as that of Alamouti space-time code using BPSK. Therefore all the analysis of the average bit error probability and diversity order done for BPSK also holds for QPSK.

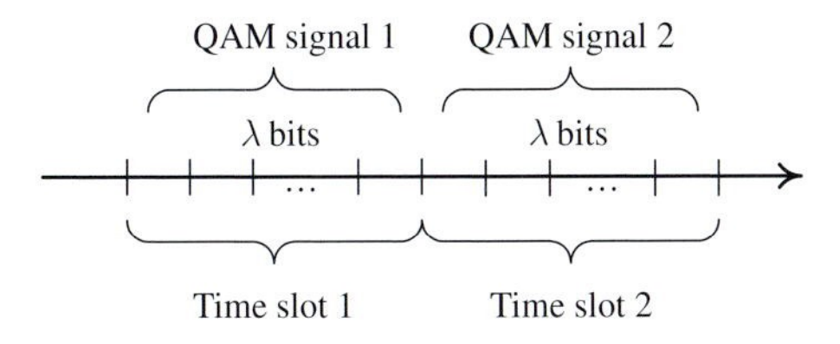

Fig. 11.52 Transmission of two consecutive QAM symbols.

Now let us generalize the Alamouti space-time code to M-QAM. Consider two QAM signals, each representing $\lambda = \log_2 M$ bits (see Figure 11.52). Write the two signals, where each occupies an interval of T_s seconds, as follows:

$$s_i(t) = s_{i,I}\sqrt{\frac{2}{T_s}}\cos(2\pi f_c t) + s_{i,Q}\sqrt{\frac{2}{T_s}}\sin(2\pi f_c t), \tag{11.126}$$

$$s_j(t) = s_{j,I}\sqrt{\frac{2}{T_s}}\cos(2\pi f_c t) + s_{j,Q}\sqrt{\frac{2}{T_s}}\sin(2\pi f_c t). \tag{11.127}$$

Similar to the case of QPSK, the transmission matrix over two symbol intervals is as follows:

$$\sqrt{\frac{2}{T_s}} \times \underbrace{\begin{bmatrix} s_{i,I}\cos(2\pi f_c t) + s_{i,Q}\sin(2\pi f_c t) \\ s_{j,I}\cos(2\pi f_c t) + s_{j,Q}\sin(2\pi f_c t) \end{bmatrix.}}_{[(k-1)T_s, kT_s]} \underbrace{\begin{bmatrix.} -s_{j,I}\cos(2\pi f_c t) + s_{j,Q}\sin(2\pi f_c t) \\ s_{i,I}\cos(2\pi f_c t) - s_{i,Q}\sin(2\pi f_c t) \end{bmatrix}}_{[kT_s, (k+1)T_s]}. \tag{11.128}$$

The demodulator is also very simple and similar to that of QPSK. Specifically, identify $s_{i,I} \to b_k^{(I)}$, $s_{i,Q} \to b_k^{(Q)}$, $s_{j,I} \to b_{k+1}^{(I)}$, $s_{j,Q} \to b_{k+1}^{(Q)}$ to obtain an expression of the form of (11.125). We then make decisions on $s_{i,I}$, $s_{i,Q}$, $s_{j,I}$, $s_{j,Q}$ if the QAM is rectangular and the bits are independently mapped onto inphase and quadrature components. The generic block diagram of the receiver is illustrated in Figure 11.53, and also applies for the cases of BPSK and QPSK.

Since the Alamouti transmission scheme yields an equivalent channel for each QAM symbol whose effective channel gain is $\sqrt{E_s}\sqrt{\alpha_1^2 + \alpha_2^2}$, where E_s is the average symbol energy (see Figure 11.54), the conditional symbol and bit error probabilities are those obtained for M-QAM over an AWGN channel in Chapter 8. The final symbol and bit error probabilities are obtained by averaging the conditional error probabilities over the pdf of the random variable $\mathbf{y} = \boldsymbol{\alpha}_1^2 + \boldsymbol{\alpha}_2^2$, given in (11.109).

Finally, it should be pointed out that space-time codes for more than two transmit antennas exist [11] and new codes are still to be found. In general, if there are N transmit antennas and one receive antenna and if the transmit antennas are placed sufficiently far apart, the maximum diversity order is N. The main criteria to design a space-time code for a given number of transmit antennas, say N, are summarized below.

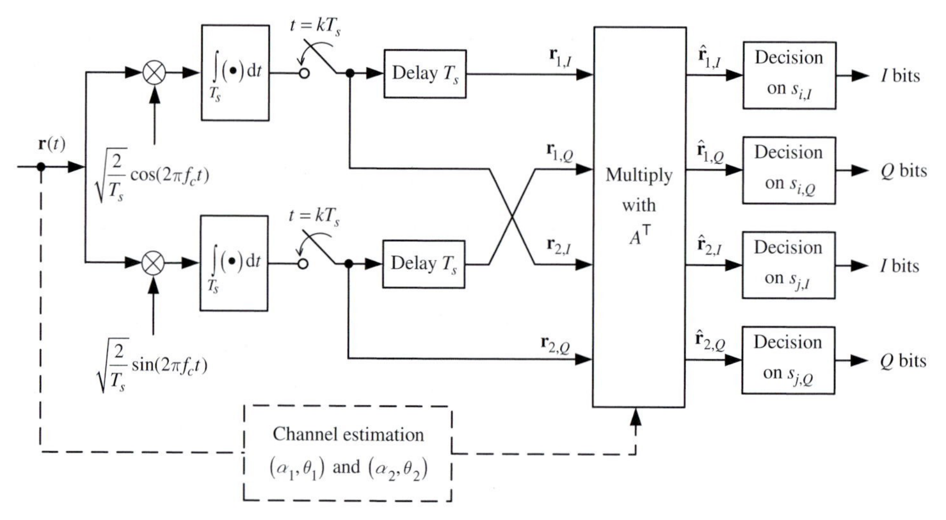

Fig. 11.53 Block diagram of the receiver for the Alamouti space-time transmission with rectangular QAM.

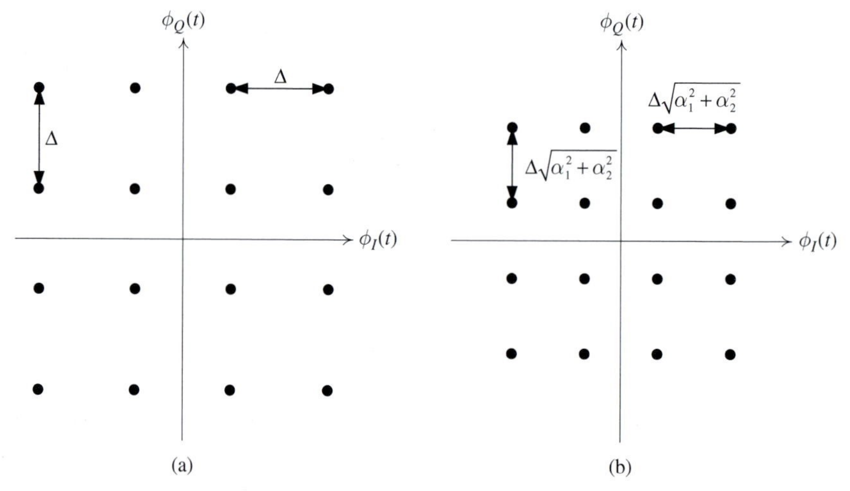

Fig. 11.54 An example of (a) a QAM constellation at the transmitter and (b) the effective constellation at the receiver (without AWGN) provided by Alamouti space-time transmission.

(1) The code should achieve a high throughput. Ideally, the code should transmit one symbol (belonging to any constellation) per one symbol duration, i.e., the maximum rate is 1 symbol/slot.
(2) The code should achieve a high diversity order. The maximum diversity order is N.
(3) The maximum likelihood receiver should be as simple as possible. The simplest maximum likelihood receiver is the one that can demodulate each symbol separately (symbol-wise ML demodulator).

Interestingly enough, Alamouti space-time code designed for two transmit antennas is the only known space-time code that optimally meets the above three criteria.

11.4 Problems

11.1 Design a shift register circuit based on the memory arrangement in Figure 11.10 and an 8-PSK constellation mapping such that the three coded bits uniquely select an 8-PSK signal according to the trellis labelings in Figure 11.11.

11.2 Design a shift register circuit based on the memory arrangement in Figure 11.13 and a 16-QAM constellation mapping such that the three coded bits together with the uncoded bit uniquely select a 16-QAM signal according to the trellis labelings in Figure 11.14.

11.3 Instead of a rectangular 32-QAM, use the 32-CROSS QAM constellation in Figure 8.18 as a reference uncoded system. Determine the coding gain achieved by the TCM scheme of Example 11.1.

11.4 Design a TCM modulator with a rectangular lattice where there is one coded bit (to choose one of two cosets) and four uncoded bits (to choose a signal point within a coset). Compare the coding gain (or loss) of the designed TCM scheme with that obtained in Example 11.1.

11.5 Justify the trellis mapping of Figure 11.14.

11.6 (*TCM demodulation with the Viterbi algorithm*) Consider the TCM scheme described by Figures 11.2, 11.6, and 11.8. The following sequence of sufficient statistics is obtained at the receiver:

$$\begin{array}{lcccccccc} r_I = & -0.173 & -0.702 & -1.166 & 1.476 & 0.838 & 0.632 & 0.472 & -0.055 \\ r_Q = & 0.334 & 0.822 & 1.183 & -0.015 & -0.637 & 0.997 & 0.166 & -0.954 \end{array} \quad \text{(P11.1)}$$

where r_I is the projection onto the inphase axis and r_Q that onto the quadrature axis. Using the Viterbi algorithm determines the most probable transmitted bit sequence. Assume that one starts in the $\boxed{00}$ state and that the average transmitted signal energy is $E_s = 1$ joule.

11.7 Verify Table 11.2.

11.8 This problem examines and compares the PSDs of typical spreading waveform and message signals.

(a) Determine the PSD of a spreading waveform, $c(t)$, produced by an m-sequence for $m = 20$ and $m = 40$ (see (11.80) and Figure 11.41). Since the autocorrelation is periodic, the spectrum is *discrete*. However, plot the "envelope" of the discrete spectrum in terms of $f_n = fT_c$.

(b) Determine also the PSD of the message signal $m(t)$ where the modulation is NRZ-L. Express the PSD in terms of the processing gain G and the chip interval T_c. On the same (horizontal) scale as the plot in (a), plot the PSD for $G = 100$.

11.9 The expression of the bit error probability in (11.41) is obtained by ignoring the usual AWGN. What happens if the AWGN is considered in the analysis.

11.10 It is of interest to get a feel for how long it takes the shift register that produces the pseudorandom sequence to repeat itself. Consider a chip rate of 1 gigahertz (not an unreasonable assumption). How long is the period of the shift register if it is of length: (a) 20 and (b) 40? If the source rate is 10^5 bits/second, how many source bits occur before the spreading sequence starts to repeat itself?

11.11 Write all possible $2^m - 1$ binary m-tuples, one to a line, and look at any column. Show that if the all zero m-tuple was included we would see that exactly half the values are zero and half are one.

11.12 Verify (11.79).

11.13 (*Multiuser detection*) Section 11.2.2 discusses *synchronous* CDMA systems where the users' signature waveforms are orthogonal and the received signals are time aligned. Here we examine the case that the system is synchronous but the users' signature waveforms are not orthogonal. Such a situation arises, for example, when the number of users exceeds the processing gain, i.e., the system is overloaded.

As the simplest example, consider a CDMA system with only two users whose signature waveforms are $c^{(1)}(t)$ and $c^{(2)}(t)$. Each waveform lasts over $[0, T_b]$ and has unit energy. Moreover, the two signature waveforms are correlated with crosscorrelation coefficient $\rho = \int_0^{T_b} c^{(1)}(t)c^{(2)}(t)\mathrm{d}t$. Also let $b_j = \pm 1, j = 1, 2$, denote the jth user's information bit in the first signaling interval.

(a) Consider the receiver in Figure 11.55 (see also Figure 11.36), known as the *correlation* or the *matched-filter* receiver. Show that the bit error probability of this receiver is

$$P^{\text{corr.}}[\text{bit error}] = \frac{1}{2}Q\left(\sqrt{\frac{2E_b}{N_0}}(1-\rho)\right) + \frac{1}{2}Q\left(\sqrt{\frac{2E_b}{N_0}}(1+\rho)\right). \quad \text{(P11.2)}$$

Hint In the detection of one user, say user 1, treat the component due to user 2's signal at the output of the integrate-and-dump circuit as interference.

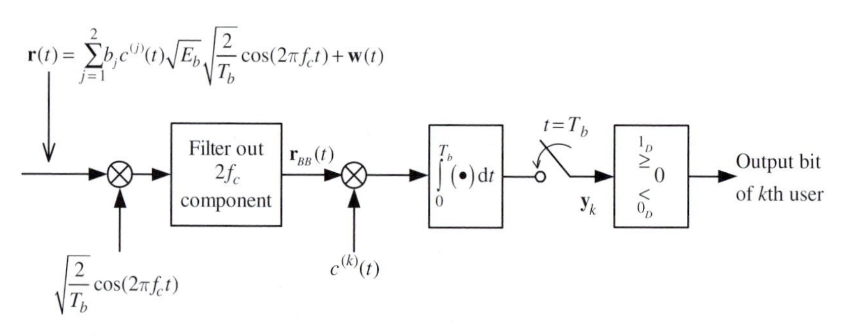

Fig. 11.55 Correlation (or matched-filter) receiver for a CDMA system.

(b) Now consider the receiver in Figure 11.56, known as the *decorrelating* receiver, where $R = \begin{bmatrix} 1 & \rho \\ \rho & 1 \end{bmatrix}$ is the correlation matrix of the signature waveforms and $\begin{bmatrix} \hat{\mathbf{y}}_1 \\ \hat{\mathbf{y}}_2 \end{bmatrix} = R^{-1} \begin{bmatrix} \mathbf{y}_1 \\ \mathbf{y}_2 \end{bmatrix}$. Show that the bit error probability of this receiver is

$$P^{\text{decorr.}}[\text{bit error}] = Q\left(\sqrt{\frac{2E_b}{N_0}}\sqrt{1-\rho^2}\right). \tag{P11.3}$$

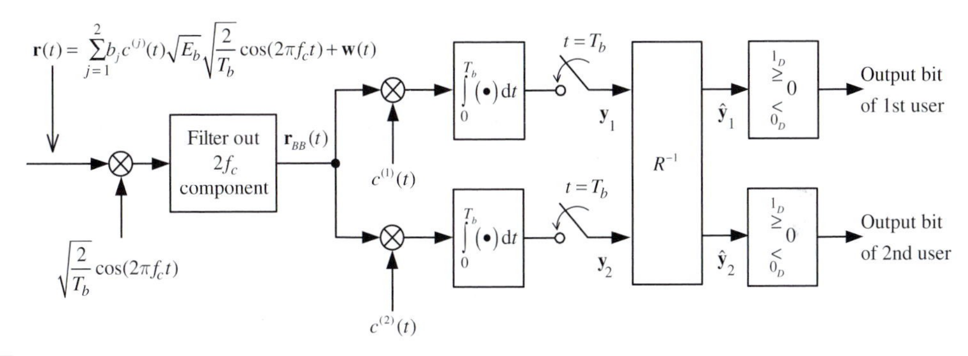

Fig. 11.56 Decorrelating receiver for a two-user CDMA system.

(c) Plot the error probabilities of the correlation and decorrelating receivers versus E_b/N_0 for $\rho = 0.2$ and $\rho = 0.8$. Comment.

(d) Develop an *optimum* multiuser receiver that *jointly* detects the two bits b_1 and b_2. *Hint* There are four combinations of b_1 and b_2 and the optimum receiver is similar to that of quaternary modulation in AWGN. Sketch a block diagram of the optimum receiver. Also clearly illustrate the signal space diagram and the decision boundary when $\rho = 0.5$.

11.14 Verify that $S_1(t)$ and $S_2(t)$ defined as the first and second rows of the matrix in (11.96) are orthogonal over $[(k-1)T_b, (k+1)T_b]$.

11.15 Show that the random variables $\hat{\mathbf{w}}_{1,I}$, $\hat{\mathbf{w}}_{2,I}$, $\hat{\mathbf{w}}_{1,Q}$, $\hat{\mathbf{w}}_{2,Q}$ in (11.102a) and (11.102b) are zero-mean statistically independent Gaussian random variables with variances given in (11.103a) and (11.103b).

11.16 Verify that $\mathbf{w}_I$ and $\mathbf{w}_Q$ in (11.105a) and (11.105b) are zero-mean statistically independent Gaussian random variables with the same variance $N_0/2$.

11.17 (*A space-time code*) We have seen that the simple repetition coding applied to two transmit antennas as described in (11.94) can achieve a diversity order 2, but at the price of a reduced (by half) transmission rate. To improve the transmission rate, your friend suggests the following space-time transmission scheme. Let $b_k\sqrt{E_b}\sqrt{2/T_b}\cos(2\pi f_c t)$ and $b_{k+1}\sqrt{E_b}\sqrt{2/T_b}\cos(2\pi f_c t)$ be two BPSK signals. First, perform the following "rotation" on these two BPSK symbols:

$$\begin{bmatrix} S_i^{(1)}(t) \\ S_i^{(2)}(t) \end{bmatrix} = \begin{bmatrix} b_k\sqrt{E_b}\sqrt{2/T_b}\cos(2\pi f_c t) \\ b_{k+1}\sqrt{E_b}\sqrt{2/T_b}\cos(2\pi f_c t) \end{bmatrix} \begin{bmatrix} \cos\theta & \sin\theta \\ -\sin\theta & \cos\theta \end{bmatrix}, \tag{P11.4}$$

where θ is some angle. Then the signals $S_i^{(1)}(t)$ and $S_i^{(2)}(t)$ are space-time transmitted as follows:

$$\underbrace{\begin{bmatrix} S_i^{(1)}(t) \\ 0 \end{bmatrix}}_{[(k-1)T_b, kT_b]} \underbrace{\begin{bmatrix} 0 \\ S_i^{(2)}(t) \end{bmatrix}}_{[kT_b, (k+1)T_b]} \begin{matrix} \text{antenna 1} \\ \text{antenna 2} \\ \text{time} \longrightarrow \end{matrix} \tag{P11.5}$$

Note that the superscripts of $S_i^{(1)}(t)$ and $S_i^{(2)}(t)$ refer to the first and second transmit antennas, while the subscript i refers to one of the four specific signals, depending on the bit pattern (b_k, b_{k+1}). Define $S_i(t) = [S_i^{(1)}(t), S_i^{(2)}(t)]$ over $(k-1)T_b \leq t \leq (k+1)T_b$.

(a) Obtain the signal space representation of the transmitted signal set $\{S_i(t)\}_{i=1}^4$.

(b) Obtain the sufficient statistics for the detection of the bits b_k and b_{k+1}. Is it possible to decouple the detections of these bits? What do you think the diversity order achieved by this space-time transmission scheme is?

(c) Let $\phi_1(t)$ and $\phi_2(t)$ be the two orthonormal basis functions used to represent the signal set $\{S_i(t)\}_{i=1}^4$ in (a). It can be shown that, over the Rayleigh fading channels, the bit error probability is minimized by maximizing the minimum *product distance*, d_P, defined as $d_P^2 = \min_{\substack{i,j \\ i \neq j}} |S_{i1} - S_{j1}||S_{i2} - S_{j2}|$. Find the angle θ that maximizes d_P.

11.18 The Alamouti scheme with BPSK modulation sends 1 bit/time slot and achieves a full diversity order of 2. We have seen that applying a repetition coding to two transmit antennas (being turned on and off alternately over two slots) can also achieve a diversity order 2. To maintain the throughput of 1 bit/time slot, 4-ASK can be employed. Assuming that Gray mapping is used with 4-ASK, obtain the exact bit error probability of the repetition coding. Compare it with that of the Alamouti/BPSK scheme by plotting (in Matlab) the two error probabilities on the same figure. Comment.

11.19 Inspired by the Alamouti scheme, your friend proposes the following space-time transmission with QPSK:

$$\sqrt{E_b'}\sqrt{2/T_s} \times \Bigg[\underbrace{\begin{matrix} b_k^{(I)}\cos(2\pi f_c t) + b_k^{(Q)}\sin(2\pi f_c t) \\ b_{k+1}^{(I)}\cos(2\pi f_c t) + b_{k+1}^{(Q)}\sin(2\pi f_c t) \end{matrix}}_{[(k-1)T_s, kT_s]} \quad \underbrace{\begin{matrix} -b_{k+1}^{(I)}\cos(2\pi f_c t) - b_{k+1}^{(Q)}\sin(2\pi f_c t) \\ b_k^{(I)}\cos(2\pi f_c t) + b_k^{(Q)}\sin(2\pi f_c t) \end{matrix}}_{[kT_s, (k+1)T_s]} \Bigg]. \tag{P11.6}$$

Note that, according to the above proposal: (i) the transmitted signal over the first antenna in the second time slot is the negative of the signal transmitted over the second antenna in the first time slot, and (ii) the transmitted signal over the second antenna in the second time slot is the same as the transmitted signal over the first antenna in the first time slot.

(a) Show that, over two time slots, the signal transmitted over the first antenna is orthogonal to the signal transmitted over the second antenna.

(b) Following the same steps as those done for the Alamouti/QPSK scheme, show that the sufficient statistics are given as:

$$\begin{bmatrix} \mathbf{r}_{1,I} \\ \mathbf{r}_{1,Q} \\ \mathbf{r}_{2,I} \\ \mathbf{r}_{2,Q} \end{bmatrix} = \sqrt{E_b'} \underbrace{\begin{bmatrix} \alpha_1 \cos\theta_1 & -\alpha_1 \sin\theta_1 & \alpha_2 \cos\theta_2 & -\alpha_2 \sin\theta_2 \\ \alpha_1 \sin\theta_1 & \alpha_1 \cos\theta_1 & \alpha_2 \sin\theta_2 & \alpha_2 \cos\theta_2 \\ \alpha_2 \cos\theta_2 & -\alpha_2 \sin\theta_2 & -\alpha_1 \cos\theta_1 & \alpha_1 \sin\theta_1 \\ \alpha_2 \sin\theta_2 & \alpha_2 \cos\theta_2 & -\alpha_1 \sin\theta_1 & -\alpha_1 \cos\theta_1 \end{bmatrix}}_{\text{“transmission matrix” } B} \begin{bmatrix} b_k^{(I)} \\ b_k^{(Q)} \\ b_{k+1}^{(I)} \\ b_{k+1}^{(Q)} \end{bmatrix} + \begin{bmatrix} \mathbf{w}_{1,I} \\ \mathbf{w}_{1,Q} \\ \mathbf{w}_{2,I} \\ \mathbf{w}_{2,Q} \end{bmatrix}. \tag{P11.7}$$

(c) Premultiply both sides of (P11.7) with $B^\top$ to obtain the following new set of the sufficient statistics:

$$\begin{cases} \hat{\mathbf{r}}_{1,I} = (\alpha_1^2 + \alpha_2^2) b_k^{(I)} + [2\alpha_1\alpha_2 \sin(\theta_1 - \theta_2)] b_{k+1}^{(Q)} + \hat{\mathbf{w}}_{1,I} \\ \hat{\mathbf{r}}_{1,Q} = (\alpha_1^2 + \alpha_2^2) b_k^{(Q)} + [-2\alpha_1\alpha_2 \sin(\theta_1 - \theta_2)] b_{k+1}^{(I)} + \hat{\mathbf{w}}_{1,Q} \\ \hat{\mathbf{r}}_{2,I} = [-2\alpha_1\alpha_2 \sin(\theta_1 - \theta_2)] b_k^{(Q)} + (\alpha_1^2 + \alpha_2^2) b_{k+1}^{(I)} + \hat{\mathbf{w}}_{2,I} \\ \hat{\mathbf{r}}_{2,Q} = [2\alpha_1\alpha_2 \sin(\theta_1 - \theta_2)] b_k^{(I)} + (\alpha_1^2 + \alpha_2^2) b_{k+1}^{(Q)} + \hat{\mathbf{w}}_{2,Q} \end{cases}. \tag{P11.8}$$

Find the statistical properties of the noise components, i.e., their mean values, variances, and crosscorrelations.

(d) Can the detection of the four information bits be decoupled? What do you think the diversity order of this scheme is?

References

[1] G. Ungerboeck, “Channel coding with multilevel/phase signals,” *IEEE Transactions on Information Theory*, vol. IT-28, pp. 55–67, Jan. 1982.

[2] G. Ungerboeck, “Trellis-coded modulation with redundant signal sets. Part I: Introduction,” *IEEE Communications Magazine*, vol. 25, pp. 5–11, Feb. 1987.

[3] G. Ungerboeck, “Trellis-coded modulation with redundant signal sets. Part II: State of the art,” *IEEE Communications Magazine*, vol. 25, pp. 12–21, Feb. 1987.

[4] S. M. Alamouti, "A simple transmit diversity technique for wireless communications," *IEEE Journal on Selected Areas in Communications*, vol. 16, pp. 1451–1458, Oct. 1998.

[5] V. Tarokh, N. Seshadri, and A. R. Calderbank, "Space-time codes for high data rate wireless communication: Performance criterion and code construction," *IEEE Transactions on Information Theory*, vol. 44, pp. 744–765, Mar. 1998.

[6] R. E. Blahut, *Digital Transmission of Information*. Addison-Wesley, 1990.

[7] J. B. Anderson and A. Svensson, *Coded Modulation Systems*. Springer, 2003.

[8] L. F. Wei, "Rotationally invariant convolutional channel coding with expanded signal space. Part I: 180°," *IEEE Journal on Selected Areas in Communications*, vol. SAC-2, pp. 659–671, Sept. 1984.

[9] L. F. Wei, "Rotationally invariant convolutional channel coding with expanded signal space. Part II: Nonlinear codes," *IEEE Journal on Selected Areas in Communications*, vol. SAC-2, pp. 672–686, Sept. 1984.

[10] S. W. Golomb, *Shift Register Sequences*. Holden-Day, 1967.

[11] V. Tarokh, H. Jafarkhani, and A. R. Calderbank, "Space-time block codes from orthogonal designs," *IEEE Transactions on Information Theory*, vol. 45, pp. 1456–1467, July 1999.

12 Synchronization

12.1 Introduction

A successful communication system must establish synchronization, in addition to utilizing the modulation and demodulation techniques discussed so far. Synchronization is required at several levels. At the physical-layer level the receiver needs to know or estimate three parameters: (i) the incoming carrier frequency, f_c (hertz); (ii) for coherent demodulation any phase shift or phase drift, $\theta(t)$ (radians), introduced during transmission; (iii) the bit (symbol) timing, i.e., where on the time axis do the kT_b (or kT_s) (seconds) ticks occur. How to obtain estimates of these parameters is the subject of this chapter.

The reader should realize, however, that one needs to establish other levels of synchronization. After detection of the transmitted bit sequence the sequence needs to be segmented or parsed into "words." The best example of this is perhaps voice where the bit sequence needs to be segmented typically into eight-bit words, each word representing a voice sample. If error coding has been used, the sequence needs to be parsed properly into codewords for error decoding. Another example occurs in time-division multiple access where the communication channel is time shared. In this case the time slots need to be properly segmented to route the information from the different users properly. Such synchronization is typically called *frame synchronization*.

Frequency, phase, symbol, and frame synchronization are done at the receiver. In a mobile cellular environment where two (or more) base stations may be involved in transmitting to (or receiving from) a mobile receiver, the transmitters need to be synchronized for satisfactory operation. Such synchronization is usually called *network synchronization*. Frame and network synchronization are normally established by insertion of special characters (special bit sequences) and by protocol. As such they are done at the network-layer level.

As mentioned, this chapter is concerned only with the fundamental level of synchronization. Many circuits and techniques are available for this synchronization. Increasingly, they are implemented by digital signal processing. However, arguably, the concepts involved are well illustrated by the discussion of the analysis and design of two basic circuits: the *phase-locked loop* (PLL) for f_c and $\theta(t)$ estimation and the *early–late gate synchronizer* for symbol timing.

But before presenting the two circuits it is of interest to obtain a feel for the effect of improper phase or symbol timing on the system's error performance. Consider the effect

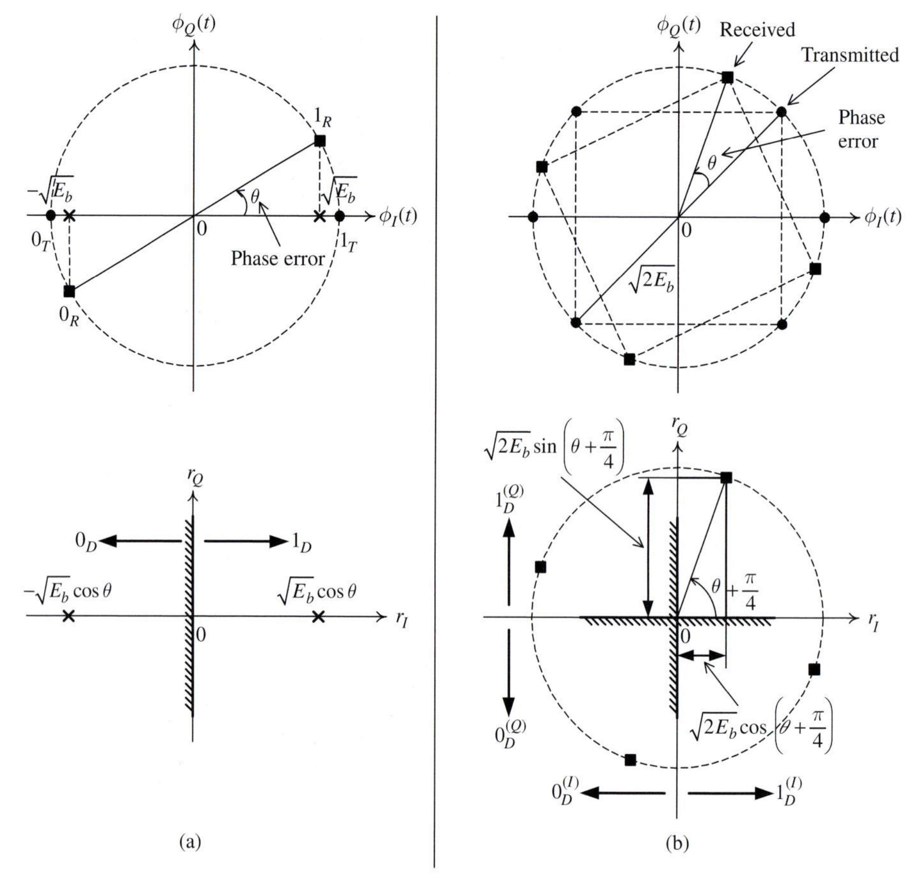

Fig. 12.1 Effect of a phase error on (a) BPSK and (b) QPSK signal constellations and received sufficient statistics.

of a phase error on the performance of BPSK and QPSK modulation. Figure 12.1 shows the transmitted and received signal constellations along with the effect on the received sufficient statistics.

It is quite straightforward to establish that (see Problems 7.8 and 7.9) the bit error probabilities in AWGN are:

$$\text{BPSK:}\quad P\,[\text{bit error}] = Q\left(\sqrt{\frac{2E_b}{N_0}}\cos\theta\right), \tag{12.1}$$

$$\begin{aligned}\text{QPSK:}\quad P\,[\text{bit error}] &= \frac{1}{2}Q\left(\sqrt{\frac{2E_b}{N_0}}\,(\cos\theta - \sin\theta)\right)\\ &\quad + \frac{1}{2}Q\left(\sqrt{\frac{2E_b}{N_0}}\,(\cos\theta + \sin\theta)\right),\end{aligned} \tag{12.2}$$

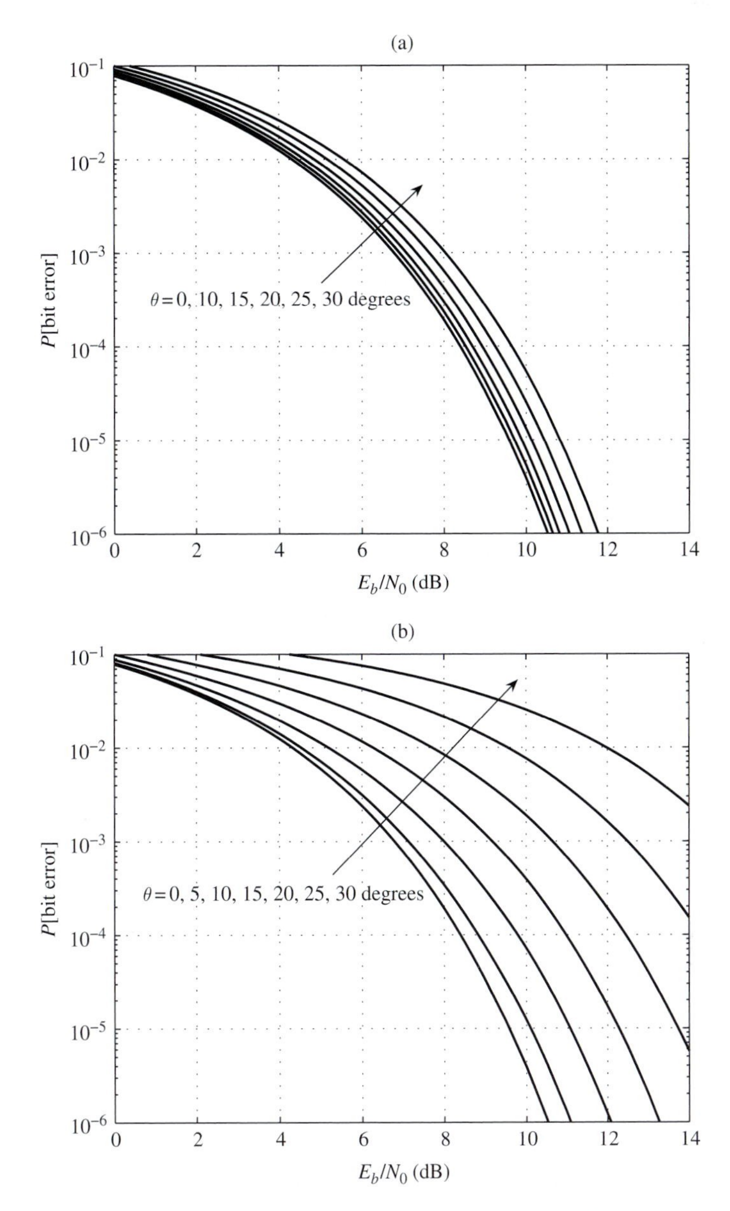

Fig. 12.2 Effect of a phase error on the bit error probabilities of (a) BPSK and (b) QPSK in AWGN.

where, as usual, the bits are assumed to be equally probable and Gray mapping is used for QPSK. Plots of these bit error probabilities are shown in Figure 12.2 for various values of phase error. They show that as long as the phase error is less than ± 5 degrees the error performance is relatively insensitive to the actual phase. Further, as intuitively expected, QPSK is more sensitive to phase error than BPSK.

To illustrate error performance when symbol timing error is encountered, consider NRZ-L modulation.[1] Figure 12.3 shows the appropriate model for the demodulator.

The sufficient statistic $\mathbf{r}$ depends on whether two consecutive bits agree or disagree. In particular, $\mathbf{r} = \pm\sqrt{E_b} + \mathbf{w}$ when they agree and $\mathbf{r} = \pm\sqrt{E_b}(1 - 2\epsilon) + \mathbf{w}$ if they disagree. The bit error probability is then given by

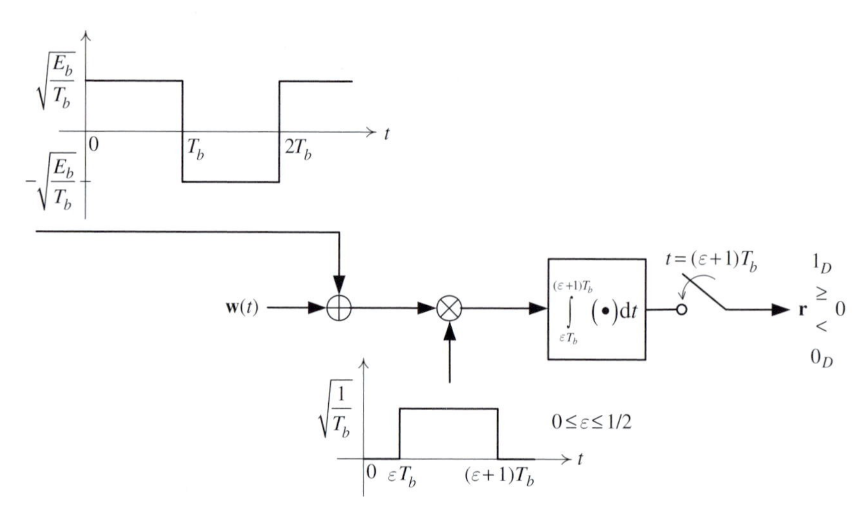

Fig. 12.3 Model to study the effect of symbol timing error on the demodulation of NRZ-L.

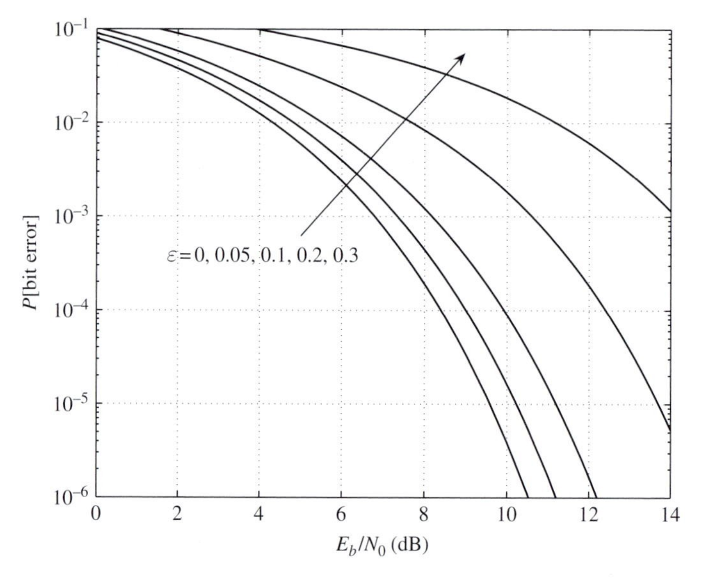

Fig. 12.4 Effect of the symbol timing error on the bit error probability of NRZ-L in AWGN.

[1] Note that this analysis is applicable for BPSK or QPSK, where we first shift the received signal to baseband and then demodulate the resulting baseband NRZ-L signal.

$$P\,[\text{bit error}] = \frac{1}{2}Q\left(\sqrt{\frac{2E_b}{N_0}}\right) + \frac{1}{2}Q\left(\sqrt{\frac{2E_b}{N_0}(1-2\epsilon)}\right). \tag{12.3}$$

A plot of error performance for various value of ϵ is shown in Figure 12.4.

Phase and symbol synchronization are similar in that both need an accurate time reference for synchronization, i.e., the $t = 0$ point. However, there are typically a very large number of carrier (f_c) cycles per symbol (T_b or T_c) period,[2] which means that symbol synchronization is a more "coarse" synchronization. The two types of synchronization are therefore done with different circuitry. Phase synchronization is done with some version of a PLL circuit and this is discussed next.

12.2 Phase offset and carrier frequency estimation

12.2.1 Phase-locked loop (PLL)

As the name suggests, the PLL is a circuit that locks onto the frequency[3] ω_c (radians/second) of a received sinusoid, $V\cos(\omega_c t + \theta)$ and estimates the phase offset, θ (radians). Loop implies a feedback circuit and this is precisely what a PLL is. It is a feedback control system whose function is to track the frequency and phase of an input sinusoid. Thus even if ω_c and θ change (slowly) over time the feedback circuit can adjust. Figure 12.5 shows a generic block diagram of a PLL. In practice, there are many different possible implementations of a PLL. To gain an understanding of the PLL we follow what could be considered the classical approach. Hopefully, this will give the reader a good grasp of the concepts underlying PLL analysis and design.

Many of the differences between PLL circuits involve the phase detector block and the loop filter block. Figure 12.6 shows what is called a sinusoid phase detector, while

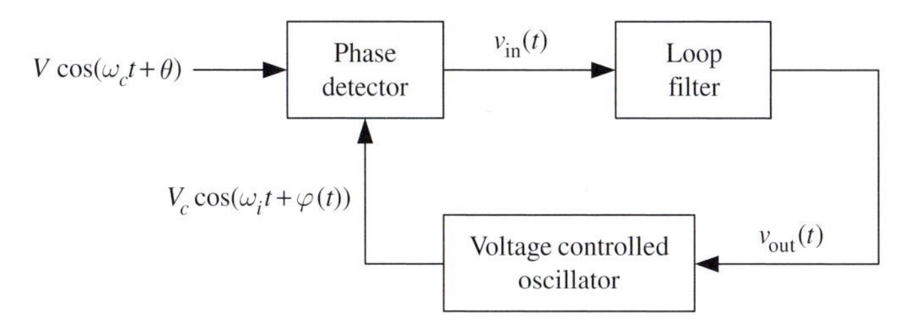

Fig. 12.5 Generic block diagram of a PLL.

[2] As a typical example take mobile voice communication: f_c is on order of gigahertz, hence $T_c = 1/f_c$ is on order of nanoseconds. On the other hand, T_c/T_b is on the order of kilohertz, which means that T_b is on the order of milliseconds, or T_b/T_c is on the order of 10^6.

[3] Though throughout the text we have thus far used f hertz or cycles/second as the frequency variable, with the PLL it is much more convenient to use ω (radians/second) as the frequency variable. This is because we are dealing with phase and the natural unit for phase is radians.

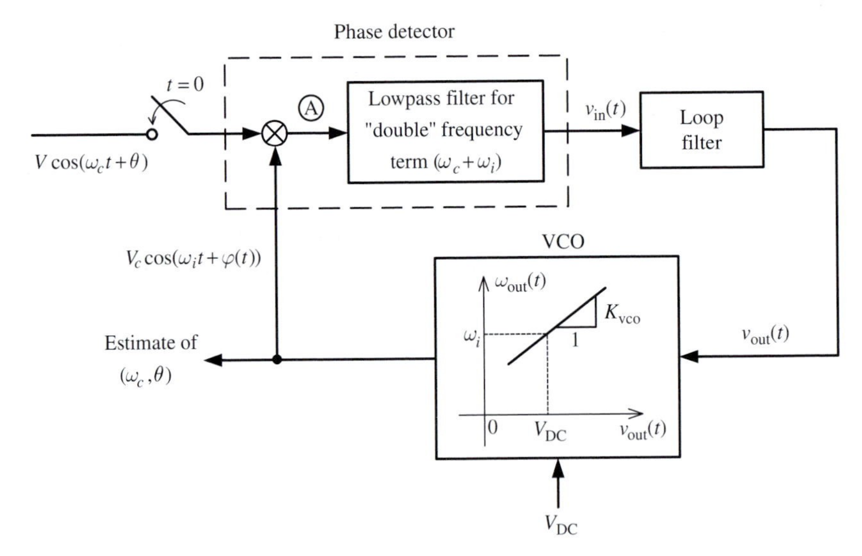

Fig. 12.6 A sinusoid phase detector. Here V_{DC} sets the bias or operating point of the voltage controlled oscillator (VCO), which determines the "free-running" frequency, ω_i, of the VCO, i.e., the frequency when $v_{out}(t)=0$.

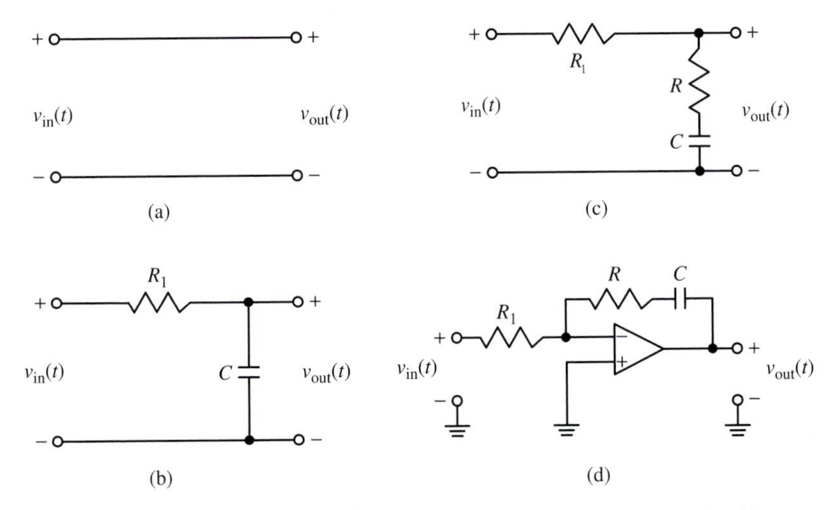

Fig. 12.7 Four possible loop filters: (a) allpass filter, (b) lowpass filter, (c) lead–lag filter, (d) active filter.

Figure 12.7 presents four possible loop filters. Because of the multiplier one would suspect that the circuit is a nonlinear one and so it is. The implication of this is that analysis of the circuit is difficult since transform methods (Laplace or Fourier) which proved to be so useful and powerful for linear circuits are no longer applicable. One must resort to analysis in the time domain.

Consider the simplest case where the loop filter is that of Figure 12.7(a), a length of copper wire which one could call an allpass filter. The governing equation for the phase error, which we now derive, is a nonlinear, first-order differential equation. The PLL is therefore called a first-order PLL. The signal at point (A) of the loop in Figure 12.6 is

$$\frac{K_m V V_c}{2} \left\{ \cos\left[(\omega_c + \omega_i) + \theta + \varphi(t) \right] + \cos\left[(\omega_c - \omega_i)t + \theta - \varphi(t) \right] \right\}, \tag{12.4}$$

where K_m is the multiplier constant, ω_i is the nominal frequency of the VCO, reasonably close to ω_c, and $\varphi(t)$ is a slow phase variation in the output of the VCO. The spectrum of the first term lies around $\omega_c + \omega_i \approx 2\omega_c$ and is filtered out. Therefore

$$v_{\text{in}}(t) = \frac{K_m V V_c}{2} \cos\left[(\omega_c - \omega_i)\, t + \theta - \varphi\,(t) \right] = \frac{K_m V V_c}{2} \cos \psi\,(t), \tag{12.5}$$

where $\psi(t) \equiv (\omega_c - \omega_i)t + \theta - \varphi(t) = (\omega_c t + \theta) - (\omega_i t + \varphi(t))$ is the *instantaneous phase error*. To obtain the differential equation for $\psi(t)$, note that for the chosen loop filter $v_{\text{out}}(t) = v_{\text{in}}(t)$. Further the VCO output is

$$\begin{aligned} \omega_{\text{out}}(t) &= \omega_i + K_{\text{vco}} v_{\text{out}}(t) = \omega_i + K_{\text{vco}} v_{\text{in}}(t) \\ &= \omega_i + \frac{K_{\text{vco}} K_m V V_c}{2} \cos \psi(t) \\ &= \omega_i + K_{\text{loop}} \cos \psi(t), \end{aligned} \tag{12.6}$$

where $K_{\text{loop}} \equiv K_{\text{vco}} K_m V V_c / 2$ is called the *loop gain*. Note that the loop gain depends on the signal levels V, V_c, a manifestation of the nonlinearity of the feedback loop. Since $\omega_{\text{out}}(t)$ is the instantaneous frequency of the VCO output, one has

$$\omega_{\text{out}}(t) = \frac{\mathrm{d}}{\mathrm{d}t}\left[\omega_i t + \varphi(t) \right] = \omega_i + \frac{\mathrm{d}\varphi(t)}{\mathrm{d}t} = \omega_i + K_{\text{loop}} \cos \psi(t). \tag{12.7}$$

It then follows that

$$\frac{\mathrm{d}\varphi(t)}{\mathrm{d}t} = K_{\text{loop}} \cos \psi(t). \tag{12.8}$$

To obtain the differential equation in $\psi(t)$, let $\Delta\omega \equiv (\omega_c - \omega_i)$, the difference in the angular frequencies[4] at $t = 0$. Therefore $\psi(t) = \Delta\omega t + \theta - \varphi(t)$. Write the left-hand side of (12.8) as

$$\frac{\mathrm{d}}{\mathrm{d}t}\left[(\varphi(t) - \theta - \Delta\omega t) + \Delta\omega t \right] = \frac{\mathrm{d}}{\mathrm{d}t}\left[-\psi(t) + \Delta\omega t \right] = -\frac{\mathrm{d}\psi(t)}{\mathrm{d}t} + \Delta\omega. \tag{12.9}$$

One then obtains the following differential equation governing the phase error, $\psi(t)$:

$$\frac{\mathrm{d}\psi(t)}{\mathrm{d}t} + K_{\text{loop}} \cos \psi(t) = \Delta\omega, \quad t \geq 0. \tag{12.10}$$

Though nonlinear, the differential equation (12.10) can be solved explicitly for $\psi(t)$. However, insight into the PLL performance can be obtained by a graphical technique used in nonlinear control system analysis, called the *phase plane* method. The phase plane is a plot of the derivative of a variable versus the variable, in this case $\mathrm{d}\psi(t)/\mathrm{d}t$ versus $\psi(t)$. This plot is shown in Figure 12.8.

[4] As time progresses the VCO frequency, ω_i, will change (hopefully) so that it becomes ω_c.

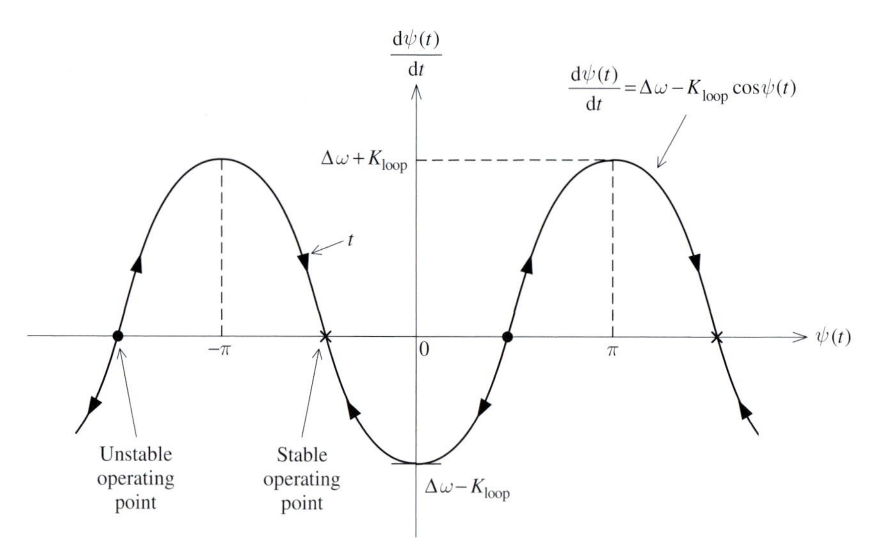

Fig. 12.8 Phase plane plot of Equation (12.10).

From the phase plane plot one can make the following observations:

(1) The phase error, $\psi(t)$, at all times has to be a point on the trajectory. Where one starts depends on $\psi(0)$, the initial condition. But since it is "inside" a sinusoid the phase value is only relevant modulo 2π.

(2) The intersection points represent equilibrium or solution points in the steady state, i.e., as $t \to \infty$ and the transient response has died out. However, some are stable and others are unstable. Note that $\mathrm{d}\psi(t)/\mathrm{d}t > 0$ implies that $\psi(t)$ increases with time, while $\mathrm{d}\psi(t)/\mathrm{d}t < 0$ means $\psi(t)$ decreases. From this one arrives at the stable and unstable operating points as shown.

(3) The stable operating points are given by

$$\psi(t)\Big|_{t\to\infty} = -\cos^{-1}\left(\frac{\Delta\omega}{K_{\text{loop}}}\right) + 2k\pi, \quad k = 0, \pm 1, \pm 2, \ldots. \tag{12.11}$$

(4) Consider the stable operating point of $-\cos^{-1}\left(\Delta\omega/K_{\text{loop}}\right)$. Since $\psi(t) = (\omega_c - \omega_i)t + \theta - \varphi(t)$ approaches a constant as $t \to \infty$, we conclude that the angular frequency ω_i approaches ω_c as $t \to \infty$, i.e., the loop locks onto the incoming frequency. The phase $\varphi(t)$, however, tends to $\left[\theta + \cos^{-1}\left(\Delta\omega/K_{\text{loop}}\right)\right]$. Therefore there is a phase error of magnitude, $\left|\cos^{-1}\left(\Delta\omega/K_{\text{loop}}\right)\right|$.

(5) The stable operating point and achievement of lock occurs only if $-K_{\text{loop}} < \Delta\omega < K_{\text{loop}}$, i.e., the trajectory must intersect the ψ axis. Otherwise the phase error either keeps increasing if $\Delta\omega > K_{\text{loop}}$, or keeps decreasing if $\Delta\omega < -K_{\text{loop}}$, with time.

(6) Observe that if $\Delta\omega = 0$ at $t = 0$, i.e., initially the incoming frequency and VCO frequency are the same, the steady-state phase error is $-\pi/2$, i.e., the VCO signal is in quadrature with the incoming signal.

In general the phase plane can usually provide information about the stable and unstable equilibrium points, the lock range, and the steady-state phase error of a PLL. What it does not provide is the transient performance of the PLL, i.e., how the phase error behaves as a function of time. Of particular interest is the acquisition time of the PLL, i.e., how long does it take (practically) to lock onto the incoming frequency. To obtain this information one must solve the differential equation, either numerically, by simulation, or by judicious approximation methods. The nonlinear differential equation of (12.10), however, can be solved explicitly.

Rewrite (12.10) as

$$\int_{\psi(t=0)}^{\psi(t)} \frac{\mathrm{d}\psi}{\Delta\omega - K_{\text{loop}}\cos\psi} = \int_{t=0}^{t} \mathrm{d}t = t. \tag{12.12}$$

The LHS can be integrated (see [1, p. 148, Eqn. 2.553.3]) to yield

$$\left[\frac{1}{\sqrt{K_{\text{loop}}^2 - (\Delta\omega)^2}} \ln\left\{\frac{\sqrt{K_{\text{loop}}^2 - (\Delta\omega)^2}\tan\frac{\psi}{2} + \Delta\omega - K_{\text{loop}}}{\sqrt{K_{\text{loop}}^2 - (\Delta\omega)^2}\tan\frac{\psi}{2} - \Delta\omega + K_{\text{loop}}}\right\}\right]_{\psi=\psi(0)}^{\psi=\psi(t)} = t, \tag{12.13}$$

where $|\Delta\omega| < K_{\text{loop}}$ (the lock-in region). Factoring out K_{loop}^2, defining $\omega_d \equiv \Delta\omega / K_{\text{loop}}$, we get

$$\ln\left\{\frac{\sqrt{1-\omega_d^2}\tan(\psi(t)/2) - (1-\omega_d)}{\sqrt{1-\omega_d^2}\tan(\psi(t)/2) + (1-\omega_d)}\right\} =$$

$$K_{\text{loop}}\left(\sqrt{1-\omega_d^2}\right)t + \ln\left\{\frac{\sqrt{1-\omega_d^2}\tan(\psi(0)/2) - (1-\omega_d)}{\sqrt{1-\omega_d^2}\tan(\psi(0)/2) + (1-\omega_d)}\right\}. \tag{12.14}$$

Without loss of generality we let $0 < \Delta\omega < K_{\text{loop}}$, which means $0 < \omega_d < 1$. Therefore the numerator of the LHS is less than the denominator of the LHS. Multiply both sides of the above equation by -1 and then take the exponential of both sides. Define parameters $a \equiv \sqrt{1-\omega_d^2}$; $b \equiv (1-\omega_d)$. The result is:

$$\frac{a\tan(\psi(t)/2) + b}{a\tan(\psi(t)/2) - b} = \left[\frac{a\tan(\psi(0)/2) + b}{a\tan(\psi(0)/2) - b}\right] \mathrm{e}^{-aK_{\text{loop}}t}, \quad t \geq 0. \tag{12.15}$$

From the above equation we infer that $\psi(t)$ approximately decays exponentially to its steady-state value with a time constant of $1/aK_{\text{loop}}$. Approximately because ψ is present only implicitly on the LHS of (12.15). Letting

$$A \equiv \left[\frac{a\tan\frac{\psi(0)}{2} + b}{a\tan\frac{\psi(0)}{2} - b}\right],$$

the explicit expression for $\psi(t)$ is

$$\psi(t) = 2\tan^{-1}\left\{\frac{-b\left(1 + Ae^{-a K_{\text{loop}} t}\right)}{a\left(1 - Ae^{-a K_{\text{loop}} t}\right)}\right\}. \tag{12.16}$$

Figure 12.9 shows plots of $\psi(t)$ for various values of ω_d and $\psi(0)$. It confirms that the decay is reasonably exponential and that steady state is reached within $3/aK_{\text{loop}}$ to $5/aK_{\text{loop}}$ seconds. Note that this holds well as long as $\omega_d \ll 1$.

12.2.2 Equivalent model of a sinusoidal PLL

In practice the loop filter is chosen so that the PLL becomes a second-order PLL, meaning that the differential equation for the phase error is a second-order, nonlinear, differential equation. The loop filter in this case is chosen from one of those shown in Figure 12.7 (b), (c), or (d).[5] A second-order PLL extends the lock range, has better performance in noise, and can achieve a steady-state phase error of zero, i.e., not only is frequency lock achieved but phase lock as well. We now develop a somewhat more general model for the sinusoidal PLL. It could be termed the equivalent baseband model or the incremental model because the frequency ω_c is suppressed. We start with the PLL as shown in Figure 12.10(a). The time reference for the incoming signal and the VCO output signal, based on observation (6) in the previous section, is chosen so that the two signals are in quadrature, i.e., one is a sine, the other a cosine. Furthermore, the VCO output is assumed to have an angular

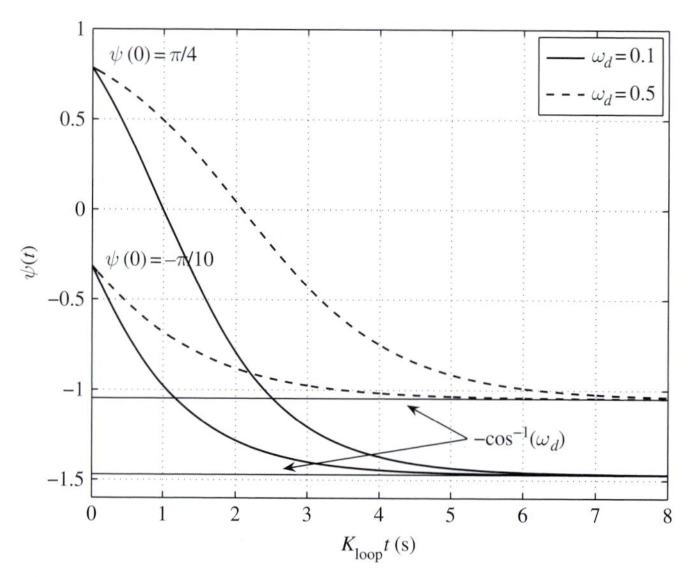

Fig. 12.9 Plots of $\psi(t)$ for different values of ω_d and $\psi(0)$.

[5] The loop filters are first-order filters but together with the VCO the PLL becomes second-order.

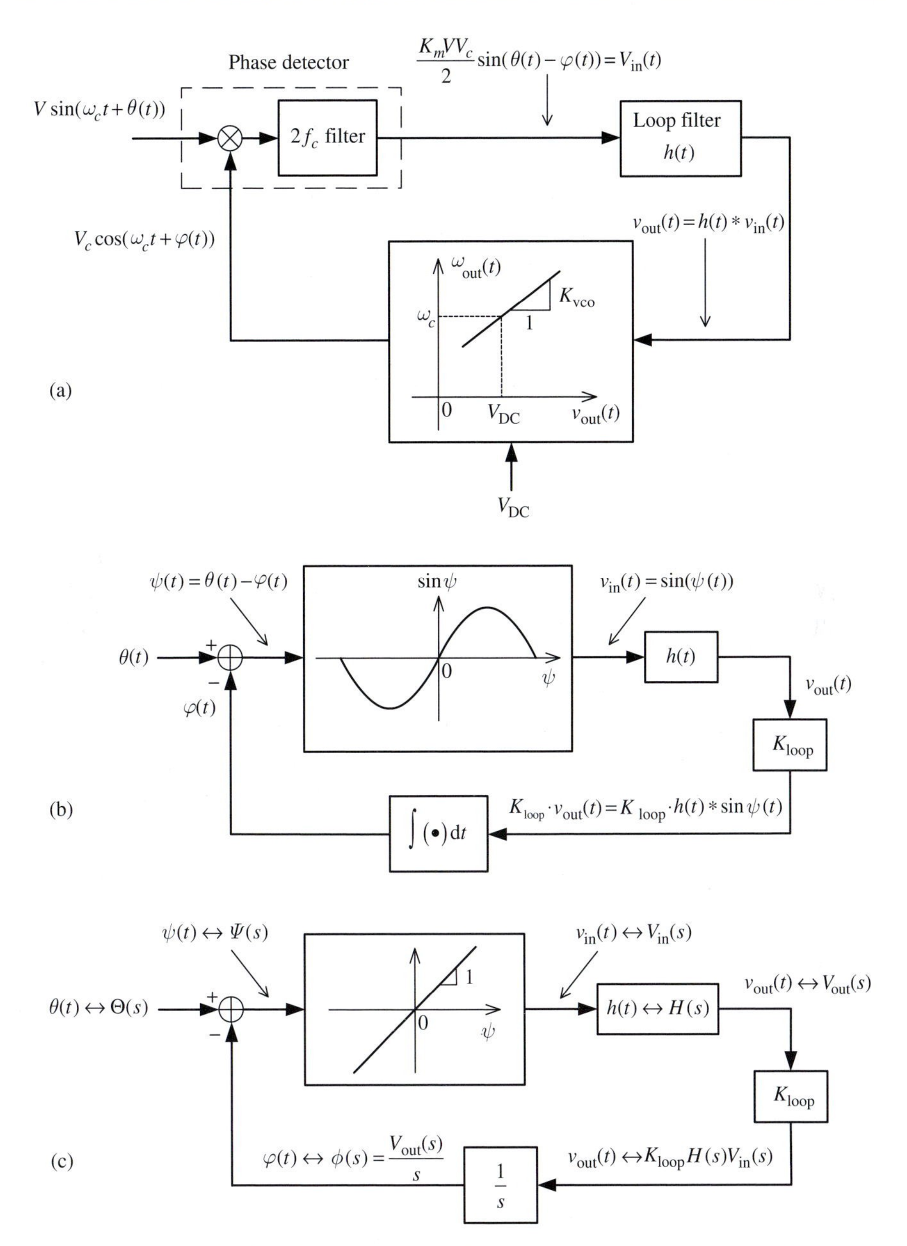

Fig. 12.10 PLL models: (a) sinusoidal PLL, (b) equivalent model of the sinusoidal PLL, (c) linear model of the sinusoidal PLL.

frequency of ω_c, any deviation from this, say $\Delta\omega t$, is subsumed in the instantaneous phase, $\varphi(t)$, i.e., $\varphi(t) = \Delta\omega t + \varphi'(t)$. The loop filter is described by its impulse response $h(t)$. Thus its output is given by $v_{out}(t) = h(t) * v_{in}(t)$. Note that though the overall loop is nonlinear the loop filter is linear and one can use convolution, a linear system operation, to describe its input–output relationship. As in the case with the first-order PLL we can establish the following relationships:

$$\frac{\mathrm{d}\varphi(t)}{\mathrm{d}t} = K_{\mathrm{loop}}h(t) * \sin(\theta(t) - \varphi(t)), \tag{12.17a}$$

$$\frac{\mathrm{d}\psi(t)}{\mathrm{d}t} + K_{\mathrm{loop}}h(t) * \sin\psi(t) = \frac{\mathrm{d}\theta(t)}{\mathrm{d}t}, \tag{12.17b}$$

where $\psi(t) \equiv \theta(t) - \varphi(t)$ is the phase error.

From Equation (12.17a) we have $\varphi(t) = K_{\mathrm{loop}} \int h(t) * \sin(\theta(t) - \varphi(t))\mathrm{d}t = K_{\mathrm{loop}} \int v_{\mathrm{out}}(t)\mathrm{d}t$. This leads to the model of the PLL shown in Figure 12.10(b). Note the sinusoidal nonlinearity (hence the name) which represents the phase detector. Other nonlinearities for the phase detector are possible, with different characteristics. Two such characteristics are shown in Figure 12.11. The operating signals for these are no longer sinusoidal but square waves. The reader is referred to the literature for more detail.

If the phase error is small enough, a situation that occurs once the PLL has locked onto the frequency of the incoming signal and now is tracking slow phase changes in it, then $\sin\psi \approx \psi$ ($\psi \ll 1$) and the PLL can be represented by a linear model as shown in Figure 12.10(c). One can now apply all the standard concepts and techniques from linear systems theory. However, before the PLL can be linearized it must achieve lock and for this one must deal with the nonlinearity. In the case of the first-order PLL we were able to achieve an explicit solution. We now consider the second-order PLL.

12.2.3 Second-order phase-locked loop dynamics

Once again we want to derive the nonlinear differential equation describing the behavior of the PLL's phase error. Consider as a specific example the filter of Figure 12.7(d). To obtain the differential equation relating the output, $v_{\mathrm{out}}(t)$, to the input, $v_{\mathrm{in}}(t)$, we can use transform methods (again only at this local level). Let the transfer

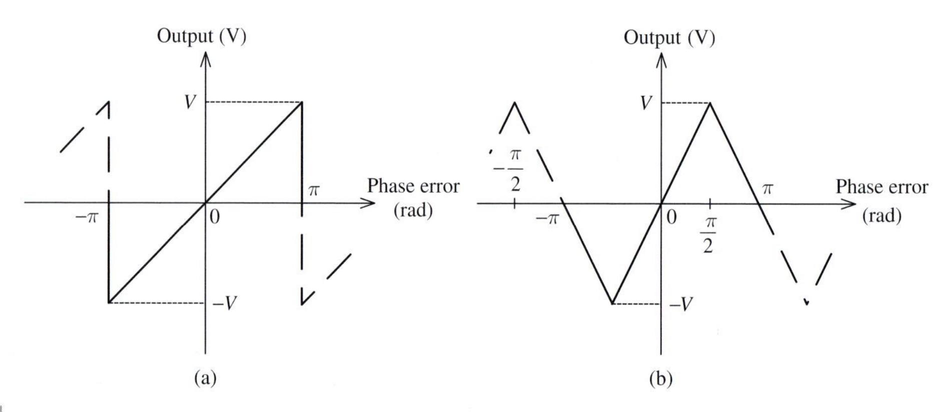

Fig. 12.11 Two commonly encountered phase characteristics: (a) sawtooth, and (b) triangular. Note these are encountered when the input signal and VCO output are (or are converted to) square waves. In this case the phase error is more appropriately thought of as a timing error.

function be[6] $H(s) = 1 + a/s$ with any possible gain (or attenuation) factor absorbed by K_{loop}. The filter output is $V_{\text{out}}(s) = ((s+a)/s)V_{\text{in}}(s)$ which in the time domain becomes $\mathrm{d}v_{\text{out}}(t)/\mathrm{d}t = av_{\text{in}}(t) + \mathrm{d}v_{\text{in}}(t)/\mathrm{d}t$. Since $\mathrm{d}\varphi(t)/\mathrm{d}t = K_{\text{loop}}v_{\text{out}}(t)$ and $v_{\text{in}}(t) = \sin\psi(t)$ (see Figure 12.10(b)), one has

$$\frac{\mathrm{d}^2\varphi(t)}{\mathrm{d}t^2} = K_{\text{loop}}\frac{\mathrm{d}v_{\text{out}}(t)}{\mathrm{d}t} = aK_{\text{loop}}\sin\psi(t) + K_{\text{loop}}\cos\psi(t)\frac{\mathrm{d}\psi(t)}{\mathrm{d}t}. \tag{12.18}$$

Again $\mathrm{d}^2\varphi(t)/\mathrm{d}t^2 = (\mathrm{d}^2/\mathrm{d}t^2)\left[\varphi(t) - \theta(t) + \theta(t)\right] = -\left(\mathrm{d}^2\psi(t)/\mathrm{d}t^2\right) + \left(\mathrm{d}^2\theta(t)/\mathrm{d}t^2\right)$ which means the differential equation for the phase error is

$$\frac{\mathrm{d}^2\psi(t)}{\mathrm{d}t^2} + K_{\text{loop}}\cos\psi(t)\frac{\mathrm{d}\psi(t)}{\mathrm{d}t} + aK_{\text{loop}}\sin\psi(t) = \frac{\mathrm{d}^2\theta(t)}{\mathrm{d}t^2}. \tag{12.19}$$

Consider now a constant frequency input, i.e., $\theta(t) = \Delta\omega t$. Equation (12.19) becomes

$$\frac{\mathrm{d}^2\psi(t)}{\mathrm{d}t^2} + K_{\text{loop}}\cos\psi(t)\frac{\mathrm{d}\psi(t)}{\mathrm{d}t} + aK_{\text{loop}}\sin\psi(t) = 0. \tag{12.20}$$

Unfortunately, no solution is available for (12.20), so one must resort to simulation, approximations, numerical techniques, or the phase plane method to obtain insight into the PLL behavior.

We shall obtain the phase plane plot for (12.20) but first make the observation that when the transients die out, which means $\mathrm{d}^2\psi(t)/\mathrm{d}t^2 \to 0$ and $\mathrm{d}\psi(t)/\mathrm{d}t \to 0$, then we have $\sin\psi(t) = 0$. This implies that $\psi(t) = 0$ (as $t \to \infty$) which in turn implies that (i) there is no steady-state phase error, i.e., the PLL is locked in both frequency and phase, (ii) the *lock-in range* is theoretically infinite, i.e., for $-\infty < \Delta\omega < \infty$ the system achieves lock. All of this is true provided the system is stable.[7] A theoretical justification of the lock range will be given shortly but first we derive the phase plane plot.

To this end, we first normalize the time axis by letting $\tau \equiv K_{\text{loop}}t$ to eliminate one of the parameters. Equation (12.20) becomes

$$K_{\text{loop}}^2\left[\frac{\mathrm{d}^2\psi(\tau)}{\mathrm{d}\tau^2} + \cos\psi(\tau)\frac{\mathrm{d}\psi(\tau)}{\mathrm{d}\tau} + \frac{a}{K_{\text{loop}}}\sin\psi(\tau)\right] = 0 \tag{12.21}$$

and upon defining $a' \equiv a/K_{\text{loop}}$ we have

$$\frac{\mathrm{d}^2\psi(\tau)}{\mathrm{d}\tau^2} + \cos\psi(\tau)\frac{\mathrm{d}\psi(\tau)}{\mathrm{d}\tau} + a'\sin\psi(\tau) = 0. \tag{12.22}$$

[6] Note that the Laplace transform, which is a generalization of the Fourier transform, is used here. For readers unfamiliar with the Laplace transform simply substitute $s = \mathrm{j}\omega$. In the time domain $s = \mathrm{j}\omega$ corresponds to the differentiation operation, i.e., to $\mathrm{d}(\cdot)/\mathrm{d}t$, $s^2 = (\mathrm{j}\omega)^2 = -\omega^2$ to $\mathrm{d}^2(\cdot)/\mathrm{d}t^2$, and in general $s^n = (\mathrm{j}\omega)^n$ to $\mathrm{d}^n(\cdot)/\mathrm{d}t^n$.

[7] Traditionally stability in a control system is defined as bounded input–bounded output. Due to the sinusoidal nonlinearity this definition is not applicable. A more suitable definition for stability would be that the PLL is able to lock onto an incoming frequency.

Now $\mathrm{d}^2\psi(\tau)/\mathrm{d}\tau^2 = (\mathrm{d}/\mathrm{d}\tau)\left[\mathrm{d}\psi(\tau)/\mathrm{d}\tau\right]$, so if we divide (12.22) by $\mathrm{d}\psi(\tau)/\mathrm{d}\tau$, upon realizing that

$$\frac{\dfrac{\mathrm{d}}{\mathrm{d}\tau}\left[\dfrac{\mathrm{d}\psi(\tau)}{\mathrm{d}\tau}\right]}{\dfrac{\mathrm{d}\psi(\tau)}{\mathrm{d}\tau}} = \frac{\mathrm{d}\left[\dfrac{\mathrm{d}\psi(\tau)}{\mathrm{d}\tau}\right]}{\mathrm{d}\psi(\tau)}$$

we get

$$\frac{\mathrm{d}\dot{\psi}}{\mathrm{d}\psi} = -\cos\psi - a'\frac{\sin\psi}{\dot{\psi}}, \tag{12.23}$$

where $\dot{\psi} = \mathrm{d}\psi(\tau)/\mathrm{d}\tau$ (we have switched to Newton's notation).

The phase plane is a plot of $\dot{\psi}$ versus ψ and (12.23) gives us the slope of the trajectories at each point $(\dot{\psi}, \psi)$ in the phase plane. This can be used to obtain the phase plane trajectories. These trajectories are shown in Figure 12.12 for two different values of a'.

To show that the lock-in range is theoretically infinite, multiply (12.23) by $\dot{\psi}$ and integrate it between $-\pi$ and π:

$$\frac{1}{2}\left[\dot{\psi}^2(\pi) - \dot{\psi}^2(-\pi)\right] = -\int_{-\pi}^{\pi} \dot{\psi}\cos\psi\,\mathrm{d}\psi - a'\int_{-\pi}^{\pi}\sin\psi\,\mathrm{d}\psi. \tag{12.24}$$

The second term on the RHS is zero. Integrate the first term by parts (where the parts are $u = \dot{\psi} \rightarrow \mathrm{d}u = \mathrm{d}\dot{\psi}$ and $\mathrm{d}v = \cos\psi\,\mathrm{d}\psi = \mathrm{d}(\sin\psi)$) to obtain

$$\frac{1}{2}\left[\dot{\psi}^2(\pi) - \dot{\psi}^2(-\pi)\right] = \int_{-\pi}^{\pi}\sin\psi\,\mathrm{d}\dot{\psi}. \tag{12.25}$$

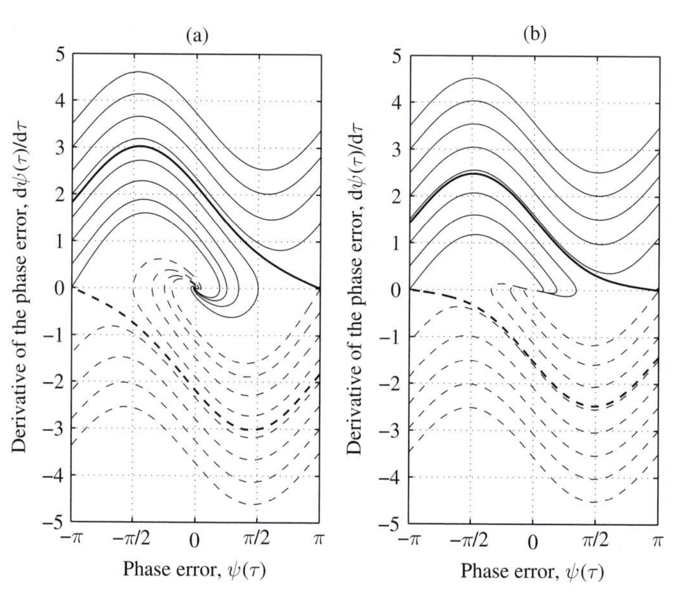

Fig. 12.12 Phase plane plot of (12.23): (a) $a' = 1/2$, (b) $a' = 1/8$.

But $d\dot{\psi} = \left[-\cos\psi - a'\sin\psi/\dot{\psi}\right]d\psi$. Substituting into (12.25) yields

$$\frac{1}{2}\left[\dot{\psi}^2(\pi) - \dot{\psi}^2(-\pi)\right] = \underbrace{-\int_{-\pi}^{\pi}\sin\psi\cos\psi\,d\psi}_{=0} - a'\int_{-\pi}^{\pi}\frac{\sin^2\psi}{\dot{\psi}}d\psi$$

$$= -a'\int_{-\pi}^{\pi}\frac{1-\cos(2\psi)}{\dot{\psi}}d\psi. \tag{12.26}$$

Therefore if $\dot{\psi} = d\psi(\tau)/d\tau$ is positive, the RHS of (12.26) is negative and if $\dot{\psi} < 0$ the RHS is positive. This, in turn, says that for any cycle of width 2π the value of $|\dot{\psi}|$ must decrease regardless of the initial value of $\dot{\psi}$, i.e., the lock range is infinite provided the integrator is perfect.

The phase plane method can be generated by the state-space approach.[8] In this approach a high-order differential equation, linear or nonlinear, is changed into a set of first-order differential equations which can then be solved numerically. Consider (12.22) and define the following two states $y = d\psi(\tau)/d\tau$ and $x = \psi(\tau)$. Then the equation is represented by the following two first-order differential equations:

$$\begin{aligned}\frac{dy}{d\tau} &= -y\cos x - a'\sin x,\\ \frac{dx}{d\tau} &= y.\end{aligned} \tag{12.27}$$

The above set can be solved numerically by a standard mathematical package such as `ode23` found in Matlab. Figure 12.13 shows several plots of $x = \psi(\tau)$ for different initial conditions.

The plots in Figure 12.12 show that there is a region in the $(\dot{\psi}, \psi)$ plane (the region inside the thicker trajectories) where the phase error converges to 0 without the phase exceeding the $[-\pi, \pi]$ range. Outside the region the PLL still converges to a stable point, but one that is a multiple of 2π. This phenomenon is known as *cycle skipping* (or slipping) and can be seen from Figure 12.13. The number of skipped cycles depends on the initial value of $d\psi(\tau)/d\tau$, i.e., $\dot{\psi}(0)$. The larger it is, the more cycles are (potentially) skipped. The phase plane plot is shown only in the range $[-\pi, \pi]$ since it is seen from (12.23) that the plot is periodic with period 2π. Note it is the plot that is periodic, not any individual trajectory. Finally, there are singular points at $(\dot{\psi} = 0, \psi = k2\pi)$ which are stable operating points and also singular points at $(\dot{\psi} = 0, \dot{\psi} = (2k+1)\pi)$. The latter, however, are unstable and are called *saddle points*. Figure 12.14 illustrates trajectory behavior in the region of a saddle point.

A similar analysis can be carried out for the other loop filters, though in detail the performances will differ. In particular the lock-in range will not be infinite as was the case for the perfect integrator. It will lie somewhere between ∞ and that of a first-order PLL. The

[8] The state-space approach is a generalization of the phase plane. It deals with differential equations of order greater than 1.

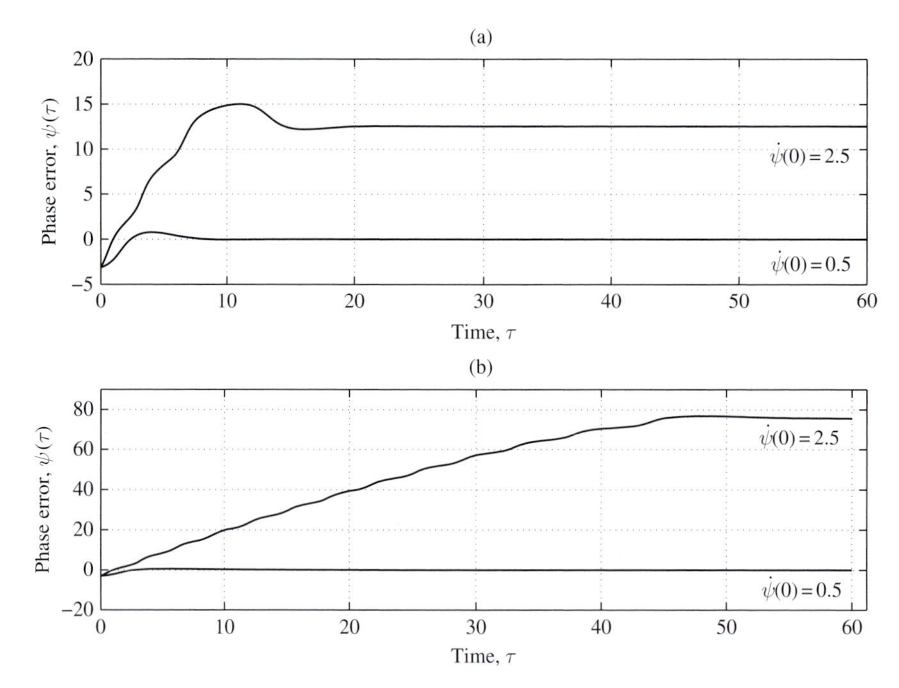

Fig. 12.13 Plots of $x = \psi(\tau)$ for different initial conditions: (a) $a' = 1/2$; (b) $a' = 1/8$.

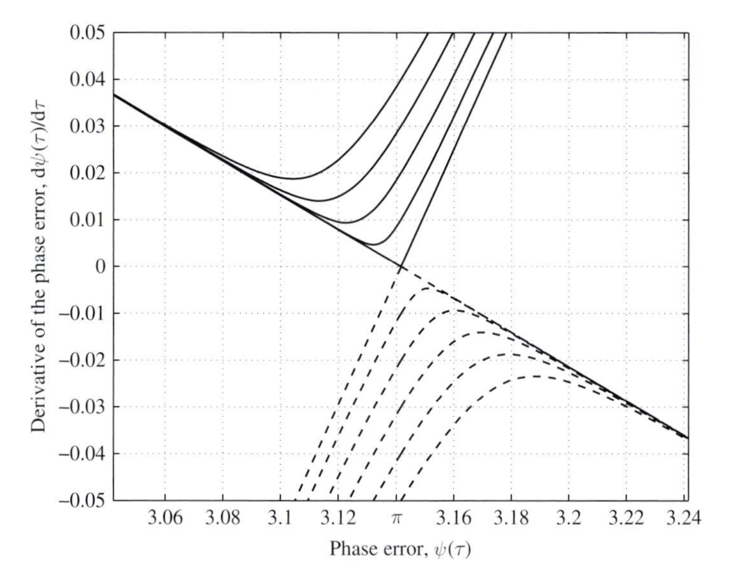

Fig. 12.14 Trajectory behavior in the region of a saddle point.

infinite lock-in range when there is a perfect integrator in the loop is somewhat intuitive. As long as there is a nonzero phase error $(\theta(t) - \varphi(t))$ and therefore a nonzero input to the integrator, it will produce an increasing (or decreasing) output which in turn drives the VCO into phase lock. Besides the lock-in frequency range, another parameter of interest is what is called the *tracking range*. This is the range of input frequencies that the PLL can stay locked onto after it has achieved frequency lock. In general, the tracking range is greater than or equal to the lock-in range. They are equal for the first-order PLL. Synonyms for the lock-in range and tracking range are pull-in and pull-out respectively.

Finally the operation of the PLL depends on the fact that the incoming signal has a spectral component (i.e., an impulse theoretically) at the carrier frequency, f_c. Modulations such as BPSK, QPSK, and M-QAM do not have one in general, i.e., they are suppressed carrier modulations. How to obtain a spectral component for these modulations is the subject of the next section.

12.3 Phase and carrier frequency acquisition for suppressed carrier modulation

Consider the simplest case of phase modulation, that of BPSK. The received signal can be written as

$$\mathbf{r}(t) = \underbrace{\left\{ \sum_{k=-\infty}^{\infty} b_k \left[u(kT_b) - u((k+1)T_b) \right] \right\}}_{I(t)} \times \sqrt{E_b}\sqrt{\frac{2}{T_b}} \cos(2\pi f_c t + \theta) + \mathbf{w}(t), \tag{12.28}$$

where $b_k = \pm 1$. The PSD of BPSK does not contain an impulse at $f = f_c$ and hence there is no spectral component for the PLL to lock on. However, we can create one by squaring $\mathbf{r}(t)$. Upon squaring one gets

$$\mathbf{r}^2(t) = I^2(t) E_b \left(\frac{2}{T_b} \right) \cos^2(2\pi f_c t + \theta) + \textbf{noise terms}. \tag{12.29}$$

But $I^2(t) = 1$ for all t and using the trigonometric identity $\cos^2 x = 1/2 + (1/2)\cos(2x)$, $\mathbf{r}^2(t)$ becomes

$$\mathbf{r}^2(t) = \frac{E_b}{T_b} + \cos(2\pi(2f_c)t + 2\theta) + \textbf{noise terms}. \tag{12.30}$$

It has a spectral component at $2f_c$ for the PLL to lock on (the DC component is easily filtered, the noise terms less so). A block diagram of the synchronizer is shown in Figure 12.15.

We now turn our attention to QPSK, which also has no spectral component at f_c. Since QPSK is conceptually two "interleaved" BPSK signals, intuitively squaring it should produce a spectral component at f_c. The received signal is

$$\mathbf{r}(t) = I(t)\sqrt{E_b}\sqrt{\frac{2}{T_s}} \cos(2\pi f_c t + \theta) + Q(t)\sqrt{E_b}\sqrt{\frac{2}{T_s}} \sin(2\pi f_c t + \theta) + \mathbf{w}(t), \tag{12.31}$$

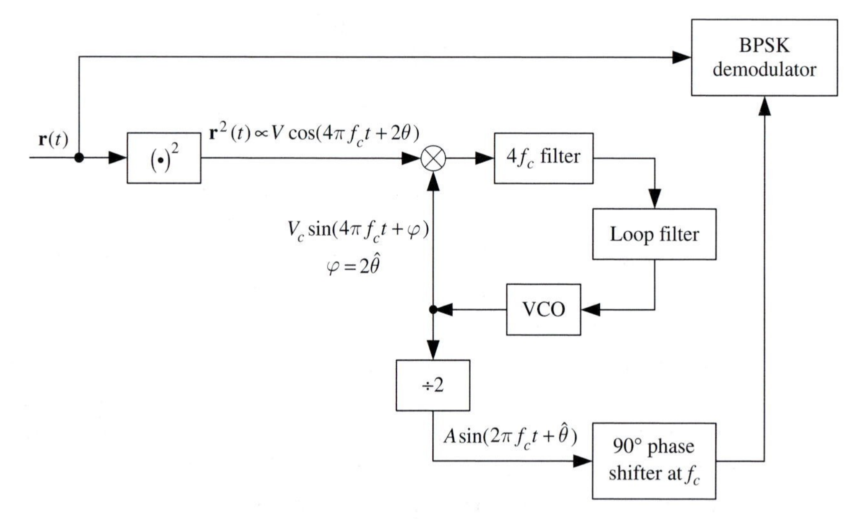

Fig. 12.15 Block diagram of the squaring (second-power) synchronizer for carrier recovery for BPSK.

where

$$I(t) \equiv \sum_{k=-\infty}^{\infty} b_k^{(I)} \left[u(kT_s) - u((k+1)T_s)\right], \tag{12.32}$$

$$Q(t) \equiv \sum_{k=-\infty}^{\infty} b_k^{(Q)} [u(kT_s) - u((k+1)T_s)], \tag{12.33}$$

$T_s = 2T_b$, and $\{b_k^{(I)}, b_k^{(Q)}\} = \pm 1$. Squaring it gives

$$\begin{aligned}\mathbf{r}^2(t) = {} & I^2(t)E_b\left(\frac{2}{T_s}\right)\cos^2(2\pi f_c t + \theta) + Q^2(t)E_b\left(\frac{2}{T_s}\right)\sin^2(2\pi f_c t + \theta) \\ & + 2I(t)Q(t)E_b\left(\frac{2}{T_s}\right)\cos(2\pi f_c t + \theta)\sin(2\pi f_c t + \theta) \\ & + \textbf{noise terms}.\end{aligned} \tag{12.34}$$

But $I^2(t) = Q^2(t) = 1$ and $\cos^2 x + \sin^2 x = 1$. Therefore

$$\mathbf{r}^2(t) = \frac{2E_b}{T_s} + \frac{2E_b}{T_s} I(t)Q(t)\sin(2\pi(2f_c)t + 2\theta) + \textbf{noise terms}. \tag{12.35}$$

Since $I(t)Q(t)$ can be treated as a zero-mean random process the term $I(t)Q(t)\sin(2\pi(2f_c)t + 2\theta)$ is a double sideband suppressed-carrier (DSB-SC) modulation, i.e., there is no spectral component at $2f_c$. To produce a spectral component, let us square $\mathbf{r}^2(t)$. Now

$$
\begin{aligned}
\mathbf{r}^4(t) = & \frac{4E_b^2}{T_s^2} + \frac{8E_b^2}{T_b} I(t)Q(t)\sin(2\pi(2f_c)t + 2\theta) \\
& + \frac{4E_b^2}{T_s^2} I^2(t)Q^2(t)\sin^2(2\pi(2f_c)t + 2\theta) \\
& + \textbf{even more noise terms}.
\end{aligned} \tag{12.36}
$$

The third term is $\left(2E_b^2/T_s^2\right)\left[1 - \cos(2\pi(4f_c)t + 4\theta)\right]$, which means that a spectral component at $4f_c$ is present, due to the term $-\left(2E_b^2/T_s^2\right)\cos(2\pi(4f_c)t + 4\theta)$. Figure 12.16 shows the synchronizer's block diagram.

BPSK requires a squarer or a second-power block to produce a spectral component (at $2f_c$), QPSK needs a fourth-power block to produce a spectral component (at $4f_c$). In general M-PSK needs an Mth-power block to produce a spectral component at Mf_c for the PLL circuit. We now consider QAM, specifically QAM with signals on a rectangular grid. Write the modulated signal part of the received signal as

$$
\begin{aligned}
s(t) = & \underbrace{\left\{ \sum_{k=-\infty}^{\infty} I_{i,k}\left[u(kT_s) - u\left((k+1)T_s\right)\right] \right\}}_{\equiv I(t)} \sqrt{\frac{2}{T_s}}\cos(2\pi f_c t + \theta) \\
& + \underbrace{\left\{ \sum_{k=-\infty}^{\infty} Q_{j,k}\left[u(kT_s) - u\left((k+1)T_s\right)\right] \right\}}_{\equiv Q(t)} \sqrt{\frac{2}{T_s}}\sin(2\pi f_c t + \theta),
\end{aligned} \tag{12.37}
$$

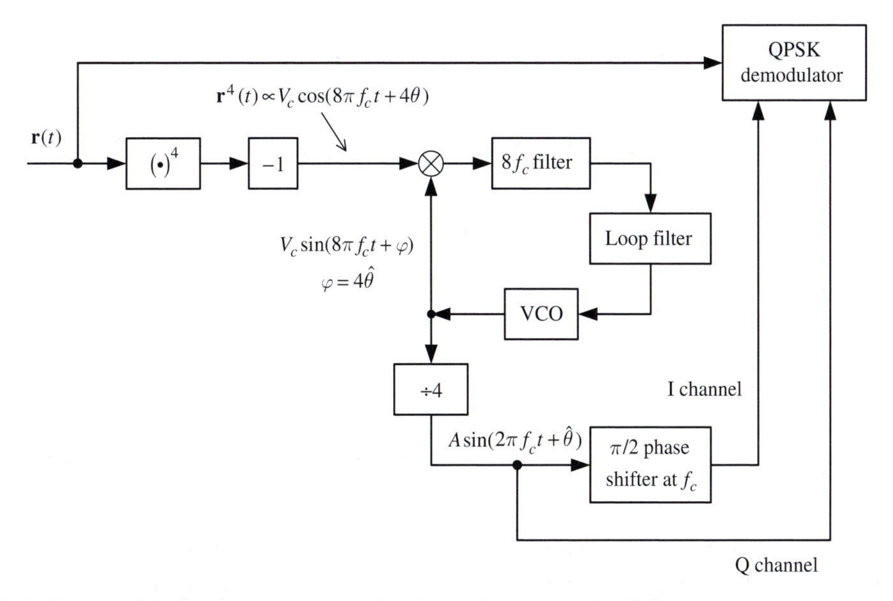

Fig. 12.16 Block diagram of the fourth-power synchronizer for carrier recovery for QPSK.

where $I_{i,k}$ is one of the amplitudes $(\pm\Delta, \pm 3\Delta, \ldots)$ on the inphase axis and $Q_{j,k}$ is one of the amplitudes $(\pm\Delta, \pm 3\Delta, \ldots)$ on the quadrature axis, $i = 1, 2, \ldots, 2^{\lambda_I}, j = 1, 2, \ldots, 2^{\lambda_Q}$.

We treat $I(t)$, $Q(t)$ as two random processes and note that $E\{\mathbf{I}(t)\} = E\{\mathbf{Q}(t)\} = 0$; $E\{\mathbf{I}(t)\mathbf{Q}(t)\} = E\{\mathbf{I}(t)\}E\{\mathbf{Q}(t)\} = 0$; $E\{\mathbf{I}^2(t)\} = \sigma_I^2$; $E\{\mathbf{Q}^2(t)\} = \sigma_Q^2$ where, as usual, the information bits are equally likely and statistically independent. This means that the amplitudes, $\mathbf{I}_{i,k}$, $\mathbf{Q}_{j,k}$ are equally likely and statistically independent.

The received signal is

$$\mathbf{r}(t) = \mathbf{I}(t)\sqrt{\frac{2}{T_s}}\cos(2\pi f_c t + \theta) + \mathbf{Q}(t)\sqrt{\frac{2}{T_s}}\sin(2\pi f_c t + \theta) + \mathbf{w}(t). \tag{12.38}$$

Upon squaring it one gets

$$\begin{aligned}\mathbf{r}^2(t) = {} & \frac{2}{T_s}\mathbf{I}^2(t)\cos^2(2\pi f_c t + \theta) + \frac{2}{T_s}\mathbf{Q}^2(t)\sin^2(2\pi f_c t + \theta) \\ & + \frac{4}{T_s}\mathbf{I}(t)\mathbf{Q}(t)\cos(2\pi f_c t + \theta)\sin(2\pi f_c t + \theta) + \textbf{noise terms}.\end{aligned} \tag{12.39}$$

If $\sigma_I^2 \neq \sigma_Q^2$ (as would be the case for rectangular 8-QAM or 32-QAM), then the mean value of squared output is

$$\begin{aligned}E\left\{\mathbf{r}^2(t)\right\} & = \frac{2}{T_s}\sigma_{\mathbf{I}}^2\sin^2(2\pi f_c t + \theta) + \frac{2}{T_s}\sigma_{\mathbf{Q}}^2\sin^2(2\pi f_c t + \theta) + A \\ & = \frac{{\sigma_I}^2 + {\sigma_Q}^2}{T_s} + \frac{{\sigma_I}^2 - {\sigma_Q}^2}{T_s}\cos\left[2\pi(2f_c)t + 2\theta + A\right]\end{aligned} \tag{12.40}$$

since $E\{\mathbf{I}(t)\mathbf{Q}(t)\} = 0$ and $E\{\textbf{noise terms}\} = A$. The above means that there is a spectral component at $2f_c$, provided $\sigma_I^2 \neq \sigma_Q^2$. To conclude, for *nonsymmetrical* QAM constellations squaring provides a spectral component and one can use the same synchronizer circuit as for BPSK.

Symmetrical QAM constellations, where $\sigma_I^2 = \sigma_Q^2 = \sigma^2$, however, need a fourth-power block to obtain a spectral component related to f_c. The output of this block is

$$\begin{aligned}\mathbf{r}^4(t) = {} & \frac{4}{T_s^2}\mathbf{I}^4(t)\cos^4(2\pi f_c t + \theta) + \frac{4}{T_s^2}\mathbf{Q}^4(t)\sin^4(2\pi f_c t + \theta) \\ & + \frac{24}{T_s^2}\mathbf{I}^2(t)\mathbf{Q}^2(t)\cos^2(2\pi f_c t + \theta)\sin^2(2\pi f_c t + \theta) \\ & + \frac{16}{T_s^2}\mathbf{I}^3(t)\mathbf{Q}(t)\cos^3(2\pi f_c t + \theta)\sin(2\pi f_c t + \theta) \\ & + \frac{16}{T_s^2}\mathbf{I}(t)\mathbf{Q}^3(t)\cos(2\pi f_c t + \theta)\sin^3(2\pi f_c t + \theta) \\ & + \textbf{more and more noise terms}.\end{aligned} \tag{12.41}$$

The mean value of $\mathbf{r}^4(t)$ is

$$E\left\{\mathbf{r}^4(t)\right\} = \frac{3}{T_s^2}\left[E\left\{\mathbf{I}^4(t)\right\} - \sigma^4\right] + \frac{1}{T_s^2}\left[E\left\{\mathbf{I}^4(t)\right\} - 3\sigma^4\right]\cos(2\pi(4f_c)t + 4\theta) + B \tag{12.42}$$

There is a spectral component at $4f_c$. Therefore the QPSK synchronizer block diagram can be used for symmetrical QAM.

A difficulty encountered in circuit implementation of the described synchronizers is the squaring or fourth-power block, especially at high frequencies. A circuit design that avoids this is one that is known as a *Costas loop*. This design is shown in block diagram form in Figure 12.17 for BPSK and in Figure 12.18 for QPSK. We describe its operation for BPSK. Ignore the AWGN and let $r(t) = a(t)\cos(2\pi f_c t + \theta)$. Initially the VCO output is a frequency that is close to the carrier frequency f_c. Any difference in frequency and initial phase difference is accounted for by the phase φ. The multipliers in the I and Q loops produce $2f_c$ terms which are filtered out and the two low-frequency terms show at the outputs of the lowpass filters. These two outputs are multiplied together and lowpass filtered to eliminate any amplitude fluctuations due to $a(t)$. The output of this lowpass filter is a phase error signal that drives the VCO to make the phase error smaller and smaller.

The working principle of the Costas loop for QPSK is similar. One difficulty, not present in BPSK, is maintaining a balance between the I and Q channels. The bipolar limiters in each arm are used to achieve this balance.

We conclude the chapter with a general discussion of symbol timing determination or what is commonly known as clock recovery.

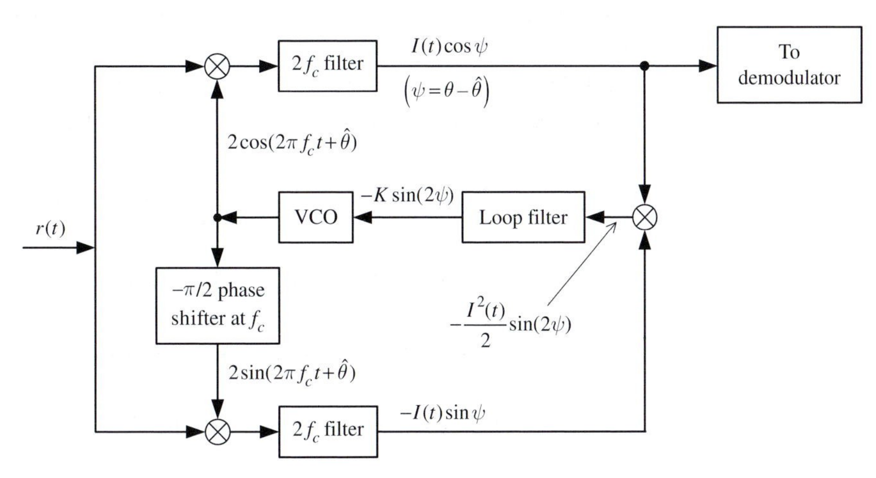

Fig. 12.17 Costas loop for carrier recovery for BPSK.

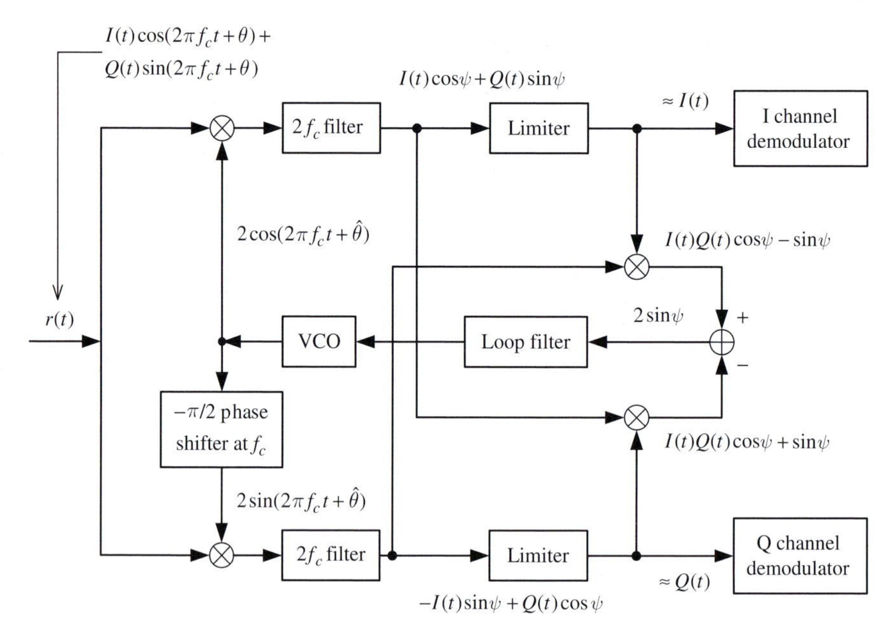

Fig. 12.18 Costas loop for carrier recovery for QPSK. Typically $\psi = \theta - \hat{\theta}$ is close to zero.

12.4 Determination (estimation) of symbol timing

Symbol timing recovery circuits can be broadly classified into open-loop symbol synchronizers and closed-loop symbol synchronizers. They may be further classified into non-data-aided (NDA) synchronizers and data-aided (DA) synchronizers which utilize either previous decisions or have a known data sequence inserted into the information sequence to aid in the synchronization process. Here we consider NDA symbol synchronizers and restrict ourselves to binary baseband signals. Extension to nonbinary baseband signals (hopefully) should be straightforward.

If a baseband modulation $m(t)$ has a spectral component at $f = 1/T_b$, then one only needs to pass the demodulated signal to a bandpass filter, centered at $1/T_b$ hertz, followed by a sgn($\cdot$) function to produce a square wave with proper timing (Figure 12.19). However, most popular baseband modulations, NRZ-L, Miller, biphase, etc., do not have a spectral component at $1/T_b$. In this case, as in the carrier–phase recovery circuits, one needs to be created. This again is accomplished by a nonlinearity such as a multiplier, squarer or full-wave rectifier.

Figure 12.20 shows two different implementations of an open-loop synchronizer for this situation. In the first one the waveform $m_1(t)$ is always positive in the second half of every bit period, T_b. It will be negative in the first half if two successive bits disagree. This produces a spectral component at the data rate $r_b = 1/T_b$ as well as at the harmonics. The bandpass filter isolates the appropriate spectral component. The second circuit is essentially an edge detector where the differentiator produces positive or negative spikes

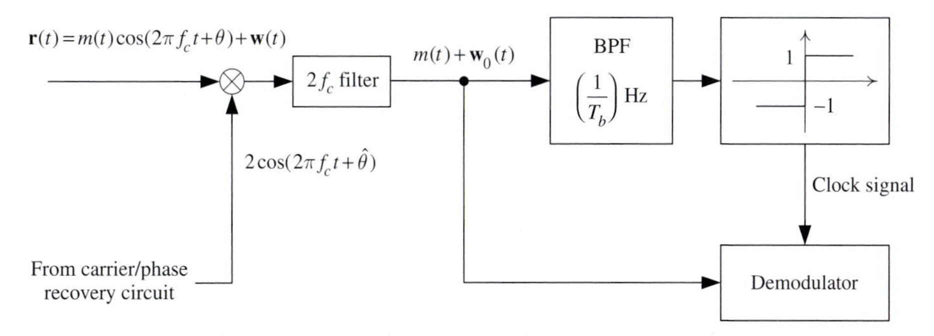

Fig. 12.19 Clock recovery circuit when $m(t)$ has a spectral component at $1/T_b$ hertz. Here BPF means bandpass filter.

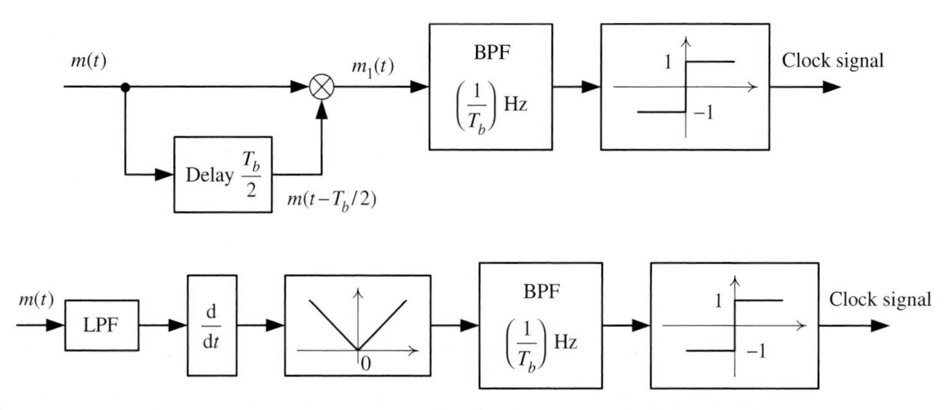

Fig. 12.20 Two different open-loop symbol synchronizers in which the baseband modulation has no spectral component at $1/T_b$ hertz. Note that, for clarity, noise component $\mathbf{w}_0(t)$ is ignored. Also LPF means lowpass filter.

at symbol transitions. Differentiation, however, is a process that is very sensitive to wideband noise, where wideband implies there are rapid changes in the time signal. Hence a lowpass filter precedes the differentiator. It, of course, causes any rapid transitions to be rounded out, i.e., there is a finite rise and fall time. The result is that rather than having an impulse at any transition point, one has a pulse-like signal that will still be of considerable amplitude.

The main disadvantage of an open-loop synchronizer is an unavoidable nonzero average tracking error. Though small for large SNR, it cannot be made zero. A closed-loop symbol synchronizer, being a feedback system, circumvents this problem. The most popular closed-loop symbol synchronizer is one called the *early–late gate synchronizer*. It is shown in block diagram form in Figure 12.21. The operation of the synchronizer is explained by referring to Figure 12.22. Let the $t = 0$ point be set by the square-wave clock generated locally by the VCO. When this clock and the incoming data, $m(t)$, are in perfect synchronization, both integrators will accumulate the same amount of signal energy over the period $(T_b - d)$, at least ideally. Therefore, the error signal is zero. If, however, the incoming data are delayed by $\Delta < d$ as shown, then the early-gate integrator accumulates signal

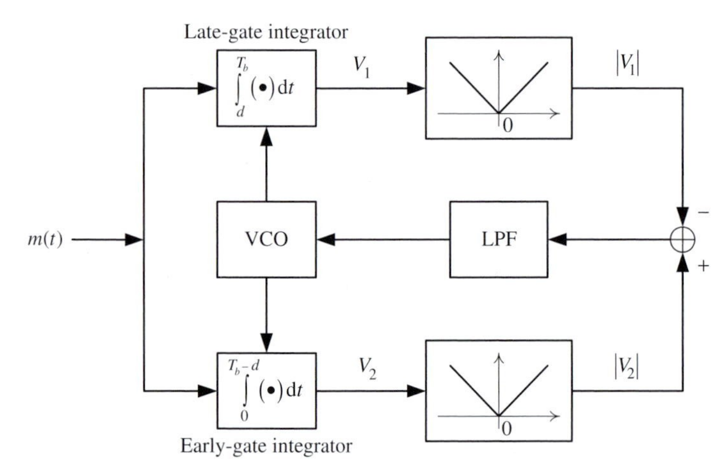

Fig. 12.21 Early–late gate clock synchronizer.

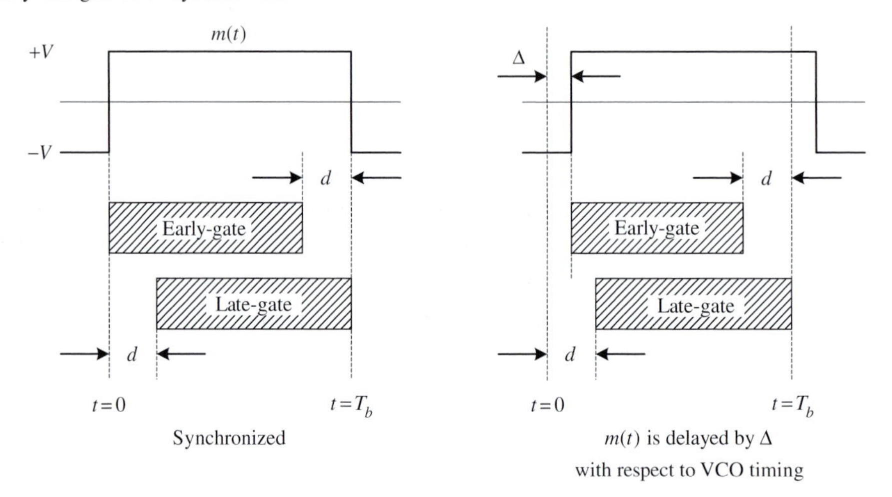

Fig. 12.22 Timing illustration of early–late gate synchronizer operation.

energy over a period of $(T_b - \Delta - d)$, while the late-gate integrator still accumulates signal energy over $(T_b - d)$ seconds. The error signal is therefore proportional to $-\Delta$, which would reduce the VCO frequency and delay its timing to bring it in synchronization with the timing of $m(t)$. The opposite happens when the data timing is in advance of the VCO timing.

12.5 Summary

As stated at the beginning of the chapter, synchronization is a vast topic. Only the most basic aspects of synchronization have been dealt with in this chapter. Important issues such as the performance of the PLL in noise, other synchronization techniques (such as sending known symbols to aid in the estimation of the "nuisance" parameters), the effect of different loop filters on the PLL's locking and tracking performance, etc., is not examined. In large part, this is because of the nonlinear nature of the circuits involved. Though some analysis is possible, as done in the chapter, eventually one needs to resort to simulation to obtain an appreciation of the circuit's behavior. If a linear approximation is justified, then standard linear system theory can be used. This approach and other issues are pursued in the problem set. To reiterate, synchronization is a very important aspect of digital communications and the reader is well advised to pursue further study of it.

12.6 Problems

The first three problems investigate further the effect of gain, phase, and frequency imperfections.

12.1 Consider BPSK and QPSK where not only is there a phase offset of θ radians in the received signal but also an attenuation of α. Derive the resultant bit error probabilities and plot the resultant error performance (see Figure 12.2 for a reference). Let the attenuation range from 0 to 20%.

12.2 As pointed out in Section 12.1, QPSK is more sensitive to phase uncertainty than BPSK. Here we investigate the sensitivity of higher-order QAM constellations, namely those on a rectangular grid with a spacing of Δ on each axis. The phase uncertainty rotates the signal constellation by θ degrees. The geometry for a general signal point is shown in Figure 12.23.

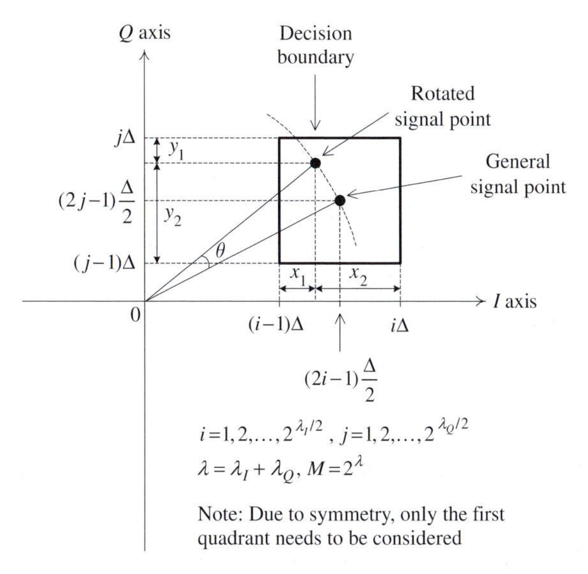

Fig. 12.23 Geometrical representation of a general QAM signal point with phase uncertainty.

(a) Show that:

$$x_1 = \frac{\Delta}{2}\sqrt{(2i-1)^2+(2j-1)^2}\cos\left[\tan^{-1}\left(\frac{2j-1}{2i-1}\right)+\theta\right]-2(i-1),$$

$$x_2 = i\Delta - \frac{\Delta}{2}\sqrt{(2i-1)^2+(2j-1)^2}\cos\left[\tan^{-1}\left(\frac{2j-1}{2i-1}\right)+\theta\right],$$

$$y_1 = j\Delta - \frac{\Delta}{2}\sqrt{(2i-1)^2+(2j-1)^2}\sin\left[\tan^{-1}\left(\frac{2j-1}{2i-1}\right)+\theta\right],$$

$$y_2 = \frac{\Delta}{2}\sqrt{(2i-1)^2+(2j-1)^2}\sin\left[\tan^{-1}\left(\frac{2j-1}{2i-1}\right)+\theta\right]-2(j-1).$$

(b) The probability of symbol error for the usual AWGN channel, where the signals are equally likely, depends on whether the signal is an *inner* signal (which has four nearest neighbors), an *outer* signal (which has three nearest neighbors), or a *corner* signal (which has two nearest neighbors). Using the geometry of the figure show that the conditional error probabilities are given by

$$P_{\text{inner}}[\text{error}] = Q\left(\frac{x_1}{\sigma}\right)+Q\left(\frac{x_2}{\sigma}\right) + \left[Q\left(\frac{y_1}{\sigma}\right)+Q\left(\frac{y_2}{\sigma}\right)\right]\left[1-Q\left(\frac{x_1}{\sigma}\right)-Q\left(\frac{x_2}{\sigma}\right)\right] \quad \text{(inner signals)}, \tag{P12.1}$$

$$P_{\text{outer-V}}[\text{error}] = Q\left(\frac{x_1}{\sigma}\right)+\left[Q\left(\frac{y_1}{\sigma}\right)+Q\left(\frac{y_2}{\sigma}\right)\right]\left[1-Q\left(\frac{x_1}{\sigma}\right)\right] \quad \text{(“vertical” outer signals)}, \tag{P12.2}$$

$$P_{\text{outer-H}}[\text{error}] = Q\left(\frac{y_2}{\sigma}\right)+\left[Q\left(\frac{x_1}{\sigma}\right)+Q\left(\frac{x_2}{\sigma}\right)\right]\left[1-Q\left(\frac{y_2}{\sigma}\right)\right] \quad \text{(“horizontal” outer signals)},$$

$$P_{\text{corner}}[\text{error}] = Q\left(\frac{x_1}{\sigma}\right)+Q\left(\frac{x_1}{\sigma}\right)\left[1-Q\left(\frac{y_2}{\sigma}\right)\right] \quad \text{(corner signals)}, \tag{P12.3}$$

where $\sigma = \sqrt{N_0/2}$.

(c) The distances x_1, x_2, y_1, y_2 have the parameter Δ in them. Express this parameter as a function of $E_b = E_s/\lambda$, the energy per bit, and $M = 2^\lambda$, the number of signals.

(d) Obtain an expression for P[symbol error]. Note that you need to consider only the first quadrant and the sum for the inner signals should index from $i = 1$ to $2^{\lambda_I/2} - 1 = N_I - 1$ and $j = 1$ to $2^{\lambda_Q/2} - 1 = N_Q - 1$; for the outer signals from $i = N_I, j = 1$ to $N_I - 1$ and $i = 1$ to $N_I - 1, j = N_I$; and for the corner signal, $i = N_I$ and $j = N_Q$.

(e) Plot P[symbol error] versus the SNR $\equiv E_b/N_0$ (in decibels) for 16-QAM and 64-QAM for phase offsets of $\theta = 0, 3, 5, 7$ degrees and $\theta = 0, 1, 2, 3$ degrees, respectively. Compare with BPSK and QPSK.

12.3 The phase offset thus far has been considered to be an unknown, but constant parameter. A more realistic model is one where it is a random variable. When a PLL is used to estimate the phase, a common model for the pdf of the phase error is what is known as Tikhonov's density. It is given as

$$f_{\boldsymbol{\theta}}(\theta) = \frac{1}{2\pi I_0(\Lambda_m)} e^{\Lambda_m \cos\theta}, \tag{P12.4}$$

where the parameter Λ_m is known as the loop SNR and changing its values results in a wide range of pdfs. In particular, $\Lambda_m = 0$ corresponds to a uniform distribution, whereas $\Lambda_m = \infty$ gives an impulse function.

(a) Plot $f_{\boldsymbol{\theta}}(\theta)$ for different values of the parameter Λ_m.

(b) Consider BPSK with this phase offset model. The error probability is given by $Q\left(\sqrt{2E_b/N_0}\cos\boldsymbol{\theta}\right)$ and is a random variable. Obtain an expression for the average error probability and plot the error performance for various values of Λ_m.

12.4 (*Frequency offset*) Consider BPSK with carrier frequency f_c hertz. At the receiver, the demodulator's local oscillator has a frequency that is slightly offset, i.e., it is $f_c + \Delta f$ hertz. Obtain the expression for the error probability. Plot it for various values of $\Delta f T_b$. One can always overcome any performance degradation by increasing the transmitted power. How accurate must the local oscillator be if this increase is to be limited to 1 decibel or less? Assume T_b is on the order of milliseconds, while f_c is on the order of gigahertz.

12.5 Consider BFSK with frequencies f_1 and f_2 hertz. The local oscillators at the demodulator are offset by Δf_1 and Δf_2, respectively. Derive an expression for the resultant bit error probability and plot it. Make whatever assumptions you deem necessary.

12.6 The phase detector characteristic discussed in the chapter was sinusoidal. In general, one can state that $v(t) = g[\psi(t)]$, where $v(t)$ is the output of the phase detector, $\psi(t) = \theta(t) - \varphi(t)$ is the phase error, and $g[\cdot]$ is the phase detector's nonlinear characteristic. Show that the general nonlinear differential equation governing the PLL's behavior is

$$\frac{\mathrm{d}\psi(t)}{\mathrm{d}t} + K_{\text{vco}} h(t) * g[\psi(t)] = \frac{\mathrm{d}\theta(t)}{\mathrm{d}t}, \tag{P12.5}$$

where $h(t)$ is the loop filter's impulse response and $*$ denotes the convolution operation.

12.7 Using the result of Problem 12.6 and the loop filter of Figure 12.7(a), derive the appropriate equation and obtain the corresponding phase plane plot for the following phase detector's characteristics:

(a) sawtooth,

(b) triangular.

The input is a unit step in frequency, i.e., $\theta(t) = \Delta\omega t u(t)$. Discuss the stable operating points, steady-state error, etc.

12.8 Consider the loop filter of Figure 12.7(b) and a sinusoidal nonlinearity for the phase detector.

(a) Derive the differential equation for the phase error.

(b) Write the state equations for the system in terms of K_{loop} and $\tau \equiv RC$.

(c) Using the `ode` function in Matlab try to obtain the phase plane plot.

12.9 Repeat Problem 12.8 for the loop filter of Figure 12.7(c). The loop filter of Figure 12.7(c) can be termed a leaky integrator. Under what condition does it act as a reasonable integrator?

12.10 Show that the loop filter of Figure 12.7(d) realizes the transfer function discussed in Section 12.2.3.

If the phase error can be assumed to be sufficiently small and the sinusoidal nonlinearity can be approximated by a linear characteristic, then linear system theory can be used to gain insight into the PLL behavior. This is explored in the next series of problems.

12.11 Show that if the phase error is less than $\pi/6$ radians then the sinusoidal nonlinearity can be approximated by a straight line to within 5%.

With the linear approximation of Problem 12.11 the PLL has the linear model shown in Figure 12.24. Problems 12.12–12.15 are concerned with such a linear PLL model.

12.12 Consider $\varphi(s)$ as the loop's output, in Figure 12.24.

(a) Show that

$$\varphi(s) = \frac{K_{\text{loop}}H(s)/s}{1 + K_{\text{loop}}H(s)/s}\Theta(s). \quad \text{(P12.6)}$$

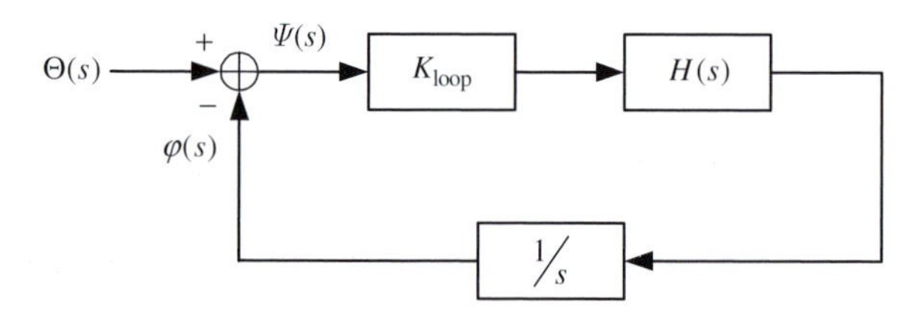

Fig. 12.24 Linear model of the PLL.

The transfer function

$$T(s) = \frac{K_{\text{loop}}H(s)/s}{1 + K_{\text{loop}}H(s)/s}$$

is known as the closed-loop transfer function.

(b) Show that $\Psi(s) = [1 - T(s)]\Theta(s)$.

12.13 Consider $H(s) = 1$, i.e., the loop filter of Figure 12.7(a). Let the input be $\theta(t) = \Delta\omega t + \theta_0$, $t \geq 0$, where θ_0 is an initial phase. Show that

$$\Psi(s) = \frac{\Delta\omega}{s(s + K_{\text{loop}})} + \frac{\theta_0}{s + K_{\text{loop}}}$$

and determine $\psi(t)$. What is the steady-state error and how does it compare with the steady-state error determined in the chapter (see page 512)? Compare the transient response given by the linear model to the exact response given in the chapter (see page 514).

12.14 Consider $H(s) = 1 + a/s$, a parallel combination of an ideal integrator with gain a and a direct connection. Again, as in Problem 12.13, let the input be $\Theta(s) = \Delta\omega/s^2 + \theta_0/s$. Determine the phase error expression for $\Psi(s)$. Though one can invert this to obtain the time-domain expression (if interested you are invited to do it) here we shall only look at the steady-state phase, i.e., $\psi(t)$ as $t \to \infty$. To do this use the final-value theorem for Laplace transforms, which states

$$\lim_{t\to\infty} \psi(t) = \lim_{s\to 0} s\Psi(s). \tag{P12.7}$$

Compare with the result obtained in the chapter.

12.15 The loop filter of Figure 12.7(c) can be used to realize the perfect integrator.

(a) Show that the transfer function $H(s)$ is of the form $K(s + a)/(s + b)$. Identify K, a, and b.

(b) Let the attenuation K be compensated for by a high-gain amplifier, gain $1/K$. Show that the closed-loop transfer function is

$$T(s) = \frac{K_{\text{loop}}(s + a)}{s^2 + (K_{\text{loop}} + b)s + aK_{\text{loop}}} \tag{P12.8}$$

and that

$$\Psi(s) = \frac{s(s + a)}{s^2 + (K_{\text{loop}} + b)s + aK_{\text{loop}}} \left(\frac{\Delta\omega}{s^2} + \frac{\theta_0}{s} \right). \tag{P12.9}$$

The input is still $\theta(t) = (\Delta\omega t + \theta_0)u(t)$.

(c) Use the final-value theorem in (P12.14) to obtain the steady-state phase error. Compare this steady-state error with that of the first-order loop and that of the perfect integrator.

In Chapter 7 (see Problems 7.14–7.20) it was shown that the bandpass noise, $\mathbf{n}(t)$, can be represented by an equivalent lowpass (or baseband) noise as

$$\mathbf{n}(t) = \mathbf{n}_I(t)\cos(\omega_c t) + \mathbf{n}_Q(t)\sin(\omega_c t), \tag{P12.10}$$

where $\mathbf{n}_I(t)$ and $\mathbf{n}_Q(t)$ have identical PSDs, which are the same as the PSD of $\mathbf{n}(t)$ but translated down by ω_c so that they are centered about zero frequency (0 hertz). Furthermore, if $\mathbf{n}(t)$ is Gaussian and zero-mean, then so are $\mathbf{n}_I(t)$ and $\mathbf{n}_Q(t)$. They are also statistically independent. With all this in mind we now proceed to develop a model of the sinusoidal PLL where there is additive white noise at the input.

12.16 Let the input to the sinusoidal PLL be

$$\begin{aligned} \mathbf{v}_{\text{in}}(t) &= V\sin(\omega_c t + \theta(t)) + \mathbf{n}(t) \\ &= V\sin(\omega_c t + \theta(t)) + \mathbf{n}_I(t)\cos(\omega_c t) + \mathbf{n}_Q(t)\sin(\omega_c t) \end{aligned} \tag{P12.11}$$

and the VCO's output be $K\cos(\omega_c t + \varphi(t))$.

(a) Show that the output of the phase detector is

$$\mathbf{v}_{\text{out}}(t) = C\sin(\theta(t) - \varphi(t)) + C_1\mathbf{n}_I(t)\cos(\omega_c t) - C_2\mathbf{n}_Q(t)\sin(\omega_c t). \tag{P12.12}$$

Identify the constants C, C_1, and C_2.

(b) Show that the PLL model when additive noise is present can be represented by the block diagram of Figure 12.25.

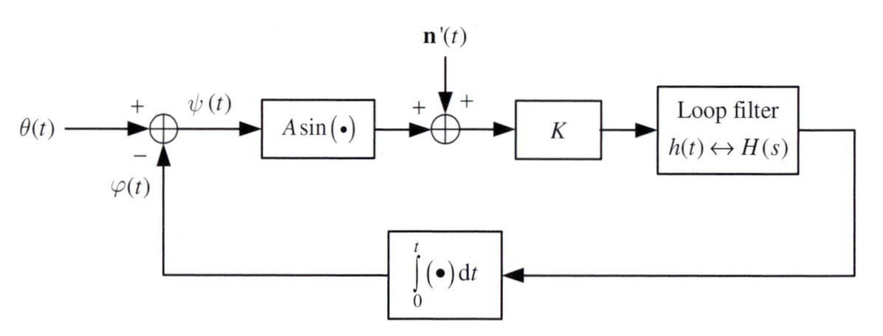

Fig. 12.25 PLL model with the presence of additive noise.

(c) Obtain the differential, more appropriately the integro-differential, equation for $\psi(t)$. It can be argued that, if $\mathbf{n}(t)$ is a stationary Gaussian process, then so is $\mathbf{n}'(t)$.

12.17 Consider the linearized model with additive noise developed in Problem 12.16. To investigate the effect of the loop filter on the noise, let $\theta(t) = 0$. Note that because the system is linear, superposition holds and considering the noise independently of the input is legitimate, as long as the linear assumption is valid.

Since $\mathbf{n}'(t)$ is a Gaussian process, then so are $\boldsymbol{\psi}(t)$ and $\boldsymbol{\varphi}(t)$. The two important parameters are the mean and variance of the processes. Note that $\boldsymbol{\psi}(t) = -\boldsymbol{\varphi}(t)$.

(a) Determine the mean values of $\boldsymbol{\psi}(t)$ and $\boldsymbol{\varphi}(t)$.

(b) Show that the PSDs of $\boldsymbol{\psi}(t)$ and $\boldsymbol{\varphi}(t)$ are

$$S_{\boldsymbol{\psi}}(\omega) = S_{\boldsymbol{\varphi}}(\omega) = \left| \frac{KH(j\omega)/(j\omega)}{1 + AKH(j\omega)/(j\omega)} \right|^2 S_{\mathbf{n}'}(\omega). \tag{P12.13}$$

(c) Let the noise be white of spectral strength $N_0/2$. Show that the PSD of the phase error due to the noise is

$$S_{\boldsymbol{\psi}}(\omega) = \frac{N_0}{2A^2} |T(j\omega)|^2, \tag{P12.14}$$

where $T(j\omega)$ is the closed-loop transfer function.

(d) The variance of the phase error due to noise, $\sigma^2_{\boldsymbol{\psi}}$, is $\int_{-\infty}^{\infty} S_{\boldsymbol{\psi}}(\omega) \mathrm{d}\omega/2\pi$. Define the loop-noise bandwidth as $B_{\text{loop}} \equiv \int_0^{\infty} |T(j\omega)|^2 \mathrm{d}\omega/2\pi$. Show that

$$\sigma^2_{\boldsymbol{\psi}} = \frac{N_0 B_{\text{loop}}}{A^2} \quad \text{(watts)}. \tag{P12.15}$$

Remark It is of interest and useful to compute B_{loop} for the various loop filters discussed in the chapter. The interested reader may pursue this. The integration is perhaps most readily accomplished by using residue theory from complex variables.

12.18 Show that to create a spectral component for M-PSK one needs to raise the received signal to the Mth power. *Hint* Express the signal point as $\sqrt{E_s} e^{j2\pi k/M}$,

$k = 1, 2, \ldots, M$. In essence you wish to create a (nonzero) constant value by the power operation.

12.19 Consider what is commonly called ring QAM. The signal points lie on concentric circles and are uniformly spaced on each circle. A general figure of the signal constellation is shown in Figure 12.26.

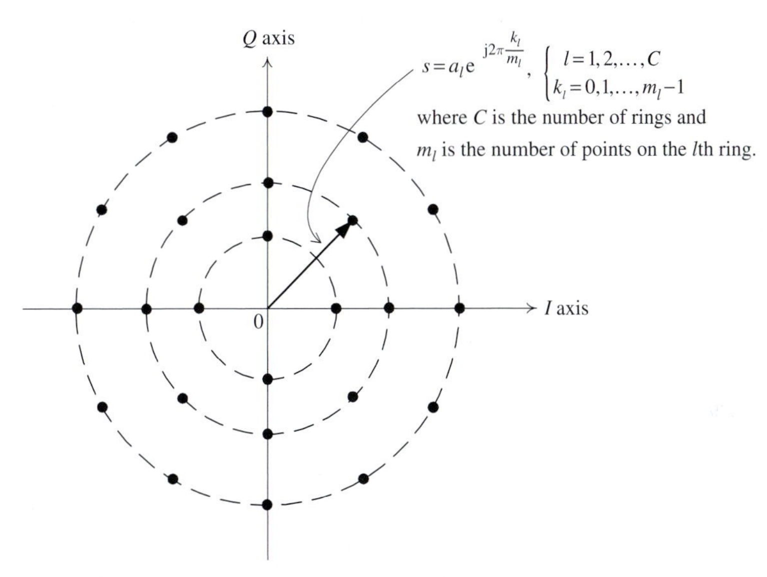

Fig. 12.26 General ring QAM constellation.

The signal constellation is obviously symmetrical and does not have a spectral component at f_c. To what power would you raise the received signal to create a spectral component?

12.20 Implement the early–late gate circuit using Matlab Simulink or Labview and investigate its performance. Note that the problem is quite open and the reader can pursue many aspects of the performance.

References

[1] L. S. Gradshteyn and L. M. Ryzhik, *Table of Integrals, Series, and Products*. Academic Press, 6th edn, 2000.

Index

RECEIVED
JUN 30 2015